T0178255

Lecture Notes in Computer Science 14519

Advanced Research in Computing and Software Science

Subline of Lecture Notes in Computer Science

More information about this series at https://link.springer.com/bookseries/558

Henning Fernau · Serge Gaspers ·
Ralf Klasing
Editors

SOFSEM 2024: Theory and Practice of Computer Science

49th International Conference on Current Trends
in Theory and Practice of Computer Science, SOFSEM 2024
Cochem, Germany, February 19–23, 2024
Proceedings

Editors
Henning Fernau
University of Trier
Trier, Germany

Serge Gaspers
UNSW Sydney
Sydney, NSW, Australia

Ralf Klasing
CNRS and University of Bordeaux
Talence, France

ISSN 0302-9743 ISSN 1611-3349 (electronic)
Lecture Notes in Computer Science
ISBN 978-3-031-52112-6 ISBN 978-3-031-52113-3 (eBook)
https://doi.org/10.1007/978-3-031-52113-3

This Springer imprint is published by the registered company Springer Nature Switzerland AG
The registered company address is: Gewerbestrasse 11, 6330 Cham, Switzerland

Paper in this product is recyclable.

Preface

A Good Tradition

The *49th International Conference on Current Trends in Theory and Practice of Computer Science* (SOFSEM 2024) was organized by the University of Trier, Germany, during February 19th–23rd, 2024. It was held in Cochem, a town on the Moselle river, about 100 km downstream from Trier and 150 km downstream from Schengen. SOFSEM (originally the SOFtware SEMinar) is an annual international winter conference, originally devoted to the theory and practice of computer science. Its aim was to present the latest developments in research to professionals from academia and industry working in leading areas of computer science. While now a well-established and fully international conference, SOFSEM also maintains the best of its original Winter School aspects, such as a high number of invited talks, in-depth coverage of selected research areas, and ample opportunity to discuss and exchange new ideas, but it has turned its focus more towards the theoretical aspects of computer science.

The series of SOFSEM conferences began in 1974 as a winter seminar for computer scientists and software engineers in former Czechoslovakia. SOFSEM soon became the foremost national seminar devoted to theoretical and practical problems of software systems. Later on, international experts were also invited, to present series of lectures on recent topics. Each SOFSEM conference consisted of several series of invited lectures, complemented by selected contributions of participants. Until 1994, the total duration of every SOFSEM conference was two weeks. Gradually, SOFSEM transformed from a mainly national seminar into an international conference. From 1995 onwards, the SOFSEM proceedings were included in the series Lecture Notes in Computer Science of Springer. The duration of SOFSEM was shortened to eight days. In 2016 the SOFSEM Steering Committee mandated that, from then on, SOFSEM conferences could be organized by colleagues anywhere in Europe, i.e., not limited to the Czech or Slovak Republics anymore. The most recent venues have been:

- 2023: Nový Smokovec, Slovak Republic
- 2021: Bozen-Bolzano, Italy (virtual)
- 2020: Limassol, Cyprus

The break during the Covid pandemic was used to renew the scope and format of SOFSEM. Now, it is focused entirely on original research and challenges in the foundations of computer science including algorithms, AI-based methods, computational complexity, and formal models.

The Newest Edition: Cochem 2024

The Program Committee of SOFSEM 2024 (with three PC Chairs) included in total 36 scientists from 22 countries and was chaired by Henning Fernau (Trier University, Germany), Serge Gaspers (UNSW Sydney, Australia), and Ralf Klasing (CNRS, University of Bordeaux, France).

This volume contains the accepted papers of SOFSEM 2024. We received 89 abstract submissions in total. Each full paper received at least three reviews per submission in a single-blind review. As a result, the PC selected 33 papers for presentation to the conference and publication in these proceedings, evaluated based on quality, originality, and relevance to the symposium. The reviewing process was supported by using the EasyChair conference system.

Highlights of the Conference

The Program Committee selected two papers to receive the Best Paper Award and one for the Best Student Paper Award, respectively. These awards were sponsored by Springer. The awardees are:

- Best Paper Award: Tesshu Hanaka, Hironori Kiya, Michael Lampis, Hirotaka Ono and Kanae Yoshiwatari. *Faster Winner Determination Algorithms for (Colored) Arc Kayles*
- Best Paper Award: Jesper Nederlof and Krisztina Szilágyi. *Algorithms and Turing Kernels for Detecting and Counting Small Patterns in Unit Disk Graphs*
- Best Student Paper Award: Christoph Grüne, Tom Janssen and Janosch Fuchs. *The Complexity of Online Graph Games*

The conference audience enjoyed five invited talks, given below in alphabetical order of the speakers:

- Edith Elkind, Univ. Oxford, UK: *Fairness in Multiwinner Voting*
- Sevag Gharibian, Univ. Paderborn, Germany: *Quantum algorithms and complexity theory: Does theory meet practice?*
- Rob van Glabbeek, Univ. Edinburgh, UK & Stanford Univ., USA & UNSW Sydney, Australia: *Modeling Time Qualitatively in Process Algebra and Concurrency Theory*
- Markus L. Schmid, HU Berlin, Germany: *The Information Extraction Framework of Document Spanners - An Overview of Concepts, Results, and Recent Developments*
- Sandra Zilles, Univ. Regina, Canada: *Machine Teaching - A Combinatorial Approach to Machine Learning from Small Amounts of Data*

Finally, Big Thanks ...

We would like to thank all invited speakers for accepting to give a talk at the conference, all Program Committee members who graciously gave their time and energy, and the more than 100 external reviewers for their expertise. Also, we are grateful to Springer for sponsoring the best (student) paper awards and for publishing the proceedings of SOFSEM 2024 in their ARCoSS subline of the LNCS series. Personal thanks go to all who helped with organizing the event.

February 2024

Henning Fernau
Serge Gaspers
Ralf Klasing

Organization

Organizing Chair

Henning Fernau	Universität Trier, Germany

Program Committee

Petra Berenbrink	Universität Hamburg, Germany
Maike Buchin	Ruhr Universität Bochum, Germany
Elisabet Burjons	York University, Canada
Maria Chudnovsky	Princeton University, USA
Sanjana Dey	National University of Singapore, Singapore
Sigrid Ewert	University of the Witwatersrand, Johannesburg, South Africa
Henning Fernau (Co-chair)	Universität Trier, Germany
Paola Flocchini	University of Ottawa, Canada
Florent Foucaud	LIMOS - Université Clermont Auvergne, Aubière, France
Robert Ganian	TU Wien, Austria
Luisa Gargano	Università di Salerno, Italy
Leszek Gąsieniec	University of Liverpool, UK
Serge Gaspers (Co-chair)	UNSW Sydney, Australia
Mingyu Guo	The University of Adelaide, Australia
Diptarama Hendrian	Tohoku University, Japan & Tokyo Medical and Dental University, Japan
Ling-Ju Hung	National Taipei University of Business, Taiwan
Tomasz Jurdziński	University of Wrocław, Poland
Philipp Kindermann	Universität Trier, Germany
Ralf Klasing (Co-chair)	CNRS and University of Bordeaux, Talence, France
Mikko Koivisto	University of Helsinki, Finland
Rastislav Královič	Comenius University, Bratislava, Slovakia
Dominik Köppl	Universität Münster, Germany & University of Yamanashi, Japan
Yaping Mao	Qinghai Normal University, China
Kitty Meeks	University of Glasgow, UK
Hirotaka Ono	Nagoya University, Japan
Marina Papatriantafilou	Chalmers University of Technology, Gothenburg, Sweden
Tomasz Radzik	King's College London, UK
Peter Rossmanith	RWTH Aachen University, Germany
Sasha Rubin	The University of Sydney, Australia
Maria Serna	Universitat Politècnica de Catalunya, Barcelona, Spain

Hadas Shachnai	Technion, Haifa, Israel
Ulrike Stege	University of Victoria, Canada
Frank Stephan	National University of Singapore, Singapore
Jan van Leeuwen	Utrecht University, The Netherlands
Jiří Wiedermann	Institute of Computer Science of the Czech Academy of Sciences, Prague, Czech Republic
Petra Wolf	University of Bergen, Norway & LaBRI - University of Bordeaux, Talence, France

Steering Committee

Barbara Catania	University of Genova, Italy
Leszek A. Gąsieniec	University of Liverpool, UK
Mirosław Kutyłowski	NASK – National Research Institute, Poland
Tiziana Margaria	University of Limerick, Ireland
Branislav Rovan	Comenius University, Bratislava, Slovakia
Petr Šaloun	Palacky University Olomouc, Czech Republic
Július Štuller (Chair)	Institute of Computer Science of the Czech Academy of Sciences, Prague, Czech Republic
Jan van Leeuwen	Utrecht University, The Netherlands

Additional Reviewers

Abu-Khzam, Faisal
Alvin, Yan Hong Yao
Ardévol Martínez, Virginia
Arrighi, Emmanuel
Arseneva, Elena
Aruleba, Kehinde
Avanzini, Martin
Baccini, Edoardo
Barsukov, Alexey
Bentert, Matthias
Berndt, Sebastian
Binucci, Carla
Biswas, Arindam
Björklund, Johanna
Bok, Jan
Bordihn, Henning
Brand, Cornelius
Brinkmann, Gunnar
Casas Torres, David Fernando
Chakraborty, Dipayan
Chalopin, Jérémie
Chen, Li-Hsuan
Chen, Po-An
Chenxu, Yang
Cooper, Linus
D'Antoni, Loris
Dailly, Antoine
Dal Lago, Ugo
Damaschke, Peter
Das, Himika
Das, Soura Sena
de Castro Mendes Gomes, Guilherme
Di Crescenzo, Antonio
Dijk, Thomas C. Van
Dobrev, Stefan
Drange, Pål Grønås
Drewes, Frank
Duvignau, Romaric
Eiben, Eduard
Elder, Murray
Faliszewski, Piotr
Ferens, Robert
Gawrychowski, Pawel
Gehnen, Matthias

Goel, Diksha
Gruner, Stefan
Guaiana, Giovanna
Hanaka, Tesshu
He, Mengmeng
Hesterberg, Adam
Hilgendorf, Martin
Hoffmann, Michael
Hoffmann, Stefan
Inamdar, Tanmay
Islam, Sk Samim
Jana, Satyabrata
Janczewski, Wojciech
Ji, Zhen
Karthik, C. S.
Kitaev, Sergey
Klawitter, Jonathan
Kleer, Pieter
Kobayashi, Koji M.
Kobayashi, Yasuaki
Kolay, Sudeshna
Kostolányi, Peter
Kunz, Pascal
Kurita, Kazuhiro
Kuske, Dietrich
Kuszmaul, John
Lambert, Dakotah
Laurenti, Luca
Lee, Chuan-Min
Lee, Troy
Leofante, Francesco
Levis, Edison
Li, Wen
Liang, Jinxia
Lin, Chuang-Chieh
Lin, Patrick
Lotze, Henri
Madathil, Jayakrishnan
Mande, Nikhil S.
Mengya, He
Mock, Daniel
Mömke, Tobias
Mráz, František
Murphy, Charlie
Nakajima, Tamio-Vesa
Nederlof, Jesper
Ngo, Quang Huy
Ortali, Giacomo
Pardubska, Dana
Pathirage Don, Thilina Chathuranga
Pellegrino, Maria Angela
Pighizzini, Giovanni
Prusa, Daniel
Pyatkin, Artem
Rebentrost, Patrick
Riquelme, Fabián
S, Taruni
Saarela, Aleksi
Sanders, Ian
Serra, Thiago
Shur, Arseny
Siemer, Stefan
Simon, Hans Ulrich
Šuppa, Marek
Szykuła, Marek
Tison, Sophie
Tsai, Meng-Tsung
Tzevelekos, Nikos
Uetz, Marc
Urbina, Cristian
von Geijer, Kåre
Wang, Xiumin
Weltge, Stefan
Yamakami, Tomoyuki
Zaborniak, Tristan
Zhang, Ayun
Zhang, Zhao
Zilles, Sandra

Contents

Invited Paper

Contributed Papers

Invited Paper

The Information Extraction Framework of Document Spanners - A Very Informal Survey

Markus L. Schmid(✉)

Humboldt-Universität zu Berlin, Berlin, Germany
MLSchmid@MLSchmid.de

Abstract. This document provides an intuitive and high-level survey of the information extraction framework of *document spanners* (Fagin, Kimelfeld, Reiss, and Vansummeren (PODS 2013, J. ACM 2015)). Originally, document spanners were presented as a formalisation of the query language AQL, which is used in IBM's information extraction engine SystemT, and over the last decade this framework is heavily investigated in the database theory community. The research topic of document spanners combines classical results from areas like formal languages, algorithms and database theory, while at the same time posing challenging new research questions.

This survey is aimed at a general theoretical computer science audience that is not necessarily familiar with database theory. Its focus are the topics of an invited talk at SOFSEM 2024.

Disclaimer

This survey particularly aims at providing an intuitive introduction to the topic of document spanners, and a list of pointers to the relevant literature. Whenever possible, we will neglect formal definitions, explain technical concepts with only examples, and discuss theoretical results in an intuitive way. We do not assume the reader to be familiar with aspects of data management that would exceed the common knowledge of most computer scientists. For more technically detailed surveys (that are particularly directed at a database theory audience), the reader is referred to [1,28].

1 Document Spanners

Document spanners are a relatively new research area that has received a lot of attention in the database theory community over the last ten years or so. An interesting fact is that the topic is motivated by practical considerations, but its theoretical foundation uses very classical and old concepts from theoretical computer science, like regular expressions, finite automata and, in general, regular languages. In order to substantiate the claim that document spanners constitute

H. Fernau et al. (Eds.): SOFSEM 2024, LNCS 14519, pp. 3–22, 2024.
https://doi.org/10.1007/978-3-031-52113-3_1

a quite relevant research area, we will just state a long list of recent papers (all of them published within the last 10 years) that all have to do with this topic. Here it comes: [2–4, 6–15, 19–24, 26, 27, 29].[1] Let us now explain what document spanners are.

Document spanners have been introduced in [9], and they are a framework for extracting information from texts (i.e., strings, sequences or words, or, as is the common term in the data management community, *documents*); it is therefore called an *information extraction framework*. Since strings are not tables as we known them from relational databases, the information they represent is usually considered by database people as being not structured, or, to use a less derogatory term, to be only *semi-structures*.[2] Therefore, we would like to process a string (or let's try to stick to the term *document* in the following (but we keep in mind that documents are nothing but strings in the sense of finite sequences of elements from a finite alphabet)), so we would like to process a document and extract (some of) its information in a structured way, i.e., as a table as found in relational databases, so with a fixed label for every column. Relevant parts of the string should then appear as entries of the cells (put into relation by the rows of the table as usual), but we are not really interested in having substrings of our document in the cells of the table. Instead we use just pointers to substrings, which are represented by the start and the endpoint of the substring. Such a 'pointer-to-substring' is called a *span*. So the span $(2, 4)$ represents the first occurrence of the substring $\mathtt{ana}$ in the document $\mathbf{D} = \mathtt{banana}$, and $(4, 6)$ represents the second occurrence of $\mathtt{ana}$.[3] Not very surprisingly, we are not just interested in the existence of substrings, but also where they occur; thus, the span representation is useful. In the following, we use the notation $\mathbf{D}[i, j]$ to refer to the content of span (i, j) of document $\mathbf{D}$, so for $\mathbf{D} = \mathtt{banana}$, we have $\mathbf{D}[2, 4] = \mathbf{D}[4, 6] = \mathtt{ana}$.

In order to say how this table of spans that we want to extract from the document looks like, we also have to label its columns (thereby also postulating how many columns we have), and we do this by simply giving a set of *variables*, e.g., $\mathcal{X} = \{\mathsf{x}, \mathsf{y}, \mathsf{z}\}$. So with respect to $\mathcal{X}$, a possible table to be extracted from a document can look like the table here to the right (such tables are also called *$\mathcal{X}$-span relations*, and their rows are called *span tuples* (or $\mathcal{X}$-span tuples)):

[1] Whether a research area should be considered important or not is always quite subjective. At the very least we can observe that many researchers like to work in the area of document spanners right now.

[2] The fate of representing data in a way that is only semi-structured is also shared by trees and graphs.

[3] In the literature, the span $(4, 6)$ is actually represented as $[4, 7\rangle$, which has some reasons, but in this survey we abstract from several such details that are not needed on this high level discussion.

$\mathbf{D} = \mathtt{abbabccabc} \quad \Longrightarrow$

x	y	z
(2, 5)	(4, 7)	(1, 10)
(3, 5)	(5, 8)	(4, 7)
(1, 3)	(3, 10)	(2, 4)
⋮	⋮	⋮

Any function that, for a fixed set $\mathcal{X}$ of variables, maps each document to a (possibly empty) $\mathcal{X}$-span relation is called a *spanner* (or *document spanner*, if there is no page limit), and actually we should add $\mathcal{X}$ somewhere and rather say $\mathcal{X}$-spanner, because the set of variables is obviously relevant. The little picture above demonstrates this scenario (and the reader should play the fun game of finding all the factors of abbabccabc the spans of the table point to).

But let's not overdo it with the informal style of this survey and maybe fix at least some more precise notation. A span of a document $\mathbf{D}$ is an element $(i, j) \in \{1, 2, \ldots, |\mathbf{D}|\} \times \{1, 2, \ldots, |\mathbf{D}|\}$ with $i \leq j$, a span tuple is a mapping from $\mathcal{X}$ to the set of spans, and a spanner is a function that maps a document to a set of span tuples. Since it is awkward to write span tuples as functions, we use a tuple notation, i.e., we write $t = ((2, 5), (4, 7), (1, 10))$ instead of $t(\mathsf{x}) = (2, 5)$, $t(\mathsf{y}) = (4, 7)$ and $t(\mathsf{z}) = (1, 10)$ (this only works if we fix some linear order for $\mathcal{X}$, but this will always be clear from the context).

This brief and informal explanation of the concept of document spanners is a sufficient basis for explaining the further concepts and results that are to follow. On the other hand, it is overly simplistic and makes the model look somewhat primitive, which does not do justice to the original paper [9], which is indeed a *seminal* paper. In particular, besides establishing some important conventions about spans and spanners and documents as data sources, the paper [9] also convincingly explains (also for researchers not too familiar with database theory), why document spanners cover relevant information extraction and therefore data management tasks. There is no need to further motivate document spanners here, since this has been done by [9] and the many papers that followed (we will, however, cite more actual literature later on).

2 Representations of Document Spanners

An important point is of course how to represent document spanners (so far, they are just abstractly defined as functions), and, as is common in database theory, we are not only interested in a mathematically rigorous formalisation, but we also want to provide a language for describing spanners that can be easily learned and applied by users (let's keep in mind that even though the theoretical research on document spanners is somewhat dominating, their original motivation was to describe a practically relevant information extraction framework, so a tool for users to tackle real-world data management tasks).

Historically, document spanners were defined by a two stage approach: First we use classes of regular language descriptors, like regular expressions and automata, to define spanners, and then we apply some relational algebra (i.e., operations on tables) on top of the span-relations that can be produced by those "regular spanners" (this term will be in quotation marks until we define it more formally, which will happen later on). This first stage makes a lot of sense from a data management perspective, because it means that if we throw a bunch of "regular spanners" at a document, then we actually turn it into a relational database. Using regular language descriptors is a great idea, since they are well understood, they have very nice algorithmic properties, they are still powerful enough to describe relevant computational tasks, and they are so simple that we can even teach them to students in the first year of their studies. Regarding the second stage: Everybody working in data management is able to manipulate relational tables with relational algebra or similar languages like SQL. So this approach is just natural.

2.1 Regular Spanners

Now how can regular language descriptors be used for describing document spanners? Well, just use a finite automaton, e.g., the following one:

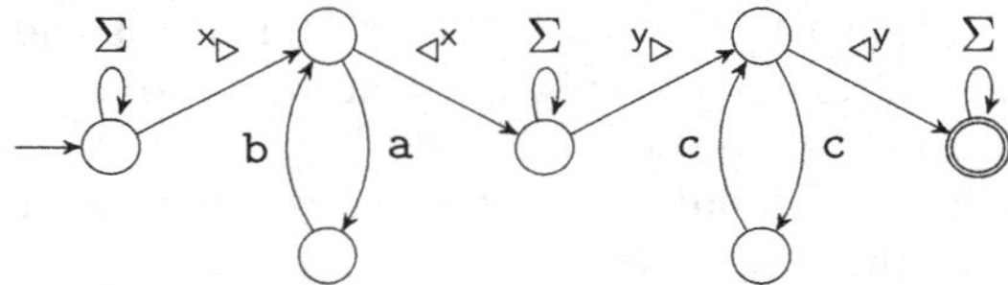

This is an automaton over the alphabet $\Sigma = \{\mathtt{a}, \mathtt{b}, \mathtt{c}\}$, but for the variable x it has a special "please start the span for variable x here"-symbol ${}^{\mathsf{x}}\!\triangleright$, and a special "please let the span for variable x end here, thank you"-symbol $\triangleleft^{\mathsf{x}}$, and analogous special symbols for variable y (it's helpful to see them as a pair of parentheses ${}^{\mathsf{x}}\!\triangleright \ldots \triangleleft^{\mathsf{x}}$). Intuitively, it is clear what's going on: The automaton reads some input over Σ and whenever it takes an ${}^{\mathsf{x}}\!\triangleright$-arc, a span for x is created that ends when a $\triangleleft^{\mathsf{x}}$-arc is traversed, and similar for y. When exactly such special-arcs are traversed obviously depends on the nondeterminism of the automaton, so the automaton can perform several different accepting runs on a fixed input, which yields several ways of extracting an $\{\mathsf{x}, \mathsf{y}\}$-span tuple from the input. As is easy to see, the automaton describes the spanner that, for any document $\mathbf{D}$, produces the table of all span-tuples $((i, j), (k, l))$, where $j \leq k$ and $\mathbf{D}[i, j] = (\mathtt{ab})^m$ and $\mathbf{D}[k, l] = (\mathtt{cc})^n$ with $m, n \geq 0$. In a similar way, the regular expression $\Sigma^* \, {}^{\mathsf{x}}\!\triangleright (\mathtt{ab})^* \triangleleft^{\mathsf{x}} \Sigma^* \, {}^{\mathsf{y}}\!\triangleright (\mathtt{cc})^* \triangleleft^{\mathsf{y}} \Sigma^*$ describes the same spanner (note that it describes the same regular language over $\Sigma \cup \{{}^{\mathsf{x}}\!\triangleright, \triangleleft^{\mathsf{x}}, {}^{\mathsf{y}}\!\triangleright, \triangleleft^{\mathsf{y}}\}$ as the automaton).

However, in order to gain a better theoretical understanding of the model, it is somewhat more convenient to refrain from thinking about specific classes of spanner representations for a moment, and establish a quite general way of how document spanners can be described (so any possible functions that maps documents to span-relations, even undecidable ones). The most relevant class of "regular spanners" can then be easily obtained by just saying the word "regular" at the right place.

Spanners are functions with Σ^* as their domain, where Σ^* is the set of all words over the alphabet Σ. Consequently, it makes sense to describe the concept of spanners in a purely language theoretic setting, which is quite convenient.[4]

A word w and an $\mathcal{X}$-span-tuple t can be merged into a single string by simply marking in w the beginning of the span for x by the symbol ${}^{\mathsf{x}}\!\rhd$ and the end of the span for x by the symbol $\lhd^{\mathsf{x}}$ (and obviously in the same way for all other variables from $\mathcal{X}$), which is then a word over the alphabet $\Sigma \cup \Gamma_{\mathcal{X}}$, where $\Gamma_{\mathcal{X}} = \{{}^{\mathsf{x}}\!\rhd, \lhd^{\mathsf{x}} \mid \mathsf{x} \in \mathcal{X}\}$. So merging the word $\mathtt{banana}$ and the span-tuple $((2,4),(4,6))$ yields $\mathtt{b}\,{}^{\mathsf{x}}\!\rhd\,\mathtt{an}\,{}^{\mathsf{y}}\!\rhd\,\mathtt{a}\,\lhd^{\mathsf{x}}\,\mathtt{na}\lhd^{\mathsf{y}}$. Words like this – i.e., words over $\Sigma \cup \Gamma_{\mathcal{X}}$ that encode a word and an $\mathcal{X}$-span-tuple for this word – are called *subword-marked words* (because that's just what they are).[5] In particular, we note that from such a subword-marked word, we can easily get the document it describes (by just deleting the special symbols from $\Gamma_{\mathcal{X}}$), and the span tuple it describes (by just looking up the positions of ${}^{\mathsf{x}}\!\rhd$ and $\lhd^{\mathsf{x}}$ for every $\mathsf{x} \in \mathcal{X}$).

The important point is now that every set L of subword-marked words (over $\mathcal{X}$), which we also call a *subword-marked language*, describes an $\mathcal{X}$-document spanner. Why? Because for any given document $\mathbf{D}$, we can simply collect all subword-marked words $w \in L$ that represent $\mathbf{D}$ and put the span-tuple represented by w in a table. So any subword-marked language L (over $\mathcal{X}$) has a natural interpretation as a function $[\![L]\!]$ that maps documents to $\mathcal{X}$-span relations, i.e., a spanner. As a concrete example, consider the subword-marked language $\{\mathtt{b}\,{}^{\mathsf{x}}\!\rhd\,\mathtt{an}\,{}^{\mathsf{y}}\!\rhd\,\mathtt{a}\,\lhd^{\mathsf{x}}\,\mathtt{na}\lhd^{\mathsf{y}}, \mathtt{b}\,{}^{\mathsf{x}}\!\rhd\,\mathtt{anana}\lhd^{\mathsf{x}}\}$, which extracts the span-relation $\{((2,4),(4,6)),((2,6),\bot)\}$ from $\mathtt{banana}$ (note that $\bot$ means *undefined*), and the empty span relation from any other document. Or the subword-marked language $\{\,{}^{\mathsf{x}}\!\rhd\, b \lhd^{\mathsf{x}} u \mid b \in \Sigma, w \in \Sigma^*\}$, which represents a spanner that extracts the first symbol of any document in a span for variable x. Or the subword-marked language $\{u_1\,{}^{\mathsf{x}}\!\rhd\,(\mathtt{ab})^m \lhd^{\mathsf{x}} u_2\,{}^{\mathsf{y}}\!\rhd\,(\mathtt{cc})^n \lhd^{\mathsf{y}} u_3 \mid u_1, u_2, u_3 \in \Sigma^*, m, n \geq 0\}$ which describes the spanner also represented by the automaton from above.

But it also works the other way around. Let S be some $\mathcal{X}$-document spanner, so some function that maps documents to $\mathcal{X}$-span relations without any further restriction. Then S maps every given document to a set $\{t_1, t_2, \ldots, t_n\}$ of $\mathcal{X}$-span tuples (note that this set is always finite, since $\mathcal{X}$ is, but for our considerations this is not even important, it's just more practically sane to have only finitely many variables). So we can simply merge $\mathbf{D}$ with each of the span-tuples $t_1, t_2, \ldots, t_n$ to obtain subword-marked words $w_1, w_2, \ldots, w_n$, and if we collect all these subword-marked words that we can obtain like this from every possible document, we have a huge subword-marked language L_S, which describes exactly the spanner S in the way explained in the previous paragraph, i.e., S equals the spanner interpretation $[\![L_S]\!]$ of L_S.

A Brief Interlude about Subword-marked Words: In the literature on spanners, subword-marked words are usually called *ref-words*. This has histori-

[4] Maybe a bit confusingly, but justified by the fact that we are now in the realm of formal languages, we use the term *word* instead of *document* for a short while.

[5] Obviously, we have to formalise when exactly a word over $\Sigma \cup \Gamma_{\mathcal{X}}$ is a proper subword-marked word, but this is not difficult.

cal reasons: Ref-words have originally been used in [25] (in the context of regular expressions with backreferences) as words that contain *references* x to some of their subwords, which are explicitly marked by brackets ${}^{\mathsf{x}}\triangleright \cdots \triangleleft^{\mathsf{x}}$. So these ref-words from [25] are strings in which subwords can be marked, but they can also contain references to marked substrings, represented by variable symbols in the string. The literature has adapted this technical tool for formalising document spanners, but for this, we only need the "subword-marking"-property, not the "subword-referencing"-property. However, the term "ref-word" has been used anyway and it stuck. Using the terms subword-marked words and ref-words synonymously is fine, as long as we only want to represent markings of subwords and no references. But in [26] – which we shall discuss in more detail below – it has been shown that the ref-words in the sense of [25] (so not only with marked subwords but also with reference-symbols) can be used for formalising a certain class of spanners. Hence, we need to distinguish between the ref-words that only mark subwords (and we call them subword-marked words here) on the one hand, and the ref-words from [25] that can also contain references to marked subwords. End of the interlude.

So we saw that subword-marked languages (over $\mathcal{X}$) and $\mathcal{X}$-document spanners are the same thing.[6] In particular, we can now conveniently define certain classes of spanners by simply stating the underlying class of subword-marked languages. Like this: The *regular spanners* are exactly the spanners $[\![L]\!]$, where L is a regular subword-marked language. The reader is encouraged to try it for herself by replacing "regular" with her favourite language class (literally any language class! No judging!).

Obviously, for the applicability of a spanner class, the algorithmic properties of the underlying class of subword-marked languages is important. So for representing regular spanners, we can use automata or regular expressions or just anything that describes regular languages (note that we can always filter out strings over $\Sigma \cup \Gamma_{\mathcal{X}}$ that are not valid subword-marked words by intersection with a regular language (note that for doing this, we do not have to parse a well-formed parenthesised expression, we only have to check that if ${}^{\mathsf{x}}\triangleright$ occurs, then it is followed by one occurrence of $\triangleleft^{\mathsf{x}}$, and that this happens at most once)).

[6] Note that this is not a one-to-one correspondence. While every subword-marked language L uniquely describes the spanner $[\![L]\!]$, there are in general several ways of representing a spanner by a subword-marked language. This is due to the fact that two subword marked words like $\mathsf{a}\ {}^{\mathsf{x}}\triangleright\ {}^{\mathsf{y}}\triangleright\ \mathsf{b} \triangleleft^{\mathsf{y}}\ \triangleleft^{\mathsf{x}}\ \mathsf{c}$ and $\mathsf{a}\ {}^{\mathsf{y}}\triangleright\ {}^{\mathsf{x}}\triangleright\ \mathsf{b} \triangleleft^{\mathsf{x}}\ \triangleleft^{\mathsf{y}}\ \mathsf{c}$ are different strings, but they nevertheless describe the same pair of document and span-tuple. Unfortunately, this can be very annoying from a technical point of view, and it can even lead to some problems for algorithms on spanners. But since there is no peer-reviewing for this article, we keep quiet and simple pretend that we did not notice this little flaw. This issue is anyway much discussed in the actual literature on document spanners and everybody is aware of it, we just neglect it in this survey because nobody stops us. For example, in order solve this issue, [7] introduces a fixed order on the consecutive occurrences of the symbols from $\Gamma_{\mathcal{X}}$, while [2,27] simply replace sequences of symbols from $\Gamma_{\mathcal{X}}$ by sets of the symbols, thus using subsets of $\Gamma_{\mathcal{X}}$ as symbols. This was a long footnote, but better than having another "interlude".

Coming back to the example automaton above: If interpreted as an NFA over $\Sigma \cup \Gamma_{\mathcal{X}}$, it obviously represents a regular subword-marked language, which is the same language represented by the regular expression $\Sigma^* \, {}^{\mathsf{x}}\!\triangleright \, (\mathtt{ab})^* \, \triangleleft^{\mathsf{x}} \, \Sigma^* \, {}^{\mathsf{y}}\!\triangleright \, (\mathtt{cc})^* \, \triangleleft^{\mathsf{y}} \, \Sigma^*$, and it is easy to see that if we interpret this subword-marked language as a spanner, it is exactly the one described above, so all span-tuples $((i,j),(k,l))$, where $j \leq k$ and $\mathbf{D}[i,j] = (\mathtt{ab})^m$ and $\mathbf{D}[k,l] = (\mathtt{cc})^n$ with $m, n \geq 0$.

Whether we interpret the ${}^{\mathsf{x}}\!\triangleright$ and $\triangleleft^{\mathsf{x}}$ transitions as special operations that trigger the construction of the span tuple, or whether we consider them as normal input symbols so that the automaton is a string-acceptor is merely a matter of taste. Although the second point of view seems to fit better to this general perspective of spanners as subword-marked languages.

Note that [9] also considers a proper subclass of regular spanners defined by so-called *regex formulas*, which is a certain class of regular expressions. The point is that regex formulas can enclose only proper sub-expressions in brackets ${}^{\mathsf{x}}\!\triangleright$ and $\triangleleft^{\mathsf{x}}$. As a result, this formalism cannot describe overlapping spans.

2.2 Core-Spanners

Recall that the regular spanners describe just the first step of the original spanner framework from [9]. So let's move on to the second stage.

Assume that we have extracted from a string a span relation or several span relations by regular spanners. We could now manipulate these tables with relational algebra operations, and in [9], we use union, natural join, projection and – let us make a dramatic pause here, because this operation is a real game changer – *string equality selection*. Union is just the set union of span relations, natural join sort of glues together tables on their common attributes (if you know how natural join is defined, you are probably annoyed by this superficial explanation and would be bored by a detailed one, if you don't know it, you can google it) and projection just deletes columns. Now these are typical operations for relational data and they are not specific to our framework of information extraction of *textual data*. The string equality selection, on the other hand, is tailored to the situation that our tables are not just any tables, but span relations, so their entries are pointers to substrings of a document. The string equality selection is an operator that is parameterised by some subset $\mathcal{Y} \subseteq \mathcal{X}$. It looks at every span tuple of the span relation and checks whether all the spans of the variables in $\mathcal{Y}$ point to an occurrence of the *same* substring. Although this is obvious, let us briefly observe that different spans might represent occurrences of the same substring, like $(2,4)$ and $(4,6)$ both represents the `ana` in `banana`. Every span tuple where this is *not* the case will be kicked out by this operator, so it selects span tuples from a span relation according to the equality of the substrings of certain spans. The next example shows what the string equality selector does with respect to $\mathcal{Y} = \{\mathsf{x}, \mathsf{y}\}$ on the following table that has been extracted from `banana`:

x	y
(1, 2)	(4, 6)
(2, 4)	(4, 6)
(3, 4)	(5, 6)
(2, 3)	(5, 6)

$\Rightarrow$

x	y
(2, 4)	(4, 6)
(3, 4)	(5, 6)

The so-called *core spanners* are those spanners that can be obtained by first extracting a span relation from a document by a regular spanner, and then apply a finite sequence of any of the relational operators from above (including the string equality selection). As shown in [9], the operators of union, natural join and projection (but not string equality selection!) can all directly be pushed into the automaton for the regular spanner, meaning that tables extracted by a regular spanner followed by any sequence of these operators can also directly be extracted by a single regular spanner. Or, putting it differently, these simpler relational operators are "regular". As a consequence, core spanners have a normal form: Every core spanner can be described by a regular spanner followed by a finite sequence of string equality selections followed by one projection.

Why the string equality selection makes such a huge difference (i.e., why core spanners are much more powerful than regular spanners) will be discussed in the next section. From an intuitive point of view, this is not surprising, since string equality selection is an inherently non-regular feature. For example, we can use a regular spanner over $\mathcal{X} = \{\mathsf{x}, \mathsf{y}\}$ that extracts from a document $\mathbf{D}$ all span tuples $((1, k), (k + 1, |\mathbf{D}|)$ for every $k \in \{1, 2, \ldots, |\mathbf{D}|\}$ (so it can arbitrarily split the document and store the two parts in the spans of the two variables), and then uses a string equality selection with respect to $\mathcal{X}$. This will turn every given document $\mathbf{D}$ into the span-relation $\{((1, |\mathbf{D}|/2), (|\mathbf{D}|/2 + 1, |\mathbf{D}|)\}$ if $\mathbf{D} = ww$ (i.e., $\mathbf{D}$ is a square) and into the span-relation $\emptyset$ if $\mathbf{D}$ is not a square. So it somehow recognises the non-regular copy language. But this is nothing! We can also get crazy and apply string equality selections to several spans that overlap each other in complicated ways to describe spanners that are not funny anymore (see [12] for further details).

The majority of the papers on document spanners is concerned with regular spanners, probably because core spanners have some issues with complexity and decidability. However, there are also several papers concerned with core spanners; see [11,12,14,23,26].

3 Problems on Regular Spanners and Core Spanners

Regular spanners are mild and core spanners are wild. Putting it more formally, regular spanners have excellent algorithmic properties (i.e., good complexities), while core spanners exhibit intractability and even undecidability for many of their relevant computational problems. As an example, let us consider some relevant problems like model checking (deciding whether a given span tuple t is in $S(\mathbf{D})$ for a given spanner S and document $\mathbf{D}$), non-emptiness (check whether $S(\mathbf{D}) \neq \emptyset$ for a given spanner S and document $\mathbf{D}$), satisfiability (for given spanner S, decide whether there is a document $\mathbf{D}$ with $S(\mathbf{D}) \neq \emptyset$), or inclusion (for given spanners S_1 and S_2, decide whether $S_1(\mathbf{D}) \subseteq S_2(\mathbf{D})$ for every document $\mathbf{D}$).

For regular spanners, algorithms for model checking, non-emptiness and satisfiability have quite good polynomial running times (see, e.g., [2,9,10,26]). The reason is that these problems reduce to problems on regular languages or finite automata (i.e., the good algorithmic properties of regular languages carry over to regular spanners). Moreover, inclusion for regular spanners is PSPACE-complete, which is not exactly tractable, but inclusion for regular languages is also PSPACE-complete, and the inclusion problem for regular spanners covers the inclusion problem for regular languages (see [19]). For core-spanners, the inclusion problem is even undecidable, and model checking, non-emptiness and satisfiability, which can be solved quite efficiently for regular spanners, are all NP-hard (see [12]).

All the aforementioned problems are typical decision problems, but in database theory, which always has an eye towards application, there is also a substantial interest in practically motivated problems. One key observation is that a computer program that merely says "yes" or "no" to Boolean database queries is of little use in the real world. Moreover, a program that computes the huge set (of potentially exponential size) of all possible answers to the query is also of questionable practical relevance. Therefore, it is common to investigate query evaluation (this term somewhat abstractly covers all scenarios where we want to evaluate a given query with respect to a given database, even if the database is just a single string) in terms of an enumeration problem. This means that we are interested in algorithms that produce a list of all answers to the queries (obviously, without repetitions). Such an algorithm is particularly worthwhile if it starts producing the list very fast, and if we do not have to wait too long to receive the next element. The optimal scenario here is therefore that the first element is produced after a running time that is only linear in the size of the data, which is called *linear preprocessing.* Note that the algorithm must somehow process the data that is queried, so assuming at least preprocessing linear in the size of the data is fair. Moreover, after one answer is produced, we would like the time we have to wait for the next element (which we call *delay*) to be completely independent from the size of the data, so the running time we need here is only a function of the size of the query (which is then called *constant* delay). We should mention here that these complexity requirements use the so-called *data-complexity* perspective, which measures running time only in the size of the data, and considers the size of the query as being constant. This is a quasi-standard in many areas of database theory, and it makes a lot of sense, since the data can be assumed to be quite large, while the query, in comparison, is tiny. The assumption that the data is large is justified by the buzzword "big data". The queries are assumed to be small since they are – in most scenarios – written by human users. Of course, linear preprocessing and constant delay is not always possible, but it is the holy-grail for enumeration algorithms of query evaluation problems. See [31,32] for surveys on the topic of enumeration algorithms in database theory, and [30] for a recent paper.

Coming back to document spanners, we are looking for an algorithm that, for some document $\mathbf{D}$ and a spanner S, makes some preprocessing that is linear in $|\mathbf{D}|$ and then enumerates all span tuples from $S(\mathbf{D})$ with constant delay. For regular spanners, this is possible, but it does not directly follow from known algorithmic results about automata (see [2,10] for details). Let us briefly discuss this on an intuitive level.

Assume that the spanner S is given by an NFA M that accepts a subword-marked language L (over Σ and $\mathcal{X}$) with $[\![L]\!] = S$. Now we are interested in all possible ways of shuffling the symbols $\Gamma_{\mathcal{X}} = \{{}^{\mathsf{x}_1}\triangleright, \triangleleft^{\mathsf{x}_1}, {}^{\mathsf{x}_2}\triangleright, \triangleleft^{\mathsf{x}_2}, \ldots\}$ into $\mathbf{D}$ such that we get a subword-marked word that is accepted by M, since these subword-marked words represent the span-tuples of $S(\mathbf{D})$. But these subword-marked words are represented by paths in M from the start state to an accepting state that are labelled with $\mathbf{D}$ (and some symbols from $\Gamma_{\mathcal{X}}$). In fact, it is better to consider the DAG of nodes (p, i), where p is a state of M and i is a position of $\mathbf{D}$, there is a $\mathbf{D}[i+1]$-labelled edge from (p, i) to $(q, i+1)$ if in M we can change from p to q by reading $\mathbf{D}[i+1]$, and there is a γ-labelled edge from (p, i) to (q, i) if in M we can change from p to q by reading some symbol $\gamma \in \Gamma_{\mathcal{X}}$ (actually, we could even drop the edge labels from Σ, since they are not relevant). In this DAG, we are interested in all paths from $(q_i, 0)$, where q_0 is the start state, to some $(q_f, |\mathbf{D}|)$, where q_f is some accepting state. The labels from $\Gamma_{\mathcal{X}}$ on such paths represents the span-tuples that we enumerate. So we simply enumerate those paths, but this is a bit tricky because, firstly, there might be different paths representing the same span-tuple and, secondly, the paths are rather long (well, of size $|\mathbf{D}|$ actually), so we cannot afford to construct them explicitly, because this would break our bound on the delay. In other words, we have to enumerate those paths, but we have to efficiently skip over the parts of the paths that are labelled with symbols from $\mathbf{D}$.

4 An Approach to Tame Core Spanners

We called core spanners *wild* earlier in this article, because with this term in mind, it is appropriate to think about how we can *tame* them. And *taming* now means to make them a bit more like regular spanners in terms of their algorithmic properties, while still maintaining the most important features of their expressive power. An approach towards this goal has been presented in [26], which we shall now briefly explain.

A nice property of regular spanners is that we can purely describe them as special regular languages (i.e., regular subword marked languages), which means that we can use classical tools like regular expressions and finite automata to handle them. So it is worth thinking about to what extent we can use the same approach also for (subclasses of) core spanners. In particular, the goal is to describe core-spanners again just by certain regular languages (obviously, the subword-marked languages are not suitable for this, since the subword-marked languages of core-spanners are not necessarily regular languages).

Let L_S be the subword-marked language for a regular spanner S. If we use on S a string equality selection (i.e., we consider a core-spanner), say with respect to variables $\mathcal{Y} = \{\mathsf{y}, \mathsf{z}\}$, then any subword-marked word of L_S can be considered irrelevant, if ${}^{\mathsf{y}}\triangleright \ldots \triangleleft^{\mathsf{y}}$ and ${}^{\mathsf{z}}\triangleright \ldots \triangleleft^{\mathsf{z}}$ enclose different factors. So with the application of the string equality selection in mind, we wish to directly get rid of such irrelevant words of L_S, and we achieve this by replacing every ${}^{\mathsf{z}}\triangleright\, u \triangleleft^{\mathsf{z}}$ factor of a subword-marked word by ${}^{\mathsf{z}}\triangleright\, \mathtt{y} \triangleleft^{\mathsf{z}}$, where the symbol y is used as a *reference* signifying that the word enclosed by ${}^{\mathsf{y}}\triangleright \ldots \triangleleft^{\mathsf{y}}$ is to be repeated here (otherwise, the subword-marked word would describe a span-tuple that would be killed by the string equality relation anyway).

As an example, consider the subword-marked language described by the regular expression

$$r := {}^{\mathsf{x}}\triangleright\, \mathtt{a}^*\mathtt{b}\; {}^{\mathsf{y}}\triangleright\, \mathtt{c} \triangleleft^{\mathsf{x}}\, \mathtt{b}^*\; {}^{\mathsf{x}'}\triangleright\, \mathtt{a}^*\mathtt{bc} \triangleleft^{\mathsf{x}'}\; \triangleleft^{\mathsf{y}}\; {}^{\mathsf{y}'}\triangleright\, \mathtt{cb}^*\mathtt{a}^*\mathtt{bc}\triangleleft^{\mathsf{y}'},$$

and assume that we want to apply a string equality selection with respect to $\{\mathsf{x}, \mathsf{x}'\}$ followed by a string equality selection with respect to $\{\mathsf{y}, \mathsf{y}'\}$. The resulting core-spanner can as well be represented by the regular expression $r' := {}^{\mathsf{x}}\triangleright\, \mathtt{a}^*\mathtt{b}\; {}^{\mathsf{y}}\triangleright\, \mathtt{c} \triangleleft^{\mathsf{x}}\, \mathtt{b}^*\; {}^{\mathsf{x}'}\triangleright\, \mathtt{x} \triangleleft^{\mathsf{x}'}\; \triangleleft^{\mathsf{y}}\; {}^{\mathsf{y}'}\triangleright\, \mathtt{y}\triangleleft^{\mathsf{y}'}$, i.e., we simply replace ${}^{\mathsf{x}'}\triangleright\, \mathtt{a}^*\mathtt{bc}\triangleleft^{\mathsf{x}'}$ and ${}^{\mathsf{y}'}\triangleright\, \mathtt{cb}^*\mathtt{a}^*\mathtt{bc}\triangleleft^{\mathsf{y}'}$ by ${}^{\mathsf{x}'}\triangleright\, \mathtt{x}\triangleleft^{\mathsf{x}'}$ and ${}^{\mathsf{y}'}\triangleright\, \mathtt{y}\triangleleft^{\mathsf{y}'}$, respectively. Now, we have represented a non-regular core-spanner "somehow" as a regular language (note that with help of the string equality selection, this core spanner checks whether some unbounded factors are repeated, which is an inherently non-regular property). But how exactly does the regular language describe the core spanner? Very easy: The regular expression r' can generate words like ${}^{\mathsf{x}}\triangleright\, \mathtt{aab}\; {}^{\mathsf{y}}\triangleright\, \mathtt{c} \triangleleft^{\mathsf{x}}\, \mathtt{b}\; {}^{\mathsf{x}'}\triangleright\, \mathtt{x} \triangleleft^{\mathsf{x}'}\; \triangleleft^{\mathsf{y}}\; {}^{\mathsf{y}'}\triangleright\, \mathtt{y}\triangleleft^{\mathsf{y}'}$ and ${}^{\mathsf{x}}\triangleright\, \mathtt{aaaab}\; {}^{\mathsf{y}}\triangleright\, \mathtt{c} \triangleleft^{\mathsf{x}}\, \mathtt{bbbb}\; {}^{\mathsf{x}'}\triangleright\, \mathtt{x} \triangleleft^{\mathsf{x}'}\; \triangleleft^{\mathsf{y}}\; {}^{\mathsf{y}'}\triangleright\, \mathtt{y}\triangleleft^{\mathsf{y}'}$ and so on. So these words are almost subword-marked words, but they have occurrences of *references* x and y, which we need to replace. Hence, we interpret each such word as the subword-marked word that we get by simply replacing all the references. Applied to ${}^{\mathsf{x}}\triangleright\, \mathtt{aab}\; {}^{\mathsf{y}}\triangleright\, \mathtt{c} \triangleleft^{\mathsf{x}}\, \mathtt{b}\; {}^{\mathsf{x}'}\triangleright\, \mathtt{x} \triangleleft^{\mathsf{x}'}\; \triangleleft^{\mathsf{y}}\; {}^{\mathsf{y}'}\triangleright\, \mathtt{y}\triangleleft^{\mathsf{y}'}$, the replacement $\mathtt{x} \mapsto \mathtt{aabc}$ gives us ${}^{\mathsf{x}}\triangleright\, \mathtt{aab}\; {}^{\mathsf{y}}\triangleright\, \mathtt{c} \triangleleft^{\mathsf{x}}\, \mathtt{b}\; {}^{\mathsf{x}'}\triangleright\, \underline{\mathtt{aabc}} \triangleleft^{\mathsf{x}'}\; \triangleleft^{\mathsf{y}}\; {}^{\mathsf{y}'}\triangleright\, \mathtt{y}\triangleleft^{\mathsf{y}'}$, and then the replacement $\mathtt{y} \mapsto \mathtt{cbaabc}$ gives us the subword-marked word ${}^{\mathsf{x}}\triangleright\, \mathtt{aab}\; {}^{\mathsf{y}}\triangleright\, \mathtt{c} \triangleleft^{\mathsf{x}}\, \mathtt{b}\; {}^{\mathsf{x}'}\triangleright\, \mathtt{aabc} \triangleleft^{\mathsf{x}'}\; \triangleleft^{\mathsf{y}}\; {}^{\mathsf{y}'}\triangleright\, \underline{\mathtt{cbaabc}}\triangleleft^{\mathsf{y}'}$. And this final subword-marked word describes a document and a span-tuple in the usual way.

So what we do is that we also allow the variables from $\mathcal{X}$ to occur in subword-marked words, and we call such words then *ref-words* (since these variables function a references to marked subwords). Then we consider *regular* languages of ref-words, and we interpret such ref-languages as spanners by first resolving all the references in all the ref-words (as sketched above) to get a subword-marked languages, which then describes a spanner as before. While the ref-language is necessarily regular (by definition), the subword-marked language that we get from it by resolving all references is not necessarily a regular language anymore (non-regularity must creep in somewhere, if we want to describe non-regular core-spanners).

Those so-called *refl-spanners* can still describe all regular spanners, but also many non-regular core-spanners.[7]

---

[7] Technically, the formalism also allows to create an unbounded number of references by using a variable symbol under a Kleene-star, which results in spanners that are not even core-spanners. But since the formalism is introduced for describing a large class of core-spanners by regular languages, we ignore this issue.

Model-checking and satisfiability for refl-spanners can be solved as efficiently as for regular spanners (while for core-spanners these problems are intractable). On the other hand, non-emptiness for refl-spanners is NP-hard as for core spanners. Moreover, refl-spanner allow a certain restriction that yields decidable inclusion (recall that inclusion is undecidable for core-spanners).

It is intuitively clear that refl-spanners cannot describe all core-spanners. Note that the construction sketched above where we replaced parts of the regular spanner by references only works in special cases.

For example, let s, t and r be some regular expressions, and consider the regular spanner described by ${}^{\mathsf{x}}\!\triangleright\, r \,\triangleleft^{\mathsf{x}}\; {}^{\mathsf{y}}\!\triangleright\, s \,\triangleleft^{\mathsf{y}}\; {}^{\mathsf{z}}\!\triangleright\, t\triangleleft^{\mathsf{z}}$. If we want to describe the core-spanner that we get by applying a string equality selection with respect to $\{\mathsf{x}, \mathsf{y}, \mathsf{z}\}$, then we cannot simply replace each of s and t by an occurrence of x, since this would also represent ref-words ${}^{\mathsf{x}}\!\triangleright\, u \,\triangleleft^{\mathsf{x}}\; {}^{\mathsf{y}}\!\triangleright\, \mathsf{x} \,\triangleleft^{\mathsf{y}}\; {}^{\mathsf{z}}\!\triangleright\, \mathsf{x}\triangleleft^{\mathsf{z}}$, where $u \in L(r)$, but $u \notin L(s)$ or $u \notin L(t)$, which gives us the subword-marked word ${}^{\mathsf{x}}\!\triangleright\, u \,\triangleleft^{\mathsf{x}}\; {}^{\mathsf{y}}\!\triangleright\, u \,\triangleleft^{\mathsf{y}}\; {}^{\mathsf{z}}\!\triangleright\, u\triangleleft^{\mathsf{z}}$, which describes a span-tuple for the document uuu that should not be in the span-relation. However, in this case, we can use the refl-spanner ${}^{\mathsf{x}}\!\triangleright\, h \,\triangleleft^{\mathsf{x}}\; {}^{\mathsf{y}}\!\triangleright\, \mathsf{x} \,\triangleleft^{\mathsf{y}}\; {}^{\mathsf{z}}\!\triangleright\, \mathsf{x}\triangleleft^{\mathsf{z}}$, where h is a regular expression that describes the intersection of $L(r)$, $L(s)$ and $L(t)$ (note that this construction causes an exponential blow-up).

Another problem arises when we want to replace some ${}^{\mathsf{x}}\!\triangleright\, s\triangleleft^{\mathsf{x}}$ by ${}^{\mathsf{x}}\!\triangleright\, \mathsf{y}\triangleleft^{\mathsf{x}}$, but s also contains symbols from $\Gamma_{\mathcal{X}}$, which we cannot afford to simply delete. Interestingly, in such cases it can help to first cut all the overlapping regions enclosed by the brackets ${}^{\mathsf{x}}\!\triangleright \ldots \triangleleft^{\mathsf{x}}$ into smaller parts (thereby introducing new variables, but only polynomially many), then, depending on the string equality selection, translating it into a ref-language, which then not quite describes the intended core spanner, but almost: It describes the core-spanner with the only difference that some spans are cut into a several smaller spans. This difference can then easily be repaired by just combining certain columns into one column and gluing the respective spans of those columns together. This result points out that refl-spanner can describe a large class of core-spanners.

5 Regular Spanners on SLP-Compressed Data

Since big data is so big, it is a good idea to compress it. The classical motivation is that compressed data can be stored in less space, or send somewhere in less time. But what about directly querying data in compressed form, i.e., without decompressing it? This would be very convenient, since then we do not have to decompress our data when we want to work with it, and algorithms for querying the data might even be faster, since their running times depend on the size of the compressed data, which might be much smaller than the uncompressed data size. Theoretically, any polynomial time algorithm that works on compressed data may outperform even a linear time algorithm for the uncompressed data, in the special case where the compressed data has size logarithmic in the uncompressed data (so measured in the uncompressed data it's polylogarithmic running time vs. linear running time).

This paradigm of *algorithmics on compressed inputs* is well-developed in the realm of string algorithms (see the explanations in [27,29] and the general survey [18]), and since spanner evaluation is a string problem, it makes sense to investigate it in this compressed setting as well. More precisely, we are interested in evaluating a document spanner over a document that is compressed, and which should not be decompressed for this purpose. Moreover, the most relevant form of evaluation problem is enumeration (as explained above), and the best running time is linear preprocessing (but now linear with respect to the compressed size of the document!) and constant delay.

Before outlining some respective results, let us discuss the underlying compression scheme in this setting. A particularly fruitful approach to algorithmics on compressed strings are so-called *straight-line programs* (SLPs). The simplicity of SLPs is very appealing: We compress a string w by a context-free grammar (for convenience in Chomsky normal form) that describes the language $\{w\}$ (i.e., it can generate exactly one string, which happens to be w). This calls for an example: The string `aabbabaabaabbab` can be described by a context-free grammar with the rules shown here on the left (we use S, A, B, C, D, E as non-terminals, where S is the start non-terminal):

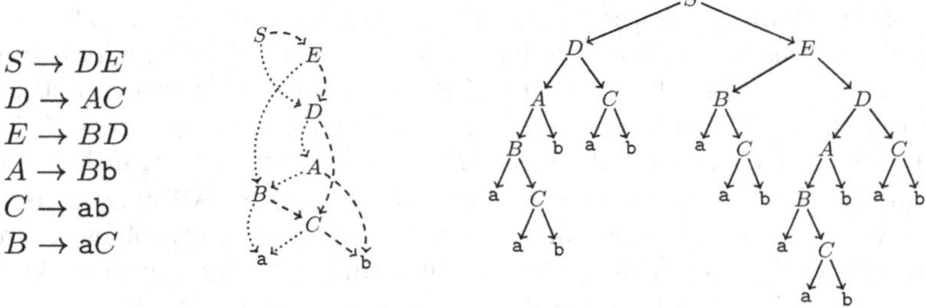

Above in the middle, it is demonstrated that we can interpret each SLP as a DAG in which each rule $A \to BC$ is represented by a node A with *left successor* B (indicated by dotted arrows) and *right successor* C (indicated by dashed arrows). Note that this is due to the fact that we assume the Chomsky normal form (without the Chomsky normal form, the outdegree of the nodes would be larger and we would need some way of expressing an order on the outgoing edges). The only possible derivation in this SLP (i.e., starting with S and then just applying the rules until we have a string over $\{\mathtt{a}, \mathtt{b}\}$) yields the compressed string `aabbabaabaabbab`. As is typical for derivations in context-free grammars, we can also consider the derivation tree, which is displayed above on the right (note, however, that the derivation tree is not a compressed representation, since it is at least as large as the string).

For the size of an SLP, we can take its number of rules (obviously, this ignores a factor of 3, but that's okay if we measure asymptotically). In order to see how the compression works, take a look at the derivation tree (recall that the derivation tree is an uncompressed representation): Here, we have to explicitly spell out each application of a rule in the construction of the string, while the actual SLP representation mentions each rule exactly once. Intuitively speaking, an SLP just tells us how to replace repeating substrings by variables in a clever way (and this is hierarchical in the sense that substrings already containing variables are again replaced by variables and so on). Obviously, this should be called clever only if it achieves a decent compression.

An SLP might be much smaller than the string it represents; in fact, even exponentially smaller (to see this, just consider an SLP which just doubles a string in every rule). There are also strings that are not well compressible, but experimental analyses have shown that the compression achieved by SLPs on natural inputs is very good. On an intuitive level, this is not surprising: Texts in natural language have many repeating words, and the same syllables occur over and over again in different words. But also for any artificial string over a finite alphabet (e.g., the alphabet $\{A, G, C, T\}$ of DNA nucleobases), we will get repeating substrings if the string is long enough.

The success of SLP-compression for strings is due to the fact that SLPs achieve good compression (already mentioned above), but also that there are very fast approximation algorithms that achieve these good compression ratios.[8] And, most importantly, there are many algorithms capable of solving basic, but important string problems directly on compressed strings (see [18] for an overview). As an example, let us recall that pattern matching (i.e., finding a given string P as substring in another string T) can be solved in time $O(|P| + |T|)$ by classical algorithms (the well-known Knuth-Morris-Pratt algorithm for example). But if T is given by an SLP S, so in SLP-compressed form, we can still solve pattern matching in linear time, but without having to decompress S, i.e., in time $O(|P| + |S|)$ (note that if $|S| \ll |T|$, then this directly translates into a faster running time); this has recently been shown in [16]. Although this is not always trivial to show, it often turns out that certain string problems can still be solved efficiently, even if we get the input string in form of an SLP. And it is the same for *regular* spanner evaluation (investigated in [20,27,29]).

For spanners (recall that we are in a data management setting), we assume that we have a whole database of documents, which is just a set of documents, e.g., $\mathcal{D} = \{\mathbf{D}_1, \mathbf{D}_2, \mathbf{D}_3\} = \{\texttt{ababbcabca}, \texttt{bcabcaabbca}, \texttt{ababbca}\}$). Such document databases can also be compressed by an SLP, for example, this one:

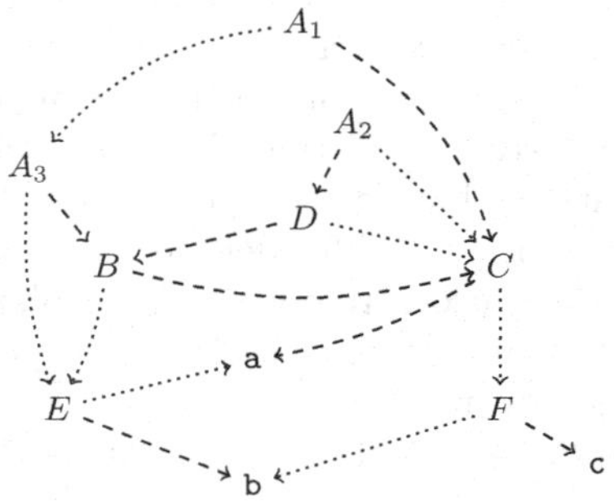

[8] Full disclosure: Computing a minimal SLP for a string is NP-complete [5]. But this is no problem at all, due to these fast practical approximate compressors.

Recall that dotted arrows point to the left successors and dashed arrows point to the right successors. Here, the non-terminals A_1, A_2 and A_3 derive exactly the documents $\mathbf{D}_1$, $\mathbf{D}_2$ and $\mathbf{D}_3$ of the example document database $\mathcal{D}$ mentioned above (the reader is welcome to verify this, although it is a bit painful).

We now get some regular spanner S and some index $i \in \{1, 2, \ldots, |\mathcal{D}|\}$ (indicating the document that we want to query), and after a preprocessing linear in the size of $\mathcal{D}$, we want to be able to enumerate the elements of $S(\mathbf{D}_i)$. This is in fact possible, but with a delay of $\log(|\mathbf{D}_i|)$. The delay is therefore not constant and depends on the size of the *uncompressed* data, but only logarithmically. An important fact is that this delay is always at most logarithmic in $|\mathbf{D}_i|$ independent of the actual compression achieved by the SLP. In order to understand where this log-factor comes from, let us sketch the general approach of this algorithm (see [27] for details).

In the uncompressed setting, the enumeration of regular spanners relies on enumerating all subword-marked versions of the document that are accepted by the automaton that represents the spanner. So in the compressed setting, we have to enumerate all subword-marked versions of the derivation tree for $\mathbf{D}_i$ that describe subword-marked documents accepted by the automaton. A subword-marked version of $\mathbf{D}_i$'s derivation tree is obtained by placing the symbols from $\Gamma_{\mathcal{X}}$ at their corresponding places into the derivation tree of $\mathbf{D}_i$. Of course, we cannot just naively construct $\mathbf{D}_i$'s derivation tree, because it is too big, but it is enough to sufficiently expand only those branches of the derivation tree that reveal positions of $\mathbf{D}_i$ where symbols from $\Gamma_{\mathcal{X}}$ have to be placed. This means that we have to expand only $O(|\mathcal{X}|)$ branches of the derivation tree (recall that, due to the data-complexity perspective, $|\mathcal{X}|$ is a constant for us), and we do not have to expand branches completely, just a long path leading to the desired position of the document. However, each of these paths may lead to a position that is buried deep, deep down in $\mathbf{D}_i$'s derivation tree. So these branches leading to the positions that must be marked with symbols from $\Gamma_{\mathcal{X}}$ may be long. To clarify this, take a look again at our first example of an SLP from above (Page 13). The fourth letter $\mathtt{b}$ of the compressed document can be reached from S by a path of length 3, i.e., the path $S, D, A, \mathtt{b}$, while the 11th letter $\mathtt{a}$ needs a path of length 6. Hence, the cost of producing the next subword-marked variant of the (still partially compressed) derivation tree, can be rather high, when we have to mark symbols deep down in the derivation tree. In fact, the length of such a path can be $\Omega(|\mathbf{D}_i|)$, which is bad.

We solve this, by first *balancing* our SLP, where an SLP is balanced if the longest path from a non-terminal to a leaf is bounded logarithmically in the size of the string derived by that non-terminal. In particular, this means that all paths of $\mathbf{D}_i$'s derivation tree would be bounded by $\log(|\mathbf{D}_i|)$. So if the SLP is balanced, then constructing these subword-marked and partially decompressed variants of $\mathbf{D}_i$'s derivation tree can be done in time $\log(|\mathbf{D}_i|)$ which explains our logarithmic delay. But what if the SLP is not balanced? In this case, we can just balance it, which is possible in linear time (see [17]).

There is also a different approach to regular spanner evaluation over SLP-compressed documents that improves the logarithmic delay to constant delay (see [20]).

5.1 Updates

In addition to the enumeration perspective, another practically motivated perspective of query evaluation problems is the so-called *dynamic case*. This is based on the observation that in practice we usually query a fixed database that is updated from time to time. Hence, the same queries are evaluated over just slightly different versions of the same database (i.e., a few tuples are added, a few tuples are deleted, but it is relatively safe to assume that the database that we query today is quite similar to the database that we query tomorrow). Consequently, it would be nice if for a query q and a database D, it is enough to run the preprocessing for q and D only once (this preprocessing provides us with the necessary data structures to evaluate q on D efficiently), and whenever we update the database, we also update directly what we have computed in the preprocessing without repeating the complete preprocessing. Obviously, since an update changes just a tiny bit of our data, the work to be done after an update should be much less in comparison to re-running the complete preprocessing.

Document spanner evaluation in the SLP-compressed setting is particularly well-suited for this dynamic setting. Let us sketch why this is the case (see [29] for details).

In an uncompressed setting, we perform an update to the database (like adding or deleting some data element), which is easy, and then we have to take care of how to maintain the preprocessing data structures under the updates. In a compressed scenario, on the other hand, we also have to update the compressed representation of our data. And, needless to say, we do not want to completely decompress the data, make the update and then compress it again. But, fortunately, an SLP that represents a document database is somehow suitable for such updates. Take a look at the example SLP from above for the document database $\mathcal{D} = \{\mathbf{D}_1, \mathbf{D}_2, \mathbf{D}_3\} = \{\mathtt{ababbcabca}, \mathtt{bcabcaabbca}, \mathtt{ababbca}\}$ (actually, take a look below, where we repeat this SLP, but now with some updates).

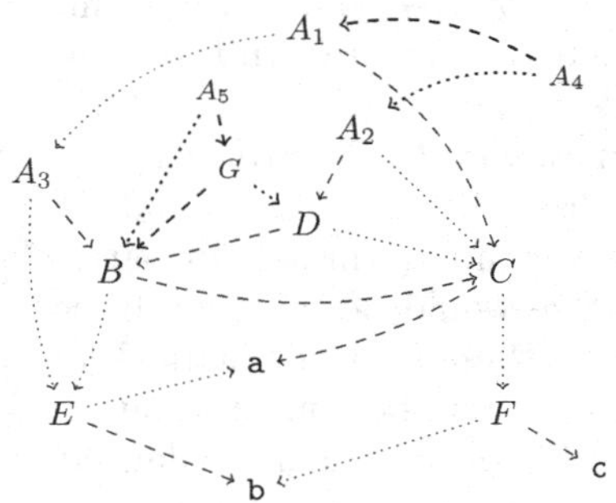

In this SLP, non-terminals A_1 and A_2 represent document $\mathbf{D}_1$ and $\mathbf{D}_2$. So by adding a new non-terminal A_4 with rule $A_4 \rightarrow A_2A_1$ (as shown above), we automatically add the new document $\mathbf{D}_2\mathbf{D}_1 =$ `bcabcaabbcaababbcabca` to our document database. Adding the non-terminals A_5 and G with rules $A_5 \rightarrow BG$ and $G \rightarrow DB$ adds the more complicated document $\mathbf{D}_B\mathbf{D}_D\mathbf{D}_B =$ `abbcabcaabbcaabbca`, where $\mathbf{D}_B =$ `abbca` is the document represented by non-terminal B and $\mathbf{D}_D =$ `bcaabbca` is the document represented by non-terminal D. So as long as we want to add documents that can be pieced together from strings already represented by non-terminals, this is simply done by adding a few more non-terminals and rules. By slightly more complicated operations, we can also cut some already present document into two parts (meaning that we construct non-terminals that derive exactly the left and right part), and with such an operation, we can realise operations like copying factors and inserting them somewhere else etc. In summary, updates that consist in adding new documents that can be created from existing documents by a sequence of copy-and-paste-like operations (note that this is how we usually work with text documents) can be easily done for SLP-compressed document databases. But how much effort is this really?

On close inspection, we can see that for implementing any of these copy-and-paste-like operations, it is sufficient to manipulate a constant number of paths in the SLP. But how long is a path? Well, if the SLP is balanced, it is not so long, i.e., logarithmic in the (uncompressed) size of the represented document database. So updates can be performed in time logarithmic in the size of the uncompressed data (multiplied by the number of individual copy-and-paste-operations, should our desired update require several of those). The only catch is that our updates may destroy the balancedness property, so we should implement them in such a way that balancedness is maintained. This is possible, at least if we require a stronger balancedness property. The weaker balancedness property that we used for the enumeration algorithm only requires that every path that starts in a non-terminal A is logarithmically bounded in the size of the string represented by A, the stronger variant requires for every rule $A \rightarrow BC$ that the longest path from B and the longest path from C differ by at most one. The advantage of the weaker property is that we can transform any SLP in an equivalent weakly balanced one of asymptotically the same size in linear time, whereas the stronger property requires time $m \log(n)$, where m is the size of the SLP and n is the (uncompressed) size of the represented document database. But the stronger property can be maintained by our updates, so that we can guarantee that all our updates only cost time logarithmic in the size of the uncompressed data.

Nice, but this only updates the SLP. However, the data structures that we have computed in the preprocessing and that we need for enumerating the regular document spanner are basically just certain information for every non-terminal of the SLP, and we can easily update those while performing our updates on the SLP. Hence, as long as we have a strongly balanced SLP (which we can always get by paying with a logarithmic factor), we can make a preprocessing linear in the size of the compressed data (which, potentially, is logarithmic in the actual data), then enumerate a regular spanner with delay that is guaranteed logarithmic in the data, then perform updates in time logarithmic in the data, then again enumerate the spanner with delay logarithmic in the data (but without re-running the preprocessing) and so forth. We only have to spend again preprocessing time linear in the compressed data if we want to evaluate a completely new query. Thus, we could also initially preprocess a finite set of document spanners (this costs time linear in the compressed data multiplied by the number of spanners) and then we can always enumerate any of these spanners with a delay logarithmic in the (current) data, which we can manipulate with updates that cost us time logarithmic in the (current) data.

References

1. Amarilli, A., Bourhis, P., Mengel, S., Niewerth, M.: Constant-delay enumeration for nondeterministic document spanners. SIGMOD Rec. **49**(1), 25–32 (2020). https://doi.org/10.1145/3422648.3422655
2. Amarilli, A., Bourhis, P., Mengel, S., Niewerth, M.: Constant-delay enumeration for nondeterministic document spanners. ACM Trans. Database Syst. **46**(1), 2:1–2:30 (2021). https://doi.org/10.1145/3436487
3. Amarilli, A., Jachiet, L., Muñoz, M., Riveros, C.: Efficient enumeration for annotated grammars. In: PODS '22: International Conference on Management of Data, Philadelphia, PA, USA, 12–17 June 2022, pp. 291–300 (2022). https://doi.org/10.1145/3517804.3526232
4. Bourhis, P., Grez, A., Jachiet, L., Riveros, C.: Ranked enumeration of MSO logic on words. In: 24th International Conference on Database Theory, ICDT 2021, 23–26 March 2021, Nicosia, Cyprus, pp. 20:1–20:19 (2021). https://doi.org/10.4230/LIPICS.ICDT.2021.20
5. Casel, K., Fernau, H., Gaspers, S., Gras, B., Schmid, M.L.: On the complexity of the smallest grammar problem over fixed alphabets. Theory Comput. Syst. **65**(2), 344–409 (2021). https://doi.org/10.1007/S00224-020-10013-W
6. Doleschal, J., Bratman, N., Kimelfeld, B., Martens, W.: The complexity of aggregates over extractions by regular expressions. In: 24th International Conference on Database Theory, ICDT 2021, 23–26 March 2021, Nicosia, Cyprus, pp. 10:1–10:20 (2021). https://doi.org/10.4230/LIPICS.ICDT.2021.10
7. Doleschal, J., Kimelfeld, B., Martens, W., Nahshon, Y., Neven, F.: Split-correctness in information extraction. In: Proceedings of the 38th ACM SIGMOD-SIGACT-SIGAI Symposium on Principles of Database Systems, PODS 2019, Amsterdam, The Netherlands, June 30–5 July 2019, pp. 149–163 (2019). https://doi.org/10.1145/3294052.3319684
8. Doleschal, J., Kimelfeld, B., Martens, W., Peterfreund, L.: Weight annotation in information extraction. In: 23rd International Conference on Database Theory, ICDT 2020, March 30–2 April 2020, Copenhagen, Denmark, pp. 8:1–8:18 (2020). https://doi.org/10.4230/LIPIcs.ICDT.2020.8
9. Fagin, R., Kimelfeld, B., Reiss, F., Vansummeren, S.: Document spanners: a formal approach to information extraction. J. ACM **62**(2), 12:1–12:51 (2015)
10. Florenzano, F., Riveros, C., Ugarte, M., Vansummeren, S., Vrgoc, D.: Efficient enumeration algorithms for regular document spanners. ACM Trans. Database Syst. **45**(1), 3:1–3:42 (2020). https://doi.org/10.1145/3351451

11. Freydenberger, D.: A logic for document spanners. Theory Comput. Syst. **63**(7), 1679–1754 (2019). https://doi.org/10.1007/s00224-018-9874-1
12. Freydenberger, D., Holldack, M.: Document spanners: from expressive power to decision problems. Theory Comput. Syst. **62**(4), 854–898 (2018)
13. Freydenberger, D.D., Kimelfeld, B., Peterfreund, L.: Joining extractions of regular expressions. In: Proceedings of the 37th ACM SIGMOD-SIGACT-SIGAI Symposium on Principles of Database Systems, Houston, TX, USA, 10–15 June 2018, pp. 137–149 (2018). https://doi.org/10.1145/3196959.3196967
14. Freydenberger, D.D., Thompson, S.M.: Dynamic complexity of document spanners. In: 23rd International Conference on Database Theory, ICDT 2020, March 30–2 April 2020, Copenhagen, Denmark, pp. 11:1–11:21 (2020). https://doi.org/10.4230/LIPIcs.ICDT.2020.11
15. Freydenberger, D.D., Thompson, S.M.: Splitting spanner atoms: a tool for acyclic core spanners. In: 25th International Conference on Database Theory, ICDT 2022, March 29 to 1 April 2022, Edinburgh, UK (Virtual Conference), pp. 10:1–10:18 (2022). https://doi.org/10.4230/LIPIcs.ICDT.2022.10
16. Ganardi, M., Gawrychowski, P.: Pattern matching on grammar-compressed strings in linear time. In: Proceedings of the 2022 ACM-SIAM Symposium on Discrete Algorithms, SODA 2022, Virtual Conference/Alexandria, VA, USA, 9–12 January 2022, pp. 2833–2846 (2022). https://doi.org/10.1137/1.9781611977073.110
17. Ganardi, M., Jez, A., Lohrey, M.: Balancing straight-line programs. J. ACM **68**(4), 27:1–27:40 (2021). https://doi.org/10.1145/3457389
18. Lohrey, M.: Algorithmics on SLP-compressed strings: a survey. Groups Complex. Cryptol. **4**(2), 241–299 (2012). https://doi.org/10.1515/gcc-2012-0016
19. Maturana, F., Riveros, C., Vrgoc, D.: Document spanners for extracting incomplete information: expressiveness and complexity. In: Proceedings of the 37th ACM SIGMOD-SIGACT-SIGAI Symposium on Principles of Database Systems, Houston, TX, USA, 10–15 June 2018, pp. 125–136 (2018). https://doi.org/10.1145/3196959.3196968
20. Muñoz, M., Riveros, C.: Constant-delay enumeration for SLP-compressed documents. In: 26th International Conference on Database Theory, ICDT 2023, 28–31 March 2023, Ioannina, Greece, pp. 7:1–7:17 (2023). https://doi.org/10.4230/LIPICS.ICDT.2023.7
21. Peterfreund, L.: The Complexity of Relational Queries over Extractions from Text. Ph.D. thesis (2019)
22. Peterfreund, L.: Grammars for document spanners. In: 24th International Conference on Database Theory, ICDT 2021, 23–26 March 2021, Nicosia, Cyprus, pp. 7:1–7:18 (2021). https://doi.org/10.4230/LIPIcs.ICDT.2021.7
23. Peterfreund, L., ten Cate, B., Fagin, R., Kimelfeld, B.: Recursive programs for document spanners. In: 22nd International Conference on Database Theory, ICDT 2019, 26–28 March 2019, Lisbon, Portugal, pp. 13:1–13:18 (2019)
24. Peterfreund, L., Freydenberger, D.D., Kimelfeld, B., Kröll, M.: Complexity bounds for relational algebra over document spanners. In: Proceedings of the 38th ACM SIGMOD-SIGACT-SIGAI Symposium on Principles of Database Systems, PODS 2019, Amsterdam, The Netherlands, June 30–5 July 2019, pp. 320–334 (2019)
25. Schmid, M.L.: Characterising REGEX languages by regular languages equipped with factor-referencing. Inf. Comput. (I&C) **249**, 1–17 (2016)
26. Schmid, M.L., Schweikardt, N.: A purely regular approach to non-regular core spanners. In: 24th International Conference on Database Theory, ICDT 2021, 23–26 March 2021, Nicosia, Cyprus, pp. 4:1–4:19 (2021). https://doi.org/10.4230/LIPIcs.ICDT.2021.4

27. Schmid, M.L., Schweikardt, N.: Spanner evaluation over SLP-compressed documents. In: PODS'21: Proceedings of the 40th ACM SIGMOD-SIGACT-SIGAI Symposium on Principles of Database Systems, Virtual Event, China, 20–25 June 2021, pp. 153–165 (2021). https://doi.org/10.1145/3452021.3458325
28. Schmid, M.L., Schweikardt, N.: Document spanners - a brief overview of concepts, results, and recent developments. In: PODS '22: International Conference on Management of Data, Philadelphia, PA, USA, 12–17 June 2022, pp. 139–150 (2022). https://doi.org/10.1145/3517804.3526069
29. Schmid, M.L., Schweikardt, N.: Query evaluation over SLP-represented document databases with complex document editing. In: PODS '22: International Conference on Management of Data, Philadelphia, PA, USA, 12–17 June 2022, pp. 79–89 (2022). https://doi.org/10.1145/3517804.3524158
30. Schweikardt, N., Segoufin, L., Vigny, A.: Enumeration for FO queries over nowhere dense graphs. J. ACM **69**(3), 22:1–22:37 (2022). https://doi.org/10.1145/3517035
31. Segoufin, L.: A glimpse on constant delay enumeration (invited talk). In: 31st International Symposium on Theoretical Aspects of Computer Science (STACS 2014), STACS 2014, 5–8 March 2014, Lyon, France, pp. 13–27 (2014). https://doi.org/10.4230/LIPICS.STACS.2014.13
32. Segoufin, L.: Constant delay enumeration for conjunctive queries. SIGMOD Rec. **44**(1), 10–17 (2015)

Contributed Papers

Generalized Distance Polymatrix Games

Alessandro Aloisio[1(✉)], Michele Flammini[2], and Cosimo Vinci[3]

[1] Universitá degli Studi Internazionali di Roma, Rome, Italy
alessandro.aloisio@unint.it
[2] Gran Sasso Science Institute, L'Aquila, Italy
michele.flammini@gssi.it
[3] Universitá del Salento, Lecce, Italy
cosimo.vinci@unisalento.it

Abstract. We consider a generalization of the distance polymatrix coordination games to hypergraphs. The classic *polymatrix coordination games* and the successive *distance polymatrix coordination games* are usually modelled by means of undirected graphs, where nodes represent agents, and edges stand for binary games played by the agents at their extremes. The utility of an agent depends at different scales on the outcome of a suitably defined subset of all binary games, plus the preference she has for her action.

We propose the new class of *generalized distance polymatrix games*, properly generalizing distance polymatrix coordination games, in which each subgame can be played by more than two agents. They can be suitably modelled by means of hypergraphs, where each hyperedge represents a subgame played by its agents. Moreover, as for distance polymatrix coordination games, the overall utility of a player x also depends on the payoffs of the subgames where the involved players are far, at most, a given distance from x. As for the original model, we discount these payoffs proportionally by factors depending on the distance of the related hyperedges.

After formalizing and motivating our model, we first investigate the existence of exact and approximate strong equilibria. Then we study the degradation of the social welfare by resorting to the standard measures of Price of Anarchy and Price of Stability, both for general and bounded-degree graphs.

Keywords: Polymatrix Games · Price of Anarchy · Price of Stability

1 Introduction

Polymatrix games [25] are well-known games that have been deeply investigated for decades. They are multi-player games and belong to the class of *graphical games* [26] since it is possible to represent player interactions using an interaction graph. In this graph, nodes correspond to players, while edges correspond to bimatrix games played by the agents at the extremes.

This work is partially supported by GNCS-INdAM and European Union, PON Ricerca e Innovazione 2014-20 TEBAKA - Fondo Sociale Europeo 2014-20.

H. Fernau et al. (Eds.): SOFSEM 2024, LNCS 14519, pp. 25–39, 2024.
https://doi.org/10.1007/978-3-031-52113-3_2

Each player chooses a pure strategy from a finite set, which she will play in all the binary games she is involved in. In the subclass of *polymatrix coordination games* [30], the interaction graph is undirected since the outcome of a binary game is the same for both players. An extension of this classic model is presented in [2], where the utility of an agent x depends not only on the games (edges) in which x is involved but also on the games (edges) played by agents that are at most a distance of d away from x.

In this paper, we present and study a new, more general model, *generalized distance polymatrix games*, where each local game can concern more than two players, and the utility of an agent x can also depend on the games at a distance bounded by d. In this new model, the interaction graph becomes an undirected hypergraph, where every hyperedge corresponds to a game played by the players it contains. Following the idea proposed in [2], the utility of an agent x is the sum of the outcomes of the games she plays plus a fraction of the outcomes of the games played by other players at a distance at most d from x. Each agent x also gets an additional payoff that is a function of her chosen strategy. Our new model is interesting both from a theoretical and a practical point of view since it is able to represent scenarios that previous models did not cover.

On the one hand, extending a local game to more than two players is reasonable because, in many natural social environments (e.g., economics, politics, sports, academia, etc.), people get a payoff from activities that involve more than two players.

As an example, in a scientific community, a project or a paper often involves more than two researchers, and its outcome depends on the choice made by each person. This can be modelled by using a hyperedge for each project/paper, with a weight (payoff) depending on the participants' strategies.

On the other hand, any individual also gets a benefit, albeit to a smaller degree, when her close colleagues succeed in some project or publish a quality paper that she is not personally a part of. This is quite obvious when considering the student-advisor relationship, but also noticeable for the collective evaluation and reputation of the department or institution where the researchers are working, for instance, in terms of increased assignment of resources and positions. We can model these indirect relationships by introducing distances and discount factors.

Our framework can also be used to key out a local economy interaction. It is well known that the businesses of a small town or an area of a city strictly depend on each other. In many cases, the interaction is positive; if a business grows, it attracts customers and also positively influences the nearby ones. When small businesses are placed throughout an area, townspeople are likelier to shop around from one business to the next. An example is provided in Sect. 3.

Once formalized our new model, we provide some conditions for the existence of β-approximate k-strong Nash equilibria. Then we focus on the degradation of the social welfare when k players can simultaneously change their strategies. We analyze the Price of Anarchy and Stability for β-approximate k-strong Nash equilibria, determining tight lower and upper bounds.

1.1 Related Work

Polymatrix games were introduced more than forty years ago [25] and have since received considerable attention in the scientific literature, as they are a very general model that can be applied to many different real-world scenarios and can be used to derive several relevant games (e.g., *hedonic games* [17], *max-cut* [22]). Some seminal papers on the topic are [18,23,24,27], and more recent studies are [4,11,16,30], where the authors showed results mainly concerning equilibria and their computational issues.

Our model is related to polymatrix coordination games [4,30] and the more recent distance polymatrix coordination games [2], where the authors introduced the idea of distances. Polymatrix coordination games are, in turn, an extension of a previously introduced model that did not include individual preferences [12]. Some preliminary results can be found in [1].

Our studies are also related to *(symmetric) additively separable hedonic games* [17] and *hypergraph hedonic games* [3], where there are no preferences, and the weights of the interaction graph are independent of the players' strategies, except being null when concerning agents playing different strategies. Our model can also be seen as a generalization of the hypergraph hedonic games [3], introducing preferences and increasing the expressiveness of the weight function (also allowing weight related to different strategies to be non-null).

Pure Nash equilibria have been studied for *synchronization games* [31], which are a generalization of polymatrix coordination games to hypergraphs. However, they do not investigate the degradation of the social welfare, and they do not consider distances.

Another closely related model is the *group activity selection problem* [10,14,15], which is positioned somewhat between polymatrix coordination games and hedonic games.

Our model is also related to the *social context games* [5,8], where the players' utilities are computed from the payoffs based on the neighbourhood graph and an aggregation function. We take into account more than just the neighbourhood of an agent, we account for the player's preference only for her utility, and we extend payoffs of local games to more than two agents.

The idea of obtaining utility from non-neighbouring players has also been analyzed for *distance hedonic games* [21], a variant of hedonic games that are non-additively separable since payoffs also depend on the size of the coalitions. They generalize *fractional hedonic games* [6,9,13,19,28] similarly as distance polymatrix games and our model do with polymatrix games.

Some negative results for our problem can be inherited from additively separable hedonic games. For instance, computing a Nash stable outcome is PLS-complete [12], while the problems of finding an optimal solution and determining the existence of a core stable, strict core stable, Nash stable, or individually stable outcome are all NP-hard [7].

2 Our Contribution

After formalizing our new model, we analyse the existence of β-approximate k-strong equilibria and investigate the degradation of social welfare when a deviation from the current strategy profile can involve up to k agents. Consequently, we compute tight bounds on the resulting Price of Anarchy and Stability. To the best of our knowledge, there are no previous results of this kind in the literature that would apply to our model.

In particular, in Sect. 4, we analyze the existence of β-approximate k-strong equilibria. In Sect. 5, we provide tight bounds on the Price of Anarchy for general hypergraphs. In Sect. 6, we prove a suitable lower bound on the Price of Stability for general hypergraphs, which is asymptotically equivalent to the upper bound of the Price of Anarchy when $\beta = 1$, meaning that the inefficiency of 1-approximate k-strong equilibria is fully characterized. Finally, in Sect. 7, we give upper and lower bounds for bounded-degree hypergraphs, with the gap being reasonably small. Some of our results are summarized in Table 1. Due to space constraints, some proofs are only sketched or omitted, while all the details are deferred to the full version.

Table 1. Summary of some of our results, where UB and LB stand for the upper and lower bound, respectively. Furthermore, Δ and r denote the maximal vertex degree and the maximum arity in the bounded-degree case, respectively, and $\alpha_h, h \in [d]$, is the discounting factor for hyperedges at distance $h-1$. The arrows denote that a result follows from an adjacent result in the table. The question mark stands for an open problem.

	general	bounded-degree
PoA_k^{β}(LB)	$\beta\frac{(n-1)_{r-1}}{(k-1)_{r-1}}(r+\alpha_2(n-r))$	$\Omega(\beta(\Delta-1)^{d/2}(r-1)^{d/2})$
PoA_k^{β}(UB)	$\beta\frac{(n-1)_{r-1}}{(k-1)_{r-1}}(r+\alpha_2(n-2))$	$\beta r\sum_{h\in[d]}\alpha_h\Delta((\Delta-1)r)^{h-1}$
PoS_k^{β}(UB)	$\downarrow$	$\downarrow$
PoS_k^{1}(LB)	$\frac{n-r}{n-1}\frac{(n-1)_{r-1}}{(k-1)_{r-1}}\frac{(r+\alpha_2(n-r))}{2(1+\alpha_2)}$	?

3 Preliminaries

Given two integers $r \geq 1$ and $n \geq 1$, let $[n] = \{1, 2, \ldots, n\}$ and $(n)_r := n \cdot (n-1) \cdot \ldots \cdot (n-r+1)$ be the falling factorial. A *weighted hypergraph* is a triple $\mathcal{H} = (V, E, w)$ consisting of a finite set $V = [n]$ of *nodes*, a collection $E \subseteq 2^V$ of *hyperedges*, and a *weight* $w : E \to \mathbb{R}$ associating a real value $w(e)$ with each hyperedge $e \in E$. For simplicity, when referring to weighted hypergraphs, we omit the term weighted.

The *arity* of a hyperedge e is its size $|e|$. An r-*hypergraph* is a hypergraph such that the arity of each hyperedge is at most r, where $2 \leq r \leq n$. A *complete* r-*hypergraph* is a hypergraph (V, E, w) such that $E := \{U \subseteq V : |U| \leq r\}$. A *uniform* r-*hypergraph* is a hypergraph such that the arity of each hyperedge is r. An *undirected graph* is a uniform 2-hypergraph. A hypergraph is said to be Δ-regular if each of its nodes is contained in exactly Δ hyperedges. It is said to be *linear* if any two of its hyperedges share at most one node. A hypergraph is called a *hypertree* if it admits a host graph T such that is a tree. Given two distinct nodes u and v in a hypergraph $\mathcal{H}$, a $u-v$ *simple path* of length l in $\mathcal{H}$ is a sequence of distinct hyperedges $(e_1, \ldots, e_l)$ of $\mathcal{H}$, such that $u \in e_1, v \in e_l$, $e_i \cap e_{i+1} \neq \emptyset$, for every $i \in [l-1]$, and $e_i \cap e_j = \emptyset$ whenever $j > i+1$. The distance from u to v, $d(u,v)$, is the length of the shortest $u-v$ simple path in $\mathcal{H}$. A *cycle* in a hypergraph $\mathcal{H}$ is defined as a simple path $(e_1, \ldots, e_l)$, but the further condition $e_1 \cap e_l \neq \emptyset$ must hold (that is, the first and the last hyperedge of the path must intersect, while in a simple path they are disjoint). This definition of cycle is originally due to Berge, and it can be also stated as an alternating sequence of $v_1, e_1, v_2, \ldots, v_n, e_n$ of distinct vertices v_i and distinct hyperedges e_i so that each e_i contains both v_i and v_{i+1}. The *girth* of a hypergraph is the length of the shortest cycle it contains.

Generalized Distance Polymatrix Games. A *generalized distance polymatrix game* (or GDPG) $\mathcal{G} = (\mathcal{H}, (\Sigma_x)_{x\in V}, (w_e)_{e\in E}, (p_x)_{x\in V}, (\alpha_h)_{h\in[d]})$, is a game based on an r-hypergraph $\mathcal{H}$, and defined as follows:

Agents: The set of agents is $V = [n]$, i.e., each node corresponds to an agent. We reasonably assume that $n \geq r \geq 2$.

Strategy profile or outcome: For any $x \in V$, Σ_x is a finite set of *strategies* of player x. A *strategy profile or outcome* $\boldsymbol{\sigma} = (\sigma_1, \ldots, \sigma_n)$ is a configuration in which each player $x \in V$ plays strategy $\sigma_x \in \Sigma_x$.

Weight function: For any hyperedge $e \in E$, let $w_e : \times_{x\in e}\Sigma_x \rightarrow \mathbb{R}_{\geq 0}$ be the *weight function* that assigns, to each subset of strategies σ_e played respectively by every $x \in e$, a *weight* $w_e(\sigma_e) \geq 0$. In what follows, for the sake of brevity, given any strategy profile $\boldsymbol{\sigma}$, we will often denote $w_e(\sigma_e)$ simply as $w_e(\boldsymbol{\sigma})$.

Preference function: For any $x \in V$, let $p_x : \Sigma_x \rightarrow \mathbb{R}_{\geq 0}$ be the *player-preference function* that assigns, to each strategy σ_x played by player x, a non-negative real value $p_x(\sigma_x)$, called *player-preference*. In what follows, for the sake of brevity, given any strategy profile $\boldsymbol{\sigma}$, we will often denote $p_x(\sigma_x)$ simply as $p_x(\boldsymbol{\sigma})$.

Distance-factors sequence: Let $(\alpha_h)_{h\in[d]}$ be the *distance-factors sequence* of the game, that is a non-negative sequence of real parameters, called *distance-factors*, such that $1 = \alpha_1 \geq \alpha_2 \geq \ldots \geq \alpha_d \geq 0$.

Utility function: For any $h \in [d]$, let $E_h(x)$ be the set of hyperedges e such that the minimum distance between x and one of the players $v \in e$ is exactly $h - 1$. Then, for any $x \in V$, the *utility function* $u_x : \times_{x\in V}\Sigma_x \rightarrow \mathbb{R}$ of player x, for any strategy profile $\boldsymbol{\sigma}$ is defined as $u_x(\boldsymbol{\sigma}) := p_x(\boldsymbol{\sigma}) + \sum_{h\in[d]} \alpha_h \sum_{e\in E_h(x)} w_e(\boldsymbol{\sigma})$.

The *social welfare* $\mathsf{SW}(\boldsymbol{\sigma})$ of a strategy profile $\boldsymbol{\sigma}$ is defined as the sum of all the agents' utilities in $\boldsymbol{\sigma}$, i.e., $\mathsf{SW}(\boldsymbol{\sigma}) := \sum_{x\in V} u_x(\boldsymbol{\sigma})$. A *social optimum* of game $\mathcal{G}$ is a strategy profile $\boldsymbol{\sigma}^*$ that maximizes the social welfare. We denote by $\mathsf{OPT}(\mathcal{G}) = \mathsf{SW}(\boldsymbol{\sigma}^*)$ the corresponding value.

β-approximate k-strong Nash Equilibrium. Given two strategy profiles $\boldsymbol{\sigma} = (\sigma_1, \ldots, \sigma_n)$ and $\boldsymbol{\sigma}^* = (\sigma_1^*, \ldots, \sigma_n^*)$, and a subset $Z \subseteq V$, let $\boldsymbol{\sigma} \xrightarrow{Z} \boldsymbol{\sigma}^*$ be the strategy profile $\boldsymbol{\sigma}' = (\sigma_1', \ldots, \sigma_n')$ such that $\sigma_x' = \sigma_x^*$ if $x \in Z$, and $\sigma_x' = \sigma_x$ otherwise. Given $k \geq 1$, a strategy profile $\boldsymbol{\sigma}$ is a *β-approximate k-strong Nash equilibrium* (or (β, k)-equilibrium) of $\mathcal{G}$ if, for any strategy profile $\boldsymbol{\sigma}^*$ and any $Z \subseteq V$ such that $|Z| \leq k$, there exists $x \in Z$ such that $\beta u_x(\boldsymbol{\sigma}) \geq u_x(\boldsymbol{\sigma} \xrightarrow{Z} \boldsymbol{\sigma}^*)$. We say that a player x *β-improves* from a deviation $\boldsymbol{\sigma} \xrightarrow{Z} \boldsymbol{\sigma}^*$ if $\beta u_x(\boldsymbol{\sigma}) < u_x(\boldsymbol{\sigma}')$. Informally, $\boldsymbol{\sigma}$ is a (β, k)-equilibrium if, for any coalition of at most k players deviating, there exists at least one player in the coalition that does not β-improve her utility by deviating. We denote the (possibly empty) set of (β, k)-equilibria of $\mathcal{G}$ by $\mathsf{NE}_k^\beta(\mathcal{G})$. Clearly, if $\beta = 1$, $\mathsf{NE}_k^\beta(\mathcal{G})$ contains all the k-strong equilibria, and when $\beta = 1$ and $k = 1$, it contains the classic Nash equilibria.

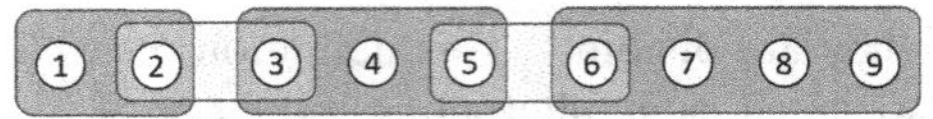

Fig. 1. Three shops in a shopping area.

(β, k)-Price of Anarchy (PoA) and (β, k)-Price of Stability (PoS). The *(β, k)-Price of Anarchy* of a game $\mathcal{G}$ is defined as $\mathsf{PoA}_k^\beta(\mathcal{G}) := \max_{\sigma \in \mathsf{NE}_k^\beta(\mathcal{G})} \frac{\mathsf{OPT}(\mathcal{G})}{\mathsf{SW}(\sigma)}$, i.e., it is the worst-case ratio between the optimal social welfare and the social welfare of a (β, k)-equilibrium. The *(β, k)-Price of Stability* of game $\mathcal{G}$ is defined as $\mathsf{PoS}_k^\beta(\mathcal{G}) := \min_{\sigma \in \mathsf{NE}_k^\beta(\mathcal{G})} \frac{\mathsf{OPT}(\mathcal{G})}{\mathsf{SW}(\sigma)}$, i.e., it is the best-case ratio between the optimal social welfare and the social welfare of a (β, k)-Nash equilibrium. Clearly, $\mathsf{PoS}_k^\beta(\mathcal{G}) \leq \mathsf{PoA}_k^\beta(\mathcal{G})$, whereas both quantities are not defined if $\mathsf{NE}_k^\beta(\mathcal{G}) = \emptyset$.

Example 1. We give here an example of GDPG applied to the local economy of a city's shopping area where the shops are one beside the other. Figure 1 schematizes three of the shops in the area, which are represented by three light blue hyperedges ($\{1, 2\}$, $\{3, 4, 5\}$, $\{6, 7, 8, 9\}$). They are positioned in the area like in Fig. 1, where the light grey hyperedges are just auxiliary hyperedges with null weights for every strategy profile. In this case, the distances are physical distances. Each node stands for the manager of a category of products sold. The manager's strategy is to choose a brand for her product category. A strategy profile $\boldsymbol{\sigma}$ corresponds to the brands the managers chose.

All the preferences are null while the weight $w_e(\boldsymbol{\sigma})$ is the number of customers visiting the shop e for a specific strategy profile $\boldsymbol{\sigma}$. It is reasonable that the number of customers $w_e(\boldsymbol{\sigma})$ strictly depends on the brand chosen by the agents.

Since the three shops are beside each other, if a person goes to $\{1, 2\}$, she will probably enter $\{3, 4, 5\}$ and enter $\{6, 7, 8, 9\}$ with less probability. This means that part of the people that visit $\{1, 2\}$ will stop at the other two shops, inversely proportional to the physical distance.

The utility of an agent is the number of received views. This number is strictly related to profit in the economy. We set $\alpha_2 = \alpha_3$ and $\alpha_4 = \alpha_5$ because of the auxiliary light grey hyperedges, which are not real shops. We can now compute the utilities of the agents. For example, agent 6 has $u_6(\boldsymbol{\sigma}) = w_{\{6,7,8,9\}}(\boldsymbol{\sigma}) + \alpha_3 \cdot w_{\{3,4,5\}}(\boldsymbol{\sigma}) + \alpha_5 \cdot w_{\{1,2\}}(\boldsymbol{\sigma})$, which equals the number of customers that shop $\{6, 7, 8, 9\}$ gets for $\sigma_{\{6,7,8,9\}}$ plus part (α_3) of the number of customers got by shop $\{3, 4, 5\}$, plus part (α_5) of the number of customers got by shop $\{1, 2\}$. Clearly, this example cannot be modelled using previous polymatrix models, i.e., without using hypergraphs and the distance factors sequence.

4 Existence of (β, k)-Equilibria

This section analyzes the existence of (β, k)-equilibria. First, we notice that (β, k)-equilibria may not exist since they cannot always exist even in polymatrix coordination games [4,30]. In the following (Theorem 1), we give some conditions on β that guarantee the existence. These results extend the ones shown in [4,30].

We say that a game $\mathcal{G}$ has a *finite* (β, k)*-improvement property* (or (β, k)-FIP for short) if every sequence of (β, k)-improving deviations is finite. In such a case, we necessarily have that any (β, k)-FIP ends in a (β, k)-equilibrium, which implies the latter's existence, too. To show that the (β, k)-FIP holds (and then the existence of a (β, k)-equilibrium), we resort to a potential function argument [29]. A function Φ that associates each strategy profile with a real number is called *potential function* if, for any strategy profile $\boldsymbol{\sigma}$ and (β, k)-improving deviation $\boldsymbol{\sigma}' = \boldsymbol{\sigma} \xrightarrow{Z} \boldsymbol{\sigma}^*$, we have that $\Phi(\boldsymbol{\sigma}') - \Phi(\boldsymbol{\sigma}) > 0$. Thus, since any (β, k)-improving deviation increases the potential function and the number of strategy profiles is finite, any sequence of (β, k)-improving deviations cannot cycle and must necessarily meet a (β, k)-equilibrium after a finite number of steps.

For a given hyperedge e and a subset $Z \subseteq V$, let $n_h^Z(e) := |\{x \in Z : e \in E_h(x)\}|$, i.e., $n_h^Z(e)$ is the number of players $x \in Z$ that are at distance $h - 1$ from e.

Theorem 1. *Let $\mathcal{G}$ be a GDPG. Then: i) $\mathcal{G}$ has the $(\beta, 1)$-FIP for every $\beta \geq 1$; ii) $\mathcal{G}$ has the (β, k)-FIP for every $\beta \geq \max_{Z \subseteq V: |Z|=k}\{\max_{e \in E}\{\sum_{h \in [d]} \alpha_h n_h^Z(e)\}\}$ and for every k.*

Proof (Proof sketch). To prove both i) and ii) we show that $\Phi(\boldsymbol{\sigma}) = \sum_{x \in V} p_x(\boldsymbol{\sigma}) + \sum_{e \in E} w_e(\boldsymbol{\sigma})$ is a potential function. Proof of i) is left to the full version.

Proof Sketch of ii). Consider a (β, k)-improving deviation $\boldsymbol{\sigma}' = \boldsymbol{\sigma} \xrightarrow{Z} \boldsymbol{\sigma}^*$. Let $\overline{Z} = V \setminus Z$. Let also $\mathsf{SW}_Z(\boldsymbol{\sigma})$ be the social welfare related to the deviating agents, that is $\mathsf{SW}_Z(\boldsymbol{\sigma}) = \sum_{x \in Z} u_x(\boldsymbol{\sigma})$. We can rewrite this social welfare as $\mathsf{SW}_Z(\boldsymbol{\sigma}) = \sum_{x \in Z} p_x(\boldsymbol{\sigma}) + \sum_{e \in E : e \not\subseteq \overline{Z}} a_e w_e(\boldsymbol{\sigma}) + \sum_{e \in E : e \subseteq \overline{Z}} a_e w_e(\boldsymbol{\sigma})$, where $a_e = \sum_{h \in [d]} \alpha_h n_h^Z(e)$. It follows that

$$\beta \sum_{x \in Z} p_x(\boldsymbol{\sigma}') + \sum_{e \in E : e \not\subseteq \overline{Z}} \beta w_e(\boldsymbol{\sigma}') + \sum_{e \in E : e \subseteq \overline{Z}} \beta a_e w_e(\boldsymbol{\sigma}') \tag{1}$$

$$\geq \mathsf{SW}_Z(\boldsymbol{\sigma}') \tag{2}$$

$$> \beta \cdot \mathsf{SW}_Z(\boldsymbol{\sigma}) \tag{3}$$

$$\geq \beta \sum_{x \in Z} p_x(\boldsymbol{\sigma}) + \sum_{e \in E : e \not\subseteq \overline{Z}} \beta w_e(\boldsymbol{\sigma}) + \sum_{e \in E : e \subseteq \overline{Z}} \beta a_e w_e(\boldsymbol{\sigma}) \tag{4}$$

where (2) is due to $\beta \geq 1$ and $\beta \geq a_e$ for any $e \in E$, (3) holds since $\boldsymbol{\sigma}'$ is a β-improving deviation; (4) is due to $a_e \geq 1$ for every $e \in E$ such that $e \not\subseteq \overline{Z}$. From (1) > (4) and $\sum_{e \subseteq \overline{Z}} w_e(\boldsymbol{\sigma}') = \sum_{e \subseteq \overline{Z}} w_e(\boldsymbol{\sigma})$, we can derive $\Phi(\boldsymbol{\sigma}') - \Phi(\boldsymbol{\sigma}) > 0$. □

The value $\sum_{h \in [d]} \alpha_h n_h^Z(e)$ strictly depends on d and $n_h^Z(e)$. When $d = 1$, we have $\sum_{h \in [d]} \alpha_h n_h^Z(e) = n_1^Z(e) \leq |e|$ for every $e \in E$ and $Z \subseteq V$, so we can assume $\beta \geq r$. When the hypergraph of a game is a hyperlist, we have $\sum_{h \in [d]} \alpha_h n_h^Z(e) \leq 2r \sum_{h \in [d]} \alpha_h$, for every $e \in E$, and $Z \subseteq V$. When the hypergraph of a game is a hyperthree of maximum degree Δ, we have $\sum_{h \in [d]} \alpha_h n_h^Z(e) \leq r \sum_{h \in [d]} \alpha_h r^{h-1} \Delta^{h-1}$, for every $e \in E$, and $Z \subseteq V$.

5 (β, k)-PoA of General Graphs

In this section, we provide tight upper and lower bounds for the (β, k)-Price of Anarchy when the hypergraph $\mathcal{H}$ of a game $\mathcal{G}$ is general, that is there is no particular assumption on it. Such bounds depend on β, k, the number of players n, the maximum arity r, and the value α_2 of the distance-factors sequence.

Theorem 2. *For any integers $\beta \geq 1$, $r \geq 2$, $k < r$, and $n \geq r$, there exists a simple GDPG $\mathcal{G}$ with n agents such that $\mathsf{PoA}_k^\beta(\mathcal{G}) = \infty$.*

Thus, in the rest of the paper, we will only take into account the estimation of the (β, k)-PoA for $k \geq r \geq 2$ since it is not possible to bound the (β, k)-PoA for $k < r$, not even for bounded-degree graphs and not even when $\Delta = 1$.

5.1 (β, k)-PoA: Upper Bound

We now provide three results that we will use to compute the upper bound of the (β, k)-Price of Anarchy (Theorem 3). The first result is an upper bound to the social welfare of any strategy profile.

Lemma 1. *For any strategy profile $\boldsymbol{\sigma}$, it holds that $\mathsf{SW}(\boldsymbol{\sigma}) \leq \sum_{x \in V} p_x(\boldsymbol{\sigma}) + (r + \alpha_2 \cdot (n-2)) \cdot \sum_{e \in E} w_e(\boldsymbol{\sigma})$.*

Before providing the other two preliminary results, we write Inequality (5), which is a necessary condition for an outcome $\boldsymbol{\sigma}$ to be a (β, k)-equilibrium. For a fixed integer $k \geq r$, let $\boldsymbol{\sigma}$ and $\boldsymbol{\sigma}^*$ be a (β, k)-equilibrium and a social optimum of $\mathcal{G}$ respectively. Since $\boldsymbol{\sigma}$ is a (β, k)-equilibrium, for every subset $Z \subseteq V$ of at most k agents, there exists an agent $z_1(Z) \in Z$ such that $\beta u_{z_1(Z)}(\boldsymbol{\sigma}) \geq u_{z_1(Z)}(\boldsymbol{\sigma} \xrightarrow{Z} \boldsymbol{\sigma}^*)$. Moreover, let $Z(2) := Z \setminus \{z_1\}$, there exists another agent $z_2(Z) \in Z(2)$ such that $\beta u_{z_2(Z)}(\boldsymbol{\sigma}) \geq u_{z_2(Z)}(\boldsymbol{\sigma} \xrightarrow{Z(2)} \boldsymbol{\sigma}^*)$. We can iterate this process for every $z_i(Z) \in Z(i) := Z \setminus \{z_1(Z), \ldots, z_{i-1}(Z)\}$, obtaining the following inequality.

$$\beta u_{z_i(Z)}(\boldsymbol{\sigma}) \geq u_{z_i(Z)}(\boldsymbol{\sigma} \xrightarrow{Z(i)} \boldsymbol{\sigma}^*) \tag{5}$$

By summing the previous inequality's left and right parts for every possible subset of Z of k players, we derive the following two results needed for Theorem (3).

Lemma 2. *For every (β, k)-equilibrium $\boldsymbol{\sigma}$ and every subset $K \subseteq V$, with $|K| = k$, it holds that $\beta \cdot \binom{n-1}{k-1} \cdot \mathsf{SW}(\boldsymbol{\sigma}) = \sum_{\substack{Z \subseteq V \\ |Z|=k}} \sum_{i \in [k]} \beta \cdot u_{z_i(Z)}(\boldsymbol{\sigma})$.*

Proof (Proof sketch). By summing $\beta u_{Z(i)}(\boldsymbol{\sigma})$ for every $Z \subseteq V$ of cardinality k and for every $i \in [k]$, it is easy to see that the utility of an agent $x \in V$ is counted exactly $\binom{n-1}{k-1}$ times, which is the number of subsets K in V containing x. □

Lemma 3. *For every (β, k)-equilibrium $\boldsymbol{\sigma}$ and every subset $K \subseteq V$, with $|K| = k$, it holds that $\sum_{\substack{Z \subseteq V \\ |Z|=k}} \sum_{i \in [k]} u_{z_i(Z)}(\boldsymbol{\sigma} \xrightarrow{Z(i)} \boldsymbol{\sigma}^*) \geq \binom{n-r}{k-r} \left(\sum_{x \in V} p_x(\boldsymbol{\sigma}^*) + \sum_{e \in E} w_e (\boldsymbol{\sigma}^*)\right)$*

Proof (Proof sketch). For every subset $K \subseteq V$, with $|K| = k$, it holds that $\sum_{i\in[k]} u_{z_i(Z)}(\boldsymbol{\sigma} \overset{Z(i)}{\rightarrow} \boldsymbol{\sigma}^*) \geq \sum_{x\in Z} p_x(\boldsymbol{\sigma}^*) + \sum_{e\subseteq Z} w_e(\boldsymbol{\sigma}^*)$. To establish this, we discarded the weights of all the hyperedges, far at least one from $z_i(Z)$, and used the fact that every hyperedge in Z is counted exactly once. By summing the previous inequality for every subset Z, we obtain $\sum_{\substack{Z\subseteq V\\|Z|=k}} \sum_{i\in[k]} u_{z_i(Z)}(\boldsymbol{\sigma} \overset{Z(i)}{\rightarrow} \boldsymbol{\sigma}^*) \geq \binom{n-1}{k-1}\sum_{x\in V} p_x(\boldsymbol{\sigma}^*) + \binom{n-r}{k-r}\sum_{e\in E} w_e(\boldsymbol{\sigma}^*)$, thus showing the claim. □

Finally, we can state the theorem for the upper bound of the (β, k)-Price of Anarchy.

Theorem 3. *For any $\beta \geq 1$, any integer $k \geq r$ and any GDPG $\mathcal{G}$ having a distance-factors sequence $(\alpha_h)_{h\in[d]}$, it holds that $\mathsf{PoA}_k^\beta(\mathcal{G}) \leq \beta\frac{(n-1)_{r-1}}{(k-1)_{r-1}}(r + \alpha_2(n-2))$.*

Proof (Proof sketch). By using the results given in Lemmas 1, 2, and 3 we obtain

$$\beta \cdot \binom{n-1}{k-1} \cdot \mathsf{SW}(\boldsymbol{\sigma}) = \sum_{Z\subseteq V:|Z|=k} \sum_{i\in[k]} \beta \cdot u_{z_i(Z)}(\boldsymbol{\sigma}) \tag{6}$$

$$\geq \sum_{Z\subseteq V:|Z|=k} \sum_{i\in[k]} u_{z_i(Z)}(\boldsymbol{\sigma} \overset{Z(i)}{\rightarrow} \boldsymbol{\sigma}^*) \tag{7}$$

$$\geq \binom{n-r}{k-r}\left(\sum_{x\in V} p_x(\boldsymbol{\sigma}^*) + \sum_{e\in E} w_e(\boldsymbol{\sigma}^*)\right) \tag{8}$$

$$\geq \binom{n-r}{k-r} \cdot (r + \alpha_2 \cdot (n-2))^{-1} \cdot \mathsf{SW}(\boldsymbol{\sigma}^*) \tag{9}$$

where (6), (7), (8), and (9) derive from Lemma 2, Eq. (5), Lemma 3, and Lemma 1, respectively. Concluding, from (6) and (9), we can get the upper bound for $\mathsf{PoA}_k^\beta(\mathcal{G})$. □

5.2 (β, k)-PoA: Lower Bound

We continue by showing the following tight lower bound.

Theorem 4. *For every $\beta \geq 1$, every integers $r \geq 2$, $k \geq r$, $d \geq 1$, $n \geq k$, and every d-distance-factors sequence $(\alpha_h)_{h\in[d]}$, there is a GDPG $\mathcal{G}$ with $\mathsf{PoA}_k^\beta(\mathcal{G}) \geq \beta\frac{(n-1)_{r-1}}{(k-1)_{r-1}}(r + \alpha_2(n-r))$.*

Proof (Proof sketch). The idea is to use a GDPG game instance $\mathcal{G}$ with n players where: (i) the underlying hypergraph $\mathcal{H}$ is a hyperstar in which all the players $x \geq 2$ are only connected to player 1; (ii) each hyperedge contains 1 and $r-1$ other players; (iii) each agent has only two strategies, $\{s, s^*\}$; (iii) $w_e(\boldsymbol{\sigma}) = \beta$ if every agent in e chooses s^* under outcome $\boldsymbol{\sigma}$, and $w_e(\boldsymbol{\sigma}) = 0$ otherwise; (iv) $p_1(\boldsymbol{\sigma}) = \binom{k-1}{r-1}$ if agent 1 chooses s under outcome $\boldsymbol{\sigma}$, and $p_1(\boldsymbol{\sigma}) = 0$ otherwise; and (v) $p_x(\boldsymbol{\sigma}) = 0$ for every $x \geq 2$ and outcome $\boldsymbol{\sigma}$. We call $\boldsymbol{\sigma}$ and $\boldsymbol{\sigma}^*$ the two outcomes where all the agents choose s and s^*, respectively. Since $\boldsymbol{\sigma}$ is a (β, k)-equilibrium, we use the ratio of the social welfare of $\boldsymbol{\sigma}^*$ and $\boldsymbol{\sigma}$ to get the result.

6 $(1, k)$-PoS of General Graphs

This section shows a lower bound for the $(1, k)$-Price of Stability asymptotically equal to the upper bound for the $(1, k)$-Price of Anarchy given in Theorem 3. This means that we can use this upper bound also for the $(1, k)$-Price of Stability and close our study for general hypergraphs for the case $\beta = 1$.

The basic idea is to start from the lower bound instance of Theorem 4, then transform it to a new instance with the property of having every outcome with social welfare different from the minimum unstable.

Theorem 5. *For any* $n \geq 6$, *there exists a GDPG* $\mathcal{G}$ *such that* $\mathsf{PoS}_k^1(\mathcal{G}) \geq \frac{n-r}{n-1} \frac{(n-1)_{r-1}}{(k-1)_{r-1}} \frac{(r+\alpha_2(n-r))}{2(1+\alpha_2)}$

Proof. Let $\mathcal{H} = (V, E, w)$ be the interaction hypergraph of $\mathcal{G}$, with $|V| = n$ and $|E| = 2\binom{n-2}{r-1} + 1$. Furthermore, let the set of hyperedges E be divided into $\{1, 2\}$, E^1, and E^2, where E^i, with $i \in \{1, 2\}$, has $\binom{n-2}{r-1}$ hyperedges of arity r, each containing node i and $r - 1$ nodes different from 1 and 2. Hypergraph $\mathcal{H}$ is a kind of hyperstar with two roots connected by an edge of arity 2. Each agent x has a set $\Sigma_x = \{1, 2, 3\}$ of three possible strategies. We call *bottom layer*, *medium layer*, and *top layer* the outcome in which every player plays strategy 3, 2, and 1, respectively.

Finally, all the non-null weights and preferences are defined as follows. For the bottom layer, $p_1(3) = p_2(3) = (1+2\epsilon) + \binom{k-1}{r-1}(1+\alpha_2)(1+\epsilon)$. For the medium layer, $w_{\{1,2\}}(2, 2) = w_{e \in E^1}(\boldsymbol{\sigma}) = w_{e \in E^2}(\boldsymbol{\sigma}) = (1 + \epsilon)$. For the top layer, $p_1(1) = p_2(1) = 1$, $w_{e \in E^1}(\boldsymbol{\sigma}) = w_{e \in E^2}(\boldsymbol{\sigma}) = 1 + \epsilon$. Non-null hyperedges between the layers are only $w_{\{1,2\}}(1, 2) = 2\epsilon$, $w_{e \in E^1}(\boldsymbol{\sigma}) = 1+\epsilon$, and $w_{e \in E^2}(\boldsymbol{\sigma}) = 1+\epsilon$, when some players play strategy 1 and all the others play strategy 2 in $\boldsymbol{\sigma}$. Please note that every hyperedge with some players in the bottom layer and all the others out of the bottom layer have a null weight. □

Lemma 4. *The bottom layer is a* $(1, k)$*-equilibrium with social welfare* $2(1 + 2\epsilon) + 2\binom{k-1}{r-1}(1 + \alpha_2)(1 + \epsilon)$.

Lemma 5. *All the* $(1, k)$*-equilibria have the same social welfare* $2(1+2\epsilon)+2\binom{n-1}{k-1}(1+\alpha_2)(1 + \epsilon)$.

Proof (Proof sketch). We only need to check the case where no one agent is in the bottom. In fact, any other outcome is either unstable or has a social welfare equal to the one given in Lemma 4. When all the players are out of the bottom, the utility of agents 1 and 2 can change only when one or both of them change layer. Now, if both 1 and 2 are in the top layer, they both prefer to go to the medium one because they get an extra ϵ each. Then, agent 1 goes back to the top layer, increasing her utility of an ϵ more. From this state, agent 2 goes back to the top layer. The last state is when agent 1 is in the medium layer and 2 is in the top layer. Then, 1 goes to the top layer. □

Lemma 6. $\mathsf{PoS}_k^1(\mathcal{G}) = \frac{n-r}{n-1} \frac{(n-1)_{r-1}}{(k-1)_{r-1}} \frac{(r+\alpha_2(n-r))}{2(1+\alpha_2)}$.

Proof (Proof sketch). We use the ratio between the social welfare of the medium and the bottom layers to get the lower bound. □

The proof of Theorem 5 is complete. □

7 (β, k)-PoA of Bounded-Degree Graphs

In this section, we analyze the (β, k)-Price of Anarchy for games whose hypergraphs have bounded-degree.[1] We also say that a game $\mathcal{G}$ is Δ-bounded degree if the degree of every node in the underlying hypergraph is at most Δ. Here, we will only focus on the cases where $k \geq r$, as observed in Theorem 2, and $\Delta \geq 2$, since the case when $\Delta = 1$ is encompassed by Sect. 5.

7.1 (β, k)-PoA: Upper Bound

As we did for general hypergraphs, we first show an upper bound on the social welfare of every outcome.

Lemma 7. *Given a Δ-bounded-degree GDPG $\mathcal{G}$, for every (β, k)-equilibrium $\boldsymbol{\sigma}$ it holds that* $\mathsf{SW}(\boldsymbol{\sigma}) \leq \sum_{x \in V} p_x(\boldsymbol{\sigma}) + r \sum_{h \in [d]} \alpha_h \cdot (\Delta - 1)^{h-1} r^{h-1} \cdot \sum_{e \in E} w_e(\boldsymbol{\sigma})$.

We can now state the main theorem on the upper bound.

Theorem 6. *For every Δ-bounded-degree GDPG $\mathcal{G}$, with distance-factor sequence $(\alpha_h)_{h \in [d]}$, and for every $k \geq r$, it holds that* $\mathsf{PoA}_k^{\beta}(\mathcal{G}) \leq \beta \cdot r \sum_{h \in [d]} \alpha_h \cdot \Delta \cdot (\Delta - 1)^{h-1} r^{h-1}$.

Proof (Proof sketch). First, we write some necessary conditions for every outcome $\boldsymbol{\sigma}$ to be a (β, k)-equilibrium. Since the maximum arity is at most equal to k, if $\boldsymbol{\sigma}$ is a (β, k)-equilibrium, then for every hyperedge e, there must exist a player $z_1(e) \in e$ such that (i): $\beta u_{z_1(e)}(\boldsymbol{\sigma}) \geq u_{z_1}(\boldsymbol{\sigma} \xrightarrow{e} \boldsymbol{\sigma}^*) \geq p_{z_1(e)}(\boldsymbol{\sigma}^*) + w_e(\boldsymbol{\sigma}^*)$. Moreover, since a (β, k)-equilibrium is also a $(\beta, 1)$-equilibrium, for every other $z_i(e) \in e$, with $z_i(e) \neq z_1(e)$, it must hold (ii): $\beta u_{z_i(e)}(\boldsymbol{\sigma}) \geq u_{z_i(e)}(\boldsymbol{\sigma} \xrightarrow{z_i(e)} \boldsymbol{\sigma}^*) \geq p_{z_i(e)}(\boldsymbol{\sigma}^*)$. By summing Eq. (i) plus all the inequalities (ii) for every hyperedge $e \in E$, and by using Lemma 7, we get

$$\beta \cdot \sum_{e \in E} \left(\sum_{z_i(e)} u_{z_i(e)}(\boldsymbol{\sigma}) \right) \geq \left(r \sum_{h \in [d]} \alpha_h \cdot (\Delta - 1)^{h-1} r^{h-1} \right)^{-1} \cdot \mathsf{SW}(\boldsymbol{\sigma}^*) \tag{10}$$

We notice now that it holds that $\sum_{e \in E} \left(\sum_{z_i(e)}, u_{z_i(e)}(\boldsymbol{\sigma}) \right) \leq \sum_{x \in V} \Delta \cdot u_x(\boldsymbol{\sigma}) = \Delta \cdot \mathsf{SW}(\boldsymbol{\sigma})$ because in the left-hand part, the utility of each player x is counted at most Δ times, which is the maximum number of edges containing x. By using both (10) and the last inequality, we obtain $\mathsf{SW}(\boldsymbol{\sigma}) \geq \beta^{-1} \Delta^{-1} \left(r \sum_{h \in [d]} \alpha_h \cdot (\Delta - 1)^{h-1} r^{h-1} \right)^{-1} \mathsf{SW}(\boldsymbol{\sigma}^*)$.

Remark 1. Please note that Theorem 6 implies that the (β, k)-price of anarchy of Δ-bounded-degree GDPG, as a function of d, grows at most as $O(\beta \cdot (\Delta - 1)^d \cdot r^d)$.

---

[1] A hypergraph $\mathcal{H}$ has degree bounded by Δ if the degree of every node x of $\mathcal{H}$ is at most Δ.

7.2 (β, k)-PoA: Lower Bound

In the following theorem, we provide a lower bound on the (β, k)-Price of Anarchy relying on a nice result from graph theory.

Theorem 7. *For every $\beta \geq 1$, any integers $k \geq r$, $\Delta \geq 3$, $d \geq 1$, and any distance-factors sequence $(\alpha_h)_{h\in[d]}$, there exists a Δ-bounded-degree GDPG $\mathcal{G}$ such that* $\mathsf{PoA}_k^\beta(\mathcal{G}) \geq \frac{\beta\cdot\sum_{h\in[d]}\alpha_h\Delta(\Delta-1)^{h-1}b^{h-1}}{1+\sum_{h=1}^{d-1}\alpha_{h+1}(2(\Delta-1)^{\lfloor(h+1)/2\rfloor}(r-1)^{\lfloor(h+1)/2\rfloor-1}+2(\Delta-1)^{\lfloor h/2\rfloor-1}(r-1)^{\lfloor h/2\rfloor})}$.

Proof. In Lemma 3 of [20], they state that for every integers Δ, r, and $\gamma_0 \geq 3$, it is always possible to find a Δ-regular r-uniform hypergraph $\mathcal{H}$ of girth at least γ_0. By using this result, given integers $k \geq r$, $\Delta \geq 3$ and $d \geq 1$, a distance-factors sequence $(\alpha_h)_{h\in[d]}$, and a Δ-regular linear hypergraph $\mathcal{H} = (V, E)$ of girth at least $\gamma_0 := \max\{2d+1, k+1\}$, we can build a Δ-bounded-degree GDPG $\mathcal{G}$, such that (i) $\mathcal{H}$ is its underlying hypergraph; (ii) $(\alpha_h)_{h\in[d]}$ is its distance-factors sequence; (iii) each player x has two strategies, s and s^*; (iv) for every hyperedge $e \in E$ and outcome $\boldsymbol{\sigma}$, $w_e(\boldsymbol{\sigma}) = \beta$ if all the nodes in e play s^* in $\boldsymbol{\sigma}$, and 0 otherwise; and (v) for every $x \in V$, $p_x(\boldsymbol{\sigma}) = 1 + \sum_{h=1}^{d-1}\alpha_{h+1}\left(1 + \Delta\frac{(\Delta-1)^p(r-1)^p-1}{(\Delta-1)(r-1)-1} + r\frac{(\Delta-1)^q(r-1)^q-1}{(\Delta-1)(r-1)-1}\right)$ if x plays s in $\boldsymbol{\sigma}$, where $p = \lfloor(h+1)/2\rfloor$ and $q = \lfloor h/2\rfloor$, otherwise $p_x(\boldsymbol{\sigma}) = 0$.

Let $\boldsymbol{\sigma}$ and $\boldsymbol{\sigma}^*$ be the strategy profiles in which all players play strategy s and s^*, respectively. First, we show that $\boldsymbol{\sigma}$ is an (β, k)-equilibrium of $\mathcal{G}$.

Lemma 8. *$\boldsymbol{\sigma}$ is a (β, k)-equilibrium.*

Proof (Proof sketch). If $\boldsymbol{\sigma}$ is not a (β, k)-equilibrium, then there must exist a subset $Z \subseteq V$, $|Z| \leq k$, such that $u_x(\boldsymbol{\sigma} \xrightarrow{Z} \boldsymbol{\sigma}^*) > \beta \cdot u_x(\boldsymbol{\sigma})$ for every $x \in Z$.

We can assume w.l.o.g. that the subhypergraph $\mathcal{T}$ induced by Z in $\mathcal{H}$, rooted in x_r, is a perfect linear hypertree of height d because y cannot get utility from hyperedges that are far more than $d-1$ from y, and each node has degree Δ. We will assume w.l.o.g. that y is one of the $r-1$ leftmost leaves, that is one of the leaves in the leftmost hyperedge.

Let $(e_0, e_1, \ldots, e_d)$ be a path $\mathcal{P}$ from the leftmost leaf y to the root, where e_0 is the leftmost hyperedge containing y, e_d is the one containing the root x_r, and $e_{l-1} \cap e_l = v_l$ for every other $l \in [d-1]$. Clearly, each hyperedge e_l is made up of v_l, v_{l+1}, and other $r-2$ nodes that we call v_l'. The distance between y and any v_l is $l-1$, while the distance between y and v_l' is l. The root is at distance $d-l-1$ from v_l and any v_l', and there is a subhypertree having v_l or any v_l' as root and height l. Let $\mathcal{T}_l$ be the subhypertree having root v_l, and all the descendant nodes without the ones in $\mathcal{P}$. Let also $\mathcal{T}_l'$ be one of the $r-2$ hypertrees with one of the nodes v_l' as root. Both $\mathcal{T}_l$ and any $\mathcal{T}_l'$ have height l. The number of hyperedges $E_{l,t}$ at level t of $\mathcal{T}_l$ is $(\Delta-2)(\Delta-1)^{t-1}(r-1)^{t-1}$, while the number of hyperedges $E_{l,t}'$ at level t of any $\mathcal{T}_l'$ is $(\Delta-1)^t(r-1)^{t-1}$.

All the hyperedges that are at distance $h \geq 1$ from y are e_h, plus $E_{h,1} \cup E_{h-1,2} \cup \ldots \cup E_{\lceil(h+1)/2\rceil,\lfloor(h+1)/2\rfloor}$, plus $E_{h-1,1}' \cup E_{h-2,2}' \cup \ldots \cup E_{\lceil h/2\rceil,\lfloor h/2\rfloor}'$ for the $(r-2)$ hypertrees $\mathcal{T}_l'$. When $h = 0$, there is only the hyperedge e_0 at distance h from y. Therefore, using the already defined p and q, the number of hyperedges at distance $h \geq 1$ from y is $1 + \Delta\frac{(\Delta-1)^p(r-1)^p-1}{(\Delta-1)(r-1)-1} + r\frac{(\Delta-1)^q(r-1)^q-1}{(\Delta-1)(r-1)-1}$.

We can conclude that the utility that a leftmost leaf y gets from the deviation $u_y(\boldsymbol{\sigma} \xrightarrow{Z} \boldsymbol{\sigma}^*)$ is $\beta \cdot p_x(\boldsymbol{\sigma})$, which is equal to $\beta \cdot u_y(\boldsymbol{\sigma})$, so y does not profit from the deviation, and $\boldsymbol{\sigma}$ is a (β, k)-equilibrium. □

Lemma 9. $u_x(\boldsymbol{\sigma}^*) = \beta \sum_{h \in [d]} \alpha_h \Delta(\Delta - 1)^{h-1}(r-1)^{h-1}$ *for any* $x \in V$.

From Lemma 8 and 9, we obtain the lower bound for $\mathsf{PoA}_k^\beta(\mathcal{G})$, which concludes the proof of the theorem. □

Remark 2. Please note that, if all the distance-factors are not lower than a constant $c > 0$, from Theorem 7 we can conclude that the (β, k)-price of anarchy of Δ-bounded-degree GDPG, as a function of d, can grow as $\Omega(\beta(\Delta - 1)^{d/2}(r-1)^{d/2})$.

8 Conclusion and Future Works

This study leaves some open problems, such as (i) closing the gap between the upper and the lower bound on the Price of Anarchy for bounded-degree hypergraphs; (ii) extending the results on the Price of Stability to values of β greater than one; and (iii) computing a lower bound on the Price of Stability for bounded-degree hypergraphs. Concerning the latter problem, we are confident that it is possible to use the same modus operandi described in Sect. 6.

Another relevant open problem we consider worth investigating concerns finding particular classes of games which guarantee the existence of equilibria. We believe the class of games with a hypertree as underlying hypergraph are a good candidate. About the existence of β-approximate k-strong equilibria, we conjecture that the condition stated in Theorem 1 is necessary, and can be proven with some ad-hoc game instances.

Another interesting research direction is studying our model with respect to different social welfare functions, e.g., using the L^p-norm for the different values of p.

References

1. Aloisio, A.: Distance hypergraph polymatrix coordination games. In: Proceedings of the 22nd Conference on Autonomous Agents and Multi-Agent Systems (AAMAS), pp. 2679–2681 (2023)
2. Aloisio, A., Flammini, M., Kodric, B., Vinci, C.: Distance polymatrix coordination games. In: Proceedings of the 30th International Joint Conference on Artificial Intelligence, IJCAI-21, pp. 3–9 (2021)
3. Aloisio, A., Flammini, M., Vinci, C.: The impact of selfishness in hypergraph hedonic games. In: Proceedings of the 34th Conference on Artificial Intelligence (AAAI), pp. 1766–1773 (2020)
4. Apt, K.R., de Keijzer, B., Rahn, M., Schäfer, G., Simon, S.: Coordination games on graphs. Int. J. Game Theory **46**(3), 851–877 (2017)
5. Ashlagi, I., Krysta, P., Tennenholtz, M.: Social context games. In: Papadimitriou, C., Zhang, S. (eds.) WINE 2008. LNCS, vol. 5385, pp. 675–683. Springer, Heidelberg (2008). https://doi.org/10.1007/978-3-540-92185-1_73
6. Aziz, H., Brandl, F., Brandt, F., Harrenstein, P., Olsen, M., Peters, D.: Fractional hedonic games. ACM Trans. Econ. Comput. **7**(2), 6:1–6:29 (2019)

7. Aziz, H., Brandt, F., Seedig, H.G.: Optimal partitions in additively separable hedonic games. In: Proceedings of the 22nd International Joint Conference on Artificial Intelligence (IJCAI), pp. 43–48 (2011)
8. Bilò, V., Celi, A., Flammini, M., Gallotti, V.: Social context congestion games. Theor. Comput. Sci. **514**, 21–35 (2013)
9. Bilò, V., Fanelli, A., Flammini, M., Monaco, G., Moscardelli, L.: Nash stable outcomes in fractional hedonic games: existence, efficiency and computation. J. Artif. Intell. Res. **62**, 315–371 (2018)
10. Bilò, V., Fanelli, A., Flammini, M., Monaco, G., Moscardelli, L.: Optimality and Nash stability in additive separable generalized group activity selection problems. In: Proceedings of the 28th International Joint Conference on Artificial Intelligence (IJCAI), pp. 102–108 (2019)
11. Cai, Y., Candogan, O., Daskalakis, C., Papadimitriou, C.H.: Zero-sum polymatrix games: a generalization of minmax. Math. Oper. Res. **41**(2), 648–655 (2016)
12. Cai, Y., Daskalakis, C.: On minmax theorems for multiplayer games. In: Proceedings of the 22nd Symposium on Discrete Algorithms (SODA), pp. 217–234 (2011)
13. Carosi, R., Monaco, G., Moscardelli, L.: Local core stability in simple symmetric fractional hedonic games. In: Proceedings of the 18th International Conference on Autonomous Agents and Multi-Agent Systems (AAMAS), pp. 574–582 (2019)
14. Darmann, A., Elkind, E., Kurz, S., Lang, J., Schauer, J., Woeginger, G.J.: Group activity selection problem. In: Proceedings of the 8th International Workshop Internet & Network Economics (WINE), vol. 7695, pp. 156–169 (2012)
15. Darmann, A., Lang, J.: Group activity selection problems. In: Trends in Computational Social Choice, pp. 385–410 (2017)
16. Deligkas, A., Fearnley, J., Savani, R., Spirakis, P.G.: Computing approximate Nash equilibria in Polymatrix games. Algorithmica **77**(2), 487–514 (2017)
17. Drèze, J.H., Greenberg, J.: Hedonic coalitions: optimality and stability. Econometrica **48**(4), 987–1003 (1980)
18. Eaves, B.C.: Polymatrix games with joint constraints. SIAM J. Appl. Math. **24**(3), 418–423 (1973)
19. Elkind, E., Fanelli, A., Flammini, M.: Price of pareto optimality in hedonic games. Artif. Intell. **288**, 103357 (2020)
20. Ellis, D., Linial, N.: On regular hypergraphs of high girth. Electron. J. Comb. **21**(1), 1 (2014)
21. Flammini, M., Kodric, B., Olsen, M., Varricchio, G.: Distance hedonic games. In: Proceedings of the 19th Conference on Autonomous Agents and Multi-Agent Systems (AAMAS), pp. 1846–1848 (2020)
22. Gourvès, L., Monnot, J.: On strong equilibria in the max cut game. In: Leonardi, S. (ed.) WINE 2009. LNCS, vol. 5929, pp. 608–615. Springer, Heidelberg (2009). https://doi.org/10.1007/978-3-642-10841-9_62
23. Howson, J.T.: Equilibria of polymatrix games. Manag. Sci. **18**(5), 312–318 (1972)
24. Howson, J.T., Rosenthal, R.W.: Bayesian equilibria of finite two-person games with incomplete information. Manag. Sci. **21**(3), 313–315 (1974)
25. Janovskaja, E.: Equilibrium points in polymatrix games. Lith. Math. J. **8**(2), 381–384 (1968)
26. Kearns, M.J., Littman, M.L., Singh, S.P.: Graphical models for game theory. In: Proceedings of the 17th International Conference on Uncertainty in Artificial Intelligence (UAI), pp. 253–260 (2001)
27. Miller, D.A., Zucker, S.W.: Copositive-plus Lemke algorithm solves polymatrix games. Oper. Res. Lett. **10**(5), 285–290 (1991)
28. Monaco, G., Moscardelli, L., Velaj, Y.: Stable outcomes in modified fractional hedonic games. Auton. Agents Multi Agent Syst. **34**(1), 4 (2020)

29. Monderer, D., Shapley, L.S.: Potential games. Games Econom. Behav. **14**(1), 124–143 (1996)
30. Rahn, M., Schäfer, G.: Efficient equilibria in polymatrix coordination games. In: Italiano, G.F., Pighizzini, G., Sannella, D.T. (eds.) MFCS 2015. LNCS, vol. 9235, pp. 529–541. Springer, Heidelberg (2015). https://doi.org/10.1007/978-3-662-48054-0_44
31. Simon, S., Wojtczak, D.: Synchronisation games on hypergraphs. In: Proceedings of the 26th International Joint Conference on Artificial Intelligence (IJCAI), pp. 402–408 (2017)

Relaxed Agreement Forests

Virginia Ardévol Martínez[2], Steven Chaplick[1], Steven Kelk[1](✉), Ruben Meuwese[1], Matúš Mihalák[1], and Georgios Stamoulis[1]

[1] Department of Advanced Computing Sciences, Maastricht University, Maastricht, The Netherlands
steven.kelk@maastrichtuniversity.nl
[2] Université Paris-Dauphine, PSL University, CNRS, LAMSADE, Paris, France

Abstract. The phylogenetic inference process can produce, for multiple reasons, conflicting hypotheses of the evolutionary history of a set X of biological entities, i.e., phylogenetic trees with the same set of leaf labels X but with distinct topologies. It is natural to wish to quantify the difference between two such trees T_1 and T_2. We introduce the problem of computing a *maximum relaxed agreement forest* (MRAF) and use this as a proxy for the dissimilarity of T_1 and T_2, which in this article we assume to be unrooted and binary. MRAF asks for a partition of the leaf labels X into a minimum number of blocks $S_1, \ldots, S_k$ such that the two subtrees induced in T_1 and T_2 by every S_i are isomorphic up to suppression of degree-2 nodes and taking the labels X into account. Unlike the earlier introduced maximum agreement forest (MAF) model, the subtrees induced by the S_i are allowed to overlap. We prove that it is NP-hard to compute MRAF, by reducing from the problem of partitioning a permutation into a minimum number of monotonic subsequences (PIMS). We further show that MRAF has a $O(\log n)$-approximation algorithm where $n = |X|$ and permits exact algorithms with single-exponential running time. When one of the trees is a caterpillar, we prove that testing whether a MRAF has size at most k can be answered in polynomial time when k is fixed. We also note that on two caterpillars the approximability of MRAF is related to that of PIMS. Finally, we establish a number of bounds on MRAF, compare its behaviour to MAF both theoretically and experimentally and discuss a number of open problems.

1 Introduction

The central challenge of phylogenetics, which is the study of phylogenetic (evolutionary) trees, is to infer a tree whose leaves are bijectively labeled by a set of species X and which accurately represents the evolutionary events that gave rise to X [23]. There are many existing techniques to infer phylogenetic trees from biological data and under a range of different objective functions [19]. The complexity of this problem arises from the fact that we typically only have indirect

R. Meuwese was supported by NWO grant *Deep kernelization for phylogenetic discordance* OCENW.KLEIN.305.

H. Fernau et al. (Eds.): SOFSEM 2024, LNCS 14519, pp. 40–54, 2024.
https://doi.org/10.1007/978-3-031-52113-3_3

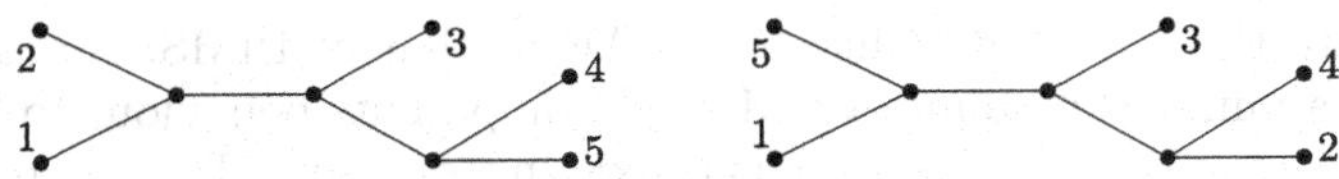

Fig. 1. The two trees, while isomorphic, are not isomorphic when taking the leaf-labeling into account, and thus both MRAF and MAF cannot be of size one. A MRAF has 2 blocks, e.g., $\{1, 2, 3\}$ and $\{4, 5\}$. A MAF has 3 blocks, e.g., $\{1, 2, 3\}$, $\{4\}$, and $\{5\}$.

data available, such as DNA sequences of the species X. Different techniques regularly yield trees with differing topologies, or the same technique constructs different trees depending on which part of a genome the DNA data is extracted from [21]. Hence, it is insightful to formally quantify the dissimilarity between (pairs of) phylogenetic trees, stimulating research into various distance measures.

Here we propose a new dissimilarity measure between unrooted phylogenetic trees T_1, T_2 which is conceptually related to the well-studied *agreement forest* abstraction. An agreement forest (AF) is a partition of X into blocks which induce, in the two input trees, non-overlapping isomorphic subtrees, modulo edge subdivision and taking the labels X into account; computing such a forest of minimum size (a MAF) is NP-hard [14] although it can be computed reasonably well in practice [26]. The AF abstraction originally derives its significance from the fact that, in unrooted (respectively, rooted) phylogenetic trees, an AF of minimum size models *Tree Bisection and Reconnection* (TBR) (respectively, *rooted Subtree Prune and Regraft*, rSPR) distance [1,6]. For background on AFs we refer to recent articles such as [7,8]. Here we propose the *relaxed agreement forest* abstraction (RAF). The only difference is that we no longer require the partition of X to induce non-overlapping subtrees; they only have to be isomorphic (see Fig. 1). We write MRAF to denote a relaxed agreement forest of minimum size. As we will observe, in the worst case MRAF can be constant while MAF grows linearly $|X|$.

The fact that RAFs are allowed to induce overlapping subtrees is potentially interesting from the perspective of biological modelling. Unlike an AF, multiple subtrees of the RAF can pass through a single branch of T_1 (or T_2). This allows us to view T_1 and T_2 as the union of several interleaved, overlapping, common evolutionary histories. It is beyond the scope of this article to expound upon this, but it is compatible with the trend in the literature of phylogenetic trees (or networks) having multiple distinct histories woven within them which sometimes evolve "in parallel" due to phenomena such as incomplete lineage sorting [9,15,21]. This greater modelling flexibility, rather than computational tractability issues, is our primary reason for studying MRAF.

Our results are as follows. First, we show that it is NP-hard to compute a MRAF. We reduce from the problem of partitioning a permutation into a minimum number of monotone subsequences (PIMS). We show that MRAF has a $O(\log n)$-approximation algorithm where $n = |X|$ and permits exact algorithms with single-exponential running time. When one of the two trees is a caterpillar, we prove that "Is there a RAF with at most k components?" can be answered in polynomial time when k is fixed, i.e., in XP parameterized by k.

We also relate the approximability of MRAF to that of PIMS. Along the way we establish a number of bounds on MRAF, compare its behaviour to MAF and undertake an empirical analysis on two existing datasets. Due to page limits some proofs/details are deferred to an appendix in a preprint of this article [2].

2 Preliminaries, Basic Properties and Bounds

Let X be a set of labels (*taxa*) representing species. An *unrooted binary phylogenetic tree* T on X is a simple, connected, and undirected tree whose leaves are bijectively labeled with X and whose other vertices all have degree 3. When it is clear from the context we will simply write (*phylogenetic*) *tree* for shorthand. For two trees T and T' both on the same set of taxa X, we write $T = T'$ if there is an isomorphism between T and T' that preserves the labels X. Tree T is a *caterpillar* if deleting the leaves of T yields a path. We say that two distinct taxa $\{a, b\} \subseteq X$ form a *cherry* of a tree T if they have a common parent. The *identity caterpillar* on n leaves is simply the caterpillar with leaves $1, \dots n$ in ascending order with the exception of the two cherries $\{1, 2\}$ and $\{n-1, n\}$ at its ends; see e.g. the tree on the left in Fig. 1. Note that caterpillars are almost total orders, but not quite: the leaves in the cherries at the ends are incomparable. Managing this subtle difference is a key aspect of our results.

A *quartet* is an unrooted binary phylogenetic tree with exactly four leaves. Let T be a phylogenetic tree on X. If $\{a, b, c, d\} \subseteq X$ are four distinct leaves, we say that quartet $ab|cd$ *is induced by (or simply 'is a quartet of')* T if in T the path from a to b does not intersect the path from c to d. Note that, for any four distinct leaves $a, b, c, d \in X$, exactly one of the three quartets $ab|cd, ac|bd, ad|bc$ will be a quartet of T. It is well-known that $T_1 = T_2$ if and only if both trees induce exactly the same set of quartet topologies [23]. For example, in Fig. 1 12|45 is a quartet of the first tree but not a quartet of the second tree. For $X' \subseteq X$, we write $T[X']$ to denote the unique, minimal subtree of T that connects all elements in the subset X'. We use $T|X'$ to denote the phylogenetic tree on X' obtained from $T[X']$ by suppressing degree-2 vertices. If $T_1|X' = T_2|X'$ then we say that the subtrees of T_1, T_2 induced by X' are *homeomorphic*.

Let T_1 and T_2 be two phylogenetic trees on X. Let $\mathcal{F} = \{S_1, \dots, S_k\}$ be a partition of X, where each block S_i, is referred to as a *component* of $\mathcal{F}$. We say that $\mathcal{F}$ is an *agreement forest* (AF) for T_1 and T_2 if these conditions hold:

1. For each $i \in \{1, 2, ..., k\}$ we have $T_1|S_i = T_2|S_i$.
2. For each pair $i, j \in \{1, 2, ..., k\}$ with $i \neq j$, we have that $T_1[S_i]$ and $T_1[S_j]$ are vertex-disjoint in T, and $T_2[S_i]$ and $T_2[S_j]$ are vertex-disjoint in T_2.

The *size* of $\mathcal{F}$ is simply its number of components, i.e., k. Moreover, an AF with the minimum number of components (over all AFs for T_1 and T_2) is called a *maximum agreement forest* (MAF) for T_1 and T_2. For ease of reading, we will also write MAF to denote the size of a MAF. This is NP-hard to compute [1,14].

A *relaxed agreement forest* (RAF) is defined similarly to an AF, except without condition 2. A RAF with a minimum number of components is a *maximum relaxed agreement forest* (MRAF). We also use MRAF for the size of a MRAF.

MAXIMUM RELAXED AGREEMENT FOREST (MRAF)
Input: Two unrooted binary phylogenetic trees T_1, T_2 on the same leaf set X, and a number k.
Task: Partition X into at most k sets $S_1, \ldots, S_k$ where $T_1|S_i = T_2|S_i$ for each i.

Observation 1 follows directly from the definitions. Observation 2 shows that MAF and MRAF can behave very differently.

Observation 1. *(a) A RAF with at most $\lceil \frac{n}{3} \rceil$ components always exists, where $n = |X|$, because if $|X'| = 3$ and $X' \subseteq X$ we have $T_1|X' = T_2|X'$ irrespective of X' or the topology of T_1 and T_2. (b) MRAF is 0 if and only if $T_1 = T_2$. (c) A partition $\{S_1, \ldots, S_k\}$ of X is a RAF of T_1, T_2 if and only if, for each S_i, the set of quartets induced by $T_1|S_i$ is identical to the set of quartets induced by $T_2|S_i$.*

Observation 2. *There are instances where MAF is arbitrarily large, $\Omega(n)$, while MRAF is constant.*

Proof. Let T be an arbitrary unrooted phylogenetic binary tree on n taxa. We create two trees T_1 and T_2, both on $4n$ taxa. We build T_1 by replacing each leaf x in T with a subtree on $\{a_x, b_x, c_x, d_x\}$ in which a_x, b_x form a cherry and c_x, d_x form a cherry. The construction of T_2 is similar except that a_x, c_x form a cherry and b_x, d_x form a cherry. Note that $T_1|\{a_x, b_x, c_x, d_x\} \neq T_2|\{a_x, b_x, c_x, d_x\}$. MRAF here is 2 because we can take one component containing all the a_x, b_x taxa and one containing all the c_x, d_x taxa. However, MAF is at least n. This is because in any AF at least one of the four taxa in $\{a_x, b_x, c_x, d_x\}$ must be a singleton component, and there are n subsets of the form $\{a_x, b_x, c_x, d_x\}$. □

Given two trees T_1, T_2 on X we say that $X' \subseteq X$ induces a *maximum agreement subtree* (MAST) if $T_1|X' = T_2|X'$ and X' has maximum cardinality ranging over all such subsets. Clearly, $\lceil \frac{n}{MAST} \rceil$ is a lower bound on MRAF, since each component of a RAF is no larger than a MAST. A MAST can be computed in polynomial time [24]. The trivial upper bound on MRAF of $\lceil \frac{n}{3} \rceil$ (see Observation 1), which already contrasts sharply with the fact that the MAF of two trees can be as large as $n(1 - o(1))$ [3], can easily be strengthened via MASTs. For example, it can be verified computationally or analytically that for any two trees on 6 or more taxa, a MAST has size at least 4. We can thus repeatedly choose and remove a homeomorphic size-4 subtree, until there are fewer than 6 taxa left, giving a loose upper bound on MRAF of $n/4 + 2$. In fact, it is known that the size of a MAST on two trees with n leaves is $\Omega(\log n)$ [20] (and that this bound is asymptotically tight). In particular, the lower bound on MAST grows to infinity as n grows to infinity. Hence, the upper bound of $n/4 + 2$ can be strengthened to $n/c + f(c)$ for any arbitrary constant $c > 1$ where f is a function that only depends on c; this is thus $n/c + O(1)$. In fact, by iteratively removing $\Omega(\log n')$ of the *remaining* number of taxa n' we obtain a (slightly) sublinear upper bound on the size of a MRAF. Namely, while $n' \geq \log n + O(1)$, each iteration removes at least $d \log n' \geq d \log \log n$ leaves for some constant d, giving an upper bound of $\frac{n}{d \log \log n} + \log n + O(1)$ which is $O(\frac{n}{\log \log n})$.

Regarding lower bounds, one can generate pairs of trees on n leaves where a MAST has $O(\log n)$ leaves [18,20]. A MRAF will thus have size $\Omega(\frac{n}{\log n})$.

3 Hardness of MRAF

We discuss a related NP-hard problem regarding partitioning permutations [25].

PARTITION INTO MONOTONE SUBSEQUENCES (PIMS)
Input: A permutation π of $\{1,\ldots,n\}$, and a number k.
Task: Partition $\{1,\ldots,n\}$ into at most k sets such that each set occurs monotonically in π, i.e., either as an increasing or a decreasing sequence.

Due to the classical Erdős Szekeres Theorem [10], for any n-element permutation there is a monotone sequence in π with at least $\sqrt{n}$ elements. This can be used to efficiently partition π into at most $2\sqrt{n}$ monotone sequences [4]. Thus, we may assume that the k in the problem statement is always at most $2\sqrt{n}$.

Theorem 1. *MRAF is NP-hard.*

Proof. Let (π, k) be an input to the PIMS problem, i.e., k is an integer greater than 1 and π is a permutation of $\{1,\ldots,n\}$, where we use π_i to denote the ith element of π. As remarked before, k is at most $2\sqrt{n}$. This will imply that our constructed instance of MRAF will have linear size in terms of the given permutation π, and as such any lower bounds, e.g., arising from the Exponential Time Hypothesis (ETH), will carry over from the PIMS problem to the MRAF problem. For each pair of integers (α, β) where $\alpha + \beta = k$ and $\alpha, \beta \geq 1$[1], we will construct an instance (T_1, T_2) of MRAF such that (T_1, T_2) has a solution consisting of k trees if and only if π can be partitioned into α increasing sequences and β decreasing sequences. The trees T_1 and T_2 are described as follows.

Recall that a *caterpillar* is a tree T where the subtree obtained by removing all leaves of T is a path. The path here is called the *spine* of the caterpillar. Note that, in the caterpillars used to construct T_1 and T_2, some spine vertices will have degree 2. However, to make proper binary trees one should contract any such vertex into one of its neighbors.

We first construct a leaf set $v_1, \ldots, v_n$ corresponding to the permutation. We create an identity caterpillar I whose spine is the n-vertex path $(x_1, \ldots, x_n)$ such that x_i is adjacent to v_i. Next, we create a caterpillar P whose spine is the n-vertex path $(y_1, \ldots, y_n)$ such that y_i is adjacent to v_{π_i}. Observe that already for the MRAF instance (I, P), any (r, s) partition of π leads to a solution to (I, P) consisting of k trees. However, the converse is not yet enforced. In particular, if the input to MRAF is (I, P), then the components in a MRAF (which are caterpillars) have cherries at their ends which, crucially, might be ordered differently in I than in P. This can violate monotonicity. To counter this we extend I and P to obtain T_1, T_2 as shown in Fig. 2. For T_1, we construct $8k$ caterpillars. First, for

[1] $\alpha = 0$ or $\beta = 0$ makes the problem easy.

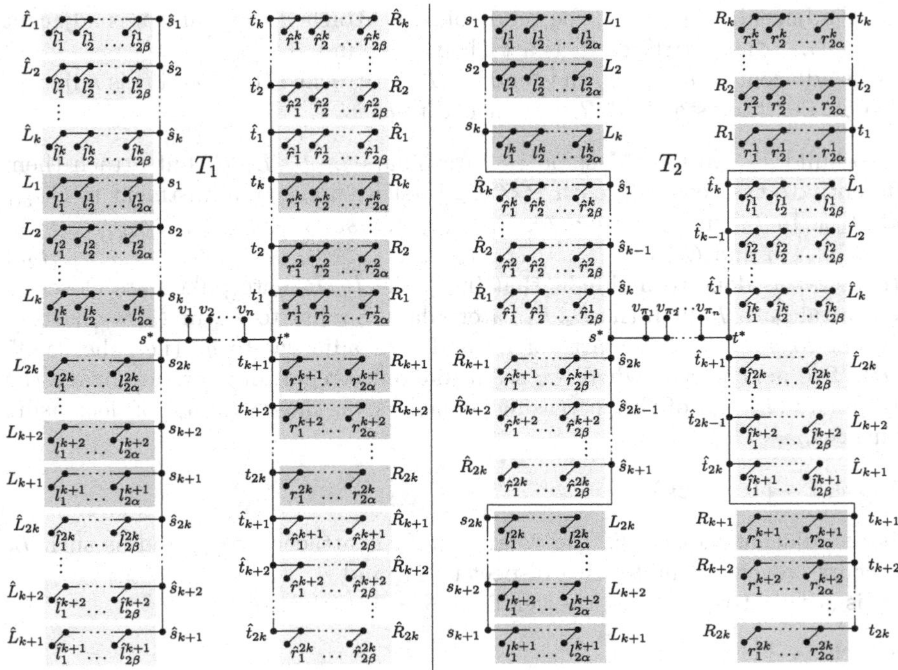

Fig. 2. The two trees T_1, T_2 constructed from an instance of PIMS in the NP-hardness proof. The dark (light) grey leaves are used to induce increasing (decreasing) subsequences in the permutation-encoding taxa in the centre of the trees.

the increasing sequences, we construct $4k$ caterpillars $L_1, \ldots, L_{2k}$, $R_1, \ldots, R_{2k}$ each having 2α leaves and 2α spine vertices. Namely, for each i,

- L_i is the caterpillar with leaf set $\{l_1^i, \ldots, l_{2\alpha}^i\}$ and spine $(w_1^i, \ldots, w_{2\alpha}^i)$ where, for each j, l_j^i is adjacent to w_j^i; and
- R_i is the caterpillar with leaf set $\{r_1^i, \ldots, r_{2\alpha}^i\}$ and spine $(z_1^i, \ldots, z_{2\alpha}^i)$ where, for each j, r_j^i is adjacent to z_j^i.

For the decreasing sequences, we similarly construct $4k$ caterpillars $\hat{L}_1, \ldots, \hat{L}_{2k}$, $\hat{R}_1, \ldots, \hat{R}_{2k}$ each having 2β leaves and 2β spine vertices. Namely, for each i,

- $\hat{L}_i$ is the caterpillar with leaf set $\{\hat{l}_1^i, \ldots, \hat{l}_{2\beta}^i\}$ and spine $(\hat{w}_1^i, \ldots, \hat{w}_{2\beta}^i)$ where, for each j, $\hat{l}_j^i$ is adjacent to $\hat{w}_j^i$; and
- $\hat{R}_i$ is the caterpillar with leaf set $\{\hat{r}_1^i, \ldots, \hat{r}_{2\beta}^i\}$ and spine $(\hat{z}_1^i, \ldots, \hat{z}_{2\beta}^i)$ where, for each j, $\hat{r}_j^i$ is adjacent to $\hat{z}_j^i$.

To form T_1, we create two $(4k+1)$-paths $Q_{\text{start}} = (\hat{s}_1, \ldots, \widehat{s_k}, s_1, \ldots, s_k, s^*, s_{2k}, \ldots, s_{k+1}, \hat{s}_{2k}, \ldots, \hat{s}_{k+1})$ and $Q_{\text{end}} = (\hat{t}_k, \ldots, \hat{t}_1)\ t_k, \ldots, t_1, t^*, t_{k+1}, \ldots, \hat{t}_{2k}\ \hat{t}_{k+1}, \ldots, \hat{t}_{2k})$ such that s^* is adjacent to x_1 (i.e., to the "start" of I) and t^* is adjacent to x_n (i.e., to the "end" of I), and for each $i \in \{1, \ldots, 2k\}$:

- s_i is adjacent to $w^i_{2\alpha}$, i.e., the "end" of L_i is attached to s_i, and t_i is adjacent to z^i_1, i.e., the "start" of R_i is attached to t_i; and
- $\hat{s}_i$ is adjacent to $\hat{w}^i_{2\alpha}$, i.e., the "end" of $\hat{L}_i$ is attached to $\hat{s}_i$, and $\hat{t}_i$ is adjacent to $\hat{z}^i_1$, i.e., the "start" of $\hat{R}_i$ is attached to $\hat{t}_i$.

To build T_2, we use the same $8k$ caterpillars $L_i, R_i, \hat{L}_i, \hat{R}_i$ but attach them differently to the "central" path P of T_2. First we make an adjustment to Q_{start} and Q_{end}. In T_2, these become: $Q_{\text{start}} = (s_1, \dots, s_k, \hat{s}_1, \dots, \hat{s_k}, s^*, \hat{s}_{2k}, \dots, \hat{s}_{k+1}$ $s_{2k}, \dots, s_{k+1})$ and $Q_{\text{end}} = (t_k, \dots, t_1, \hat{t}_k, \dots, \hat{t}_1, t^*, \hat{t}_{k+1}, \dots, \hat{t}_{2k}, t_{k+1}, \dots, \hat{t}_{2k})$ – this swap is done to highlight that in T_2 the $\hat{L}_i, \hat{R}_i$ caterpillars are closer to the central path P than the L_i, R_i caterpillars. Similar to T_1, in T_2, we have s^* adjacent to y_1 (i.e., the "start" of P) and t^* is adjacent to y_n (i.e., the "end" of P). The next part is where we see a difference regarding how we attach the caterpillars (L_i, R_i) of the increasing sequences vs. those $(\hat{L}_i, \hat{R}_i)$ of decreasing sequences.

For each $i \in \{1, \dots, 2k\}$:

- s_i is adjacent to w^i_1, i.e., the "start" of L_i is attached to s_i and as such L_i occurs "reversed" in T_2 with respect to T_1, and
- t_i is adjacent to $z^i_{2\alpha}$, i.e., the "end" of R_i is attached to t_i.

For each $i \in \{1, \dots, k\}$:

- $\hat{s}_{k-i+1}$ $(\hat{s}_{2k-i+1})$ is adjacent to $\hat{z}^i_{2\beta}$ $(\hat{z}^{k+i}_{2\beta})$, i.e., the "end" of $\hat{R}_i$ $(\hat{R}_{i+k})$ is attached to $\hat{s}_{k-i+1}$ (and $\hat{s}_{2k-i+1}$) and as such $\hat{R}_i$ $(\hat{R}_{k+i})$ occurs "on the opposite side" in T_2 with respect to its location in T_1, and
- $\hat{t}_{k-i+1}$ $(\hat{t}_{2k-i+1})$ is adjacent to $\hat{w}^i_1$ $(\hat{w}^{k+i}_1)$, i.e., the "start" of $\hat{L}_i$ $(\hat{L}_{k+i})$ is attached to $\hat{t}_{k-i+1}$ $(\hat{t}_{2k-i+i})$.

This completes the construction of T_1 and T_2 from π. It is easy to see that this construction can be performed in polynomial time and that our trees contain precisely $16k^2 + 8k + 4 + 4n$ vertices, i.e., since $k \le 2\sqrt{n}$, our instance of MRAF has $O(n)$ size.

Suppose π can be partitioned into α increasing sequences $\tau_1, \dots, \tau_\alpha$ and β decreasing sequences $\sigma_1, \dots, \sigma_\beta$. The leaf set corresponding to τ_i consists of $\{v_p : p \in \tau_i\}$ together with two leaves from each of L_j and R_j ($j \in \{1, \dots, 2k\}$), i.e., $l^j_{2i-1}, l^j_{2i}, r^j_{2i-1}, r^j_{2i-1}$. Similarly, the leaf set corresponding to σ_i consists of $\{v_p : p \in \sigma_i\}$ together with two leaves from each of $\hat{L}_j$ and $\hat{R}_j$ ($j \in \{1, \dots, 2k\}$), i.e., $\hat{l}^j_{2i-1}, \hat{l}^j_{2i}, \hat{r}^j_{2i-1}, \hat{r}^j_{2i}$. It can be verified that this is a valid solution to MRAF.

Now suppose that we have a solution $S_1, \dots, S_k$ to MRAF (T_1, T_2). We need to show that this leads to a solution to the PIMS problem on π consisting of (at most) α increasing sequences and (at most) β decreasing sequences. The proof of the following lemma is in the appendix.

Lemma 1. *If some S_j uses three leaves of any caterpillar $C \in \{L_i, R_i, \hat{L}_i, \hat{R}_i : i \in \{1, \dots, 2k\}\}$ then all elements of S_j are leaves of C.*

A consequence of this lemma is that if some S_j uses more than two leaves from any single one of our left/right caterpillars, then S_j can contain at most $\max\{2\alpha, 2\beta\} < 2k$ elements. In the next part we will see that every S_j must contain precisely $8k$ leaves from our left/right caterpillars in order to cover them all. In particular, this means that no S_j contains more than two leaves from any single left/right caterpillar. Note that, the total number of leaves is $n + 4k \cdot 2\alpha + 4k \cdot 2\beta = n + 8k^2$ where the set of n leaves is $\{v_1, \ldots, v_n\}$ (i.e., corresponding to the permutation) and the $8k^2$ leaves are the leaves of the left/right caterpillars. We now define the following eight leaf sets related to our caterpillars $L_i, R_i, \hat{L}_i, \hat{R}_i$.

- $\mathcal{L}_1 = \{l : l \text{ is a leaf of some } L_i,\ i \in \{1, \ldots, k\}\}$,
- $\mathcal{L}_2 = \{l : l \text{ is a leaf of some } L_i,\ i \in \{k+1, \ldots, 2k\}\}$,
- $\mathcal{R}_1 = \{r : r \text{ is a leaf of some } R_i,\ i \in \{1, \ldots, k\}\}$,
- $\mathcal{R}_2 = \{r : r \text{ is a leaf of some } R_i,\ i \in \{k+1, \ldots, 2k\}\}$.

The definition of $\hat{\mathcal{L}}_1, \hat{\mathcal{L}}_2, \hat{\mathcal{R}}_1, \hat{\mathcal{R}}_2$ is analogous. The proof of the following is also deferred to the appendix.

Lemma 2. *No S_j can contain five elements where each one belongs to a different set among: $\mathcal{L}_1, \mathcal{L}_2, \mathcal{R}_1, \mathcal{R}_2, \hat{\mathcal{L}}_1, \hat{\mathcal{L}}_2, \hat{\mathcal{R}}_1, \hat{\mathcal{R}}_2$.*

Now, observe that a component S_j can contain at most $2k$ taxa from each of the 8 sets listed above. That is because each of the 8 sets is formed from k caterpillars (e.g., $\mathcal{L}_1$ is formed from the caterpillars $L_1, ..., L_k$) and each of these k caterpillars contributes at most 2 taxa to a RAF component. (If one of the k caterpillars contributed more than 2 taxa, we would automatically be limited to at most $2k$ taxa, by Lemma 1.) It follows from this that a component S_j can in total intersect with at most $4 \times 2k = 8k$ taxa ranging over all the 8 sets: intersecting with more would require intersecting with at least 5 of the 8 sets, which as we have shown in Lemma 2 is not possible.

Given that there are k components in the RAF, and T_1, T_2 have $n + 8k^2$ taxa, each of the k components must therefore contain *exactly* $8k$ taxa from the 8 sets, and each component intersects with *exactly* 4 of the 8 sets (as this is the only way to achieve $8k$). In the appendix we prove that the only way for S_j to intersect with four sets *and* a permutation-encoding taxon v_i, is if the four sets are $\{\mathcal{L}_1, \mathcal{L}_2, \mathcal{R}_1, \mathcal{R}_2\}$ or $\{\hat{\mathcal{L}}_1, \hat{\mathcal{L}}_2, \hat{\mathcal{R}}_1, \hat{\mathcal{R}}_2\}$. The permutation-encoding taxa v_i contained in components of the first type, necessarily induce increasing subsequences, and those contained in the second type are descending. There can be at most α components of the first type, and at most β of the second, which means that the permutation π can be partitioned into at most α increasing and β decreasing sequences. This concludes the proof. □

4 Exact Algorithms

We now observe a single-exponential exact algorithm for MRAF and then show that when one input tree is a caterpillar, MRAF is in XP parameterized by k.

Recall that the NP-hard Set Cover problem (U, F), where F consists of subsets of U, is to compute a minimum-size subset of F whose union is U.

Observation 3. *Let T_1, T_2 be two unrooted binary phylogenetic trees on X. Let $U = X$ and let F be the set of all subsets of X that induce homeomorphic trees in T_1, T_2. Each RAF of T_1, T_2 is a set cover of (U, F), and each set cover of (U, F) can be transformed in polynomial time into a RAF of T_1, T_2 with the same or smaller size, by allocating each element of X to exactly one of the selected subsets. In particular, any optimum solution to the set cover instance (U, F) can be transformed in polynomial time to yield a MRAF of T_1, T_2 of the same size.*

Lemma 3. *MRAF can be solved in time $O(c^n)$, $n = |X|$, for some constant c.*

Proof. The construction in Observation 3 yields $|U| = n$ and $|F| \leq 2^n$. Minimum set cover can be solved in time $O(2^{|U|} \cdot (|U| + |F|)^{O(1)})$ thanks to [5]. □

Lemma 3 concerns general instances. When one of the given trees is a caterpillar, we can place MRAF into XP (parameterized by the solution size k). We use dynamic programming for this. We will assume that $n > 3k$, as otherwise an arbitrary partition $S_1, \ldots, S_k$ where each S_i has at most three taxa is a MRAF. For $n > 3k$ it follows that if there is a MRAF for T_1 and T_2, then there always is a MRAF $S_1, \ldots, S_k$ where no S_i is a singleton. To see this, observe that for any MRAF with a singleton S_i, it must contain a component S_j with $|S_j| \geq 3$ (since $n > 3k$), and moving any element from S_j to S_i gives another MRAF where S_i is not a singleton.

We let T_1 be the caterpillar, and T_2 an arbitrary tree. Similarly to our hardness result, we consider, without loss of generality, T_1 to consist of a spine (a path) $(y_1, \ldots, y_n)$ and leaves $v_1, \ldots, v_n$, where leaf v_i, $i = 1, \ldots, n$ is adjacent to vertex y_i. See Fig. 3 for an illustration. The spine naturally orders the leaves (up to arbitrarily breaking ties on the end cherry taxa) and this will guide our dynamic-programming approach. We write $u \prec v$ for two leaves u and v, if u appears before v in the considered ordering along the spine of T_1. We decide whether a MRAF $S_1, \ldots, S_k$ of T_1 and T_2 exists as follows: we enumerate over all possible pairs of vertices l_i, r_i, $i = 1, \ldots, k$, and check (compute) whether there exists a MRAF where the first leaf of S_i, $i = 1, \ldots, k$, is l_i and the last leaf of S_i is r_i. We call such MRAF a *MRAF constrained by* l_i, r_i, $i = 1, \ldots, k$, or simply a *constrained MRAF* if l_i and r_i are clear from the context. If for one of the guesses (enumerations) we find a constrained MRAF, we output YES, and otherwise (if for all guesses we do not find a MRAF) we output NO.

We now present our algorithm to decide, for input T_1, T_2, and pairs l_i, r_i, $i = 1, \ldots, k$, whether a constrained MRAF exists. We define $L := \{l_1, \ldots, l_n\}$ and $R := \{r_1, \ldots, r_n\}$. We view the process of computing constrained MRAF $S_1, \ldots, S_k$ as an iteration over v_i, $i = 1, \ldots, n$, and assigning $v_i \notin (L \cup R)$ to one of the components $S_1, \ldots, S_k$ (every taxon $v_i \in (L \cup R)$ is already assigned). Figure 3 illustrates this by the gray arrows from each taxon to one of the sets S_i. In the constrained MRAF, taxon v_i can only be assigned to component S_j if and only if $l_j \prec v_i \prec r_j$.

Tree T_2 further limits how taxon v_i can be assigned to components S_j (because we want that $T_1|S_j = T_2|S_j$). Clearly, for any $S_j \subset X$, $T_1|S_j$ is a caterpillar of maximum degree at most three. Thus, since l_j and r_j are the first

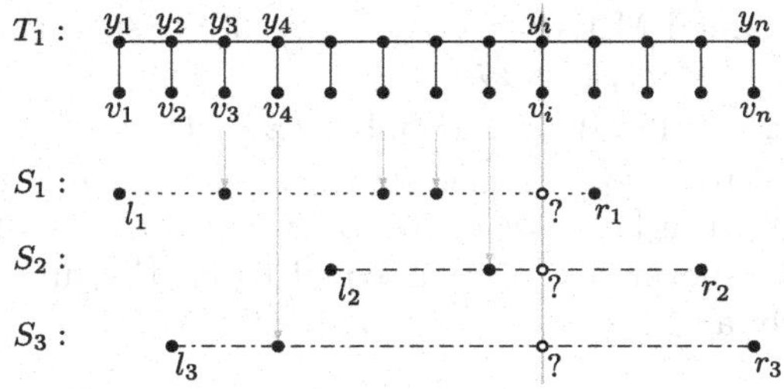

Fig. 3. Caterpillar T_1 induces a natural ordering on the taxa (leaves). The gray vertical arrows assign each taxa to one of the sets S_1, S_2, S_3. At iteration i, the question marks denote possible assignment of v_i.

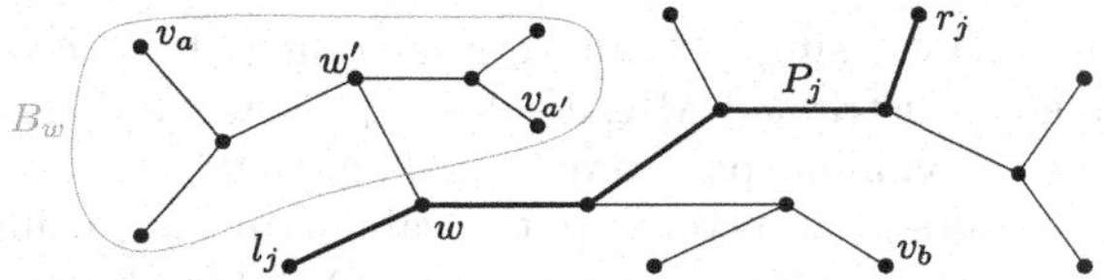

Fig. 4. A bag B_w on the (l_j, r_j)-path P_j. At most one of $v_a, v_{a'}$ can occur in S_j.

and last leaf in $T_1|S_j$, they also need to be first and last in $T_2|S_j$. Hence, the inner vertices of the unique path P_j from l_j to r_j in T_2 is the subdivision of the spine of $T_2|S_j$. For a vertex $w \in P_j$ that has a neighbor $w' \notin P_j$ we define a *bag* B_w *of* P_2 to be the maximal subtree of T_2 rooted at w' that does not include w. See Fig. 4 for an illustration. Observe that for any bag B_w of P_j, at most one taxon from B_w can be assigned to S_j. (Because if two taxa $v_a, v_{a'} \in B_w$, $a < a'$, are assigned to S_j then $l_j v_a | v_{a'} r_j$ will not be a quartet of $T_2|S_j$, while it is a quartet of $T_1|S_j$, and thus $T_1|S_j \neq T_2|S_j$.) The path P_j of $T_2|S_j$ naturally orders all bags of P_j. It follows that for two bags B_w and $B_{w'}$ where B_w appears before $B_{w'}$ in the ordering along P_j, we can select taxa $v_a \in B_w$ and $v_b \in B_{w'}$ into S_j if and only if $v_a \prec v_b$, i.e., if v_a appears before v_b in the caterpillar T_1. We write $v \prec_{P_j} v'$ for taxa v, v' such that v is from a bag B_w and v' is from a bag $B_{w'}$, and B_w appears before bag $B_{w'}$ along path P_j. Relation $\prec_{P_j}$ is thus a partial ordering of X, where any two taxa from the same bag are uncomparable. Observe now that any assignment of taxa to S_j that satisfies the above conditions, i.e., (i) for every $v_i \in S_j$, $l_j \prec v_i r_i$, (ii) for every bag B_w of P_j there is at most one vertex $v_i \in B_w \cap S_j$, and (*iii*) for any two taxa $v_p, v_q \in S_j$, $p < q$, $v_p \prec_{P_j} v_q$, we have $T_1|S_j = T_2|S_j$.

We can thus assign taxon v_i to component S_j whenever the previously assigned taxon $v_{i'}$ to S_j satisfies $v_{i'} \prec_{P_j} v_i$. We thus do not need to know all previously assigned taxa to S_j, only the last assigned. We compute a (partial) restricted MRAF for taxa $X_i := \{1, 2, \ldots, i\} \cup E$ iteratively for $i = 1, 2, \ldots, k$. We set $X_0 := L \cup R$. For $\boldsymbol{z} = (z_1, \ldots, z_k) \in (X \setminus (L \cup R))^k$ and $i = 0, 1, \ldots, k$ we define a boolean function $\mathrm{craf}^{(i)}(\boldsymbol{z})$ as follows: $\mathrm{craf}^{(i)}(\boldsymbol{z}) :=$ TRUE if and only if there exists a constrained MRAF $S_1^i, S_2^i, \ldots, S_k^i$ of X_i such that the last taxon from $X_i \setminus R$ in S_ℓ^i, $\ell = 1, \ldots, k$, is z_ℓ.

Clearly, $\text{craf}^{(0)}(\boldsymbol{z}) = \text{TRUE}$ if and only if $\boldsymbol{z} = (l_1, l_2, \ldots, l_k)$. Also observe that if no z_j is equal to taxon v_i, then $\text{craf}^{(i)}(z_1, \ldots, z_k)$ is FALSE, because in every partition of X_i, the last element v_i of $X_i \setminus R$ needs to be last in one of the sets S_j. Now, whenever one of z_j is equal to v_i, the function craf^i can be computed recursively as follows:

$$\text{craf}^{(i)}(z_1, \ldots, z_{j-1}, z_j = v_i, z_{j+1} \ldots, z_k) = \bigvee_{\substack{z \in X_{i-1} \setminus R \\ z \prec_{P_j} = v_i}} \text{craf}^{(i-1)}(z_1, \ldots, z_{j-1}, z, z_{j+1}, \ldots, z_k) \tag{1}$$

This recurrence follows simply because removing v_i from every constrained MRAF of X_i gives a constrained MRAF of X_{i-1}. Now we can compute $\text{craf}^{(i)}$ bottom-up using the dynamic programming. For every value $i = 1, \ldots, k$ we enumerate $O(n^k)$ vectors $\boldsymbol{z}$, and compute the value $\text{craf}^{(i)}(\boldsymbol{z})$ using the recursive relation from Eq. (1), thus looking at at most $O(n)$ different entries of $\text{craf}^{(i-1)}$. This thus leads to the overall runtime of $O(k \cdot n^k \cdot n)$. Accounting for the enumeration of the $O(n^{2k})$ pairs l_i, r_i, $i = 1, \ldots, k$ results in the following theorem.

Theorem 2. *MRAF can be computed in time $O(k \cdot n^{3k+1})$ whenever one of the trees is a caterpillar.*

5 Approximation Algorithms

We now provide a polytime approximation algorithm for MRAF (Lemma 4) and relate the approximability of PIMS to that of MRAF on caterpillars (Lemma 5).

Lemma 4. *There is a $O(\log n)$-approximation algorithm for computing MRAF, where $n = |X|$. This algorithm cannot be better than a $(4/3)$ approximation.*

Proof. Given an instance (U, F) of Set Cover, the natural greedy algorithm yields a $O(\log |U|)$ approximation. Recall the encoding of MRAF as a Set Cover instance in Observation 3. We cannot construct this directly, since $|F|$ is potentially exponential in n, but this is not necessary to simulate the greedy algorithm. Let X' be the set of currently uncovered elements of X, initially $X = X'$. We compute a MAST of $T_1|X'$ and $T_2|X'$ in polynomial time [24]. Let S be the leaf-set of this MAST; we add this to our RAF. We then delete S from X' and iterate this process until X' is empty. Figure 6 (in the appendix) shows that this algorithm cannot be better than a $(4/3)$ approximation. □

Lemma 5. *Let π be a permutation of $\{1, \ldots, n\}$ and let T_1 and T_2 be two caterpillars on leaves $\{1, \ldots n\}$ where T_1 is the identity caterpillar and the ith leaf of T_2 is $\pi(i)$. For any solution to the MRAF problem of size k, there is a corresponding solution to the PIMS problem of size at most $k + 2\sqrt{2k}$.*

Proof. We start with an agreement forest for the two caterpillars; each component is itself a caterpillar. We "cut off" one leaf from each end of the components in this forest. (This is because the "interior" of each component induces a monotonic subsequence, but the cherries at the end of each component potentially violate this). This leaves behind a subpermutation of π of length $2k$, which can always be partitioned into at most $2\sqrt{2k}$ monotone subsequences. □

We can create an instance of PIMS from a caterpillar instance of MRAF by treating one caterpillar as the identity and the other as the permutation. Any solution for this PIMS instance yields a feasible MRAF solution. Hence:

$$\text{MRAF} \leq \text{PIMS} \leq \text{MRAF} + 2\sqrt{2 \cdot \text{MRAF}}.$$

Recall that MRAF on caterpillars is in XP by Theorem 3. PIMS is also in XP. Specifically, the PIMS problem is equivalent to the *co-chromatic number* problem on permutation graphs, i.e., partitioning the vertices of a permutation graph into cliques and independent sets. When a graph can be partitioned into r cliques and s independent sets it is sometimes called an (r, s)*-split graph.* It is known that the perfect (r, s)-split graphs can be characterized by a finite set of forbidden induced subgraphs [17]. This implies that their recognition is in XP parameterized by r and s, i.e., when r and s are fixed, (r, s)-split graphs can be recognized in polynomial time—this was later improved to FPT [13]. These XP results are relevant here because they mean that if one of the problems has a polynomial time c-approximation, c constant, then for each fixed constant $\epsilon > 0$ the other has a polynomial-time $(c+\epsilon)$-approximation. For example, given a polynomial-time c-approximation for MRAF, and $\epsilon > 0$, we first use the XP algorithm for PIMS to check in polynomial time whether PIMS $\leq \frac{8c}{\epsilon^2}$. If so, we are done. Otherwise, the described transformation of MRAF solutions to PIMS solutions yields a $(c+\epsilon)$-approximation. The direction from PIMS to MRAF is similar. PIMS has a polynomial-time 1.71-approximation [11]. Hence, for every constant $\epsilon > 0$ MRAF on caterpillars has a polynomial-time $(1.71+\epsilon)$-approximation.

6 Implementation and Experimental Observations

MRAF can be modelled as the *weak chromatic number* of hypergraph: the minimum number of colours assigned to vertices, such that no hyperedge is monochromatic. The set of vertices is X and there is a hyperedge $\{a, b, c, d\}$ whenever the two trees have a different quartet topology on $\{a, b, c, d\}$, leveraging Observation 1. We implemented this as a constraint program (CP) using MiniZinc [22]. For trees with ≤ 30 leaves the CP solves quickly. Code is available at https://github.com/skelk2001/relaxed_agreement_forests. We used this to extend the analysis of [16] on the grass dataset of [12], consisting of fifteen pairs of trees. See Table 1; as expected MRAF grows more slowly than MAF. An FPT algorithm parameterized by MRAF, if it exists, might therefore scale well in practice. FPT algorithms for MAF struggle for MAF ≥ 25 [26]. In fact, MRAF seems more comparable to the *treewidth* of the *display graph* of the input tree pair (obtained by identifying

Table 1. Comparison of MAF and MRAF for the fifteen tree pairs in the data set [12] analysed in [16]. We also include MAST, the lower bound on MRAF given by $\lceil \frac{n}{MAST} \rceil$, and $tw(D)$ which is the treewidth of the display graph obtained from the tree pair.

tree pair	$\|X\| = n$	MAF	MRAF	tw(D)	MAST	$\lceil n/MAST \rceil$
00_rpoC_waxy.txt	10	2	**2**	3	8	2
01_phyB_waxy.txt	14	3	**2**	3	11	2
02_phyB_rbcL.txt	21	5	**3**	3	14	2
03_rbcL_waxy.txt	12	4	**2**	3	9	2
04_phyB_rpoC.txt	21	5	**2**	3	15	2
05_waxy_ITS.txt	15	6	**3**	4	10	2
06_phyB_ITS.txt	30	8	**4**	4	17	2
07_ndhF_waxy.txt	19	5	**3**	4	11	2
08_ndhF_rpoC.txt	34	9	**3**	5	20	2
09_rbcL_rpoC.txt	26	7	**4**	5	14	2
10_ndhF_rbcL.txt	36	7	**4**	3	20	2
11_rbcL_ITS.txt	29	11	**4**	5	17	2
12_ndhF_phyB.txt	40	7	**3**	3	30	2
13_rpoC_ITS.txt	31	11	**4**	6	16	2
14_ndhF_ITS.txt	46	16	**5**	6	20	3

vertices with the same leaf label: the treewidth of this graph is bounded by a function of MAF [16]). We obtained similar results on a more challenging dataset comprising the 163 tree pairs from the dataset in [26] that had at ≤ 50 leaves after pre-processing. See Table 2 in the appendix.

7 Discussion and Open Problems

It remains unclear whether it is NP-hard to compute MRAF on caterpillars, although it seems likely. Can the finite forbidden obstructions that characterize solutions to PIMS be mapped to MRAF on caterpillars and then generalized to general trees? Indeed, how far can MRAF be viewed as a generalization of the PIMS problem to partial orders? Is MRAF on caterpillars FPT? Does it (or PIMS) have a polynomial kernel? What should reduction rules look like, given that rules for MAF seem of limited use (see Appendix A.2)? Strikingly, we do not know whether it is NP-hard to determine whether MRAF ≤ 2 for two general trees, so the FPT landscape is also unclear. How far can the logarithmic approximation for MRAF on general trees, and the 1.71 approximation for MRAF on caterpillars (equivalently, PIMS) be improved? Finally, it will be instructive to elucidate the biological interpretation of this model and to explore MRAF on multiple and/or non-binary trees; such generalisations exist for MAF [7].

References

1. Allen, B., Steel, M.: Subtree transfer operations and their induced metrics on evolutionary trees. Ann. Comb. **5**, 1–15 (2001)
2. Ardevol Martinez, V., Chaplick, S., Kelk, S., Meuwese, R., Mihalák, M., Stamoulis, G.: Relaxed agreement forests. arXiv:2309.01110 [cs.DS] (2023)
3. Atkins, R., McDiarmid, C.: Extremal distances for subtree transfer operations in binary trees. Ann. Comb. **23**, 1–26 (2019)
4. Bar-Yehuda, R., Fogel, S.: Partitioning a sequence into few monotone subsequences. Acta Inform. **35**(5), 421–440 (1998)
5. Björklund, A., Husfeldt, T., Koivisto, M.: Set partitioning via inclusion-exclusion. SIAM J. Comput. **39**(2), 546–563 (2009)
6. Bordewich, M., Semple, C.: On the computational complexity of the rooted subtree prune and regraft distance. Ann. Comb. **8**(4), 409–423 (2005)
7. Bulteau, L., Weller, M.: Parameterized algorithms in bioinformatics: an overview. Algorithms **12**(12), 256 (2019)
8. Chen, J., Shi, F., Wang, J.: Approximating maximum agreement forest on multiple binary trees. Algorithmica **76**, 867–889 (2016)
9. Degnan, J., Rosenberg, N.: Gene tree discordance, phylogenetic inference and the multispecies coalescent. Trends Ecol. Evol. **24**(6), 332–340 (2009)
10. Erdős, P., Szekeres, G.: A combinatorial problem in geometry. Compos. Math. **2**, 463–470 (1935)
11. Fomin, F., Kratsch, D., Novelli, J.C.: Approximating minimum cocolorings. Inf. Process. Lett. **84**(5), 285–290 (2002)
12. Grass Phylogeny Working Group, et al.: Phylogeny and subfamilial classification of the grasses (Poaceae). Ann. Missouri Botanical Garden **88**(3), 373–457 (2001)
13. Heggernes, P., Kratsch, D., Lokshtanov, D., Raman, V., Saurabh, S.: Fixed-parameter algorithms for cochromatic number and disjoint rectangle stabbing via iterative localization. Inf. Comput. **231**, 109–116 (2013)
14. Hein, J., Jiang, T., Wang, L., Zhang, K.: On the complexity of comparing evolutionary trees. Discret. Appl. Math. **71**(1–3), 153–169 (1996)
15. Iersel, L.V., Jones, M., Scornavacca, C.: Improved maximum parsimony models for phylogenetic networks. Syst. Biol. **67**(3), 518–542 (2018)
16. Kelk, S., van Iersel, L., Scornavacca, C., Weller, M.: Phylogenetic incongruence through the lens of monadic second order logic. J. Graph Algorithms Appl. **2**, 189–215 (2016)
17. Kézdy, A., Snevily, H., Wang, C.: Partitioning permutations into increasing and decreasing subsequences. J. Comb. Theory Ser. A **73**(2), 353–359 (1996)
18. Kubicka, E., Kubicki, G., McMorris, F.: On agreement subtrees of two binary trees. Congressus Numerantium 217 (1992)
19. Lemey, P., Salemi, M., Vandamme, A.M.: The Phylogenetic Handbook: A Practical Approach to Phylogenetic Analysis and Hypothesis Testing. Cambridge University Press (2009)
20. Markin, A.: On the extremal maximum agreement subtree problem. Discret. Appl. Math. **285**, 612–620 (2020)
21. Nakhleh, L.: Computational approaches to species phylogeny inference and gene tree reconciliation. Trends Ecol. Evol. **28**(12), 719–728 (2013)
22. Nethercote, N., Stuckey, P.J., Becket, R., Brand, S., Duck, G.J., Tack, G.: MiniZinc: towards a standard CP modelling language. In: Bessière, C. (ed.) CP 2007. LNCS, vol. 4741, pp. 529–543. Springer, Heidelberg (2007). https://doi.org/10.1007/978-3-540-74970-7_38

23. Steel, M.: Phylogeny: Discrete and Random Processes in Evolution. SIAM (2016)
24. Steel, M., Warnow, T.: Kaikoura tree theorems: computing the maximum agreement subtree. Inf. Process. Lett. **48**(2), 77–82 (1993)
25. Wagner, K.: Monotonic coverings of finite sets. J. Inf. Process. Cybern. **20**(12), 633–639 (1984)
26. van Wersch, R., Kelk, S., Linz, S., Stamoulis, G.: Reflections on kernelizing and computing unrooted agreement forests. Ann. Oper. Res. **309**(1), 425–451 (2022)

On the Computational Complexity of Generalized Common Shape Puzzles

Mutsunori Banbara[1], Shin-ichi Minato[2], Hirotaka Ono[1], and Ryuhei Uehara[3(✉)]

[1] Nagoya University, Furocho, Chikusa-ku, Nagoya, Aichi 464-8601, Japan
banbara@nagoya-u.ac.jp, ono@i.nagoya-u.ac.jp
[2] Kyoto University, Kyoto, Japan
minato@i.kyoto-u.ac.jp
[3] School of Information Science, Japan Advanced Institute of Science and Technology, Asahidai, Nomi, Ishikawa 923-1292, Japan
uehara@jaist.ac.jp

Abstract. In this study, we investigate the computational complexity of some variants of generalized puzzles. We are provided with two sets $\mathcal{S}_1$ and $\mathcal{S}_2$ of polyominoes. The first puzzle asks us to form the same shape using polyominoes in $\mathcal{S}_1$ and $\mathcal{S}_2$. We demonstrate that this is polynomial-time solvable if $\mathcal{S}_1$ and $\mathcal{S}_2$ have constant numbers of polyominoes, and it is strongly NP-complete in general. The second puzzle allows us to make copies of the pieces in $\mathcal{S}_1$ and $\mathcal{S}_2$. That is, a polyomino in $\mathcal{S}_1$ can be used multiple times to form a shape. This is a generalized version of the classical puzzle known as the common multiple shape puzzle. For two polyominoes P and Q, the common multiple shape is a shape that can be formed by many copies of P and many copies of Q. We show that the second puzzle is undecidable in general. The undecidability is demonstrated by a reduction from a new type of undecidable puzzle based on tiling. Nevertheless, certain concrete instances of the common multiple shape can be solved in a practical time. We present a method for determining the common multiple shape for provided tuples of polyominoes and outline concrete results, which improve on the previously known results in the puzzle community.

Keywords: Common shape puzzle · shape logic · least common multiple shape · NP-completeness · polyform compatibility · polypolyomino · SAT-based solver · undecidability

1 Introduction

Research on the computational complexity of puzzles and games has become increasingly important in theoretical computer science (see [13] for a comprehensive survey). Since the 1990s, numerous puzzles have been demonstrated to be NP-complete in general. These results provide a certain amount of some common intuition of the properties of the NP class. However, it has not been possible to

H. Fernau et al. (Eds.): SOFSEM 2024, LNCS 14519, pp. 55–68, 2024.
https://doi.org/10.1007/978-3-031-52113-3_4

capture certain puzzles, among which the sliding block puzzle is representative. The complexity of these kinds of puzzles had remained an open problem since Martin Gardner pointed out in the 1960s that some certain theory is required to understand such puzzles. However, after 40 years, Hearn and Demaine proposed a framework known as constraint logic, and demonstrated that these puzzles are PSPACE-complete [8,9] (some related work was also done by Flake and Baum [5]). Combinatorial reconfiguration problems have been investigated towards an understanding of the PSPACE class [10].

With the developments in theoretical computer science in the past decade, new series of puzzles have been developed in the puzzle community. In comparison to classical packing puzzles, one major property of these puzzles is that the target shape is not explicitly stated. The first example is the *symmetric shape puzzle*. This puzzle asks us to form a symmetric shape using a given set of pieces. It is extremely challenging to solve such a puzzle because we cannot be sure whether or not we are approaching the goal. This property makes the puzzle very difficult, and in fact, only a few pieces are sufficient to cause this difficulty [4]. The second example is the *anti-slide puzzle*. This puzzle asks us to interlock a given set of pieces. A typical instance asks us to pack the given pieces into a frame so that no piece can be slid in the frame. This puzzle is also difficult because the goal is not explicitly stated. The computational complexity of this puzzle was recently investigated by [11].

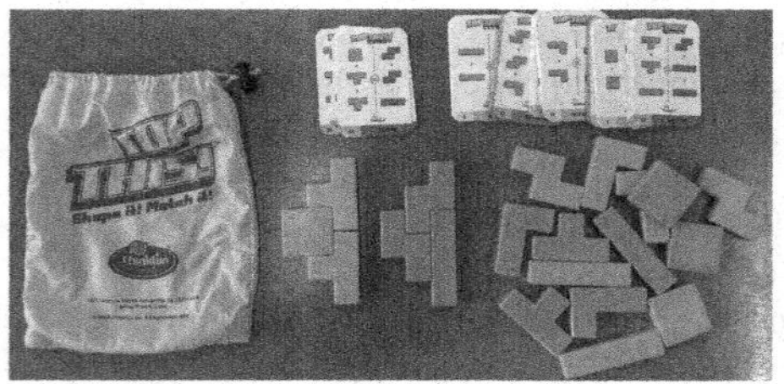

Fig. 1. Shape Logic (commercial product by ThinkFun).

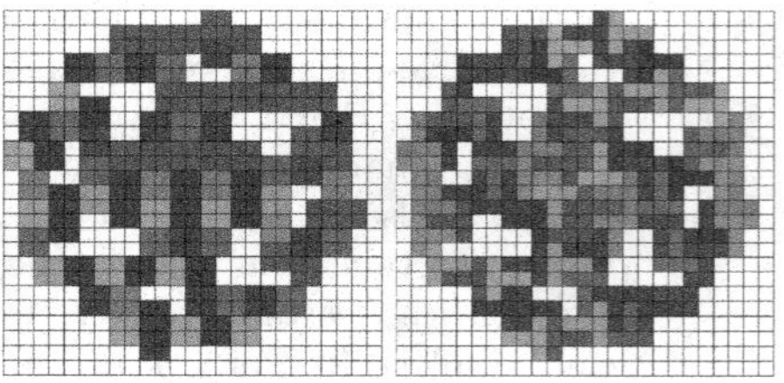

Fig. 2. Copies of a pentomino and a tetromino share a large common shape.

In this study, we focus on such a puzzle that is known as the *common shape puzzle*. Many instances of this puzzle are available in the puzzle community, and a commercial product named "Shape Logic" exists. (The authors confirmed that this puzzle was named "Top This!" in 2008 (Fig. 1), "ShapeOmetry" in 2012, and "Shape Logic" more recently by the same puzzle maker. The puzzle "Top This!" won three awards in 2008.[1] However, in this paper, we use the most recent name.) In the shape logic puzzle, we are provided with two sets $\mathcal{S}_1$ and

[1] https://www.thinkfun.com/about-us/awards/.

$\mathcal{S}_2$ of polygons. We must find a polygon X that can be formed by the pieces in $\mathcal{S}_1$ and $\mathcal{S}_2$, respectively, as in the classic silhouette puzzle Tangram. The main difference between the Tangram and the shape logic puzzle is that the target shape X is not provided, which drastically increases the difficulty of the puzzles.

Hereafter, we suppose that $\max\{\,|\mathcal{S}_1|\,/\,|\mathcal{S}_2|\,,\,|\mathcal{S}_2|\,/\,|\mathcal{S}_1|\,\}$ is bounded from above by a constant. We note that if $\mathcal{S}_1$ contains only one piece, the target shape X is fixed to it. Therefore, it is equivalent to the classic puzzle Tangram for $\mathcal{S}_2$, and it is NP-complete even if all pieces in $\mathcal{S}_1$ and $\mathcal{S}_2$ are rectangles [3].

We first demonstrate that it is polynomial-time solvable if $|\mathcal{S}_1| + |\mathcal{S}_2|$ is a constant. Subsequently, we show that the shape logic puzzle is strongly NP-complete even if all the pieces in $\mathcal{S}_1$ and $\mathcal{S}_2$ are small rectangles. We state that a rectangle in $\mathcal{S}_1 \cup \mathcal{S}_2$ is small if its size is polynomial in $(\,|\mathcal{S}_1| + |\mathcal{S}_2|\,)$.

Next, we focus on a similar puzzle named the *common multiple shape puzzle*, which has been investigated in the puzzle community for a long time under a few different names such as "polypolyomino"[2] and "polyform compatibility"[3]. We propose the term "the (least) common multiple shape" based on the term "the least common multiple," which is the corresponding Japanese name used in Japanese puzzle community.[4] An instance of this puzzle is as follows: We are provided with two polygons P and Q. The puzzle asks us to find the (smallest area) shape X that can be tiled by P, and also tiled by Q. That is, we can use any number of copies of P and Q, and find the common shape X that can be filled by only copies of P, as well as only copies of Q. This puzzle aims at finding the minimum shape, however, it is known that some pairs result in a huge solution (e.g., Fig. 2). The problem of finding a small common multiple shape of the T-pentomino and O-tetromino (shown in Fig. 2) was first proposed by Robert Wainright as a problem at the conference of games and puzzles competitions on computers[5] in 2005 and 2011. A solution with an area of 600 was found in 2011, and it was improved to 340 in 2011.[6] However, it remains open whether or not this shape with an area of 340 in Fig. 2 is the smallest.

We naturally consider the (least) common multiple shape variant of the shape logic puzzle. That is, for given sets $\mathcal{S}_1$ and $\mathcal{S}_2$ of polygons, the puzzle asks us to find a small common shape X that can be filled by copies of pieces in $\mathcal{S}_1$ (and $\mathcal{S}_2$, respectively). We show that this puzzle is undecidable even if each set of $\mathcal{S}_1$ and $\mathcal{S}_2$ contains small polyominoes. As a corollary, we also demonstrate that the following problem is undecidable: For a given set $\mathcal{S}$ of small polyominoes, determine whether a rectangle can be formed using copies of the pieces in $\mathcal{S}$.

In this study, we also present a formulation of these puzzles and verify the feasibility with a computer. We recently discovered that such puzzles can be solved by SAT-based solvers with sophisticated modeling far more efficiently

---

2 https://www.iread.it/Poly/.

3 https://sicherman.net/polycur.html.

4 In Japan, we use 最小公倍図形 (least common multiple shape) following 最小公倍数 (least common multiple number).

5 http://hp.vector.co.jp/authors/VA003988/gpcc/gpcc.htm.

6 http://deepgreen.game.coocan.jp/MCFG/MCFG_index.htm.

than when using other methods [1]. By determining an efficient formulation of this puzzle and using a SAT-based solver, we also improve several known instances of the common multiple shapes that have been investigated in the puzzle community.

2 Preliminaries

A *polyomino* is a polygon that can be obtained by joining one or more unit squares edge to edge [7]. In this study, we only consider simple polyominoes (without holes) as polygons. (We note that even if all pieces are simple, the solution may have holes, as indicated in Fig. 2.) If a polyomino P is formed by k unit squares, we refer to P as a *k-omino*. For a specific k, we also refer to it as a *monomino*, *domino*, *tromino*, *tetromino*, *pentomino*, and *hexomino* for $k = 1, 2, 3, 4, 5, 6$, respectively.

A set $\mathcal{S}_1$ of polyominoes is said to be a set of *small* polyominoes when the maximum polyomino in $\mathcal{S}_1$ is a k-omino for $k = O(|\mathcal{S}_1|^c)$ for a positive constant c. In this case, we assume that the input size of the problem is bounded by $O(p(n))$ for a polynomial function p, where $n = |\mathcal{S}_1| + |\mathcal{S}_2|$.

In this study, we consider two problems on polyominoes. The first problem is the *shape logic puzzle*. Given two sets of $\mathcal{S}_1$ and $\mathcal{S}_2$ of small polyominoes, the puzzle asks us to form a common polyomino X using all pieces in $\mathcal{S}_1$ and all pieces in $\mathcal{S}_2$, respectively. The goal shape X is not provided. Clearly, the shape logic puzzle is in NP when all pieces are small as we can guess X and verify the feasibility of the given packing of $\mathcal{S}_1$ and $\mathcal{S}_2$ on X in polynomial time.

The second problem is the *common multiple shape puzzle*. Given two finite sets $\mathcal{S}_1$ and $\mathcal{S}_2$ of polyominoes, the puzzle asks us to form a common polyomino X with a positive area using copies of the pieces in $\mathcal{S}_1$ and copies of the pieces in $\mathcal{S}_2$, respectively. This puzzle generalizes both of the shape logic puzzle and the puzzle known as the *polypolyomino* (also referred to as *polyform compatibility*). The latter puzzle is the case in which $|\mathcal{S}_1| = |\mathcal{S}_2| = 1$. It can be extended from two sets to three or more sets naturally. (See Sect. 5 in this case).

For a finite set $\mathcal{S}$ of polyominoes, we define a set $\hat{\mathcal{S}}$ of polyominoes P such that P can be formed by copies of the pieces in $\mathcal{S}$. Clearly, $\hat{\mathcal{S}}$ is infinite and countable. That is, the common multiple shape puzzle asks whether or not $\hat{\mathcal{S}}_1 \cap \hat{\mathcal{S}}_2 \neq \emptyset$. When $\hat{\mathcal{S}}_1 \cap \hat{\mathcal{S}}_2 \neq \emptyset$, we refer to an element in $\hat{\mathcal{S}}_1 \cap \hat{\mathcal{S}}_2$ as a *common multiple shape*. Among the common multiple shapes, the smallest one is the *least common multiple shape*.

3 Complexity of Shape Logic Puzzle

In this section, we focus on the generalized shape logic puzzle. That is, given two sets $\mathcal{S}_1$ and $\mathcal{S}_2$ of polyominoes, we need to decide if all pieces in $\mathcal{S}_1$ (and in $\mathcal{S}_2$) can form a common polyomino X.

Observation 1. *When $|\mathcal{S}_1| + |\mathcal{S}_2|$ is a constant k, the generalized shape logic puzzle can be solved in time polynomial in n, where n is the total number of vertices in $\mathcal{S}_1 \cup \mathcal{S}_2$.*

Proof (Outline). We solve the problem by brute force using the same technique as in [4, Section 3]. In [4, Section 3], they presented a method for solving the symmetric assembly puzzle, which asks us to form a symmetric shape by using the pieces in a set of (general) simple polygons in polynomial time.

The generalized shape logic puzzle can be reduced to the symmetric shape puzzle in polynomial time as follows: Suppose that the generalized shape logic puzzle has a solution and the pieces in $\mathcal{S}_1$ forms a polygon P, which can also be formed by the pieces in $\mathcal{S}_2$. Then, without loss of generality, P can be placed so that its rightmost vertex v is on ∂P; that is, any point in P is not right of v. At this point, we obtain a symmetric shape by joining P and its mirror image P^R at vertex v with its mirror image on ∂P^R. That is, when the shape logic puzzle with $\mathcal{S}_1$ and $\mathcal{S}_2$ has a solution, the symmetric shape puzzle also has a solution for $\mathcal{S}_1 \cup \mathcal{S}_2$ such that the left half of the symmetric shape consists of the pieces in $\mathcal{S}_1$ and the right half of the symmetric shape consists of the pieces in $\mathcal{S}_2$. The proof in [4, Section 3] is based on brute force. Therefore, we can restrict our search to a symmetric shape that also provides a solution for the shape logic puzzle in $\mathcal{S}_1 \cup \mathcal{S}_2$. As the original brute force algorithm for the symmetric shape puzzle runs in time polynomial in n, so does our algorithm. □

We note that the brute force algorithm in the proof of Observation 1 runs in $O(n^{f(k)})$ for some polynomial function f. That is, the generalized shape logic puzzle problem is fixed parameter tractable.

Theorem 2. *The shape logic puzzle is strongly NP-complete even if all pieces in $\mathcal{S}_1$ and $\mathcal{S}_2$ are small rectangles.*

Proof. According to the definition of a small polyomino, the problem is in NP. Thus, we demonstrate the NP-hardness using a reduction from the 3-partition problem. In the 3-partition problem, we are provided with a multiset of $3m$ positive integers $A = \{a_1, a_2, \ldots, a_{3m}\}$, where the a_is are bounded from above by a polynomial of m. The goal is to partition the multiset A into m triples such that every triple has the same sum $B = (\sum_{i=1}^{3m} a_i)/m$. It is known that the 3-partition problem is strongly NP-complete even if every a_i satisfies $B/4 < a_i < B/2$ [6, SP16]. Without loss of generality, we assume that $a_i > 3$ for each i and $B = 3mB'$ for a positive integer B'. Then, the set $\mathcal{S}_1$ consists of $3m$ rectangles of size $1 \times (a_i + 3m^2)$ for each $i = 1, 2, \ldots, 3m$. Furthermore, $\mathcal{S}_2$ consists of $3m$ congruent rectangles of size $m \times (B/(3m) + 3m)$. The construction can be computed in time polynomial in m.

Subsequently, we observe that $3m < B/(3m) + 3m < B/4 + 3m^2 < a_i + 3m^2$ for each i, as $3m$ pieces exist in $\mathcal{S}_1$, $a_i > 3$, and $a_i > B/4$. That is, (1) $B/(3m) + 3m$ is larger than $3m$, which is the number of long and slender rectangles in $\mathcal{S}_1$, and (2) each length $a_i + 3m^2$ cannot fit into any rectangle that is formed by the pieces in $\mathcal{S}_2$ except if a rectangle with a width of m is created. Therefore, the only means of forming the same shape using the pieces in $\mathcal{S}_1$ and the pieces in $\mathcal{S}_2$ is to form a rectangle with a size of $m \times (B + 9m^2)$ using the pieces in $\mathcal{S}_2$ and to pack long rectangles with a size of $1 \times (a_i + 3m^2)$ into this frame. The arrangement of pieces in $\mathcal{S}_1$ directly provides the solution to the original instance of the 3-partition problem. □

4 Undecidability of Common Multiple Shape Puzzle

In this section, we demonstrate that the common multiple shape puzzle is undecidable. We first show that a generalized jigsaw puzzle is undecidable. In this puzzle, each edge is colored, which will be modified to polygonal shapes without color.

4.1 Undecidability of a Generalized Jigsaw Puzzle

We first consider a generalized jigsaw puzzle. We borrow several notions from [2]. Each piece is a square with four edges and has its own color. We denote the set of colors as $C = \{0, 1, 2, \ldots, c, \bar{1}, \bar{2}, \ldots, \bar{c}\}$. In our jigsaw puzzle, we tile the pieces into a rectangular frame so that each edge is shared by two adjacent pieces with colors i and $\bar{i}$, except for the boundary of the frame. A special color 0 exists, which should match to the frame. That is, when we tile the pieces, the outer boundary has the color 0, and no inside edge has the color 0.

In our jigsaw puzzle, we are allowed to use copies of a piece in $\mathcal{S}$ multiple times, which is the significant difference between our puzzle and that in [2]. Therefore, for a given finite set $\mathcal{S}$, we have infinitely countable means of tiling the pieces. Subsequently, the jigsaw puzzle problem is defined as follows:

Input: A set $\mathcal{S}$ of unit square pieces such that each piece has four colors in C on its four edges.

Output: Decide if there is a polyomino region R such that R can be tiled by copies of pieces in $\mathcal{S}$ in which each inner edge is shared by two adjacent pieces of colors i and $\bar{i}$ (with $i > 0$), and each edge on the boundary ∂R has the color 0.

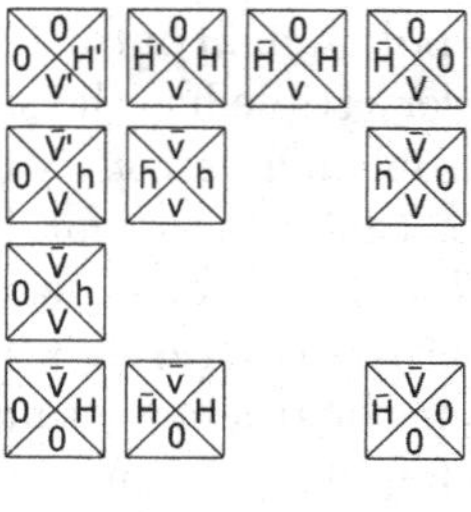

We first present the following lemma. Intuitively, the rectangle in the following lemma will be used as a template for the other set of pieces.

Lemma 1. *There exists a finite set $\mathcal{S}$ of jigsaw puzzle pieces such that the area R is tiled by copies of the pieces in $\mathcal{S}$ with the boundary color 0 along ∂R if and only if R is a rectangle with a size of at least 3×3.*

Fig. 3. Jigsaw puzzle in a rectangle.

Proof. We consider the set of 11 jigsaw puzzle pieces depicted in Fig. 3. For ease of reference, we use some letters such as H, V, etc. instead of the numbers 1, 2, etc. in the figure. As every piece contains H,

$\bar{H}$, H', $\bar{H}'$, h, or $\bar{h}$, without loss of generality, we can assume that one piece is placed so that its $\bar{H}$, $\bar{H}'$, or $\bar{h}$ is on its left, and all the pieces are then aligned in the same direction, as indicated in the figure. The boundary of the jigsaw puzzle is labeled by 0. We observe that we cannot form a rectangle with a size of $2 \times n$ (and $n \times 2$) for any n because V' and $\bar{V}$ (and H' and $\bar{H}$) do not match. Furthermore, we observe that an edge is colored by h or v if and only if it is not incident to a vertex on the boundary. In particular, we cannot create a corner with the angle 270°; to achieve this, we must place one "corner boundary" piece inside, which results in the color 0 being inside the shape, and this is not permitted. □

We show that our jigsaw puzzle problem is undecidable.

Lemma 2. *There exists a finite set $\mathcal{S}$ of pieces of the jigsaw puzzle such that the jigsaw puzzle is undecidable.*

Proof (Outline). We present a polynomial-time reduction from the following Post Correspondence (PC) problem:

Input: A sequence of pairs of strings $s_1 = (t_1; b_1), s_2 = (t_2; b_2), \ldots, s_n = (t_n, b_n)$. We define $T(s_i) = t_i$,$B(s_i) = b_i$ for a pair $s_i = (t_i; b_i)$.

Question: Decide if there exists a sequence of pairs $s_{i_1}, s_{i_2}, s_{i_3}, \ldots, s_{i_k}$ of strings such that $T(s_{i_1})T(s_{i_2})T(s_{i_3}) \cdots T(s_{i_k}) = B(s_{i_1})B(s_{i_2})B(s_{i_3}) \cdots B(s_{i_k})$.

Let Σ be an alphabet, namely the set of letters that is used in the sequence. We note that we can use each pair s_i can be used any number of times. It is well known that the PC problem is undecidable even if $|\Sigma|$ is a constant [12].

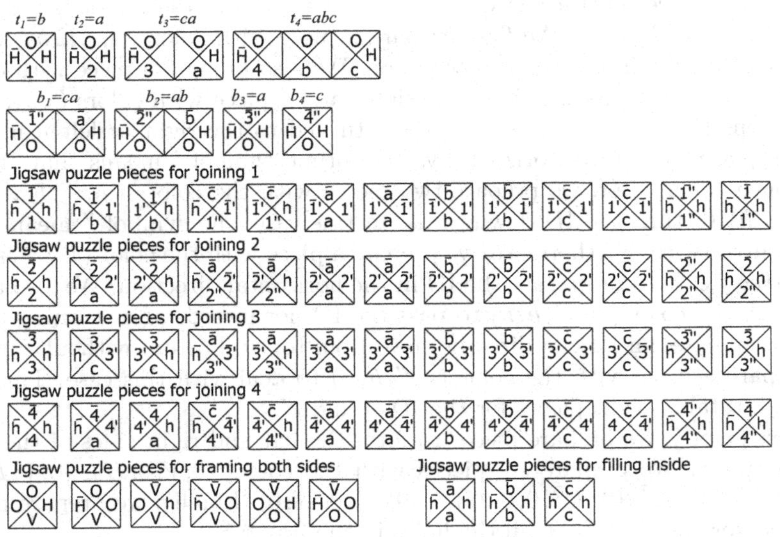

Fig. 4. A reduction from the PC problem to the jigsaw puzzle problem.

We demonstrate the reduction by using a concrete example $\Sigma = \{a, b, c\}$, $s_1 = (b; ca), s_2 = (a; ab), s_3 = (ca; a), s_4 = (abc; c)$ (Fig. 4). We prepare one piece, one piece, two pieces, and three pieces of jigsaw puzzle for each string $t_1 = b$, $t_2 = a$, $t_3 = ca$, and $t_4 = abc$, respectively. We set each string to be uniquely constructible: two pieces for $t_3 = ca$ have their own color (distinct from any other color) between them, and three pieces for $t_4 = abc$ have their own two colors between a and b, and b and c (these are blank in Fig. 4). The top color is 0, the left color is $\bar{H}$, and the right color is H for each piece. (As in the proof of Lemma 1, we regard certain letters as numbers greater than n). Hereafter, we consider these strings as rectangular pieces that are represented by sizes of 1×1, 1×1, 2×1, and 3×1, respectively. The leftmost bottom color of the rectangular piece t_i is the color i, which is referred to as the *ID* of this rectangular piece. The color of the other edge corresponds to the letter in the string. That is, the second and the third pieces of the rectangle representing $t_4 = abc$ have the colors b and the color c, respectively. (We regard these letters as unique numbers in the color set C. As the size of the alphabet Σ is a constant, regarding these letters as numbers has no influence on our arguments).

Subsequently, we prepare two, two, one, and one pieces for the strings $b_1 = ca$, $b_2 = ab$, $b_3 = a$, and $b_4 = c$, respectively. As with the strings t_i, b_1, b_2, b_3, b_4 each corresponds to a rectangular piece with a size of 2×1, 2×1, 1×1, and 1×1, respectively. The bottom color is 0, the left color is $\bar{H}$, and the right color is H for these rectangular pieces. The top colors of the rectangular piece are represented by the letters, except for the leftmost edge, which has the color ID $\bar{i''}$ for the string b_i.

Next, we prepare to join two pieces with the IDs i and i''. Hereafter, we use (c_u, c_b, c_l, c_r) to denote the top color c_u, bottom color c_b, left color c_l, and right color c_r of a piece. Furthermore, we assume that the top letter of t_i is x_i and the top letter of b_i is y_i. We first prepare a piece with colors $(\bar{i}, i, \bar{h}, h)$, which is a wire of the ID in the vertical direction. We also prepare a piece with colors $(\bar{i}, x_i, \bar{h}, i')$, which turns the ID to the right, and a piece with colors $(\bar{i}, x_i, \bar{i'}, h)$, which turns the ID to the left. The ID is turned to the right or left using one of these pieces and runs horizontally. The prime symbol $'$ means that the ID turns once. Thereafter, we prepare two pieces with the colors $(\bar{y_i}, i'', \bar{h}, i')$ and $(\bar{y_i}, i'', \bar{i'}, h)$ to turn the ID downwards. In this case, the symbol $''$ means that the ID turns twice. Furthermore, we prepare a piece with the color $(\bar{j}, j, \bar{i'}, i')$ for each letter $j \in \Sigma$ to propagate the ID in the horizontal direction. We also add a piece with the color $(\bar{i''}, i'', \bar{h}, h)$ to pass the ID downwards after turning twice. In a special case, an ID can directly move from top to bottom without turning. We prepare a piece with the color $(\bar{i}, i'', \bar{h}, h)$ to deal with this case. Thus, we prepare a total of eight pieces for the IDs i and i''.

Subsequently, we prepare pieces to form the left and right sides of the rectangular frame. We prepare six pieces with the colors $(0, V, 0, H)$, $(0, V, \bar{H}, 0)$, $(\bar{V}, V, 0, h)$, $(\bar{V}, V, \bar{h}, 0)$, $(\bar{V}, 0, 0, H)$, and $(\bar{V}, 0, \bar{H}, 0)$. Finally, we prepare pieces $(\bar{j}, j, \bar{h}, h)$ for each $j \in \Sigma$ to fill the holes in the frame.

We prepare $\sum_{i=1}^{n}(\,|t_i|\, + \,|b_i|\,) + 8n + 6 + |\Sigma|$ pieces in total. Therefore, the jigsaw puzzle can be constructed in time polynomial in the size of a given instance of the PC problem.

We demonstrate that the instance $s_1 = (t_1; b_1), s_2 = (t_2; b_2), \ldots, s_n = (t_n, b_n)$ of the PC problem has a solution if and only if the jigsaw puzzle has a solution such that a rectangular area R is filled with copies of the pieces with the color 0 only on ∂R.

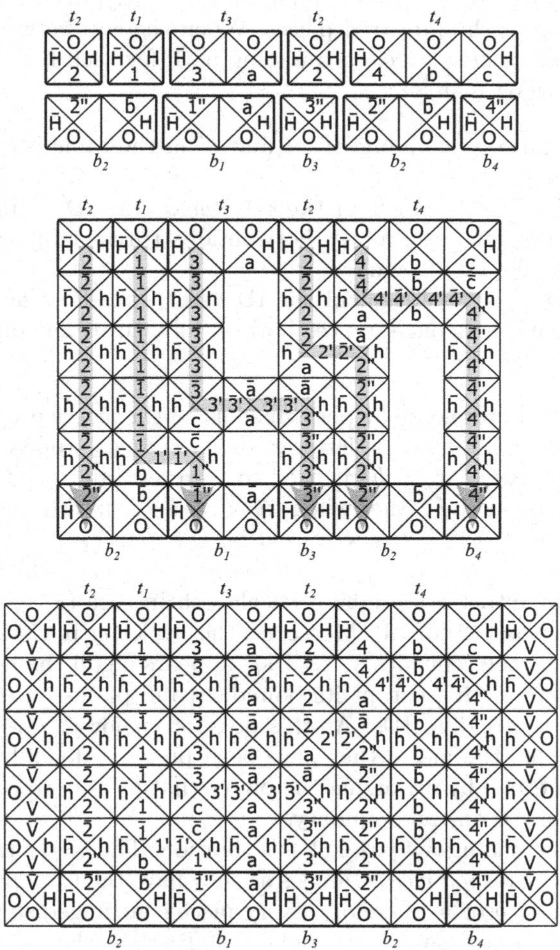

Fig. 5. Construction of a solution of the jigsaw puzzle from a solution of the instance of the PC problem.

We first assume that the sequence $s_{i_1}, \ldots, s_{i_k}$ is a solution, from which we construct a rectangular shape using the set of pieces of the jigsaw puzzle. We use $s_1 = (b; ca), s_2 = (a; ab), s_3 = (ca; a), s_4 = (abc; c)$ as an example (Fig. 5). Intuitively, we verify that two corresponding IDs are joined by a zig-zag path with two (or zero) turns, these zig-zag paths do not cross one another, and the corresponding letters are joined by vertical matching pieces.

We first align the rectangular pieces $t_{i_1}, t_{i_2}, \ldots, t_{i_k}$ on the top row and $b_{i_1}, b_{i_2}, \ldots, b_{i_k}$ on the bottom row following the solution $s_{i_1}, \ldots, s_{i_k}$ of the PC problem. (As a reminder, each string t_i (and s_i) produces unique rectangular pieces.) Thus, we obtain 0s on the top and bottom boundaries, and we can join all rectangular pieces by matching h and $\bar{h}$. Subsequently, we join all corresponding pairs of IDs using the prepared pieces. When the gap between the top and bottom rows is sufficiently large, each joining path for each ID i can be created in one of the following manners:

(1) The pair of the corresponding IDs i and i'' in the same column is directly joined vertically,
(2) When the ID i'' of b_i is left of the ID i of t_i, the ID i first moves down vertically, turns left once, moves horizontally, turns right once, and moves downwards to the ID i'', and
(3) When the ID i'' of b_i is right of the ID i of t_i, the ID i first moves down vertically, turns right once, moves horizontally, turns left once, and moves downwards to the ID i''.

Any of these procedures can be performed using the pieces prepared as above. Note that in the case (2), the first letter x_i of the string t_i appears at the corner when we use the piece $(\bar{i}, x_i, \bar{i'}, h)$ is used to turn left, and the first letter y_i of the string b_i appears at the corner when $(\bar{y}_i, i'', \bar{h}, i')$ is used. In the case (3), the pieces $(\bar{i}, x_i, \bar{h}, i')$ and $(\bar{y}_i, i'', \bar{i'}, h)$ are used for this purpose.

Following all the above steps, we can observe that each corresponding pair of IDs is joined by either a straight vertical path in case (1) or a zig-zag path with two turns in cases (2) or (3). Moreover, the ith letter in the common string that is produced by the sequence $s_{i_1}, \ldots, s_{i_k}$ appears on all the horizontal edges of the ith piece (from left), except its boundary and pieces on the vertical line that join two corresponding IDs. At this holds even for the holes, we can fill all of the holes using the pieces that have been prepared for filling. Finally, we can complete the frame by arranging the pieces that have been prepared for the frame with the color 0 on the boundary.

We assume that the jigsaw puzzle has a solution. The pieces that correspond to t_i and b_i form the respective rectangles as they have their unique colors. As all pieces have the color h or $\bar{h}$, therefore, every piece is arranged so that $\bar{h}$ appears on the left side and h appears on the right side. Because the color 0 matches no other colors, the rectangles for t_i and the corner pieces of color 0 on the upper edges are arranged on the top row, as are the rectangles for b_i and the corner pieces of color 0 on the lower edges. We need to form a rectangle using the pieces of the color 0 on the left or right side. (Although it may appear that we can form any polyomino other than rectangles, we cannot create any concave corner of 270° because an edge with the color 0 cannot be placed inside the polyomino).

According to the color properties, the ID color of each rectangle corresponding to t_i should be connected to the ID color of each rectangle for b_i, and these k paths cannot cross. If a path has no turn, it is necessary to use some copies of the piece with the color $(\bar{i}, i, \bar{h}, h)$, one copy of the piece $(\bar{i}, i'', \bar{h}, h)$, and some copies of the piece $(\bar{i''}, i'', \bar{h}, h)$. If a path has turns, the only possible solution is that the color i of t_i starts vertically, is changed to i' after one 90° turn, moves horizontally, is changed to i'' after one 90° turn, and moves down to the piece in the rectangle corresponding to b_i. The colors of t_i and b_i appear at each turn on a horizontal edge. Thus, the remaining holes should be packed using the pieces with the color $(\bar{j}, j, \bar{h}, h)$ for the matching color j, and each pair of IDs of t_i and b_i should match.

Therefore, when the jigsaw puzzle has a solution, the pieces form a rectangle, the pieces for t_i are arranged on the top, the pieces for b_i are arranged on the bottom, the corresponding pairs of IDs of t_i and b_i match, and a consistent letter is obtained along each vertical line of the pieces. Thus, s_i can be arranged following the sequence, and the same sequence of letters that is produced by the sequences of t_i and b_i can be obtained, which provides a solution to the PC problem, thereby completing the proof. □

4.2 Undecidability of the Common Multiple Shape Puzzle

We now turn to the common multiple shape puzzle. Lemmas 1 and 2 imply the following Theorem.

Theorem 3 *For two finite sets $\mathcal{S}_1$ and $\mathcal{S}_2$ of small polyominoes, the common multiple shape puzzle for $\mathcal{S}_1$ and $\mathcal{S}_2$ is undecidable.*

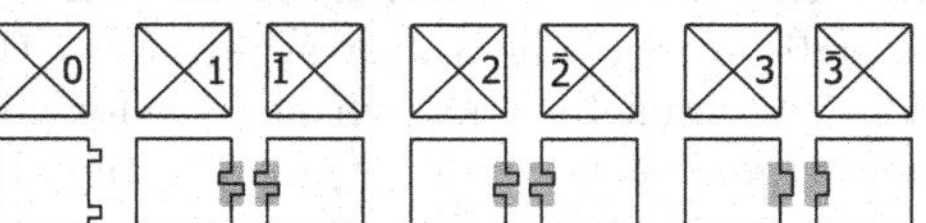

Fig. 6. Colored jigsaw piece for a polyomino. Each color i corresponds to a zigzag pattern that represents the integer i in the binary system. The color $\bar{i}$ is its negative.

Proof (Outline). We first demonstrate how to represent each piece of the jigsaw puzzle in Lemmas 1 and 2 using a small polyomino. The basic concept is explained in [3, Fig. 7]. Each color is represented by its original zig-zag pattern. See Fig. 6 for an example of the representation. Using the binary system, the size of the polyomino is $O((\log |C|)^2)$, where C is the set of colors.

We consider the set $\mathcal{S}_1$ of jigsaw pieces in Lemma 1 and the set $\mathcal{S}_2$ of jigsaw pieces in Lemma 2. Different colors can be used for each set, except the common color 0. Subsequently, according to Lemma 1, a solution to the common multiple shape puzzle is a shape that corresponds to a rectangle. Moreover, according to Lemma 2, whether it can be constructed using the pieces in $\mathcal{S}_2$ is undecidable. The number of colors used in $\mathcal{S}_1 \cup \mathcal{S}_2$ is linear with the size n of the input. Thus, each polyomino has an area of $O((\log n)^2)$, which means that it is small. This completes the proof. □

5 Improved Solutions for Common Multiple Shapes

In this section, we provide a brief formulation of generalized common shape puzzles. The rep-tile problem,[7] which is a type of packing puzzle on polyominoes, has been formulated and examined using several different computer methods [1] recently. In [1], the authors demonstrated that the rep-tile problem can be formulated in a natural form that can be handled using various methods. They compared a well-known puzzle solver, a few algorithms based on dancing links, an MIP solver, and a SAT-based solver with respect to for solving the packing puzzles. In [1], the authors concluded that the SAT-based solver is significantly faster than the other methods. The common shape puzzle has similar properties to the rep-tile problem. Therefore, we examined several instances of the common shape puzzle that are available online,[8] and improved some of the known results by using the SAT-based solver used in [1].

For example, the previous best known shape for F5Q4T4 on https://www.iread.it/Poly/ was a 760-omino, and our new shape is only 160-omino (Fig. 7). (Here, "F5Q4T4" means that this problem asks for finding a common shape of using copies of F-pentomino, Q-tetromino, and T-tetromino, respectively, which are commonly used in the puzzle society. See https://www.iread.it/lz/pttomini.html for details.) The previous best known shape for T5L4Q4 on https://sicherman.net/n445com/n445com.html, which was a 560-omino, is improved to 480-omino (Fig. 8). In addition to them, we improved the following cases: The previous best known shapes for I5P5T5, I5P5Z5, L5P5X5, and P5U5V5 on https://sicherman.net/rosp/triplep.html were 120-omino, 200-omino, 400-omino, and 160-omino, respectively. We obtain new better shapes of 110-omino for I5P5T5, 150-omino for I5P5Z5, 360-omino for L5P5X5, 120-omino for P5U5V5, respectively.

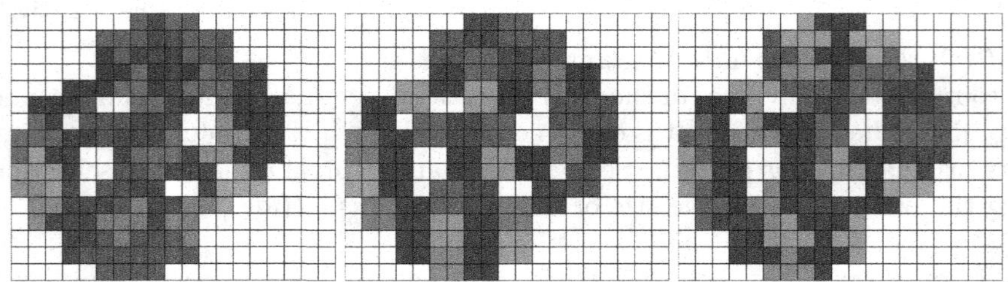

Fig. 7. Tiling patterns for F5Q4T4 improved from 760-omino to 160-omino.

Two main differences exist between the formulations of the common shape puzzle and the packing puzzle in [1]. The first one is that the goal shape is not

[7] A polygon P is called a *rep-tile* if it can be divided into congruent polygons with each other similar to P.

[8] https://www.iread.it/Poly/ and https://sicherman.net/polycur.html.

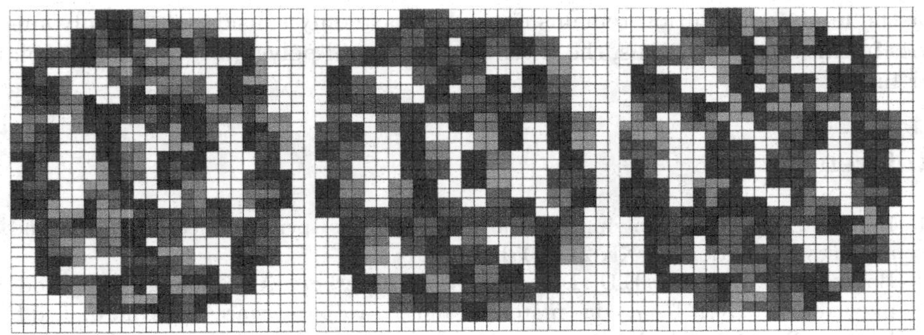

Fig. 8. Tiling patterns for T5L4Q4 improved from 560-omino to 480-omino.

provided in the common shape puzzle, whereas it is provided in the packing puzzle. The second is that we must create a common (or congruent) shape using two different sets $\mathcal{S}_1$ and $\mathcal{S}_2$ of pieces in the common shape puzzle, whereas we have only one set of pieces in the packing puzzle.

To address the first point, we fix the bounding box of the goal shape. We first fix the number of pieces (or $|\mathcal{S}_1|$ and $|\mathcal{S}_2|$), and we attempt to create possible bounding boxes that contains these pieces.

In the packing puzzle, we can assume that each unit square of a goal shape is covered exactly once by a piece. However, in the common shape puzzle, each unit square of a bounding box is covered by either 0 or 2 pieces. Moreover, when the square is covered by 2 pieces, these should be in $\mathcal{S}_1$ and $\mathcal{S}_2$.

We can modify the formulation of the packing puzzle in [1] to that for the common shape puzzle using these concepts. Furthermore, it is straightforward to extend the problem from two sets $\mathcal{S}_1$ and $\mathcal{S}_2$ to three sets $\mathcal{S}_1$, $\mathcal{S}_2$, and $\mathcal{S}_3$ (and more).

6 Concluding Remarks

We have considered the computational complexities of generalized common shape puzzles, in which the goal shapes are not provided. The puzzle is tractable when the number of pieces is a constant; however, it is strongly NP-complete even if the piece sets consist of small rectangles. Moreover, if we are allowed to use the copies of the pieces repeatedly, the problem becomes undecidable. It is possible to formulate the puzzle for several different solvers in a natural form, and we improved some known records for concrete instances using a SAT-based solver. However, we have not yet succeeded in confirming that the results are the minimum solutions. For example, we verified the pattern in Fig. 2 for each boundary box with a size of $i \times \lfloor 625/i \rfloor$ using $1 \leq i \leq 25$ and confirmed that there are no smaller patterns in these boundary boxes. However, this does not imply that the pattern in Fig. 2 is the smallest area pattern. Thus, efficient searching for the minimum solution remains open. We have only considered the polyominoes in this study, and thus, the extension to general polygons is a natural topic for future work.

References

1. Banbara, M., et al.: Solving rep-tile by computers: performance of solvers and analyses of solutions. arXiv:2110.05184 (2021)
2. Bosboom, J., Demaine, E.D., Demaine, M.L., Hesterberg, A., Manurangsi, P., Yodpinyanee, A.: Even $1 \times n$ edge matching and jigsaw puzzles are really hard. J. Inf. Process. **25**, 682–694 (2017)
3. Demaine, E.D., Demaine, M.L.: Jigsaw puzzles, edge matching, and polyomino packing: connections and complexity. Graphs Combin. **23**(Suppl 1), 195–208 (2007). https://doi.org/10.1007/s00373-007-0713-4
4. Demaine, E.D., et al.: Symmetric assembly puzzles are hard, beyond a few pieces. Comput. Geom.: Theory Appl. **90**, 101648, 1–11 (2020). https://doi.org/10.1016/j.comgeo.2020.101648
5. Flake, G.W., Baum, E.B.: Rush hour is PSPACE-complete, or "why you should generously tip parking lot attendants". Theoret. Comput. Sci. **270**, 895–911 (2002)
6. Garey, M.R., Johnson, D.S.: Computers and Intractability—A Guide to the Theory of NP-Completeness. Freeman (1979)
7. Golomb, S.W.: Polyominoes. Princeton University Press, Princeton (1994)
8. Hearn, R.A., Demaine, E.D.: Games, Puzzles, and Computation. A K Peters Ltd. (2009)
9. Hearn, R.A., Demaine, E.D.: PSPACE-completeness of sliding-block puzzles and other problems through the nondeterministic constant logic model of computation. Theoret. Comput. Sci. **343**(1–2), 72–96 (2005)
10. Ito, T., et al.: On the complexity of reconfiguration problems. Theoret. Comput. Sci. **412**, 1054–1065 (2011)
11. Minamisawa, K., Uehara, R., Hara, M.: Mathematical characterizations and computational complexity of anti-slide puzzles. Theor. Comput. Sci. **939**, 216–226 (2023). https://doi.org/10.1016/j.tcs.2022.10.026
12. Post, E.L.: A variant of a recursively unsolvable problem. Bull. Amer. Math. Soc. **52**, 264–268 (1946). https://doi.org/10.1090/S0002-9904-1946-08555-9
13. Uehara, R.: Computational complexity of puzzles and related topics. Interdisc. Inf. Sci. **29**(2), 119–140 (2023). https://doi.org/10.4036/iis.2022.R.06

Fractional Bamboo Trimming and Distributed Windows Scheduling

Arash Beikmohammadi[1], William Evans[2], and Seyed Ali Tabatabaee[2(✉)]

[1] Department of Computer Science, Simon Fraser University, Burnaby, Canada
arash_beikmohammadi@sfu.ca

[2] Department of Computer Science, University of British Columbia, Vancouver, Canada
{will,salitaba}@cs.ubc.ca

Abstract. This paper studies two related scheduling problems: fractional bamboo trimming and distributed windows scheduling. In the fractional bamboo trimming problem, we are given n bamboos with different growth rates and cut fractions. At the end of each day, we can cut a fraction of one bamboo. The goal is to design a perpetual schedule of cuts to minimize the height of the tallest bamboo ever. For this problem, we present a 2-approximation algorithm. In addition, we prove upper bounds on the approximation factors of well-known algorithms Reduce-Max and Reduce-Fastest(x) for this problem. In the closely related windows scheduling problem, given a multiset of positive integers $W = \{w_1, ..., w_n\}$, we want to schedule n pages on broadcasting channels such that the time interval between any two consecutive appearances of the i-th page ($1 \leq i \leq n$) is at most w_i. The goal of this problem is to minimize the number of channels. We provide an algorithm for the windows scheduling problem that uses at most $\left\lceil \frac{d(W)+1}{0.75} \right\rceil$ channels, where $d(W) = \sum_{i=1}^{n} \frac{1}{w_i}$. When $d(W) \leq 46$, our algorithm guarantees a smaller upper bound on the number of channels than the best-known algorithm in the literature. We also describe the first approximation algorithm for the windows scheduling problem in a distributed setting, where input data is partitioned among a set of m machines. Furthermore, we introduce patterns of some multisets with $d(W) \leq 1$ for which windows scheduling on one channel (i.e., pinwheel scheduling) is impossible.

Keywords: Scheduling problems · Bamboo trimming · Windows scheduling · Distributed algorithms

1 Introduction

Scheduling problems have numerous applications across various areas of computer science, including operating systems, computer networks, and parallel processing. The bamboo trimming [19] and windows scheduling [6] problems are two

This work was partially funded by NSERC Discovery Grants.

H. Fernau et al. (Eds.): SOFSEM 2024, LNCS 14519, pp. 69–81, 2024.
https://doi.org/10.1007/978-3-031-52113-3_5

prominent and closely related examples of scheduling problems. The bamboo trimming problem and its generalizations have applications in cloud systems [2], network-switch buffer management [20], and quality of service guarantees [1]. The windows scheduling problem can be utilized in media-on-demand [33] and push systems [3].

In the bamboo trimming problem (BT), we are given n bamboos with growth rates $r_1, ..., r_n$ whose initial heights are 0. At the end of each day, we can cut one bamboo whose height becomes 0 while the height of every uncut bamboo i increases by r_i. The goal is to design a perpetual schedule of cuts to minimize the height of the tallest bamboo ever. We introduce a natural generalization of BT that considers cut fractions $q_1, ..., q_n$ (positive numbers less than or equal to 1), meaning that cutting the i-th bamboo ($1 \leq i \leq n$) at any time will multiply its height by $(1 - q_i)$. We refer to this problem as the *fractional bamboo trimming problem* (FBT). In this problem, we let $q = \min_{1 \leq i \leq n} q_i$. Two natural greedy algorithms for BT and FBT are Reduce-Max and Reduce-Fastest(x). Reduce-Max cuts the bamboo with the maximum height at the end of each day. Reduce-Fastest(x), on the other hand, cuts the bamboo with the highest growth rate among the ones that have height at least the threshold $x \cdot \sum_{i=1}^{n} r_i$.

In the windows scheduling problem (WS), we are given a multiset of positive integers $W = \{w_1, ..., w_n\}$, where each element of the multiset is called a window. We want to schedule n pages on broadcasting channels (one page per channel at each time slot) such that the interval between any two consecutive appearances of the i-th page ($1 \leq i \leq n$) is at most w_i slots. The goal is to minimize the number of channels. The density of W is defined as $d(W) = \sum_{i=1}^{n} \frac{1}{w_i}$. WS for a single broadcasting channel is called the pinwheel scheduling problem (PS). WS and BT are closely related. If we modify BT (where all cut fractions are equal to 1) to allow more than one bamboo to be cut at the end of each day and aim to minimize the number of cuts per day while preventing the height of any bamboo from exceeding a certain given amount, we obtain WS. In the distributed windows scheduling problem (DWS), the input multiset W is partitioned among a set of m machines, where the k-th machine ($1 \leq k \leq m$) has a subset W_k (which is itself a multiset) of the input, and the goal is to design collaboratively a perpetual schedule of pages on broadcasting channels for the whole multiset W.

In this paper, we provide the following results:

- We present a 2-approximation algorithm for FBT.
- We prove the upper bound of $\frac{6}{q}$ on the approximation factor of Reduce-Max for FBT. For the same problem, we also prove the upper bound of $x+1$ on the approximation factor of Reduce-Fastest(x) for all $x \geq \frac{3-q}{q}$. This shows that $\frac{3}{q}$ is an upper bound on the approximation factor of Reduce-Fastest$(\frac{3-q}{q})$. Thus, if q is fixed, both algorithms can have constant approximation factors for FBT.
- We provide an algorithm for WS that uses at most $\left\lceil \frac{d(W)+1}{0.75} \right\rceil$ channels. When $d(W) \leq 46$, our algorithm provides a better guarantee than the best-known algorithm in the literature.

- We present an algorithm for DWS that uses at most $\min(\frac{d(W)}{0.75} + \frac{7m}{3}, d(W) + e \cdot \ln(d(W)) \cdot m + 7.3595 \cdot m, (2+\epsilon) \cdot d(W) + 1)$ channels. The communication complexity of the algorithm is $O(m \cdot \log(\frac{1}{\epsilon}))$.
- We describe some infinite patterns of instances of PS with $d(W) \leq 1$ that do not have a pinwheel schedule, augmenting those of Holte et al. [22].

The rest of this paper is organized as follows. Section 2 provides background information on BT and WS. Section 3 studies different algorithms for FBT. Section 4 presents algorithms for WS and DWS. Finally, Sect. 5 concludes the results provided in this paper.

2 Background

BT has been the subject of several research projects in recent years [13,18,19, 21,26,32]. The best known algorithm for BT has an approximation factor of $\frac{10}{7}$ [21]. Reduce-Max has been shown to provide the approximation guarantee of 4 for this problem [26]. The same work proved the upper bound of $x+1$ on the approximation factor of Reduce-Fastest(x) for all $x \geq 2$ [26]. The best known approximation guarantee achieved by Reduce-Fastest(x) is $\frac{3+\sqrt{5}}{2} < 2.62$ for $x = 1 + \frac{1}{\sqrt{5}}$ [13]. There is a close relationship between BT and PS [14,17,22], and this relationship has been used to design algorithms for BT [19,32]. BT has also been viewed as a variation of the cup game [1,10–12,23,27–29]. Kuszmaul [26] explained this connection in detail. Anily et al. [4] considered a problem that shares similarities with BT but has a different objective function: to minimize the long-run average height of the bamboos, whereas in BT, the goal is to minimize the height of the tallest bamboo ever.

WS [6,8] is classified as a variant of periodic scheduling problems [30]. A restricted version of WS has been proved to be NP-hard, but it is unknown whether or not the main version of the problem is also NP-hard [8]. It has been shown that $\lceil d(W) \rceil$ is a lower bound on the minimum number of required broadcasting channels [6]. The best known algorithm for WS uses $d(W) + O(\ln d(W))$ channels [6]. WS for a single broadcasting channel is known as PS [14,17,22]. A necessary condition for having a pinwheel schedule is $d(W) \leq 1$ [22]. On the other hand, $d(W) \leq 0.75$ is a sufficient condition for having a pinwheel schedule [17]. The harmonic windows scheduling problem (HWS) [6] is another related problem where for a given number c, the goal is to maximize n such that $W = \{1, ..., n\}$ can be scheduled on c channels. HWS has applications in media-on-demand [7]. Furthermore, the relationship between WS and a special case of bin packing [15], known as the unit fractions bin packing problem (UFBP) [8], has been investigated in the literature. In UFBP, the size of each input item is a reciprocal of an integer.

Distributed computation has been used to process massive datasets. The Massively Parallel Computation (MPC) model is frequently used to study such computation [5,9,25]. The MPC model provides an abstraction of frameworks such as MapReduce [16] for processing datasets in a distributed manner. In

MPC, data is processed in multiple rounds carried out by several machines of strongly sublinear space in the input size. In every round, each machine performs polynomial-time computations on its given data, and the results are communicated with other machines to be used by those machines in the next round of the algorithm. The total size of messages sent or received by each machine in each round should not exceed its memory. Moreover, it is desired to keep the number of rounds low, ideally constant. Distributed algorithms have been devised for various problems in the MPC model and techniques such as using composable coresets have been proposed to address some of those problems [24,31]. The idea of using composable coresets is to extract a small subset of data, called a coreset, from each machine in a way that the union of the coresets represents a summary of the whole data.

3 Fractional Bamboo Trimming

In this section, we present an algorithm with an approximation factor of 2 for FBT. We then provide upper bounds on the approximation factors of Reduce-Max and Reduce-Fastest(x) for this problem.

Theorem 1. *There exists a 2-approximation algorithm for FBT.*

Proof. Let M be the height of the tallest bamboo ever for the optimal algorithm. Moreover, let $C_{i,t}$ denote the number of times the optimal algorithm cuts the i-th bamboo ($1 \leq i \leq n$) within the first t days ($t \geq 1$). Every time the optimal algorithm cuts the i-th bamboo, the height of the bamboo decreases by at most $q_i M$. Hence, for any $t \geq 1$, the optimal algorithm must maintain $t \cdot r_i - C_{i,t} \cdot q_i M \leq M$, which is a necessary condition for preventing the height of the i-th bamboo from exceeding M within the first t days. As a result, we have $\frac{r_i}{q_i M} - \frac{1}{q_i t} \leq \frac{C_{i,t}}{t}$. Thus, we deduce that $\sum_{i=1}^{n} \frac{r_i}{q_i M} - \sum_{i=1}^{n} \frac{1}{q_i t} \leq \sum_{i=1}^{n} \frac{C_{i,t}}{t} = 1$ (for any $t \geq 1$). Therefore, considering that $\lim_{t \to \infty} \sum_{i=1}^{n} \frac{1}{q_i t} = 0$, we must have $\sum_{i=1}^{n} \frac{r_i}{q_i M} \leq 1$, which means that $\sum_{i=1}^{n} \frac{r_i}{q_i} \leq M$.

Let us define $R = \sum_{i=1}^{n} \frac{r_i}{q_i}$ and $W = \{w_1, ..., w_n\}$ such that for $1 \leq i \leq n$, w_i is the smallest power of 2 greater than or equal to $\frac{q_i R}{r_i}$. Since $d(W) = \sum_{i=1}^{n} \frac{1}{w_i} \leq \sum_{i=1}^{n} \frac{r_i}{q_i R} = 1$ and all elements of W are powers of 2, W has a pinwheel schedule [22]. If we cut the i-th bamboo ($1 \leq i \leq n$) exactly once every w_i days, the height of that bamboo will never exceed $H_i = \frac{w_i r_i}{q_i}$. We can easily prove this by arguing that the height of the i-th bamboo at time 0 or after a cut is at most $(1 - q_i)H_i$ and the height increases by $w_i r_i = q_i H_i$ between two consecutive cuts, which means that it will never exceed $(1 - q_i)H_i + q_i H_i = H_i$. We have $H_i = \frac{w_i r_i}{q_i} \leq \frac{2 q_i R}{r_i} \cdot \frac{r_i}{q_i} = 2R \leq 2M$. Consequently, a pinwheel schedule for W indicates a perpetual schedule of cuts to keep the bamboos' heights at most $2M$ by cutting the i-th bamboo ($1 \leq i \leq n$) once every w_i days. □

We note that the $(\frac{12}{7})$-approximation algorithm for BT [32] (the best approximation algorithm known at the time of writing this article) cannot be used to achieve the same approximation guarantee for FBT. That solution uses the fact that in BT with more than one bamboo, the lowest possible height of the tallest bamboo ever is at least twice the maximum growth rate among the bamboos (in the notation of the proof of Theorem 1, $M \geq 2\max_{1 \leq i \leq n} r_i$), and thus no bamboo needs to be cut more frequently than every two days. However, this is not the case in FBT. Let us consider an example with two bamboos where r_1 is a large integer (at least 7), $r_2 = 1$, $q_1 = q_2$, and cut fractions are approaching 0. Assume we cut the second bamboo exactly once every $r_1 + 1$ days and cut the first bamboo every day except on the days we cut the second bamboo. This way, the height of the second bamboo will never exceed $H_2 = \frac{r_1+1}{q_2} = \frac{r_1}{q_1} + \frac{r_2}{q_2} = R$, which as we saw in the proof of Theorem 1 is a lower bound for the height of the tallest bamboo ever. Moreover, for the maximum height H_1 of the first bamboo, we have $H_1 = r_1 + r_1 \sum_{i=0}^{r_1-1}(1-q_1)^i + (1-q_1)^{r_1} H_1$ and thereby, $H_1 = \frac{r_1}{q_1} + \frac{r_1}{1-(1-q_1)^{r_1}}$. Consequently, we have $\frac{H_1}{R} = \frac{r_1 + r_1/\sum_{i=0}^{r_1-1}(1-q_1)^i}{r_1+1}$, which means that $\lim_{q_1 \to 0} \frac{H_1}{R} = 1$. However, if we cut the first bamboo every two days, its height will approach $\frac{2r_1}{q_1} > R$ as the number of days approaches infinity. This example shows that in FBT with more than one bamboo, we might need to cut a bamboo more frequently than every two days to minimize the height of the tallest bamboo ever.

Now, we show that $\frac{6}{q}$ is an upper bound on the approximation factor of Reduce-Max for FBT. The proof of the following proposition is inspired by the proof of Theorem 2.1 of [26].

Proposition 1. *Reduce-Max is a* $(\frac{6}{q})$*-approximation algorithm for FBT.*

Proof. Without loss of generality, we assume that $r_1 \geq \ldots \geq r_n$ and $\sum_{i=1}^{n} r_i = 1$. We let $h_{k,t}$ be the height of the k-th bamboo ($1 \leq k \leq n$) at the end of the t-th day ($t \geq 0$), immediately after a bamboo has been cut (note that $h_{k,0} = 0$). For $1 \leq i \leq n$ and $t \geq 0$, we define the volume function $V(i,t) = \sum_{k=1}^{i} h_{k,t}$ and the potential function

$$\Phi(i,t) = \sum_{\substack{1 \leq k \leq i \\ 3(k-1)/q < V(i,t)/2}} r_k \cdot \min\left(\frac{3}{q}, \frac{V(i,t)}{2} - \frac{3(k-1)}{q}\right).$$

The potential function is a weighted sum of $r_1, \ldots, r_i$. Each weight is less than or equal to $\frac{3}{q}$. To maximize the weighted sum, the higher growth rates are given as much weight as possible. Since $\sum_{k=1}^{n} r_k = 1$, we have $0 \leq \Phi(i,t) \leq \frac{3}{q}$ for all $1 \leq i \leq n$ and $t \geq 0$.

We use induction to show that for $1 \leq i \leq n$ and $t \geq 0$, we have $h_{i,t} \leq \frac{6}{q} - \Phi(i,t)$. The base case of $t = 0$ holds because for every i, $h_{i,0} = 0 \leq \frac{6}{q} - \Phi(i,0)$. The induction hypothesis states that for some $t \geq 0$, and every $1 \leq i \leq n$, we have $h_{i,t} \leq \frac{6}{q} - \Phi(i,t)$, and we want to prove that the inequality $h_{i,t+1} \leq \frac{6}{q} - \Phi(i,t+1)$ holds for every i. Let us consider j to be the index of the bamboo that Reduce-Max cuts at the end of the $(t+1)$-th day. If the height of the j-th bamboo right before that cut is less than $\frac{3}{q}$, then for every i we have $h_{i,t+1} < \frac{3}{q} \leq \frac{6}{q} - \Phi(i,t+1)$. Therefore, we consider the cases where $h_{j,t} + r_j \geq \frac{3}{q}$.

For the i-th bamboo, where $i \geq j$, we have

$$V(i,t+1) = V(i,t) + \sum_{k=1}^{i} r_k - q_j(h_{j,t} + r_j) \leq V(i,t) + 1 - q\frac{3}{q} = V(i,t) - 2.$$

Consequently, we obtain that $\Phi(i,t+1) \leq \Phi(i,t) - r_i$ (since we know that $r_i = \min_{1 \leq k \leq i} r_k$). As a result, we have

$$h_{i,t+1} \leq h_{i,t} + r_i \leq \frac{6}{q} - \Phi(i,t) + r_i \leq \frac{6}{q} - \Phi(i,t+1) - r_i + r_i = \frac{6}{q} - \Phi(i,t+1).$$

Otherwise, if $i < j$, we have

$$\begin{aligned} V(i,t+1) &= V(i,t) + \sum_{k=1}^{i} r_k \leq V(i,t) + 1 - r_j \\ &\leq V(i,t) + \frac{3}{q} - r_j - 2 \leq V(i,t) + h_{j,t} - 2 \leq V(j,t) - 2. \end{aligned}$$

Hence, we deduce that $\Phi(i,t+1) \leq \Phi(j,t) - r_j$ (since $r_j = \min_{1 \leq k \leq j} r_k$). Thus, we have

$$h_{i,t+1} \leq h_{j,t} + r_j \leq \frac{6}{q} - \Phi(j,t) + r_j \leq \frac{6}{q} - \Phi(i,t+1) - r_j + r_j = \frac{6}{q} - \Phi(i,t+1).$$

Therefore, the inequality $h_{i,t+1} \leq \frac{6}{q} - \Phi(i,t+1)$ holds for $1 \leq i \leq n$. In addition, we can derive $h_{j,t} + r_j \leq \frac{6}{q} - \Phi(j,t+1)$ from the inequalities related to the case $i \geq j$. Thus, Reduce-Max keeps the heights of the bamboos from exceeding $\frac{6}{q}$ at any time. On the other hand, the height of the tallest bamboo ever for the optimal algorithm is at least $R = \sum_{i=1}^{n} \frac{r_i}{q_i} \geq 1$ and this completes the proof. □

We now prove an upper bound on the approximation factor of Reduce-Fastest(x) for FBT. This proof is inspired by the proof of Theorem 3.1 of [26].

Proposition 2. *For all $x \geq \frac{3-q}{q}$, Reduce-Fastest(x) is an $(x+1)$-approximation algorithm for FBT.*

Proof. Suppose, towards a contradiction, that for some $1 \leq k \leq n$, the k-th bamboo is the first to achieve the height of $h_2 = (x+1)\sum_{i=1}^{n} r_i$ during the day d_2. Moreover, let d_1 be the closest day before d_2 during which the height of the k-th bamboo reaches the threshold $h_1 = x\sum_{i=1}^{n} r_i$. This means that the algorithm does not cut the k-th bamboo at the end of any day between d_1 and $d_2 - 1$, as otherwise the height of the bamboo would fall below $(1-q_k)h_2 \leq (1-q)(x+1)\sum_{i=1}^{n} r_i \leq (x+1-q\frac{3}{q})\sum_{i=1}^{n} r_i < h_1$. For $1 \leq j \leq n$, let m_j denote the number of times Reduce-Fastest(x) cuts the j-th bamboo between days d_1 and $d_2 - 1$.

For $1 \leq j \leq n$, we show that $r_j \geq m_j r_k$. If $m_j = 0$, it is trivial that $r_j \geq 0 \cdot r_k$. If $m_j = 1$, we have $r_j \geq 1 \cdot r_k$ because the algorithm cuts the j-th bamboo at a time when the height of the k-th bamboo is at least h_1. Otherwise, if $m_j \geq 2$, we have $\frac{h_2 - h_1}{r_k} \geq \frac{(h_1 - (1-q_j)h_2)(m_j - 1)}{r_j}$ because the j-th bamboo must regrow from a height of at most $(1 - q_j)h_2$ to a height of at least h_1 at least $m_j - 1$ times before the height of the k-th bamboo grows from h_1 to h_2. We deduce that

$$\frac{\sum_{i=1}^{n} r_i}{r_k} \geq \frac{(x \sum_{i=1}^{n} r_i - (1 - q_j)(x + 1) \sum_{i=1}^{n} r_i)(m_j - 1)}{r_j}$$

and consequently, $r_j \geq (x - (1 - q_j)(x + 1))(m_j - 1)r_k$. Considering that $x - (1 - q_j)(x + 1) = q_j(x + 1) - 1 \geq q\frac{3}{q} - 1 = 2$ and $m_j \geq 2$, we have

$$r_j \geq (x - (1 - q_j)(x + 1))(m_j - 1)r_k \geq 2(m_j - 1)r_k \geq m_j r_k.$$

As a result, we have

$$d_2 - d_1 = \sum_{i=1}^{n} m_i \leq \frac{1}{r_k} \sum_{\substack{1 \leq i \leq n \\ i \neq k}} r_i = \frac{\sum_{i=1}^{n} r_i}{r_k} - 1.$$

Since $d_2 - d_1$ is an integer, we have $d_2 - d_1 \leq \lfloor \frac{\sum_{i=1}^{n} r_i}{r_k} \rfloor - 1$. The height of the k-th bamboo at the end of the day d_1 is less than $h_1 + r_k$. Thus, the height of that bamboo at the end of the day d_2 is less than

$$h_1 + r_k + (d_2 - d_1)r_k \leq h_1 + r_k + \left(\left\lfloor \frac{\sum_{i=1}^{n} r_i}{r_k} \right\rfloor - 1 \right) r_k \leq h_1 + \sum_{i=1}^{n} r_i = h_2.$$

For this reason, the k-th bamboo does not achieve the height of h_2 during the day d_2, which is a contradiction. On the other hand, the height of the tallest bamboo ever for the optimal algorithm is at least $R = \sum_{i=1}^{n} \frac{r_i}{q_i} \geq \sum_{i=1}^{n} r_i$; thereby, Reduce-Fastest(x) is an $(x + 1)$-approximation algorithm for FBT. □

As a result of the preceding proposition, Reduce-Fastest($\frac{3-q}{q}$) achieves the approximation guarantee of $\frac{3}{q}$ for FBT.

4 Distributed Windows Scheduling

In this section, we provide approximation algorithms for WS and DWS. In addition, we present some instances of PS with densities at most 1 that do not have a pinwheel schedule.

First, we give an algorithm for WS that uses at most $\left\lceil \frac{d(W)+1}{0.75} \right\rceil$ channels. When $d(W) \leq 46$, our algorithm achieves a better guarantee than the best known algorithm in the literature, which uses at most $d(W) + e \cdot \ln(d(W)) + 7.3595$ channels [6]. The proof of the upper bound on the number of channels used by our algorithm is inspired by the proof of Theorem 3.1 of [8].

Proposition 3. *There exists an algorithm for WS that uses at most* $\left\lceil \frac{d(W)+1}{0.75} \right\rceil$ *channels.*

Proof. We begin by describing our algorithm. First, we sort the pages in non-descending order of their windows ($w_1 \leq ... \leq w_n$). Initially, no channel is open. Starting with the first page, we assign the pages to channels, one by one, and open new channels along the way. For every page, if we can assign it to an open channel such that the density of the windows of the pages assigned to that channel remains less than or equal to 0.75, we do so. Else, if we can assign the page to an open channel whose assigned pages all have windows equivalent to the window of this page and the density of the windows of the pages assigned to the channel remains less than or equal to 1, we do so. Otherwise, we open a new channel and assign the page to that channel. In the end, it is possible to pinwheel schedule both types of channels: those with pages of equivalent windows and density at most 1 [22] and those with pages of windows of density at most 0.75 [17].

Now, we prove that our algorithm uses at most $\left\lceil \frac{d(W)+1}{0.75} \right\rceil$ channels. We first use induction to show that for $k \geq 1$, after including all pages whose corresponding windows are equivalent to k, we have at most $k-1$ channels with density less than 0.75. The base case of $k = 1$ holds because for each page with a unit window, we need to open a new channel only for that page, and the density assigned to that channel will be exactly 1. By the induction hypothesis, for some $k \geq 1$, we know that after including all pages with windows equivalent to k, there are at most $k-1$ channels with density less than 0.75. When including the pages with windows equivalent to $k+1$ (in case there are any), we will never have more than one channel with density less than 1, such that the windows of all pages assigned to it are equivalent to $k+1$ (because until this channel has $k+1$ pages and density equal to 1, the algorithm will not open another one). Therefore, after including all pages with windows equivalent to $k+1$, there are at most k channels with density less than 0.75, which completes the induction step.

Let i be the index of the last page for which the algorithm opened a new channel ($1 \leq i \leq n$). If $w_i = 1$, then the total number of channels that the algorithm uses is $d(W) \leq \left\lceil \frac{d(W)+1}{0.75} \right\rceil$. Otherwise, the number of channels with density less than 0.75 that were opened before including any pages with windows equivalent to w_i is at most $w_i - 2$. Moreover, all those channels have density more than $0.75 - \frac{1}{w_i}$ because the algorithm could not assign the i-th page to any of them. As a consequence, the total number of channels that the algorithm uses is less than or equal to $\left\lceil \frac{d(W)+(w_i-2)/w_i}{0.75} \right\rceil \leq \left\lceil \frac{d(W)+1}{0.75} \right\rceil$. □

We now leverage the idea of composable coresets to devise an algorithm for DWS in the MPC model, where each of the m machines uses sublinear memory. Our algorithm requires three MPC rounds. As well as acting as a regular machine, the first machine in our algorithm acts as the central machine. We show that our algorithm uses at most $\min(\frac{d(W)}{0.75} + \frac{7m}{3}, d(W) + e\ln(d(W))m +$

$7.3595m, (2+\epsilon)d(W)+1)$ channels. We also argue that the communication complexity of the algorithm is $O(m \log(\frac{1}{\epsilon}))$.

Theorem 2. *There exists an MPC algorithm for DWS that uses at most* $\min\left(\frac{d(W)}{0.75} + \frac{7m}{3}, d(W) + e\ln(d(W))m + 7.3595m, (2+\epsilon)d(W)+1\right)$ *channels.*

Proof. The algorithm runs in three rounds. In the first round, for $1 \leq k \leq m$, the k-th machine runs the algorithms from Proposition 3 of this paper and Sect. 3.4 of [6] on W_k and chooses the algorithm that would use fewer channels (note that both algorithms schedule all appearances of each page on the same channel). The k-th machine opens each channel, suggested by the chosen algorithm, with a density of at least 0.5 and then sorts the remaining pages in non-descending order of their corresponding windows. The k-th machine creates a new channel (but does not open it right away) and assigns the remaining pages (one by one, starting from the first page in the sorted order) to that channel until the density assigned to the channel becomes at least 0.5 or no more page is left. If at any point, the density assigned to the channel becomes greater than or equal to 0.5, the k-th machine finds a pinwheel schedule for the pages assigned to that channel, opens that channel, creates a new channel, and continues assigning the remaining pages to the new channel. It can be easily seen that the density assigned to each channel opened during this process is less than or equal to 0.75, and hence finding a pinwheel schedule is indeed possible.

If the density assigned to the last created channel is less than 0.5, the k-th machine does not open that channel and instead sends a summary of the remaining pages to the central machine (which is actually the first machine). Let W'_k be a multiset containing the windows of the remaining pages. We know that $d(W'_k) < 0.5$. Also, let $\hat{W}_k$ be a multiset that has the same size as W'_k such that for each $w' \in W'_k$, there exists a corresponding $\hat{w} \in \hat{W}_k$, where $\hat{w}$ is the greatest power of 2 less than or equal to w'. Furthermore, consider a positive integer p as a parameter of the algorithm, which is universally known by all machines. We define the summary S_k to be a set of at most p unique positive integers such that all elements of the set are powers of 2, $d(S_k) \geq d(\hat{W}_k)$, and $d(S_k)$ is the lowest possible. From this, we deduce that $d(S_k) - d(\hat{W}_k) \leq \frac{d(\hat{W}_k)}{2^{p-1}}$. The k-th machine sends S_k to the central machine.

In the second round of the algorithm, if $\lceil \sum_{k=1}^{m} d(S_k) \rceil$ is greater than the number of machines that sent non-empty summaries, the central machine instructs every machine to open a channel for its own remaining pages in the subsequent round. Otherwise, the central machine finds a perfect schedule with $\lceil \sum_{k=1}^{m} d(S_k) \rceil$ channels for the imaginary pages with corresponding windows in summaries received from all machines (according to Lemma 4 of [6], this is possible). In a perfect schedule, the time interval between any two consecutive appearances of each page is equivalent to its corresponding window. The central machine then sends the information of slots allocated for the imaginary pages with corresponding windows in S_k to the k-th machine, for $1 \leq k \leq m$. Finally, in the third round, the k-th machine $(1 \leq k \leq m)$ schedules its remaining pages, one by one and in a non-descending order of their corresponding windows, on the allocated slots. It is easy to check that using the first-fit strategy, the k-th machine can schedule the remaining pages in a way that the time interval between any two consecutive appearances of each page is equivalent to the greatest power of 2 less than or equal to the window of that page.

Next, we give upper bounds on the number of channels used by our algorithm. This number is less than or equal to the number of channels used if all machines separately ran the algorithm from Proposition 3 of this paper or if they all ran the algorithm from Sect. 3.4 of [6]. Hence, the number of channels used by our algorithm is less than or equal to

$$\sum_{k=1}^{m} \left\lceil \frac{d(W_k)+1}{0.75} \right\rceil \leq \sum_{k=1}^{m} \frac{d(W_k)+1.75}{0.75} = \frac{d(W)}{0.75} + \frac{7m}{3}.$$

Furthermore, this number is less than or equal to

$$\sum_{k=1}^{m} (d(W_k) + e\ln(d(W_k)) + 7.3595) \leq d(W) + e\ln(d(W))m + 7.3595m.$$

We know that $d(S_k) \leq (1+\frac{1}{2^{p-1}})d(\hat{W}_k) \leq 2(1+\frac{1}{2^{p-1}})d(W_k')$. Therefore, by setting $\epsilon = \frac{1}{2^{p-2}}$, we have $d(S_k) \leq (2+\epsilon)d(W_k')$. In addition, if $\tilde{W}$ denotes the multiset of the windows of all pages assigned to the channels that were opened in the first round across all machines, then the number of channels that contain such pages is at most $2d(\tilde{W})$. This is because each of those channels has a density of at least 0.5. Consequently, the number of channels used by our algorithm is at most

$$\left\lceil \sum_{k=1}^{m} d(S_k) \right\rceil + 2d(\tilde{W}) \leq \left\lceil (2+\epsilon)\sum_{k=1}^{m} d(W_k') \right\rceil + (2+\epsilon)d(\tilde{W}) \leq (2+\epsilon)d(W) + 1.$$

Finally, we analyze the communication complexity of our algorithm. The size of the data that each machine sends to the central machine at the end of the first round is $O(p)$. Also, the size of the data that the central machine sends to each machine at the end of the second round is $O(p)$. Thus, the communication complexity of the algorithm is $O(mp) = O(m\log(\frac{1}{\epsilon}))$. □

Next, we introduce some instances of PS with density at most 1 for which there exists no pinwheel schedule. These provide further examples, beyond those of Holte et al. [22], of unschedulable low-density instances. Before proceeding, we need to establish some notation. For a positive integer c and a multiset of positive integers $A = \{a_1, ..., a_n\}$, we define $M(c, A) = \{ca_1, ..., ca_n\}$. For two multisets of positive integers $A = \{a_1, ..., a_n\}$ and $B = \{b_1, ..., b_m\}$, we define $A + B = \{a_1, ..., a_n, b_1, ..., b_m\}$. Furthermore, for a positive integer c and a multiset of positive integers $A = \{a_1, ..., a_n\}$, we define $R(c, A) = \underbrace{A + ... + A}_{c \text{ times}}$.

Proposition 4. *Let k and g be integers such that $k \geq 2$ and g is arbitrarily large (much larger than k). Furthermore, let U be a multiset of positive integers with $d(U) \leq 1$ that does not have a pinwheel schedule. The following instances of PS do not have a pinwheel schedule, even though their densities are less than or equal to 1:*

- $W_1 = \{k\} + R(k-1, \{k+1\}) + \{g\}$
- $W_2 = R(k-1, \{k\}) + \{2k-1, g\}$
- $W_3 = R(k-1, \{k\}) + M(k, U)$

Proof. We prove by contradiction that W_1, W_2, and W_3 do not have pinwheel schedules. If there exists a pinwheel schedule for W_1, then for some $t_1 > k$, the page with window g appears in slot t_1. In addition, the page with window k must appear in some slot t_2, where $t_1 < t_2 \leq t_1 + k - 1$. Furthermore, the page with window k must appear in some other slot t_0, where $t_2 - k \leq t_0 < t_2$. Consequently, the pages with windows k and g must appear at least twice and once, respectively, in the interval $[t_2 - k, t_2]$. Hence, at least one of the $k-1$ pages whose windows are equivalent to $k+1$ cannot appear in that interval. This contradicts our assumption that W_1 has a pinwheel schedule.

If W_2 has a pinwheel schedule then for some $t_1 > k-1$, the page with window g appears in slot t_1. Each of the $k-1$ pages whose windows are equivalent to k must appear in the interval $[t_1 - k + 1, t_1 - 1]$. Similarly, each of those pages must appear in the interval $[t_1 + 1, t_1 + k - 1]$. Hence, the page with window $2k-1$ cannot appear in the interval $[t_1 - k + 1, t_1 + k - 1]$, which contradicts the existence of a pinwheel schedule for W_2.

If W_3 has a pinwheel schedule, consider any slot t_1 such that a page whose window is in $M(k, U)$ appears in that slot ($t_1 \geq 1$). Therefore, each of the $k-1$ pages whose windows are equivalent to k must appear in the interval $[t_1 + 1, t_1 + k - 1]$. This means that the interval between any two consecutive appearances of the pages with windows in $M(k, U)$ is at least k slots. On the other hand, for any $u \in U$, the interval between any two consecutive appearances of the page with window ku is at most ku slots. Consequently, there are at most $u-1$ appearances of the other pages with windows in $M(k, U)$ between any two consecutive appearances of the page with window ku. As a result, a pinwheel schedule for W_3 implies a pinwheel schedule for U. This is a contradiction as U does not have a pinwheel schedule. □

5 Conclusion

In this paper, we introduced FBT and presented a 2-approximation algorithm for it. We also proved upper bounds on the approximation factors of well-known algorithms Reduce-Max and Reduce-Fastest(x) for FBT. For WS, we designed an algorithm that uses at most $\left\lceil \frac{d(W)+1}{0.75} \right\rceil$ channels. We argued that our algorithm works better than the best-known algorithm in the literature when $d(W) \leq 46$. In addition, we designed the first approximation algorithm for DWS in the MPC model. Our algorithm can be implemented in $O(1)$ rounds of MapReduce and leads to a streaming algorithm with $O(1)$ passes. Finally, we introduced patterns of some instances of PS with $d(W) \leq 1$ that do not have a pinwheel schedule.

It remains open whether it is possible to find algorithms with better approximation guarantees for FBT, WS, and DWS. It is also interesting to see if there exist better upper bounds on the approximation factors of Reduce-Max and Reduce-Fastest(x) for FBT. Another open problem is to find other instances of PS with densities of at most 1 for which there exists no pinwheel schedule. We believe finding such instances can help us determine the requirements for having a pinwheel schedule.

References

1. Adler, M., Berenbrink, P., Friedetzky, T., Goldberg, L.A., Goldberg, P., Paterson, M.: A proportionate fair scheduling rule with good worst-case performance. In: Proceedings of the Fifteenth Annual ACM Symposium on Parallel Algorithms and Architectures, pp. 101–108 (2003)
2. Alshamrani, S.: How reduce max algorithm behaves with symptoms appearance on virtual machines in clouds. In: 2015 International Conference on Cloud Computing (ICCC), pp. 1–4. IEEE (2015)
3. Ammar, M.H., Wong, J.W.: The design of teletext broadcast cycles. Perform. Eval. **5**(4), 235–242 (1985)
4. Anily, S., Glass, C.A., Hassin, R.: The scheduling of maintenance service. Discrete Appl. Math. **82**(1–3), 27–42 (1998)
5. Assadi, S.: Combinatorial optimization on massive datasets: streaming, distributed, and massively parallel computation. University of Pennsylvania (2018)
6. Bar-Noy, A., Ladner, R.E.: Windows scheduling problems for broadcast systems. SIAM J. Comput. **32**(4), 1091–1113 (2003)
7. Bar-Noy, A., Ladner, R.E., Tamir, T.: Scheduling techniques for media-on-demand. In: SODA, pp. 791–80 (2003)
8. Bar-Noy, A., Ladner, R.E., Tamir, T.: Windows scheduling as a restricted version of bin packing. ACM Trans. Algorithms (TALG) **3**(3), 28-es (2007)
9. Beame, P., Koutris, P., Suciu, D.: Communication steps for parallel query processing. J. ACM (JACM) **64**(6), 1–58 (2017)
10. Bender, M.A., Farach-Colton, M., Kuszmaul, W.: Achieving optimal backlog in multi-processor cup games. In: Proceedings of the 51st Annual ACM SIGACT Symposium on Theory of Computing, pp. 1148–1157 (2019)
11. Bender, M.A., et al.: The minimum backlog problem. Theor. Comput. Sci. **605**, 51–61 (2015)
12. Bender, M.A., Kuszmaul, W.: Randomized cup game algorithms against strong adversaries. In: Proceedings of the 2021 ACM-SIAM Symposium on Discrete Algorithms (SODA), pp. 2059–2077. SIAM (2021)
13. Bilò, D., Gualà, L., Leucci, S., Proietti, G., Scornavacca, G.: Cutting bamboo down to size. Theoret. Comput. Sci. **909**, 54–67 (2022)
14. Chan, M.Y., Chin, F.Y.L.: General schedulers for the pinwheel problem based on double-integer reduction. IEEE Trans. Comput. **41**(06), 755–768 (1992)
15. Garey, M.R., Johnson, D.S.: Approximation algorithms for bin packing problems: a survey. In: Ausiello, G., Lucertini, M. (eds.) Analysis and Design of Algorithms in Combinatorial Optimization. ICMS, vol. 266, pp. 147–172. Springer, Vienna (1981). https://doi.org/10.1007/978-3-7091-2748-3_8

16. Dean, J., Ghemawat, S.: MapReduce: simplified data processing on large clusters. Commun. ACM **51**(1), 107–113 (2008)
17. Fishburn, P.C., Lagarias, J.C.: Pinwheel scheduling: achievable densities. Algorithmica **34**(1), 14–38 (2002)
18. Gasieniec, L., et al.: Perpetual maintenance of machines with different urgency requirements. J. Comput. Syst. Sci. **139**, 103476 (2024)
19. Gąsieniec, L., Klasing, R., Levcopoulos, C., Lingas, A., Min, J., Radzik, T.: Bamboo garden trimming problem (perpetual maintenance of machines with different attendance urgency factors). In: Steffen, B., Baier, C., van den Brand, M., Eder, J., Hinchey, M., Margaria, T. (eds.) SOFSEM 2017. LNCS, vol. 10139, pp. 229–240. Springer, Cham (2017). https://doi.org/10.1007/978-3-319-51963-0_18
20. Goldwasser, M.H.: A survey of buffer management policies for packet switches. ACM SIGACT News **41**(1), 100–128 (2010)
21. Höhne, F., van Stee, R.: A 10/7-approximation for discrete bamboo garden trimming and continuous trimming on star graphs. In: Approximation, Randomization, and Combinatorial Optimization. Algorithms and Techniques (APPROX/RANDOM 2023). Schloss Dagstuhl-Leibniz-Zentrum für Informatik (2023)
22. Holte, R., Mok, A., Rosier, L., Tulchinsky, I., Varvel, D.: The pinwheel: a real-time scheduling problem. In: Proceedings of the 22nd Hawaii International Conference of System Science, pp. 693–702 (1989)
23. Im, S., Moseley, B., Zhou, R.: The matroid cup game. Oper. Res. Lett. **49**(3), 405–411 (2021)
24. Indyk, P., Mahabadi, S., Mahdian, M., Mirrokni, V.S.: Composable core-sets for diversity and coverage maximization. In: Proceedings of the 33rd ACM SIGMOD-SIGACT-SIGART Symposium on Principles of Database Systems, pp. 100–108 (2014)
25. Karloff, H., Suri, S., Vassilvitskii, S.: A model of computation for MapReduce. In: Proceedings of the Twenty-First Annual ACM-SIAM Symposium on Discrete Algorithms, pp. 938–948. SIAM (2010)
26. Kuszmaul, J.: Bamboo trimming revisited: Simple algorithms can do well too. In: Proceedings of the 34th ACM Symposium on Parallelism in Algorithms and Architectures, pp. 411–417 (2022)
27. Kuszmaul, W.: Achieving optimal backlog in the vanilla multi-processor cup game. In: Proceedings of the Fourteenth Annual ACM-SIAM Symposium on Discrete Algorithms, pp. 1558–1577. SIAM (2020)
28. Kuszmaul, W.: How asymmetry helps buffer management: achieving optimal tail size in cup games. In: Proceedings of the 53rd Annual ACM SIGACT Symposium on Theory of Computing, pp. 1248–1261 (2021)
29. Kuszmaul, W., Westover, A.: The variable-processor cup game. arXiv preprint arXiv:2012.00127 (2020)
30. Liu, C.L., Layland, J.W.: Scheduling algorithms for multiprogramming in a hard-real-time environment. J. ACM (JACM) **20**(1), 46–61 (1973)
31. Mirjalali, K., Tabatabaee, S.A., Zarrabi-Zadeh, H.: Distributed unit clustering. In: CCCG, pp. 236–241 (2019)
32. van Ee, M.: A 12/7-approximation algorithm for the discrete bamboo garden trimming problem. Oper. Res. Lett. **49**(5), 645–649 (2021)
33. Viswanathan, S., Imielinski, T.: Metropolitan area video-on-demand service using pyramid broadcasting. Multimed. Syst. **4**, 197–208 (1996)

New Support Size Bounds and Proximity Bounds for Integer Linear Programming

Sebastian Berndt[1], Matthias Mnich[2(✉)], and Tobias Stamm[2]

[1] University of Lübeck, Institute for Theoretical Computer Science, Lübeck, Germany
s.berndt@uni-luebeck.de

[2] Hamburg University of Technology, Institute for Algorithms and Complexity, Hamburg, Germany
{matthias.mnich,tobias.stamm}@tuhh.de

Abstract. Integer linear programming (ILP) is a fundamental research paradigm in algorithms. Many modern algorithms to solve structured ILPs efficiently follow one of two main approaches. The first one is to prove a small upper bound on the support size of the ILP, which is the number of variables taking non-zero values in an optimal solution, and then to only search for ILP solutions of small support. The second one is to apply an augmentation algorithm using Graver elements to an initial feasible solution obtained from a small proximity bound for the ILP, which is the distance between an optimal solution of the ILP and that of its LP relaxation.

Our first contribution are new lower bounds for the support size of ILPs. Namely, we discover a connection between support sizes and an old number-theoretic conjecture by Erdős on subset-sum distinct sets. Further, we improve the previously best lower bounds on the support size of ILPs with m constraints and largest absolute value Δ of any coefficient in the constraint matrix from $\Omega(m \log(\Delta))$ to $\Omega(m \log(\sqrt{m}\Delta))$. This new lower bound asymptotically matches the best-known upper bounds.

Our second contribution are new bounds on the size of Graver elements and on the proximity for ILPs. We first show nearly tight lower and upper bounds for $g_1(A)$, the largest 1-norm $\|\boldsymbol{g}\|_1$ of any Graver basis element $\boldsymbol{g}$ of the constraint matrix A. Then we show that the proximity of any ILP in standard form with support size s is bounded by $s \cdot c_1(A)$, where $c_1(A)$ is the largest 1-norm $\|\boldsymbol{c}\|_1$ of any circuit $\boldsymbol{c}$ of A. This improves over the known proximity bound of $n \cdot g_1(A)$, as s and $c_1(A)$ can be much smaller than n and $g_1(A)$, respectively.

Keywords: Integer linear programming · proximity bounds · Graver bases

1 Introduction

Integer linear programming (ILP) is one of the most studied approaches in modelling and solving combinatorial optimization problems, both from a theoretical and a practical perspective. ILPs model these problems through a system of n

H. Fernau et al. (Eds.): SOFSEM 2024, LNCS 14519, pp. 82–95, 2024.
https://doi.org/10.1007/978-3-031-52113-3_6

variables linked through m constraints and an objective function, such that an optimal assignment of integer values to the variables which satsfies all constraints recovers amn optimal solution to the optimization problem. Due to it's strong expressivity, it allows to formulate problems from a wide range of domains in a schematic and succinct way.

On the one hand, there are ILP solvers for which one observes practically fast run times for ILPs even for very large n and m, which thus allows to solve the optimization problem fast on real-world instances. However, these solvers usually do not come with (good) theoretical guarantees with regard to their run time in terms of n and m. On the other hand, ILP solvers with theoretical guarantees in terms of n and m are known for several decades, but their practical run time can be very high [23, 25]. This gap between theoretical and practical performance motivates the study of solving of ILPs through *parameterized algorithms* [11]; these algorithms discover structures in the problem instances which crucially influence the problem complexity and therefore the run time of algorithms. These structures are measured in terms of parameters, which thus lead to a fine-grained analysis of run times that has the potential to narrow gap between theoretically proven and practically observed run times.

Parameterized algorithms for ILPs have been studied intensively, and for structured ILPs—i.e., ILPs for which good parameter bounds on the structures are known—the resulting algorithms allowed for drastic run-time improvements [15, 24] when compared to algorithms focusing on general ILPs. Many of these algorithms solve structured ILPsfollow one of two possible approaches. The first approach uses the fact that the *support size* s of structured ILPs is typically bounded by some function of the parameters, i.e., there exists an optimal solution that assigns non-zero values to only few ($s \ll n$) variables and assigns zero to all other variables. These bounds thus allows to restrict the search space heavily by only considering solutions with small support size [2, 21]. Clearly, the smaller the support size is, the more restricted the search space becomes, and thus the faster algorithms can be designed. In the second approach, one starts by some feasible, but not necessarily optimal, solution, and gradually augments it by so-called *Graver elements* [14]. Such elements are special integral elements of the kernel of the underlying ILP. The two major bottlenecks of this approach are thus to (i) find the augmenting Graver elements and (ii) find the initial solution. As in the first approach, the existence of small Graver elements allows to restrict the search space, and thus to speed up the ILP solver. To find the initial solution, a phenomenon called *proximity* is often used that guarantees that solutions of ILPs are close to solutions of the fractionally relaxed linear program (LP). Hence, to find a feasible initial solution, one can solve the LP in polynomial time and then search for neighboring integral solutions. Equivalently, this allows to reduce the right-hand side of the ILP significantly.

Our Results. In this work, we give a fine-grained analysis of the limits of the two approaches mentioned above for solving structured ILPs. To do so, we study the three important parameters of ILPs: their support sizes, bounds on the sizes of their Graver elements, and bounds on their proximity. On the one hand, we improve the best-known *upper bounds* on all of these quantities in terms of the parameters of the structured ILPs. This allows for more efficient algorithms as the search space can be shrunk dramatically. On the other hand, we prove strong lower bounds on these quantities in terms of the parameters, thereby establishing limitations of the aforementioned algorithmic approaches for solving ILPs. While this does not rule out faster algorithms based on other approaches, it identifies the bottlenecks of the algorithms and, simultaneously, shows the tightness of the upper bounds. Finally, we provide results based on extensive computational experiments to show the optimality of our bounds for small parameter values.

SSD sets: To improve the bounds on the support size of ILPs, we reveal a close relation to largest elements in certain *subset sum distinct (SSD) sets*, which are sets where the sum of any subset of items is unique. In the one-dimensional case, bounds on the largest elements of SSD sets have been well-studied due to a long-standing (and still open) conjecture by Erdős from 1931. However, multi-dimensional SSD sets have received less attention. We remedy this situation by giving upper and lower bounds on the largest element Δ and the largest 1-norm $\|A\|_1 := \max_{\boldsymbol{a} \in A} \|\boldsymbol{a}\|_1$ of any vector of m-dimensional SSD sets of size n. Our upper bounds show that $\|A\|_1 \leq (1/2) \cdot 2^{\lceil n/m \rceil}$ and $\Delta \leq (1/\sqrt{2m})2^{\lceil n/m \rceil}$, and our lower bounds show that $\|A\|_1 \geq (1/2) \cdot 2^{n/m}/\sqrt{n/m}$ and $\Delta \geq (1/\sqrt{2m})2^{n/m}$.

Support size bounds: Based on our analysis of SSD sets, we first show that a long-standing conjecture of Erdős would imply an optimal upper bound of $\log(c\Delta)$ on the support size of any single-constraint ILP with largest absolute coefficient Δ in their constraint matrix A, for some universal constant c. Using the best-known bounds on the size of SSD sets, we then improve a best known upper bound for single-constraint ILPs to $1.1\log(2.02\Delta)$, unconditionally. For ILPs with $m > 1$ constraints, we improve the lower bound on their support size from $m\log(\Delta)$ from Berndt et al. [5] to $m\log(\sqrt{m}\Delta)$. We also improve the upper bound on the support size from $2m\log(1.46\|A\|_1)$, by Berndt et al. [4], to $2m\log(\sqrt{2}\|A\|_1)$.

Graver bases: For single constraint ILPs (e.g., KNAPSACK), tight bounds on the size of the largest Graver basis element were found by Diaconis, Graham and Sturmfels [12] in the 1990s. We derive lower and upper bounds on the size of the largest Graver basis element for ILPs with $m > 1$ constraints. On the one hand, we give constructions yielding lower bounds $(2\Delta - 1)(\Delta + 1)^{m-1}$ and $(2(\|A\|_1 - 1)^{m+1} - \|A\|_1)/(\|A\|_1 - 2)$. On the other hand, we show the upper bounds $(2m\Delta - 1)^m$ and $(2e \cdot (\|A\|_1 + 1))^m$. These bounds depend exponentially on m. For $\|A\|_1 = 2$, we show a *linear* upper bound of $2m + 1$.

Proximity: For the proximity we first show an upper bound of $s \cdot c_1(A)$ on the proximity of ILPs with support size s and largest circuit 1-norm $c_1(A)$. For $s \ll n$, this significantly improves the bound $n \cdot g_1(A)$ given by Hemmecke et al. [22]. Secondly we show how the Steinitz lemma gives a bound of $m \cdot g_1(A)$ on the proximity of any m-constraint ILP.

We defer all proofs of statements to a full version of this paper.

2 Subset Sum Distinct Sets

Subset sum distinct (SSD) sets are sets of vectors such that no two subsets have the same sum of elements. In this work, we only consider vectors of integers.

Definition 1. *A set $S \in \mathbb{Z}^m$ is* subset sum distinct (SSD) *if the sum function is injective on 2^S, i.e., if $\sum_{s \in S_1} s = \sum_{s \in S_2} s$ implies $S_1 = S_2$ for all $S_1, S_2 \subseteq S$.*

Equivalently, we can think of SSD sets as linear equations $S \cdot \{-1, 0, 1\}^{|S|} = \mathbf{0}$ with a unique solution $\mathbf{0}$, as the $\{-1, 0, 1\}^{|S|}$ vectors are bijective to disjoint subset pairs (S_1, S_2). Consequently, the SSD property is invariant under negation of any element, and $\mathbf{0}$ is never part of an SSD set. Additionally, the SSD property is invariant under any permutation applied to all vectors, i.e., swapping rows.

2.1 One-Dimensional Subset Sum Distinct Sets

In dimension $m = 1$, the problem of constructing SSD sets reduces to studying subsets of positive integers, $S \subset \mathbb{N}$. As such, it has been a long-standing subject of number theory. On the one hand, for any $n \in \mathbb{N}$ there is an SSD set S of size n with $\max_{a \in S} a \leq 2^n/2$, because the powers of two form SSD sets, as any subset then gives a distinct number in binary. On the other hand, for any SSD set S of size n it holds that $\max_{a \in S} a \geq (2^n - 1)/n$, as the n numbers in S have to reach 2^n distinct points. Such constructions naturally raise the question about the smallest possible maximum element in an n-element SSD set.

Definition 2. *Let $(a_n)_{n \in \mathbb{N}}$ be the sequence $a_n := \min_{S \subset \binom{\mathbb{N}}{n}, \mathrm{SSD}(S)} \max_{a \in S} a$ of smallest numbers for which there is an SSD set of size n with largest entry a_n.*

The sequence $(a_n)_n$ is denoted as A276661 in the online encyclopedia of integer sequences [27]. For $n \leq 8$, the values a_n were found by Lunnon [26], the value a_9 was found by Borwein and Mossinghof [9]. By computer search, we found $a_{10} = 309$, with the unique minimal SSD set $\{309, 308, 307, 305, 302, 296, 285, 265, 225, 148\}$. The asymptotic growth of a_n has been an open problem for almost a century, with Erdős [16] calling it his "first serious conjecture" and offering \$500 for a proof or disproof:

Erdős' Conjecture (1931). *There is a constant $\tilde{c} \in \mathbb{R}$ such that $a_n \geq 2^{n+\tilde{c}}$.*

In 1967, Conway and Guy [10] (cf. [27, A005318]) constructed a series of sets, which they conjectured to be SSD. Yet, only in 1996 did Bohman [6] prove that all Conway-Guy sets are SSD. The Conway-Guy sets give an upper bound of $a_n < 0.23513 \cdot 2^n$ for $n \geq 352$. Surprisingly, for $n \leq 10$ the Conway-Guy construction gives SSD sets with optimal largest entry a_n. For $n = 11$, our computer search did not yield a better set. For $n = 12$, the Conway-Guy set has largest entry 1164, whereas a construction by Bohmann [7] gives the SSD set $S_{2,12} = \{1159, 1157, 1156, 1155, 1151, 1145, 1134, 1112, 1073, 995, 845, 556\}$. Currently, the best construction [7] has $a_n \leq 0.22002 \cdot 2^n$, for sufficiently large n.

The best known lower bounds are all of the form $a_1 \geq (\tilde{c} - o(1)) \cdot 2^n/\sqrt{n}$, with a long history of improvements on c, well summarized by Steinerberger [31]. In such a lower bound, any $\tilde{c} > 0.5$ can only hold asymptotically, because of the value $a_1 = 1$. A particularly elegant, simple and short proof of the currently best constant shows:

Proposition 1 (Dubroff, Fox, Xu [13]). *For any one dimensional SSD-set of size n the largest entry Δ satisfies $\Delta \geq \binom{n}{\lfloor n/2 \rfloor} \geq (\sqrt{2/\pi} - o(1)) \cdot 2^n/\sqrt{n}$.*

2.2 Higher-Dimensional Subset Sum Distinct Sets

For higher-dimensional SSD sets, there is only little research. There are three common ways to define the largest "number" in a set S of vectors: (i) the largest absolute value $\Delta := \max_{\boldsymbol{a} \in S} \|\boldsymbol{a}\|_\infty$ of any entry; (ii) the largest sum of absolute values $\|A\|_1 := \max_{\boldsymbol{a} \in S} \|\boldsymbol{a}\|_1$ of any vector; and (iii) the largest subdeterminant δ of any square subset of vectors. In this work, we mainly consider Δ and $\|A\|_1$.

Upper Bounds. To obtain upper bounds on the size of SSD sets, we consider constructions which iteratively generate larger SSD sets from smaller SSD sets. For starters, we show that any SSD set can be enlarged by doubling $\Delta/\|A\|_1$.

Observation 1. *For any SSD set S the set $2S + I$ is also SSD.*

Consequently, any SSD set of size n with largest entry Δ can be used to create a set of size $n + k$ with largest entry $\Delta \cdot 2^k$. As there is no construction for a_n in $o(2^n)$ so far, all known one-dimensional construction converge to this method of set enlargement. If we double the dimension m instead, then interestingly the behavior of the parameters Δ and $\|A\|_1$ differs. For the parameter Δ, doubling the dimension m allows for a construction analogous to Hadamard matrices, with the addition of further set elements.

Observation 2. *For any SSD set S, the set $\begin{pmatrix} S & S & I \\ S & -S & 0 \end{pmatrix}$ is also SSD.*

Hence, an SSD set of size n in dimension m generates one of size $2n + m$ in dimension $2m$. Thus, the doubling of m allows for a superlinear gain in the size of the SSD set for Δ. For $\|A\|_1$, a different construction works:

Observation 3. *For any two SSD sets S and T, the set $\begin{pmatrix} S & 0 \\ 0 & T \end{pmatrix}$ is also SSD.*

Obviously, Observation 2 only works for Δ and doubling the dimension. In contrast, Observation 3 works for both Δ and $\|A\|_1$ and also allows for combining sets of distinct dimensions. These constructions directly give upper bounds for the smallest Δ or $\|A\|_1$ needed for an n-element SSD set in m dimensions.

Corollary 1. *For all $n, m \in \mathbb{N}$ there is an SSD set with $\|A\|_1 \leq (1/2) \cdot 2^{\lceil n/m \rceil}$.*

Corollary 2. *For all $n \in \mathbb{N}$ and $m = 2^i, i \in \mathbb{N}$ there is an SSD set with largest entry $\Delta \leq \max(c/(2\sqrt{m})) \cdot 2^{\lceil n/m \rceil}, 1)$ and $c = 1$ for i even, $c = \sqrt{2}$ for i odd.*

However, in general these constructions do not achieve optimal SSD sets. For example, $\begin{pmatrix} 2 & 2 & 2 & 1 & 0 \\ 1 & -1 & 0 & 0 & 3 \end{pmatrix}$ with $\|A\|_1 = 3$ and $m = 2$ is an SSD set of size 5.

Lower Bounds. In multiple dimensions, any single dimension does *not* need to form an SSD set by itself, as a duplicate value in one dimension might be distinguished by another dimension. We adapt the approach of Aliev et al. [1] to use Siegel's lemma for the derivation of our bounds.

Proposition 2 (Siegel's lemma, Bombieri and Vaaler [8]). *The linear system $A \cdot \boldsymbol{x} = \mathbf{0}$ with $A \in \mathbb{Z}^{m \times n}, n > m$ and the rows of A linearly independent has a non-zero integer solution $\boldsymbol{y} \in \mathbb{Z}^n$ with $\|\boldsymbol{y}\|_\infty \leq \sqrt{\det(AA^T)}^{1/(n-m)}$.*

The equation interpretation of SSD sets directly bounds their size:

Lemma 1. *Any SSD set $S \subset \mathbb{Z}^m$ of size n satisfies $2^n \leq 2^m \sqrt{\det(SS^T)}$.*

Because the matrix SS^T is symmetric and positive semidefinite, its determinant is smaller than the product of the diagonal entries of SS^T. Using this fact, we obtain the following lower bounds of the form $c \cdot 2^{n/m} / \sqrt{n/m}$:

Lemma 2. *Any SSD set $S \subset \mathbb{Z}^m$ of size n satisfies $\Delta \geq 1/(2\sqrt{m}) \cdot 2^{n/m} / \sqrt{n/m}$.*

Lemma 3. *Any SSD set $S \subset \mathbb{Z}^m$ of size n satisfies $\|A\|_1 \geq (1/2) \cdot 2^{n/m} / \sqrt{n/m}$.*

The asymptotic difference between the known lower and upper bounds is a factor of $1/\sqrt{n/m}$, just like in the one-dimensional case.

For $\|A\|_1 = 1$, the identity matrix I is obviously the largest SSD set. For $\|A\|_1 = 2$, we show that, surprisingly, $\begin{pmatrix} 2I & I \end{pmatrix}$ is still a largest SSD set.

Theorem 1. *If $n/m > 2$ then $\|A\|_1 \geq 3$ for any SSD set $S \subset \mathbb{Z}^m$ of size n.*

2.3 Numerical Results

Our approach to find SSD sets S in dimension $m = 11$ bears similarity to that of Lunnon [26]. We performed a depth first search for the elements S_i of the set. The search is trivially parallelizable, which we did. To keep track of the SSD property, we maintained a bitvector of the currently reachable points. That way duplicates are easily detected by the bitwise AND operation on the bitvector and its shift by the current number. If there is no duplicate, the new bitvector is simply the bitwise OR of the original and shifted bitvector. We obtained faster implementations by searching for decreasing sets with $S_i \geq S_j$ for $i \leq j$, instead of increasing. The smaller values of a_n were used as lower bounds $S_{n-i} \geq a_i$ for the elements which needed to be considered. In our implementation, the use of bounds like Borwein and Mossinghoff [9] or number theoretical results about the sums of reciprocals of SSD sets resulted in no significant speedup. Searching numbers descendingly gave all solutions relatively quickly, with the majority of the compute time spend checking that no more solutions exist. Hence, if one could prove that in an SSD set for a_n most elements are almost as large as a_n, as suggested by Steinerberger [31], the search could be sped up significantly.

For dimensions $m > 1$, we kept an explicit list of the reachable points, instead of implementing multi-dimensional bit vectors. We also did not implement symmetry reductions, which would give a speedup of $m!$.

We list the smallest possible parameters $\|A\|_1$ and Δ for an SSD set $S \subset \mathbb{Z}^m$ of size n in Table 1 and Table 2 respectively.

Table 1. Smallest possible $\|A\|_1$, given m and n.

$m\backslash n$	1	2	3	4	5	6	7	8	9	10
1	1	2	4	7	13	24	44	84	161	309
2	1	1	2	2	3	4	6	7	10	13
3	1	1	1	2	2	2	3	4	4	5
4	1	1	1	1	2	2	2	2	3	3
5	1	1	1	1	1	2	2	2	2	2

Table 2. Smallest possible Δ, given m and n.

$m\backslash n$	1	2	3	4	5	6	7	8	9	10
1	1	2	4	7	13	24	44	84	161	309
2	1	1	1	2	2	3	4	5	7	10
3	1	1	1	1	1	2	2	2	3	3
4	1	1	1	1	1	1	1	1	2	2
5	1	1	1	1	1	1	1	1	1	1

3 Support Size Bounds for Integer Linear Programming

For matrix $A \in \mathbb{Z}^{m \times n}$ and vectors $\boldsymbol{b} \in \mathbb{Z}^m, \boldsymbol{c} \in \mathbb{Z}^n$, the ILP in standard form is

$$\max\ \boldsymbol{c}^T \cdot \boldsymbol{x} \quad \text{subject to} \quad A \cdot \boldsymbol{x} = \boldsymbol{b}, \quad \boldsymbol{x} \in \mathbb{Z}^n_{\geq 0}. \tag{1}$$

For simplicity, we only consider ILPs with optimal solutions, i.e., we assume that our ILPs are feasible and bounded in the direction of the objective function and denote such ILPs by $(A, \boldsymbol{b}, \boldsymbol{c})$. We are interested in the *support size* of the ILP (1), which we define as the smallest $s := |\text{supp}(\boldsymbol{x})| := |\{i \in \{1, \dots, n\} \mid \boldsymbol{x}_i > 0\}|$ for any optimal solution $\boldsymbol{x}$. In other words, we want to know the smallest number of necessary non-zero variables among all optimal solutions. For ILPs in standard form there exists a vertex solution with equal or larger objective value, and equal or smaller support size, than any other solution. Hence, we only need to consider vertex solutions. However, vertex solutions of ILPs have an important property.

Proposition 3 (Berndt, Jansen, Klein [5]). *For any ILP $(A, \boldsymbol{b}, \boldsymbol{c})$ with vertex solution $\boldsymbol{x}$, the $\mathbf{0}$-vector is the only feasible solution for $A_{\text{supp}(x)} \cdot \boldsymbol{y} = 0$ with $\boldsymbol{y} \in \{-1, 0, 1\}^s$ and $A_{\text{supp}(x)}$ the matrix of columns A_i with non-zero $\boldsymbol{x}_i$.*

The crucial connection between SSD sets and minimal support optimal solution, which lets us use the results of Sect. 2, is the following theorem.

Theorem 2. *For any ILP $(A, \boldsymbol{b}, \boldsymbol{c})$ a minimal support optimal solution $\boldsymbol{x}$ with support size s forms an SSD set $A_{\text{supp}(x)}$ of size s with the columns of A.*

3.1 Single-Constraint ILPs

ILPs with a single constraint ($m = 1$) already model important combinatorial optimization problems, such as KNAPSACK. Our goal is to bound the support size s of single-constraint ILPs by $s \leq \log(c\Delta)$. From our study of SSD sets, we can relate this task to Erdős' conjecture and give a conditional answer.

Corollary 3. *If Erdős' conjecture holds with constant $\tilde{c}$, then any ILP with constraint matrix $A \in \mathbb{Z}^{1 \times n}$ and largest absolute value Δ of any coefficient in A has support size $s \leq \log(\Delta / 2^{\tilde{c}})$.*

For $m = 1$, Proposition 1 can be used to give a strong support size bound.

Theorem 3. *Any single-constraint ILP with largest absolute value Δ of any coefficient in the constraint matrix has support size $s \leq 1.1\log(2.02\Delta)$.*

3.2 ILPs with Multiple Constraints

Another indicator of the close connection between SSD sets and support solutions is that the exact same constructions from Sect. 2 also work for the support.

We first establish a set of lower bounds.

Theorem 4. *For any ILP $(A, \boldsymbol{b}, \boldsymbol{c})$ with support size s there is a vector $\tilde{\boldsymbol{c}}$ such that the ILP $\left(\begin{pmatrix}2A & I\end{pmatrix}, 2\boldsymbol{b} + \boldsymbol{1}, \tilde{\boldsymbol{c}}\right)$ has support size $s + m$.*

Theorem 5. *For any ILP $(A, \boldsymbol{b}, \boldsymbol{c})$ with support size s there is a vector $\tilde{\boldsymbol{c}}$ such that the ILP $\left(\begin{pmatrix}A & A & I\\ A & -A & 0\end{pmatrix}, \begin{pmatrix}2\boldsymbol{b}+\boldsymbol{1}\\ \boldsymbol{0}\end{pmatrix}, \tilde{\boldsymbol{c}}\right)$ has support size $2s + m$.*

Theorem 6. *For any ILPs $(A, \boldsymbol{b}, \boldsymbol{c})$ and $(A', \boldsymbol{b}', \boldsymbol{c}')$ with support sizes s and t, respectively, the ILP $\left(\begin{pmatrix}A & 0\\ 0 & A'\end{pmatrix}, \begin{pmatrix}\boldsymbol{b}\\ \boldsymbol{b}'\end{pmatrix}, \begin{pmatrix}\boldsymbol{c} & \boldsymbol{c}'\end{pmatrix}\right)$ has support size $s + t$.*

For the parameter $\|A\|_1$, Berndt et al. [4] showed the asymptotically optimal lower bound $s \geq m \cdot \log(\|A\|_1)$ by using the matrix which results from the combination of Theorem 4 and Theorem 6.

Corollary 4. *For m a power of two there is an m-constraint ILP $(A, \boldsymbol{b}, \boldsymbol{c})$ with support size $s \geq m(\log(2\sqrt{m}) + \lfloor\log(\Delta)\rfloor) \geq m\log(\sqrt{m}\Delta)$.*

Upper Bounds. Gribanov et al. [19] gave a support size bound of the form $s \leq m(\log(\sqrt{m}\Delta) + \mathcal{O}(\log(\log(\sqrt{m}\Delta)))$. They were thereby the first to achieve a leading coefficient of 1. Our lower bounds show that a leading coefficient of 1 is asymptotically optimal and that $\sqrt{m}\Delta$ is indeed the best attainable result inside the logarithm. Berndt et al. [4] gave a parametric bound of the form $s \leq m(\log(\sqrt{m}\Delta) + \mathcal{O}(\sqrt{\log(\sqrt{m}\Delta)}))$ with leading coefficient 1 and raised the question, whether a bound of the form $s \leq m\log(c\sqrt{m}\Delta)$ for a constant c is possible. We made progress towards this question with Corollary 3 in dimension one. In terms of bounds with greater leading coefficient, Berndt et al. [4] showed $s \leq 2m\log(1.46 \cdot \sqrt{m}\Delta)$ and $s \leq 1.1m\log(3.42 \cdot \sqrt{m}\Delta)$. For a leading coefficient of 2 we show an optimal factor of $\sqrt{2}$ inside the logarithm.

Theorem 7. *Any ILP $(A, \boldsymbol{b}, \boldsymbol{c})$ with m constraints and largest absolute value Δ of any coefficient in A has support size $s \leq 2m \cdot \log(\sqrt{2} \cdot \sqrt{m}\Delta)$.*

Theorem 8. *Any ILP $(A, \boldsymbol{b}, \boldsymbol{c})$ with m constraints has support size $s \leq 2m \cdot \log(\sqrt{2} \cdot \|A\|_1)$.*

3.3 Numerical Results

As elaborated before, the columns of minimal support solutions form SSD sets. The following insight is useful for finding SSD sets with minimal parameter.

Lemma 4. *Consider any ILP* $(A, \boldsymbol{b}, \boldsymbol{c})$ *and let* $\boldsymbol{x}$ *be an optimal solution with minimum support. Then for the ILP* $(A_{\text{supp}(\boldsymbol{x})}, A_{\text{supp}(\boldsymbol{x})} \cdot \mathbf{1}, \boldsymbol{c}_{\text{supp}(\boldsymbol{x})})$*, the vector* $\mathbf{1}$ *is an optimal solution with minimum support.*

Hence, we can find an optimal solution with minimum support by testing SSD sets. Note that ILPs are *not* invariant under sign change of columns. So for every SSD set S, we need to test $2^{|S|-1}$ many sign combinations (-1 because $-\boldsymbol{x}$ is optimal for $(-A, -\boldsymbol{b}, -\boldsymbol{c})$). To test a matrix A resulting from a sign configuration of an SSD set, we set up the following LP and ILP:

$$\{\boldsymbol{c} \in \mathbb{R}^n \mid G \cdot \boldsymbol{c} \leq -1\} \qquad \{\boldsymbol{g} \in \mathbb{Z}^n \mid A \cdot \boldsymbol{g} = 0,\ \boldsymbol{c}^T \cdot \boldsymbol{g} \geq 0,\ \boldsymbol{g} \neq 0\}$$

The LP tries to find a direction such that no known augmentation improves the objective. The ILP tries to find an augmentation relative to the current objective. An objective $\boldsymbol{c}$ found by the LP is used in the ILP and an augmentation step $\boldsymbol{g}$ is added to the set G used in the LP. If the ILP becomes infeasible, then $\mathbf{1}$ is an optimal solution of minimal support. If the LP becomes infeasible, it is not. Some care needs to be taken to avoid an infinite loop for unbounded ILPs. We used GuRoBi [20] to successively solve the LPs and ILPs. For any SSD set, the augmentation steps were stored and reused for other sign combinations, if applicable, which significantly sped up the process. We found performance decreases from GuRoBi multithreading or multithreading on more than half of the physical cores, suggesting memory speed as the bottleneck of the computation.

We list the smallest possible parameters $\|A\|_1$ and Δ for an ILP with m constraints and support size s in Table 3 and Table 4 respectively.

Table 3. Smallest possible $\|A\|_1$, given m and s.

$m \backslash s$	1	2	3	4	5	6	7	8	9	10
1	1	2	4	7	13	24	46	90	176	345
2	1	1	2	2	3	4	6	7	10	13
3	1	1	1	2	2	2	3	4	4	5
4	1	1	1	1	2	2	2	2	3	3
5	1	1	1	1	1	2	2	2	2	2

Table 4. Smallest possible Δ, given m and s.

$m \backslash s$	1	2	3	4	5	6	7	8	9	10
1	1	2	4	7	13	24	46	90	176	345
2	1	1	1	2	2	3	4	5	7	10
3	1	1	1	1	1	2	2	2	3	3
4	1	1	1	1	1	1	1	1	2	2
5	1	1	1	1	1	1	1	1	1	1

4 Bounds on Largest Graver Basis Elements

Graver [18] introduced the concept now known as Graver bases to study augmentation in integer programs. For an integer matrix A, the *Graver basis* $G(A)$ is the set of inclusion-minimal non-zero integer solutions $\boldsymbol{x}$ to $A \cdot \boldsymbol{x} = 0$. A solution to an ILP is optimal if and only if it cannot be improved by addition of a vector in the Graver basis of the constraint matrix [18]. An equivalent characterization of Graver bases as primitive partition identities was given by Diaconis, Graham and Sturmfels [12]. A (multi-)set $A \in \mathbb{Z}^m$ is a primitive partition identity if $\sum_{\boldsymbol{a} \in A} \boldsymbol{a} = \boldsymbol{0}$ and for all non-empty $B \subsetneq A$ it holds $\sum_{\boldsymbol{a} \in B} \boldsymbol{a} \neq \boldsymbol{0}$. Another interpretation of Graver bases is as cycles on the integer points $\mathbb{Z}^m$, where each edge is a column vector of A and there is no rearrangement into two cycles [15]. Hence, if we take any column vector for a Graver basis element $\boldsymbol{g}$, all other column vectors for $\boldsymbol{g}$ have to be necessary to return to the origin $\boldsymbol{0}$. We are interested in bounds on the size of the largest Graver basis elements:

Definition 3. *For any matrix A let $g_1(A) := \max_{\boldsymbol{g} \in G(A)} \|\boldsymbol{g}\|_1$ be the largest 1-norm of any Graver basis element $\boldsymbol{g}$ in the Graver basis $G(A)$ of A.*

For a matrix A, let $\delta(A)$ be the absolute value of its largest subdeterminant. Using the third interpretation of Graver bases, Diaconis et al. [12] showed $g_1(A) \leq (2m)^m(m+1)^{m+1} \cdot \delta(A)$. This extended the earlier result $g_1(A) \leq n \cdot (n-m) \cdot \delta(A)$ by Sturmfels [32]. Onn [28] gave the bound $g_1(A) \leq (n-m) \cdot \delta(A)$. Sturmfels and Onn's bounds use the number n of distinct columns, which can be as large as $(2\Delta+1)^m$. Recently, Eisenbrand et al. [14] showed $g_1(A) \leq (2\Delta m + 1)^m$.

We first consider single-constraint ILPs that model problems like KNAPSACK. For dimension $m = 1$, Diaconis et al. [12] characterized $g_1(A)$ as follows:

Proposition 4. *Any matrix $A \in \mathbb{Z}^{1 \times n}$ with largest absolute entry Δ satisfies $g_1(A) \leq 2\Delta - 1$ for $\Delta > 1$ and $g_1(A) \leq 2$ for $\Delta = 1$.*

Next, we deal with ILPs with multiple constraints. Regarding lower bounds, Graver bases elements in dimension m directly allow for the construction of larger Graver bases elements in dimension $m+1$.

Theorem 9. *For any matrix $A \in \mathbb{Z}^{m \times n}$ with largest absolute entry Δ and any Graver basis element $\boldsymbol{g} \in G(A)$ there is a matrix $A' \in \mathbb{Z}^{(m+1)\times(n+1)}$ with largest absolute entry Δ and Graver basis element $\tilde{\boldsymbol{g}} \in G(A')$ with $\|\tilde{\boldsymbol{g}}\|_1 = \|\boldsymbol{g}\|_1(1+\Delta)$.*

Even for very restricted matrices, Theorem 9 already implies exponential growth in the size of Graver basis elements.

Corollary 5. *For all $m \in \mathbb{N}$ there is a matrix $A \in \{-1,0,1\}^{m \times m+1}$ with $g_1(A) \geq 2^m$.*

Starting at $m = 3$, the construction of Corollary 5 is no longer optimal, as $1 \cdot (1\ {-1}\ {-1})^T + 2 \cdot (1\ 1\ 0)^T + 3 \cdot (-1\ 1\ {-1})^T + 4 \cdot (0\ {-1}\ 1)^T = \boldsymbol{0}$ is a Graver basis element of length $10 > 8$. Our best lower bounds do not have m in the basis when considering the parameter Δ, a gap we closed in the support case.

Corollary 6. *For all $m \in \mathbb{N}$ and all $\Delta \geq 2$ there is an m-row matrix A with largest absolute entry Δ such that $g_1(A) \geq (2\Delta - 1)(\Delta + 1)^{m-1}$.*

However, for $\|A\|_1$ we are able to construct an almost matching lower bound:

Theorem 10. *For all $m \in \mathbb{N}$ and all $\|A\|_1 > 2$ there is an m-row matrix A such that $g_1(A) \geq (2(\|A\|_1 - 1)^{m+1} - \|A\|_1)/(\|A\|_1 - 2)$, and with $g_1(A) = 2m+1$ for $\|A\|_1 = 2$.*

For $\|A\|_1$, there is no black-box extension method like Theorem 9, as already for $m = 2$ and $\|A\|_1 = 3$ the only largest Graver element, up to symmetry and sign changes, is $\begin{pmatrix} 3 & -2 & -1 \\ 0 & 1 & -2 \end{pmatrix} \cdot (5\ 6\ 3)^T = \begin{pmatrix} 0 \\ 0 \end{pmatrix}$. Hence there are optimal solutions to which we can not attach anything without increasing the norm, because all column vectors already have the largest allowed 1-norm.

Upper Bounds. The third interpretation of Graver bases suggests using Steinitz type rearrangement lemmas to derive bounds on $g_1(A)$, as shown by Diaconis et al. [12]. The idea of the Steinitz lemma is that any cycle of vectors of limited size admits a rearrangement such that each partial sum is also of limited size.

Proposition 5 (Steinitz Lemma). *Given $\boldsymbol{x}_1, \ldots, \boldsymbol{x}_n \in \mathbb{R}^m$ with $\sum_{i=1}^n \boldsymbol{x}_i = \boldsymbol{0}$ and $\|\boldsymbol{x}_i\| \leq \Delta$, there is a permutation $\pi \in S_n$ such that $\|\sum_{i=1}^k \boldsymbol{x}_{\pi(i)}\| \leq C(m) \cdot \Delta$ holds for all $k = 1, \ldots, n$.*

Eisenbrand et al. [14] used the bound $C(m) \leq m$ proved by Sevast'janov [29]. Banaszczyk [29] is credited with the improved bound $C(m) \leq m - 1 + 1/m$. Sevast'janov [30] gave a constructive proof (cf. de Gelder [17] for a non-constructive proof). As a consequence, there is a slightly better bound on Graver basis elements. Using the techniques of Eisenbrand et al. [14], we show:

Theorem 11. *For any matrix $A \in \mathbb{Z}^{m \times n}$ with largest absolute entry Δ it holds that $g_1(A) \leq (2m\Delta - 1)^m$.*

Interestingly, Bárány [3] conjectured $C(m) \in \mathcal{O}(\sqrt{m})$, which would directly improve the Graver bounds. Further, every Hadamard matrix of dimension $m+1$ gives $C(m) \geq \sqrt{m}/2(1 + 1/m)$, as shown by Bárány [3] and de Gelder [17]. The achievable bounds in terms of $\|A\|_1$ are asymptotically smaller than $m^{\mathcal{O}(m)}$ for $\|A\|_1 \in m^{o(1)}$, which can occur for sparse constraint matrices.

Theorem 12. *For any matrix $A \in \mathbb{Z}^{m \times n}$, its largest Graver basis element size is bounded by $g_1(A) \leq \sum_{k=0}^m 2^k \binom{m}{k} \binom{\lfloor \|A\|_1 (m-1+1/m) \rfloor}{k} \leq (2e \cdot (\|A\|_1 + 1))^m$.*

The case $\|A\|_1 = 2$ is special, similar to Theorem 10, and only gives a linear bound, whereas $\|A\|_1 = 1$ is trivial and $\|A\|_1 > 2$ is exponential. In fact, we can show that $g_1(A) \leq 2m+1$ for matrices A with $\|A\|_1 \leq 2$, and in the process also classify all Graver elements for such matrices.

Theorem 13. *For any matrix A with $\|A\|_1 = 2$, any Graver basis element $\boldsymbol{g} \in G(A)$ either uses no $\pm \boldsymbol{e}_i$ vectors with $\|\boldsymbol{g}\|_1 \leq 2m$, or uses two $\pm \boldsymbol{e}_i$ vectors with $\|\boldsymbol{g}\|_1 \leq 2m + 1$. The non-zero entries in a row of the partition identity of $\boldsymbol{g}$ are, up to sign, $\{2, -1, -1\}, \{1, 1, -1, -1\}, \{2, -2\}, \{1, -1\}$ or $\emptyset$.*

4.1 Numerical Results

To find the largest Graver basis element for any parameter bound, we calculated the Graver bases for the matrix of all admissible vectors with 4ti2 [33]. This approach quickly became computationally infeasible, used large amounts of RAM, is single-threaded, and does not utilize symmetry reductions. We thus generated longer Graver basis elements by letting GuRoBi generate solutions to $A \cdot \boldsymbol{g} = 0$ of specified length. Then we used GuRoBi to check that no $\boldsymbol{g}' \subsetneq \boldsymbol{g}$ is already a Graver basis element. If $\boldsymbol{g}'$ exists, we add the smaller of $\boldsymbol{g}'$ and $\boldsymbol{g} - \boldsymbol{g}'$ as a forbidden subset to the original model. We list the largest found $g_1(A)$ for given dimension m and parameters $\|A\|_1$ and Δ in Table 5 and Table 6 respectively.

Table 5. Largest possible $g_1(A)$, given m and Δ.

$m\backslash\Delta$	1	2	3	4	5
1	2	3	5	7	9
2	4	13	34	≥61	≥98
3	10	≥79	≥160		
4	≥30				
5	≥90				

Table 6. Largest possible $g_1(A)$, given m and $\|A\|_1$.

$m\backslash\|A\|_1$	1	2	3	4	5
1	2	3	5	7	9
2	2	5	14	25	46
3	2	7	38	≥89	≥206
4	2	9	≥86		
5	2	11			

4.2 Proximity

We show two types of results regarding the proximity of optimal fractional solutions to optimal integral solutions of linear programs. First, we improve the bound of $n \cdot g_1(A)$ by Hemmecke et al. [22] via a bound on the support size of optimal integer solutions. Second, we show that a Graver basis element bound in the style of Eisenbrand et al. [14] can be used to obtain a proximity bound in the style of Eisenbrand and Weismantel [15] with a small additional factor of m. The circuits of a matrix A are the elements of its Graver basis of minimum support. Let $c_1(A) := \max_{c \in \mathcal{C}(A)} \|c\|_1$ be the largest circuit 1-norm of A. The inequality $c_1(A) \leq (m+1)\delta(A)$ is useful for deriving bounds.

Theorem 14. *For any m-constraint ILP $(A, \boldsymbol{b}, \boldsymbol{c})$ with an optimal solution and any optimal vertex fractional solution $\tilde{\boldsymbol{x}}$ there is, for any support size bound $f(A)$, an optimal integer solution $\tilde{\boldsymbol{z}} \in \mathbb{Z}^n$ satisfying $\|\tilde{\boldsymbol{z}} - \tilde{\boldsymbol{x}}\|_1 \leq f(A) \cdot c_1(A)$.*

For matrices A with bounded Δ or bounded 1-norm $\|A\|_1$ this directly yields:

Corollary 7. *For any m-constraint ILP $(A, \boldsymbol{b}, \boldsymbol{c})$ with an optimal solution and any optimal vertex fractional solution $\tilde{\boldsymbol{x}}$ there is an optimal integer solution $\tilde{\boldsymbol{z}} \in \mathbb{Z}^n$ satisfying $\|\tilde{\boldsymbol{z}} - \tilde{\boldsymbol{x}}\|_1 \leq 2m \log(\sqrt{2m}\Delta) \cdot (m+1) \cdot (\sqrt{m}\Delta)^m$.*

Corollary 8. *For any m-constraint ILP $(A, \boldsymbol{b}, \boldsymbol{c})$ with an optimal solution and any optimal vertex fractional solution $\tilde{\boldsymbol{x}}$ there is an optimal integer solution $\tilde{\boldsymbol{z}} \in \mathbb{Z}^n$ satisfying $\|\tilde{\boldsymbol{z}} - \tilde{\boldsymbol{x}}\|_1 \leq 2m \log(\sqrt{2}\|A\|_1) \cdot (m+1) \cdot (\|A\|_1)^m$.*

For small or fixed values of m better bounds can be obtained with an approach based on the Steinitz lemma.

Theorem 15. *For any m-constraint ILP $(A, \boldsymbol{b}, \boldsymbol{c})$ with an optimal solution and any optimal vertex fractional solution $\tilde{\boldsymbol{x}}$ there is an optimal integer solution $\tilde{\boldsymbol{z}} \in \mathbb{Z}^n$ satisfying $\|\tilde{\boldsymbol{z}} - \tilde{\boldsymbol{x}}\|_1 \leq m \cdot (2m\Delta - 1)^m$.*

Theorem 16. *For any m-constraint ILP $(A, \boldsymbol{b}, \boldsymbol{c})$ with an optimal solution and any optimal vertex fractional solution $\tilde{\boldsymbol{x}}$ there is an optimal integer solution $\tilde{\boldsymbol{z}} \in \mathbb{Z}^n$ satisfying $\|\tilde{\boldsymbol{z}} - \tilde{\boldsymbol{x}}\|_1 \leq m \cdot (2e(\|A\|_1 + 1))^m$.*

Contrarily, the proximity does not bound the circuit length:

Observation 4. *For all $m \in \mathbb{N}$ there is a matrix $A \in \mathbb{Z}^{m \times m+1}$ with proximity* 0 *and largest circuit 1-norm $c_1(A) = m + 1$.*

Naturally, a proximity bound directly implies bounds the integrality gap:

Corollary 9. *For any ILP $(A, \boldsymbol{b}, \boldsymbol{c})$ the absolute integrality gap is bounded by*

$$\boldsymbol{c}^T \cdot (\tilde{\boldsymbol{x}} - \tilde{\boldsymbol{z}}) \leq \|\boldsymbol{c}\|_\infty \cdot \|\tilde{\boldsymbol{z}} - \tilde{\boldsymbol{x}}\|_1.$$

References

1. Aliev, I., De Loera, J.A., Oertel, T., O'Neill, C.: Sparse solutions of linear Diophantine equations. SIAM J. Appl. Algebra Geom. **1**(1), 239–253 (2017)
2. Bansal, N., Oosterwijk, T., Vredeveld, T., van der Zwaan, R.: Approximating vector scheduling: almost matching upper and lower bounds. Algorithmica **76**(4), 1077–1096 (2016). https://doi.org/10.1007/s00453-016-0116-0
3. Bárány, I.: On the power of linear dependencies. In: Grötschel, M., Katona, G.O.H., Sági, G. (eds.) Building Bridges. BSMS, vol. 19, pp. 31–45. Springer, Heidelberg (2008). https://doi.org/10.1007/978-3-540-85221-6_1
4. Berndt, S., Brinkop, H., Jansen, K., Mnich, M., Stamm, T.: New support size bounds for integer programming, applied to makespan minimization on uniformly related machines. In: Proceedings of ISAAC 2023 (2023). https://drops.dagstuhl.de/entities/document/10.4230/LIPIcs.ISAAC.2023.13
5. Berndt, S., Jansen, K., Klein, K.M.: New bounds for the vertices of the integer hull. In: Proceedings of SOSA 2021, pp. 25–36 (2021)
6. Bohman, T.: A sum packing problem of Erdős and the Conway-Guy sequence. Proc. Am. Math. Soc. **124**(12), 3627–3636 (1996)
7. Bohman, T.: A construction for sets of integers with distinct subset sums. Electron. J. Comb. **5**, 14 (1998). Research Paper 3
8. Bombieri, E., Vaaler, J.: On Siegel's lemma. Inventiones Math. **73**, 11–32 (1983). https://doi.org/10.1007/BF01393823
9. Borwein, P., Mossinghoff, M.J.: Newman polynomials with prescribed vanishing and integer sets with distinct subset sums. Math. Comput. **72**(242), 787–800 (2003)
10. Conway, J.H., Guy, R.K.: Sets of natural numbers with distinct subset sums. Not. Am. Math. Soc. **15**, 345 (1968)
11. Cygan, M., et al.: Parameterized Algorithms. Springer, Cham (2015). https://doi.org/10.1007/978-3-319-21275-3

12. Diaconis, P., Graham, R.L., Sturmfels, B.: Primitive partition identities. In: Combinatorics, Paul Erdős is Eighty, vol. 2 (Keszthely, 1993), Bolyai Soc. Math. Stud., vol. 2, pp. 173–192 (1996)
13. Dubroff, Q., Fox, J., Xu, M.W.: A note on the Erdős distinct subset sums problem. SIAM J. Discrete Math. **35**(1), 322–324 (2021)
14. Eisenbrand, F., Hunkenschröder, C., Klein, K.M.: Faster algorithms for integer programs with block structure. In: Proceedings of ICALP 2018. Leibniz International Proceedings in Informatics, vol. 107, p. 13 (2018). Article No. 49
15. Eisenbrand, F., Weismantel, R.: Proximity results and faster algorithms for integer programming using the Steinitz lemma. ACM Trans. Algorithms **16**(1), 14 (2020). Article 5
16. Erdős, P.: Problems and results on extremal problems in number theory, geometry, and combinatorics. In: Proceedings of the 7th Fischland Colloquium, I, Wustrow, no. 38, pp. 6–14 (1989)
17. de Gelder, M.: Investigating various upper and lower bounds of the Steinitz constant. Bachelor thesis, TU Delft, Delft Institute of Applied Mathematics (2016)
18. Graver, J.E.: On the foundations of linear and integer linear programming. I. Math. Program. **9**(2), 207–226 (1975). https://doi.org/10.1007/BF01681344
19. Gribanov, D., Shumilov, I., Malyshev, D., Pardalos, P.: On Δ-modular integer linear problems in the canonical form and equivalent problems. J. Glob. Optim. 1–61 (2022). https://doi.org/10.1007/s10898-022-01165-9
20. Gurobi Optimization, LLC: Gurobi Optimizer Reference Manual (2023)
21. Haase, C., Zetzsche, G.: Presburger arithmetic with stars, rational subsets of graph groups, and nested zero tests. In: Proceedings of LICS 2019, pp. 1–14 (2019)
22. Hemmecke, R., Köppe, M., Weismantel, R.: Graver basis and proximity techniques for block-structured separable convex integer minimization problems. Math. Program. **145**(1), 1–18 (2014). https://doi.org/10.1007/s10107-013-0638-z
23. Kannan, R.: Minkowski's convex body theorem and integer programming. Math. Oper. Res. **12**(3), 415–440 (1987)
24. Knop, D., Koutecký, M., Mnich, M.: Combinatorial n-fold integer programming and applications. Math. Program. **184**(1–2), 1–34 (2020). https://doi.org/10.1007/s10107-019-01402-2
25. Lenstra, H.W., Jr.: Integer programming with a fixed number of variables. Math. Oper. Res. **8**(4), 538–548 (1983)
26. Lunnon, W.F.: Integer sets with distinct subset-sums. Math. Comput. **50**(181), 297–320 (1988)
27. OEIS Foundation Inc.: The On-Line Encyclopedia of Integer Sequences (2023). Published electronically at http://oeis.org
28. Onn, S.: Nonlinear discrete optimization. In: Zurich Lectures in Advanced Mathematics (2010)
29. Sevast'janov, S.: On the approximate solution of some problems of scheduling theory. Metody Diskretnogo Analiza **32** (1978)
30. Sevast'janov, S.: On the compact summation of vectors. Diskret. Mat. **3**(3), 66–72 (1991)
31. Steinerberger, S.: Some remarks on the Erdős distinct subset sums problem. Int. J. Number Theory **19**(08), 1783–1800 (2023)
32. Sturmfels, B.: Gröbner bases of toric varieties. Tohoku Math. J. **43**(2), 249–261 (1991)
33. 4ti2 team: 4ti2–a software package for algebraic, geometric and combinatorial problems on linear spaces. https://4ti2.github.io/

On the Parameterized Complexity of Minus Domination

Sriram Bhyravarapu[1(✉)], Lawqueen Kanesh[2(✉)], A Mohanapriya[3(✉)], Nidhi Purohit[4(✉)], N. Sadagopan[3(✉)], and Saket Saurabh[1,4(✉)]

[1] The Institute of Mathematical Sciences, HBNI, Chennai, India
{sriramb,saket}@imsc.res.in
[2] Indian Institute of Technology, Jodhpur, Jodhpur, India
lawqueen@iitj.ac.in
[3] Indian Institute of Information Technology, Design and Manufacturing, Kancheepuram, Chennai, India
{coe19d003,sadagopan}@iiitdm.ac.in
[4] University of Bergen, Bergen, Norway
Nidhi.Purohit@uib.no

Abstract. Dominating Set is a well-studied combinatorial problem. Given a graph $G = (V, E)$, a dominating function $f : V(G) \to \{0, 1\}$ is a labeling of the vertices of G such that $\sum_{w \in N[v]} f(w) \geq 1$ for each vertex $v \in V(G)$, where $N[v] = \{v\} \cup \{u \mid uv \in E(G)\}$. We study a generalization of Dominating Set called Minus Domination (in short, MD) where $f : V(G) \to \{-1, 0, 1\}$. Such a function is said to be a *minus dominating function* if for each vertex $v \in V(G)$, we have $\sum_{w \in N[v]} f(w) \geq 1$. The objective is to minimize the weight of a minus domination function, which is $f(V) = \sum_{u \in V(G)} f(u)$. The problem is NP-hard even on bipartite, planar, and chordal graphs.

In this paper, we study MD from the perspective of parameterized complexity. After observing the complexity of the problem with the natural parameters such as the number of vertices labeled 1, -1 and 0, we study the problem with respect to structural parameters. We show that MD is fixed-parameter tractable when parameterized by twin-cover number, neighborhood diversity or the combined parameters component vertex deletion set and size of the largest component. In addition, we give an XP-algorithm when parameterized by distance to cluster number.

Keywords: Minus Domination · fixed-parameter tractability · twin-cover · neighborhood diversity · disjoint paths deletion · cluster vertex deletion

1 Introduction

Given a graph $G = (V, E)$, a *dominating function* $f : V(G) \to \{0, 1\}$ is a labeling of $V(G)$ from $\{0, 1\}$ such that for each vertex $v \in V(G)$ we have $\sum_{w \in N[v]} f(w) \geq 1$, where $N[v] = \{v\} \cup \{u \mid uv \in E(G)\}$. The weight of f is denoted by $f(V) = \sum_{u \in V(G)} f(u)$. The Dominating Set (in short, DS)

H. Fernau et al. (Eds.): SOFSEM 2024, LNCS 14519, pp. 96–110, 2024.
https://doi.org/10.1007/978-3-031-52113-3_7

problem asks to find a dominating function of minimum weight. Several variants of DS have been studied in literature, some of which include independent, total, global, perfect and k-dominating [1,12,13]. In this paper, we study another variant of domination called MINUS DOMINATION (in short, MD) which was introduced by Dunbar et al. in 1996 [6]. Given a graph $G = (V, E)$, a *minus dominating function* $f : V(G) \to \{-1, 0, 1\}$ is an assignment of labels to the vertices of G such that for each $v \in V(G)$, the sum of labels assigned to the vertices in the closed neighborhood of v (denoted by $N[v]$) is at least one, i.e., $\sum_{w \in N[v]} f(w) \geq 1$. The *weight* of a minus dominating function f denoted by $f(V)$ is $\sum_{v \in V(G)} f(v)$. Given a graph G, MINUS DOMINATION asks to compute the minimum weight of a minus dominating function of G. The decision version of the problem takes as input a graph G and an integer k, and outputs whether there exists a minus dominating function of weight at most k. MD has applications in electrical networks, social networks, voting, etc. [6,19].

The weight of a minus dominating function can be negative. For example, consider a clique on n vertices and for each edge uv in the clique, add a private vertex adjacent to only u and v. Consider the minus dominating function f that assigns all the clique vertices the label 1 and all the private vertices the label -1. The clique vertices have as many private neighbors as they have clique neighbors while the private vertices have exactly two clique neighbors. Thus, for each vertex v, we have $f(N[v]) \geq 1$ and $f(V) = n - n(n-1)/2 < 0$ for a large n. The authors in [6] show that given a positive integer k there exists a bipartite, chordal and outer-planar graphs with weight at most $-k$.

MINUS DOMINATION is NP-complete in general [8] and NP-complete even on chordal bipartite graphs, split graphs, and bipartite planar graphs of degree at most 4 [3,6,8,17]. The problem is polynomial-time solvable on trees, graphs of bounded rank-width, cographs, distance hereditary graphs and strongly chordal graphs [6,8]. Given the hardness results for the problem, it is natural to ask for ways to confront this hardness. Parameterized complexity is an approach towards solving NP-hard problems in "feasible" time. Parameterized problems that admit such an algorithm are called fixed-parameter tractable (in short, FPT). For more details, we refer the reader to the book by Cygan et al. [2] and Downey and Fellows [4].

MINUS DOMINATION has been studied from the realm of parameterized complexity. On subcubic graphs, MD is FPT when parameterized by weight [18]. As far as near-optimal solutions are concerned, the minimum weight of a minus dominating function cannot be approximated in polynomial time within $(1+\epsilon)$, for some $\epsilon > 0$, unless $\mathsf{P} \neq \mathsf{NP}$ [3]. The problem is APX-hard on graphs of maximum degree 7 [18]. Several combinatorial bounds for the problem on regular graphs, and small degree graphs ($\Delta \leq 3$ or $\Delta \leq 4$) have been studied [3,7].

A parameter may originate from the formulation of the problem itself (called natural parameters) or it can be a property of the input graph (called structural parameters). DOMINATING SET when parameterized by solution size is W[2]-hard [4]; however, when parameterized by structural parameters such as tree-width [2], modular-width, or distance to cluster (size of the cluster vertex deletion set) [11], the problem is fixed-parameter tractable. MD has various natural parameters such as n_{-1}, n_0, n_1 and $f(V)$ (where $n_{-1} = |f^{-1}(-1)|$, $n_0 = |f^{-1}(0)|$, and $n_1 = |f^{-1}(1)|$ for a minus dominating function f) and it was shown in [8] that the problem is para-NP-hard when parameterized by $f(V)$.

Domination vs Minus Domination: One may think that the ideas used for solving DS can be extended to MD. But this is not the case always. There are graphs for example connected cographs, wheel graphs, windmill graphs, chain graphs, etc. where dominating set size is constant and the corresponding set can be found trivially. However, it is not the case with MD.

On graphs of bounded tree-width, the dynamic programming based FPT algorithm for DS when parameterized by tree-width (tw) focuses on guessing vertices from each bag that are in the dominating set. However, for MD, just guessing the labels 1, −1 and 0 for the vertices of a bag does not suffice. We may also need to store the information about the sum that each vertex in the bag receives from its subtree to be able to extend to the rest of the graph. Since the degree of a vertex can be unbounded, the sum it receives from the subtree can be unbounded. This gives us an $n^{\mathcal{O}(\mathsf{tw})}$ time algorithm. Notice that this gives us an FPT algorithm when parameterized by maximum degree and tree-width. The authors of [8] believe that MD is not FPT when parameterized by tree-width or rank-width. To the best of our knowledge, the FPT status of minus domination with respect to tree-width is still open.

Our Contribution: First, we analyse the problem on natural parameters. We obtain the following result, the proof of which is not hard and follows from Dominating Set and its well-known variants.

Theorem 1 (⋆).[1] Minus Domination *when parameterized by* n_{-1} *or* n_0 *is para-*NP*-hard, when parameterized by* n_1 *or* $n_{-1} + n_1$ *is* W[2]*-hard, and when parameterized by* $n_0 + n_1$ *is FPT.*

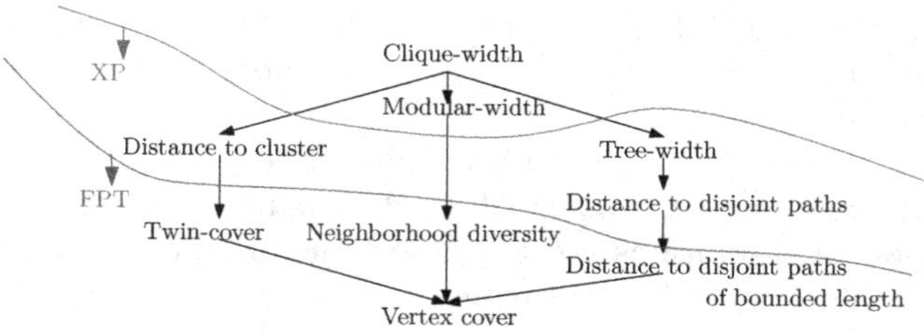

Fig. 1. Hasse diagram of graph parameters for MD. A directed edge from the parameter a to the parameter b indicates that $a \leq g(b)$ for some computable function g. The parameters below the blue curve are those for which MD is FPT while the parameters between the red and blue curves are those for which XP algorithms are known for MD. (Color figure online)

[1] Due to space constraints, all the proofs of the results marked (⋆) will be presented in the full version of the paper.

We shift our focus to various structural graph parameters. In Sect. 3, we show that MD is FPT when parameterized by twin-cover number. The next parameter to consider is distance to cluster number, which is a generalization of twin-cover number. In Sect. 4, we obtain an XP-algorithm when parameterized by distance to cluster number. Then we move our attention to a more general parameter: the size of component vertex deletion set. In Sect. 5, we study the problem on this parameter and obtain an FPT algorithm when parameterized by the size of component vertex deletion set and the size of a largest component. This implies an FPT algorithm for MD when parameterized by (i) distance to cluster number and the size of a largest clique, (ii) distance to disjoint paths and the size of a largest path, or (iii) feedback vertex set number and the size of a largest tree component. We also show that MD is FPT when parameterized by the parameter neighborhood diversity. An illustration of the results is given in Fig. 1. We now state the theorems of the above discussed results.

Theorem 2. Minus Domination *can be solved in* $2^{\mathcal{O}(k\cdot 2^k)}n^{\mathcal{O}(1)}$ *time, where* k *is the twin cover number of the graph.*

Theorem 3. Minus Domination *can be solved in time* $g(k)\cdot n^{2k+6}$*, where* k *is the distance to cluster number.*

Theorem 4 **(⋆).** *Let* G *be a graph and* $S\subseteq V(G)$ *of size* k *be such that* $G-S$ *is a disjoint union of components where each component has at most* d *vertices. Then,* Minus Domination *is FPT when parameterized by* k *and* d.

Theorem 5 **(⋆).** Minus Domination *can be solved in time* $t^{\mathcal{O}(t)}n^{\mathcal{O}(1)}$*, where* t *is the neighborhood diversity of the graph.*

Open Question: What is the parameterized complexity of MD when parameterized by distance to cluster, tree-width or feedback vertex set, or distance to disjoint paths?

2 Preliminaries

In this paper, we consider finite, undirected and connected graphs. If the graph is disconnected, then we apply our algorithms on each of the components independently. Given a graph $G=(V,E)$, we use $V(G)$ and $E(G)$ to denote the vertex and the edge sets of G. For a vertex $v\in V(G)$, we use $N(v)$ (open neighborhood of v) to denote the neighbors of v in G. The closed neighborhood of v is denoted by $N[v]=N(v)\cup\{v\}$. For a vertex $v\in V(G)$ and a set $C\subseteq V(G)$, we denote $N_C(v)=N(v)\cap C$. For a pair of vertices $u,v\in V(G)$, we say u and v are *true twins*, if and only if $N[u]=N[v]$. We say that a vertex v *satisfies the sum property*, if $\sum_{u\in N[v]}f(u)\geq 1$. For a set $X\subseteq V(G)$ and a vertex w, we say w *receives the sum* s from X if $\sum_{v\in N[w]\cap X}f(v)=s$.

The *size* of a minus dominating function is the number of vertices assigned the label 1. For a nonempty subset $S\subseteq V(G)$, we denote by $G[S]$ the subgraph of G induced by S. Let $f : X \to Y$ be a function. If $A \subseteq X$ then the restriction of f to A is the function $f|_A : A \to Y$ given by $f|_A(x) = f(x)$, for $x \in A$. We say a labeling $f : X \to \{-1, 0, 1\}$ *extends* $g : Y \to \{-1, 0, 1\}$ if $Y \subseteq X$ and for each $w \in Y$, we have $f(w) = g(w)$. We use $\mathcal{O}^*$ notation to hide factors that are polynomial in the input size.

Definition 1 (Twin-cover [10]). *Given a graph G, a set $S \subseteq V(G)$ is called a* twin cover *of G if the following conditions hold: (i) $G[V \setminus S]$ is a disjoint union of cliques, and (ii) each pair of vertices of a clique in $G[V \setminus S]$ are true twins in G. We then say that G has twin cover number k if k is the minimum possible size of a twin cover of G.*

Definition 2 (Distance to cluster [14]). *A* cluster graph *is a disjoint union of cliques. Given a graph G, a set of vertices $S \subseteq V(G)$ is called a* cluster vertex deletion set *of G if $G - S$ is a cluster graph. The size of the smallest set S for which $G - S$ is a cluster graph is referred to as* distance to cluster number.

Definition 3 (Neighborhood diversity [16]). *Let $G = (V, E)$ be a graph. Two vertices $u, v \in V(G)$ are said to have the* same type *if and only if $N(u) \setminus \{v\} = N(v) \setminus \{u\}$. A graph G has neighborhood diversity at most t, if there exists a partition of $V(G)$ into at most t sets $V_1, V_2, \ldots, V_t$ such that all the vertices in each set have the same type.*

Ganian [10] showed that a twin-cover of size k can be found in time $\mathcal{O}^*(1.2738^k)$. Hüffner et al. [14] showed that showed that a cluster vertex deletion set of size k can be computed in $\mathcal{O}^*(1.811^k)$ time. Lampis [16] showed that the neighborhood diversity of a graph can be found in polynomial time. Thus we will assume that a twin-cover, a cluster vertex deletion set and a partition of vertex set into types of vertices are given as input, in the respective sections.

We use Integer Linear Programming (ILP) feasibility problem, stated in [15] as subroutine in some of our results.

Theorem 6 ([15]). *An integer linear programming instance of size L with p variables can be solved using*

$$\mathcal{O}(p^{2.5p+o(p)} \cdot (L + \log M_x) \log(M_x M_c))$$

arithmetic operations and space polynomial in $L + \log M_x$, where M_x is an upper bound on the absolute value a variable can take in a solution, and M_c is the largest absolute value of a coefficient in the vector c.

Lemma 1. *Let $f : V(G) \to \{-1, 0, 1\}$ be a minus dominating function and $u, v \in V(G)$ be true twins such that $f(u) = 1$ and $f(v) = -1$. Then there exists a minus dominating function $g : V(G) \to \{-1, 0, 1\}$ of weight $f(V)$ and $g(u) = g(v) = 0$.*

Proof. We construct a function $g : V(G) \to \{-1, 0, 1\}$ as follows: $g(u) = g(v) = 0$ and $g(z) = f(z)$, for each $z \in V(G) \setminus \{u, v\}$. We claim that g is the minus dominating function of weight $f(V)$. It is easy to see that $g(V) = f(V)$ because $g(u) + g(v) = f(u) + f(v) = 0$ and the remaining vertices are assigned the same labels in both the labelings. Now we show that g is a minus dominating function. It is easy to see that for each vertex $w \in V(G)$ that is not adjacent to either u or v, $\sum_{y \in N[w]} g(y) = \sum_{y \in N[w]} f(y) \geq 1$, as u and v are the only vertices whose labels are changed. Since u and v are true twins, for each $w \in N(u) \cup N(v)$,

$$\begin{aligned}\sum_{y \in N[w]} g(y) &= g(u) + g(v) + \sum_{y \in N[w] \setminus \{u,v\}} g(y) = f(u) + f(v) + \sum_{y \in N[w] \setminus \{u,v\}} f(y) \\ &= \sum_{y \in N[w]} f(y) \geq 1.\end{aligned}$$

Thus, g is a minus dominating function. □

3 Twin-Cover

Let G be a graph and $S \subseteq V(G)$ be a twin cover of G of size k. Let $C_1, C_2, \ldots, C_\ell$ be the set of maximal cliques in $G - S$. From the definition of twin cover, we have that each vertex of a clique C in $G[V \backslash S]$ has the same neighborhood in S. We denote the neighborhood of a clique C in S by $N_S(C)$, i.e., $N_S(C) = N(C) \cap S$. We partition the cliques in $G[V \backslash S]$ based on its neighborhood in S. For each non-empty subset $A \subseteq S$, we denote by $T_A = \{C \mid C \text{ is a maximal clique in } G[V \backslash S] \text{ and } N_S(C) = A \subseteq S\}$ the set of cliques where each clique is adjacent to every vertex in A. We call T_A a *clique type*. Notice that the number of clique types is at most 2^k. Consider the case when $A = \emptyset$. If $T_A \neq \emptyset$ then G is disconnected. Since we are considering connected graphs, we can assume that $A \neq \emptyset$ throughout this section. In addition, we only consider $A \subseteq S$ for which $|T_A| \geq 1$, or T_A is non-empty.

Next, we show that for each $A \subseteq S$, the vertices of cliques in set T_A receive their labeling from a fixed set of labels.

Lemma 2 (⋆). *Let $f : V(G) \to \{-1, 0, 1\}$ be any minus dominating function with minimum weight. Then there exists a minus dominating function $g : V(G) \to \{-1, 0, 1\}$ with weight $f(V)$ such that for each $A \subseteq S$, vertices in T_A receives labels from exactly one of the three sets $\{-1, 0\}, \{0, 1\}$ or $\{0\}$.*

Proof (Proof of Theorem 2). Let G be a graph and $S \subseteq V(G)$ be a twin cover of G of size k. Let $f : V(G) \to \{-1, 0, 1\}$ be a minus dominating function of minimum weight. WLOG we can assume that f is a minus dominating function such that for each $A \subseteq S$, the vertices in T_A receive the labels from exactly one of the three sets $\{-1, 0\}$, $\{0, 1\}$ or $\{0\}$, as any minus dominating function can be converted to the function satisfying the above said conditions using Lemma 2. Recall that there are $2^k - 1$ many clique types.

For each $A \subseteq S$, we denote $b_A \in \{\{-1, 0\}, \{0, 1\}, \{0\}\}$ to be the set of labels that the vertices of the cliques in T_A are labeled with. We say

$\widehat{b} = \{b_A \colon A$ is non-empty and $|T_A| \geq 1\}$ *respects* f if for each non-empty $A \subseteq S$ with $|T_A| \geq 1$, the vertices of the cliques in T_A are assigned the labels from b_A.

We guess the tuple $(f|_S, \widehat{b}, \widehat{s})$, where $\widehat{s} = f(V) = \sum_{v \in V(G)} f(v)$ is the minimum weight and $-n < \widehat{s} \leq n$. For each guess, we formulate the problem as an ILP feasibility problem. Let $c_A = \sum_{v \in A} f|_S(v)$ denote the sum that each vertex in the clique type T_A receives from $A \subseteq S$. Let ℓ_A represent the number of cliques in each clique type T_A and $d_A = \min\{|D| \colon D$ is a clique in $T_A\}$ represent the minimum size of a clique in T_A. Let m_A denote the total number of vertices over all the cliques in the clique type T_A.

From now on, we work with a fixed guess $(f|_S, \widehat{b}, \widehat{s})$. Observe that for each $A \subseteq S$, c_A, ℓ_A, d_A and m_A are constants. We say a guess $(f|_S, \widehat{b}, \widehat{s})$ is *invalid* if for any $A \subseteq S$, one of the following holds: (i) $c_A \leq 1$ and $b_A = \{-1, 0\}$ (i.e., the vertices of T_A are assigned the labels from $\{-1, 0\}$), (ii) $c_A \leq 0$ and $b_A = \{0\}$, and (iii) $b_A = \{0, 1\}$ and $c_A \leq -d_A$. Otherwise, we call $(f|_S, \widehat{b}, \widehat{s})$ a *valid* guess.

For a valid guess $(f|_S, \widehat{b}, \widehat{s})$, we formulate an instance of the ILP problem. The goal of ILP is to obtain an assignment of the variables such that each vertex $v \in V(G)$ satisfies its sum property respecting $\widehat{b}$ and $\widehat{s}$. For each $A \subseteq S$, let $n_{-1,A}$, $n_{0,A}$ and $n_{1,A}$ be the variables of the ILP instance that denote the number of vertices assigned -1, 0 and 1 respectively in the clique type T_A. Notice that the number of variables is at most $3 \cdot 2^k$. We now describe the constraints of ILP for each $A \subseteq S$ as follows.

(C1) The number of vertices from T_A assigned the labels from $\{-1, 0, 1\}$ is m_A. That is,

$$n_{-1,A} + n_{0,A} + n_{1,A} = m_A.$$

(C2) If $b_A = \{0\}$ and $c_A \geq 1$, then

$$n_{-1,A} = n_{1,A} = 0, \text{ and } n_{0,A} = m_A.$$

(C3) If $b_A = \{0, 1\}$ and $c_A > -d_A$, then

$$n_{-1,A} = 0, n_{0,A} \geq 0 \text{ and } (1 - c_A)\ell_A \leq n_{1,A} \leq m_A.$$

(C4) If $b_A = \{-1, 0\}$ and $c_A > 1$, then

$$n_{1,A} = 0, n_{0,A} \geq 0, \text{ and}$$

$$0 \leq n_{-1,A} \leq \sum_{j=1}^{\ell_A} \min\{c_A - 1, |B_j|\}.$$

where $B_1, B_2, \ldots, B_{\ell_A}$ are the cliques of the type T_A.

(C5) For each $v \in S$, the sum property is satisfied. That is,

$$\sum_{A : v \in A} (n_{1,A} - n_{-1,A}) + \sum_{w : w \in N[v] \cap S} f(w) \geq 1.$$

(C6) Weight of our desired minus dominating function is $\widehat{s}$,

$$\sum_A n_{1,A} - n_{-1,A} + \sum_{v \in S} f(v) = \widehat{s}.$$

We next show a one-to-one correspondence between feasible assignments of ILP and minus dominating functions of G.

Lemma 3. *ILP has a feasible assignment if and only if there exists a minus dominating function with weight $\widehat{s}$.*

Proof. Suppose that there is a feasible assignment returned by ILP. We show that there exists a minus dominating function $f : V(G) \to \{-1, 0, 1\}$ respecting $f|_S$, $\widehat{b}$ and $\widehat{s}$. For each $A \subseteq S$, we assign the labels to the vertices of the clique type T_A in the following manner. Let $V(T_A)$ be the vertices of the cliques in T_A.

- **Case 1:** $b_A = \{0\}$.
  Assign $f(v) = 0$ for each vertex v in the cliques of T_A.
- **Case 2:** $b_A = \{0, 1\}$ and $c_A > -d_A$.
  We have the following subcases.
  - $c_A \geq 1$.
    Choose $n_{1,A}$ vertices arbitrarily from the cliques of T_A and assign the label 1 to them. Rest of the vertices (if any) are assigned the label 0.
  - $c_A \leq 0$.
    Choose $(1 - c_A)$ vertices arbitrarily from each clique of T_A and assign the label 1 to them. From the constraint (C3), we have that $n_{1,A} \geq (1 - c_A)\ell_A$. The remaining $n_{1,A} - (1 - c_A)\ell_A$ vertices are picked arbitrarily from the unassigned vertices of the cliques of T_A.
- **Case 3:** $b_A = \{-1, 0\}$ and $c_A \geq 2$.
  For each clique $B_j \in T_A$, choose $\min\{c_A - 1, |B_j|\}$ many vertices and assign the label -1 to each of them. Remaining vertices (if any) are assigned the label 0.

Clearly, the labeling $f : V(G) \to \{-1, 0, 1\}$ respects $\widehat{b}$, $\widehat{s}$ and the fact that the vertices in each clique type T_A receive the set of labels from exactly one of the three sets $\{-1, 0\}$, $\{0, 1\}$ or $\{0\}$ from the constraints (C1), (C2), (C3), (C4) and (C6). The sum property for each vertex $v \in S$ is satisfied because of Constraint (C5). Therefore it is sufficient to show that for each $v \in V(G)$, the sum property is satisfied.

Consider a vertex v from T_A. If T_A falls under Case 1, then v receives its positive sum from A from constraint (C2). If T_A falls under Case 2 and $c_A \geq 1$, then v receives its positive sum from A irrespective of whether v is assigned 0 or 1, because of constraint (C3). Else if T_A falls under Case 2 and $c_A \leq 0$, then we ensured that each clique in T_A has at least $(1 - c_A)$ vertices assigned the label 1, from constraint (C3), making the total sum in neighborhood of v to be at least 1. If T_A falls under Case 3, then we ensured that the number of vertices assigned -1 in each clique B_j of T_A is $\min\{c_A - 1, |B_j|\}$ from constraint (C4), making the closed neighborhood sum to be at least 1 for v. Notice that the above argument works irrespective of the label assigned to v.

Conversely, let $f : V(G) \to \{-1, 0, 1\}$ be a minus dominating function respecting $f|_S$, $\widehat{b}$ and $\widehat{s}$. Thus the constraint (C6) is satisfied. The variables are assigned as follows depending on the labeling of T_A.

- $b_A = \{0\}$.
  Assign $n_{0,A} = m_A$ and $n_{-1,A} = n_{1,A} = 0$.
- $b_A = \{0, 1\}$ and $c_A \geq -d_A$.
  Assign $n_{-1,A} = 0$, $n_{0,A} = f^{-1}(0) \cap V(T_A)$ and $n_{1,A} = f^{-1}(1) \cap V(T_A)$.
- $b_A = \{-1, 0\}$ and $c_A \geq 2$.
  Assign $n_{1,A} = 0$, $n_{0,A} = f^{-1}(0) \cap V(T_A)$ and $n_{-1,A} = f^{-1}(-1) \cap V(T_A)$.

Each vertex from T_A is assigned a label from $\{-1, 0, 1\}$ and hence constraint (C1) is satisfied. The assignment of labels to vertices in $V \backslash S$ is such that every vertex in V satisfies sum property. Thus the constraints (C2), (C3) and (C4) are satisfied. Each vertex in S satisfied the sum property in f. Thus the constraint (C5) is satisfied. □

We run ILP over all the valid guesses and check whether there exists an assignment leading to a minus dominating function. Over all such assignments we pick the assignment that has the minimum $\widehat{s}$.

Running Time: Guessing a labeling of S, the set of labels the vertices in each T_A can receive, and the weight $\widehat{s}$, takes time $\mathcal{O}(3^k \cdot 3^{2^k} \cdot n)$. For each of the above guesses, we run the ILP feasibility problem where the number of variables is at most $3 \cdot 2^k$. Thus from Lemma 3 and Theorem 6, the total time taken is $2^{\mathcal{O}(k \cdot 2^k)} \cdot n^{\mathcal{O}(1)}$. □

4 Cluster Vertex Deletion Set

Let G be a graph and $S \subseteq V(G)$ be a cluster vertex deletion set of size k. Also let $C_1, C_2, \ldots, C_\ell$ be the maximal cliques of $G - S$. We partition the vertices of each clique C_i, $i \in [\ell]$, based on its neighborhood in S. For each $A \subseteq S$, we use $C_{i,A} = \{v \mid v \in C_i \text{ and } N(v) \cap S = A\}$ to denote the set of vertices from C_i that are adjacent to each vertex in A. Next, we show that for each clique C_i, the vertices in $C_{i,A}$ receive their labels from a fixed set of labels, for each $A \subseteq S$. Notice that A could be an empty set.

Lemma 4. *Let $f : V(G) \to \{-1, 0, 1\}$ be a minus dominating function. Then there exists a minus dominating function of weight $f(V)$ such that in each clique C_i, for each $A \subseteq S$ and a non-empty $C_{i,A}$, the vertices of $C_{i,A}$ receive labels from exactly one of the three sets $\{-1, 0\}$, $\{0, 1\}$ or $\{0\}$.*

Proof. If all vertices in $C_{i,A}$ are assigned the label 0 then the claim is trivially satisfied. For each $i \in \{1, \ldots, \ell\}$ and $A \subseteq S$, if f assigns the vertices of $C_{i,A}$ from the labels $\{1, 0\}$ or $\{-1, 0\}$, then we conclude that f is the desired function.

Otherwise, there exists a clique C_i and an $A \subseteq S$ with vertices u and v in $C_{i,A}$ such that $f(u) = 1$ and $f(v) = -1$ or $f(u) = -1$ and $f(v) = 1$. WLOG let $f(u) = 1$ and $f(v) = -1$ (similar arguments apply for the other case). Since u and v are true twins, we apply Lemma 1 and obtain a minus dominating function of weight $f(V)$ with u and v assigned the label 0.

After repeated application of Lemma 1 on each $C_{i,A}$, where $A \subseteq S$ and $i \in [t]$, all vertices in $C_{i,A}$ are either assigned labels from exactly one of the three sets $\{-1,0\}$, $\{0,1\}$ or $\{0\}$. □

From now on, we can assume that in any minus dominating function, for each $A \subseteq S$ and a clique C_i, the vertices in $C_{i,A}$ are assigned labels from exactly one of the three sets $\{-1,0\}$, $\{0,1\}$ or $\{0\}$.

We now look at the following lemma. Suppose that we are given a labeling of S, the number of vertices in a clique C_i assigned the labels -1 and 1, and the sum each vertex in S receives from C_i. Then we can decide whether there exists a assignment of labels to C_i extending the labeling of S and satisfying the assumptions.

Lemma 5. *Let $f : S \to \{-1,0,1\}$ be a labeling of S and C_i be a clique in $G-S$. Let $a_i, b_i \in \mathbb{N} \cup \{0\}$. Also let $X^i = (x_1^i, \dots, x_k^i)$ be a tuple where x_j^i corresponds to $v_j \in S$. Then there is an algorithm that runs in $\mathcal{O}^*(2^{O(k \cdot 2^k)})$ and either returns a labeling $g : S \cup C_i \to \{-1,0,1\}$ that extends f with the following properties,*

- *$a_i = |g^{-1}(1) \cap C_i|$, $b_i = |g^{-1}(-1) \cap C_i|$,*
- *$\forall v_j \in S$, $x_j^i = \sum_{u \in N_{C_i}(v_j)} g(u)$,*
- *for each $A \subseteq S$ and a non-empty $C_{i,A}$, the vertices of $C_{i,A}$ receive the labels from exactly one of the three sets $\{-1,0\}$, $\{0,1\}$ or $\{0\}$, and*
- *for each $v \in C_i$, $\sum_{w \in N_{C_i}[v]} g(w) \geq 1$,*

or returns that there is no labeling g extending f satisfying the properties.

Proof. Given a labeling $f : S \to \{-1,0,1\}$, a clique C_i, $X^i = (x_1^i, \dots, x_k^i)$, and two integers a_i and b_i, the goal is to find a labeling $g : S \cup C_i \to \{-1,0,1\}$ extending f satisfying some constraints. We formulate this as an ILP feasibility problem with the variables: $n_{1,A}$, $n_{-1,A}$ and $n_{0,A}$ that denote the number of vertices in $C_{i,A}$, for each $A \subseteq S$, that are assigned the labels -1, 0 and 1 respectively. The number of variables is at most $3 \cdot 2^k$. We now present the constraints.

(C1) For each $A \subseteq S$, the number of vertices assigned the labels -1, 0, and 1 in $C_{i,A}$ is at least 0 and at most $|C_{i,A}|$. In addition, each vertex is assigned some label.

$$0 \leq n_{-1,A}, n_{1,A}, n_{0,A} \leq |C_{i,A}| \text{ and } n_{-1,A} + n_{1,A} + n_{0,A} = |C_{i,A}|.$$

(C2) For each $v_j \in S$, the sum it receives from C_i is x_j^i.

$$\sum_{A:\, v_j \in A} n_{1,A} - n_{-1,A} = x_j^i.$$

(C3) For each $A \subseteq S$, the vertices in $C_{i,A}$ are assigned the labels from exactly one of the three sets $\{-1, 0\}$, $\{0, 1\}$ or $\{0\}$.

$$n_{1,A} > 0 \implies n_{-1,A} = 0,$$
$$n_{-1,A} > 0 \implies n_{1,A} = 0.$$

(C4) Total sum of vertices assigned the labels 1 and -1 are a_i and b_i respectively.

$$\sum_A n_{1,A} = a_i, \text{ and } \sum_A n_{-1,A} = b_i.$$

(C5) For each $A \subseteq S$, the sum property for vertices in $C_{i,A}$ is satisfied.

$$\sum_{A:\, v \in A} f(v) + a_i - b_i \geq 1.$$

We now have to show that there is a feasible assignment of ILP if and only if there is a labeling g that extends f and satisfying the properties.

Feasibility Implies Labeling: Let there be a feasible assignment of values to variables returned by the ILP. In each $C_{i,A}$, choose $n_{1,A}$, $n_{0,A}$ and $n_{-1,A}$ many vertices arbitrarily and assign them the label 1, 0 and -1, respectively. Notice that each vertex is assigned a label because of Constraint (C1). For each $v_j \in S$, the sum it receives from C_i is x^i_j, which is ensured from Constraint (C2). For each $A \subseteq S$, the vertices in $C_{i,A}$ are assigned labels from exactly one of the three sets $\{-1, 0\}$, $\{0, 1\}$ or $\{0\}$ which is ensured by Constraint (C3) and Constraint (C1). The number of vertices assigned 1 and -1 is a_i and b_i respectively and is ensured by Constraint (C4). Every vertex in C_i satisfies the sum property and this is ensured by Constraint (C5).

Labeling Implies Feasibility: Let $g : S \cup C_i \to \{-1, 0, 1\}$ be an extension of f satisfying the properties of a_i, b_i, x^i_j, the set of labels used in each $C_{i,A}$ and the sum property for each vertex in C_i with respect to g. Using Lemma 4, we convert g to be a labeling such that each $C_{i,A}$ receives the labels from exactly one of the three sets $\{-1, 0\}$, $\{1, 0\}$ or $\{0\}$. We obtain a feasible assignment for variables in ILP as follows. For each $A \subseteq S$, we set $n_{-1,A} = |g^{-1}(-1) \cap C_{i,A}|$, $n_{0,A} = |g^{-1}(0) \cap C_{i,A}|$, and $n_{1,A} = |g^{-1}(1) \cap C_{i,A}|$. By definition of g, the constraints (C1), (C2), (C3), (C4), and (C5) are satisfied.

Since the number of variables is at most $3 \cdot 2^k$, from Theorem 6, the running time of our algorithm is $2^{\mathcal{O}(k \cdot 2^k)} n^{\mathcal{O}(1)}$. □

We say a tuple (a_i, b_i, X^i) is *feasible* for C_i if Lemma 5 returns a feasible labeling of C_i extending the labeling of S. Else we call it infeasible. We now proceed to the proof of Theorem 3.

Proof (Proof of Theorem 3*).* Let G be a graph and $S \subseteq V(G)$ of size k be such that $G - S$ is a disjoint union of cliques. Let $C_1, C_2, \ldots, C_\ell$ be the cliques of $G - S$. Let $g : V(G) \to \{-1, 0, 1\}$ be a minus dominating function of minimum weight. The first step of the algorithm is to guess the labeling $g|_S$ of S. For each clique C_i in $G - S$, we try to obtain a labeling of C_i (if one exists) that extends $g|_S$ using Lemma 5. Towards this, we guess the following: a_i and b_i, which are the number of vertices assigned the labels 1 and -1 respectively in C_i, and $X^i = (x^i_1, \ldots, x^i_k)$ the tuple where x^i_p corresponds to the sum that the vertex $v_p \in S$ receives from C_i. Thus for each of the guesses, we should able to decide whether there exists a labeling of $C_i \cup S$ extending $g|_S$ that satisfies the sum property for each vertex in C_i.

We give a bottom-up dynamic programming based algorithm to find g.

An entry $T[i, a, b, X]$ in the table is set to 1 if there exists a labeling of the vertices in $C_1, C_2, \cdots, C_i$ such that

- a is the number of vertices labelled 1 over the cliques from C_1 to C_i,
- b is the number of vertices labelled -1 over the cliques from C_1 to C_i,
- $X = (x_1, x_2, \cdots, x_k)$ is the tuple where x_j corresponds to the sum the vertex $v_j \in S$ receives from the cliques C_1 to C_i, and
- each vertex in $C_1 \cup C_2 \cup \cdots \cup C_i$ satisfies the sum property.

Otherwise, we store $T[i, a, b, X] = 0$.

We now define the recurrence. For an entry $T[i, a, b, X]$, we go over all feasible tuples of C_i and look at the corresponding subproblem over the cliques C_1 to C_{i-1}. That is,

$$T[i, a, b, X] = \bigvee_{\text{feasible tuples } (a_i, b_i, X^i) \text{ of } C_i} T[i-1, a-a_i, b-b_i, (x_1 - x^i_1, \ldots, x_k - x^i_k)].$$

Notice that in a feasible tuple we have $n \geq a \geq a_i \geq 0$, $n \geq b \geq b_i \geq 0$ and $n \geq x_j \geq x^i_j \geq -n$ for all j. The base case of the recurrence is obtained at the clique C_1, which is computed as follows:

$$T[1, a, b, X] = \begin{cases} 1 & (a, b, X) \text{ is a feasible tuple for } C_1, \\ 0 & \text{otherwise.} \end{cases}$$

The correctness of the algorithm follows from the description. We now compute the running time of the algorithm. The number of labelings $g|_S$ is at most 3^k and the number of feasible tuples for a clique C_i is at most n^{k+2}. Using Lemma 5, we can decide if a tuple is feasible or not in time $2^{\mathcal{O}(k \cdot 2^k)} n^{\mathcal{O}(1)}$. The number of entries in T is at most n^{k+3}. For each of the entry in T, we go over all feasible tuples of C_i and thus the total running time is $2^{\mathcal{O}(k \cdot 2^k)} n^{2k+6}$. □

5 Distance to Disjoint Components and Component Size

Let G be a graph and $k, d \in \mathbb{N}$ be two integers. We consider the problem of computing a set $S \subseteq V(G)$ of size at most k such that $G - S$ is a disjoint union of connected components where each connected component has at most d vertices. This problem is known in literature as d-COMPONENT ORDER CONNECTIVITY (in short, d-COC) [9]. Notice that when $d = 1$, d-COC is the VERTEX COVER problem. There is a $\mathcal{O}(\log d)$-approximation algorithm for d-COC [5,9].

In this section, we consider MD when parameterized by k and d where the input is a graph G, an integer ℓ and a set $S \subseteq V(G)$ of size k such that $G - S$ is a disjoint union of components each of size at most d vertices. The objective is to check whether there exists a minus dominating function of weight at most ℓ.

Towards this, we consider the solution set obtained from the approximation algorithm of d-COC as our modulator set S. Notice that $|S| \leq \mathcal{O}(k \log d)$, if a solution exists for d-COC. We now provide a proof sketch of Theorem 4.

Proof Sketch of Theorem 4. Let $\mathcal{C} = \{C \mid C \text{ is a component in } G-S\}$ denote the set of components in $G - S$. Notice that for each $C \in \mathcal{C}$, we have $1 \leq |C| \leq d$. Let $V(C) = \{u_1, u_2, \ldots, u_{d'}\}$, $d' \leq d$. For each component $C \in \mathcal{C}$, we apply the following procedure.

- We find the *equivalence class* of C based on its neighborhood in $S \cup V(C)$. Let $\mathcal{T} = \{T_A \mid A \subseteq S \cup V(C)\}$. Since $|S| = k$ and $|C| \leq d$, we have $|\mathcal{T}| = 2^{k+d}$. An equivalence class is defined by the function $g : \mathcal{C} \to \mathcal{T}^d$. Note that the number of equivalence classes is at most $2^{(k+d)d}$.
  The equivalence class $g(C)$ is denoted by $(T_{A_1}, T_{A_2}, \ldots, T_{A_{|d'|}})$ where for each $u_i \in V(C)$ we have $N(u_i) \cap S = A_i$.
- We now consider all possible labelings $h : V(C) \to \{-1, 0, 1\}$ of C. We say a labeling h is *feasible* for C, when the vertices of C are assigned the labels from h and each vertex v of C satisfies the sum property (i.e., the sum that v receives from $C \cup S$ is at least one).
  The set of feasible labelings $\mathcal{H}_{g(C)} = \{h \mid h \text{ is feasible for } C \text{ and } C \text{ belongs to the equivalence class } g(C)\}$ is constructed. Notice that all the components that belong to an equivalence class $g(C)$ have the same set of feasible labelings.
- We formulate an ILP feasibility instance using the above information. The variables of ILP represent the number of components in $G - S$ belonging to an equivalence class that receive a particular labeling from the feasible list of labelings (of that equivalence class). The number of variables for each equivalence class $g(C)$ is equal to $|\mathcal{H}_{g(C)}|$.

The running time of the algorithm is majorly dependent on the running time of the ILP which in turn depends on the number of variables which is $2^{(k+d)d}3^d$. □

Using Theorem 4, we get the following results when the components are cliques and trees respectively.

Corollary 1. MD *is FPT when parameterized by (i) cluster vertex deletion number and size of a largest clique, or (ii) feedback vertex set number and size of a largest tree.*

Acknowledgement. We would like to thank the anonymous reviewers for their helpful comments. The first author acknowledges SERB-DST for supporting this research via grant PDF/2021/003452. The fifth author acknowledges NBHM for supporting this research via project NBHM-02011/24/2023/6051. The fifth author would also like to acknowledge DST for supporting this research via project CRG/2023/007127.

References

1. Chang, G.J.: Algorithmic aspects of domination in graphs. In: Pardalos, P.M., Du, D.-Z., Graham, R.L. (eds.) Handbook of Combinatorial Optimization, pp. 221–282. Springer, New York (2013). https://doi.org/10.1007/978-1-4419-7997-1_26
2. Cygan, M., et al.: Parameterized Algorithms. Springer, Cham (2015). https://doi.org/10.1007/978-3-319-21275-3
3. Damaschke, P.: Minus domination in small-degree graphs. Discrete Appl. Math. **108**(1), 53–64 (2001). Workshop on Graph Theoretic Concepts in Computer Science
4. Downey, R.G., Fellows, M.R.: Fundamentals of Parameterized Complexity. Texts in Computer Science, Springer, London (2013). https://doi.org/10.1007/978-1-4471-5559-1
5. Drange, P.G., Dregi, M., van 't Hof, P.: On the computational complexity of vertex integrity and component order connectivity. Algorithmica **76**(4), 1181–1202 (2016). https://doi.org/10.1007/s00453-016-0127-x
6. Dunbar, J., Goddard, W., Hedetniemi, S., McRae, A., Henning, M.A.: The algorithmic complexity of minus domination in graphs. Discrete Appl. Math. **68**(1), 73–84 (1996)
7. Dunbar, J., Hedetniemi, S., Henning, M.A., McRae, A.A.: Minus domination in regular graphs. Discrete Math. **149**(1), 311–312 (1996)
8. Faria, L., Hon, W.-K., Kloks, T., Liu, H.-H., Wang, T.-M., Wang, Y.-L.: On complexities of minus domination. Discrete Optim. **22**, 6–19 (2016). SI: ISCO 2014
9. Fomin, F.V., Gaspers, S., Kratsch, D., Liedloff, M., Saurabh, S.: Iterative compression and exact algorithms. Theor. Comput. Sci. **411**(7), 1045–1053 (2010)
10. Ganian, R.: Twin-cover: beyond vertex cover in parameterized algorithmics. In: Marx, D., Rossmanith, P. (eds.) IPEC 2011. LNCS, vol. 7112, pp. 259–271. Springer, Heidelberg (2012). https://doi.org/10.1007/978-3-642-28050-4_21
11. Goyal, D., Jacob, A., Kumar, K., Majumdar, D., Raman, V.: Structural parameterizations of dominating set variants. In: Fomin, F.V., Podolskii, V.V. (eds.) CSR 2018. LNCS, vol. 10846, pp. 157–168. Springer, Cham (2018). https://doi.org/10.1007/978-3-319-90530-3_14
12. Haynes, T.W., Hedetniemi, S., Slater, P.: Fundamentals of Domination in Graphs. CRC Press, Boca Raton (1998)
13. Hedetniemi, S.T., Laskar, R.C.: Bibliography on domination in graphs and some basic definitions of domination parameters. Discrete Math. **86**(1), 257–277 (1990)
14. Hüffner, F., Komusiewicz, C., Moser, H., Niedermeier, R.: Fixed-parameter algorithms for cluster vertex deletion. Theory Comput. Syst. **47**(1), 196–217 (2010). https://doi.org/10.1007/s00224-008-9150-x
15. Kannan, R.: Minkowski's convex body theorem and integer programming. Math. Oper. Res. **12**(3), 415–440 (1987)
16. Lampis, M.: Algorithmic meta-theorems for restrictions of treewidth. Algorithmica **64**(1), 19–37 (2012). https://doi.org/10.1007/s00453-011-9554-x

17. Lee, C.-M., Chang, M.-S.: Variations of Y-dominating functions on graphs. Discret. Math. **308**(18), 4185–4204 (2008)
18. Lin, J.-Y., Liu, C.-H., Poon, S.-H.: Algorithmic aspect of minus domination on small-degree graphs. In: Xu, D., Du, D., Du, D. (eds.) COCOON 2015. LNCS, vol. 9198, pp. 337–348. Springer, Cham (2015). https://doi.org/10.1007/978-3-319-21398-9_27
19. Zheng, Y., Wang, J., Feng, Q.: Kernelization and lower bounds of the signed domination problem. In: Fellows, M., Tan, X., Zhu, B. (eds.) AAIM/FAW -2013. LNCS, vol. 7924, pp. 261–271. Springer, Heidelberg (2013). https://doi.org/10.1007/978-3-642-38756-2_27

Exact and Parameterized Algorithms for Choosability

Ivan Bliznets[1(✉)] and Jesper Nederlof[2]

[1] University of Groningen, Nijenborgh 9, 9747 AG Groningen, The Netherlands
i.bliznets@rug.nl

[2] Utrecht University, Princetonplein 5, 3584 CC Utrecht, The Netherlands
j.nederlof@uu.nl

Abstract. In the Choosability problem (or list chromatic number problem), for a given graph G, we need to find the smallest k such that G admits a list coloring for any list assignment where all lists contain at least k colors. The problem is tightly connected with the well-studied Coloring and List Coloring problems. However, the knowledge of the complexity landscape for the Choosability problem is pretty scarce. Moreover, most of the known results only provide lower bounds for its computational complexity and do not provide ways to cope with the intractability. The main objective of our paper is to construct the first non-trivial exact exponential algorithms for the Choosability problem, and complete the picture with parameterized results.

Specifically, we present the first single-exponential algorithm for the decision version of the problem with fixed k. This result answers an implicit question from Eppstein on a stackexchange thread discussing upper bounds on the union of lists assigned to vertices. We also present a $2^{n^2} poly(n)$ time algorithm for the general Choosability problem.

In the parameterized setting, we give a polynomial kernel for the problem parameterized by vertex cover, and algorithms that run in FPT time when parameterized by clique-modulator and by the dual parameterization $n - k$. Additionally, we show that Choosability admits a significant running time improvement if it is parameterized by cutwidth in comparison with the parameterization by treewidth studied by Marx and Mitsou [ICALP'16]. On the negative side, we provide a lower bound parameterized by a modulator to split graphs under assumption of the Exponential Time Hypothesis.

Keywords: choosability · cutwidth · exact exponential

1 Introduction

For an undirected graph G, we call k a choice number of G (or choosability number) if under any list assignment of length at least k to the vertices, it is

Supported by the project CRACKNP that has received funding from the European Research Council (ERC) under the European Union's Horizon 2020 research and innovation programme (grant agreement No. 853234).

H. Fernau et al. (Eds.): SOFSEM 2024, LNCS 14519, pp. 111–124, 2024.
https://doi.org/10.1007/978-3-031-52113-3_8

possible to find a proper list coloring. In the CHOOSABILITY problem, the goal is to find a choice number of an input graph. Sometimes the choosability number is also called the list chromatic number. CHOOSABILITY was introduced by Vizing [24] and independently by Erdős, Rubin, and Taylor [9]. It is known that the problem is Π_2^p-complete even for very restricted classes of graphs. For example, k-CHOOSABILITY (the version of CHOOSABILITY where k is a fixed parameter) is Π_2^p-complete in bipartite graphs for any k [16], 4-CHOOSABILITY is Π_2^p-complete in planar graphs [15], 3-CHOOSABILITY is Π_2^p-complete in planar triangle-free graphs [15] and in graphs that are a union of two forests [15]. Moreover, 3-CHOOSABILITY is one of the first natural problems for which a double exponential lower bound parameterized by treewidth was established [20]. There is an enormous amount of combinatorial results for CHOOSABILITY [1,2,6,9,19,21,22]. Generally, these results provide bounds on the choice number or describe situations where a graph is k-choosable. However, the number of algorithmic results is surprisingly small. Besides the mentioned lower bounds on the computational complexity of the problem, the following algorithmic results were established: k-CHOOSABILITY is Fixed Parameter Tractable (for short FPT, recall this means that instances (x, k) of the problem can be solved $f(k)|x|^{O(1)}$ time for some function f) in P_5-free graphs parameterized by k [12,13], k-CHOOSABILITY admits a linear algorithm for fixed k on H-free graphs where H is a disjoint union of paths [14], CHOOSABILITY admits an FPT-algorithm parameterized by treewidth [10,20].

One might expect that the CHOOSABILITY problem has a similar behavior to COLORING or LIST COLORING problems, and that results for these problems can be easily carried over to CHOOSABILITY. Surprisingly, this is not the case. For example, LIST COLORING is $W[1]$-hard parameterized by treewidth (even parameterized by vertex cover) [7,10], while CHOOSABILITY admits an FPT-algorithm [10,20]. The addition of a universal vertex always increases the chromatic number but this is not true for the choice number. Moreover, even the choice number of bipartite graphs is unbounded. k-COLORING and LIST COLORING with lists' length bounded by k are trivially solvable in $\mathcal{O}^*(k^n)$ time. However, it is not trivial to construct a single-exponential algorithm for k-CHOOSABILITY for a fixed k. The existence of a single-exponential algorithm for k-CHOOSABILITY implicitly was asked by Eppstein in a stackexchange post.[1] There, it was noted that it is easy to construct such an algorithm if we know that the total number of different colors used in lists is bounded by some function $f(k)$. Unfortunately, it was shown that there are no such functions [19], and our algorithm cannot build on this argument. Nevertheless, we show:

Theorem 1. *Let G be an undirected graph on n vertices and k be an integer given as input.*

1. k-CHOOSABILITY *can be solved in* $f(k)^{O(n)}$ *for some function* f,
2. CHOOSABILITY *can be solved in* $\mathcal{O}^*(2^{n^2})$ *time.*

[1] https://cstheory.stackexchange.com/questions/2661/how-many-distinct-colors-are-needed-to-lower-bound-the-choosability-of-a-graph.

The constructed single-exponential algorithm is based on the fact that in k-choosable graphs the average degree is bounded by some function $f(k)$. Our second exact exponential algorithm for the choice number with running time $2^{n^2} poly(n)$ provides an improvement upon a slightly naive $2^{\mathcal{O}(n^2 \log n)}$ time algorithm. It should be noted that this run time does not show up often for graph problems; one exception we are aware of is the LIST EDGE COLORING problem [18]. A key combinatorial ingredient of our algorithm is the fact that in some sense we can bound the total number of colors by n. We note that the obtained running time is significantly worse than the running time of the Subset Convolution algorithm for the COLORING and LIST COLORING problems from [3]. That is why one can consider the question whether there exists an improvement on the $2^{n^2} poly(n)$ time algorithm or a lower bound that excludes such improvement based on the Exponential Time Hypothesis (ETH) to be an interesting open question.

In this paper, we also consider different parameterizations of CHOOSABILITY. The most important such result is for CHOOSABILITY parameterized by cutwidth. Specifically, we prove the following theorem.

Theorem 2. *Let G be an undirected graph on n vertices given along with a cutwidth ordering of width w. Then* CHOOSABILITY *can be solved in time $2^{2^{O(w)}} n$ time.*

We note that parameterization by cutwidth allows us to obtain a significant improvement over the parameterization by treewidth for which only a $2^{k^{O(tw)}} poly(n)$ time algorithm is known [10,20]. We know that for parameterization by cutwidth, there is a matching lower bound under the assumption of the ETH (however, the result is not published yet [5]). To prove Theorem 2, we build on the method from [17] that computes the chromatic number of a graph in $O(2^w n)$ time, when given a cutwidth ordering of width w. That algorithm used an argument based on matrix rank to show the table size of the natural dynamic programming algorithm can be sparsified (i.e. we view a dynamic programming table as a family of partial solutions, and with sparsifying we mean that some partial solutions are omitted) in such a way that global solutions remain detected. We show that this approach can also be employed for the CHOOSABILITY problem.

We also provide a polynomial *kernel* for k-CHOOSABILITY parameterized by vertex cover if k is fixed. Recall a $g(p)$-kernel is a polynomial time algorithm that, given an instance x with parameter p, outputs an equivalent instance of the same problem of size (measured here in terms of the number of vertices) $g(p)$. Formally we prove the following.

Theorem 3. k-CHOOSABILITY *admits a kernel with $O(w^k)$ vertices for a fixed integer k, where w is the size of a vertex cover of the input graph G.*

Using Ohba's Conjecture (proved by Noel et al. in [22]) that connects the chromatic number with the choice number in case of a very large chromatic number, we construct an FPT-algorithm for CHOOSABILITY parameterized by *clique-modulator* and a *linear compression* for k-Choosability parameterized by $n - k$. Here the term $\mathcal{G}$-modulator of a graph G, for some graph class $\mathcal{G}$, refers to a vertex subset X of G such that $G - X$ is in $\mathcal{G}$. A $g(p)$-compression is a polynomial time algorithm that, given an instance x with parameter p, outputs an equivalent instance of a possibly different computational problem of size (measured here in terms of the number of vertices) $g(p)$.

Precisely, we prove the following FPT algorithms that show some instances of Choosability *can* be solved efficiently:

Theorem 4. k-Choosability *under the dual parametrization by $n - k$ can be compressed into an instance of k-*Choosability *or into an instance of* Dual-Coloring. *Moreover, number of vertices in the compressed instance is at most $3(n - k)$.*

Theorem 5. Choosability *admits an FPT-algorithm parameterized by clique-modulator d.*

Given Theorem 5 and the easy observation that Choosability is solvable in polynomial time on split graphs (recall, split graphs are graphs whose vertex set can be partitioned in one clique and one independent set), one may wonder whether Choosability also has an efficient algorithm when parameterized with the size of a modulator to a split graph.

We employ Ohba's Conjecture to show a lower bound for Choosability assuming ETH, to temper expectation here:

Theorem 6. *Unless ETH fails there is no constant θ such that* Choosability *is solvable in $\mathcal{O}^*(\theta^d)$ time where d denotes size of a modulator to split graphs.*

Organization of the Paper. The rest of this paper is organized as follows. In Sect. 2 we set up notation. In Sect. 3 we present exact exponential algorithms for k-Choosability and Choosability (Theorem 1). Section 3 is dedicated to k-Choosability parameterized by cutwidth (Theorem 2). In Sect. 5 we provide a linear compression for k-Choosability parameterized by $n - k$ (Theorem 4), an FPT-algorithm for Choosability parameterized by clique-modulator (Theorem 5), and lower bound for Choosability parameterized by modulator to split graphs (Theorem 6). Proofs of Lemma's and Theorem's marked with ($\star$) are deferred to the full version of this paper.

2 Preliminaries

In this paper, we consider only simple graphs. Used notation for graphs is standard and can be found in [8]. We also use standard definitions (FPT-algorithm, kernel, compression) and notation from parameterized complexity, for details we refer the interested reader to the textbook [7]. For a natural number k, we denote by $[k]$ the set $\{1, \dots, k\}$. Generally by n we denote number of vertices in a graph unless stated otherwise and if the graph is clear from the context. Similarly, by

m we denote number of edges in a graph. $N_G(v)$ denotes the open neighborhood of vertex v in a graph G, we omit subscript G if the graph is clear from the context.

For a graph G, function $c : V(G) \to C$ for some set C is called a proper coloring if for any edge $uw \in E(G)$ we have $c(u) \neq c(w)$. If a proper coloring exists for the set $C = [k]$, then we call graph k-colorable. For a graph G, we call a function $\mathcal{L}$ a *list assignment* if for each vertex $v \in V(G)$ the function assigns a set/list of colors. For a list assignment $\mathcal{L}$, we call graph $\mathcal{L}$-colorable if there is a proper coloring c of vertices such that for any vertex v we have $c(v) \in \mathcal{L}(v)$. A graph G is called k*-choosable* if it is $\mathcal{L}$-colorable for any list assignment $\mathcal{L}$ such that $|\mathcal{L}(v)| = k$ for any $v \in V(G)$. *The choosability number (choice number, list chromatic number)* of a graph G is the smallest integer k such that G is k-choosable. By $ch(G)$ we denote the choosability number of the graph G. Similarly, by $\chi(G)$ we denote chromatic number of the graph G, i.e. the smallest integer k such that G is k-colorable. As one of the central studied parameters is cutwidth we provide its definition below.

Definition 1. *The cutwidth of a graph G is the smallest k such that its vertices can be arranged in a sequence $v_1, \ldots, v_n$ such that for every $i \in [n-1]$, there are at most k edges between $\{v_1, \ldots, v_i\}$ and $\{v_{i+1}, \ldots, v_n\}$.*

A *split graph* is a graph G such that $V(G) = C \cup I$, $C \cap I = \emptyset$, $G[C]$ is a clique, $G[I]$ is an independent set. We call a subset of vertices D in a graph G *a modulator to split graphs* if subgraph $G \backslash D$ is a split graph.

Formal definitions of DUAL-COLORING, LIST COLORING, k-CHOOSABILITY, CHOOSABILITY can be found below.

DUAL-COLORING
Input: A graph G and an integer k.
Question: Is graph G $(n-k)$-colorable?

LIST COLORING
Input: A graph G and a list assignment $\mathcal{L}$.
Question: Is graph G $\mathcal{L}$-colorable?

k-CHOOSABILITY
Input: A graph G and an integer k.
Question: Is graph G k-choosable?

CHOOSABILITY
Input: A graph G.
Question: What is the smallest k such that G is k-choosable?

For the construction of a polynomial kernel, we are using the fact that bipartite graph K_{k,k^k} is not k-choosable. It is a well known fact, but nevertheless we present a proof in order to provide some intuition about the CHOOSABILITY problem.

Lemma 1 (⋆, Folklore). *The bipartite graph K_{k,k^k} is not k-choosable.*

Lemma 1 shows that choice number and chromatic number can be very different. However, in some situations values of these numbers coincide. In these cases we can determine the choice number by finding the chromatic number. One of such cases is described by Ohba's conjecture, proved by Noel, Reed, Wu [22].

Theorem 7 (Ohba's conjecture, [22]). *If $|V(G)| \leq 2\chi(G) + 1$, then G is chromatic-choosable, i.e. $ch(G) = \chi(G)$.*

3 Exact Algorithms

In this section we provide exact exponential algorithms for k-Choosability and Choosability. Before we proceed we provide some definitions and lemmas mainly used only in this section.

For a list assignment $\mathcal{L}$, we denote by $\mathcal{L}[c]$ the induced subgraph of G that contains all vertices v such that $c \in \mathcal{L}(v)$.

Lemma 2. *If a graph G is not $\mathcal{L}$-colorable for some list assignment $\mathcal{L}$, then there is a list assignment $\mathcal{L}'$ such that $|\mathcal{L}'(v)| = |\mathcal{L}(v)|$ for all $v \in V(G)$ and for any color c the induced subgraph $\mathcal{L}'[c]$ is a connected subgraph in G.*

Proof. If for some c the subgraph $\mathcal{L}[c]$ contains $p > 1$ connected components $C_1, C_2, \dots, C_p$, then we create new colors $c^1, \dots, c^p$, and for $v \in C_i$ in $\mathcal{L}'$ we use color c^i instead of color c like it is in the assignment $\mathcal{L}$. Clearly, for any color c', the induced subgraph $\mathcal{L}'[c']$ is connected. If the graph G admits a list-coloring under assignment $\mathcal{L}'$, then it also admits list-coloring under the list assignment $\mathcal{L}$. Namely, it is enough to re-color vertices colored in c^i into color c. Adjacent vertices u, v cannot be colored in c^i and c^j so we have a proper $\mathcal{L}$-coloring.

We need the following theorem bounding the average degree in a k-choosable graph [1].

Theorem 8 ([1]). *Let G be a simple graph with average degree at least d. If $d > 4\binom{k^4}{k}\log(2\binom{k^4}{k})$ and k is an integer, then $ch(G) > k$.*

Now we have all ingredients to prove main result of this section. We will first restate it for convenience:

Theorem 1. *Let G be an undirected graph on n vertices and k an integer given as input.*

1. *k-*Choosability*can be solved in $f(k)^{O(n)}$ for some function f,*
2. Choosability*can be solved in $\mathcal{O}^*(2^{n^2})$ time.*

Proof. Throughout the proof we identify colors with integers. We assume that $V(G) = \{v_1, v_2, \dots, v_n\}$.

First of all we present an algorithm for k-CHOOSABILITY with $f(k)^{O(n)}$ running time. Recall that if a graph G is not k-choosable, then there is a list assignment $\mathcal{L}'$ with $|\mathcal{L}'(v_i)| = k, \forall i \in [n]$ such that the graph G is not $\mathcal{L}'$-colorable. We show how to find such an assignment in $2^{O(n)}$ running time. By Lemma 2 we can assume that for each color c the graph $\mathcal{L}'[c]$ is connected. Without loss of generality we can consider only assignments that satisfy the following property: if $p \in \mathcal{L}'(v_i)$, then for any positive integer $q < p$ there is $j \leq i$ such that $q \in \mathcal{L}'(v_j)$ (otherwise we can simply rename colors). Under this assumption, for example, we have that $\mathcal{L}'(v_1) = \{1, 2, 3, \dots, k\}$. Moreover, $\mathcal{L}'$ with this property satisfy the following: for any l we have $\mathcal{L}'(v_1) \cup \mathcal{L}'(v_2) \cup \dots \cup \mathcal{L}'(v_l) = \{1, 2, 3, \dots, c'\}$ for some positive integer c'. For each i, we order colors in $\mathcal{L}'(v_i)$ in increasing order. Denote by $\mathcal{L}'(v_i)[\ell]$ the ℓ-th element in the list $\mathcal{L}'(v_i)$. Taking into account that list $\mathcal{L}'(v_i)$ is ordered we have $\mathcal{L}'(v_i)[1] < \mathcal{L}'(v_i)[2] < \dots < \mathcal{L}'(v_i)[k]$.

Taking into account Theorem 8 we may assume that our input graph has at most $4\binom{k^4}{k}\log(2\binom{k^4}{k})n$ edges, otherwise G is not k-choosable. Note that the number $4\binom{k^4}{k}\log(2\binom{k^4}{k})n$ is $\mathcal{O}(n)$ for a fixed integer k.

For a list assignment $\mathcal{L}$ and edge $e = v_iv_j \in E(G)$, we denote by $S_{ij}^{\mathcal{L}}$ a set of pairs $\{(a_1, b_1), (a_2, b_2), \dots, (a_\ell, b_\ell)\}$ such that $\mathcal{L}(v_i)[a_q] = \mathcal{L}(v_j)[b_q]$ for all $q \in [\ell]$. Note that without loss of generality we can assume that $a_1 < a_2 < \dots < a_\ell$ and $b_1 < b_2 < \dots < b_\ell$. This means that the number of such sets of pairs is bounded by $\sum_{\ell=1}^{k} \binom{k}{\ell}^2 \leq 2^{2k}$. In order to determine $\mathcal{L}'$ for each edge v_iv_j we consider all possible values of $S_{ij}^{\mathcal{L}'}$. It creates at most $(2^{2k})^{|E(G)|}$ sub-branches which is $2^{O(n)}$ for a fixed k. Knowing the $S_{ij}^{\mathcal{L}'}$ for each edge v_iv_j and the fact that for each $l \in [n]$ we have $\mathcal{L}'(v_1) \cup \mathcal{L}'(v_2) \cup \dots \cup \mathcal{L}'(v_l) = \{1, 2, 3, \dots, c'\}$ for some positive integer c' we can recover $\mathcal{L}'$. Indeed, having this information we can identify induced subgraphs $\mathcal{L}'[1], \mathcal{L}'[2], \dots, \mathcal{L}'[k]$. After that we can find the smallest i such that $k+1 \in \mathcal{L}'(v_i)$. Having information on edges about common colors on endpoints we find $\mathcal{L}'[k+1]$, since subgraph $\mathcal{L}'[k+1]$ is connected. Then we repeat the process for color $k+2$ and so on. Finally, we realize that either our branch is not valid and such list assignment do not exists or we obtain a list assignment. We check that the obtained assignment $\mathcal{L}'$ contains exactly k colors for each vertex. Recall that list coloring can be solved in $\mathcal{O}^*(2^n)$. It is enough to use a subset convolution method [4] for this. Therefore, after recovering $\mathcal{L}'$ from S_{ij} we run the algorithm for list-coloring and check if G is $\mathcal{L}'$-colorable or not. If G is not $\mathcal{L}'$-colorable, then G is not k-choosable. However, if in all our subbranches we either failed to recover $\mathcal{L}'$ or G was $\mathcal{L}'$-colorable, then we conclude that G is k-choosable. So, the overall running time is at most $\mathcal{O}^*((2^{2k})^{|E(G)|} \cdot 2^n)$, which is sufficient since $|E(G)| = O(n)$ for fixed k.

Now we provide an algorithm for CHOOSABILITY with $\mathcal{O}^*(2^{n^2})$ running time. It is known [23] that if a graph G is not k-choosable for some k, then there is a list assignment $\mathcal{L}$ containing at most $n-1$ colors in the union $\mathcal{L}(v_1) \cup \mathcal{L}(v_2) \cup \dots \cup \mathcal{L}(v_n)$ such that for each vertex v we have $|\mathcal{L}(v)| = k$ and G is not $\mathcal{L}$-colorable. Therefore, in order to find such $\mathcal{L}$ it is enough to brute-force all 2^n possibilities of $\mathcal{L}[i]$ for each color $i \in [n-1]$. This step creates $2^{n(n-1)}$ subcases, we eliminate those that have unequal length of lists assigned to vertices and then run the $\mathcal{O}^*(2^n)$ time algorithm for list coloring in each subcase. Overall, the running time will be at most $\mathcal{O}^*(2^{n^2})$.

4 Cutwidth

The goal of this section is to prove the following theorem.

Theorem 2. *Let G be an undirected graph on n vertices given along with a cutwidth ordering of width w. Then* CHOOSABILITY *can be solved in time $2^{2^{O(w)}}n$ time.*

We want to solve k-choosability on graphs of cutwidth w. Let q be the maximum assigned number showing up in a list. In Lemma 1 from [10] it was shown that it is enough to consider $q \leq (2tw+1)k < 4k \cdot w$ (tw is the treewidth of the graph) which is at most $4kw$ since cutwidth is larger than treewdith. We can assume $k \leq w$ since otherwise the answer is automatically yes (because a greedy algorithm can always find a list coloring if each list is of size at least $w+1$). Hence, we can assume that q is at most $4w^2$. Let $\{v_1, \dots, v_n\} = V(G)$ be an ordering of cutwidth at most w, and let G_i be the prefix graph $G[\{v_1, \dots, v_i\}]$. Denote $V_i = \{v_1, \dots, v_i\}$ for the set of vertices of G_i. Consider the i'th cut and the corresponding bipartite graph H_i with sides L_i and R_i. Thus, $|L_i|, |R_i| \leq |E(H_i)| \leq w$, and $E(H_i) = \{\{v_h, v_j\} : h \leq i \leq j\}$. Note that $L_i \subseteq L_{i-1} \cup \{v_i\}$ and $R_{i-1} \subseteq R_i \cup \{v_i\}$.

We will use the following standard vector and function notation: For sets A, B, we let A^B denote the set of vectors with values from A indexed by B. We will interchangeably address these vectors as functions $f : B \to A$ and use function notation $f(\cdot)$ and its inverse $f^{-1}(\cdot)$ (if it exists). For $f \in A^B$ (or thus, equivalently, $f : B \to A$), we denote the *projection* $f_{|B'}$ for the unique vector $f' \in A^{B'}$ that agrees with f on all its values, i.e. $f'(b) = f(b)$ for all $b \in B'$. If $f : B_1 \to A_1$ and $g : B_2 \to A_2$ and B_1 and B_2 are disjoint we denote $f \cup g$ for the unique vector with domain $B_1 \cup B_2$ that agrees with both f and g on all values. We use the shorthand $b \mapsto a$ for the vector f with co-domain $\{b\}$ with $f(b) = a$.

The standard dynamic programming algorithm that solves k-choosability in $2^{q^w} n^{O(1)}$ time defines for each $i = 1, \dots, n$ and list assignment $l : L_i \to \binom{[q]}{k}$ the table entry $T_i[l]$ to be the set

$$\Big\{\{f_{|L_i} : f \text{ is a list coloring of } (G_i, l')\} \;\Big|\; l' : V(G_i) \to \binom{[q]}{k} \text{ s.t. } \forall v \in L_i,\, l'(v) = l(v)\Big\}. \quad (1)$$

Then we have that $T_0[\emptyset] = \{\emptyset\}$, where $\emptyset$ denotes the empty (zero-dimensional) vector. For $i > 0$, we have that $T_i[l]$ equals to

$$\bigcup_{\hat{l}:L_{i-1}\setminus L_i \to \binom{[q]}{k}} \bigcup_{\mathcal{A}\in T_{i-1}[(l\cup\hat{l})_{|L_i}]} \Big\{ \bigcup_{c\in l(v_i)} \{f_{|L_i} \cup (v_i \mapsto c) | f \in \mathcal{A}, f^{-1}(c) \cap N(v_i) = \emptyset\} \Big\}. \tag{2}$$

Intuitively, this expression first iterates over all list assignments $\hat{l}$ of $L_{i-1}\setminus L_i$ and subsequently over all sets of colorings $\mathcal{A}$ that are the result of some list assignment of V_i consistent with this list assignment. For each such combination, we need to extend $\mathcal{A}$ with all possible colorings $c \in l(v_i)$ of the new vertex v_i (where v_i is "new" in the sense that it is the only vertex in L_i but not in L_{i-1}).

Since $T_i[l] \subseteq [q]^{L_i}$ we have that $|T_i[l]| \leq 2^{[q]^{L_i}}$, and the number of possibilities for l is at most $\binom{q}{k}^w \leq 2^{O(w^3)}$. Thus number of table entries is at most $2^{O(q^w)}$. Computing table entries by following (2) in the naïve way, this results in a $2^{O(q^w)}$ time algorithm. We will now show how to improve this using the idea of representation.

Definition 2 (Definition 8 from [17], paraphrased). *Fix a bipartite graph H with parts L and R. A family $\mathcal{F}' \subseteq [q]^L$* H-represents *another family $\mathcal{F} \subseteq [q]^L$ if $\mathcal{F}' \subseteq \mathcal{F}$ and for each coloring $b \in [q]^R$ the following holds: there exists $a \in \mathcal{F}$ such that $a \cup b$ is a proper coloring of H if and only if there exists an $a' \in \mathcal{F}'$ such that $a' \cup b$ is a proper coloring of H.*

Theorem 11 (Lemma 10 from [17], paraphrased). *For every $\mathcal{F} \subseteq [q]^L$, there exists an $\mathcal{F}' \subseteq \mathcal{F}$ that H-represents $\mathcal{F}$ such that $|\mathcal{F}'| \leq 2^{|E(H)|}$. Moreover, given $|\mathcal{F}|$, such an $\mathcal{F}'$ can be computed in time $|\mathcal{F}|2^{O(w)}$, where w is an upper bound on the number of edges of H.*

We refer to a *table* as an element from $2^{[q]^{L_i}}$.

Definition 3 (Representation of tables). *Fix a bipartite graph H with parts L and R. A table $T' \subseteq 2^{[q]^L}$* H-represents *another table $T \subseteq 2^{[q]^L}$ if*

1. *for each $\mathcal{A} \in T$ there is an $\mathcal{A}' \in T'$ that H-represents $\mathcal{A}$,*
2. *for each $\mathcal{A}' \in T'$ there is an $\mathcal{A} \in T$ such that $\mathcal{A}'$ H-represents $\mathcal{A}$.*

If H is clear from the context, we simply say that T' *represents* T. It is easy to see (and recorded as Observation 9 in [17]) that the H-representation relation of families of colorings is *transitive*: If $\mathcal{A}''$ represents $\mathcal{A}'$ and $\mathcal{A}'$ represents $\mathcal{A}$, then $\mathcal{A}''$ also represents $\mathcal{A}$. Using this, it is also easy to see that the H-representation relation of tables is *transitive*: If T'' represents T' and T' represents T, then T'' also represents T.

Theorem 12. *Given T, we can compute a table T' in time $\sum_{\mathcal{A}\in T} |\mathcal{A}|2^{O(w)}$ such that $|T'| \leq \sum_{i=0}^{2^w} \binom{q^w}{i} \leq 2^w \binom{q^w}{2^w}$ and T' H-represents T, where w is an upper bound on the number of edges of H.*

Proof. We can simply apply Theorem 11 on each member of T, and remove copies.

Lemma 3. *Let $0 < i \leq n$. Suppose that $T'_{i-1}[l]$ H_{i-1}-represents $T_{i-1}[l]$ for each l. We define $T'_i[l]$ as*

$$\bigcup_{\hat{l}:L_{i-1}\setminus L_i \to \binom{[q]}{k}} \quad \bigcup_{\mathcal{A}\in T'_{i-1}[(l\cup\hat{l})_{|L_i}]} \left\{ \bigcup_{c\in l(v_i)} \{f_{|L_i} \cup (v_i \mapsto c) | f \in \mathcal{A}, f^{-1}(c) \cap N(v_i) = \emptyset\} \right\}. \tag{3}$$

Then $T'_i[l]$ H_i-represents $T_i[l]$ for each l.

Proof. We prove the statement by induction on i. For $i = 0$, we define $T_0[\emptyset] = T'_0[\emptyset] = \{\emptyset\}$, and therefore the base case trivially holds.

Suppose that $\mathcal{A} \in T_i[l]$. Thus by the definition in (1) we have that

$$\mathcal{A} = \{f_{|L_i} : f \text{ is a list coloring of } (G_i, l')\},$$

for some $l' : V(G_i) \to \binom{[q]}{k}$ that agrees with l on all vertices in L_i. Consider the corresponding family $\mathcal{A}_0 \in T_{i-1}[l'_{|L_{i-1}}]$:

$$\mathcal{A}_0 = \{f_{|L_{i-1}} : f \text{ is a list coloring of } (G_{i-1}, l'_{|V_{i-1}})\}.$$

Since $T'_{i-1}[l'_{|L_{i-1}}]$ by assumption H_{i-1}-represents $T_{i-1}[l'_{|L_{i-1}}]$, there is some l^* that agrees with l' on all vertices in L_{i-1} such that $T'_{i-1}[l'_{|L_{i-1}}]$ contains a set

$$\mathcal{A}_0^* = \{f_{|L_{i-1}} : f \text{ is a list coloring of } (G_{i-1}, l^*)\}.$$

that H_{i-1}-represents $\mathcal{A}_0$. We claim that the set $\mathcal{A}^*$ defined below, which is in $T'_i[l]$ by (3),

$$\mathcal{A}^* = \bigcup_{c\in l(v_i)} \{f_{|L_i} \cup (v_i \mapsto c) | f \in \mathcal{A}_0^*, f^{-1}(c) \cap N(v_i) = \emptyset\}$$

H_i-represents $\mathcal{A}$.

First note that $\mathcal{A}_0^* \subseteq \mathcal{A}_0$, and since $\mathcal{A}^*$ is obtained from $\mathcal{A}_0^*$ by assigning all possible colors not in $\{l(u) : u \in N(v_i)\}$ to v_i it also assigns to $l(v_i)$ at some point. Thus $\mathcal{A}^* \subseteq \mathcal{A}$. Now, suppose there exists an $a \in \mathcal{A}$ and a $b \in [q]^{R_i}$ such that $a \cup b$ is a proper coloring of H_i. That means there is an f that is a list coloring of (G, l') such that $f_{|L_i} = a$ and l' agrees with l on all vertices in L_i. Then $f_{|L_{i-1}} \cup (f \cup b)_{|R_{i-1}}$ forms a proper coloring of the bipartite graph H_{i-1}. Since $f_{|L_{i-1}} \in \mathcal{A}_0$ and $\mathcal{A}_0^*$ is H_{i-1}-representing $\mathcal{A}_0$, there also is a $g \in \mathcal{A}_0^*$ that forms a proper coloring of H_{i-1} together with $(b \cup f)_{|R_{i-1}}$. Now $g \cup (v_i \mapsto l(v_i)) \in \mathcal{A}^*$ since $\mathcal{A}^*$ is constructed in (3) by mapping v to each allowed color, and $(g \cup (v_i \mapsto l(v_i)))_{L_i}$ forms a proper coloring with b of H_i. Thus $\mathcal{A}^*$ indeed represents $\mathcal{A}$.

It remains to prove item 2 of Definition 3. Suppose $\mathcal{A}' \in T'_i[l]$. Then

$$\mathcal{A}' = \bigcup_{c\in l(v_i)} \{f_{|L_i} \cup (v_i \mapsto c) | f \in \mathcal{A}'_0, f^{-1}(c) \cap N(v_i) = \emptyset\},$$

for some $\hat{l}$ and $\mathcal{A}'_0 \in T'_{i-1}[(l \cup \hat{l})_{|L_i}]$. We know that there exists an $\mathcal{A}_0 \in T_{i-1}[(l \cup \hat{l})_{|L_i}]$ with the property that $\mathcal{A}'_0$ H_{i-1}-represents $\mathcal{A}_0$. Then, similarly as before, we have that $\mathcal{A}'$ H_i-represents

$$\mathcal{A} = \bigcup_{c \in l(v_i)} \{f_{|L_i} \cup (v_i \mapsto c) | f \in \mathcal{A}_0, f^{-1}(c) \cap N(v_i) = \emptyset\}.$$

Now we are ready to combine all parts:

Proof (Proof of Theorem 2). The algorithm is as follows: First, set $T'_0[\emptyset] = \{\emptyset\}$. Then, for $i = 1, \ldots, n$, compute $T'_i[l]$ for each $l \in [q]^{L_i}$ from the table entries T'_{i-1} using Lemma 3 and before increasing i in each round, replace $T'_i[l]$ with a table that H_i-represents it and with the property that $|T'_i[l]| \leq 2^w\binom{q^w}{2^w}$ using Theorem 12.

Afterwards, we can conclude that the choice number is at least k if $T'_n[v_n] \neq \emptyset$, since it H_n-represents T_n. Doing this for every k, we solve the CHOOSABILITY problem.

For the running time, the number of table entries we work with is at most $n \cdot \binom{q}{k}^w \leq n2^{q \cdot w} \leq n2^{O(w^3)}$. Since we apply Theorem 12 each step before we compute new table entries, using (3) takes time at most $2^w\binom{q^w}{2^w}2^{O(w^3)}$, even when done in the naïve way. Using $\binom{n}{k} \leq n^k$ and $q \leq 4w^2$, we get that

$$2^w\binom{q^w}{2^w}2^{O(w^3)}n \leq q^{w2^w}2^{O(w^3)}n \leq 2^{(\log q)w2^w}2^{O(w^3)}n \leq 2^{2^{O(w)}}n.$$

5 Other Structural Parametrizations

In algorithms of this section we use the following simple lemma.

Lemma 4. *Let G be a graph and v a vertex in the graph such that $deg(v) < k$. We claim that G is k-choosable if and only if $G \backslash v$ is k-choosable.*

Proof. It is obvious that if $G \backslash v$ is not k-choosable, then G is also not k-choosable.

We show that if $G \backslash v$ is k-choosable, then G is also k-choosable. First of all we assign colors to all vertices except v since $G \backslash v$ is k-choosable. After this we assign color to vertex v. It is possible to do so since the list of vertex v contains k different colors and $deg(v) \leq k - 1$. Hence, there is at least one color in which vertex v can be colored.

5.1 Polynomial Kernel Parameterized by Vertex Cover

Theorem 3. k-CHOOSABILITY*admits a kernel with $O(w^k)$ vertices for a fixed integer k, where w is the size of a vertex cover of the input graph G. For convenience, we assume that vertex cover C' is provided with the input.*

Proof. Since a 2-approximation of a vertex cover can be found in polynomial time assumption that a vertex cover is provided with the input is not very restrictive.

We need the following simple reduction rule.

Reduction Rule 1. *If $v \in V(G)$ and $deg(v) < k$, then replace graph G by graph $G \backslash v$.*

The correctness of the above rule follows from Lemma 4. It is easy to see that exhaustive application of the rule requires at most polynomial time. We note that the size of the vertex cover can only decrease after application of Reduction Rule 1 and vertex cover of the final graph equals to C' without deleted vertices.

After exhaustive application of Reduction Rule 1 we partition vertices of the graph into vertex cover $C \subseteq C'$ and independent set I. We claim that if $|I| \geq \binom{w}{k} \cdot k^k$, then G is a No-instance, which immediately leads to $O(w^k)$ polynomial kernel since $|C| \leq w$ and k is a fixed integer. In order to prove the claim we note that $deg(u) \geq k$ for all $u \in I$. We prove that if $|I| \geq \binom{w}{k} \cdot k^k$, then G contains K_{k,k^k} as an induced subgraph which is not k-choosable by Lemma 1. In order to do this, for each subset $S \subseteq C$ of size k we assign a subset $N_S = \{u | u \in I, S \subseteq N(u)\}$. If for some S' of size k we have that $N_{S'} \geq k^k$, then we have found an induced subgraph K_{k,k^k} (take S' as one part and k^k arbitrary vertices from $N_{S'}$ as a second part). Since $\sum_{S \subseteq C} |N_S| \geq |I| \geq \binom{w}{k} \cdot k^k$ we conclude that such S' exists. Hence we obtained the desired kernel with $O(w^k)$ vertices.

We note that by trivial AND-composition (see [11]) to itself we get the following result.

Note 1. If $NP \not\subseteq coNP/poly$, then k-Choosability does not admit a polynomial kernel under parametrizations by treewidth, pathwidth, cutwidth, treedepth and all parameters p that satisfy $p(H_1 \sqcup H_2 \sqcup \cdots \sqcup H_\ell) \leq max_{i=1}^{\ell} p(H_i) + const$ where $\sqcup$ denotes disjoint union.

5.2 Dual Parameterization

Theorem 4 (⋆). *k-Choosability under the dual parametrization by $n - k$ can be compressed into an instance of k-Choosability or into an instance of Dual-Coloring. Moreover, the number of vertices in the compressed instance is at most $3(n - k)$.*

5.3 Clique-Modulator Parameterization

Theorem 5. *Choosability admits an FPT-algorithm parameterized by clique-modulator d. For convenience, we assume that a clique-modulator of size d is provided with the input.*

Proof. We note that even though we assume that a clique-modulator D is provided together with the input this fact does no lead to significant restriction. Since a clique-modulator is a vertex cover in the complement graph $\bar{G}$ and 2-approximation of a vertex cover can be found in polynomial time.

Denote by $C = G\backslash D$. Since D is a clique-modulator we have that C is a clique. We consider two cases: (i) $|C| \leq d$, (ii) $|C| > d$.

In the first case we have $|V(G)| = |C| + |D| \leq 2d$. Therefore using algorithm from Theorem 1 we find the choosability number in $\mathcal{O}^*(2^{4d^2})$ time.

In the second case we have $\chi(G) \geq |C|$ and $2\chi(G) > V(G)$. So by Theorem 7, we know that $ch(G) = \chi(G)$. Hence, it is enough to find the chromatic number of G. Note that $n \geq \chi(G) \geq |C|$. So in order to determine the $ch(G)$ we simply check for each $p \in [|C|, n]$ whether $\chi(G) = p$.

For each vertex $v \in V(G)$, we create a list of potential colors $\mathcal{L}v = \{1, 2, \ldots, p\}$. Without loss of generality we may assume that vertices from the clique C are colored in colors $\{1, 2, \ldots, |C|\}$. After that for each vertex $u \in D$ we delete color i from the list $\mathcal{L}(u)$ if there is a vertex $v \in C \cap N(u)$ with color i. If after these modifications a vertex w has a list of length larger than d, then we delete vertex w. Namely, such a vertex can be easily colored after we color other vertices. We also remove all vertices from the clique C, as a result we get a graph G' with list of length at most d for each vertex. It is easy to see that the graph G' admits a list coloring if and only if the chromatic number of G was at most p. List Coloring can be solved in $\mathcal{O}^*(2^d)$ by subset convolution method [4]. So, we obtain that $ch(G)$ can be found in $\mathcal{O}^*(2^{4d^2})$.

5.4 Split Graphs

Lemma 5 ($\star$). Choosability *on split graphs can be solved in polynomial time.*

Since Choosability is solvable in polynomial time on split graphs, it is natural to investigate the parameterization by distance to split graphs. We denote by d the size of modulator to split graphs. We prove the following theorem.

Theorem 6 **($\star$).** *Unless ETH fails there is no constant θ such that* Choosability*is solvable in $\mathcal{O}^*(\theta^d)$ time where d denotes size of modulator to split graphs.*

6 Conclusion

In this paper we constructed a $\mathcal{O}^*(2^{n^2})$-time algorithm for Choosability. We think that existence of algorithm with running time $\mathcal{O}^*(2^{n^{2-\epsilon}})$ for some $\epsilon > 0$ for the Choosability problem is an interesting open question.

References

1. Alon, N.: Restricted colorings of graphs. Surv. Comb. **187**, 1–33 (1993)
2. Alon, N., Tarsi, M.: Colorings and orientations of graphs. Combinatorica **12**, 125–134 (1992). https://doi.org/10.1007/BF01204715
3. Björklund, A., Husfeldt, T., Kaski, P., Koivisto, M.: Fourier meets möbius: fast subset convolution. In: Johnson, D.S., Feige, U. (eds.) Proceedings of the 39th Annual ACM Symposium on Theory of Computing, San Diego, California, USA, 11–13 June 2007, pp. 67–74. ACM (2007)

4. Björklund, A., Husfeldt, T., Kaski, P., Koivisto, M.: Fourier meets möbius: fast subset convolution. In: Proceedings of the Thirty-Ninth Annual ACM Symposium on Theory of Computing, pp. 67–74 (2007)
5. Bliznets, I., Hecher, M.: Private communication (2023)
6. Bonamy, M., Kang, R.J.: List coloring with a bounded palette. J. Graph Theory **84**(1), 93–103 (2017)
7. Cygan, M., et al.: Parameterized Algorithms, vol. 5. Springer, Cham (2015). https://doi.org/10.1007/978-3-319-21275-3
8. Diestel, R.: Graph Theory. Electronic Library of Mathematics, Springer, Heidelberg (2006)
9. Erdos, P., Rubin, A.L., Taylor, H.: Choosability in graphs. In: Proceedings of the West Coast Conference on Combinatorics, Graph Theory and Computing, Congressus Numerantium, vol. 26, pp. 125–157 (1979)
10. Fellows, M.R., et al.: On the complexity of some colorful problems parameterized by treewidth. Inf. Comput. **209**(2), 143–153 (2011)
11. Fomin, F., Lokshtanov, D., Saurabh, S., Zehavi, M.: Kernelization: Theory of Parameterized Preprocessing. Cambridge University Press, Cambridge (2019). Publisher Copyright: Fedor V. Fomin, Daniel Lokshtanov, Saket Saurabh, and Meirav Zehavi 2019
12. Golovach, P.A., Heggernes, P.: Choosability of P_5-free graphs. In: Královič, R., Niwiński, D. (eds.) MFCS 2009. LNCS, vol. 5734, pp. 382–391. Springer, Heidelberg (2009). https://doi.org/10.1007/978-3-642-03816-7_33
13. Golovach, P.A., Heggernes, P.: Choosability of P_5-free graphs. Bull. Syktyvkar Univ. Ser. 1. Math. Mech. Comput. Sci. (11), 126–139 (2010)
14. Golovach, P.A., Heggernes, P., van't Hof, P., Paulusma, D.: Choosability on H-free graphs. Inf. Process. Lett. **113**(4), 107–110 (2013)
15. Gutner, S.: The complexity of planar graph choosability. Discrete Math. **159**(1–3), 119–130 (1996)
16. Gutner, S., Tarsi, M.: Some results on (a: b)-choosability. Discrete Math. **309**(8), 2260–2270 (2009)
17. Jansen, B.M., Nederlof, J.: Computing the chromatic number using graph decompositions via matrix rank. Theor. Comput. Sci. **795**, 520–539 (2019)
18. Kowalik, L., Socala, A.: Tight lower bounds for list edge coloring. In: Eppstein, D. (ed.) 16th Scandinavian Symposium and Workshops on Algorithm Theory, SWAT 2018. LIPIcs, Malmö, Sweden, 18–20 June 2018, vol. 101, pp. 28:1–28:12. Schloss Dagstuhl - Leibniz-Zentrum für Informatik (2018)
19. Král', D., Sgall, J.: Coloring graphs from lists with bounded size of their union. J. Graph Theory **49**(3), 177–186 (2005)
20. Marx, D., Mitsou, V.: Double-exponential and triple-exponential bounds for choosability problems parameterized by treewidth (2016)
21. Molloy, M.: The list chromatic number of graphs with small clique number. J. Comb. Theory Ser. B **134**, 264–284 (2019)
22. Noel, J.A., Reed, B.A., Wu, H.: A proof of a conjecture of Ohba. J. Graph Theory **79**(2), 86–102 (2015)
23. Reed, B., Sudakov, B.: List colouring when the chromatic number is close to the order of the graph. Combinatorica **25**(1), 117–123 (2004)
24. Vizing, V.G.: Coloring the vertices of a graph in prescribed colors. Diskret. Analiz **29**(3), 10 (1976)

Parameterized Algorithms for Covering by Arithmetic Progressions

Ivan Bliznets[1], Jesper Nederlof[2], and Krisztina Szilágyi[2](✉)

[1] University of Groningen, Groningen, The Netherlands
i.bliznets@rug.nl
[2] Utrecht University, Utrecht, The Netherlands
{j.nederlof,k.szilagyi}@uu.nl

Abstract. An *arithmetic progression* is a sequence of integers in which the difference between any two consecutive elements is the same. We investigate the parameterized complexity of two problems related to arithmetic progressions, called Cover by Arithmetic Progressions (CAP) and Exact Cover by Arithmetic Progressions (XCAP). In both problems, we are given a set X consisting of n integers along with an integer k, and our goal is to find k arithmetic progressions whose union is X. In XCAP we additionally require the arithmetic progressions to be disjoint. Both problems were shown to be NP-complete by Heath [IPL'90].

We present a $2^{O(k^2)}poly(n)$ time algorithm for CAP and a $2^{O(k^3)}$ $poly(n)$ time algorithm for XCAP. We also give a fixed parameter tractable algorithm for CAP parameterized below some guaranteed solution size. We complement these findings by proving that CAP is Strongly NP-complete in the field $\mathbb{Z}_p$, if p is a prime number part of the input.

Keywords: Arithmetic Progressions · Set Cover · Parameterized Complexity · Number Theory

1 Introduction

In the Set Cover problem one is given a universe U and a set system $\mathcal{S} \subseteq 2^U$ of subsets of U, along with an integer k. The challenge is to detect whether there exist sets $S_1, \ldots, S_k \in \mathcal{S}$ such that $\bigcup_{i=1}^{k} S_i = U$. Unfortunately, the problem is $W[2]$-hard parameterized by k [11, Theorem 13.28], and thus we do not expect an algorithm with Fixed Parameter Tractable (FPT) runtime, i.e. runtime $f(k)|x|^{O(1)}$ for some function f and input size $|x|$.

However, Set Cover is incredibly expressive, and it contains many well-studied parameterized problems such as d-Hitting Set (for a constant d), Vertex Cover and Feedback Vertex Set (see [11]), albeit the last problem requires an exponential number of elements. While all these mentioned special

Supported by the project CRACKNP that has received funding from the European Research Council (ERC) under the European Union's Horizon 2020 research and innovation programme (grant agreement No. 853234).

The original version of the chapter has been revised. The name of the author Jesper Nederlof has been corrected. A correction to this chapter can be found at https://doi.org/10.1007/978-3-031-52113-3_35

H. Fernau et al. (Eds.): SOFSEM 2024, LNCS 14519, pp. 125–138, 2024.
https://doi.org/10.1007/978-3-031-52113-3_9

cases are FPT parameterized by k, many other special cases of SET COVER remain $W[1]$-hard, and the boundary between special cases that are solvable in FPT and $W[1]$-hard has been thoroughly studied already (see e.g. Table 1 in [6]). An especially famous special case is the POINT LINE COVER problem, in which one is given points $U \subseteq \mathbb{Z}^2$ and asked to cover them with at most k line segments. While it is a beautiful and commonly used exercise to show this problem is FPT parameterized by k,[1] many slight geometric generalizations (such as generalizing it to arbitrary set systems of VC-dimension 2 [6]) are already $W[1]$-hard.

In this paper we study another special case of SET COVER, related to *Arithmetic Progressions (APs)*. Recall that an AP is a sequence of integers of the form $a, a+d, a+2d, \ldots, a+xd$, for some integers x, *start value* a and *difference* d. We study two computational problems: COVER BY APS (abbreviated with CAP) and EXACT COVER BY APS (abbreviated with XCAP). In both problems we are given a set of integers $X = \{x_1, x_2, \ldots, x_n\}$, and our goal is to find the smallest number of APs $s_1, s_2, \ldots, s_k$ consisting of only elements in X such that their union covers[2] exactly the set X. In the XCAP problem, we additionally require that the APs do not have common elements. While CAP and XCAP are already known to be weakly NP-complete since the 90's [14], the problems have been surprisingly little studied since then.

The study of the parameterized complexity of CAP and XCAP can be motivated from several perspectives.

- **The SET COVER perspective:** CAP and XCAP are natural special cases of SET COVER for which the parameterized complexity is unclear. While the problem looks somewhat similar to POINT LINE COVER, the crucial insight[1] towards showing it is FPT in k does not apply since an AP can be covered with 2 other AP's. In order to understand the jump in complexity from FPT to $W[1]$-hardness of restricted SET COVER problems better, it is natural to wonder whether properties weaker than the one of POINT LINE COVER[1] also are sufficient for getting FPT algorithms.
- **The practical perspective:** There is a connection between these problems and some problems that arise during the manufacture of VLSI chips [13]. The connection implies the NP-hardness of the latter problems. Bast and Storandt [3,4] used heuristics for these problems to compress bus timetables and speed up the process of finding the shortest routes in public transportation networks.
- **The additive number theory perspective**: The extremal combinatorics of covers with (generalized) APs is a very well studied in the field of additive combinatorics. This study already started in the 50's with conjectures made by among others Erdös (see [9] and the references therein), and recently results in spirit of covering sets of integers with sets of low additive energy (of which APs are a canonical example) such as Freiman's Theorem and the Balog-Szemerédi-Gowers Theorem also found algorithmic applications [7,8].

[1] A crucial insight is that any line containing at least $k+1$ points must be in a solution..

[2] We frequently interpret APs as sets by omitting the order, and "covers" can be read as "contains".

- **The "not about graphs" perspective:** Initially, applications of FPT algorithms were mostly limited to graph problems.[3] More recently, FPT algorithms have significantly expanded the realm of their applicability. It now includes geometry, computational social choice, scheduling, constraint satisfiability, and many other application domains. However, at this stage the interaction of number theory and FPT algorithms seems to be very limited.

Our Contributions. Our main results are FPT algorithms for CAP and XCAP:

Theorem 1. CAP *admits an algorithm running in time* $2^{O(k^2)}poly(n)$.

On a high level, this theorem is proved with a bounded search tree technique (similar to POINT LINE COVER): In each recursive call we branch on which AP to use. Since k is the number of APs in any solution the recursion depth is at most k. The difficulty however, is to narrow down the number of recursive calls made by each recursive call. As mentioned earlier, the crucial insight[1] does not apply since an AP can be covered with 2 other AP's. We achieve this by relying non-trivially on a result (stated in Theorem 5) by Crittenden and Vanden Eynden [9] about covering an interval of integers with APs, originally conjectured by Erdös. The proof of this theorem is outlined in Sect. 3.

Theorem 2. XCAP *admits an algorithm running in time* $2^{O(k^3)}poly(n)$.

On a high level, the proof of this theorem follows the proof of Theorem 1. However, to accomodate the requirement that the selected APs are disjoint we need a more refined recursion strategy. The proof of this theorem is outlined in Sect. 4.

We complement these algorithms with a new hardness result. Already in the 1990s, the following was written in the paper that proved weak NP-hardness of CAP and XCAP:

> "*Because the integers used in our proofs are exponentially larger than $|X|$, we have not shown our problems to be NP-complete in the strong sense [..]. Therefore, there is hope for a pseudopolynomial time algorithm for each problem.*"
> — HEATH [14]

While we do not directly make progress on this question, we show that two closely related problems *are* strongly NP-hard. Specifically, if p is an integer, we define an AP in $\mathbb{Z}_p$ as a sequence of the form

$$a, a+d \pmod p, a+2d \pmod p, \ldots, a+xd \pmod p.$$

In the COVER BY ARITHMETIC PROGRESSIONS IN $\mathbb{Z}_p$ problem one is given as input an integer p and a set $X \subseteq \mathbb{Z}_p$ and asked to cover X with APs in $\mathbb{Z}_p$ that are contained in X that cover X. In EXACT COVER BY ARITHMETIC PROGRESSIONS IN $\mathbb{Z}_p$ we additionally require the APs to be disjoint. We show the following:

[3] There has even been a series of workshops titled "Parameterized Complexity: Not-About-Graphs" (link) to extend the FPT framework to other fields.

Theorem 3. Cover by Arithmetic Progressions in $\mathbb{Z}_p$ *and* Exact Cover by Arithmetic Progressions in $\mathbb{Z}_p$ *are strongly NP-complete.*

While this may hint at strong NP-completeness for CAP and XCAP as well, since often introducing a (big) modulus does not incur big jumps in complexity in number theoretic computational problems (confer e.g. k-SUM, Partition, etc.), we also show that our strategy cannot directly be used to prove CAP and XCAP to be strongly-NP. Thus this still leaves the mentioned question of Heath [14] open. This result is proven in Sect. 5.

Finally, we illustrate that CAP is generalized by a variant of Set Cover that allows an FPT algorithm for a certain below guarantee parameterization. In particular, CAP always has a solution consisting of $|X|/2$ APs that cover all sets.

Theorem 4. *There is an* $2^{O(k)}n^{O(1)}$ *time algorithm that detects if a given set* X *of integers can be covered with at most* $|X|/2 - k$ *APs.*

This result is proved in Sect. 6. Proofs that are omitted due to space restrictions (indicated with †) can be found in the full version on arXiv.

2 Preliminaries

For integers a, b we denote by $[a, b]$ a set $\{a, a+1, \dots, b\}$, for $a = 1$ instead of $[1, b]$ we use a shorthand $[b]$, i.e. $[b] = \{1, 2, \dots, b\}$. For integers a, b we write $a|b$ to show that a divides b. For integers $a_1, \dots, a_n$ we denote their greatest common divisor by $gcd(a_1, \dots, a_n)$.

An *arithmetic progression* (AP) is a sequence of numbers such that the difference of two consecutive elements is the same. While AP is a sequence, we will often identify an AP with the set of its elements. We say an AP *stops in between* l and r if it largest element is in between l and r. We say it *covers* a set A if all integers in A occur in it. Given an AP $a, a+d, a+2d, \dots$ we call d the *difference*. We record the following easy observation:

Observation. The intersection of two APs is an AP.

If X is a set of integers, we denote

$$X^{>c} = \{x | x \in X, x > c\}, \quad X^{\geq c} = \{x | x \in X, x \geq c\},$$
$$X^{<c} = \{x | x \in X, x < c\}, \quad X^{\leq c} = \{x | x \in X, x \leq c\}.$$

Given a set X and an AP $A = \{a, a+d, a+2d, \dots\}$, we denote by $A \sqcap X$ the longest prefix of A that is contained in X. In other words, $X \sqcap A = \{a, a+d, \dots, a+\ell d\}$, where ℓ is the largest integer such that $a + \ell' d \in X$ for all $\ell' \in \{0, \dots, \ell\}$.

For a set of integers X and integer p we denote by $X_p = \langle x \bmod p | x \in X \rangle$. Here the $\langle\rangle$ symbols indicate that we build a *multiset* instead of a set (so each number is replaced with its residual class mod p and we do not eliminate copies).

We call an AP s *infinite* if there are integers a, d such that $s = (a, a+d, a+2d, \dots)$. Note that under this definition, the constant AP containing only one number and difference 0 is also infinite.

The following result by Crittenden and Vanden Eynden will be crucial for many of our results:

Theorem 5 [9]. *Any k infinite APs that cover the integers $\{1, \dots, 2^k\}$ cover the whole set of positive integers.*

3 Algorithm for Cover By Arithmetic Progressions (CAP)

Before describing the algorithm, we introduce an auxiliary lemma. This lemma will be crucially used to narrow down the number of candidates for an AP to include in the solution to at most 2^k.

Lemma 1 (†). *Let s_0 be an AP with at least $t+1$ elements. Let $s_1, \dots, s_k$ be APs that cover the elements $s_0(0), \dots, s_0(t-1)$, but not $s_0(t)$. The APs $s_1, \dots, s_k$ may contain other elements. Suppose that each AP $s_1, \dots, s_k$ has an element larger than $s_0(t)$. Then we have $t < 2^k$.*

Equipped with Lemma 1 we are ready to prove our first main theorem:

Theorem 1. CAP *admits an algorithm running in time $2^{O(k^2)} poly(n)$.*

Proof. Denote the set of integers given in the input by X. Without loss of generality we can consider only solutions where all APs are inclusion-maximal, i.e. solutions where none of the APs can be extended by an element of X. In particular, given an element a and difference d, the AP is uniquely determined: it is equal to $\{a - \ell d, \dots, a - d, a, a + d, \dots, a + rd\}$, where ℓ, r are largest integers such that $a - \ell' d \in X$ for all $\ell' \in [\ell]$ and $a + r'd \in X$ for all $r' \in [r]$.

Our algorithm consists of a recursive function COVERING(C, k_1, k_2). The algorithm takes as input a set C of elements and assumes there are APs $s_1, \dots, s_{k_1}$ whose union equals C, so the elements of C are 'covered' already. With this assumption, it will detect correctly whether there exist k_2 additional APs that cover all remaining elements $X \setminus C$ from the input. Thus COVERING$(\emptyset, 0, k)$ indicates whether the instance is a Yes-instance. At Line 4 we solve the subproblem, in which set C is already covered, in $2^{k^2} poly(n)$ time if $|X \setminus C| \leq k^2$. This can be easily done by writing the subproblem as an equivalent instance of SET COVER (with universe $U = X \setminus C$ and a set for each AP in U) and run the $2^{|U|} poly(n) = 2^{k^2} poly(n)$ time algorithm for SET COVER from [5].

For larger instances, we consider the $k^2 + 1$ smallest uncovered elements $u_1, \dots, u_{k^2+1}$ in Line 6, and guess (i.e. go over all possibilities) u_i and u_j such that u_i and u_j both belong to some AP s in a solution and none of the AP's $s_1, \dots, s_{k_1}$ stops (i.e., has its last element) in between u_i and u_j. In order to prove correctness of the algorithm, we will show later in Claim 1 such u_i and u_j exist.

Note that u_i and u_j are not necessarily consecutive in s, so we can only conclude that the difference of s divides $u_j - u_i$. We use Lemma 1 to lower bound the difference of s with $(u_j - u_i)/2^k$, and after we guessed the difference we recurse with the unique maximal AP with the guessed difference containing u_i (and u_j). In pseudocode, the algorithm works as follows:

Algorithm 1. Algorithm for CAP

```
 1: Algorithm Covering(C, k1, k2)
 2: Let k = k1 + k2
 3: if |X \ C| ≤ k^2 then
 4:     Use the algorithm for Set Cover from [5]
 5: else
 6:     Let u1, ..., u_{k^2} be the k^2 + 1 smallest elements of X \ C
 7:     for i = 1 ... k^2 do
 8:         for j = i + 1 ... k^2 + 1 do
 9:             Let D = uj − ui
10:             for ℓ = 1 ... 2^k do
11:                 if ℓ divides D then
12:                     Let s = MakeAP(ui, D/ℓ)
13:                     if Covering(C ∪ s, k1 + 1, k2 − 1) then
14:                         return true
15:     return false
```

The procedure MakeAP(a, d) returns the AP $\{a - \ell_1 d, \ldots, a - d, a, a + d, \ldots, a + \ell_2 d\}$, where ℓ_1 (respectively, ℓ_2) are the largest numbers such that $a - \ell' d \in X$ for all $\ell' \leq \ell_1$ (respectively, $a + \ell' d \in X$ for all $\ell' \leq \ell_2$). It is easy to see that if Algorithm 1 outputs **true**, we indeed have a covering of size k: Since we only add elements to C if they are indeed covered by an AP, and each time we add elements to C because of an AP we decrease our budget k_2.

For the other direction of correctness, we first claim that our algorithm will indeed at some moment consider an AP of the solution.

Claim 1. Suppose there exist an AP-covering $s_1, \ldots, s_k$, and let $k_1 \in \{0, \ldots, k\}$ be an integer and let $\cup_{i=1}^{k_1} s_i = C$. Then in the recursive call Covering$(C, k_1, k - k_1)$ we have in some iteration of the for-loops $s = s_h$ at Line 12, for some $h \in \{k_1 + 1, \ldots, k\}$.

Proof of Claim. Consider the set B consisting of the $k^2 + 1$ smallest elements of $X \setminus C$. By the pigeonhole principle, there is an $h \in \{j + 1, \ldots, k\}$ such that s_h covers at least $k + 1$ elements of B. Thus, there are two consecutive elements of $B \cap s_h$, $s_h(\alpha)$ and $s_h(\beta)$, between which no AP $s_1, \ldots, s_{k_1}$ ends. Without loss of generality, we may assume that $s_h(\alpha)$ and $s_h(\beta)$ are the closest such pair (i.e. the pair such that $|s_h(\alpha) - s_h(\beta)|$ is minimal). By applying Lemma 1, we conclude that there are at most $2^k - 1$ elements of s_h between $s_h(\alpha)$ and $s_h(\beta)$. In other words, the difference of s_h is a divisor of $s_h(\beta) - s_h(\alpha)$ and at least $(s_h(\beta) - s_h(\alpha))/2^k$. Therefore, in some iteration of the for-loops we get $s = s_h$. □

Using the above claim, it directly follows by induction on $k_2 = 0, \ldots, k$ that if there is a covering $s_1, \ldots, s_k$ of X such that $\cup_{i=1}^{k_1} s_i = C$, then the function COVERING$(C, k - k_2, k_2)$ returns true.

Let us now analyse the running time of the above algorithm. The recursion tree has height k (since we reduce k by one on every level). The maximum degree is $2^k k^4$, so the total number of nodes is $2^{O(k^2)}$. The running time at each node is at most $2^{O(k^2)} poly(n)$. Therefore, the overall running time is $2^{O(k^2)} poly(n)$. □

4 FPT Algorithm for Exact Cover by Arithmetic Progressions

Now we show that XCAP is Fixed Parameter Tractable as well. Note that this problem is quite different in character than CAP. For example, as opposed to CAP, in XCAP we cannot describe an AP with only one element and its difference. Namely, it is not always optimal to take the longest possible AP: for example, if $X = \{0, 4, 6, 7, 8, 9\}$, the optimal solution uses the progression $0, 4$ rather than $0, 4, 8$. While we still apply a recursive algorithm, we significantly need to modify our structure lemma (shown below in Lemma 2) and the recursive strategy.

Before we proceed with presenting the FPT algorithm for EXACT COVER BY ARITHMETIC PROGRESSIONS (XCAP) we state an auxiliary lemma.

Lemma 2 (†). *Let $s_1, s_2, \ldots, s_k$ be a solution of an instance of* XCAP *with input set X. Let s be an AP that is contained in X. For each $i \in [k]$, we denote by t_i the intersection of s and s_i. Suppose that for some i, t_i has at least $k+1$ elements. Then there are at most $2^{k-1} - 1$ elements of s between any two consecutive elements of t_i.*

Now we have all ingredients to prove the main theorem of this section.

Theorem 2. XCAP *admits an algorithm running in time $2^{O(k^3)} poly(n)$.*

Proof. Let $X = \{a_1, \ldots, a_n\}$ be the input set. Without loss of generality, we may assume that $a_1 < \cdots < a_n$. Assume that the input instance has a solution with k APs: $o_1, \ldots, o_k$. Let d_i' be the difference of o_i. Our algorithm PARTITION recursively calls itself and is described in Algorithm 2. The algorithm has the following parameters: $T_1, \ldots, T_k, P_1, \ldots, P_k, d_1, d_2, \ldots, d_k$. The sets T_i describe the elements that are in o_i (i.e. the elements that are definitely covered by o_i). The integer d_i is either 0 or equal to the guessed value of the difference of the AP o_i. Once we assign a non-zero value to d_i, we never change it within future recursive calls. We denote by P_i the set of "potentially covered" elements. Informally, P_i consists of elements that could be covered by o_i unless o_i is interrupted by another AP. Formally, if for an AP o_i we know two elements $a, b \in o_i$ $(a < b)$, and the difference d_i then $P_i = \{b + d_i, b + 2d_i, \ldots b + \ell d_i\}$ where ℓ is the largest number such that: $\{b + d_i, b + 2d_i, \ldots b + \ell d_i\} \subset X$ and none of the elements of the set $\{b + d_i, b + 2d_i, \ldots b + \ell d_i\}$ belong to T_j for some $j \neq i$, $j \in [k]$.

Algorithm 2. Algorithm for XCAP

```
Algorithm PARTITION(X, T_1, ... T_k, P_1, ..., P_k, d_1, ..., d_k)
if there are i, j ∈ [k] s.t. i ≠ j and P_i ∩ P_j ≠ ∅ then
    Let c be the smallest element s.t. ∃i, j, c ∈ P_i ∩ P_j, and i < j
    PARTITION(X, T_1, ..., T_k, P_1, ..., P_{i-1}, P_i^{<c}, P_{i+1}, ..., P_k, d_1, ..., d_k)
    PARTITION(X, T_1, ..., T_k, P_1, ..., P_{j-1}, P_j^{<c}, P_{j+1}, ..., P_k, d_1, ..., d_k)
else if there is an i such that T_i = {a_α, a_β} and d_i = 0 then
    for (b_1, ..., b_k) ∈ {0, 1, ..., 2^k + 1}^k do
        g ← gcd(a_β − a_α, b_1 d_1, ... b_k d_k)
        D_i ← {g, g/2, ..., g/(k(k+1))}
        for d ∈ D_i and d is integer do
            d_i ← d
            T_i ← {a_α, a_α + d_i, ..., a_β}
            C^∞ ← {a_β + d_i, a_β + 2d_i, ...}
            P_i ← C^∞ ⊓ X
            if for all j ∈ [k] \ {i} we have T_i ∩ T_j = ∅ then
                for j ∈ [k] \ {i} do
                    P_j ← UPDATE(P_j, T_i)    ▷ remove ints larger than min(P_j ∩ T_i)
                PARTITION(X, T_1, ..., T_k, P_1, ..., P_k, d_1, ..., d_k)
else if X \ (P ∪ T) = ∅ then
    return o_1 = T_1 ∪ P_1, ..., o_k = T_k ∪ P_k
else
    a_β ← min(X \ (T ∪ P))
    if there is an i such that |T_i| < 2 then
        J ← {i | i ∈ [k] and |T_i| < 2}
        for i ∈ J do
            T_i ← T_i ∪ {a_β}
            PARTITION(X, T_1, ..., T_k, P_1, ..., P_k, d_1, ..., d_k)
    else
        abort                                ▷ this branch does not generate a solution
```

Initially, we call the algorithm PARTITION with parameters $T_1 = \{a_1\}$, $T_2 = \cdots = T_k = P_1 = \cdots = P_k = \emptyset, d_1 = d_2 = \cdots = d_k = 0$. We denote by $T = \cup_i T_i$ and $P = \cup_i P_i$. Let a_β be the smallest element of the input sequence that does not belong to $T \cup P$. For each $i \in [k]$ such that $|T_i| \leq 1$, we recursively call PARTITION$(T_1, \ldots, T_{i-1}, T'_i, T_{i+1}, \ldots, T_k, P_1, \ldots, P_k)$, where $T'_i = T_i \cup \{a_\beta\}$. In other words, we consider all possible APs that cover a_β. We do not assign a_β to i-th AP with $|T_i| \geq 2$ as for such APs we know that $a_\beta \notin o_i$ (since $a_\beta \notin P_i$).

If in the input for some i we have $|T_i| = 2$ and $d_i = 0$ then we branch on the value of the difference of the i-th AP. Let $T_i = \{a_\alpha, a_\beta\}$, where $a_\beta > a_\alpha$. By construction (specifically, by the choice of a_β), all elements of the sequence $B = \{a_{\alpha+1}, a_{\alpha+2}, \ldots, a_{\beta-1}\}$ belong to $P \cup T$, i.e. can be covered by at most k APs. Note that purely based on the knowledge of two elements $a_\alpha, a_\beta \in o_i$ we cannot determine immediately the difference of o_i, because a_α, a_β might not be consecutive elements of o_i. In other words, it could happen that some elements of B belong to o_i. Therefore we need to consider cases when potentially covered elements from some P_j actually belong to o_i instead of o_j. We note that the number of elements between a_α and a_β can be very large, so a simple consideration of all cases where the difference of o_i is a divisor of $a_\beta - a_\alpha$ will not provide an FPT-algorithm.

Instead of considering all divisors we use Lemma 2 and bound the number of candidates for value of the difference of o_i. Note that we allow integers to be 0 here, and define that all integers are a divisor of 0. Hence $a|0$ is always true and $gcd(0, y_1, y_2, \dots) = gcd(y_1, y_2, \dots)$.

Informally, for all known differences d_j up to this point (i.e. all $d_j \neq 0$) we branch on the smallest positive value b_j such that $d'_i | b_j d_j$, where d'_i is the difference of the i-th AP in the solution that we want to find. We treat all cases when $b_j > 2^k + 1$ at once, and instead of the actual value we assign 0 to a variable responsible for storing value of b_j. Intuitively, b_j describes the number of elements of o_j between two consecutive interruptions by o_i. By Lemma 2, if b_j is large (larger than $2^k + 1$), each of these interruptions implies that an AP stops.

Formally, for each k-tuple $(b_1, \dots, b_k) \in \{0, \dots, 2^k + 1\}^k$, we do the following. Let $g = gcd(a_\beta - a_\alpha, b_1 d_1, \dots, b_k d_k)$. From the definition of b_j it follows that $d'_i | g$.

Claim 2. If all previous branchings are consistent with a solution $o_1, \dots, o_k$ then $d'_i \geq \frac{g}{k(k+1)}$.

Proof of Claim. Indeed, if $d'_i = g/t$ and $t > k(k+1)$ then between a_α and $a_\alpha + g$ there are at least $k(k+1)$ elements from o_i. All these elements are covered by sets $P_1, \dots, P_{i-1}, P_{i+1}, \dots, P_k$. Therefore, by the pigeonhole principle there is an index q such that P_q contains at least $k+1$ elements. This means that P_q and o_i have at least $k+1$ common elements (P_q is intersected by o_i at least $k+1$ times). Denote these common elements by $c_1, \dots, c_t$ and let e be the number of elements of P_q between c_ℓ and $c_{\ell+1}$ (e does not depend on ℓ as P_q and o_i are APs). First of all, recall that $c_\ell, c_{\ell+1}$ are from interval $(a_\alpha, a_\alpha + g)$. Therefore, $(e+1)d_q = c_{\ell+1} - c_\ell < g$. Moreover, by Lemma 2 (applied with $s = P_q$ and $s_i = o_i$), we have $e \leq 2^{k-1} - 1$. Taking into account that $d'_i | (e+1)d_q$ and $e \leq 2^{k-1} - 1$ we have that g divides $(e+1)d_q$ which contradicts $(e+1)d_q < g$. □

From the previous claim it follows that $d'_i \in \{g, g/2, \dots, g/k(k+1)\}$ and we have the desired bound on the number of candidates for the value of the difference of o_i. We branch on the value of d'_i, i.e. in each branch we assign to d_i some integer value from the set $\{g, g/2, \dots, g/k(k+1)\}$. In other words, for each $d \in \{g, g/2, \dots, g/k(k+1)\}$ we call

$$\text{PARTITION}(T_1, \dots, T'_i, \dots, T_k, P'_1, \dots, P'_k, d_1, \dots, d_{i-1}, d, d_{i+1}, \dots, d_k),$$

where $C_d = \{a_\alpha, a_\alpha + d, \dots, a_\beta - d, a_\beta\}$, $C_d^\infty = \{a_\beta + t, a_\beta + 2t, \dots\}$, $T'_i = T_i \cup C_d$, $P'_i = C_d^\infty \sqcap X$ and for $j \neq i$ we set $P'_j = \text{UPDATE}(P_j, T_i)$. The function $\text{UPDATE}(A, B)$ returns $A^{<x}$, where $x = \min(A \cap B)$ (if $A \cap B = \emptyset$, it returns A). If in some branch the sequence $a_\alpha, a_\alpha + d_i, \dots, a_\beta - d_i, a_\beta$ intersects T, we abort

this branch. Overall, in order to determine the difference of o_i after discovering $a_\alpha, a_\beta \in o_i$ we create at most $(2^k+2)^k \cdot k(k+1) = 2^{O(k^2)}$ branches.

Let us now compute the number of nodes in the recursion tree. Observe that we never remove elements from T. Consider a path from the root of the tree to a leaf. It contains at most $2k$ nodes of degree at most k (adding two first elements to T_i), at most k^2 nodes of degree 2 (resolving the intersections of two APs, lines 2–5 in pseudocode) and at most k nodes of degree $2^{O(k^2)}$ (determining the difference of an AP that contains two elements). Hence, the tree has at most $k^{2k} \cdot 2^{k^2} \cdot (2^{O(k^2)})^k = 2^{O(k^3)}$ leaves. Therefore, the overall number of nodes in the recursion tree and the running time of the algorithm is $2^{O(k^3)} poly(n)$.

Claim 2 proves that we iterate over all possibilities. Hence, our algorithm is correct. □

5 Strong NP-Hardness of Cover by Arithmetic Progressions in $\mathbb{Z}_p$

A natural question to ask, in order to prove Strong NP-hardness for CAP, is whether we can replace an input set X with an equivalent set which has smaller elements. Specifically, could we replace the input number with numbers polynomial in $|X|$, while preserving the set of APs? This intuition can be further supported by result on Simultaneous Diophantine approximation that exactly achieve results in this spirit (though with more general properties and larger upper bounds) [12]. However, it turns out that this is not always the case, as we show in Lemma 3. This means that one of the natural approaches for proving strong NP-hardness of CAP does not work: namely, not all sets X can be replaced with set X' of polynomial size which preserves all APs in X.

Lemma 3 (†). *Let $x_1 = 0$, $x_i = 2^{i-2}$ for $i \geq 2$ and $X_n = \{x_1, \ldots, x_{n+2}\}$. Then for any polynomial p there exists an integer n such that there is no set $A_n = \{a_1, \ldots, a_{n+2}\}$ that satisfies the following criteria:*

- *$a_i \leq p(n)$ for all $i \in [n+2]$,*
- *For all $i, j, k \in [n+2]$, the set $\{x_i, x_j, x_k\}$ forms an AP if and only if $\{a_i, a_j, a_k\}$ forms an AP.*

Unfortunately, we do not know how to directly circumvent this issue and improve the weak NP-hardness proof of Heath [14] to *strong* NP-hardness. Instead, we work with a small variant of the problem in which we work in $\mathbb{Z}_p$. The definition of APs naturally carries over to $\mathbb{Z}_p$. It is easy to see that APs are preserved:

Claim 3 (†). Let p be a prime and let X be a set of integers that forms an AP. Then the multiset X_p generates AP in the field $\mathbb{Z}_p$.

However, the converse does not hold. Formally, if for some p the multiset X_p is an AP in $\mathbb{Z}_p$ it is not necessarily true that X is an AP in $\mathbb{Z}$.

We now show strong NP-completeness for the modular variants of CAP and XCAP. In the Cover by Arithmetic Progressions in $\mathbb{Z}_p$ problem one is given as input an integer p and a set $X \subseteq \mathbb{Z}_p$ and asked to cover X with APs in $\mathbb{Z}_p$ that are contained in X that cover X. In Exact Cover by Arithmetic Progressions in $\mathbb{Z}_p$ we additionally require the APS to be disjoint.

Theorem 3. Cover by Arithmetic Progressions in $\mathbb{Z}_p$ *and* Exact Cover by Arithmetic Progressions in $\mathbb{Z}_p$ *are strongly NP-complete.*

Proof. We recall that Heath [14] showed that CAP and XCAP are weakly NP-complete via reduction from Set Cover. Moreover, the instances of CAP and XCAP, obtained after reduction from Set Cover, consist of numbers that are bounded by $2^{q(n)}$ for some polynomial $q(n)$. To show that Cover by Arithmetic Progressions in $\mathbb{Z}_p$ and Exact Cover by Arithmetic Progressions in $\mathbb{Z}_p$ are strongly NP-complete we take a prime p and convert instances of CAP and XCAP with set S into instances of Cover by Arithmetic Progressions in $\mathbb{Z}_p$ and Exact Cover by Arithmetic Progressions in $\mathbb{Z}_p$ respectively with a set S_p (we can guarantee that the multiset S_p contains no equal numbers) and modulo p.

As shown in Claim 3, under such transformation a Yes-instance is converted into a Yes-instance. However, if we take an arbitrary p then a No-instance can be mapped to a Yes-instance or S_p can become a multiset instead of a set. In order to prevent this, we carefully pick the value of p.

We need to guarantee that if X is not an AP then X_p also does not generate an AP in $\mathbb{Z}_p$. Suppose $X_p = \{y_1, y_2, \ldots, y_k\}$ generates an AP in $\mathbb{Z}_p$ exactly in this order. We assume that x_i maps into y_i, i.e. $x_i \equiv_p y_i$. Since X_p is an AP in $\mathbb{Z}_p$, we have

$$y_2 - y_1 \equiv_p y_3 - y_2 \equiv_p \cdots \equiv_p y_k - y_{k-1}.$$

Since X is not an AP there exists an index j such that $x_j - x_{j-1} \neq x_{j+1} - x_j$. Therefore, we have that $2x_j - x_{j-1} - x_{j+1} \neq 0$ and $2y_j - y_{j-1} - y_{j+1} \equiv_p 0$. Since $x_i \equiv_p y_i$ we conclude that p divides $2x_j - x_{j-1} - x_{j+1}$. Hence if we want to choose p that does not transform a No-instance into a Yes-instance, it is enough to choose p such that p is not a divisor of $2x - y - z$ where x, y, z are any numbers from the input and $2x - y - z \neq 0$. Similarly, if we want S_p to be a set instead of a multiset, then for any different x, y the prime p should not be a divisor of $x - y$. Note that the number of different non-zero values of $2x - y - z$ and $x - y$ is at most $O(n^3)$. Since all numbers are bounded by $2^{q(n)}$ the values of $2x - y - z \neq 0$ and $x - y$ are bounded by $2^{q(n)+2}$. Note that any integer N has at most $\log N$ different prime divisors. Therefore at most $O(n^3)(q(n)+2)$ prime numbers are not suitable for our reduction. In order to find a suitable prime number we do the following:

- for each number from 2 to $n^6(q(n)+2)^2$ check if it is prime (it can be done in polynomial time [1]),
- for each prime number $p' \leq n^6(q(n)+2)^2$ check if there are integers x, y, z from the input such that ($2x - y - z \neq 0$ and $2x - y - z \equiv_p 0$) or $x \equiv_p y$ if such x, y, z exist go to the next prime number,

– when the desired prime p' is found output instance $S_{p'}$ with modulo p'.

Note that number of primes not exceeding N is at least $\frac{N}{2\log N}$ for large enough N. Hence by pigeonhole principle we must find the desired p' as we consider all numbers smaller than $n^6(q(n)+2)^2$ and $\frac{n^6(q(n)+2)^2}{\log(n^6(q(n)+2)^2)} > O(n^3)(q(n)+2)$ for sufficiently large n.

Therefore, in polynomial time we can find p' that is polynomially bounded by n and COVER BY ARITHMETIC PROGRESSIONS IN $\mathbb{Z}_p$ with input $(S_{p'}, p')$ is equivalent to CAP with input S (similarly for EXACT COVER BY ARITHMETIC PROGRESSIONS IN $\mathbb{Z}_p$ and XCAP). Hence, COVER BY ARITHMETIC PROGRESSIONS IN $\mathbb{Z}_p$ and EXACT COVER BY ARITHMETIC PROGRESSIONS IN $\mathbb{Z}_p$ are strongly NP-complete. □

6 Parameterization Below Guarantee

In this section we present an FPT algorithm parameterized below guarantee for a problem that generalizes CAP, namely t-UNIFORM SET COVER. The t-UNIFORM SET COVER is a special case of the SET COVER problem in which all instances $\mathcal{S}, U$ satisfy the property that $\{A \subseteq U : |A| = t\} \subseteq \mathcal{S}$. Clearly the solution for the t-UNIFORM SET COVER problem is at most $\lceil \frac{n}{t} \rceil$ where n is the size of universe. Note that CAP is a special case of 2-UNIFORM SET COVER since any pair of element forms an AP. Thus we focus on presenting a fixed parameter tractable algorithm for the t-UNIFORM SET COVER problem parameterized below $\lceil \frac{n}{t} \rceil$ (we consider t to be a fixed constant). We note that for a special case with $t = 1$ the problem was considered in works [2,10].

We will use the deterministic version of color-coding, which uses the following standard tools:

Definition 1. *For integers n, k a (n, k)-*perfect hash family *is a family $\mathcal{F}$ of functions from $[n]$ to $[k]$ such that for each set $S \subseteq [n]$ of size k there exists a function $f \in \mathcal{F}$ such that $\{f(v) : v \in S\} = [k]$.*

Lemma 4 [15]. *For any $n, k \geq 1$, one can construct an (n, k)-perfect hash family of size $e^k k^{O(\log k)} \log n$ in time $e^k k^{O(\log k)} n \log n$.*

Now we are ready to state and prove the main result of this section:

Theorem 6. *There is an $2^{O(k)}poly(n)$-time algorithm that for constant t and a given instance of t-UNIFORM SET COVER determines the existence of a set cover of size at most $\lceil \frac{n}{t} \rceil - k$ where k is an integer parameter and n is the size of the universe.*

Proof. In the first stage of our algorithm we start picking sets greedily (i.e. in each step, we pick the set that covers the largest number of previously uncovered elements) until there are no sets that cover at least $t+1$ previously uncovered elements. If during this stage we pick s sets and cover at least $st + tk$ elements then our instance is a YES-instance. Indeed, we can cover the remaining elements using $\lceil \frac{n-st-tk}{t} \rceil$ sets. In total, such a covering has at most $\lceil \frac{n-st-tk}{t} \rceil + s = \lceil \frac{n}{t} \rceil - k$ subsets. Intuitively, each set picked during this greedy stage covers at least one additional element. Therefore, if we pick more than tk subsets then our input instance is a YES-instance. This means that after the greedy stage we either immediately conclude that our input is a YES-instance or we have used at most tk subsets and covered at most $tk \cdot t + tk = O(k)$ elements, for a fixed t. Let us denote the subset of all covered elements by G. Note that there is no subset that covers more than t elements from $U \setminus G$.

If there is a covering of size $\lceil \frac{n}{t} \rceil - k$ then there are $s' \leq |G| \leq tk \cdot t + tk$ subsets that cover G and at least $s't + tk - |G|$ elements of $U \setminus G$. Moreover, if such subsets exist then our input is a YES-instance.

Hence, it is enough to find s' such subsets. For each $s'' \in [|G|]$ we attempt to find subsets $S_1, S_2, \ldots, S_{s''}$ such that $G \subset S_1 \cup S_2 \cup \cdots \cup S_{s''}$ and $S_1 \cup S_2 \cup \cdots \cup S_{s''}$ contains at least $s''t + tk - |G|$ elements from $U \setminus G$. Assume that for some s'' such sets exist. Let H be an arbitrary subset of $(S_1 \cup S_2 \cup \cdots \cup S_{s''}) \setminus G$ of size $s''t + tk - |G|$. Note that we do not know the set H. However, we employ the color-coding technique, and construct a $(n, |H|)$-perfect hash family $\mathcal{F}$. We iterate over all $f \in \mathcal{F}$. Recall that $|H| = s''t + tk - |G|$. Now using dynamic programming we can find H in time $2^{O(k)}$ in a standard way. In order to do that, we consider a new universe U' which contains elements from the set G and elements corresponding to $|H|$ colors corresponding to values assigned by f to $U \setminus G$. Moreover, if a subset P was a subset that can be used for covering in t-UNIFORM SET COVER then we replace it with $(P \cap G) \cup \{\text{all values } f \text{ assign to elements in } P \cap (U \setminus G)\}$. We replace our t-UNIFORM SET COVER instance with an instance of SET COVER with a universe of size $|G| + |H| \leq |G| + s''t + tk - |G| \leq |G| \cdot t + tk \leq (tk \cdot t + tk) \cdot t + tk = O(k)$. It is easy to see that our original instance is a YES-instance if and only if the constructed instance of SET COVER admits a covering by at most s'' sets (under the assumption that f indeed assigns distinct values to elements of H). If f does not assign distinct values then a YES-instance can be become a NO-instance. However, a NO-instance cannot become a YES-instance. Since $|H| = O(k)$, the overall running time is $2^{O(k)} poly(n)$. □

As a corollary of the previous theorem we get the following result.

Theorem 4. *There is an $2^{O(k)} n^{O(1)}$ time algorithm that detects if a given set X of integers can be covered with at most $|X|/2 - k$ APs.*

References

1. Agrawal, M., Kayal, N., Saxena, N.: Primes is in P. Ann. Math. **160**, 781–793 (2004)
2. Basavaraju, M., Francis, M.C., Ramanujan, M., Saurabh, S.: Partially polynomial kernels for set cover and test cover. SIAM J. Discret. Math. **30**(3), 1401–1423 (2016)
3. Bast, H., Storandt, S.: Frequency data compression for public transportation network algorithms. In: Proceedings of the International Symposium on Combinatorial Search, vol. 4, pp. 205–206 (2013)

4. Bast, H., Storandt, S.: Frequency-based search for public transit. In: Proceedings of the 22nd ACM SIGSPATIAL International Conference on Advances in Geographic Information Systems, pp. 13–22 (2014)
5. Björklund, A., Husfeldt, T., Koivisto, M.: Set partitioning via inclusion-exclusion. SIAM J. Comput. **39**(2), 546–563 (2009)
6. Bringmann, K., Kozma, L., Moran, S., Narayanaswamy, N.S.: Hitting set in hypergraphs of low VC-dimension. CoRR abs/1512.00481 (2015). http://arxiv.org/abs/1512.00481
7. Bringmann, K., Nakos, V.: Top-k-convolution and the quest for near-linear output-sensitive subset sum. In: Makarychev, K., Makarychev, Y., Tulsiani, M., Kamath, G., Chuzhoy, J. (eds.) Proceedings of the 52nd Annual ACM SIGACT Symposium on Theory of Computing, STOC 2020, Chicago, IL, USA, 22–26 June 2020, pp. 982–995. ACM (2020). https://doi.org/10.1145/3357713.3384308
8. Chan, T.M., Lewenstein, M.: Clustered integer 3SUM via additive combinatorics. In: Servedio, R.A., Rubinfeld, R. (eds.) Proceedings of the Forty-Seventh Annual ACM on Symposium on Theory of Computing, STOC 2015, Portland, OR, USA, 14–17 June 2015, pp. 31–40. ACM (2015). https://doi.org/10.1145/2746539.2746568
9. Crittenden, R.B., Vanden Eynden, C.: Any n arithmetic progressions covering the first 2^n integers cover all integers. Proc. Am. Math. Soc. **24**(3), 475–481 (1970)
10. Crowston, R., Gutin, G., Jones, M., Raman, V., Saurabh, S.: Parameterized complexity of MaxSat above average. Theoret. Comput. Sci. **511**, 77–84 (2013)
11. Cygan, M., et al.: Parameterized Algorithms. Springer, Cham (2015). https://doi.org/10.1007/978-3-319-21275-3
12. Frank, A., Tardos, E.: An application of simultaneous Diophantine approximation in combinatorial optimization. Combinatorica **7**(1), 49–65 (1987). https://doi.org/10.1007/BF02579200
13. Grobman, W., Studwell, T.: Data compaction and vector scan e-beam system performance improvement using a novel algorithm for recognition of pattern step and repeats. J. Vac. Sci. Technol. **16**(6), 1803–1808 (1979)
14. Heath, L.S.: Covering a set with arithmetic progressions is NP-complete. Inf. Process. Lett. **34**(6), 293–298 (1990)
15. Naor, M., Schulman, L.J., Srinivasan, A.: Splitters and near-optimal derandomization. In: 36th Annual Symposium on Foundations of Computer Science, Milwaukee, Wisconsin, USA, 23–25 October 1995, pp. 182–191. IEEE Computer Society (1995). https://doi.org/10.1109/SFCS.1995.492475

Row-Column Combination of Dyck Words

Stefano Crespi Reghizzi[1], Antonio Restivo[2], and Pierluigi San Pietro[1(✉)]

[1] DEIB - Politecnico di Milano, Milan, Italy
{stefano.crespireghizzi,pierluigi.sanpietro}@polimi.it
[2] Dipartimento di Matematica e Informatica, Università di Palermo, Palermo, Italy
antonio.restivo@unipa.it

Abstract. We lift the notion of Dyck language from words to 2-dimensional arrays of symbols, i.e., pictures. We define the Dyck crossword language DC_k as the row-column combination of Dyck word languages, which prescribes that each column and row is a Dyck word over an alphabet of size $4k$. The standard relation between matching parentheses is represented in DC_k by an edge of the matching graph situated on the picture array. Such edges form a circuit, of path length multiple of four, where row and column matches alternate. Length-four circuits are rectangular patterns, while longer ones exhibit a large variety of patterns. DC_k languages are not recognizable by the Tiling Systems of Giammarresi and Restivo. DC_k contains pictures where circuits of unbounded length occur, and where any Dyck word occurs in a row or in a column. We prove that the only Hamiltonian circuits of the matching graph of DC_k have length four. A proper subset of DC_k, called quaternate, includes only the rectangular patterns; we define a proper subset of quaternate pictures that (unlike the general ones) preserves a characteristic property of Dyck words: availability of a cancellation rule based on a geometrical partial order relation between rectangular circuits. Open problems are mentioned.

1 Introduction

The Dyck language is a fundamental concept in formal language theory. Its alphabet $\{a_1, \dots, a_k, b_1, \dots, b_k\}$, for any $k \geq 1$, is associated with the pairs $[a_1, b_1], \dots, [a_k, b_k]$. The language is the set of all words that can be reduced to the empty word by cancellations of two coupled letters: $a_i b_i \to \varepsilon$. Dyck words represent the last-in-first-out order of events, a fundamental concept for theoretical computer science and especially for formal language and automata theory, where the Chomsky-Schützenberger theorem [1] states that any context-free language is the homomorphic image of the intersection of a Dyck language and a regular one.

Motivated by our interest in the theory of two-dimensional (2D) or picture languages (from now on simply "languages"), we investigate the possibility to transport the Dyck concept from one dimension to 2D. When moving from 1D to 2D, most formal language concepts and relationships drastically change. In particular, in 2D the Chomsky hierarchy of languages is blurred because the notions of regularity and context-freeness cannot be formulated for pictures without giving up some characteristic properties that hold for words. In fact, it is known [7] that the three equivalent definitions of

H. Fernau et al. (Eds.): SOFSEM 2024, LNCS 14519, pp. 139–153, 2024.
https://doi.org/10.1007/978-3-031-52113-3_10

regular languages by means of finite-state recognizers, by regular expressions, and by the homomorphism of local languages, produce in 2D three distinct language families. The third one gives the family of *tiling system recognizable languages* (REC) [7], that is perhaps the best known definition for regularity in 2D.

The situation is less satisfactory for context-free languages, of which Dyck languages are a notable example, where a transposition in 2D remains problematic. None of the existing proposals of "context-free" picture grammars ([3,5,9–12], a survey is [2]) match the expressiveness and richness of formal properties of 1D context-free grammars. In particular, we are not aware of any existing definitions of 2D Dyck languages,[1] and we hope that the present one will open a new direction of research on (picture) languages.

It is time to describe our proposal. We consider the picture languages obtained by the *row-column combination*, also known as *crossword*, of two Dyck word languages over the same alphabet. In such a combination, all rows and all columns are Dyck words. Crosswords have been studied for regular languages (e.g., in [6,8]) but not, to our knowledge, for context-free ones. In particular it is known [8] that the REC family coincides with the projection of the crosswords of two regular languages.

The family of Dyck crosswords over an alphabet of size $4k$, denoted by DC_k, $k \geq 1$, represents, for reasons later explained, a rather general case. It includes a spectrum of pictures where a surprising variety of complex patterns may occur. To analyze them, we introduce the *matching graph* of a picture, where the array cells are the nodes and the matching relation defines the edges. The graph is partitioned into simple (disjoint) circuits, made by alternating horizontal and vertical edges, representing a Dyck match on a row and on a column. A circuit label is a word of length multiple of 4. The edges of a circuit path may cross each other–the case of zero crossings is the length 4 circuit or rectangle. Pictures containing just such rectangular circuits may present quite evident geometrical analogies with the Dyck word case.

We prove that DC_k is not in REC and we positively answer the question whether each Dyck word can occur in DC_k pictures. We present some interesting types of Dyck crosswords that contain multiple circuits including complex ones, but much remains to be understood about the general patterns that are possible and the trade-off between circuit length and the number of circuits that cover a picture. We show that the only pictures covered by one circuit (i.e., Hamiltonian) have size 2×2; furthermore, we prove that for any $h \geq 0$ there exist pictures in DC_k featuring a circuit of length $4 + 8h$, i.e., the circuit length is unbounded.

As said, the structure of pictures, called *quaternate*, such that their circuits are rectangular, is intuitively similar to the structure of Dyck words since the vertexes of a rectangle delimit a subpicture much as two coupled parentheses delimit a substring. To formalize such an intuition, we introduce a further subset of Dyck crosswords characterized by a variant of the Dyck cancellation rule. First, we transform cancellation into a *neutralization* rule that maps the four vertex letters of a rectangle on a new neutral

---

[1] We just know of a particular example, the *Chinese box language* in [3], that intuitively consists of embedded or concatenated rectangles, and was proposed to illustrate the expressiveness of the grammars there introduced. But that language is not a satisfactory proposal, since it is in the family REC, hence "regular" rather than "context-free".

(i.e., non-coupled) letter N. Then a quaternate picture is *neutralizable* if it reduces to a picture over alphabet $\{N\}$ by applying neutralization steps. We prove that neutralizable pictures are a subset of quaternate ones. The analogy between Dyck words and neutralizable pictures is thus substantiated by the fact that both use neutralization rules for recognition, but there is a difference. The partial order of neutralization is a tree order for words, while for pictures it is a directed acyclic graph that represents the geometric relation of partial containment between rectangles.

Section 2 lists basic concepts of picture languages and Dyck word languages. Section 3 introduces the DC_k languages, exemplifies the variety of circuits they may contain, proves formal properties, and defines the quaternate subclass. Section 4 studies the neutralizable case. Section 5 mentions open problems.

2 Preliminaries

All the alphabets considered are finite. The concepts and notations for picture languages follow mostly [7]. We assume some familiarity with the basic theory of the family REC of tiling system languages, defined as the projection of a local 2D language; the relevant properties of REC will be reminded when needed. A *picture* is a rectangular array of letters over an alphabet. The set of all non-empty pictures over Σ is denoted by Σ^{++}.

A *domain* d of a picture p is a quadruple (i, j, i', j'), with $1 \leq i \leq i' \leq |p|_{row}$, and $1 \leq j \leq j' \leq |p|_{col}$, where $|p|_{row}$ and $|p|_{col}$ denote the number of rows and columns, respectively. The *subpicture of* p with domain $d = (i, j, i', j')$, denoted by $subp(p, d)$ is the (rectangular) portion of p defined by the top-left coordinates (i, j) and by the bottom-right coordinates (i', j').

Let $p, q \in \Sigma^{++}$. The *horizontal concatenation* of p and q is denoted as $p \oplus q$ and it is defined when $|p|_{row} = |q|_{row}$. Similarly, the *vertical concatenation* $p \ominus q$ is defined when $|p|_{col} = |q|_{col}$. We also use the power operations $p^{\ominus k}$ and $p^{\oplus k}$, $k \geq 1$, their closures $p^{\ominus +}$, $p^{\oplus +}$ and the closure under both concatenations $p^{\ominus +, \oplus +}$; concatenations and closures are extended to languages in the obvious way.

The notation $N^{m,n}$, where N is a symbol and $m, n > 0$, stands for a homogeneous picture in N^{++} of size m, n. For later convenience, we extend this notation to the case where either m or n are 0, to introduce identity elements for vertical and horizontal concatenations: given a picture p of size (m, n), by definition $p \oplus N^{m,0} = N^{m,0} \oplus p = p$ and $p \ominus N^{0,n} = N^{0,n} \ominus p = p$.

Dyck Alphabet and Language. The definition and properties of Dyck word languages are basic concepts in formal language theory. Let Γ_k, $k \geq 1$, be an alphabet of cardinality $2k$. Γ_k is called a Dyck alphabet if it is associated with a partition into two sets Γ', Γ'' of cardinality k and with a one-to-one total mapping, called *coupling*, from Γ' into Γ''. If the pair $[a, b]$ is in the coupling, $a \in \Gamma'$, $b \in \Gamma''$, then it is called *coupled* pair and the *coupled letters* a, b are called, respectively, *open* and *closed*. The Dyck language D_k over alphabet Γ_k is the set of words congruent to ε, via the *cancellation rule* $a_i b_i \to \varepsilon$ that erases two adjacent coupled letters. A pair of coupled letters occurring in a word is called *matching* if it is erased by the same cancellation rule application. Notice that the number of letters between the two letters of a matching pair is always even.

3 Row-Column Combination of Dyck Languages

In this section we define the languages, called simple Dyck Crosswords (DC), such that their pictures have Dyck words in rows and in columns. They may be viewed as analogous in 2D of Dyck 1D languages. Following [7] we introduce the row-column combination operation that takes two word languages and produces a picture language.

Definition 1 (row-column combination a.k.a. crossword). *Let $S', S'' \subseteq \Sigma^*$ be two word languages, resp. called* row and column component languages. *The* row-column combination *or* crossword *of S' and S'' is the language L such that a picture $p \in \Sigma^{++}$ belongs to L if, and only if, the words corresponding to each row (in left-to-right order) and to each column (in top-down order) of p belong to S' and S'', respectively.*

The crossword of regular languages has received attention in the past since its alphabetic projection coincide with the REC family [7]; some complexity issues for this case are recently addressed in [6] where the crosswords are called "regex crosswords".

Remark 1. Given two regular languages S', S'', it is undecidable to establish whether their crossword is empty. This implies that in general there are crosswords that do not *saturate* their components, i.e., such that the set of all rows (or the set of all columns) occurring in pictures of the crossword is a proper subset of the row component language (or of the column component language).

We investigate the properties of the crossword of a fundamental type of context-free, non-regular languages, the Dyck ones. First, we discuss the alphabet size and couplings.

Theorem 1 (Alphabet size of crosswords of Dyck languages). *Let D', D'' be two Dyck languages over the same alphabet Δ (with possibly distinct couplings over Δ).*

i) If Δ has fewer than four letters, then the crossword of D', D'' is empty.
ii) If Δ has four letters, then (up to isomorphism) there is one and only one coupling for D' and for D'' such that the crossword of D', D'' is not empty.
iii) If the number of letters of Δ is a multiple of four, then there is a coupling for D' and for D'' such that the crossword of D', D'' is not empty.

Proof. Part (i): a Dyck alphabet has an even number of letters, hence the only relevant case is the binary alphabet, e.g., $\{a, b\}$. If the coupling for the row language is, say, $[a, b]$, then $[a, b]$ or $[b, a]$ is the coupling for the columns. Given a picture with an occurrence of a, say, in the leftmost column, a letter b must occur in the same column, which would require a coupling $[b, a]$ for rows, a contradiction.
Part (ii): let $\{a, b, c, d\}$ be a Dyck alphabet of four letters. As in Part (i), in the top left corner of any picture there is a letter, say, a, which is an open letter for both rows and columns. Hence, the row language has a coupled pair, say, $[a, b]$ and the column language has a coupled pair, say, $[a, c]$–we proved above that the couplings $[a, b]$ or $[b, a]$ for columns would lead to an empty language. The letter b is thus on the first row, hence it is an open letter for the column language, therefore the latter must include the coupled pair $[b, d]$ and similarly the row language must include the coupled pair $[c, d]$: there is no other letter left and any other choice than d for the closed letter in either case

would again lead to the empty language. The corresponding crossword is not empty since, among others, it includes all pictures of the form: $\begin{pmatrix} a & b \\ c & d \end{pmatrix}^{\Phi+,\ominus+}$.

Part (iii): The cardinality of Δ is $4k$, for some $k \geq 1$. It is enough to partition Δ in k subsets of four elements and then for each subset use the same coupling of Part (ii). □

In particular, the alphabet used in the proof of Part (ii) of Theorem 1 can be denoted as $\Delta_1 = \{a, b, c, d\}$, with the coupling $\{[a, b][c, d]\}$ for the rows and $\{[a, c][b, d]\}$ for the columns. The corresponding (unique) crossword is denoted as DC_1. A simple example of a picture in DC_1 is in Fig. 1.

We now generalize the definition of DC_1 to alphabets of any cardinality multiple of 4 (as in the proof of Part (iii) of Theorem 1).

Definition 2 (Dyck crossword alphabet and language). *The Dyck crossword alphabet Δ_k is a set of quadruplets, namely $\{a_i, b_i, c_i, d_i \mid 1 \leq i \leq k\}$, together with the following couplings of the Dyck row alphabet Δ_k^{Row} for the row component language D_k^{Row}, and of the column alphabet Δ_k^{Col} for the column component language D_k^{Col}:*

$$\begin{cases} \textit{for } \Delta_k^{Row} : \{[a_i, b_i] \mid i \leq 1 \leq k\} \cup \{[c_i, d_i] \mid 1 \leq i \leq k\} \\ \textit{for } \Delta_k^{Col} : \{[a_i, c_i] \mid i \leq 1 \leq k\} \cup \{[b_i, d_i] \mid 1 \leq i \leq k\} \end{cases}. \tag{1}$$

The simple[2] Dyck crossword *DC_k is the row-column combination of D_k^{Row} and D_k^{Col}.*

For brevity, we later drop "simple" when referring to Dyck crosswords.

It is easy to notice that, for every $k \geq 1$, the language DC_k is closed under horizontal and vertical concatenation and their closures, and that for every $n, m \geq 1$ there exist pictures of DC_k of size $(2n, 2m)$.

We prove that DC_k is not recognizable by a tiling system, hence it is not in REC.

Theorem 2 (Comparison with REC). *For every $k \geq 1$, the language DC_k is not in the REC family.*

Proof. By contradiction, assume that DC_k is in REC. Without loss of generality, we consider only the case $k = 1$. Consider the following picture p in DC_1: $\begin{matrix} a & b \\ c & d \end{matrix}$. From closure properties of REC, the language $p^{\Phi+}$ is in REC, hence also the language:

$$R = \left(a^{\Phi+} \oplus b^{\Phi+}\right) \ominus \left((a \ominus c) \oplus p^{\Phi+} \oplus (b \ominus d)\right) \ominus \left(c^{\Phi+} \oplus d^{\Phi+}\right).$$

A picture in R has a^+b^+ in the top row and c^+d^+ in the bottom row. Let $T = DC_1 \cap R^{\ominus+}$. By closure properties of REC, both T and $T^{\ominus+}$ are in REC. The first row of every picture in $T^{\ominus+}$ has the form $a^n b^n$, since it is the intersection of Dyck word language over $\{a, b\}$ with the regular language a^+b^+. By applying the Horizontal Iteration Lemma of [7] (Lemma 9.1) to $T^{\ominus+}$, there exists a (suitably large) picture t in $T^{\ominus+}$ which can be written as the horizontal concatenation of the three (non empty) pictures x, q, y, i.e., $t = x \oplus q \oplus y$, such that $x \oplus q^{i\Phi} \oplus y$ is also in $T^{\ominus+}$, thus contradicting the fact that the top row of the pictures in $T^{\ominus+}$ must have the form $a^n b^n$. □

[2] More general definitions of Dyck crosswords are possible if the component languages have different alphabets.

A question, related to Remark 1, to be positively answered, is whether the row and column languages of DC_k, respectively, saturate the row and column components D_k^{Row}, D_k^{Col}. Let $P \subseteq \Delta^{++}$ be a language over an alphabet Δ; the *row language* of P is: $\text{ROW}(P) = \{w \in \Delta^+ \mid \text{ there exist } p \in P, p', p'' \in \Delta^{++} \text{ such that } p = p' \ominus w \ominus p''\}$. The column language of P, $\text{COL}(P)$, is analogously defined.

Theorem 3 (Saturation of components). *$ROW(DC_k) = D_k^{Row}$, $COL(DC_k) = D_k^{Col}$.*

Proof. It is enough to prove that $D_k^{Row} \subseteq \text{ROW}(DC_k)$, since the other inclusion is obvious and the case for columns is symmetrical. Without loss of generality, we consider only the case $k = 1$. We prove by induction on $n \geq 2$, that for every word $w \in D_1^{Row}$ of length n there exists a picture $p \in DC_1$ of the form $w_1 \ominus w_2 \ominus w \ominus w_3$ for $w_1, w_2, w_3 \in D_1^{Row}$. There are two base cases, the words ab and cd. The word ab is (also) the third row in the DC_1 picture $ab \ominus cd \ominus ab \ominus cd$, while cd is (also) the third row in the DC_1 picture $ab \ominus ab \ominus cd \ominus cd$. The induction step has three cases: a word $w \in D_1^{Row}$ of length $n > 2$ has the form $w'w''$, or the form $aw'b$ or the form $cw'd$, for some $w', w'' \in D_1^{Row}$ of length less than n. Let p', p'' be the pictures verifying the induction hypothesis for w' and w'', respectively. The case of concatenation $w'w''$ is obvious (just consider the picture $p' \oplus p''$). The case $aw'b$ can be solved by considering the picture $(a \ominus c \ominus a \ominus c) \oplus p' \oplus (b \ominus d \ominus b \ominus d)$, which is in DC_1. Similarly, for the case $cw'd$ just consider the DC_1 picture $(a \ominus a \ominus c \ominus c) \oplus p' \oplus (b \ominus b \ominus d \ominus d)$. □

3.1 Matching-Graph Circuits

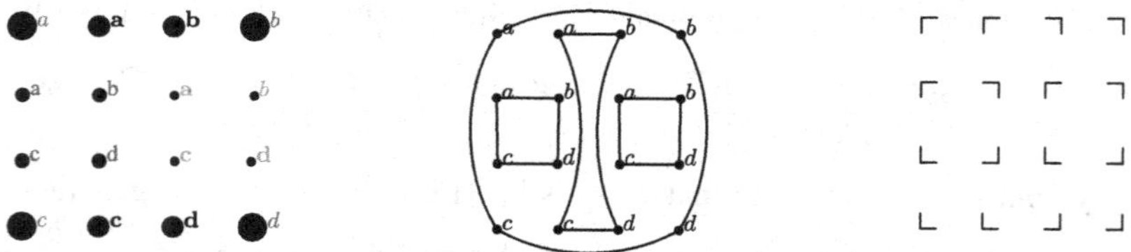

Fig. 1. (Left) A DC_1 picture whose cells are partitioned into 4 quadruples of matching symbols, identified by the same node size (color). (Middle) An alternative visualization by a graph using edges that connect matching symbols (see Definition 3). (Right) The use of corner symbols instead of letters highlights the row and column couplings of rectangle vertexes. (Color figure online)

Indeed, some interesting and surprising patterns may occur in DC_k pictures. The simplest patterns are found in pictures that are partitioned into rectangular circuits connecting four elements, see, e.g., Fig. 1, middle, where an edge connects two symbols on the same row (or column) which match in the row (column) Dyck word. Notice that the graph made by the edges contains four disjoint circuits of length four, called *rectangles* for brevity. Three of the circuits are nested inside the outermost one.

We formally define the graph, situated on the picture grid, made by such circuits.

Definition 3 (Matching graph). *The* matching graph *associated with a picture* $p \in DC_k$*, of size* (m, n)*, is a pair* (V, E) *where the set* V *of nodes is the set* $\{1, \ldots n\} \times \{1 \ldots m\}$ *with the obvious labeling over* D_k*, and the set* E *of edges is partitioned in two sets of* row *and* column *edges defined as follows, for all* $1 \leq i \leq n, 1 \leq j \leq m$*:*

- *for all pairs of matching letters* $p_{i,j}, p_{i,j'}$ *in* Δ_k^{Row}*, with* $j < j' \leq m$*, there is a row (horizontal) edge connecting* (i, j) *and* (i, j')*,*
- *for all pairs of matching letters* $p_{i,j}, p_{i',j}$ *in* Δ_k^{Col}*, with* $i < i' \leq n$*, there is a column (vertical) edge connecting* (i, j) *and* (i', j)*.*

Therefore, there is a horizontal edge connecting two matching letters a_i, b_i or c_i, d_i that occur in the same row; analogously, there is a vertical edge connecting two matching letters a_i, c_i or b_i, d_i, that occur in the same column.

Theorem 4 (Matching circuits). *Let* p *be a picture in* DC_k*. Then:*

1. *its matching graph is partitioned into simple circuits, called* matching circuits*;*
2. *for all* $1 \leq j \leq k$*, the clockwise visit of a matching circuit, starting from any of its nodes with label* a_j*, yields a word in* $(a_j b_j d_j c_j)^+$*, called the* circuit label*.*

Proof. Part (1): By Definition 3, every node of G has degree 2, with one row edge and one column edge, since its corresponding row and column in picture p are Dyck words. Every node must be on a circuit, otherwise there would be a node of degree 1. Each circuit must be simple and the sets of nodes on two circuits are disjoint, else one of the nodes would have degree greater than 2. Part (2) is obvious, since from a node labeled a_j there is a row edge connecting with a node labeled b_j, for which there is a column edge connecting with a d_j, then a row edge connecting d_j with c_j, etc., finally closing the circuit with a column edge connecting a c_j with the original a_j. □

Notice that when a picture on Δ_1 is represented by its matching graph, the node labels are redundant since they are uniquely determined on each circuit.

Theorem 4 has a simple interpretation in the case of Dyck words: the associated matching graph of a Dyck word is the well-known, so-called rainbow representation, e.g.,

$$a \; a \; b \; a \; b \; b$$

of the syntax tree of the word. A matching graph then corresponds to the binary relation induced by the rainbow arcs and a matching circuit just to an arc.

Remark 2. The following elementary property of Dyck words immediately generalizes to crosswords. Let $x \; a_i \; y \; b_i \; w$ be a Dyck word, where a_i, b_i match; then, for any coupled pair a_j, b_j, $1 \leq j \leq k$, the string $x \; a_j \; y \; b_j \; w$ is a Dyck word. For crosswords, the statement is that, by replacing a matching circuit labeled $a_i \; b_i \; d_i \; c_i$ in a picture in DC_k with a matching circuit labeled $a_j \; b_j \; d_j \; c_j$, the result is still in DC_k.

A natural question is whether there are pictures with more complex matching circuits than rectangular ones. It is maybe unexpected that moving from 1D to 2D the circuit length is not just 2×2, but may increase without an upper bound. Two examples of pictures in DC_1 with matching circuits longer than four are in Fig. 2: (left), with a circuit of length 12 labeled by the word $(abdc)^3$, and (right) with a circuit of length 36.

The pictures of DC_k, like the ones in Figs. 1 and 4, that are devoid of circuits longer than four make a proper subset that we define for later convenience.

Definition 4 (Quaternate DC_k). *A Dyck crossword picture such that all its matching circuits are of length 4 is called* quaternate*; the corresponding language, denoted by DQ_k, is the* quaternate Dyck language.

Corollary 1. *Quaternate Dyck languages DQ_k are strictly included in Dyck crosswords DC_k for all $k \geq 1$.*

The structure of quaternate pictures having only rectangular circuits is made more evident by an alternative typography for the Dyck alphabet, using so-called *corner symbols* instead of Latin letters. Let Δ_1 be the alphabet {┌, ┐, └, ┘} with the correspondence: a = ┌, b = ┐, c = └, d = ┘. Thus, the picture ┌ ┐┌ ┐ / └ ┘└ ┘ is the same as $\begin{smallmatrix} a & b & a & b \\ c & d & c & d \end{smallmatrix}$. Another example is in Fig. 1, right.

Section 4 studies the quaternate pictures and defines a sublanguage where the containment relation of rectangles defines a partial order.

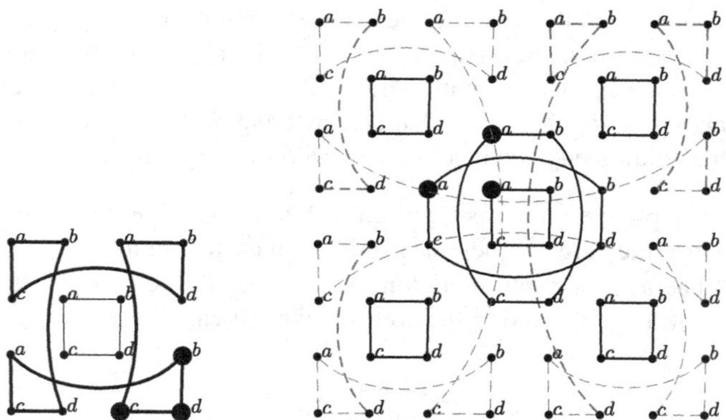

Fig. 2. Two pictures in DC_1. (Left) The picture is partitioned into two circuits of length 12 and 4. (Right) The picture includes a circuit of length 36 and seven rectangular circuits. Its pattern embeds four partial copies (direct or rotated) of the left picture; in, say, the NW copy of the evidenced "triangle", the letters b, d, c have been changed to a, a, a, also evidenced by larger dots. Such a transformation can be reiterated to grow a series of pictures.

We continue with the study of longer circuits.

Theorem 5 (Unbounded circuit length). *For all $h \geq 0$ there exist a picture $p_{(h)}$ in DC_k that contains a matching circuit of length $4 + 8h$.*

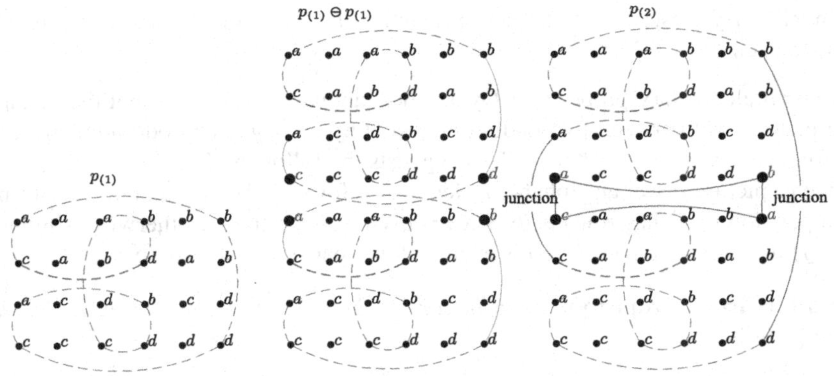

Fig. 3. (Left) Picture $p_{(1)}$ used as induction basis of Theorem 5. It is covered by a circuit of length $4 + 8 \cdot 1 = 12$ and by 3 rectangles (not shown). (Middle) Picture $p_{(1)} \ominus p_{(1)}$, the four arcs to be deleted are in green (solid lines), and the four nodes to be relabeled are in blue (also larger dots). (Right) Inductive step: picture $p_{(2)}$ is obtained from $p_{(1)} \ominus p_{(1)}$ by canceling the four green arcs, relabeling the four larger blue nodes as shown (the corresponding rectangle is in blue) and finally adding two solid arcs (blue) that join the double-noose circuits. A circuit of length $4+8 \cdot 2$ results. All length 4 circuits of $p_{(h-1)}$ and $p_{(1)}$ (not shown for clarity) are unchanged in $p_{(h)}$. (Color figure online)

Proof. We prove the statement for DC_1, since $DC_1 \subseteq DC_k$. The case $h = 0$ is obvious. The case $h > 0$ is proved by induction on a sequence of pictures $p_{(1)}, \dots p_{(h)}$ using as basis the DC_1 picture $p_{(1)}$ in Fig. 3 (left), that has size $(m_{(1)}, 6)$, where $m_{(1)} = 4$, and contains a circuit of length $12 = 4 + 8$, referred to as double-noose.

Induction step. It extends picture $p_{(h-1)}$, $h > 1$, by appending a copy of $p_{(1)}$ underneath and making a few changes essentially defined in Fig. 3 (right). It is easy to see that the result is a picture $p_{(h)}$ of size $(m_{(h-1)} + 4, 6)$ such that: $p_{(h)} \in DC_1$ and $p_{(h)}$ contains a circuit of length $4 + 8h$. □

Another series of pictures that can be enlarged indefinitely is the one in Fig. 2, where the first two terms of the series are shown.

An examination of Fig. 3 in the next example shows that there are subsets of DC_k that are in REC, yet they contain quite complex matching circuits.

Example 1. The language L composed of all pictures $p_{(h)}$, for all $h \geq 1$, of Theorem 5 is in the REC family. We first extend the alphabet of L to $\{a, b, c, d, a_1, b_1, c_1, d_1\}$ so that the circuits longer than 4 are over the alphabet $\{a_1, b_1, c_1, d_1\}$ and the remaining circuits are over $\{a, b, c, d\}$. The resulting pictures $p'_{(h)}$, constituting a language L', have only 6 distinct rows, here identified (from top to bottom) with the letters $1, \dots, 6$:

$$1\text{: } a_1aa_1b_1bb_1,\ 2\text{: } c_1abd_1ab,\ 3\text{: } a_1cdb_1cd,\ 4\text{: } acc_1d_1db,\ 5\text{: } caa_1b_1bd,\ 6\text{: } c_1cc_1d_1dd_1\,.$$

It is clear from the construction of the pictures p_h for $h > 1$ that L' can be defined as $1 \ominus (2 \ominus 3 \ominus 4 \ominus 5)^{*\ominus} 2 \ominus 3 \ominus 6$. Since each of rows $1, \dots, 6$ can be seen as a finite language (thus, in REC) and tiling systems are closed by vertical concatenation and closure, also

L' is in REC. By closure of REC under projection, also L is in REC (by projecting a_1 to a, b_1 to b, etc.)

From an elementary property of Dyck word languages it follows that the distance on the picture grid between two nodes connected by an edge is an odd number, to let room for an even number of letters. This suggests the following Lemma 1.

Given a picture p over an alphabet Γ, let $x{=}p_{i,j}$, for $x \in \Gamma$. We say that the occurrence of x in position (i, j) has *row parity* 1 if i is even, row parity -1 otherwise; similarly, x in (i, j) has *column parity 1* if j is even, column parity -1 otherwise.

Lemma 1 (Circuit property). *Let γ be a matching circuit of a picture in DC_k, with label in $(a_i b_i d_i c_i)^+$.*

i) All occurrences of a_i and b_i have the same row parity, but they have opposite row parity to every occurrence of c_i and d_i;
ii) All occurrences of a_i and c_i have the same column parity, but they have opposite column parity to every occurrence of b_i and d_i.

Proof. Without loss of generality, let $k = 1$. Let an occurrence of a in γ be in a position row (r, s). The vertical matching symbol c of a (in the same column s) must occur in a row of the form $2n + 1 + r$, for some $n \geq 0$, since there must be an even number of positions in p between the occurrence of a and c. The same happens for the symbol d matching c and for the b matching the above occurrence of a. The circuit γ continue alternating between odd and even rows, and between odd and even columns, without modifying the row and column parity of each occurrence of the same letter. □

An application of Lemma 1 follows.

Let p be a picture in DC_k and G its matching graph. A matching circuit that visits all the nodes of G is called *Hamiltonian.*

Theorem 6 (Hamiltonian circuits). *The only existing DC_k pictures with a Hamiltonian matching circuit are defined by the set* $\left\{ \begin{matrix} a_i & b_i \\ c_i & d_i \end{matrix} \mid 1 \leq i \leq k \right\}$.

Proof. Without loss of generality, let $k = 1$. By contradiction, assume that a picture $p \in DC_1$, of size greater than $(2, 2)$, has a Hamiltonian circuit. The first row of any picture is a Dyck word over $\{a, b\}$ and the leftmost column is a (vertical) word over $\{a, c\}$. By Lemma 1, the second row must be a word over $\{c, d\}$ and the second column from the left is over $\{b, d\}$. Therefore, the subpicture $(p(1, 1) \ominus p(2, 1))$ must be $a \ominus c$ (a cannot occur in the second row, therefore the row must begin with the open letter c for rows) and similarly the subpicture $(p(1, 2) \ominus p(2, 2))$ is $b \ominus d$.

Therefore p contains the subpicture $\begin{matrix} a & b \\ c & d \end{matrix}$ in the top, left corner, i.e., it has a matching circuit of length 4, a contradiction with the existence of a Hamiltonian circuit for a picture of size greater than $(2, 2)$. □

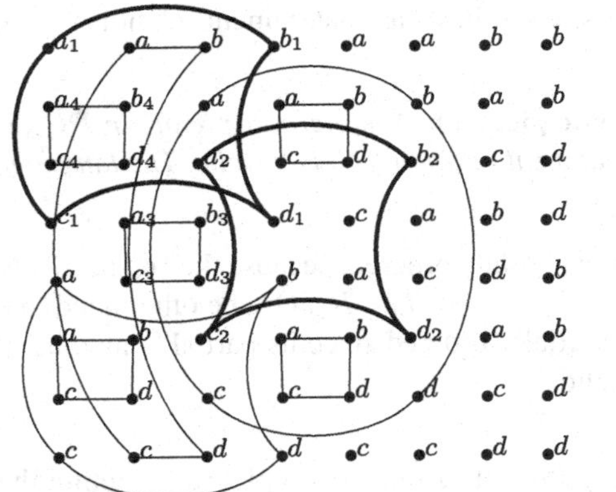

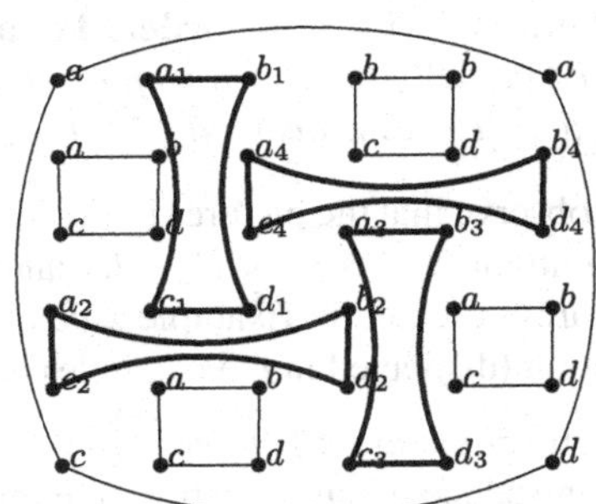

Fig. 4. Both pictures are quaternate but not partially ordered, hence not neutralizable by Theorem 7. (Left) To avoid clogging, the rectangles in the specular right half of the left picture are not drawn. (Right) The vertexes relevant for illustrating partial containment are indexed.

4 A Sublanguage Preserving Characteristic Dyck Words Properties

This section only deals with quaternate pictures, whose circuits we call "rectangles". We show that the standard definition of Dyck words by means of the cancellation rule[3] can be extended to a sublanguage of quaternate pictures that is characterized by a geometrical relation of containment between the rectangles.

The absence of long and intricate circuits will permit to define a partial containment relation between the rectangles present in a picture, and then to define a partial order if such a relation is acyclic. The corresponding language is called *partially ordered quaternate*, DPO_k. We also define a subset of Dyck crosswords, named *neutralizable* (DN_k), by means of a cancellation rule suitably transformed into a neutralization operation. At last we prove that the partially ordered and the neutralizable languages are the same, and we list some of their properties.

Preliminarily, we transform the cancellation rule for words $a_i b_i \to \varepsilon$, which erases innermost matching letters, into a length preserving rule, since in 2D the erasure of an internal subpicture would create a "hole", producing an object that no longer qualifies as a picture. The Dyck cancellation rule is rephrased as the *neutralization rule* $a_i b_i \to NN$, where N is a new "neutral" (i.e., not coupled) letter; in this way a Dyck word is mapped to a word in N^+ by a series of neutralization steps.

Geometrical Containment Relation. Consider two rectangles R_1 and R_2 with vertexes, resp., a_1, b_1, c_1, d_1 and a_2, b_2, c_2, d_2 (the letters are distinct to simplify reference).

We say that R_1 is *partially contained* in R_2, writing $R_1 < R_2$, if some vertexes of R_1 are inside or on a side of R_2. The partial containment relation of a picture is the set of all such relations. Notice that the number of vertexes contained in R_2 may be 1, 2 or 4, but not 3 which is geometrically impossible.

[3] In [4] the property of well nesting of parentheses is also reformulated for quaternate pictures.

Figure 4, left, illustrates (among others) the following containment relations: $R_1 < R_2, R_2 < R_1, R_3 < R_1, R_3 < R_2, R_4 < R_1$.

Definition 5 (Partially ordered quaternate picture). *A quaternate picture in DQ_k is called* partially ordered *if its partial containment relation "<" is acyclic. The language of such pictures is denoted by DPQ_k.*

We observe that the pictures in Fig. 4 are not partially ordered, because they respectively contain the cycles $R_1 < R_2 < R_1$ and $R_1 < R_2 < R_3 < R_4 < R_1$. On the other hand, the picture presented in Example 2 below is partially ordered since its partial containment relation (displayed in the example) is acyclic.

Neutralizable Dyck Languages
We introduce a neutralization rule mapping the letters in a quadruple, representing the corners of a subpicture, to a new neutral letter N. The *neutralizable Dyck* language DN_k is obtained by iterating neutralization, starting from 2-by-2 subpictures, until the picture is wholly neutralized. Given a picture p, all subpictures of the form $\begin{smallmatrix}\ulcorner & \urcorner\\ \llcorner & \lrcorner\end{smallmatrix}$ are neutralized, i.e., replaced in p by the subpicture $\begin{smallmatrix}N & N\\ N & N\end{smallmatrix}$. If p includes a subpicture with four matching corners and having its interior and sides completely neutralized, then also the four corners are neutralized. This is schematized by the subpicture $\begin{smallmatrix}\ulcorner & N \dots N & \urcorner\\ N & \dots & N\\ \vdots & N \dots N & \vdots\\ N & \dots & N\\ \llcorner & N \dots N & \lrcorner\end{smallmatrix}$ that is replaced by a subpicture of the same size having only N as letters. The procedure successfully terminates when the resulting picture is in N^{++}.

Definition 6 (Neutralizable Dyck language). *Let N be a new symbol not in Δ_k. The* neutralization relation $\xrightarrow{\nu} \subseteq (\{N\} \cup \Delta_k)^{++} \times (\{N\} \cup \Delta_k)^{++}$, *is the smallest relation such that for every pair of pictures p, p' in $(\{N\} \cup \Delta_k)^{++}$, $p \xrightarrow{\nu} p'$ if there are $m, n \geq 2$ and $1 \leq i \leq k$, such that p' is obtained from p by replacing a subpicture of p of the form:*

$$(a_i \ominus N^{m-2,1} \ominus c_i) \oplus N^{m,n-2} \oplus (b_i \ominus N^{m-2,1} \ominus d_i). \tag{2}$$

with the picture (of the same size) $N^{m,n}$.

The neutralizable Dyck language, *denoted by $DN_k \subseteq \Delta_k^{++}$, is the set of pictures p such that there exists $p' \in N^{++}$ with $p \xrightarrow{\nu}^{+} p'$.*

To sum up, a DN_k picture is transformed into a picture in N^{++} by a series of neutralizations, applied in any order. Clearly, every neutralizable picture is a quaternate.

Example 2 (Neutralizations). The following picture p on the alphabet Δ_1 is in DN_1 since it reduces to the neutral one by means of a sequence of six neutralization steps:

$$p = \begin{matrix}\ulcorner & \ulcorner & \urcorner & \ulcorner & \urcorner & \urcorner\\ \ulcorner & \ulcorner & \urcorner & \llcorner & \lrcorner & \urcorner\\ \llcorner & \llcorner & \lrcorner & \ulcorner & \urcorner & \lrcorner\\ \llcorner & \llcorner & \lrcorner & \llcorner & \lrcorner & \lrcorner\end{matrix} \xrightarrow{\nu} \begin{matrix}\ulcorner & \ulcorner & \urcorner & \ulcorner & \urcorner & \urcorner\\ \ulcorner & N & N & \llcorner & \lrcorner & \urcorner\\ \llcorner & N & N & \ulcorner & \urcorner & \lrcorner\\ \llcorner & \llcorner & \lrcorner & \llcorner & \lrcorner & \lrcorner\end{matrix} \xrightarrow{\nu} \begin{matrix}\ulcorner & N & N & \ulcorner & \urcorner & \urcorner\\ \ulcorner & N & N & \llcorner & \lrcorner & \urcorner\\ \llcorner & N & N & \ulcorner & \urcorner & \lrcorner\\ \llcorner & N & N & \llcorner & \lrcorner & \lrcorner\end{matrix}$$

┌ N N N N ┐		┌ N N N N ┐		┌ N N N N ┐		N N N N N N
┌ N N N N ┐	$\overset{\nu}{\rightarrow}$	┌ N N N N ┐	$\overset{\nu}{\rightarrow}$	N N N N N N	$\overset{\nu}{\rightarrow}$	N N N N N N
└ N N ┌ ┐ ┘		└ N N N N ┘		N N N N N N		N N N N N N
└ N N └ ┘ ┘		└ N N N N ┘		└ N N N N ┘		N N N N N N

$\overset{\nu}{\rightarrow}$ (applied before the first picture)

Neutralizations have been applied in a left to right order.

We show the partial containment relation "<", with the numbering below.

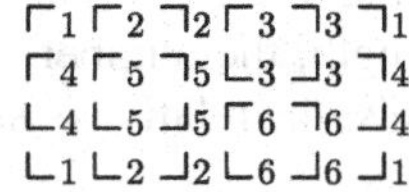

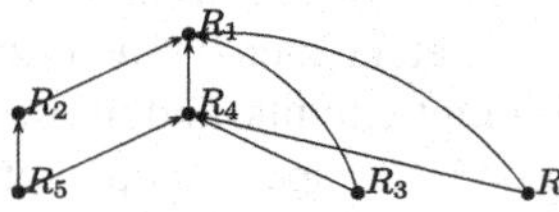

The relation represented by the graph is acyclic and defines a partial order on the set of rectangles, thus proving that this picture is in DPO_1.

It is no coincidence that the picture of Example 2 is both neutralizable and partially ordered; the next theorem proves that the two definitions define the same set of pictures.

Theorem 7 (Partially ordered equals neutralizable). *A quaternate picture is neutralizable if, and only if, it is partially ordered, i.e.,* $DN_k = DPO_k$.

Proof. Let relation < be acyclic. Then sort the rectangles in topological order and apply neutralization starting from a rectangle without predecessors. When a rectangle is checked, all of its predecessors have already been neutralized, and neutralization can proceed until all rectangles are neutralized. The converse is obvious: if relation < has a cycle, no rectangle in the cycle can be neutralized. □

This result supports the analogy between the neutralization rule for Dyck words and the rule of the same name for pictures: both rely on a partial order relation such that any topological sorting order can be applied to perform neutralization. For Dyck words, the order is a tree partial order, whereas for pictures it is a directed acyclic graph.

Properties of Neutralizable Picture Languages. The result on row/column language saturation (Theorem 3) remains valid, i.e., $\text{ROW}(DN_k) = D_k^{Row}$, $\text{COL}(DN_k) = D_k^{Col}$, since the languages used in the proof of that theorem are also in DN_k.

Similarly, by a proof almost identical to the one of Theorem 2, since the language $T^{\ominus+}$ can be obtained from DN_k by intersection with a recognizable language, we have:

Theorem 8 (Comparison with REC). *The languages* DN_k *and* DQ_k *are not in REC for every* $k \geq 1$.

From Theorem 7 and from the examples of Fig. 4, we have the inclusions:

Theorem 9 (Hierarchy). $DN_k \subsetneq DQ_k \subsetneq DC_k$.

5 Conclusion

In our opinion, the mathematical study of the properties of 2D Dyck languages is a promising research area, where much remains to be understood, for the general case of (simple) Dyck crosswords containing matching circuits of any length. Very diverse patterns may occur in such crosswords, that we have been able to start classifying just in the simpler case of rectangular circuits. In fact, the variety of patterns depends on quite a few circuit parameters such as the circuit length, the number of crossings in a circuit or between different circuits, and, more vaguely, the relative positions of circuits on the grid. We mention a few specific open problems.

First, by Theorem 4 the length of circuits in DC_1 pictures is unbounded, of the form 4 + $8h$ for all values $h \geq 0$. The question is whether, for each $n \geq 1$, there is a DC_1 picture containing a circuit of length $4n$.

Second, it seems that every picture in DC_k has at least one circuit of length 4.

Third, the number of circuits present in a picture is unbounded for the picture series used in the proof of Theorem 4. This raises the more general question whether, by bounding the number of circuits present in a picture, the number of such pictures is also bounded. (Theorem 6 bounds the number of pictures with only one circuit.)

Another question concerns the properties of those DC_k sublanguages that are in REC. For instance, Example 1, though visually complex, satisfies such a hypothesis.

At last, we mention a related future research direction on context-free crosswords, having as baseline the present work on Dyck crosswords and the variant of the Chomsky-Schützenberger Theorem [1] that characterizes the context-free word languages as the non-erasing homomorphism of the intersection of a Dyck language and a regular one.

References

1. Chomsky, N., Schützenberger, M.: The algebraic theory of context-free languages. In: Brafford, H. (ed.) Computer Programming and Formal Systems, pp. 118–161. North-Holland, Amsterdam (1963)
2. Crespi Reghizzi, S., Giammarresi, D., Lonati, V.: Two-dimensional models. In: Pin, J. (ed.) Handbook of Automata Theory, pp. 303–333. European Mathematical Society Publishing House (2021)
3. Crespi-Reghizzi, S., Pradella, M.: Tile rewriting grammars and picture languages. Theor. Comput. Sci. **340**(1), 257–272 (2005). https://doi.org/10.1016/j.tcs.2005.03.041
4. Crespi Reghizzi, S., Restivo, A., San Pietro, P.: Two-dimensional Dyck words. CoRR abs/2307.16522 (2023)
5. Drewes, F.: Grammatical Picture Generation: A Tree-Based Approach. Springer, Heidelberg (2006). https://doi.org/10.1007/3-540-32507-7
6. Fenner, S.A., Padé, D., Thierauf, T.: The complexity of regex crosswords. Inf. Comput. **286**, 104777 (2022). https://doi.org/10.1016/j.ic.2021.104777
7. Giammarresi, D., Restivo, A.: Two-dimensional languages. In: Rozenberg, G., Salomaa, A. (eds.) Handbook of Formal Languages, vol. 3, pp. 215–267. Springer, Heidelberg (1997). https://doi.org/10.1007/978-3-642-59126-6_4
8. Latteux, M., Simplot, D.: Recognizable picture languages and domino tiling. Theor. Comput. Sci. **178**(1–2), 275–283 (1997). https://doi.org/10.1016/S0304-3975(96)00283-6

9. Matz, O.: Regular expressions and context-free grammars for picture languages. In: Reischuk, R., Morvan, M. (eds.) STACS 1997. LNCS, vol. 1200, pp. 283–294. Springer, Heidelberg (1997). https://doi.org/10.1007/BFb0023466
10. Nivat, M., Saoudi, A., Subramanian, K.G., Siromoney, R., Dare, V.R.: Puzzle grammars and context-free array grammars. Int. J. Pattern Recogn. Artif. Intell. **5**, 663–676 (1991)
11. Průša, D.: Two-dimensional Languages. Ph.D. thesis, Charles University, Faculty of Mathematics and Physics, Czech Republic (2004)
12. Siromoney, R., Subramanian, K.G., Dare, V.R., Thomas, D.G.: Some results on picture languages. Pattern Recogn. **32**(2), 295–304 (1999)

Group Testing in Arbitrary Hypergraphs and Related Combinatorial Structures

Annalisa De Bonis(✉)

Dipartimento di Informatica, Università di Salerno, Fisciano, SA, Italy
adebonis@unisa.it

Abstract. We consider a generalization of *group testing* where the potentially contaminated sets are the members of a given hypergraph $\mathcal{F} = (V, E)$. This generalization finds application in contexts where contaminations can be conditioned by some kinds of social and geographical clusterings. We study non-adaptive algorithms, two-stage algorithms, and three-stage algorithms. Non-adaptive group testing algorithms are algorithms in which all tests are decided beforehand and therefore can be performed in parallel, whereas two-stage and three-stage group testing algorithms consist of two stages and three stages, respectively, with each stage being a completely non-adaptive algorithm. In *classical* group testing, the potentially infected sets are all subsets of up to a certain number of elements of the given input set. For classical group testing, it is known that there exists a correspondence between non-adaptive algorithms and superimposed codes, and between two-stage group testing and disjunctive list-decoding codes and selectors. Bounds on the number of tests for those algorithms are derived from the bounds on the dimensions of the corresponding combinatorial structures. Obviously, the upper bounds for the classical case apply also to our group testing model. In the present paper, we aim at improving on those upper bounds by leveraging on the characteristics of the particular hypergraph at hand. In order to cope with our version of the problem, we introduce new combinatorial structures that generalize the notions of *classical* selectors and superimposed codes.

1 Introduction

Group testing consists in detecting the defective members of a set of objects O by performing tests on properly chosen subsets (*pools*) of the given set O. A test yields a "yes" response if the tested pool contains one or more defective elements, and a "no" response otherwise. The goal is to find all defectives by using as few tests as possible. Group testing origins date back to World War II when it was introduced as a possible technique for mass blood testing [4]. Since then, group testing has become a fundamental problem in computer science and has been widely investigated in the literature both for its practical implications and for the many theoretical challenges it poses. Applications of group testing span a wide variety of situations ranging from conflict resolution algorithms for

H. Fernau et al. (Eds.): SOFSEM 2024, LNCS 14519, pp. 154–168, 2024.
https://doi.org/10.1007/978-3-031-52113-3_11

multiple-access systems [19], fault diagnosis in optical networks [9], quality control in product testing [17], failure detection in wireless sensor networks [13], data compression [10], and many others. Among the modern applications of group testing, some of the most important ones are related to the field of molecular biology, where group testing is especially employed in the design of screening experiments. Du and Hwang [5] provide an extensive coverage of the most relevant applications of group testing in this area. In classical group testing, the set of defectives is any of the possible subsets of size less than or equal to a certain parameter d. In the present paper, we consider a more general version of group testing parameterized by a hypergraph $\mathcal{F} = (V, E)$, with the contaminated set being one of the hyperedges of E.

Related Work and Our Contribution. Classical group testing has been studied very thoroughly both with respect to *adaptive strategies*, i.e. algorithms that at each step decide which group to test by looking at the responses of previous tests, and with respect to *non-adaptive* strategies, i.e., strategies in which all tests are decided beforehand. It is well-know that the best adaptive strategies achieve the information theoretic lower bound $\Omega(d \log(n/d))$, where n is the total number of elements and d is the upper bound on the number of defectives. Further, it is known that non-adaptive strategies for classical group testing are much more costly than their adaptive counterparts. The minimum number of tests used by the non-adaptive procedures is estimated by the minimum length of certain combinatorial structures known under the name of *d-cover free families*, or equivalently, *d-superimposed codes* and *strongly selective families* [1], [6,7,12]. The known bounds for these combinatorial structures [6,16] imply that the number of tests of any non-adaptive group testing algorithm is lower bounded by $\Omega((d^2/\log d)\log n)$ and that there exist non-adaptive group testing algorithms that use $O(d^2 \log n)$ tests. Of particular practical interest in the field of biological screening are two-stage group testing procedures, i.e., procedures consisting of two stages each of which is a completely non-adaptive algorithm. As far as it concerns classical group testing, the best two-stage algorithms [3] are asymptotically as efficient as the best completely adaptive strategy in that they achieve the information theoretic lower bound $\Omega(d \log(n/d))$.

In this paper, we focus on non-adaptive group testing and two-stage group testing. In particular, we are interested in *trivial* two-stage group testing algorithms, i.e., two-stage algorithms that in the second stage are allowed only to perform tests on individual elements. These algorithms are very useful in practice since in many applications confirmatory tests on individual elements are needed anyway to ensure that they are really defective. We also give existential results for three-stage group testing algorithms, i.e., algorithms that consist of three stages, each of which is a non-adaptive algorithm.

The version of group testing considered in the present paper has been initiated only recently in [8] and continued in [18]. Similar search models were previously considered by the authors of [15] who assumed a known community structure in virtue of which the population is partitioned into separate families and the defective hyperedges are those that contain elements from a certain number of families. While in that paper the information on the structure of potentially infected groups is used to improve on the efficiency of the group testing algorithms, other papers, [11,20], exploit this knowledge to improve on the efficiency of decoding the tests' responses. A formal definition of group testing in general hypergraphs has been given in [8]. The authors of [8] consider both adaptive and non-adaptive group testing. For the adaptive setting, when hyperedges in E are of size exactly d, they give an $O(\log|E| + d\log^2 d)$ algorithm that is close to the $\Omega(\log|E| + d)$ lower bound they also prove in the paper. In the non-adaptive setting, they exploit a random coding technique to prove an $O(\frac{d}{p}\log|E|)$ bound on the number of tests, where d is the maximum size of a set $e \in E$ and p is a lower bound on the size of the difference $e' \setminus e$ between any two hyperedges $e, e' \in E$. In [18] the author presents a new adaptive algorithm for generalized group testing, which is asymptotically optimal if $d = o(\log_2|E|)$ and, for $d = 2$, gives an asymptotically optimal algorithm that works in three stages.

In this paper, we formally define combinatorial structures that are substantially equivalent to non-adaptive algorithms for group testing in general hypergraphs. The combinatorial structures introduced in this paper extend the above-mentioned classical superimposed codes [12] in a way such that the desired property is not enforced for all subsets of up to a certain number of codewords but only for subsets of codewords associated to the hyperedges of the given hypergraph. Constructions for these generalized superimposed codes allow to achieve, in the non-adaptive setting, the same $O(\frac{d}{p}\log|E|)$ upper bound obtained in [8]. In order to design our algorithms, we introduce a notion of selectors, also parameterized by the set of hyperedges E, that generalizes the notion of (k, m, n)-selector introduced in [3]. For particular values of the involved parameters, our selectors correspond to non-adaptive algorithms for our group testing problem, These selectors are also at the basis of our two-stage and three-stage algorithms. In particular, we give a trivial two-stage algorithm that uses $O(\frac{qd}{\chi}\log|E| + dq)$ tests, where χ is a lower bound on $|\bigcup_{i=1}^{q} e'_i \setminus e|$, for any $q+1$ distinct hyperedges $e, e'_1 \ldots, e'_q$. If for a given constant $q \geq 1$, it holds that $|\bigcup_{i=1}^{q} e'_i \setminus e| = \Omega(d)$, then this algorithm asymptotically achieves the above-mentioned $\Omega(\log|E| + d)$ lower bound. Further, we give an $O(\sqrt{d}\log|E| + d)$ three-stage group testing algorithm.

2 Notations and Terminology

For any positive integer m, we denote by $[m]$ the set of integers $\{1, \ldots, m\}$. A hypergraph is a pair $\mathcal{F} = (V, E)$, where V is a finite set and E is a family of subsets of V. The elements of E will be called hyperedges. If all hyperedges of E have the same size d then the hypergraph is said to be d-uniform.

In the present paper, the hypergraph specifying the set of potentially contaminated sets is assumed to have $V = [n]$. Let $\mathcal{F} = ([n], E)$ be a given hypergraph, and let $\chi = \min\{|\bigcup_{i=1}^{q} e'_i \setminus e|$, for any $q+1$ distinct $e, e'_1, \ldots, e'_q \in E\}$. For any hyperedge $e \in E$ and any q distinct hyperedges $e'_1, \ldots, e'_q \in E \setminus \{e\}$, we denote by $S_{e,(e'_1,\ldots,e'_q)}$ the set of $d+\chi$ integers in $[n]$ such that d of these integers are the elements of e whereas the remaining are the χ smallest elements in $\bigcup_{i=1}^q e'_i \setminus e$. In order to properly define the combinatorial structures related to our group testing problem, we extend the definition of $S_{e,(e'_1,\ldots,e'_q)}$ to arbitrary hypergraphs without assuming any lower bound on the size of the unions $\bigcup_{i=1}^q e'_i \setminus e$. To this aim, we augment the set of vertices $[n]$ with χ dummy vertices $i_{n+1}, \ldots i_{n+\chi}$, with χ this time being any positive integer smaller than or equal to $n-d$, and define $S_{e,(e'_1,\ldots,e'_q),\chi}$ as follows: $S_{e,(e'_1,\ldots,e'_q),\chi}$ is defined exactly as $S_{e,(e'_1,\ldots,e'_q)}$ if $|\bigcup_{i=1}^q e'_i \setminus e| \geq \chi$; otherwise it is defined as the subset of $d+\chi$ integers in $[n]$ such that d of these integers are the vertices of e and the remaining are the integers in $(\bigcup_{i=1}^q e'_i \setminus e) \cup \{i_{n+1}, \ldots, i_{n+\chi-|\bigcup_{i=1}^q e'_i \setminus e|}\}$.

For $q=1$, we will denote $S_{e,(e'_1,\ldots,e'_q)}$ and $S_{e,(e'_1,\ldots,e'_q),\chi}$ with $S_{e,e'}$ and $S_{e,e',\chi}$, respectively. Notice that given two hyperedges $e, e' \in E$, the set $S_{e,e',\chi}$ contains one or more dummy vertices if and only if $|e' \setminus e| < \chi$.

Notice that there are at most $|E|\binom{|E|-1}{q}$ distinct sets $S_{e,(e'_1,\ldots,e'_q)}$ since for a given $(q+1)$-tuple of distinct hyperedges $(e, e'_1, \ldots, e'_{q+1})$ any permutation of $e'_1, \ldots, e'_{q+1}$ does not affect $S_{e,(e'_1,\ldots,e'_q)}$. Moreover, for any positive χ, this estimate holds also for the number of distinct sets $S_{e,(e'_1,\ldots,e'_q),\chi}$ since the indices of the dummy vertices that are eventually added to $S_{e,(e'_1,\ldots,e'_q)}$, in order to obtain $S_{e,(e'_1,\ldots,e'_q,\chi)}$, are fixed once the hyperedges $e, e'_1, \ldots, e'_q$ have been chosen. More precisely, they are the smallest $\chi - |\bigcup_{i=1}^q e'_i \setminus e|$ integers in $\{i_{n+1}, \ldots i_{n+\chi}\}$.

We remark that in our group testing problem, given an input hypergraph $\mathcal{F} = ([n], E)$, every vertex of $[n]$ is contained in at least one hyperedge of E. If otherwise, one could remove the vertex from the hypergraph without changing the collection of potentially defective hyperedges. As a consequence, for a given hypergraph $\mathcal{F} = ([n], E)$ we need only to specify its set of hyperedges E to characterize both the input of the problems and the related combinatorial tools.

3 Non-adaptive Group Testing for General Hypergraphs

In this section, we illustrate the correspondence between non-adaptive group testing algorithms for an input set of size n and families of n subsets. Indeed, given a family $\mathbb{F} = \{F_1, \ldots, F_n\}$ with $F_i \subseteq [t]$, we design a non-adaptive group testing strategy as follows. We denote the elements in the input set by the integers in $[n] = \{1, \ldots, n\}$, and for $i = 1, \ldots, t$, we define the group $T_i = \{j : i \in F_j\}$. Obviously, $T_1, \ldots, T_t$ can be tested in parallel and therefore the resulting algorithm is non-adaptive. Conversely, given a non-adaptive group testing strategy for an input set of size n that tests $T_1, \ldots, T_t$, we define a family $\mathbb{F} = \{F_1, \ldots, F_n\}$ by setting $F_j = \{i \in [t] : j \in T_i\}$, for $j = 1, \ldots, n$. Alternatively, any non-adaptive group testing algorithm for an input set of size n that performs t tests can be represented by a binary code of size n, with each codeword being a binary vector of length t. This is due to the fact that any family of size n on the ground set $[t]$ is associated with the binary code of length t whose codewords are the characteristic vectors of the members of the family. Given such a binary code $\mathcal{C} = \{\mathbf{c}_1, \ldots, \mathbf{c}_n\}$, one has that j belongs to pool T_i if and only if the i-th entry $\mathbf{c}_j(i)$ of $\mathbf{c}_j$ is equal to 1. Such a code can be represented by a $t \times n$ binary matrix M such that $M[i, j] = 1$ if and only if element j belongs to T_i. We represent the responses to tests on $T_1, \ldots, T_t$ by a binary vector whose i-th entry is equal to 1 if and only if T_i tests positive. We call this vector the *response vector*. For any input set of hyperedges E on $[n]$, the response vector is the bitwise OR of the columns associated with the vertices of the defective hyperedge $e \in E$. It follows that a non-adaptive group testing strategy successfully detects the defective hyperedge in E if and only if for any two distinct hyperedges $e, e' \in E$ we obtain two different response vectors. In terms of the associated binary matrix M, this means that the bitwise OR of the columns with indices in e and the OR of the columns with indices in e' are distinct. As for the two-stage algorithm, the non-adaptive algorithm used in the first stage should guarantee to "separate" at least a certain number of non-defective hyperedges from the defective one.

4 Combinatorial Structures for Group Testing in Arbitrary Hypergraphs

The following definition provides a combinatorial tool which is essentially equivalent to a non-adaptive group testing algorithm in our search model.

Definition 1. *Given a hypergraph $\mathcal{F} = (V, E)$ with $V = [n]$ with hyperedges of size at most d, we say that a $t \times n$ matrix M with entries in $\{0, 1\}$ is an E-separable code if for any two distinct hyperedges $e, e' \in E$, it holds that $\bigvee_{j \in e} c_j \neq \bigvee_{j \in e'} c_j$, where c_j denotes the column of M with index j. The integer t is the length of the E-separable code.*

Having in mind the correspondence between binary codes and non-adaptive algorithms illustrated in Sect. 3, one can see that a non-adaptive algorithm successfully determines the contaminated hyperedge in E if and only if the binary code associated to the algorithm is E-separable. Therefore, the minimum number of tests of such an algorithm coincides with the minimum length of an E-separable code. Our existential results for E-separable codes are in fact based on existential results for the combinatorial structure defined below.

Definition 2. *Given a d-uniform hypergraph $\mathcal{F} = (V, E)$ with $V = [n]$ and two integers p and m with $1 \le p \le n - d$ and $1 \le m \le d + p$, we say that a $t \times n$ matrix M with entries in $\{0, 1\}$ is an (E, p, m)-selector of size t if for any two distinct hyperedges $e, e' \in E$, the submatrix of M consisting of the columns with indices in $S_{e,e',p}$ contains at least m distinct rows of the identity matrix I_{d+p}.*

We will see in Sect. 5.1 that for $m = d+1$ and $p = \min\{|e' \setminus e| : e, e' \in E \text{ and } e \neq e'\}$, an (E, p, m)-selector is indeed an E-separable code. The following definition generalizes Definition 2 and is at the basis of our two-stage algorithm.

Definition 3. *Given a d-uniform hypergraph* $\mathcal{F} = (V, E)$ *with* $V = [n]$ *and integers* q, m *and* χ *such that* $1 \leq q \leq |E| - 1$ *and* $1 \leq m \leq \chi + d \leq n$, *we say that a* $t \times n$ *matrix* M *with entries in* $\{0, 1\}$ *is an* (E, q, m, χ)*-selector of size* t *if, for any* $e \in E$ *and any other* q *distinct hyperedges* $e'_1, \ldots, e'_q \in E$, *the submatrix of* M *consisting of the columns with indices in* $S_{e,(e'_1,\ldots,e'_q),\chi}$ *contains at least* m *distinct rows of the identity matrix* $I_{d+\chi}$.

For $q = 1$ and $\chi = p$, an (E, q, m, χ)-selector is indeed an (E, p, m)-selector.

We remark that even if the combinatorial objects in the present section are defined for uniform hypergraphs, the algorithms we build upon them work also for non-uniform hypergraphs.

5 Upper Bound on the Size of (E, q, m, χ)-Selectors

Let E be a set of hyperedges, each consisting of d vertices in $[n]$ and, for a given q, $1 \leq q \leq |E| - 1$, let χ and m be positive integers with $\chi \leq n - d$ and $m \leq d + \chi$. We will show how to define a hypergraph $\mathcal{H} = (X, \mathcal{B})$, with X being a set of vectors of length n, in such a way that the vectors of any *cover* of $\mathcal{H} = (X, \mathcal{B})$ are the rows of an (E, q, m, χ)-selector. We recall that, given a hypergraph $\mathcal{H} = (X, \mathcal{B})$, a cover of $\mathcal{H}$ is a subset $T \subseteq X$ such that for any hyperedge $B \in \mathcal{B}$ we have $T \cap B \neq \emptyset$. In order to avoid confusion with the hyperedges of the input hypergraph E, here we denote the hyperedges of the above said hypergraph $\mathcal{H} = (X, \mathcal{B})$ by using the letter B. The following upper bound on the minimum size $\tau(\mathcal{H})$ of a cover of a hypergraph $\mathcal{H} = (X, \mathcal{B})$ is due to Lovász [14]:

$$\tau(\mathcal{H}) < \frac{|X|}{\min_{B \in \mathcal{B}} |B|}(1 + \ln \Delta), \tag{1}$$

where $\Delta = \max_{x \in X} |\{B\colon B \in \mathcal{B} \text{ and } x \in B\}|$.

In the following, for $e, e'_1, \ldots, e'_q \in E$, we write $(i_1, \ldots, i_{d+\chi}) = S_{e,(e'_1,\ldots,e'_q),\chi}$ to refer to the tuple of the elements in $S_{e,(e'_1,\ldots,e'_q),\chi}$ arranged in increasing order. We will show that (E, q, m, χ)-selectors are covers of the hypergraph defined below. Let X be the set of all binary vectors $\mathbf{x} = (x_1, \ldots, x_n)$ of length n containing $\lfloor \frac{n}{d+\chi} \rfloor$ entries equal to 1. For any integer i, $1 \leq i \leq d + \chi$, let us denote by $\mathbf{a}_i$ the binary vector of length $d + \chi$ having all components equal to zero but that in position i, that is, $\mathbf{a}_1 = (1, 0, \ldots, 0)$, $\mathbf{a}_2 = (0, 1, \ldots, 0)$, ..., $\mathbf{a}_{d+\chi} = (0, 0, \ldots, 1)$. For any $(q + 1)$-tuple $(e, e'_1, \ldots, e'_q)$ of $q + 1$ distinct hyperedges in E and, for any binary vector $\mathbf{a} = (a_1, \ldots, a_{d+\chi}) \in \{\mathbf{a}_1, \ldots, \mathbf{a}_{d+\chi}\}$, let us define the set of binary vectors $B_{\mathbf{a},e,(e'_1,\ldots,e'_q)} = \{\mathbf{x} = (x_1, \ldots, x_n) \in X : x_{i_1} = a_1, \ldots, x_{i_{d+\chi}} = a_{d+\chi}$, with $(i_1, \ldots, i_{d+\chi}) = S_{e,(e'_1,\ldots,e'_q),\chi}\}$. For any set $A \subseteq \{\mathbf{a}_1, \ldots, \mathbf{a}_{d+\chi}\}$ of size s, $1 \leq s \leq d + \chi$, and any set $S_{e,(e'_1,\ldots,e'_q),\chi}$, for $e, e'_1, \ldots, e'_q \in E$, let us define $B_{A,e,(e'_1,\ldots,e'_q)} = \bigcup_{\mathbf{a} \in A} B_{\mathbf{a},e,(e'_1,\ldots,e'_q)}$. For any $s = 1, \ldots, d + \chi$, we define $\mathcal{B}_s = \{B_{A,e,(e'_1,\ldots,e'_q)} : A \subset \{\mathbf{a}_1, \ldots, \mathbf{a}_{d+\chi}\}, |A| = s$, and $e, e'_1, \ldots, e'_q \in E\}$ and the hypergraph $\mathcal{H}_s = (X, \mathcal{B}_s)$. We claim that *any* cover T of $\mathcal{H}_{d+\chi-m+1}$ is an (E, q, m, χ)-selector, that is, for any $e, e'_1, \ldots, e'_q \in E$, the submatrix of $d + \chi$ columns of T with indices in $S_{e,(e'_1,\ldots,e'_q),\chi}$ contains at least m distinct rows of the identity matrix $I_{d+\chi}$. The proof is by contradiction. Assume that there exist $q+1$ hyperedges $e, e'_1, \ldots, e'_q \in E$ such that the submatrix of the columns of T with indices in the $(d+\chi)$-tuple $(i_1, \ldots, i_{d+\chi}) = S_{e,(e'_1,\ldots,e'_q),\chi}$ contains *at most* $m-1$ distinct rows of $I_{d+\chi}$. Let such rows be $\mathbf{a}_{j_1}, \ldots, \mathbf{a}_{j_g}$, with $g \leq m-1$, and let A be *any* subset of $\{\mathbf{a}_1, \ldots, \mathbf{a}_{d+\chi}\} \setminus \{\mathbf{a}_{j_1}, \ldots, \mathbf{a}_{j_g}\}$ of cardinality $|A| = d+\chi-m+1$. By definition of $\mathcal{H}_{d+\chi-m+1}$, the hyperedge $B_{A,e,(e'_1,\ldots,e'_q)}$ of $\mathcal{H}_{d+\chi-m+1}$ is such that $T \cap B_{A,e,(e'_1,\ldots,e'_q)} = \emptyset$, thus contradicting the fact that T is a cover for $\mathcal{H}_{d+\chi-m+1}$.

To prove the following theorem, we exploit Lovász's result (1) to derive an upper bound on the minimum size of a cover of the hypergraph $\mathcal{H}_{d+\chi-m+1}$.

Theorem 1. *Let E be a set of hyperedges of size d with vertices in $[n]$. Moreover, let q, m and χ be integers such that $1 \leq q \leq |E|-1$ and $1 \leq m \leq \chi+d \leq n$. There exists an (E, q, m, χ)-selector of size t with*

$$t < \frac{2e(d+\chi)}{d+\chi-m+1}\left(1+\ln\left(\binom{d+\chi-1}{d+\chi-m}\alpha\right)\right), \tag{2}$$

where $\alpha = \min\left\{e^q|E|\left(\frac{|E|-1}{q}\right)^q, e^{d+\chi-1}\left(\frac{n}{d+\chi-1}\right)^{d+\chi}\right\}$ *and $e = 2.7182...$ is the base of the natural logarithm.*

Proof. Let $\mathcal{H}_{d+\chi-m+1} = (X, \mathcal{B}_{d+\chi-m+1})$ be the hypergraph defined in the discussion preceding the statement of the theorem. Inequality (1) implies that

$$\tau(\mathcal{H}_{d+\chi-m+1}) < \frac{|X|}{\min\{|B| : B \in \mathcal{B}_{d+\chi-m+1}\}}(1+\ln \Delta), \tag{3}$$

where $\Delta = \max_{x \in X} |\{B\colon B \in \mathcal{B}_{d+\chi-m+1} \text{ and } x \in B\}|$.

In order to derive the stated upper bound on t, we estimate the quantities that appear on the right hand side of (3), that is, we evaluate the quantities

$$|X|, \quad \min\{|B| : B \in \mathcal{B}_{d+\chi-m+1}\}, \quad \text{and } \Delta,$$

for the hypergraph $\mathcal{H}_{d+\chi-m+1}$. Since X is the set of all binary vectors of length n containing $\lfloor \frac{n}{d+\chi} \rfloor$ entries equal to 1, it holds that $|X| = \binom{n}{\lfloor \frac{n}{d+\chi} \rfloor}$. Each $B_{\mathbf{a},e,(e'_1,\ldots,e'_q)}$ has size $\binom{n-d-\chi}{\lfloor \frac{n}{d+\chi} \rfloor - 1}$ since this is the number of vectors, with $\lfloor \frac{n}{d+\chi} \rfloor$ entries equal to 1, for which one has that the entries with indices in $(i_1, \ldots, i_{d+\chi}) = S_{e,(e'_1,\ldots,e'_q),\chi}$ form the vector $\mathbf{a}$. Moreover, each hyperedge $B_{A,e,(e'_1,\ldots,e'_q)}$ of $\mathcal{H}_{d+\chi-m+1}$ is the union of $d+\chi-m+1$ disjoint sets $B_{\mathbf{a},e,(e'_1,\ldots,e'_q)}$, and therefore it has cardinality

$$|B_{A,e,(e'_1,\ldots,e'_q)}| = (d+\chi-m+1) \cdot |B_{\mathbf{a},e,(e'_1,\ldots,e'_q)}| = (d+\chi-m+1)\binom{n-d-\chi}{\lfloor \frac{n}{d+\chi} \rfloor - 1}.$$

Therefore, one has that $\frac{|X|}{\min\{|B|:B\in\mathcal{B}_{d+\chi-m+1}\}}=\frac{\binom{\lfloor\frac{n}{d+\chi}\rfloor}{}}{(d+\chi-m+1)\binom{n-d-\chi}{\lfloor\frac{n}{d+\chi}\rfloor-1}}$, and simple calculations imply that

$$\frac{|X|}{\min\{|B|:B\in\mathcal{B}_{d+\chi-m+1}\}}\leq\frac{2e(d+\chi)}{d+\chi-m+1}. \quad (4)$$

We refer the reader to [2] for details of calculations that lead to (4). To compute Δ, we notice that each $\mathbf{x}\in X$ belongs to at most

$$\min\left\{|E|\binom{|E|-1}{q},\binom{\lfloor\frac{n}{d+\chi}\rfloor}{1}\binom{n-\lfloor\frac{n}{d+\chi}\rfloor}{d+\chi-1}\right\} \quad (5)$$

distinct sets $B_{\mathbf{a},e,(e'_1,\ldots,e'_q)}$. Indeed, as observed in Sect. 2, the number of distinct sets $S_{e,(e'_1,\ldots,e'_q),\chi}$ is at most $|E|\binom{|E|-1}{q}$. Notice that, for a vector $\mathbf{a}=(a_1,\ldots,a_{d+\chi})$, the actual number of $(d+\chi)$-tuples $(i_1,\ldots,i_{d+\chi})=S_{e,(e'_1,\ldots,e'_q),\chi}$ for which it holds that $x_{i_1}=a_1,\ldots,x_{i_{d+\chi}}=a_{d+\chi}$ might be significantly smaller than the total number of $(d+\chi)$-tuples $(i_1,\ldots,i_{d+\chi})=S_{e,(e'_1,\ldots,e'_q),\chi}$, but this estimate is sufficient to obtain the first value in the "min" in the expression of α that appears in the statement of the theorem. However, if $|E|$ is close to $\binom{n}{d}$, i.e., it contains almost all hyperedges of size d on $[n]$, then it might be convenient to upper bound the number of distinct hyperedges $B_{\mathbf{a},e,(e'_1,\ldots,e'_q)}$ containing a given vector $\mathbf{x}\in X$ by $\binom{\lfloor\frac{n}{d+\chi}\rfloor}{1}\binom{n-\lfloor\frac{n}{d+\chi}\rfloor}{d+\chi-1}$ which is obtained by considering all possible ways of choosing $(d+\chi)$ positions in $\mathbf{x}$ so that exactly one of those positions corresponds to a 1-entry of $\mathbf{x}$, whereas the others correspond to 0-entries.

Now observe that each $B_{\mathbf{a},e,(e'_1,\ldots,e'_q)}$ belongs to $\binom{d+\chi-1}{d+\chi-m}$ distinct hyperedges $B_{A,e,(e'_1,\ldots,e'_q)}$. Therefore, (5) implies that the maximum degree of vertices in $\mathcal{H}_{d+\chi-m+1}$ is

$$\Delta\leq\binom{d+\chi-1}{d+\chi-m}\min\left\{|E|\binom{|E|-1}{q},\binom{\lfloor\frac{n}{d+\chi}\rfloor}{1}\binom{n-\lfloor\frac{n}{d+\chi}\rfloor}{d+\chi-1}\right\}. \quad (6)$$

In order to estimate our upper bound, we resort to the following well known inequality:

$$\binom{a}{b}\leq(ea/b)^b. \quad (7)$$

Applying inequality (7) to $\binom{|E|-1}{q}$ and to $\binom{n-\lfloor\frac{n}{d+\chi}\rfloor}{d+\chi-1}$ in the upper bound (6) on Δ, we get that

$$\Delta\leq\binom{d+\chi-1}{d+\chi-m}\min\left\{e^q|E|\left(\frac{|E|-1}{q}\right)^q,e^{d+\chi-1}\left(\frac{n}{d+\chi-1}\right)^{d+\chi}\right\}. \quad (8)$$

Please see [2] for detailed calculations that lead to (8). By plugging into (3) the upper bound (8) on Δ and the upper bound (4) on $\frac{|X|}{\min\{|B|:B\in\mathcal{B}_{d+\chi-m+1}\}}$, we obtain the stated upper bound on t.

5.1 A Non-adaptive Group Testing Algorithm for General Hypergraph

In order to prove the upper bound on the minimum number of tests of non-adaptive algorithms, we prove a more general result that can be exploited also to prove existential results for the three-stage algorithms.

Theorem 2. *Let $\mathcal{F} = (V, E)$ be a hypergraph with $V = [n]$ with all hyperedges in E of size* at most d. *For any positive integer $p \leq d$, there exists a non-adaptive algorithm that allows to discard all but those hyperedges e such that $|e \setminus e^*| < p$, where e^* is the defective hyperedge, and uses $t = O\left(\frac{d}{p} \log E\right)$ tests.*

Proof. Let E denote the input set of hyperedges. First we consider the case when all hyperedges of E have size exactly d and show that in this case an $(E, p, , d+1)$-selector corresponds to a non-adaptive algorithm that allows to discard all but those hyperedges $e \in E$ such that $|e \setminus e^*| < p$. Then we will consider the case of hypergraphs that are not necessarily uniform. Let M be an $(E, p, d+1)$-selector. By Definition 2, for any two distinct hyperedges e and e', the submatrix M' of M formed by the $d+p$ columns with indices in $S_{e,e',p}$ contains at least $d+1$ rows of the identity matrix I_{d+p}. At least one of these rows has 0 at the intersection with each of the d columns with index in e and an entry equal to 1 at the intersection with one of the p columns with indices in $S_{e,e',p} \setminus e$. By definition of $S_{e,e',p}$, if $|e' \setminus e| \geq p$ then $S_{e,e',p} \setminus e \subseteq e' \setminus e$, and consequently the above said entry equal to 1 is at the intersection with one column with index in $e' \setminus e$. It follows that the OR of the d columns associated to vertices in e is different from that of the d columns associated to vertices in e'. Now suppose that e^* be the defective hyperedge and e' be an arbitrary hyperedge such that $|e' \setminus e^*| \geq p$. From the above argument, there exists a row index, say r, such that at least one column associated to a vertex in e' has an entry equal to 1 in position r, whereas all columns associated to vertices in e^* have 0 in that position. It follows that the response vector has an entry equal to 0 in position r whereas at least one element of $e' \setminus e^*$ is in the pool associated to the row of M with index r, thus showing that e' is not the defective hyperedge. By inspecting each column of the $(E, p, d+1)$-selector and comparing it with the response vector, one can identify those columns that have an entry equal to 1 in a position where the response vector has 0 and get rid of all hyperedges containing one of the vertices associated to those columns. By the above argument one gets rid of all hyperedges e' such that $|e' \setminus e^*| \geq p$.

Now let us consider a not necessarily uniform hypergraph E with hyperedges of size at most d. We add dummy vertices to each hyperedge of size smaller than d, so as to obtain a set $\tilde{E}$ of hyperedges all having size d. We will see that a non-adaptive group testing based on an $(\tilde{E}, p, d+1)$-selector allows to discard all non-defective hyperedges e such that $|e \setminus e^*| \geq p$. Let $\{i_{n+1}, \ldots, i_{n+d-1}\}$ be the dummy vertices. Notice that these dummy vertices do not need to be different from those we have used in the definition of the sets $S_{e,e',p}$'s, as we will see below. For each $e \in E$, we denote with $\tilde{e}$ the corresponding hyperedge in $\tilde{E}$. The hyperedge $\tilde{e}$ is either equal to e, if $|e| = d$, or equal to $e \cup \{i_{n+1}, \ldots, i_{n+d-|e|}\}$, if

$|e| < d$. Obviously, $\tilde{E}$ has all hyperedges of size d and it holds that $|\tilde{E}| = |E|$. Let M be an $(\tilde{E}, p, d+1)$-selector. By Definition 2, for any two distinct hyperedges $\tilde{e}$ and $\tilde{e}'$, the submatrix M' of M formed by the $d+p$ columns with indices in $S_{\tilde{e},\tilde{e}',p}$ contains at least $d+1$ rows of the identity matrix I_{d+p}. At least one of these rows, say the row with index r, has 0 at the intersection with all the columns with index in $\tilde{e}$ and an entry equal to 1 at the intersection with one of the remaining columns of M'. If $|\tilde{e}' \setminus \tilde{e}| \geq p$, it holds that $S_{\tilde{e},\tilde{e}',p} = S_{\tilde{e},\tilde{e}}$. In other words, all column indices in $S_{\tilde{e},\tilde{e}',p}$ that are not in $\tilde{e}$, are in fact all contained in $\tilde{e}' \setminus \tilde{e}$ and consequently the above said entry equal to 1 in the row with index r is at the intersection with one of the columns in $\tilde{e}' \setminus \tilde{e}$. Now, let e^* denote the defective hyperedge in the original set of hyperedges E and let us replace e with e^* in the above discussion. Suppose that a hyperedge e' of the original hypergraph is such that $|e' \setminus e^*| \geq p$. Since for each $e \in E$ it holds that $e \subseteq \tilde{e}$ and $e \cap \{i_{n+1}, \ldots i_{n+\chi}\} = \emptyset$, it follows that $\tilde{e} \setminus \tilde{e}' \supseteq e \setminus e'$. This implies that $|\tilde{e}' \setminus \tilde{e}^*|$ is larger than or equal to p, and consequently, from the above discussion, it follows that the submatrix M' formed by the columns of M with index in $S_{\tilde{e},\tilde{e}'}$ contains a row that has 0 at the intersection with each of the columns with index in $\tilde{e}^*$ and an entry equal to 1 at the intersection with one of columns with index in $\tilde{e}' \setminus \tilde{e}^*$. We need to show that this index does not correspond to a dummy vertex, i.e., we need to show that it belongs to $e' \setminus e^*$. To see this, we recall that $S_{\tilde{e}^*,\tilde{e}} \cap (\tilde{e} \setminus \tilde{e}^*)$ consists of the smallest p vertices in $\tilde{e}' \setminus \tilde{e}^*$ and since $|e' \setminus e^*| \geq p$, it holds that these p vertices belong all to $e' \setminus e^*$. Therefore, we have proved that there exists a row in M', that contains 0 at the intersection with all the columns with index in $\tilde{e}^*$ and an entry equal to 1 at the intersection with a column associated to a vertex in $e' \setminus e^*$. Since $\tilde{e}^* \supseteq e^*$, this row has all 0's at the intersection with the columns with index in e^*. As a consequence, for any non-defective hyperedge e' with $|e' \setminus e^*| \geq p$, there exists a row index r such that at least one column with index in e' has a 1-entry in position r, whereas the response vector has a 0-entry in that position. In other words there is an element x in e' that is contained in a pool that tests negative, and consequently, e' is not the defective hyperedge.

Notice that decoding can be achieved by inspecting all columns of the selector that are associated with non-dummy vertices. Indeed, as we have just seen, any non-defective hyperedge e' such that $|e' \setminus e^*| \geq p$ contains a vertex that is included in a pool that tests negative, i.e., a vertex associated with a column with a 1 in a position that corresponds to a 0 in the response vector. By inspecting all columns of the selectors, the decoding algorithm finds all columns with a 1 in a position corresponding to a negative response and gets rid of all hyperedges that contain one or more vertices associated with those columns. In this way, the decoding algorithm discards any non-defective hyperedge e' such that $|e' \setminus e^*| \geq p$.

Finally, we observe that the dummy vertices added to the hyperedges of E do not need to be different from those used in the definition of the sets $S_{e,e',p}$ in Sect. 2. The fact that some of the dummy vertices used to define the sets $S_{e,e',p}$'s might be contained in some of the hyperedges of $\tilde{E}$ might lead to a fault in the above discussion only if $|\tilde{e}' \setminus \tilde{e}| < p$ and we need to include dummy vertices in $S_{\tilde{e},\tilde{e}',p}$. Indeed, some dummy vertices might be also included in $\tilde{e}$ and we could end up at adding some vertices twice to the set $S_{\tilde{e},\tilde{e}',p}$. However, this cannot happen since in the above discussion we have exploited the property of $(\tilde{E}, p, d+1)$-selectors only in reference to pairs of hyperedges $\tilde{e}$ and $\tilde{e}'$ associated to vertices e and e' of E such that $|e \setminus e'| \geq p$. We have seen that this implies not only that $S_{\tilde{e},\tilde{e}',p}$ is equal to $S_{\tilde{e},\tilde{e}'}$ but also that $S_{\tilde{e},\tilde{e}'} \cap (\tilde{e}' \setminus \tilde{e}) \subseteq e' \setminus e$. In other words, the only dummy vertices possibly involved in the definition of $S_{\tilde{e},\tilde{e}',p}$ are those in $\tilde{e}$.

The upper bound in the statement of the theorem then follows from the upper bound Theorem 1 with $\chi = p$, $q = 1$, and $m = d+1$. Notice that we have replaced n with $n+d-1$ in that bound, since we have added up to $d-1$ dummy vertices to $[n]$ in order to obtain the set of hyperedges $\tilde{E}$.

The following corollary is an immediate consequence of Theorem 2.

Corollary 1. *Let d and n be integers with $1 \leq d \leq n$, and let E be a set of hyperedges of size* at most *d on $[n]$. Moreover, let p be an integer such that $1 \leq p \leq min\{|e' \setminus e| : e, e' \in E\}$. There exists a non-adaptive group testing algorithm that finds the defective hyperedge in E and uses at most $t = O\left(\frac{d}{p}\log|E|\right)$ tests.*

Proof. If for any two distinct hyperedges $e, e' \in E$ it holds that $min\{|e' \setminus e| : e, e' \in E\} \geq p$, then the non-adaptive algorithm of Theorem 2 discards all non-defective hyperedges.

5.2 A Two-Stage Group Testing Algorithm for General Hypergraphs

Theorem 3. *Let $\mathcal{F} = (V, E)$ be a hypergraph with $V = [n]$ and all hyperedges in E of size at most d. Moreover, let q and χ be positive integers such that $1 \leq q \leq |E| - 1$ and $\chi = \min\{|\bigcup_{i=1}^{q} e'_i \setminus e|$, for any $q+1$ distinct $e, e'_1, \ldots, e'_q \in E\}$. There exists a (trivial) two-stage algorithm that uses a number of tests t with*

$$t < \frac{2e(d+\chi)}{\chi}\left(1 + \ln\left(\binom{d+\chi-1}{d+\chi-d-1}\beta\right)\right) + dq,$$

where $\beta = \min\left\{e^q|E|\left(\frac{|E|-1}{q}\right)^q, e^{d+\chi-1}\left(\frac{n+d-1}{d+\chi-1}\right)^{d+\chi}\right\}$ and $e = 2.7182...$ is the base of the natural logarithm.

Proof. Let E be a set of hyperedges each consisting of at most d vertices in $[n]$. As in the proof of Theorem 2, let us denote with $\tilde{E}$ the set of hyperedges obtained by replacing in E any hyperedge e of size smaller than d by $e \cup \{i_{n+1}, \ldots, i_{n+d-|e|}\}$. For each $e \in E$, we denote with $\tilde{e}$ the corresponding hyperedge in $\tilde{E}$. This hyperedge is either equal to e, if $|e| = d$, or is equal to $e \cup \{i_{n+1}, \ldots, i_{n+d-|e|}\}$, if $|e| < d$. Let us consider the non-adaptive algorithm that tests the pools having as characteristic vectors the rows of an $(\tilde{E}, q, d+1, \chi)$-selector M. We will show that after executing this non-adaptive algorithm, one is left with at most q hyperedges that are candidate to be the defective hyperedge. Let e^* be the defective hyperedge and let us suppose by contradiction that, after executing the algorithm associated with M, there are at least q hyperedges, in addition to e^*, that are still candidate to be the defective one. Let us consider q such hyperedges, say $e'_1, \ldots, e'_q$.

By Definition 3, it holds that the submatrix M' of M consisting of the columns with indices in $S_{\tilde{e}^*,(\tilde{e}'_1,\ldots,\tilde{e}'_q),\chi}$ contains at least $d+1$ distinct rows of the identity matrix $I_{d+\chi}$. This implies that at least one of these rows, say the row with index r, has a 0 at the intersection with each column with index in $\tilde{e}^*$ and 1 at the intersection with one of the remaining columns of $S_{\tilde{e}^*,(\tilde{e}'_1,\ldots,\tilde{e}'_q),\chi}$. Since for each $e \in E$ it holds that $e \subseteq \tilde{e}$ and $e \cap \{i_{n+1}, \ldots i_{n+\chi}\} = \emptyset$, it follows that $\bigcup_{i=1}^q \tilde{e}'_i \setminus \tilde{e}^* \supseteq \bigcup_{i=1}^q e'_i \setminus e^*$. Consequently, the hypothesis $|\bigcup_{i=1}^q e'_i \setminus e^*| \geq \chi$ implies that $|\bigcup_{i=1}^q \tilde{e}'_i \setminus \tilde{e}^*| \geq \chi$. This means that $S_{\tilde{e}^*,(\tilde{e}'_1,\ldots,\tilde{e}'_q),\chi} = S_{\tilde{e}^*,(\tilde{e}'_1,\ldots,\tilde{e}'_q)}$, i.e., that the columns with index in $S_{\tilde{e}^*,(\tilde{e}'_1,\ldots,\tilde{e}'_q),\chi}$ that are not in $\tilde{e}^*$, are in fact all contained in $\bigcup_{i=1}^q \tilde{e}'_i \setminus \tilde{e}^*$. Therefore, the above said entry equal to 1 in the row with index r is at the intersection with one of the columns in $\bigcup_{i=1}^q \tilde{e}'_i \setminus \tilde{e}^*$. Now we need to prove that this entry equal to 1, in fact, intersects a column associated with a non-dummy vertex, i.e., a vertex in $\bigcup_{i=1}^q e'_i \setminus e^*$. By definition of $S_{\tilde{e}^*,(\tilde{e}'_1,\ldots,\tilde{e}'_q),\chi}$, the set $S_{\tilde{e}^*,(\tilde{e}'_1,\ldots,\tilde{e}'_q),\chi}$ consists of the smallest χ integers in $\bigcup_{i=1}^q \tilde{e}'_i \setminus \tilde{e}^*$. Moreover, since $\bigcup_{i=1}^q e'_i \setminus e^* \subseteq \bigcup_{i=1}^q \tilde{e}'_i \setminus \tilde{e}^*$ and $|\bigcup_{i=1}^q e'_i \setminus e^*| \geq \chi$, one has that $S_{\tilde{e}^*,(\tilde{e}'_1,\ldots,\tilde{e}'_q),\chi} \cap (\bigcup_{i=1}^q \tilde{e}'_i \setminus \tilde{e}^*) \subseteq \bigcup_{i=1}^q e'_i \setminus e^*$, thus implying that the above said entry equal to 1 of the row with index r is at the intersection with a column associated to a vertex in $\bigcup_{i=1}^q e'_i \setminus e^*$. We also observe that since $e^* \subseteq \tilde{e}^*$, all columns in e^* have zeros at the intersection with the row with index r. Let T_r be the pool having this row as characteristic vector. From what we have just said, it holds that $|T_r \cap e^*| = 0$ and $|T \cap \bigcup_{i=1}^q e'_i \setminus e^*| \geq 1$. As a consequence, T_r contains at least one element belonging to one of the hyperedges $e'_1, \ldots, e'_q$ whereas the result of the test on T_r is negative, thus indicating that at least one of $e'_1, \ldots, e'_q$ is non-defective.

Let us consider now a two-stage algorithm whose first stage consists in the non-adaptive group testing algorithm associated with the rows of the $(\tilde{E}, q, d+1, \chi)$-selector M. From the above argument, after the first stage, we are left with at most q hyperedges that are still candidate to be the defective hyperedge. Therefore, in order to determine the defective hyperedge, one needs only to perform individual tests on at most dq elements. Notice that these tests can be performed in parallel.

As for the decoding algorithm that identifies the non-defective hyperedges from the responses to the tests performed in stage 1, the algorithm needs only to inspect each column of the selector and every time finds a column with a 1-entry in a position corresponding to a 0-entry in the response vector, discards the hyperedges that contain the vertex associated with that column. The stated upper bound follows by the upper bound of Theorem 1 by replacing m with $d+1$ and n with $n+d-1$ since the set of vertices of E has been augmented with at most $d-1$ dummy vertices.

Corollary 2. *Let d and n be integers with $1 \leq d \leq n$, and let E be a set of hyperedges of size at most d on $[n]$. If there exists a constant $q \geq 1$ such that for* *any $q+1$ distinct hyperedges $e, e'_1, \ldots, e'_q \in E$, $|\bigcup_{i=1}^{q} e'_i \setminus e| = \Omega(d)$, then there exists a trivial two-stage algorithm that finds the defective hyperedge in E and uses $t = O(\log|E|)$ tests.*

5.3 A Three-Stage Group Testing Algorithm for General Hypergraphs

The following theorem furnishes an upper bound on the minimum number of tests of three-stage algorithms. An interesting feature of this upper bound is that it holds independently of the size of the pairwise intersections of the hyperedges.

Theorem 4. *Let $\mathcal{F} = (V, E)$ be a hypergraph with $V = [n]$ with hyperedges of size* at most *d and let b be any positive integer smaller than d. There exists a three-stage algorithm that finds the defective hyperedge in E and uses $t = O(\frac{d}{b}\log|E| + b\log|E|)$ tests.*

Proof. Let us describe the three stage algorithm that achieves the stated upper bound. The first two stages aim at restricting the set of potentially defective hyperedges to hyperedges with at most $b-1$ vertices. From Theorem 2, one has that there is an $O(\frac{d}{b}\log|E|)$ non-adaptive algorithm $\mathcal{A}$ that discards all but the hyperedges e such that $|e \setminus e^*| < b$, where e^* is the defective hyperedge. Stage 1 consists in running algorithm $\mathcal{A}$. If all hyperedges that have not be discarded by $\mathcal{A}$ have size smaller than b, then the algorithm skips stage 2 and proceeds to stage 3. If otherwise, the algorithm chooses a hyperedge e of maximum size among those that have not been discarded by $\mathcal{A}$ and proceeds to stage 2 where it performs in parallel individual tests on the vertices of e. Since e has not been discarded, it means that $|e \setminus e^*| < b$ and therefore at least $|e| - b + 1 \geq 1$ of the individual tests yield a positive response. Let $i_1, \ldots, i_f$ denote the vertices of e that have been tested positive. The algorithm looks at the intersection between e and any other not yet discarded hyperedge e' and discards e' if it either contains one or more of the vertices of e that have been tested negative, or if $\{i_1, \ldots, i_f\} \not\subseteq e'$. In other words, after this second stage the algorithm is left only with the hyperedges e' such that $e \cap e' = \{i_1, \ldots, i_f\}$. The vertices $i_1, \ldots, i_f$ are removed from all these hyperedges since it is already known that they are defective. Each hyperedge e' that has not been discarded in stage 2 is therefore replaced by a hyperedge of size $|e'| - f \leq |e'| - |e| + b - 1 \leq b - 1$. The last inequality is a consequence of having chosen e as a hyperedge of maximum size among those that have not been discarded in stage 1. In stage 3, the algorithm is left with a hypergraph with hyperedges of at most $b-1$ vertices and, by Corollary 1 with d being replaced by $b-1$, the defective hyperedge can be detected non-adaptively using $O(b\log|E|)$ tests. In applying Corollary 1, we do not make any assumption on the size of the set differences between hyperedges and take p in the bound of that corollary as small as 1. Notice that the algorithm of Corollary 1 might end up with more that one hyperedge if there exist hyperedges that are proper subsets of the defective one. However, in this eventuality, the algorithm would choose the largest hyperedge among those that have not be discarded.

By setting $b = \sqrt{d}$ in Theorem 4, we get the following corollary.

Corollary 3. *Let E be a hypergraph with $V = [n]$ with hyperedges of size at most d. There exists a three-stage algorithm that finds the defective hyperedge in E and uses $t = O(\sqrt{d} \log |E|)$ tests.*

References

1. Clementi, A.E.F., Monti, A., Silvestri, R.: Selective families, superimposed codes, and broadcasting on unknown radio networks. In: Twelfth Annual ACM-SIAM Symposium on Discrete Algorithms, pp. 709–718 (2001)
2. De Bonis, A.: Group testing in arbitrary hypergraphs and related combinatorial structures (2023). https://arxiv.org/abs/2307.09608
3. De Bonis, A., Gasieniec, L., Vaccaro, U.: Optimal two-stage algorithms for group testing problems. SIAM J. Comput. **34**(5), 1253–1270 (2005)
4. Dorfman, R.: The detection of defective members of large populations. Ann. Math. Statist. **14**, 436–440 (1943)
5. Du, D.Z., Hwang, F.K.: Pooling Design and Nonadaptive Group Testing. Series on Applied Mathematics, vol. 18. World Scientific (2006)
6. D'yachkov, A.G., Rykov, V.V.: A survey of superimposed code theory. Probl. Control Inform. Theory **12**, 229–242 (1983)
7. Erdös, P., Frankl, P., Füredi, Z.: Families of finite sets in which no set is covered by the union of r others. Israel J. Math. **51**, 79–89 (1985)
8. Gonen, M., Langberg, M., Sprintson, A.: Group testing on general set-systems. IEEE Trans. Inf. Theory **2022**, 874–879 (2022)
9. Harvey, N.J.A., Patrascu, M., Wen, Y., Yekhanin, S., Chan, V.W.S.: Non-adaptive fault diagnosis for all-optical networks via combinatorial group testing on graphs. In: 26th IEEE International Conference on Computer Communications, pp. 697–705 (2007)
10. Hong, E.S., Ladner, R.E.: Group testing for image compression. IEEE Trans. Image Process. **11**(8), 901–911 (2002)
11. Goenka, R., Cao, S.J., Wong, C.W., Rajwade, A., Baron, D.: Contact tracing enhances the efficiency of COVID-19 group testing. In: ICASSP 2021, pp. 8168–8172 (2021)
12. Kautz, W.H., Singleton, R.C.: Nonrandom binary superimposed codes. IEEE Trans. Inf. Theory **10**, 363–377 (1964)
13. Lo, C., Liu, M., Lynch, J.P., Gilbert, A.C.: Efficient sensor fault detection using combinatorial group testing. In: 2013 IEEE International Conference on Distributed Computing in Sensor Systems, pp. 199–206 (2013)
14. Lovàsz, L.: On the ratio of optimal integral and fractional covers. Discrete Math. **13**, 383–390 (1975)
15. Nikolopoulos, P., Srinivasavaradhan, S.R., Guo T., Fragouli, C., Diggavi S.: Group testing for connected communities. In: The 24th International Conference on Artificial Intelligence and Statistics, vol. 130, pp. 2341–2349. PMLR (2021)
16. Ruszinkó, M.: On the upper bound of the size of the r-cover-free families. J. Combin. Theory Ser. A **66**, 302–310 (1994)
17. Sobel, M., Groll, P.A.: Group testing to eliminate efficiently all defectives in a binomial sample. Bell Syst. Tech. J. **38**, 1179–1252 (1959)
18. Vorobyev, I.: Note on generalized group testing (2022). https://doi.org/10.48550/arXiv.2211.04264

19. Wolf, J.: Born again group testing: multiaccess communications. IEEE Trans. Inf. Theory **31**, 185–191 (1985)
20. Zhu, J., Rivera, K., Baron, D.: Noisy pooled PCR for virus testing (2020). https://doi.org/10.48550/arXiv.2004.02689

On the Parameterized Complexity of the Perfect Phylogeny Problem

Jorke M. de Vlas[1,2(✉)]

[1] Utrecht University, Utrecht, The Netherlands
jorkedevlas@gmail.com
[2] Linköping Universitet, Linköping, Sweden

Abstract. This paper categorizes the parameterized complexity of the algorithmic problems Perfect Phylogeny and Triangulating Colored Graphs when parameterized by the number of genes and colors, respectively. We show that they are complete for the parameterized complexity class XALP using a reduction from Tree-chained Multicolor Independent Set and a proof of membership. We introduce the problem Triangulating Multicolored Graphs as a stepping stone and prove XALP-completeness for this problem as well. We also show that, assuming the Exponential Time Hypothesis, there exists no algorithm that solves any of these problems in time $f(k)n^{o(k)}$, where n is the input size, k the parameter, and f any computable function.

Keywords: Perfect phylogeny · Triangulated graphs · XALP · Parameterized complexity · W-hierarchy

1 Introduction

A phylogeny is a tree that describes the evolution history of a set S of species. Every vertex corresponds to a species: leafs correspond to species from S, and internal vertices correspond to hypothetical ancestral species. Species are characterized by their gene-variants, and the quality of a phylogeny is determined by how well it represents those variants. In particular, a phylogeny is *perfect* if each gene-variant was introduced at exactly one point in the tree. That is, the subset of vertices that contain the variant is connected. Perfect Phylogeny is the algorithmic problem of determining the existence of a perfect evolutionary tree. It has large implications on determining the evolutionary history of genetic sequences and is therefore of major importance. This application is not limited to biology: it can also be used to determine the history of languages or cultures.

The concept of phylogenies as an algorithmic problem has been well researched since the 60s. The first formal definition of Perfect Phylogeny was given by Estabrook [10]. In 1974, Buneman showed that the problem can be reduced to the more combinatorial Triangulating Colored Graphs [6] which by itself has also become an important, well-studied problem. An inverse

H. Fernau et al. (Eds.): SOFSEM 2024, LNCS 14519, pp. 169–182, 2024.
https://doi.org/10.1007/978-3-031-52113-3_12

reduction, and thus equivalence, was given by Kannan and Warnow [11]. In 1992, Bodlaender et al. showed that PERFECT PHYLOGENY is NP-complete [3].

After Downey and Fellows introduced parameterized complexity [8], people have tried to determine the complexity of PERFECT PHYLOGENY when seen as a parameterized problem. There are two main ways to parameterize the problem: either by using the number of genes or by using the maximum number of variants for each gene. In the second case, the problem becomes FPT [12]. In the first case, the parameterized complexity was unknown. There are some partial results: On one hand, it was shown that the problem is $W[t]$-hard for every t [2]. On the other hand, there exists an algorithm that runs in $\mathcal{O}(n^{k+1})$ time and space (where n is the input size and k the parameter) which implies that the problem is contained in XP [13].

In this paper we will close this gap and show that PERFECT PHYLOGENY is complete for the complexity class XALP, which is a relatively new parameterized complexity class that was introduced by Bodlaender et al. in [5]. We will show XALP-completeness by giving a reduction from the XALP-complete problem TREE-CHAINED MULTICOLOR INDEPENDENT SET, using TRIANGULATING MULTICOLORED GRAPHS as a stepping stone. This makes PERFECT PHYLOGENY the first example of a "natural" problem that is XALP-complete and allows it to be used as a starting point for many other XALP-hardness proofs. Finally, we use the same reduction to give some lower bounds dependent on the Exponential Time Hypothesis.

2 Definitions and Preliminary Results

All problems in this paper are parameterized. This means that the input contains a parameter separate from the rest of the input which allows us to analyze the runtime as a function of both the input and the parameter. If a parameterized problem with input size n and parameter k can be solved in $\mathcal{O}(f(k)n^c)$ time (with f any computable function and c any constant), we say that it is Fixed Parameter Tractable (FPT). A *parameterized reduction* is an algorithm that transforms instances of one parameterized problem into instances of another parameterized problem, runs in FPT time, and whose new parameter is only dependent on the old parameter. A *log-space reduction* is a parameterized reduction that additionally only uses $\mathcal{O}(f(k)\log(n))$ space. These reductions form the base of all parameterized complexity classes: all classes are defined up to equivalence under one of these reductions.

We use the following definition of PERFECT PHYLOGENY, which is a parameterized version of the original definition from Estabrook [10].

PERFECT PHYLOGENY (PP)
Input: A set G of genes, for each gene $g \in G$ a set V_g of variants, and a set S of species, where each species is defined as a tuple of gene-variants (exactly one per gene)
Parameter: The number of genes
Question: Does there exist a tree T of species (not necessarily from S) that contains all species from S and where the subtree of species containing a specific gene-variant is connected?

Triangulated and Colored Graphs. A graph is *colored* if every vertex is assigned a color. The graph is *properly colored* if there are no edges between vertices of the same color. For any cycle C in a graph, a *chord* is an edge between two vertices of C that are not neighbors on C. A graph is *triangulated* if every cycle of length at least four contains a chord. A *triangulation* of a graph is a supergraph that is triangulated. We now define the problem TRIANGULATING COLORED GRAPHS, which was first given by Buneman [6].

TRIANGULATING COLORED GRAPHS (TCG)
Input: A colored graph G
Parameter: The number of colors used
Question: Does there exist a properly colored triangulation of G?

We now introduce a multicolored variant of this problem. A graph is *multicolored* if every vertex is assigned a (possibly empty) set of colors. The graph is *properly multicolored* if there are no edges between vertices which share a color. This gives us the following problem:

TRIANGULATING MULTICOLORED GRAPHS (TMG)
Input: A multicolored graph G
Parameter: The number of colors used
Question: Does there exist a properly multicolored triangulation of G?

This problem is equivalent to TRIANGULATING COLORED GRAPHS under parameterized reductions. The general idea is to replace every multicolored vertex with a clique of normally colored vertices. A full proof is given in the appendix of the ArXiv version. We now define a tree decomposition and state some well-known properties of triangulated colored graphs.

Definition 1 (Tree Decomposition). *Given a graph $G = (V, E)$, a tree decomposition is a tree T where each vertex (bag) is associated with a subset of vertices from T. This tree must satisfy three conditions:*

- *For each vertex $v \in V$, there is at least one bag that contains v.*
- *For each edge $e \in V$, there is at least one bag that contains both endpoints of e.*
- *For each vertex $v \in V$, the subgraph of bags that contain v is connected.*

Proposition 1. *Let G be a (multi)colored graph and C be a cycle.*

(i) Suppose there exist two colors such that every vertex from C is colored with at least one of these colors. Then G admits no properly (multi)colored triangulation.
(ii) Let v be any vertex from C. In every triangulation of G there is either an edge between v's neighbors (in C) or a chord between v and some non-neighbor vertex from C.
(iii) G admits a properly colored triangulation if and only if G admits a tree decomposition where each bag contains each color at most once.

Proof. Omitted. Included as an appendix in the ArXive version of this paper. □

XALP. A new complexity class in parameterized complexity theory is XALP [5]. Intuitively, it is the natural home of parameterized problems that are $W[t]$-hard for every t and contain some hidden tree-structure. For PERFECT PHYLOGENY, this tree-structure is the required phylogeny. For TRIANGULATING COLORED GRAPHS, it is the tree decomposition arising from Proposition 1(iii).

Formally, XALP is the class of parameterized problems that are solvable on an alternating Turing machine using $\mathcal{O}(f(k)\log(n))$ memory and at most $\mathcal{O}(f(k) + \log(n))$ co-nondetermenistic computation steps, where n is the input size and k is the parameter. It is closed under log-space reductions. On Downey and Fellows' W-hierarchy, it lies between $W[t]$ and XP: XALP-hardness implies $W[t]$-hardness for every t and XALP membership implies XP-membership.

An example of an XALP-complete problem is TREE-CHAINED MULTICOLOR INDEPENDENT SET [5]. It is defined as a tree-chained variant of the well-known MULTICOLOR INDEPENDENT SET problem.

MULTICOLOR INDEPENDENT SET (MIS)
Input: A colored graph G
Parameter: The number of colors used
Question: Does G contain an independent set consisting of exactly one vertex of each color?

TREE-CHAINED MULTICOLOR INDEPENDENT SET (TCMIS)
Input: A binary tree T, for each vertex (bag) $B \in T$ a colored graph $G_B = (V_B, E_B)$ which we view as an instance of MULTICOLOR INDEPENDENT SET, and for each edge $e \in T$ a set of extra edges E_e between the graphs corresponding to the endpoints of e.
Parameter: The maximum number of colors used in each instance of MIS
Question: Does there exist a solution to each instance of MIS such that for each of the extra edges at most one of the endpoints is contained in the solution?

3 Main Results

In this section we will state the main result and explore some of its corollaries. We postpone the proof to the next sections.

Theorem 1. TRIANGULATING COLORED GRAPHS *is contained in XALP.*

Theorem 2. *There exists a log-space reduction from* TREE-CHAINED MULTICOLOR INDEPENDENT SET *to* TRIANGULATING MULTICOLORED GRAPHS. *This reduction has a linear change of parameter.*

We will prove Theorem 1 in Sect. 4. For Theorem 2, we describe the reduction in Sect. 6 together with an intuitive reasoning on the correctness. The full proof is omitted and can be seen in the ArXive version of this paper.

Theorem 3 (Main Result). *The problems* PERFECT PHYLOGENY, TRIANGULATING COLORED GRAPHS *and* TRIANGULATING MULTICOLORED GRAPHS *are all XALP-complete.*

Proof. Combine Theorem 1, Theorem 2 and the equivalences between these three problems. □

We now use these complexity results to show some lower bounds on the space and time usage of PERFECT PHYLOGENY. In the remainder of this section, let n be the input size, k the parameter, f any computable function and c any constant. We start with a bound on the runtime based on the Exponential Time Hypothesis.

Proposition 2. *Assuming ETH, the problems* PERFECT PHYLOGENY, TRIANGULATING COLORED GRAPHS *and* TRIANGULATING MULTICOLORED GRAPHS *cannot be solved in $f(k)n^{o(k)}$ time.*

Proof. We use as a starting point that, assuming ETH, the problem MULTICOLOR INDEPENDENT SET cannot be solved in $f(k)n^{o(k)}$ time [7]. A trivial reduction to TREE-CHAINED MULTICOLOR INDEPENDENT SET using a single-vertex tree then shows the same for that problem. Since the reduction given in Theorem 2 has a linear change in parameter we obtain the same lower bound for TRIANGULATING MULTICOLORED GRAPHS. Finally, using the equivalences between TMG, TCG and PP (all with no change in parameter), the result follows. □

We now bound the space usage based on the Slice-wise Polynomial Space Conjecture (SPSC). This conjectures that LONGEST COMMON SUBSEQUENCE cannot be solved in both $n^{f(k)}$ time and $f(k)n^c$ space [14].

Corollary 1. *Assuming SPSC, the problems* PERFECT PHYLOGENY, TRIANGULATING COLORED GRAPHS *or* TRIANGULATING MULTICOLORED GRAPHS *cannot be solved in both $n^{f(k)}$ time and $f(k)n^c$ space.*

Proof. This proof uses the parameterized complexity class XNLP, which is defined as the class of parameterized problems that are solvable on a deterministic Turing machine using $\mathcal{O}(f(k)\log(n))$ memory. Comparing this with the definition of XALP shows that XALP-hardness implies XNLP-hardness. Since LARGEST COMMON SUBSEQUENCE is XNLP-complete [9], SPSC applies to all XNLP-hard problems and consequently also to all XALP-hard problems such as the three problems from this corollary. □

Compared with the existing algorithm that runs in $\mathcal{O}(n^{k+1})$ time and space [13], these are close but not tight gaps.

4 XALP Membership of Triangulating Colored Graphs

In this section we will prove Theorem 1.

Recall that Triangulating Colored Graphs asks us to determine whether a colored graph can be triangulated. Because of Proposition 1(iii), this is equivalent to finding a tree decomposition where each bag contains each color at most once. We claim that it is equivalent to find a tree decomposition where each bag contains each color *exactly* once.

Lemma 1. *A colored graph admits a tree decomposition where each bag contains each color at most once, if and only if it admits a tree decomposition where each bag contains each color exactly once.*

Proof. Omitted. Included as an appendix in the ArXive version of this paper. □

We can now prove XALP membership.

Proof (of Theorem 1). We construct an alternating Turing machine (ATM) that, given an instance of Triangulating Colored Graphs, determines whether there exists a tree decomposition that contains each color exactly once. As a refresher, an ATM is a Turing machine that has access to both nondetermenistic and co-nondeterministic branching steps. A nondetermenistic step leads to ACCEPT if at least one successor state leads to ACCEPT and a co-nondetermenistic step leads to ACCEPT if all successor states lead to ACCEPT.

Our Turing machine is based on the XP-time algorithm we mentioned before [13]. We use the following claim without proof: given a graph G, a deterministic Turing machine can determine the whether two vertices belong to the same connected component in logarithmic space and polynomial time [15]. Repeated application of this result allows us to branch on all connected components of a graph using several co-nondeterministic steps.

Let G be any colored graph. The Turing machine will use nondeterministic steps to determine how to modify each bag compared to its parent and co-nondeterministic steps to simultaneously verify all subtrees. A precise formulation is given below:

- Using k nondeterministic steps, determine an initial bag S which contains one vertex of each color. During computations that lead to ACCEPT, each S will be a bag from the tree decomposition.
- Keep track of some vertex i that is initially NULL. This will signify the parent of the current bag S.
- Repeat the following until an ACCEPT or REJECT state is reached:
    - Determine all components of $G \setminus S$. Using a co-nondeterministic step, we branch into every component except the one that contains i. If this results in zero branches (e.g. when there are no other components), ACCEPT.

- Let C be the component our current branch is in. We determine a vertex $v \in C$ with a nondeterministic step.
- Determine the vertex $w \in S$ that has the same color as v. Since S contains one vertex of every color, w exists.
- If w is adjacent to any vertex from C, REJECT. This means that the current guess for how to modify S is incorrect.
- Modify S by adding v and removing w. Set i to w.

Overall, this alternating Turing machine constructively determines a rooted tree decomposition if one exists and thus solves TRIANGULATING COLORED GRAPHS. It also satisfies the memory requirement: the only memory usage is the set S, a constant number of extra vertices, and the memory needed to branch on connected components. Since memory of a vertex uses $\mathcal{O}(\log(n))$ space and $|S| = k$, we need $\mathcal{O}(k \log(n))$ space. We also use polynomial time: the time usage in the computation of each bag is a constant plus the time needed to find the connected components which results in polynomial time overall. Finally, we require at most $\mathcal{O}(n)$ co-nondeterministic computation steps: each co-nondeterministic step corresponds to branching into a subtree of the eventual (rooted) tree decomposition. Since each subtree introduces at least one vertex that is used nowhere else in the tree, there are at most $\mathcal{O}(n)$ subtrees.

Overall, we conclude that TRIANGULATING COLORED GRAPHS is contained in XALP. □

5 Zipper Chains and Gadgets

In this section we will introduce two multicolored graph components, the *zipper chain* and the *zipper gadget*. Their most important property is Proposition 4 which says that a zipper gadget has a fixed number of triangulations. This will be used in the XALP-hardness proof to represent a choice.

Definition 2. *A zipper chain is a multicolored graph that consists of two paths P and Q, not necessarily of the same length. The vertices of P and Q are respectively labeled as $p_1, p_2, \dots$ and $q_1, q_2, \dots$.*

The vertices are colored in 7 colors, with 2 colors per vertex. For ease of explanation, the colors are grouped in three groups with sizes 1, 2, and 4. The first group contains one color a which is added to odd-labeled vertices from P and even-labeled vertices from Q. The second group contains the color b_P which is added to even-labeled vertices of P and the color b_Q which is added to odd-labeled vertices of Q. The third group contains four colors c_1, c_2, c_3 and c_4 where c_i is added to vertices in P whose index is equivalent to i (mod 4) and vertices in Q whose index is equivalent to $i + 2$ (mod 4).

To summarize: the colors on path P are $ac_1, b_Pc_2, ac_3, b_Pc_4, ac_1, \dots$ and those of Q are $b_Qc_3, ac_4, b_Qc_1, ac_2, b_Qc_3, \dots$. This is visualized in Fig. 1.

This color pattern repeats every four vertices. We call such a repetition a *tooth* of the zipper chain. If a triangulation of the zipper chain contains an edge between some tooth of P and some tooth of Q and at least one endpoint of this edge contains the color a, we say that these two teeth are *locked together*.

Proposition 3. *Let G be a graph containing a zipper chain (P, Q) and assume that there is a cycle that fully contains both P and Q. Any triangulation of G satisfies the following properties. Because of symmetry, all properties also hold with P and Q reversed.*

(i) There is no edge between two non-adjacent vertices of P.
(ii) If there exist two edges between P and Q which share an endpoint in P, then the common endpoint in P is connected to all vertices of Q that lie between the other two endpoints.
(iii) If (p_i, q_j) is an edge, then either (p_{i+1}, q_j) or (p_i, q_{j+1}) is also an edge (as long as either p_{i+1} or q_{j+1} exists).
(iv) If (p_i, q_j) is an edge and p_i contains the color a, then (p_{i+1}, q_{j+1}) is also an edge (as long as both p_{i+1} and q_{j+1} exist). Here, q_{j+1} contains the color a.
(v) If the i-th tooth of P and the j-th tooth of Q are locked together, then the $i+1$-th tooth of P and the $j+1$-th tooth of Q are also locked together.
(vi) Each tooth from P is locked together with at most one tooth from Q.

Proof. We prove the statements in order.

(i) If, to the contrary, such an edge does exist, then this edge together with the rest of P forms a cycle whose vertices alternate between the colors a and b_P. Because of Proposition 1(i) such a cycle cannot be triangulated.
(ii) Because of part (i), the cycle formed by these two edges and the path between the two endpoints on Q can only be triangulated by adding edges with an endpoint in P.
(iii) If (p_{i+1}, q_j) is not an edge then Proposition 1(ii) shows that p_i must be connected to another vertex in the cycle. Because of part (i) this neighbor is a vertex from Q. Because of part (ii) p_i must then also be connected to q_{j+1}.
(iv) Without loss of generality, say that p_i also contains the color c_1. Then, q_j must have the colors b_Q and c_3: all other color combinations share a color with p_i. Since q_{j+1} and p_i both contain the color a there is no edge between them so part (iii) implies that there is one between p_{i+1} and q_j. Since p_{i+2} and q_j share the color c_3, the same argument implies that there is an edge between p_{i+1} and q_{j+1} (Fig. 2).

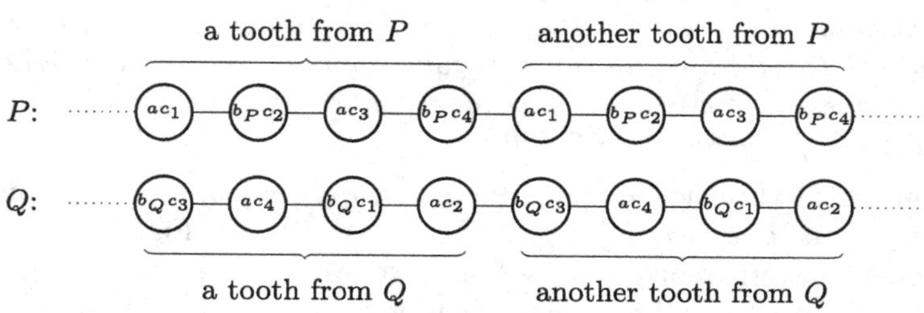

Fig. 1. A zipper chain.

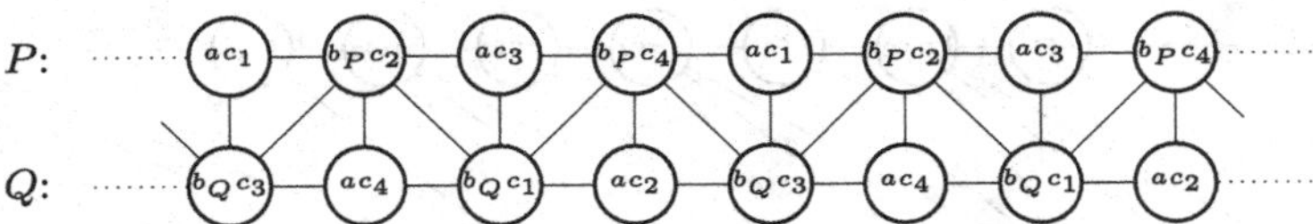

Fig. 2. A possible triangulation of a zipper chain.

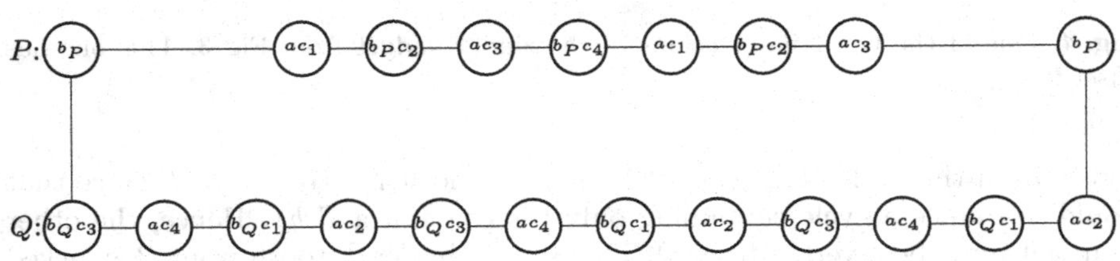

Fig. 3. A zipper gadget of size 2 and skew 1.

(v) Apply part (iv) four times to the edge connecting the i-th and j-th teeth of P and Q (respectively) to obtain an edge connecting the $i+1$-th and $j+1$-th teeth of P and Q (respectively).

(vi) Suppose to the contrary that a tooth from P is locked together with two teeth from Q. After some applications of parts (iii) and (iv) we find that the last vertex from the tooth from P (which has colors b_P and c_3) is connected to the last vertex from both teeth from Q. Because of part (ii), it is connected to all four vertices of the tooth from Q with the higher index. At least one of these also contains the color c_3 so this is a contradiction. □

Zipper Gadgets. We now introduce the *zipper gadget*. It is a zipper chain with a specific length and a head and tail. An example is given in Fig. 3.

Definition 3. *A zipper gadget of size n and skew s (satisfying $n > 0, s \geq 0$) is a zipper chain with the following modifications:*

- *The path P contains $4n - 1$ vertices, and thus n teeth. The last tooth misses one vertex.*
- *The path Q contains $4(n + s)$ vertices, and thus $n + s$ teeth.*
- *There are two additional vertices with just the color b_P: a head h and a tail t. The head is connected to the first vertices of P and Q and the tail to the last vertices of P and Q.*

Proposition 4. *There are exactly $s+1$ ways to triangulate a zipper gadget with skew s. These ways are identified by the offset at which the teeth lock together.*

Proof. Observe that the entire gadget forms a cycle, so Proposition 3 applies. Consider a vertex from P that contains the color a. Its neighbors share the color b_P, so Proposition 1(ii) shows that this vertex must be connected to some other

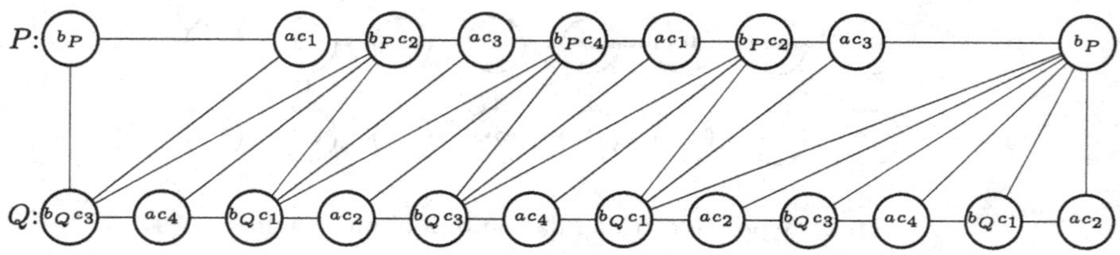

Fig. 4. One of the two triangulations of the zipper gadget from Fig. 3. This one has offset 0.

vertex from the cycle. This cannot be h, t or another vertex from P since that would introduce a cycle containing only the colors a and b_P. Hence, the other endpoint must be a vertex from Q. This shows that each tooth from P is locked together with at least one tooth from Q.

Proposition 3(vi) now shows that each tooth from P is locked together with exactly one tooth from Q. Let Δ be the index of the tooth locked together with the first tooth of P. Proposition 3(v) now shows that any tooth with index i must be connected to tooth $i + \Delta$. Since Q has s more teeth than P, the offset Δ must be between 0 and s. We conclude that there are at most $s + 1$ ways to triangulate a zipper gadget with offset s and that these ways are identified by the offset.

To complete the proof, we now show that each case can actually be extended into a triangulation of the zipper gadget. Let Δ be the target offset. We add the following edges:

- An edge between the head h and every vertex from the first Δ teeth from Q.
- Edges between the i-th tooth from path P and the $i + \Delta$-th tooth from Q according to the pattern described in parts (iii) and (iv) of Proposition 3. This includes one overlap edge between the last vertex of each tooth from P and the first vertex from the next tooth from Q.
- An edge between the tail t and every vertex from the last $s - \Delta$ teeth from Q.

An example of such a triangulation is given in Fig. 4. One can observe that this construction indeed triangulates the zipper gadget. □

6 XALP-Hardness of Triangulating Multicolored Graphs

In this section we describe the reduction from Theorem 2. The intuition is as follows. We want to reduce from TREE-CHAINED MULTICOLOR INDEPENDENT SET, which comes down to selecting a vertex from each color for each instance of MULTICOLOR INDEPENDENT SET. These choices must be compatible: we may not choose two vertices which share an edge. The selection of a vertex will be done by creating zipper gadgets and interpreting each possible triangulation as a choice of a vertex. The compatibility checks will be done by combining two zipper gadgets in a way that makes it impossible to simultaneously triangulate both zipper gadgets in the respective choices. This construction borrows a technique, namely on how to create and combine gadgets from the TCMIS tree, from the XALP-completeness proof for Tree Partition Width from [4]. The actual gadgets and their combination procedure are new.

Let an instance of TCMIS be given. Let T and k be the (binary) tree and parameter from this instance. For any node $n \in T$, we have an associated instance of Multicolor Independent Set consisting of a set of vertices S_c for each color c. Without loss of generality, we can assume that all sets S_c have the same size, say $r + 1$: if not, then we can add extra vertices to S_c that are connected to all other vertices and thus never occur in an independent set. We also assume that S_c is ordered in some way. This allows us to refer to vertices as $v_{n,c,i}$ where n is the node from T, c is the color, and i is the index in S_c (which, for ease of explanation, is zero-based). We also have a set of edges E, which we again assume to be ordered in some way. Each edge connects two vertices v_{n_1,c_1,i_1} and v_{n_2,c_2,i_2} where n_1 and n_2 are either the same node or neighbors in T and where c_1 and c_2 are distinct if $n_1 = n_2$. Let $m := |E|$ be the total number of edges.

First, we transform T into a rooted tree T' by choosing any node $u \in T$, adding two new nodes v and w and two edges (u, v) and (v, w), and setting w as the root. This way, each node from the original tree T has a parent and a grandparent in T'. We now construct a graph G which will be an instance of TMG. It will consist of several zipper gadgets in which some vertices have been identified with each other: that is, where some vertices with distinct colors are merged into one vertex with the combined set of colors. For each node n in T and each color c in its associated instance of Multicolor Independent Set, we add a zipper gadget $z_{n,c}$ of size $2mr + 1$ and skew r. The middle tooth of path P (with index $mr + 1$) is special: we call it the *middle*. We now say that this zipper gadget starts in n, passes through the parent of n and ends in the grandparent of n. This is supported with some vertex identifications: for each node n in T', we identify the heads of all zipper gadgets starting at n, the tails of all zipper gadgets ending at n, and the last vertex of the middles (with colors c_4 and b_P) of all zipper gadgets that pass through n. Observe that the path P of each zipper gadget now consists of m sets of r teeth between its head and middle, and also m sets of r teeth between its middle and tail.

Each zipper gadget is assigned its own set of 7 colors such that no two zipper gadgets which start, pass through, or end in a common node share a color. We claim that this can be done using at most $7k$ sets of 7 colors. Assign colors to nodes in order of distance to the root of T' (closest to the root first). Let n be the current node. All zipper gadgets that have already been assigned colors and intersect with zipper gadgets starting from n are those that start at either: n's parent, the other child of n's parent (n's sibling), n's grandparent, the other child of n's grandparent (n's uncle), or any of the two children from that vertex (n's cousins). In total, this is at most $6k$ other zipper gadgets. To color the k zipper gadgets starting at n, we can thus use the remaining $7k - 6k = k$ sets of 7 colors.

In a triangulation of G, each zipper gadget will represent a choice of a vertex from S_c: if the zipper gadget is triangulated with offset Δ, then we choose the vertex with index Δ from S_c. Each of the m sets of r teeth between head and middle or between middle and tail will represent a restriction regarding one of the edges. For each edge e_i (with index i) with endpoints v_{n_1,c_1,i_1} and v_{n_2,c_2,i_2} we want to exclude the possibility of simultaneously triangulating the zipper gadget z_{n_1,c_1} with offset i_1 and the zipper gadget z_{n_2,c_2} with offset i_2. This is done as follows.

Let (P_1, Q_1) and (P_2, Q_2) be the paths which form the zipper gadgets. We now identify two vertices from P_1 and P_2 and add a new color d to some vertices from Q_1 and Q_2. This is visualized in Fig. 5. The idea is that if we would triangulate both zipper gadgets in a way that adds edges between the vertices with color d and the merged vertex, then any triangulation of both zipper gadgets together forces an edge between the vertices with the color d which is impossible. We now describe exactly which vertices should be modified.

We consider two cases: either n_1 and n_2 are the same node or they are neighbors in T. In the first case, we consider the tooth with index ir from both P_1 and P_2 and identify the first vertex from these teeth with each other. We also consider tooth $ir + i_1$ from Q_1 and tooth $ir + i_2$ from Q_2 and add a new color d to the first vertex of these teeth. In the second case, we assume without loss of generality that n_1 is the parent of n_2. We do almost the same as in the first case, except that we use the second half of the zipper gadget z_{n_2,c_2}: we identify the first vertex of tooth ir from P_1 and tooth $mr + 1 + ir$ from P_2, and we add color d to the first vertex of tooth $ir + i_1$ from Q_1 and tooth $mr + 1 + ir + i_2$ from Q_2.

This completes the construction. Observe that this construction uses $49k + 1$ colors ($7k$ sets of 7 colors for the zipper gadgets and one for the extra color d) and thus that the change in parameter is linear. Also observe that the construction can be performed in logarithmic working space since the creation and merging of the zipper gadgets only require local information from the original TCMIS instance. This shows that we indeed have a logspace reduction.

The proof that this TMG instance admits a triangulation if and only if the original TCMIS instance admits a solution is a direct result of the intuitive insights mentioned during the construction and thus omitted from the main text. A full proof is given in the appendix of the ArXive version of the paper.

7 Future Research

Let n be the input size, k the parameter, f any computable function, c any constant, and ϵ any small positive constant. We have shown that Perfect Phylogeny and Triangulating Colored Graphs are XALP-complete and that (assuming ETH) there exist no algorithms that solve any of them in $f(k)n^{o(k)}$ time. This increases the number of "natural" problems in the complexity class XALP and gives more reason to determine properties of this complexity class. Additionally, these problems can be used as a starting point for XALP-hardness reductions for other parameterized problems.

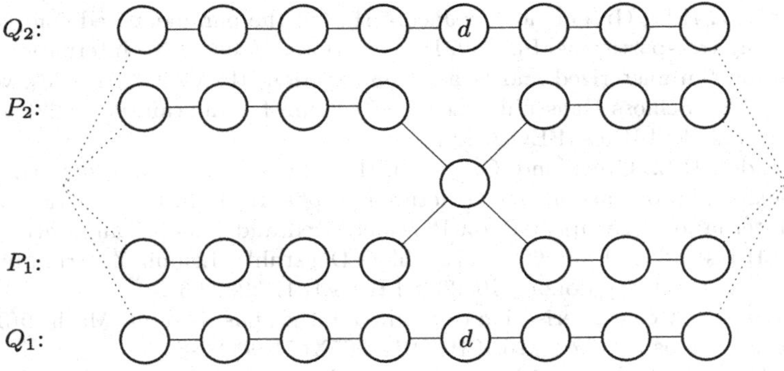

Fig. 5. How two zipper gadgets are combined: two vertices from P_1 and P_2 are merged into one vertex and two vertices from Q_1 and Q_2 are given the extra color d.

Another future research direction might be to close or reduce the gaps between the current upper and lower bounds on space and time usage. For the time gap, there is a lower bound of $f(k)n^{o(k)}$ (assuming ETH) and an upper bound of $\mathcal{O}(n^{k+1})$ [13]. For the space gap on algorithms that run in $n^{f(k)}$ time, there is a lower bound of $f(k)n^c$ (assuming SPSC) and an upper bound of again $\mathcal{O}(n^{k+1})$ [13]. One way to close the time gap could be by assuming the Strong Exponential Time Hypothesis (SETH). We expect that, assuming SETH, a lower bound like $f(k)n^{k-\epsilon}$ should be possible.

We also rule out a research direction. Triangulating a colored graph comes down to finding a tree decomposition where each bag contains each color at most once. A similar problem would be to instead look for a *path* decomposition where each bag contains each color at most one. This problem, known as INTERVALIZING COLORED GRAPHS, is already NP-complete for the case $k = 4$ [1].

Acknowledgements. This paper was written as a master thesis at Utrecht University. I wish to thank my supervisors Hans L. Bodlaender and Carla Groenland for the discussions and guidance.

References

1. Bodlaender, H.L., de Fluiter, B.: On intervalizing k-colored graphs for DNA physical mapping. Discret. Appl. Math. **71**(1), 55–77 (1996). https://doi.org/10.1016/S0166-218X(96)00057-1
2. Bodlaender, H.L., Fellows, M.R., Hallett, M.T., Wareham, H., Warnow, T.J.: The hardness of perfect phylogeny, feasible register assignment and other problems on thin colored graphs. Theoret. Comput. Sci. **244**(1), 167–188 (2000). https://doi.org/10.1016/S0304-3975(98)00342-9
3. Bodlaender, H.L., Fellows, M.R., Warnow, T.J.: Two strikes against perfect phylogeny. In: Kuich, W. (ed.) ICALP 1992. LNCS, vol. 623, pp. 273–283. Springer, Heidelberg (1992). https://doi.org/10.1007/3-540-55719-9_80

4. Bodlaender, H.L., Groenland, C., Jacob, H.: On the parameterized complexity of computing tree-partitions. In: Dell, H., Nederlof, J. (eds.) 17th International Symposium on Parameterized and Exact Computation, IPEC 2022. LIPIcs, vol. 249, pp. 7:1–7:20. Schloss Dagstuhl – Leibniz-Zentrum für Informatik (2022). https://doi.org/10.4230/LIPIcs.IPEC.2022.7
5. Bodlaender, H.L., Groenland, C., Jacob, H., Pilipczuk, M., Pilipczuk, M.: On the complexity of problems on tree-structured graphs. In: Dell, H., Nederlof, J. (eds.) 17th International Symposium on Parameterized and Exact Computation, IPEC 2022. LIPIcs, vol. 249, pp. 6:1–6:17. Schloss Dagstuhl – Leibniz-Zentrum für Informatik (2022). https://doi.org/10.4230/LIPIcs.IPEC.2022.6
6. Buneman, P.: A characterisation of rigid circuit graphs. Discret. Math. **9**(3), 205–212 (1974). https://doi.org/10.1016/0012-365X(74)90002-8
7. Chen, J., et al.: Tight lower bounds for certain parameterized NP-hard problems. In: 19th IEEE Annual Conference on Computational Complexity, vol. 19, pp. 150–160 (2004). https://doi.org/10.1109/CCC.2004.1313826
8. Downey, R.G., Fellows, M.R.: Fixed-parameter tractability and completeness I: basic results. SIAM J. Comput. **24**(4), 873–921 (1995). https://doi.org/10.1137/S0097539792228228
9. Elberfeld, M., Stockhusen, C., Tantau, T.: On the space and circuit complexity of parameterized problems: classes and completeness. Algorithmica **71**(3), 661–701 (2015). https://doi.org/10.1007/s00453-014-9944-y
10. Estabrook, G., Johnson, C., McMorris, F.: A mathematical foundation for the analysis of cladistic character compatibility. Math. Biosci. **29**(1), 181–187 (1976). https://doi.org/10.1016/0025-5564(76)90035-3
11. Kannan, S., Warnow, T.: Inferring evolutionary history from DNA sequences. In: 31st Annual Symposium on Foundations of Computer Science, FOCS 1990, vol. 1, pp. 362–371 (1990). https://doi.org/10.1109/FSCS.1990.89555
12. Kannan, S., Warnow, T.: A fast algorithm for the computation and enumeration of perfect phylogenies. SIAM J. Comput. **26**(6), 1749–1763 (1997). https://doi.org/10.1137/S0097539794279067
13. McMorris, F.R., Warnow, T.J., Wimer, T.: Triangulating vertex-colored graphs. SIAM J. Discret. Math. **7**(2), 296–306 (1994). https://doi.org/10.1137/S0895480192229273
14. Pilipczuk, M., Wrochna, M.: On space efficiency of algorithms working on structural decompositions of graphs. ACM Trans. Comput. Theory **9**(4), 18:1–18:36 (2018). https://doi.org/10.1145/3154856
15. Reingold, O.: Undirected connectivity in log-space. J. ACM **55**(4), 1–24 (2008). https://doi.org/10.1145/1391289.1391291

Data Reduction for Directed Feedback Vertex Set on Graphs Without Long Induced Cycles

Jona Dirks[1], Enna Gerhard[1], Mario Grobler[1], Amer E. Mouawad[2], and Sebastian Siebertz[1(✉)]

[1] University of Bremen, Bremen, Germany
{dirks2,gerhard,grobler,siebertz}@uni-bremen.de
[2] American University of Beirut, Beirut, Lebanon
aa368@aub.edu.lb

Abstract. We study reduction rules for Directed Feedback Vertex Set (DFVS) on instances without long cycles. A DFVS instance without cycles longer than d naturally corresponds to an instance of d-Hitting Set, however, enumerating all cycles in an n-vertex graph and then kernelizing the resulting d-Hitting Set instance can be too costly, as already enumerating all cycles can take time $\Omega(n^d)$. To the best of our knowledge, the kernelization of DFVS on graphs without long cycles has not been studied in the literature, except for very restricted cases, e.g., for tournaments, in which all induced cycles are of length three. We show how to compute a kernel with at most $2^d k^d$ vertices and at most $d^{3d} k^d$ induced cycles of length at most d (which however, cannot be enumerated efficiently). We then study classes of graphs whose underlying undirected graphs have bounded expansion or are nowhere dense; these are very general classes of sparse graphs, containing e.g. classes excluding a minor or a topological minor. We prove that for such classes without induced cycles of length greater than d we can compute a kernel with $\mathcal{O}_d(k)$ and $\mathcal{O}_{d,\varepsilon}(k^{1+\varepsilon})$ vertices for any $\varepsilon > 0$, respectively, in time $\mathcal{O}_d(n^{\mathcal{O}(1)})$ and $\mathcal{O}_{d,\varepsilon}(n^{\mathcal{O}(1)})$, respectively, where k is the size of a minimum directed feedback vertex set. The most restricted classes we consider are planar graphs without any (induced or non-induced) long cycles. We show that strongly connected planar graphs without long cycles have bounded treewidth and hence DFVS on such graphs can be solved in time $2^{\mathcal{O}(d)} \cdot n^{\mathcal{O}(1)}$. We finally present a new data reduction rule for general DFVS and prove that the rule together with a few standard rules subsumes all the rules applied by Bergougnoux et al. to obtain a polynomial kernel for DFVS[FVS], i.e., DFVS parameterized by the feedback vertex set number of the underlying (undirected) graph.

1 Introduction

A directed feedback vertex set of a directed n-vertex graph G is a subset $S \subseteq V(G)$ of vertices such that every directed cycle of G intersects with S. In the

H. Fernau et al. (Eds.): SOFSEM 2024, LNCS 14519, pp. 183–197, 2024.
https://doi.org/10.1007/978-3-031-52113-3_13

Directed Feedback Vertex Set (DFVS) problem, we are given a directed graph G and an integer k, and the objective is to determine whether G admits a directed feedback vertex set of size at most k. In what follows, unless stated otherwise, when we speak of a graph we always mean a directed graph, and when we speak of a cycle we mean a directed cycle.

DFVS is one of Karp's 21 NP-complete problems [28]. Its NP-completeness follows easily by a reduction from Vertex Cover, which is a special case of DFVS where every edge lies on an induced cycle of length two. The fastest known exact algorithm for DFVS, due to Razgon [37], runs in time $\mathcal{O}(1.9977^n \cdot n^{\mathcal{O}(1)})$. Chen et al. [7] proved that the problem is fixed-parameter tractable when parameterized by solution size k; providing an algorithm running in time $\mathcal{O}(k!4^k k^4 nm) = 2^{\mathcal{O}(k \log k)} \cdot nm$, for digraphs with n vertices and m edges. The dependence on the input size has been improved to $\mathcal{O}(k!4^k k^5(n+m))$ by Lokshtanov et al. [32]. The problem has also been studied under different parameterizations. Bonamy et al. [6] proved that one can solve the problem in time $2^{\mathcal{O}(t \log t)} \cdot n^{\mathcal{O}(1)}$, where t denotes the treewidth of the underlying undirected graph. They also proved that this running time is tight assuming the exponential-time hypothesis (ETH). On planar graphs the running time can be improved to $2^{\mathcal{O}(t)} \cdot n^{\mathcal{O}(1)}$. A natural question is whether these results extend to directed width measures, e.g., whether the problem is fixed-parameter tractable when parameterized by directed treewidth. This is not the case: DFVS remains NP-complete even on very restricted classes of graphs such as graphs of cycle rank at most four (which in particular have bounded directed treewidth), as shown by Kreutzer and Ordyniak [29], and hence the problem is not even in XP when parameterized by cycle rank.

The question whether DFVS parameterized by solution size k admits a polynomial kernel, i.e., an equivalent polynomial-time computable instance of size polynomial in k, remains one of the central open questions in the area of kernelization. Bergougnoux et al. [4] showed that the problem admits a kernel of size $\mathcal{O}(f^4)$ in general graphs and $\mathcal{O}(f)$ in graphs embeddable on a fixed surface, where f denotes the size of a minimum undirected feedback vertex set in the underlying undirected graph. Note that f can be arbitrarily larger than k. More generally, for an integer η, a subset $M \subseteq V(G)$ of vertices is called a *treewidth η-modulator* if $G - M$ has treewidth at most η. Lokshtanov et al. [33] showed that when given a graph G, an integer k, and a treewidth η-modulator of size ℓ, one can compute a kernel with $(k \cdot \ell)^{\mathcal{O}(\eta^2)}$ vertices. This result subsumes the result of Bergougnoux et al. [4], as the parameter $k + \ell$ is upper bounded by $\mathcal{O}(f)$ and can be arbitrarily smaller than f. On the other hand, unless NP $\subseteq$ coNP/poly, for $\eta \geq 2$, there cannot exist a polynomial kernel when we parameterize by the size of a treewidth-η modulator alone, as even Vertex Cover cannot have a polynomial kernel when parameterized by the size of a treewidth-2 modulator [8]. Polynomial kernels are known for several restricted graph classes, see e.g. [2,11,20].

From the viewpoint of approximation, the best known algorithms for DFVS are based on integer linear programs whose fractional relaxations can be solved efficiently. It was shown by Seymour [38] that the integrality gap for DFVS is at most $\mathcal{O}(\log k^* \log\log k^*)$, where k^* denotes the optimal value of a fractional directed feedback vertex set. Note that the linear programming formulation of DFVS may contain an exponential number of constraints. Even et al. [16] circumvented this obstacle and provided a related combinatorial polynomial-time algorithm yielding an $\mathcal{O}(\log k^* \log\log k^*) \subseteq \mathcal{O}(\log k \log\log k)$-approximation. Assuming the Unique Games Conjecture (UGC), the problem does not admit a polynomial-time computable constant-factor approximation algorithm [25,26,39]. Lokshtanov et al. [31] showed how to compute a 2-approximation in time $2^{\mathcal{O}(k)} \cdot n^{\mathcal{O}(1)}$.

This work was initiated after successfully participating in the PACE 2022 programming challenge [24]. In the scope of a student project at the University of Bremen, we participated in the competition and our solver ranked second in the exact track [3]. In this paper we present our theoretical findings, whereas an empirical evaluation of the implemented rules will be presented in future work.

We first study DFVS instances without long cycles, which is intimately linked to the study of the Hitting Set problem. Many of the known data reduction rules for DFVS are special cases of general reduction rules for Hitting Set. A hitting set in a set system $\mathcal{G}$ with ground set $V(\mathcal{G})$ and edge set $E(\mathcal{G})$, where each $S \in E(\mathcal{G})$ is a subset of $V(\mathcal{G})$, is a subset $H \subseteq V(\mathcal{G})$ such that $H \cap S \neq \varnothing$ for all $S \in E(\mathcal{G})$. Given a graph G, a directed feedback vertex set in G corresponds one-to-one to a hitting set for the set system $\mathcal{G}$ where $V(\mathcal{G}) = V(G)$ and $E(\mathcal{G}) = \{V(C) \mid C \text{ is a cycle in } G\}$. The main difficulty in applying reduction rules designed for Hitting Set is that we first need to efficiently convert an instance of DFVS to an instance of Hitting Set. However, in general, we want to avoid computing $\mathcal{G}$ from G, as $|E(\mathcal{G})|$ may be super-polynomial in the size of the vertex set, i.e., super-polynomial in $|V(G)| = |V(\mathcal{G})|$. One simple reduction rule for Hitting Set is to remove all sets $S \in E(\mathcal{G})$ such that there exists $S' \in E(\mathcal{G})$ with $S' \subseteq S$. Instances of Hitting Set that do not contain such pairs of sets are called *vertex induced.* The remaining minimal sets in the corresponding DFVS instance are the induced cycles of G. It follows that in a DFVS instance it suffices to hit all induced cycles. Unfortunately, it is NP-complete to detect if a vertex or an edge lies on an induced cycle [18] even on planar graphs, implying that it is not easy to exploit this property for DFVS directly. Overcoming this obstacle requires designing data reduction rules based on sufficient conditions guaranteeing that a vertex or an edge does not lie on an induced cycle and can therefore be safely removed.

An instance of DFVS without cycles of length greater than d naturally corresponds to an instance of d-Hitting Set. As shown in [1], d-Hitting Set admits a kernel with $k+(2d-1)k^{d-1}$ vertices, which can be efficiently computed when the d-Hitting Set instance is explicitly given as input. This is known to be near optimal, as d-Hitting Set does not admit a kernel of size $\mathcal{O}(k^{d-\varepsilon})$ unless the polynomial hierarchy collapses [9]; note that here size refers to the total size of the instance and not to the number of vertices. The question of whether there exists a kernel for d-Hitting Set with fewer elements is considered to be one of the most important open problems in kernelization [5,11,21,22,42]. However, even in this restricted case we cannot efficiently generate the d-HITTING SET instance from the DFVS instance, as even testing if a vertex lies on an induced cycle of length at most d is W[1]-hard [27] when parameterized by d. We hence have to avoid computing a HITTING SET instance explicitly but must rather work on the implicit graph representation of a DFVS instance. To the best of our knowledge the kernelization of DFVS on graphs without long cycles has not been studied in the literature, except for some very restricted cases, e.g., on tournaments in which all induced cycles are of length three [5,11,20]. We remark that due to the hardness of deciding whether a graph contains an induced cycle of length greater than d, we need to be given the promise that the input graph does not contain such induced cycles.

We show that after applying standard reduction rules we can compute in polynomial time a superset W of the vertices that lie on induced cycles of length at most d and which is of size at most $2^d k^d$. As it suffices to hit all induced cycles, $G[W]$ is an equivalent instance. Up to a factor k and constants depending only on d this matches the best known bounds (stated above) for the kernelization of d-HITTING SET. Potentially in the kernelized instance we could have $(2^d k^d)^d = 2^{d^2} k^{d^2}$ induced cycles. Based on the classical sunflower lemma, we prove however, that kernelized instances contain at most $d^{3d} k^d$ induced cycles of length at most d, for any fixed $d \geq 2$. In light of the question whether DFVS admits a polynomial kernel, we pose as a question whether it admits a kernel of size $\mathcal{O}_d(k^{\mathcal{O}(1)})$ computable in time $\mathcal{O}_d(n^{\mathcal{O}(1)})$ on instances without induced cycles of length greater than d.

We then turn our attention to restricted graph classes for which we can efficiently test whether a vertex lies on an induced cycle of length at most d, e.g., by efficient algorithms for first-order model-checking [13,23]. We study classes of graphs whose underlying undirected graphs have bounded expansion or are nowhere dense. We show that DFVS on classes of bounded expansion admits a kernel with $\mathcal{O}_d(k)$ vertices, and a kernel with $\mathcal{O}_{d,\varepsilon}(k^{1+\varepsilon})$ vertices, for any $\varepsilon > 0$, on nowhere dense classes, respectively, computable in time $\mathcal{O}_d(n^{\mathcal{O}(1)})$ and $\mathcal{O}_{d,\varepsilon}(n^{\mathcal{O}(1)})$, respectively. This answers our above question for very general classes of sparse graphs. Our method is based on the approach of [12,14] for the kernelization of the DISTANCE-r DOMINATING SET problem on bounded expansion and nowhere dense classes. We conclude our study of restricted graph classes by observing that a strongly connected planar graph without any long (induced or non-induced) cycles has bounded treewidth. We observe that after the application of some reduction rules, weak components are equal to strong components. Hence, the DAG of strong components in fact is a tree. Then, if each strong component has bounded treewidth, we can combine the tree decompositions of the strong components with the tree of strong components to derive that the whole graph after application of the rules has bounded treewidth and solve it efficiently.

We proceed by designing a new data reduction rule that provides a sufficient condition for a vertex or edge to lie on an induced cycle. The new rule conveniently generalizes many of the complicated rules presented by Bergougnoux et al. [4] to establish a kernel of size $\mathcal{O}(f^4)$, where f is the size of a minimum feedback vertex set for the underlying undirected graph. In addition to being simpler, our rule does not require the initial computation of a feedback vertex set for the underlying undirected graph.

Due to space constraints not all proofs can be presented in this conference version. They can be found in the full version of the paper [10].

2 Preliminaries

A *graph* G consists of a (non-empty) *vertex set* $V(G)$ and *edge set* $E(G) \subseteq V(G) \times V(G)$. For vertices $u, v \in V(G)$ we write uv for the edge directed from u to v. An edge vv is called a *loop*. We denote the *in-* and *out-neighborhood* of $v \in V(G)$ by $N_G^-(v) = \{u \mid uv \in E(G)\}$ and $N_G^+(v) = \{u \mid vu \in E(G)\}$, respectively. The *neighborhood* of v is denoted by $N_G(v) = N_G^-(v) \cup N_G^+(v)$. A *cycle* C in a graph G is a sequence of vertices $v_1 v_2 \dots v_{\ell+1}$ such that $v_1 = v_{\ell+1}$, $v_i \neq v_j$ for all $i \neq j \leq \ell$, and $v_i v_{i+1} \in E(G)$ for all $i \leq \ell$. We denote the set of vertices that appear in C by $V(C) = \{v_1, \dots, v_\ell\}$. We denote by ℓ the *length* of C. A u-v-path is a sequence of vertices $v_1 v_2 \dots v_{\ell+1}$ of pairwise distinct vertices such that $v_1 = u$ and $v_{\ell+1} = v$. Likewise, we define $V(P) = \{v_1, \dots, v_{\ell+1}\}$ and call ℓ the *length* of the path, that is, the number of edges of P. For a u-v-path P and a v-w-path Q we write PQ for the *u-w-walk* obtained by concatenating P and Q (removing the repetition of v in the middle). Recall that a u-w-walk in a graph G implies the existence of a u-w-path in G using a subset of the vertices and edges of the walk. By a slight abuse of notation, we sometimes use PQ to denote the u-w-path. For a set $S \subseteq V(G)$, we write $N_G^{d+}[S]$ to denote the *d-out-neighborhood* of S in G and $N_G^{d-}[S]$ to denote the *d-in-neighborhood* of S in G. That is, $N_G^{d+}[S]$ contains all vertices of G that are reachable from some vertex in S via a path of length at most $d \geq 0$, and $N_G^{d-}[S]$ contains all vertices of G that can reach some vertex in S via a path of length at most d. Note that since paths of length zero are allowed, we have $S \subseteq N_G^{d+}[S] \cap N_G^{d-}[S]$.

For a vertex subset $U \subseteq V(G)$, we denote by $G[U]$ the graph *induced by* U, that is the graph obtained from G where we only keep the vertices in U and the edges incident on them. We write $G - U$ for the graph $G[V(G) \setminus U]$ and for a singleton vertex set $\{v\}$ we write $G - v$ instead of $G - \{v\}$. For an edge uv we write $G + uv$ and $G - uv$ for the graph obtained by adding or removing the edge uv, respectively. A cycle C is an induced cycle if the graph $G[V(C)]$ is isomorphic to a cycle and a path P is an induced path if the graph $G[V(P)]$ is a path. We call a u-v-path *almost induced* if P is an induced path in $G - vu$. Slightly abusing notation, when u, v are distinct vertices on an induced cycle C, we will say that C decomposes into an induced u-v-path and an induced v-u-path, even though this is not true if $uv \in E(G)$, in which case the u-v-path is almost induced.

We call a set $S \subseteq V(G)$ a *directed feedback vertex set*, dfvs for short, if $G - S$ does not contain any (directed) cycle. An input of the Directed Feedback

Vertex Set problem consists of a graph G and a positive integer k. The goal is to determine whether G admits a dfvs of size at most k. We will constantly make use of the following simple lemma.

Lemma 2.1. *Let G be a graph without induced cycles of length greater than d and let k be a positive integer. Then, we can compute in polynomial time either a dfvs of size at most dk or decide that there is no dfvs of size at most k in G.*

3 DFVS in Graphs Without Long Induced Cycles

We begin our study of DFVS in graphs without induced cycles of length greater than d. Unfortunately, it is NP-complete to determine if a vertex lies on an induced cycle [18]. In fact, this is even W[1]-hard when parameterized by d [27]. By Lemma 2.1 we can approximate a small dfvs S. As a first rule we can delete all vertices that do not lie in $N_G^{d+}[S] \cap N_G^{d-}[S]$. It would be even better to delete all vertices that do not lie on an induced path of length at most d between two vertices $u, v \in S$ (making a copy of u (and its incident edges) when dealing with the case $u = v$). Since $G - S$ is acyclic, one could hope that this is possible in time $\mathcal{O}_d(n^{\mathcal{O}(1)})$, however, as we show next even this is not possible. The Directed Chordless (s, v, t)-Path problem asks, given a graph G, vertices s, v, t, and integer d, whether there exists an induced s-t-path in G of length at most d containing v. The W[1]-hardness of the problem on general (directed and undirected) graphs was proved in [27]. We show hardness on directed acyclic graphs via a reduction from Grid Tiling.

Lemma 3.1. *The* Directed Chordless (s, v, t)-Path *problem parameterized by the length d of a path is W[1]-hard even restricted to directed acyclic graphs.*

By inspecting the proof of Lemma 3.1 we obtain the following corollaries.

Corollary 3.1. *It is W[1]-hard to decide if a vertex lies on an induced cycle of length at most d even on graphs that become acyclic after the deletion of a single edge.*

Corollary 3.2. *It is W[1]-hard to decide if a graph contains an induced cycle of length at least d even on graphs that become acyclic after the deletion of a single edge.*

We will rely on the following sunflower-like rule which was presented as Rule 3 in [4] and the special case of $u = v$ as Rule 6 in [19].

Rule 1. *Let $u, v \in V(G)$. If u and v are connected by more than k internally vertex-disjoint u-v-paths, then insert the edge uv.*

We assume that Rule 1 has been applied exhaustively, i.e., reiterated after any successful application of the rule. We slightly abuse notation and use G to denote the resulting graph, which we call a *reduced graph*.

We now consider graphs that have no induced cycles of length greater than d. We start with a high-level description of our strategy as well as the obstacles that we need to overcome. Given a reduced graph G, we first compute a dfvs S of size at most dk as guaranteed by Lemma 2.1. Since we assume that G has no induced cycles of length greater than d, all vertices of G that are at distance $d+1$ or more from every vertex in S can be discarded as they cannot belong to induced cycles of length at most d that intersect with S. Hence, in what follows, we let $G = G[N_G^{d+}[S] \cap N_G^{d-}[S]]$ (which can be easily computed in polynomial time by standard breadth-first searches). The vertex set of $G = G[N_G^{d+}[S] \cap N_G^{d-}[S]]$ is partitioned into S and $R = V(G)\backslash S$, where $|S| \leq dk$ and every vertex in R is at distance at most d to some vertex in S. Note that we would like to check for each $w \in R$ whether there exists an induced path of length at most d from some $u \in S$ to w and back. However, this is not possible due to Lemma 3.1, since it implies that we cannot efficiently iterate through the vertices of R one by one and decide if they belong to some induced path. Our solution consists of adopting a "relaxed approach". That is, for $u \in S$ let $I_u^d \subseteq V(G)$ denote the set of all vertices that belong to some induced cycle of length at most d containing u. We shall compute, for each vertex $u \in S$, a set $W_u^d \supseteq I_u^d$. In other words, we compute a superset, which we call W_u^d, of the vertices that share an induced cycle of length at most d with u. We call W_u^d the set of *d-weakly relevant vertices for u*. Most crucially, we show that each W_u^d can be computed efficiently and will be of bounded size. We let $W_S^d = \bigcup_{u \in S} W_u^d$ and we call W_S^d the set of *d-weakly relevant vertices for S*. It is not hard to see that $G[S \cup W_S^d]$ is indeed an equivalent instance (to G) as it includes all vertices that participate in induced cycles of length at most d.

We describe the construction of W_u^d for a single vertex. That is, we fix a non-reducible directed graph G, an integer $k \geq 2$, a constant $d \geq 2$, a dfvs S of size at most dk, and a vertex $u \in S$. We first construct a graph H_u^d as follows:

- We begin by setting $H_u^d = G[N_G^{d+}[u]]$.
- Then, we add a new vertex v to H_u^d and make all the in-neighbors of u become in-neighbors of v instead, i.e., u will only have out-neighbors and v will only have in-neighbors.
- Next, we delete all vertices in H_u^d that do not belong to some directed path from u to v of length at most d.

Note that H_u^d can be computed in polynomial time. Moreover, there exists an induced cycle of length at most d containing u in G if and only if there exists an induced u to v path of length at most d in H_u^d. By a slight abuse of notation, we also denote the graph H_u^d by $H_{u,v}^d$ to emphasize the source and sink vertices. We call a directed graph *k-nice* whenever any two vertices x, z are either connected by the directed edge xz or by a set of at most k pairwise internally vertex-disjoint (directed) paths. In particular, either xz is an edge or there exists a set Y (disjoint from $\{x, z\}$) of at most k vertices that hits every directed path from x to z. Observe that H_u^d is indeed k-nice (since Rule 1 has been exhaustively applied on G). Given a k-nice graph H_u^d, two vertices

Algorithm 1. Algorithm for computing weakly relevant vertices for $u \in S$

```
procedure WeaklyRelevant(G, S, u, d)
    return Recurse(H^d_{u,v}, u, v, {}, d)                ▷ Returns W^d_u
end procedure

procedure Recurse(H, x, z, W, d)
    if |V(H)\{x, z}| ≤ k  or  d == 2 then
        return W ∪ (V(H)\{x, z})
    end if
    Y ← VertexSeparator(H, x, z)          ▷ Recall that |Y| ≤ k and x, z ∉ Y
    W ← W ∪ Y
    for y ∈ Y do
        W ← W ∪ Recurse(H^{d-1}_{x,y}, x, y, W, d − 1) ∪ Recurse(H^{d-1}_{y,z}, y, z, W, d − 1)
    end for
    return W
end procedure
```

$x, z \in V(H_u^d)$, and $2 \leq d' < d$, we let $H_{x,z}^{d'}$ denote the (k-nice) graph obtained from H_u^d by deleting all incoming edges of x, deleting all outgoing edges of z, and deleting all vertices that do not belong to a path of length at most d' from x to z. We are now ready to compute W_u^d, for $u \in S$, recursively as described in Algorithm 1. Recall that since Rule 1 is not applicable in G, there do not exist k internally vertex-disjoint (directed) paths between any two non-adjacent vertices of G (and any $H_{x,z}^{d'}$ resulting from the recursive calls). Hence, whenever we compute (via a flow algorithm) a set Y separating two non-adjacent vertices we know that Y will be of size at most k.

Lemma 3.2. *For $u \in S$, every induced cycle C_u of length at most d including u only includes vertices that are d-weakly relevant for u, i.e., $V(C_u) \subseteq W_u^d$.*

Lemma 3.2 immediately implies the safeness of the following rule.

Rule 2. *If a vertex $w \notin S$ is not d-weakly relevant for some vertex $u \in S$ then remove w from G.*

It remains to prove that the rule can be efficiently implemented and that its application leads to a small kernel.

Lemma 3.3. *For $u \in S$ and $2 < \ell \leq d$ we have $|W_u^\ell| \leq k(2|W_u^{\ell-1}| + 1) \leq 2^{\ell-1}k^{\ell-1}$.*

Lemma 3.4. *Rule 2 is safe and, if $2^d k^d \leq n^{\mathcal{O}(1)}$, it can be implemented in polynomial time.*

Theorem 3.1. DFVS *parameterized by solution size k and restricted to graphs without induced cycles of length greater than d admits a kernel with $2^d k^d$ vertices computable in polynomial time.*

We further study the structure of kernelized instances and count how many induced cycles we can find. Our key tool is the classical sunflower lemma. A *sunflower* with ℓ *petals* and a *core* Y is a collection of sets $S_1, \ldots, S_\ell \in E(\mathcal{G})$ such that $S_i \cap S_j = Y$ for all $i \neq j \leq \ell$. The sets $S_i \setminus Y$ are called *petals* and we require none of them to be empty (while the core Y may be empty). Erdös and Rado [15] proved in their famous sunflower lemma that every hypergraph with edges of size at most d with at least $\text{sun}(d,k) = d!k^d$ edges contains a sunflower with at least $k+1$ petals. Kernelization for d-Hitting Set based on the sunflower lemma yields a kernel with at most $\mathcal{O}(d!k^d)$ sets on hypergraphs with hyperedges of size at most d, see e.g. [17,40]. We can prove the following lemma.

Lemma 3.5. *Kernelized instances of* DFVS *contain at most* $d^{3d}k^d$ *induced cycles of length at most* d.

4 Nowhere Dense Classes Without Long Induced Cycles

We now improve the general kernel construction for DFVS on graphs without induced cycles of length greater than d by additionally restricting the class of (the underlying undirected) graphs. We obtain a kernel with $\mathcal{O}_d(k)$ vertices on classes with bounded expansion and $\mathcal{O}_{d,\varepsilon}(k^{1+\varepsilon})$ vertices, for any $\varepsilon > 0$, on nowhere dense classes of graphs (when we say G belongs to a class $\mathscr{C}$ of graphs we in fact mean that the underlying undirected graph belongs to $\mathscr{C}$). We present the proof for nowhere dense classes since it subsumes the bounded expansion case. To keep the presentation clean we omit the details for the latter case since the required modifications are negligible. We refer the reader to [34,35] for formal definitions of bounded expansion and nowhere dense classes of graphs. We only need the following properties, which will also motivate our additional reduction rule. Recall that every class of bounded expansion is also nowhere dense. For every nowhere dense class of graphs $\mathscr{C}$ there exists a positive integer $t > 0$ such that $K_{t,t}$ (the complete biparite graph with t vertices in each part) is not a subgraph of any $G \in \mathscr{C}$.

Let us fix an approximate solution S as described in Lemma 2.1. Let $X \subseteq V(G)$ and let $u \in V(G) \setminus X$. The *undirected d-projection* of u onto X is defined as the set $\Pi_d(u, X)$ of all vertices $w \in X$ for which there exists an undirected path P of length at most d in G that starts in u, ends in w, with internal vertices not in X.

Lemma 4.1 ([14]). *Let* $\mathscr{C}$ *be a nowhere dense class of graphs. There exists a polynomial time algorithm that given a graph* $G \in \mathscr{C}$, d, $\varepsilon > 0$ *and* $X \subseteq V(G)$, *computes the d-projection-closure of* X, *denoted by* X°, *with the following properties:*

1. $X \subseteq X^\circ$,
2. $|X^\circ| \leq \kappa_{d,\varepsilon} \cdot |X|^{1+\varepsilon}$ *for a constant* $\kappa_{d,\varepsilon}$ *depending only on* d *and* ε,
3. $|\Pi_d(u, X^\circ)| \leq \kappa_{d,\varepsilon} \cdot |X|^\varepsilon$ *for each* $u \in V(G) \setminus X^\circ$, *and*

4. $|\{\Pi_d(u, X) \ : \ u \in V(G)\backslash X^\circ\}| \leq \kappa_{d,\varepsilon} \cdot |X|^{1+\varepsilon}$.

We need the following strengthening for ℓ-tuples [36]. For a set $X \subseteq V(G)$ and an ℓ-tuple $\bar{x}$ of vertices we call the tuple $(N[\bar{x}_1] \cap X, \ldots, N[\bar{x}_\ell] \cap X)$ the *undirected projection of* $\bar{x}$ *onto* X. We say that $\bar{X} = (X_1, \ldots, X_\ell)$ is realized as a projection if there is a tuple $\bar{x}$ whose projection is equal to $\bar{X}$.

Lemma 4.2 ([36]). *Let* $\mathscr{C}$ *be a nowhere dense class of graphs and let* ℓ *be a natural number. Let* $G \in \mathscr{C}$ *and* $X \subseteq V(G)$. *Then, for every* $\varepsilon > 0$ *there exists a constant* $\tau_{\ell,\varepsilon}$ *such that there are at most* $\tau_{\ell,\varepsilon} \cdot |X|^{\ell+\varepsilon}$ *different realized undirected projections of* ℓ*-tuples.*

Let $X \subseteq V(G)$ and let $x, y \in X$. Let $P = u_1, \ldots, u_\ell$ be an almost induced x-y-path with $|V(P)| = \ell \leq d$ and let $u_i \in V(P)$. Then, the *X-path-projection profile* of (P, u) is the tuple $(i, N^-(u_1) \cap X, N^+(u_1) \cap X, \ldots, N^-(u_\ell) \cap X, N^+(u_\ell) \cap X)$. The *X-path-projection profile* of vertex u is the set of all X-path-projection profiles (P, u), where P is any almost induced x-y-paths on at most d vertices and $x, y \in X$ are any two vertices in X. Two vertices u, v are *equivalent over* X if they have the same X-path-projection profiles.

Lemma 4.3. *Let* $\mathscr{C}$ *be a nowhere dense class of graphs and let* $t > 0$ *be some fixed positive integer such that* $K_{t,t} \not\subseteq G$, *for all* $G \in \mathscr{C}$. *Let* $G \in \mathscr{C}$ *and* $X \subseteq V(G)$. *Then, for every* $\varepsilon > 0$, *there exists a constant* $\chi_{d,t,\varepsilon}$ *such that the number of X-path-projection profiles for* $u \in V(G)\backslash X$ *is bounded by* $\chi_{d,t,\varepsilon} \cdot |X|^{d+\varepsilon}$.

Rule 3. *If we can find in polynomial time sets* $B, X \subseteq V(G)$ *such that the following holds:*

1. *the* $d+1$*-neighborhoods in* $G - X$ *of distinct vertices from* B *are disjoint,*
2. *every induced cycle using a vertex of* $N_G^d[B]$ *also uses a vertex of* X,
3. *vertices in* B *are pairwise equivalent over* X, *i.e., they have the same X-path-projection profile (in particular, if one vertex of* B *lies on an* x-y*-path of length* $\ell \leq d$, *then all vertices of* B *do as well), and*
4. $|B| > c + d + 1$ *and* $|X| \leq c$, *for some fixed constant* c.

Then choose an arbitrary vertex of B *and delete it from* G.

Lemma 4.4. *Rule 3 is safe.*

Lemma 4.5. *Given a graph* G, $X \subseteq V(G)$ *and* $u \in V(G)$, *we can test in time* $\mathcal{O}_{d,\varepsilon}(n^{\mathcal{O}(1)})$ *whether every induced cycle (on at most* d *vertices) using a vertex of* $N_G^d[u]$ *also uses a vertex of* X. *Furthermore, we can decide in the same time if two vertices* u, v *have the same X-path-projection profile.*

We now use a property of nowhere dense classes of graphs called *uniformly quasi-wideness* [30,35,36] to prove our main theorem.

Theorem 4.1. *Rule 3 can be efficiently applied until the reduced graph has* $\mathcal{O}_{d,\varepsilon}(k^{1+\varepsilon})$ *vertices.*

5 DFVS in Planar Graphs Without Long Cycles

One may wonder whether the stronger assumption that a graph does not contain long cycles, induced or non-induced, leads to even more efficient algorithms. We show that this is indeed the case when considering planar graphs. We show that strongly connected planar graphs without cycles of length d have treewidth $\mathcal{O}(d)$. We observe in Corollary 6.1 that after the application of Rule 7 (stated below), weak components are equal to strong components. Hence, the DAG of strong components in fact is a tree. Then, if each strong component has bounded treewidth, we can combine the tree decompositions of the strong components with the tree of strong components to derive that the whole graph after application of Corollary 6.1 has bounded treewidth. We can then use the algorithm of Bonamy et al. [6] to solve the instance in time $2^{\mathcal{O}(d)} \cdot n^{\mathcal{O}(1)}$.

Theorem 5.1. *Let G be a strongly connected planar graph without cycles of length greater than d. Then, G has treewidth at most $30d$.*

6 Long Induced Cycles

In this section we prove that Rule 1, Rule 4, Rule 5, and Rule 7 lead to a kernel of size $\mathcal{O}(f^4)$, where f is the size of a minimum feedback vertex set in the underlying undirected graph. In fact, we prove the stronger bound of $\mathcal{O}(f^3 k)$. Our analysis is based on the analysis of Bergnouxnoux et al. [4]. Essentially, we prove that all complicated rules of Bergnouxnoux et al. are subsumed by Rule 7. In the full version we also present and attribute special cases that may be more efficiently computable.

Rule 4 is presented, e.g., as Rule 1 in [19].

Rule 4. *If $v \in V(G)$ lies on a loop then add v to the solution, remove v from G, and decrease the parameter by one.*

Rule 5 is based on a folklore rule for Hitting Set, which in the literature is often attributed to [41]. If there are two vertices $u, v \in V(\mathcal{G})$ such that u appears in every hyperedge in which v appears then remove v. We say that u *dominates* v. Removing an element from the universe of $\mathcal{G}$ *almost* corresponds to the following operation in G. For $v \in V(G)$, we write $G \ominus v$ for the graph obtained by connecting all in-neighbors of v with all out-neighbors of v and then removing v, or simply removing v if it has no in- or out-neighbors. We say that $G \ominus v$ is obtained from G by *shortcutting* v. Shortcutting may introduce new cycles that cannot be recovered in G by re-inserting v. However, shortcutting cannot introduce new *induced* cycles (that did not exist in G).

Rule 5. *If there are distinct vertices $u, v \in V(G)$ such that u appears on every cycle on which v appears then shortcut v in G.*

Finally, we state the following conditional rule and then present a special case that is efficiently implementable.

Rule 6. *Remove all vertices and edges for which we can decide in polynomial time that do not lie on induced cycles.*

We now formulate a modified depth-first search rule. We say that a cycle $C = v_1, \dots, v_\ell$ for $\ell \geq 4$ is *induced on an initial segment of length* i (for some $3 \leq i < \ell$) if $v_1, \dots v_i$ induce a path in $G[V(C)]$ (or a cycle in case $i = 3$) and only v_i can have out-neighbors among $v_{i+1}, \dots, v_{\ell-1}$ and only v_1 can have in-neighbors among $v_{i+1}, \dots, v_{\ell-1}$. Note that this definition depends on the vertex that we distinguish as v_1 on the cycle. Note also that by definition every cycle of length three is induced on an initial segment of length three. We say that an edge $uv \in E(G)$, which does not lie on an induced cycle of length 2, *lies on a cycle that is induced on an initial segment of length* i if there exists a cycle $C = v_1, \dots, v_\ell$ that is induced on an initial segment of length i such that $(v_1, v_2) = (u, v)$.

Lemma 6.1. *If* $uv \in E(G)$, *which does not lie on an induced path of length* 2, *does not lie on a cycle that is induced on an initial segment of length* i *for some* $i \geq 3$, *then* uv *does not lie on an induced cycle of* G. *Furthermore, we can test this property in time* $\mathcal{O}(n^{i-2}(n+m))$.

Proof. The first statement is immediate from the fact that an induced cycle of length ℓ is induced on an initial segment of length $\ell - 1$ (independent of the choice of initial vertex v_1). Moreover, if a cycle is induced on an initial segment of length i, then it is induced on an initial segment of length j for every $3 \leq j \leq i$.

For the running time we consider an algorithm that non-deterministically guesses vertices $v_3, \dots, v_i$ such that $v_1, \dots, v_i$ is an induced path in G. We remove all vertices that are out-neighbors of one of the v_j for $1 \leq j \leq i-1$ and all vertices that are in-neighbors of one of the v_j for $2 \leq j \leq i$ and carry out a regular depth-first search from v_i. If we find v_1 in this search, say by visiting the vertices $v_{i+1}, \dots, v_\ell = v_1$ we return the cycle $v_1, \dots, v_\ell$, which is induced on the initial segment of length i by construction. Otherwise, we return that uv does not lie on such a cycle. A deterministic version of the algorithm iterates through all possible sets $v_3, \dots, v_i$ in time $\mathcal{O}(n^{i-2})$. For each set, the algorithm constructs the graph with deleted vertices and carries out a depth-first search in time $\mathcal{O}(n+m)$.

We prove that testing for containment in cycles that are induced on an initial segment of length three subsumes many non-trivial reduction rules of Bergougnoux et al. [4]. Hence, we state the following reduction rule, though rules with larger i may be interesting to consider.

Rule 7. *If an edge* $uv \in E(G)$, *which does not lie on an induced cycle of length* 2, *does not lie on a cycle that is induced on an initial segment of length three, then remove* uv *from* G.

Lemma 6.2. *Rule 7 is safe and can be implemented in time* $\mathcal{O}(nm(n+m))$.

From the proof we note the following corollary that we used in the previous section.

Corollary 6.1. *After exhaustive application of Rule 7 every weak component is strongly connected.*

Finally, we prove that Rule 1, Rule 4, Rule 5, and Rule 7 lead to a kernel of size $\mathcal{O}(f^3k)$, where f is the size of a minimum feedback vertex set in the underlying undirected graph. Our analysis is based on the analysis of Bergnouxnoux et al. [4]. Essentially, we prove that all complicated rules of Bergnouxnoux et al. are subsumed by Rule 7.

Theorem 6.1. *After the exhaustive application of Rule 4, Rule 5, Rule 1 and Rule 7 we obtain a kernel with $\mathcal{O}(f^3k)$ vertices.*

References

1. Abu-Khzam, F.N.: A kernelization algorithm for d-hitting set. J. Comput. Syst. Sci. **76**(7), 524–531 (2010)
2. Bang-Jensen, J., Maddaloni, A., Saurabh, S.: Algorithms and kernels for feedback set problems in generalizations of tournaments. Algorithmica **76**(2), 320–343 (2016)
3. Bergenthal, M., et al.: Pace solver description: Grapa-java. In: IPEC 2022. Schloss Dagstuhl-Leibniz-Zentrum für Informatik (2022)
4. Bergougnoux, B., Eiben, E., Ganian, R., Ordyniak, S., Ramanujan, M.: Towards a polynomial kernel for directed feedback vertex set. Algorithmica **83**(5), 1201–1221 (2021)
5. Bessy, S., et al.: Kernels for feedback arc set in tournaments. JCSS **77**(6), 1071–1078 (2011)
6. Bonamy, M., Kowalik, Ł, Nederlof, J., Pilipczuk, M., Socała, A., Wrochna, M.: On directed feedback vertex set parameterized by treewidth. In: Brandstadt, A., Kohler, E., Meer, K. (eds.) WG 2018. LNCS, vol. 11159, pp. 65–78. Springer, Cham (2018). https://doi.org/10.1007/978-3-030-00256-5_6
7. Chen, J., Liu, Y., Lu, S., O'sullivan, B., Razgon, I.: A fixed-parameter algorithm for the directed feedback vertex set problem. In: STOC 2008, pp. 177–186 (2008)
8. Cygan, M., Lokshtanov, D., Pilipczuk, M., Pilipczuk, M., Saurabh, S.: On the hardness of losing width. Theory Comput. Syst. **54**(1), 73–82 (2014)
9. Dell, H., Van Melkebeek, D.: Satisfiability allows no nontrivial sparsification unless the polynomial-time hierarchy collapses. JACM **61**(4), 1–27 (2014)
10. Dirks, J., Gerhard, E., Grobler, M., Mouawad, A.E., Siebertz, S.: Data reduction for directed feedback vertex set on graphs without long induced cycles. arXiv preprint arXiv:2308.15900 (2023)
11. Dom, M., Guo, J., Hüffner, F., Niedermeier, R., Truß, A.: Fixed-parameter tractability results for feedback set problems in tournaments. J. Discrete Algorithms **8**(1), 76–86 (2010)
12. Drange, P.G., et al.: Kernelization and sparseness: the case of dominating set. In: STACS 2016, LIPIcs, vol. 47, pp. 31:1–31:14. Schloss Dagstuhl - Leibniz-Zentrum für Informatik (2016)
13. Dreier, J., Mählmann, N., Siebertz, S.: First-order model checking on structurally sparse graph classes. In: STOC 2023, pp. 567–580. ACM (2023)
14. Eickmeyer, K., et al.: Neighborhood complexity and kernelization for nowhere dense classes of graphs. In: ICALP 2017, LIPIcs, vol. 80, pp. 63:1–63:14. Schloss Dagstuhl - Leibniz-Zentrum für Informatik (2017)

15. Erdös, P., Rado, R.: Intersection theorems for systems of sets. J. London Math. Soc. **1**(1), 85–90 (1960)
16. Even, G., Schieber, B., Sudan, M.: Approximating minimum feedback sets and multicuts in directed graphs. Algorithmica **20**(2), 151–174 (1998)
17. Fafianie, S., Kratsch, S.: A shortcut to (sun)flowers: kernels in logarithmic space or linear time. In: Italiano, G.F., Pighizzini, G., Sannella, D.T. (eds.) MFCS 2015. LNCS, vol. 9235, pp. 299–310. Springer, Heidelberg (2015). https://doi.org/10.1007/978-3-662-48054-0_25
18. Fellows, M.R., Kratochvíl, J., Middendorf, M., Pfeiffer, F.: The complexity of induced minors and related problems. Algorithmica **13**(3), 266–282 (1995)
19. Fleischer, R., Wu, X., Yuan, L.: Experimental study of FPT algorithms for the directed feedback vertex set problem. In: Fiat, A., Sanders, P. (eds.) ESA 2009. LNCS, vol. 5757, pp. 611–622. Springer, Heidelberg (2009). https://doi.org/10.1007/978-3-642-04128-0_55
20. Fomin, F.V., Le, T., Lokshtanov, D., Saurabh, S., Thomassé, S., Zehavi, M.: Subquadratic kernels for implicit 3-hitting set and 3-set packing problems. TALG **15**(1), 1–44 (2019)
21. Fomin, F.V., Le, T., Lokshtanov, D., Saurabh, S., Thomasse, S., Zehavi, M.: Lossy kernelization for (implicit) hitting set problems. arXiv preprint arXiv:2308.05974 (2023)
22. Fomin, F.V., Lokshtanov, D., Saurabh, S., Zehavi, M.: Kernelization: Theory of Parameterized Preprocessing. Cambridge University Press, Cambridge (2019)
23. Grohe, M., Kreutzer, S., Siebertz, S.: Deciding first-order properties of nowhere dense graphs. JACM **64**(3), 1–32 (2017)
24. Großmann, E., Heuer, T., Schulz, C., Strash, D.: The pace 2022 parameterized algorithms and computational experiments challenge: directed feedback vertex set. In: IPEC 2022. Schloss Dagstuhl-Leibniz-Zentrum für Informatik (2022)
25. Guruswami, V., Håstad, J., Manokaran, R., Raghavendra, P., Charikar, M.: Beating the random ordering is hard: Every ordering CSP is approximation resistant. SICOMP **40**(3), 878–914 (2011)
26. Guruswami, V., Lee, E.: Simple proof of hardness of feedback vertex set. Theory Comput. **12**(1), 1–11 (2016)
27. Haas, R., Hoffmann, M.: Chordless paths through three vertices. TCS **351**(3), 360–371 (2006)
28. Karp, R.M.: Reducibility among combinatorial problems. In: Miller, R.E., Thatcher, J.W., Bohlinger, J.D. (eds.) Complexity of Computer Computations. The IBM Research Symposia Series, pp. 85–103. Springer, Heidelberg (1972). https://doi.org/10.1007/978-1-4684-2001-2_9
29. Kreutzer, S., Ordyniak, S.: Digraph decompositions and monotonicity in digraph searching. TCS **412**(35), 4688–4703 (2011)
30. Kreutzer, S., Rabinovich, R., Siebertz, S.: Polynomial kernels and wideness properties of nowhere dense graph classes. ACM Trans. Algorithms **15**(2), 24:1–24:19 (2019)
31. Lokshtanov, D., Misra, P., Ramanujan, M., Saurabh, S., Zehavi, M.: FPT-approximation for FPT problems. In: SODA 2021, pp. 199–218. SIAM (2021)
32. Lokshtanov, D., Ramanujan, M.S., Saurabh, S.: A linear time parameterized algorithm for directed feedback vertex set. CoRR abs/1609.04347 (2016)
33. Lokshtanov, D., Ramanujan, M.S., Saurabh, S., Sharma, R., Zehavi, M.: Wannabe bounded treewidth graphs admit a polynomial kernel for DFVS. In: Friggstad, Z., Sack, J.-R., Salavatipour, M.R. (eds.) WADS 2019. LNCS, vol. 11646, pp. 523–537. Springer, Cham (2019). https://doi.org/10.1007/978-3-030-24766-9_38

34. Nešetřil, J., Ossona de Mendez, P.: Grad and classes with bounded expansion i. decompositions. Eur. J. Comb **29**(3), 760–776 (2008)
35. Nešetřil, J., de Mendez, P.O.: On nowhere dense graphs. Eur. J. Comb. **32**(4), 600–617 (2011)
36. Pilipczuk, M., Siebertz, S., Toruńczyk, S.: On the number of types in sparse graphs. In: LICS 2018, pp. 799–808. ACM (2018)
37. Razgon, I.: Computing minimum directed feedback vertex set in $o*(1.9977^n)$. In: TCS, pp. 70–81. World Scientific (2007)
38. Seymour, P.D.: Packing directed circuits fractionally. Combinatorica **15**(2), 281–288 (1995)
39. Svensson, O.: Hardness of vertex deletion and project scheduling. In: Gupta, A., Jansen, K., Rolim, J., Servedio, R. (eds.) APPROX/RANDOM -2012. LNCS, vol. 7408, pp. 301–312. Springer, Heidelberg (2012). https://doi.org/10.1007/978-3-642-32512-0_26
40. Van Bevern, R.: Towards optimal and expressive kernelization for d-hitting set. Algorithmica **70**(1), 129–147 (2014)
41. Weihe, K.: Covering trains by stations or the power of data reduction. ALEX, 1–8 (1998)
42. You, J., Wang, J., Cao, Y.: Approximate association via dissociation. Discret. Appl. Math. **219**, 202–209 (2017)

Visualization of Bipartite Graphs in Limited Window Size

William Evans[1], Kassian Köck[2], and Stephen Kobourov[3](✉)

[1] Department of Computer Science, University of British Columbia, Vancouver, Canada
will@cs.ubc.ca

[2] Department of Computer Science, University of Passau, Passau, Germany
koeckk@fim.uni-passau.de

[3] Department of Computer Science, University of Arizona, Tucson, AZ, USA
kobourov@cs.arizona.edu

Abstract. Bipartite graphs are commonly used to visualize objects and their features. An object may possess several features and several objects may share a common feature. The standard visualization of bipartite graphs, with objects and features on two (say horizontal) parallel lines at integer coordinates and edges drawn as line segments, can often be difficult to work with. A common task in visualization of such graphs is to consider one object and all its features. This naturally defines a drawing window, defined as the smallest interval that contains the x-coordinates of the object and all its features. We show that if both objects and features can be reordered, minimizing the average window size is NP-hard. However, if the features are fixed, then we provide an efficient polynomial time algorithm for arranging the objects, so as to minimize the average window size. Finally, we introduce a different way of visualizing the bipartite graph, by placing the nodes of the two parts on two concentric circles. For this setting we also show NP-hardness for the general case and a polynomial time algorithm when the features are fixed.

Keywords: bipartite graphs · NP-hardness · two-layer drawing · circle layout

1 Introduction

Bipartite graphs arise in many applications and are usually visualized with 2-layer drawings, where vertices are drawn as points at integer coordinates on two distinct parallel lines, and edges are straight-line segments between their endpoints. Such drawings occur as components in layered drawings of directed graphs [19] and also as final drawings, e.g., in tanglegrams for phylogenetic trees [6,11].

A common task in the exploration of such bipartite graphs $G = (P \cup C, E)$, where P is the set of *parent* (object) vertices and C is the set of *child* (feature) vertices, is to identify the neighbors (children) of a parent vertex of interest. A

H. Fernau et al. (Eds.): SOFSEM 2024, LNCS 14519, pp. 198–210, 2024.
https://doi.org/10.1007/978-3-031-52113-3_14

typical approach is to click on this parent and highlight the edges to its children, while hiding/shading the rest of the graph. Naturally, it is desirable that the highlighted edges fit in the display. This motivates work on placing the vertices at integer coordinates so as to minimize the maximum *window*, i.e. the smallest x-interval that contains a parent and all its children, over all parents. Bekos et al. [4] show that minimizing the maximum window size can be solved efficiently when the children C are fixed and the parents P can be placed, and is NP-hard when the parents are fixed and the children can be placed. Note that this asymmetry is due to the windows being defined only for parents P. As a side effect of the underlying greedy approach, the algorithm of Bekos et al. [4] often results in a much larger than optimal average window size in order to minimize the max window. So if the max window exceeds the display size, many parents may have windows that exceed that size. In this paper, we consider the problem of minimizing the average window size directly.

Unlike our approach, methods for constructing 2-layer drawings often try to minimize the number of edge crossings, which is an NP-hard problem even when the order of one layer is fixed [10]. Vertex splitting provides another alternative approach to reduce the number of crossings, by replacing some vertices on one layer by multiple copies and distributing incident edges among these copies [9]. In bipartite graphs arising in domain applications, such as visualizing relationships between anatomical structures and cell types in the human body [1], vertex splitting makes sense only on one side of the layout. Several variants of optimizing such layouts have been recently studied [2,3,17].

Formally, the input consists of a bipartite graph $G = (P \cup C, E)$. The output is a **2-layer drawing** of G in which the vertices in P and in C are located at distinct integer coordinates on two parallel lines ℓ_P and ℓ_C, respectively (w.l.o.g., $\ell_P : y = 1$ is the *parent layer* and $\ell_C : y = 0$ is the *child layer*). The drawing is specified by a function $x : P \cup C \to \mathbb{Z}_{\geq 0}$ which defines the x-coordinate of each vertex in the drawing. No two parents can have the same x-coordinate and neither can two children; so x is injective when restricted to P or C. The objective is to minimize the average window size of the parents in the drawing (defined by) x. The *window* $w(p)$ of a parent $p \in P$ is the smallest x-interval that contains the locations of p and its neighbors (children) in G. Its size is its length, i.e., $\max_{a,b \in S} |x(a) - x(b)|$ where $S = \{c \mid (p, c) \in G\} \cup \{p\}$. Note that the smallest window size is 0 for a parent with no children or one child that shares its parent's x-coordinate. The *span* $s(p)$ of a parent $p \in P$ is the smallest x-interval that contains the children of p ($s(p) \subseteq w(p)$). Motivated by common assumptions in layered graph drawing [8,15] we consider two variants: one when we can choose the x-coordinates of the vertices of both P and C and the other when the x-coordinates of one of them is fixed. The Minimum Average Window Problem takes as input a graph $G = (P \cup C, E)$ and a value σ and determines if G has a drawing in which the average window size of the parents is at most σ. We focus on the equivalent Minimum Window Sum Problem (MWS) which determines if the *sum* of the window sizes can be at most some given Σ.

We also consider the same problem when the drawing maps parents and children to two concentric circles: parents to the inner circle and children to the outer circle. A **2-ring drawing** of G is specified by an integer size $r \geq 0$ and a function $x : P \cup C \rightarrow \mathbb{Z}_r$ (the integers mod r) which determines the locations of the vertices: The polar coordinates for $p \in P$ are $(1, \frac{2\pi}{r}x(p))$ and for $c \in C$ are $(2, \frac{2\pi}{r}x(c))$. Again, we require x to be injective when restricted to P or C (so $|P|, |C| \leq r$). The distance between two vertices a and b with locations specified by x is $d(a, b) = \min\{|x(a) - x(b)| \bmod r, |x(b) - x(a)| \bmod r\}$, i.e., the smallest of the clockwise or counter-clockwise distances from $x(a)$ to $x(b)$. The *window* $w(p)$ of parent $p \in P$ is the smallest interval of the circle that contains the angle locations of p and its children. Its size is measured in units of $\frac{2\pi}{r}$ radians. The *span* $s(p)$ of parent $p \in P$ is the smallest circle interval that contains the angle locations of the children of p ($s(p) \subseteq w(p)$). See Fig. 1 for an example of 2-layer and 2-ring drawing (with size $r = 7$) of the same bipartite graph with parents $A, B, \ldots, H$ and children $a, b, \ldots, h$.

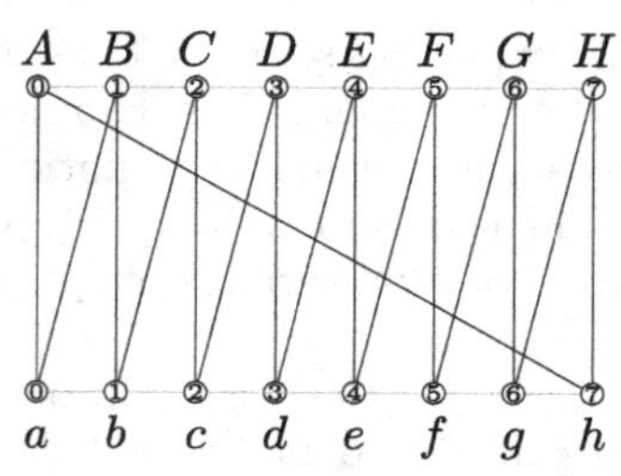

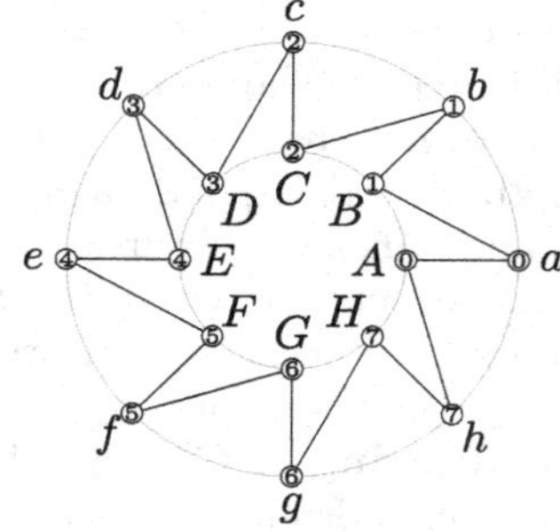

Fig. 1. An even cycle represented using a 2-layer drawing with minimum total window sum 14 and using a 2-ring drawing with minimum total window sum 8.

Our Contributions

In this paper we present the following results.
In the 2-layer setting:

- Minimizing the average window size is NP-hard when we can choose the locations of both parents and children.
- When the children are fixed, determining whether every parent can be placed in its span can be done in polynomial time.
- When the children are fixed, placing the parents to minimize the average window size can be done in polynomial time.

In the 2-ring setting:

- Minimizing the average window size is NP-hard when we can choose the locations of both parents and children.
- When the children are fixed, placing the parents to minimize the average window size can be done in polynomial time.

2 The Two Layer Setting

We begin by showing the NP-hardness of minimizing the average window size when both parents and children can be placed, and then provide two polynomial time algorithms for the case when the children are fixed.

2.1 Hardness of Minimizing Average Window Size.

The Linear Arrangement Problem (LAP) takes as input an undirected graph $G = (V, E)$ and an integer W and decides whether there is a bijection $x : V \to \{0, 1, \ldots, |V| - 1\}$ such that $\sum_{(u,v)\in E} |x(u) - x(v)| \leq W$. In other words, LAP decides whether there is a straight-line drawing of G with vertices at integer coordinates on the x-axis so that the sum of the edge lengths is at most W. LAP is a classic NP-complete problem [13].

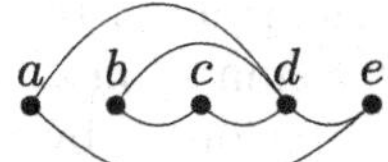

(a) A sample input graph to LAP.

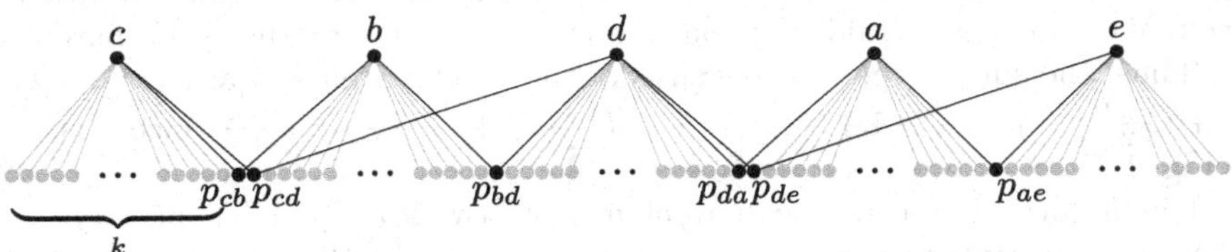

(b) The corresponding bipartite graph instance of MWS, drawn to minimize the window sum.

(c) The optimal linear arrangement of the original graph.

Fig. 2. Example illustrating the reduction from the linear arrangement problem to the problem of minimizing the window sum.

Theorem 1. *Deciding whether a given a bipartite graph $G = (P \cup C, E)$ has a 2-layer drawing with average window size σ when vertices in both P and C can be placed is NP-complete.*

Proof. We show the equivalent statement that Minimum Window Sum (MWS) is NP-complete. MWS is in NP since it takes time polynomial in the size of the input graph to verify that, for a given drawing $x : P \cup C \to \mathbb{Z}_{\geq 0}$, the sum of the window sizes of $p \in P$ is at most W.

To show that MWS is NP-hard, we reduce LAP to it. We construct an input $G' = (P \cup C, E')$, Σ to MWS from an input $G = (V, E)$, W to LAP as follows:

- Let k be an odd number with $k \geq 3|E|^2 + |E| + 2|V|$.
- The children C in G' are the vertices V in G.
- The parents P in G' are one *edge parent* p_{uv} (which is shorthand for $p_{\{u,v\}}$) for each $(u, v) \in E$ with edges (p_{uv}, u) and (p_{uv}, v) in G'; and k *block parents* $v_1, v_2, \ldots, v_k$ for each vertex $v \in V$, with edges $(v_1, v), (v_2, v), \ldots, (v_k, v)$ in G'.
- Let $\Sigma = |V|\frac{k^2-1}{4} + |E|^2 + kW$.

This construction takes time polynomial in the size of G. It remains to show that $G = (V, E)$ has a linear arrangement of total edge length at most W if and only if $G' = (P \cup C, E')$ has a drawing with window size sum at most Σ.

We first show that if G has a linear arrangement $v(0), v(1), \ldots, v(|V|)$ (where $v(i)$ is the ith vertex in the arrangement) of total edge length at most W then G' has a drawing with window size sum at most Σ.

We construct a bipartite drawing of G' by placing the k block parents of $v(i)$, starting with $i = 0$, consecutively (on line ℓ_P) followed (in any order) by the edge parents $p_{v(i)v(j)}$ for all edges $(v(i), v(j)) \in E$ with $j > i$, and then repeating this process for $i = i + 1$ until all block parents and edge parents are placed. We place each child $v(i)$ (on line ℓ_C) below the middle block parent of $v(i)$. Thus, the sum of the window sizes of the block parents of any child $v(i)$ is $\frac{k-1}{2} + (\frac{k-1}{2} - 1) + \cdots + 1 + 0 + 1 + \cdots + (\frac{k-1}{2} - 1) + \frac{k-1}{2} = \frac{k^2-1}{4}$ since k is odd. See Fig. 2.

The window size for an edge parent $p_{v(i)v(j)}$ (with $i < j$) is at most $(j - i)k$ for the block parents in the window plus at most $|E|$ for the edge parents in the window. The total window sum is at most:

$$|V|\frac{k^2-1}{4} + \sum_{\substack{\{v(i),v(j)\}\in E \\ i<j}} ((j-i)k + |E|) \leq |V|\frac{k^2-1}{4} + |E|^2 + kW = \Sigma$$

since the sum of $(j - i)$ over all edges is the total edge length of the linear arrangement $v(0), v(1), \ldots, v(|V|)$, which we assumed to be at most W.

We next show that if G' has a 2-layer drawing with total window sum at most Σ then G has a linear arrangement with total edge length at most W. This requires establishing several properties of any optimal drawing x of G':

Property 1. We may assume, by relabeling block parents if necessary, that if $x(u) < x(v)$ for $u, v \in C$ then $x(u_i) < x(v_j)$ for all block parents for u and v where $i, j \in [k]$.

Proof. Suppose $x(u_i) > x(v_j)$ for some $i, j \in [k]$. If we switch u_i and v_j, the sum of the window sizes of these two block parents remains the same if u_i and v_j are both to the left of u or both to the right of v, and decreases otherwise.

Property 2. Child v lies strictly within the x-interval of its block parents, i.e., for all $v \in C$, $x(v_1) < x(v) < x(v_k)$, where we have assumed (by renumbering) that v_i is the ith leftmost block parent of v in the realization.

Proof. Suppose $x(v) \leq x(v_1)$ (a symmetric argument applies when $x(v_k) \leq x(v)$). There must be an empty spot s for v with $s \in \{x(v_1)+1, x(v_1)+2, \ldots, x(v_1)+|V|\}$ since there are only $|V|$ children in G' and $x(v)$ is not in that set. Moving v from $x(v)$ to s can increase the window size of any edge parent by at most $s - x(v)$. It can increase the window size of any block parent p with $x(v_1) \leq x(p) < s$ by at most $s - x(v)$ as well. It *decreases* the window size of any block parent p with $s \leq x(p) \leq x(v_k)$ by at least $s - x(v)$. Since $k > |E| + 2|V|$, there are more than $|E| + |V|$ block parents whose windows decrease (by at least $s - x(v)$) and at most $|E|$ edge parents plus at most $|V|$ block parents whose windows increase (by at most $s - x(v)$) as a result of moving child v to s. Thus the sum of the window sizes decreases and the original realization cannot be optimal.

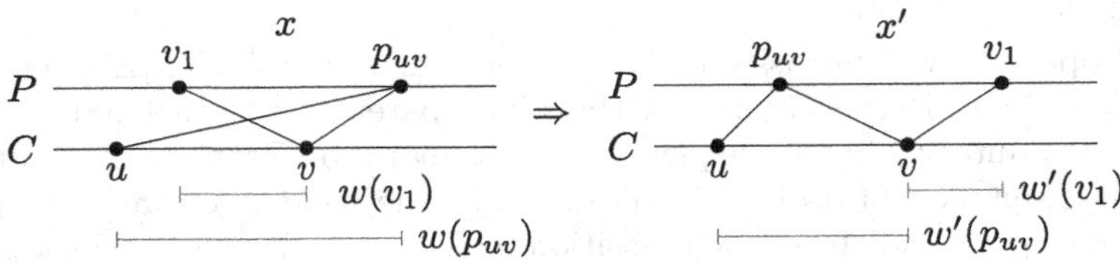

Fig. 3. Switching edge parent p_{uv} with leftmost block parent v_1 of v so that p_{uv} lies between u and v.

Property 3. The location $x(p_{uv})$ of edge parent p_{uv} lies strictly between the locations of u and v.

Proof. We may assume by renaming that $x(u) < x(v)$. Suppose $x(v) \leq x(p_{uv})$ (a similar symmetric argument applies when $x(p_{uv}) \leq x(u)$). Switch the locations of the leftmost block parent v_1 of v with p_{uv} to obtain a new drawing x' with window sizes w' (see Fig. 3). Note that x' differs from x only in that $x'(v_1) = x(p_{uv})$ and $x'(p_{uv}) = x(v_1)$. Since (by Properties 1 and 2) v_1 lies strictly between u and v in realization x, the new location of p_{uv} in realization x' lies strictly between u and v, decreasing the window size of p_{uv} by $x(p_{uv}) - x(v)$. The new window size of v_1 is at most $x(p_{uv}) - x(v)$ and since the original window size of v_1 is at least 1 (by Property 2), the sum of window sizes in realization x' is less than in x. Thus, x cannot be optimal.

Property 4. The parents P occupy $k|V| + |E|$ consecutive locations.

Proof. Suppose there is an empty spot s with parents to the left and right of s. No child v has $x(v) = s$ otherwise we could move a block parent of v to location s and decrease the window size sum. Thus we can decrement the location of all parents and children located to the right of s by 1 to create a new realization (no two parents or two children at the same location) with no increase in window size sum, and a strict decrease if G is connected.

Property 5. The block parents of v are consecutive, i.e., $x(v_{1+i}) = x(v_1) + i$ for all $i \in [k-1]$ and $v \in V$.

Proof. Suppose an edge parent p_{uv} is in the x-interval of the block parents of a vertex t, i.e., $x(t_1) < x(p_{uv}) < x(t_k)$. It can't be in the x-interval of the block parents of more than one vertex by Property 1. By Property 3, $x(u) \leq x(p_{uv}) \leq x(v)$. Swapping p_{uv} with t_1 or t_k thus keeps p_{uv} in its span (since at least one of t_1 or t_k is in the span of p_{uv}), but out of the x-interval of any vertex's block parents. It also decreases the window size of the swapped block parent (t_1 or t_k) by at least one.

With these properties established, we continue with the proof of the theorem. Let x be an optimal 2-layer drawing of G' with window sum at most Σ. We claim that the order $v(0), v(1), \ldots, v(|V|-1)$ of the children on line ℓ_C (where $x(v(i)) < x(v(j))$ for all $i < j$) is a linear arrangement of the vertices of G with total edge length at most W.

By Property 4, we may assume that the parents occupy $k|V|+|E|$ consecutive locations and, by Property 5, that the block parents of each child v form a consecutive x-interval in this sequence. Furthermore, by Property 2, child v lies within the x-interval of its block parents. Thus the window of any edge parent $p_{v(i)v(j)}$ $(i<j)$ extends from some position in the x-interval of the block parents of $v(i)$ to some position in the x-interval of the block parents of $v(j)$. To minimize the sum of the window sizes of the block parents of child v requires placing v in the middle of the x-interval of its block parents, resulting in a total contribution of v's block parents to the window sum of at least $\frac{k^2-1}{4}$. To minimize the sum of all window sizes, including the windows of edge parents, an optimal drawing may place v at a location that is not in the middle of the x-interval of its block parents. However, if v is placed at distance d from the middle, the sum of the window sizes of the block parents of v increases by d^2 (by symmetry, the edge lengths after v is moved are the same as the lengths before the move, except for the lengths to the furthest d block parents after the move, which each increase by d), while the decrease in the window sum of all $|E|$ edge parents is at most $d|E|$. Thus, v must be placed at distance $d \leq |E|$ from the middle of its block parents in an optimal drawing. Even with this placement of v, the window size of $p_{v(i)v(j)}$ is at least

$$(j-i-1)k + 2\left(\frac{k-1}{2} - d\right) \geq (j-i)k - (2|E|+1).$$

The total contribution of block parents and edge parents to the window sum is thus at least

$$|V|\frac{k^2-1}{4} + \sum_{\substack{\{v(i),v(j)\}\in E\\ i<j}} ((j-i)k - 2|E| - 1)\,.$$

Since this is at most $\Sigma = |V|\frac{k^2-1}{4} + |E|^2 + kW$, we know

$$|V|\frac{k^2-1}{4} + \sum_{\substack{\{v(i),v(j)\}\in E\\ i<j}} ((j-i)k - 2|E| - 1) \leq |V|\frac{k^2-1}{4} + |E|^2 + kW$$

and thus

$$\sum_{\substack{\{v(i),v(j)\}\in E\\ i<j}} (j-i)k \leq 3|E|^2 + |E| + kW.$$

Since $k \geq 3|E|^2 + |E| + 2|V|$,

$$\sum_{\substack{\{v(i),v(j)\}\in E\\ i<j}} (j-i) \leq W + 1 - \frac{2|V|}{k}$$

and since the sum and W are integers,

$$\sum_{\substack{\{v(i),v(j)\}\in E\\ i<j}} (j-i) \leq W.$$

So G has a linear arrangement with total edge length at most W.

2.2 Minimizing Average Window Size for Fixed Children

When the children are fixed, the best possible solution to the minimum average window problem is to place every parent in the span of its children. We first observe how to efficiently test if such a solution exists. After that, we show that even when it is not possible to place every parent in the span of its children we can still find a solution that minimizes the average window size in polynomial time.

Placing Parents in Their Span. If the children C of a bipartite graph $G = (P \cup C, E)$ have already been placed at distinct integer x-coordinates then the span of every parent p is fixed: $s(p) = [\mathrm{lo}(p), \mathrm{hi}(p)]$ where $\mathrm{lo}(p)$ is the smallest x-coordinate of a child of p and $\mathrm{hi}(p)$ is the largest x-coordinate of a child of p. Our problem is to determine if it is possible to place each parent at an integer coordinate within its span without placing two parents at the same location. This is an instance of a matching problem in a *convex* bipartite graph $A = (P \cup S, F)$ where $F = \{(p, \ell) | p \in P, \ell \in s(p)\}$ and $S = \bigcup_{p\in P} s(p)$ (so S is a set of integers). The bipartite graph $A = (P \cup S, F)$ is convex (in S) since there is an ordering of S (the integer order "$<$" in this case) such that if $(p, a) \in F$ and $(p, c) \in F$ then $(p, b) \in F$ for $b \in S$ with $a < b < c$. Graph A is defined by the parents P and the pairs $\mathrm{lo}(p)$, $\mathrm{hi}(p)$ for $p \in P$, which in our case can be calculated from G in $O(|E|)$ time.

The problem of finding a maximum matching in a convex (in S) bipartite graph $(P \cup S, F)$ has a long history starting with Glover's $O(|P||S|)$ time algorithm from 1967 [14] and ending with the $O(|P|)$ time algorithm of Steiner and Yeomans [18]. Using the latter algorithm, we observe that:

Observation 1. *Given a bipartite graph $G = (P \cup C, E)$ where the x-coordinates of the children C are fixed, finding a placement of parents $p \in P$ where every parent is in its span, or determining that no such placement exists, takes $O(|P| + |E|)$ time.*

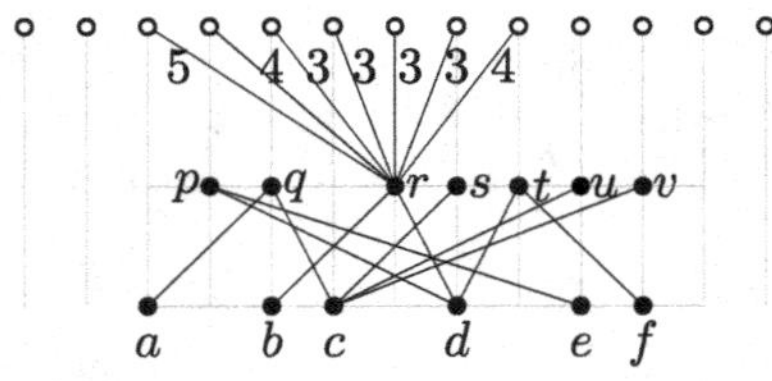

(a) Shown above the graph G are the edges matching parent vertex r to its possible locations ∘ in an optimal drawing of G. The weight of edge (r, ℓ) is the size of r's window if r were at location ℓ.

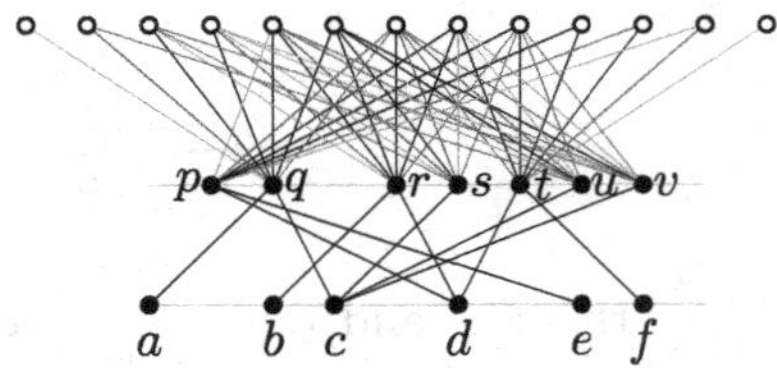

(b) The entire bipartite matching graph shown above the graph G. Black edges have weight equal to the parent's window size. Blue have that weight + 1. Orange: + 2. Red: + 3.

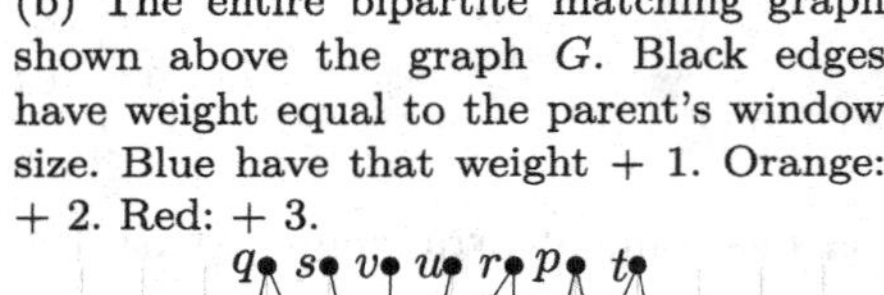

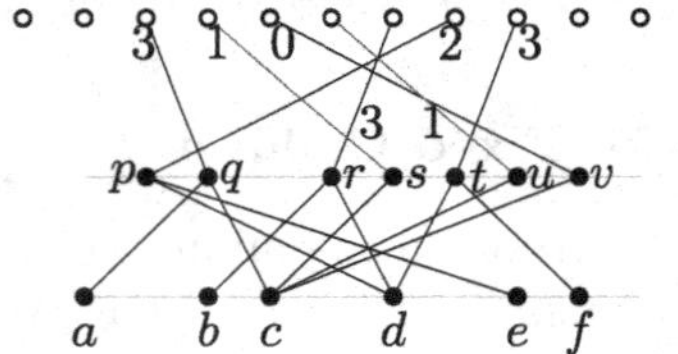

(c) A min-weight maximum size matching of parents to locations.

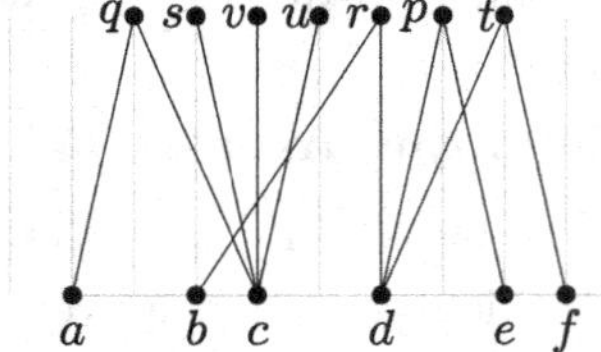

(d) The resulting optimal drawing with window sum 13.

Fig. 4. Using min-weight maximum size matching to find an optimal drawing when children are fixed.

Minimizing Average Window Size. If we cannot place every parent in its span, we would like to place parents as close to their span as possible to minimize their average window size. We adopt a similar technique of finding a matching in a convex bipartite graph to determine the parents' locations. However, in this case the edges in the matching are weighted and we ask for a minimum weight maximum matching.

Theorem 2. *Minimizing the average window size of a 2-layer drawing of a given bipartite graph $G = (P \cup C, E)$, when the locations of the children C are fixed, but the parents P can be placed, can be accomplished in $O(|P|^{2+o(1)} + |E|)$ time.*

Proof. Given the fixed locations of the children C, for each parent $p \in P$, compute the x-coordinate of the midpoint between its children's smallest and largest locations, and let $M(p)$ be the set of $|P|$ potential parent locations closest to this midpoint. We do not need to place p outside of $M(p)$ when minimizing the average window size for p since $M(p)$ contains the locations that result in the $|P|$ smallest window sizes for p and we're guaranteed to find at least one empty spot among them since there are only $|P| - 1$ other parents.

We construct a weighted bipartite graph B between parents P and their possible locations $M(P) = \bigcup_{p \in P} M(p)$ with edges (p, ℓ) for every parent p and $\ell \in M(p)$. (See Fig. 4.) There are exactly $|P|^2$ edges in the graph B and each edge (p, ℓ) has weight corresponding to the size of parent p's window if p were placed at location ℓ. The construction of B takes $O(|E| + |P|^2)$ time. We then find a min cost matching in B using the algorithm of Chen et al. [7] in time $O(|P|^{2+o(1)})$.

3 The Two Ring Setting

In this setting we are given two concentric circles, with the parents on the inner circle and the children on the outer circle.

The problems here are similar to the two layer setting, but sufficiently different that neither result in the previous section directly works here. It might appear that we can extract an optimal 2-layer drawing x of $G = (P \cup C, E)$ from an optimal 2-ring drawing x' of G by simply setting $x(v) = x'(v)$ for all $v \in P \cup C$. However, Fig. 1 illustrates a difficulty: the 2-ring setting allows us to measure the distance from a parent to its child in two ways, clockwise or counter-clockwise around the inner ring. This creates the possibility of a smaller window sum in the 2-ring setting than in the 2-layer setting. We can do such an extraction if the 2-ring drawing has a point at which we can "split" the two circles and straighten them into two parallel lines.

Definition 1. *A 2-ring drawing x of a graph $G = (P \cup C, E)$ of size r is* splittable *at $s \in \mathbb{Z}_r$ if $s + 1/2$ (and thus the open interval $(s, s + 1 \bmod r)$) is not in the window $w(p)$ for any $p \in P$.*

Remark 1. A 2-ring drawing x' of $G = (P \cup C, E)$ of size r that is splittable at s can be converted into a 2-layer drawing x with the same window sum by setting $x(v) = (x'(v) - s) \bmod r$ for all $v \in P \cup C$.

Theorem 3. *Deciding whether a given bipartite graph $G = (P \cup C, E)$ has a 2-ring drawing with average window size σ when vertices in both P and C can be placed is NP-complete.*

Proof. We show the equivalent statement that the window sum version of the problem, with target sum Σ, is NP-complete. The problem is in NP since it takes polynomial time to verify that, for a given drawing $x : P \cup C \to \mathbb{Z}_r$, the sum of the window sizes of $p \in P$ is at most Σ. We reduce the 2-layer version, shown to be NP-complete in the proof of Theorem 1, to this setting by making every non-splittable 2-ring drawing of G prohibitively expensive. We do this by adding $k = |P|(|P| + |C|) + 1$ independent edges to G.

Given an input $G = (P \cup C, E)$, Σ to the 2-layer minimum sum problem, we create a new graph $G' = (P' \cup C', E')$ where $P' = P \cup \{u_1, u_2, \ldots, u_k\}$, $C' = C \cup \{v_1, v_2, \ldots, v_k\}$, and $E' = E \cup \{(u_i, v_i) | i \in [k]\}$. We claim that G has a 2-layer drawing with window sum at most Σ if and only if G' has a 2-ring drawing with window sum at most Σ.

If G has a 2-layer drawing x with window sum at most Σ then it is safe to assume by translating that the drawing places vertices at x-coordinates in $[0, 1, \ldots, |P| + |C| - 1]$. Otherwise, if the drawing x exceeds this interval then there is a common empty spot s in both the parents and children in x and decreasing the positions of parents and children to the right of s by one results in a drawing whose window size sum is at most the original sum. We construct a 2-ring drawing x' from x by setting $r = |P| + |C| + k$ and $x'(v) = x(v)$ for all $v \in P \cup C$ and $x'(u_i) = x'(v_i) = |P| + |C| + i - 1$ for all $i \in [k]$. Since the window sum for all parents u_i is zero the window sum of the 2-ring drawing is at most Σ.

If G' has a 2-ring drawing x' of size r with window sum at most Σ then if x' is splittable at s, Remark 1 implies we can extract a 2-layer drawing x of G from it with window sum at most Σ by setting $x(v) = (x'(v) - s) \bmod r$ for all $v \in P \cup C$. The addition to G of the (many) independent edges (u_i, v_i) to create G' ensures that the optimal drawing of G' when restricted to the vertices of G is splittable. Let x^* be this optimal 2-ring drawing of G. Suppose, for the sake of contradiction, that x^* is not splittable, then every interval between adjacent children in x^* is contained in the window $w^*(p)$ for some parent $p \in P$. Since the children of the independent edges lie in these intervals, the sum of the window sizes in the 2-ring drawing x^* is at least k. Since x^* is a subdrawing of x', the window sum of drawing x' of G' is also at least k. However, k is chosen to be larger than the window sum of the optimal 2-ring drawing of any graph G with $|P|$ parents and $|C|$ children along with k independent edges. To see this, note that one possible 2-ring drawing of such a graph starts with a 2-layer drawing of G in which every parent has a window of size at most $|P| + |C|$; followed by k independent edges each with window size 0. Placing these vertices, in order around two concentric circles of size $|P| + |C| + k$ creates a 2-ring drawing with window sum of $|P|(|P| + |C|) < k$. Thus the 2-ring drawing x' of G' is splittable and by Remark 1 G has a 2-layer drawing of size Σ.

Theorem 4. *Minimizing the average window size of a 2-ring drawing of size r of a given bipartite graph $G = (P \cup C, E)$, when the locations of the children C are fixed, but the parents can be placed, takes $O(|P|^{2+o(1)})$ time.*

Proof. This is identical to the proof of Theorem 2 with the understanding that ring size r is large enough to place the parents P and that the window size associated with a potential location of $p \in P$ is measured in the ring setting, i.e. as the smallest angle of a sector containing that potential location and p's children measured in units of $\frac{2\pi}{r}$ radians.

4 Conclusions and Open Problems

The visualization of bipartite graphs on displays of limited size is motivated by real-world applications. Minimizing the largest window size of any parent provides a reasonable solution but only if that largest size is at most the display size. We consider minimizing the average window size. Our solution may not obey the display size limit for all parents (even if such a solution exists), but it produces smaller parent windows on average, which might be preferable. We showed that in the general setting the problem is NP-hard, while for more restricted settings we provide efficient polynomial time algorithms.

The matching re-formulation of the problem is broad enough to be applied to several related problems in both the 2-layer and 2-ring settings. For example, in the 2-ring setting when the children are fixed, we can find an optimal solution using the same graph matching formulation as in Theorem 4 but instead of using a solution to the min cost matching problem on graph B, we use a solution technique for the bottleneck matching problem. This, in effect, replaces the min-average computation with a min-max computation.

Observation 2. *In the 2-ring setting, minimizing the maximum window size over all parents for a given bipartite graph* $G = (P \cup C, E)$*, when the size of the ring* r *and the children* C *are fixed, but parents* P *can be placed, takes* $O(|P|\log|P| + |E|)$ *time.*

Proof. Let B be the graph defined in the proof of Theorem 2 for the ring size r. If we restrict this graph to edges of weight at most w, we obtain a *circular convex bipartite graph* $B(w)$ for which the neighbors of a parent $p \in P$ form a contiguous interval around the ring. For such graphs, Liang and Blum [16] describe an algorithm to find a maximum matching based on two runs of a maximum matching algorithm for a regular convex bipartite subgraph of this graph. Using a binary search technique credited to Bhat [5] by Gabow and Tarjan [12], we can find the smallest $w \in [0, |P|]$ so that $B(w)$ contains a maximum matching (of size $|P|$ in our case). After a single $O(|E|)$-time preprocessing step, the running time is $O(|P|)$ for each test of whether the restricted graph $B(w)$ contains a maximum matching [16,18]. The total time is thus $O(|P|\log|P| + |E|)$.

We conclude with two natural open problems. First, is there a faster algorithm for minimizing the average window size when children are fixed but parents can be placed? Second, what is the complexity of minimizing the average window size when parents are fixed, but children can be placed?

References

1. https://hubmapconsortium.github.io/ccf-asct-reporter/
2. Ahmed, R., et al.: Splitting vertices in 2-layer graph drawings. IEEE Comput. Graph. Appl. (2023)
3. Baumann, J., Pfretzschner, M., Rutter, I.: Parameterized complexity of vertex splitting to pathwidth at most 1. arXiv preprint arXiv:2302.14725 (2023)

4. Bekos, M.A., et al.: On the 2-layer window width minimization problem. In: Gasieniec, L. (ed.) SOFSEM 2023. LNCS, vol. 13878, pp. 209–221. Springer, Cham (2023)
5. Bhat, K.V.S.: An $O(n^{2.5} \log_2 n)$ time algorithm for the bottleneck assignment problem. unpublished. AT&T Bell Laboratories, Napiendle, IL (1984)
6. Buchin, K., et al.: Drawing (complete) binary tanglegrams. Algorithmica **62**(1–2), 309–332 (2012). https://doi.org/10.1007/s00453-010-9456-3
7. Chen, L., Kyng, R., Liu, Y.P., Peng, R., Gutenberg, M.P., Sachdeva, S.: Maximum flow and minimum-cost flow in almost-linear time. In 2022 IEEE 63rd Annual Symposium on Foundations of Computer Science (FOCS), pp. 612–623. IEEE (2022)
8. Battista, G.D., Eades, P., Tamassia, R., Tollis, I.G.: Graph Drawing: Algorithms for the Visualization of Graphs. Prentice-Hall, Hoboken (1999)
9. Eades, P., de Mendonça N, C.F.X.: Vertex splitting and tension-free layout. In: Brandenburg, F.J. (ed.) GD 1995. LNCS, vol. 1027, pp. 202–211. Springer, Heidelberg (1996). https://doi.org/10.1007/BFb0021804
10. Eades, P., Wormald, N.C.: Edge crossings in drawings of bipartite graphs. Algorithmica **11**(4), 379–403 (1994)
11. Fernau, H., Kaufmann, M., Poths, M.: Comparing trees via crossing minimization. J. Comput. Syst. Sci. **76**(7), 593–608 (2010). https://doi.org/10.1016/j.jcss.2009.10.014
12. Gabow, H.N., Tarjan, R.E.: Algorithms for two bottleneck optimization problems. J. Algorithms **9**, 411–417 (1988)
13. Garey, M.R., Johnson, D.S., Stockmeyer, L.: Some simplified NP-complete graph problems. Theoret. Comput. Sci. **1**(3), 237–267 (1976)
14. Glover, F.: Maximum matching in convex bipartite graphs. Naval Res. Logistic Q. **14**, 313–316 (1967)
15. Kaufmann, M., Wagner, D.: Drawing Graphs, Methods and Models, vol. 2025. Springer, Cham (2001). https://doi.org/10.1007/3-540-44969-8
16. Liang, Y.D., Blum, N.: Circular convex bipartite graphs: maximum matching and Hamiltonian circuits. Inf. Process. Lett. **56**(4), 215–219 (1995)
17. Nöllenburg, M., Sorge, M., Terziadis, S., Villedieu, A., Wu, H.Y., Wulms, J.: Planarizing graphs and their drawings by vertex splitting. In: Angelini, P., von Hanxleden, R. (eds.) GD 2022. LNCS, vol. 13764, pp. 232–246. Springer, Cham (2022). https://doi.org/10.1007/978-3-031-22203-0_17
18. Steiner, G., Yeomans, J.S.: A linear time algorithm for determining maximum matchings in convex, bipartite graphs. Comput. Math. Appl. **31**, 91–96 (1996)
19. Sugiyama, K., Tagawa, S., Toda, M.: Methods for visual understanding of hierarchical system structures. IEEE Trans. Syst. Man Cybern. **11**(2), 109–125 (1981). https://doi.org/10.1109/TSMC.1981.4308636

Outerplanar and Forest Storyplans

Jiří Fiala[1], Oksana Firman[2(✉)], Giuseppe Liotta[3], Alexander Wolff[2], and Johannes Zink[2]

[1] Charles University, Prague, Czech Republic
[2] Universität Würzburg, Würzburg, Germany
oksana.firman@uni-wuerzburg.de
[3] Università degli Studi di Perugia, Perugia, Italy

Abstract. We study the problem of gradually representing a complex graph as a sequence of drawings of small subgraphs whose union is the complex graph. The sequence of drawings is called *storyplan*, and each drawing in the sequence is called a *frame*. In an outerplanar storyplan, every frame is outerplanar; in a forest storyplan, every frame is acyclic.

We identify graph families that admit such storyplans and families for which such storyplans do not always exist. In the affirmative case, we present efficient algorithms that produce straight-line storyplans.

1 Introduction

A possible approach to the visual exploration of large and complex networks is to gradually display them by showing a sequence of *frames*, where each frame contains the drawing of a portion of the graph. When going from one frame to the next, some vertices and edges appear while others disappear. To preserve the mental map, the geometric representation of vertices and edges that are shared by two consecutive frames must remain the same. Informally speaking, a *storyplan* for a graph consists of a sequence of frames such that every vertex and edge of the graph appears in at least one frame. Moreover, there is a consistency requirement (as for the labels in a zoomable digital map [2]): once a vertex disappears, it may not re-appear. Hence, after a vertex appears, it remains visible until all its incident edges are represented; then it disappears in the transition to the next frame. See Fig. 1 for a storyplan.

Since edge crossings are a natural obstacle to the readability of a graph layout [10], Binucci at al. [4] introduced and studied the *planar storyplan problem* that asks whether a graph G admits a storyplan such that every frame is a crossing-free drawing and in every frame a single new vertex appears. Binucci et al. showed that the problem is NP-complete in general and fixed-parameter tractable w.r.t. the vertex cover number. They also proved that every graph of treewidth at most 3 admits a planar storyplan.

Motivated by the research of Binucci et al., we forward the idea of representing a graph with a storyplan such that each frame is a drawing whose visual inspection is as simple as possible. Specifically, we study the *outerplanar storyplan problem* and the *forest storyplan problem*, which are defined analogously

H. Fernau et al. (Eds.): SOFSEM 2024, LNCS 14519, pp. 211–225, 2024.
https://doi.org/10.1007/978-3-031-52113-3_15

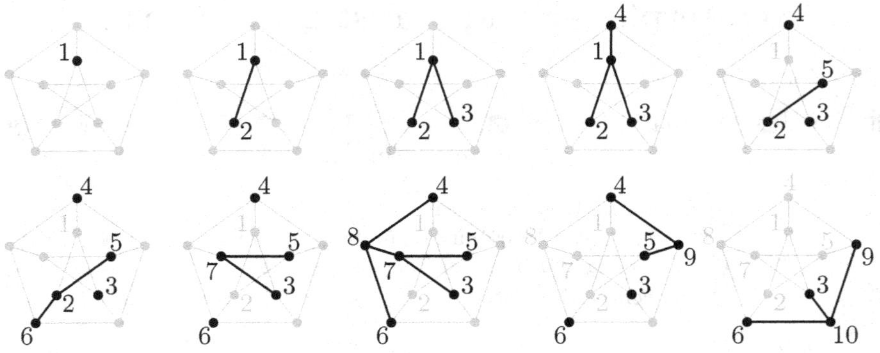

Fig. 1. A forest storyplan of the Petersen graph.

to the planar storyplan problem (see Definition 1 below). We let the classes of graphs that admit planar, outerplanar and forest storyplans be denoted by $\mathcal{G}_{\text{planar}}$, $\mathcal{G}_{\text{outerpl}}$, and $\mathcal{G}_{\text{forest}}$, respectively. Clearly, $\mathcal{G}_{\text{forest}} \subseteq \mathcal{G}_{\text{outerpl}} \subseteq \mathcal{G}_{\text{planar}} \subseteq \mathcal{G}$, where $\mathcal{G}$ is the class of all graphs. To further simplify visual inspection, our algorithms draw all frames with straight-line edges. We call storyplans with this property *straight-line storyplans*.

Beside the work of Binucci et al., our research relates to the graph drawing literature that assumes either dynamic or streaming models (see, e.g., [1,3,6,7]) and to recent work about graph stories (see, e.g., [5,9]). The key difference to our work is that these papers (except [4]) assume that the order of the vertices is given as part of the input. We now summarize our contribution, using *△-free* as shorthand for triangle-free.

- We establish the chain of strict containment relations $\mathcal{G}_{\text{forest}} \subsetneq \mathcal{G}_{\text{outerpl}} \subsetneq \mathcal{G}_{\text{planar}} \subsetneq \mathcal{G}$ (see Fig. 2) by showing that
  - there is a △-free 6-regular graph that does not admit a planar storyplan;
  - there is a K_4-free 4-regular planar graph that (trivially) admits a planar storyplan, but does not admit an outerplanar storyplan; and
  - there is a △-free 4-regular (nonplanar) graph that admits an outerplanar storyplan, but does not admit a forest storyplan.

  Recall that a *triangulation* is a maximal planar graph; it admits a planar drawing where every face is a triangle. We show that no triangulation (except for K_3) admits an outerplanar storyplan; see Sect. 3.
- We show that every partial 2-tree and every subcubic graph except K_4 admits an outerplanar straight-line storyplan (in linear time); see Sect. 4. In our construction for subcubic graphs, every frame contains at most five edges.
- A graph must be △-free in order to admit a forest storyplan. We show that △-free subcubic graphs (as the Petersen graph in Fig. 1), and △-free planar graphs admit straight-line forest storyplans (which we can compute in linear and polynomial time, respectively); see Sect. 5.

We start with some preliminaries in Sect. 2 and close with open problems in Sect. 6. The full proofs of statements with a "⋆" are in the full version of this paper [12]. Given a positive integer n, we use $[n]$ as shorthand for $\{1, 2, \ldots, n\}$.

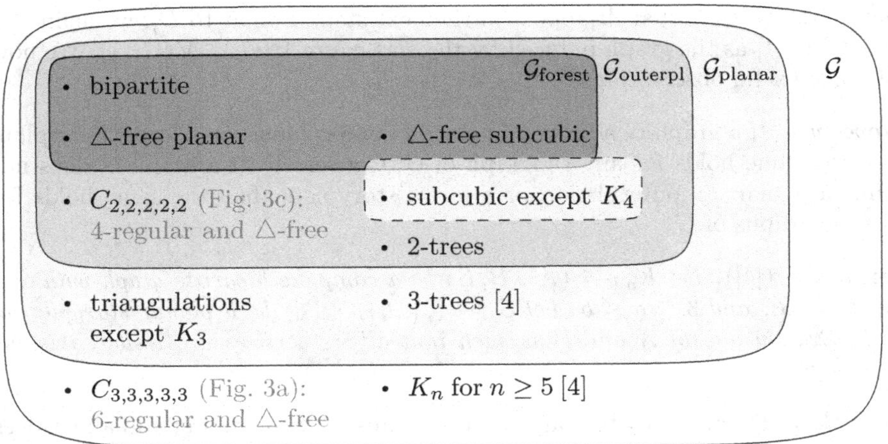

Fig. 2. Overview: existing [4] and new storyplan results, implying $\mathcal{G}_{\text{forest}} \subsetneq \mathcal{G}_{\text{outerpl}} \subsetneq \mathcal{G}_{\text{planar}} \subsetneq \mathcal{G}$. (For simplicity, we mention 2-/3-trees rather than partial 2-/3-trees.)

2 Preliminaries

Our definitions of a planar, an outerplanar, and a forest storyplan are based on the definition of a planar storyplan of Binucci et al. [4].

Definition 1. *A* planar storyplan $\mathcal{S} = \langle \tau, (D_i)_{i \in [n]} \rangle$ *of* G *is a pair defined as follows. The first element is a bijection* $\tau \colon V \to [n]$ *that represents a total order of the vertices of* G*. For a vertex* $v \in V$*, let* $i_v = \tau(v)$ *and let* $j_v = \max_{u \in N[v]} \tau(u)$*, where* $N[v]$ *is the set containing* v *and its neighbors. The interval* $[i_v, j_v]$ *is the* lifespan *of* v*. We say that* v appears *at step* i_v*, is* visible *at step* i *for each* $i \in [i_v, j_v]$*, and* disappears *at step* $j_v + 1$*. Note that a vertex disappears only when all its neighbors have appeared. The second element of* $\mathcal{S}$ *is a sequence of drawings* $(D_i)_{i \in [n]}$*, called* frames *of* $\mathcal{S}$*, such that, for* $i \in [n]$*: (i)* D_i *is a drawing of the graph* G_i *induced by the vertices visible at step* i*, (ii)* D_i *is planar, (iii) the point representing a vertex* v *is the same over all drawings that contain* v*, and (iv) the curve representing an edge* e *is the same over all drawings that contain* e*.*

We emphasize that though for the definition of a storyplan we allow that edges could be represented by curves, our constructions use only straight-line segments. For an *outerplanar storyplan* and a *forest storyplan*, we strengthen requirement (ii) to D_i being outerplanar and D_i being a crossing-free drawing of a forest, respectively. In what follows, we will sometimes use a slight variant of Definition 1, in which we enrich the sequence $(D_i)_{i \in [n]}$ of frames by explicitly representing the portions of the drawings that consecutive frames have in common. More precisely, for $i \in [n-1]$, let $D'_i = D_i \cap D_{i+1}$. Then, a storyplan is a sequence of drawings $\langle D_1, D'_1, \ldots, D_{n-1}, D'_{n-1}, D_n \rangle$, where in each step $i < n$, we first introduce a vertex (in D_i) and then remove all *completed* vertices (in D'_i), that is, the vertices that disappear in the next step. Similar to D'_i, we define G'_i for $i \in [n-1]$ as the graph induced by the vertices of $V(G_i) \cap V(G_{i+1})$. We now list some useful observations.

Property 1. If a graph G admits a planar, an outerplanar, or a forest storyplan, then the same holds for any subgraph of G. Conversely, if a graph G does not admit a planar, an outerplanar, or a forest storyplan, then the same holds for all supergraphs of G.

Lemma 1 ([4]). *Let $K_{a,b} = (A \cup B, E)$ be a complete bipartite graph with $a = |A|$, $b = |B|$, and $3 \leq a \leq b$. Let $\mathcal{S} = \langle \tau, \{D_i\}_{i \in [a+b]} \rangle$ be a planar storyplan of $K_{a,b}$. Exactly one of A and B is such that all its vertices are visible at some $i \in [a+b]$.*

Example 1. Every bipartite graph admits a forest storyplan: first add all vertices of one set of the bipartition and then, one by one, the vertices of the other set. Note that each vertex of the second set is visible in only one frame.

3 Separation of Graph Classes

Trivially, triangulations admit planar storyplans, but as we show now, no triangulation (except for K_3) admits an outerplanar storyplan.

Theorem 1. *No triangulation (except for K_3) admits an outerplanar storyplan.*

Proof. For a triangulation, the closed neighborhood of each vertex induces a wheel, which is not outerplanar. For the first vertex that disappears according to a given storyplan, however, its whole closed neighborhood, which is not outerplanar, must be visible. □

Example 2 (Platonic graphs). According to Theorem 1, the tetrahedron, the octahedron, and the icosahedron do not admit outerplanar storyplans because they are triangulations. The cube is bipartite; hence, it admits a forest storyplan due to Example 1. The dodecahedron is △-free and cubic; hence, it admits a forest storyplan due to Theorem 5.

We now separate the graph classes $\mathcal{G}_{\text{forest}}$, $\mathcal{G}_{\text{outerpl}}$, $\mathcal{G}_{\text{planar}}$, and $\mathcal{G}$; see Fig. 2.

Theorem 2. *The following statements hold:*

1. *There is a △-free 6-regular graph that does not admit a planar storyplan; hence $\mathcal{G}_{\text{planar}} \subsetneq \mathcal{G}$.*
2. *There is a K_4-free 4-regular planar graph that does not admit an outerplanar storyplan; hence $\mathcal{G}_{\text{outerpl}} \subsetneq \mathcal{G}_{\text{planar}}$.*
3. *There is a △-free 4-regular (nonplanar) graph that admits an outerplanar storyplan, but does not admit a forest storyplan; hence $\mathcal{G}_{\text{forest}} \subsetneq \mathcal{G}_{\text{outerpl}}$.*

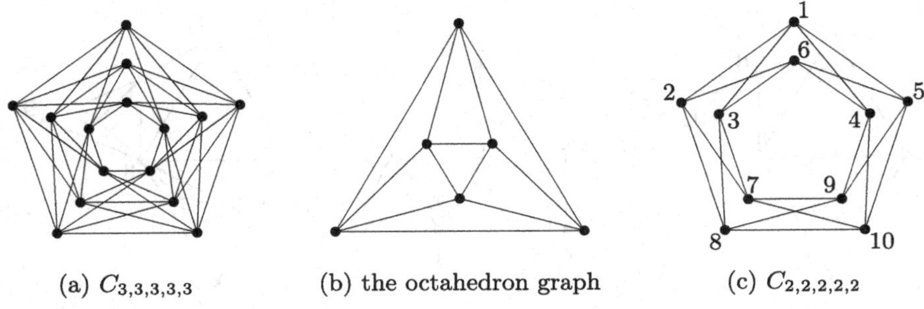

(a) $C_{3,3,3,3,3}$ (b) the octahedron graph (c) $C_{2,2,2,2,2}$

Fig. 3. Three graphs from the proof of Theorem 2. The graph in (a) is $\triangle$-free and does not admit any planar storyplan. The octahedron graph in (b) does not admit any outerplanar storyplan. The graph in (c) is $\triangle$-free and does not admit any forest storyplan (but the vertex numbering corresponds to an outerplanar storyplan – if vertex 8 is placed at the position of vertex 6, which will have disappeared by then).

Proof. 1. The graph $C_{3,3,3,3,3}$ (see Fig. 3a) is $\triangle$-free and 6-regular, but does not admit a planar storyplan as we will now show. Let $V(G) = V_1 \cup \cdots \cup V_5$ be the partition of the vertex set into independent sets of size 3. Note that, for $i \in \{1, 2, 3, 4, 5\}$, $G[V_i \cup V_{(i \bmod 5)+1}]$ is isomorphic to $K_{3,3}$. For $K_{3,3} = G[V_1 \cup V_2]$, we know by Lemma 1 that, in any planar storyplan, either all vertices of V_1 or all vertices of V_2 are shown simultaneously, say, those of V_1. Hence, for a frame to be planar, the vertices of V_2 and V_5 cannot be shown simultaneously. This, in turn, means that the vertices of V_3 and V_4 must be shown simultaneously. But then there must be a frame with a drawing of the non-planar graph $G[V_3 \cup V_4] = K_{3,3}$.
2. Observe that the octahedron (see Fig. 3b) is planar, 4-regular, and K_4-free, but does not admit an outerplanar storyplan due to Example 2.
3. The graph $C_{2,2,2,2,2}$ (see Fig. 3c) is $\triangle$-free and 4-regular, but does not admit a forest storyplan. The proof is analogous to the one above. There needs to be a frame with a drawing of $K_{2,2}$, which is not a tree. On the other hand, the order of the vertices shown in Fig. 3c yields an outerplanar storyplan. Note that we cannot use the vertex positions exactly as in the figure, but if we place vertex 8 at the position of vertex 6 (which will have disappeared by then), every frame is crossing-free. □

4 Outerplanar Storyplans

In this section we present families of graphs that admit outerplanar storyplans.

Theorem 3. *Every partial 2-tree admits a straight-line outerplanar storyplan, and such a storyplan can be computed in linear time.*

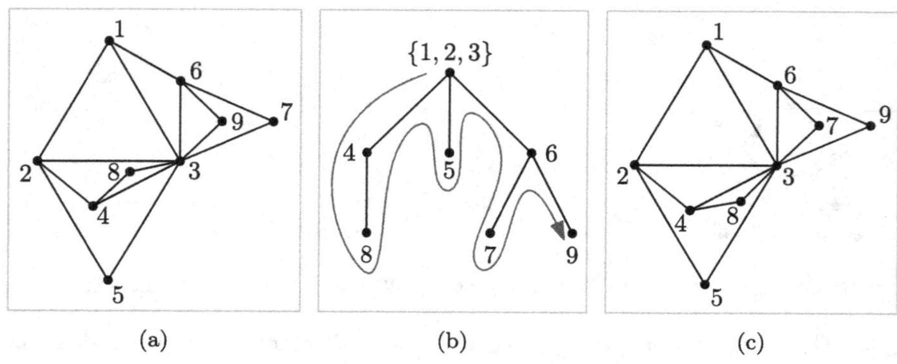

Fig. 4. A 2-tree G with a stacking order (a); its tree decomposition yields a vertex order $\sigma = \langle 1, 2, 3, 4, 8, 5, 6, 7, 9 \rangle$ (b); and an embedding of G that together with σ defines an outerplanar storyplan (c).

Proof. Due to Property 1, it suffices to prove the statement for 2-trees.

Let G be a 2-tree. Hence, there exists a *stacking order* $\sigma = \langle v_1, \ldots, v_n \rangle$ of the vertex set $V(G)$. In other words, G can be constructed as follows: we start with v_1, v_2, v_3 forming a K_3 and then, for $i \geq 4$, v_i is *stacked* on an edge $v_k v_\ell$ with $k, \ell < i$, that is, v_i is connected to v_k and v_ℓ by edges. We claim that we can choose a vertex order σ' and an embedding $\mathcal{E}$ of G such that σ' (together with $\mathcal{E}$) defines an outerplanar storyplan. Moreover, we can obtain a straight-line drawing of G with embedding $\mathcal{E}$ in linear time [8,14]. Let Γ be such a drawing. For the outerplanar storyplan that we construct we use the positions of vertices and edges as in Γ. This yields a straight-line storyplan. Figure 4(a) shows a 2-tree with a stacking order (that is not an outerplanar storyplan).

To show that an outerplanar storyplan always exists, we create a tree decomposition $T_{G,\sigma}$ of G. The root of $T_{G,\sigma}$ represents the triangle $\Delta v_1 v_2 v_3$ given by the first three vertices of σ. For $i = 4, 5, \ldots$, let v_i of σ be stacked onto the edge $v_k v_\ell$ with $k < \ell < i$. We add a node to $T_{G,\sigma}$ that represents v_i and is a child of the node representing v_ℓ. Note that if $\ell \leq 3$, then this new node is a child of the root. Figure 4(b) shows a tree decomposition of the 2-tree in Fig. 4(a).

From $T_{G,\sigma}$, we obtain a vertex order $\sigma' = \langle v'_1, v'_2, \ldots, v'_n \rangle$ being an outerplanar storyplan as follows; see Fig. 4(c). Let $v'_1 = v_1$, $v'_2 = v_2$, and $v'_3 = v_3$. Now, we traverse the nodes of $T_{G,\sigma}$ in (depth-first) pre-order and add the represented vertices of G to σ'. We claim that for σ', we can choose an embedding $\mathcal{E}$ (defined implicitly next) of G such that all frames are outerplanar. Note that the first three vertices form a triangle, which always admits an outerplanar drawing. Now consider v'_i for $i = 4, 5, \ldots$. Our invariant is that, before the i-th frame starts, the parent p of v'_i in $T_{G,\sigma}$ has degree 2 in the current outerplanar drawing and lies on the outer face. This implies that v'_i can be added to the outer face because it is stacked onto an edge of the outer face resulting again in an outerplanar drawing. Of course, for $i = 4$, our invariant is satisfied. If $p = v'_{i-1}$, then our invariant is trivially satisfied. Otherwise, let $p = v'_j$ for some $j < i - 1$. Observe that, for

$k \in \{j+1, \ldots, i-1\}$, each v'_k will have disappeared by the end of the $(i-1)$-th frame. This is due to the fact that v'_k is not an ancestor of v_i, which means that all of the neighbors of v'_k have already been introduced to the storyplan due to the depth-first pre-order traversal. Essentially, every frame given by σ' shows a subpath of $T_{G,\sigma}$, which is a sequence of stacked triangles admitting an outerplanar drawing. □

Theorem 4 ($\star$). *Every subcubic graph except K_4 admits a straight-line outerplanar storyplan with at most five edges in each frame, and such a storyplan can be computed in linear time.*

Proof. Due to Property 1, it suffices to prove the statement for cubic graphs.

We can assume that the given cubic graph G (which is not K_4) is connected; otherwise we consider each connected component separately. For an outerplanar storyplan, we will order the vertices $v_1, \ldots, v_n$ of G such that the resulting sequence of graphs $\langle G_1, G'_1 \ldots, G_{n-1}, G'_{n-1}, G_n \rangle$ has the following property: for $4 \leq i \leq n-1$, G'_i has at most two edges. Only for $i = 3$, G'_i may be a triangle and would thus contain three edges. Then we show how to obtain outerplanar drawings $D_1, D'_1 \ldots, D_{n-1}, D'_{n-1}, D_n$ of the graphs $G_1, G'_1 \ldots, G_{n-1}, G'_{n-1}, G_n$, respectively. For $i \in [n]$, let $H_i = G[\{v_1, \ldots, v_i\}]$.

We pick the first vertex v_1 arbitrarily. For $1 < i \leq n$, let v denote a vertex of G'_{i-1} with maximum degree in H_{i-1}. If there are more choices, let v additionally have maximum degree in G'_{i-1}. We then select $v_i \in V(G) \setminus \{v_1, \ldots, v_{i-1}\}$ as a neighbor of v in G. Note that v always has such a neighbor, otherwise v would already be completed and, hence, would not be in G'_{i-1}. The intuition behind this choice is that we want to remove v from the drawing as soon as possible.

We claim that, for $4 \leq i \leq n-1$, the graph G'_i contains at most two edges. In addition, if G'_i contains two edges, then these edges are both incident with v_i. This would mean that, for $i \in [n]$, G_i contains at most five edges. Indeed, even if G'_3 has three edges (that is, G'_3 is a triangle), then G_4 still has at most five edges since G is not K_4. Clearly, D_1 and D_2 have at most two edges.

We consider three cases depending on the degree of v in G'_{i-1}; see Fig. 5.

(C1) Vertex v does not have any neighbors in G'_{i-1}. By the choice of v, this implies that there are no edges in G'_{i-1} because H_{i-1} is connected and, for an edge in G'_{i-1}, H_{i-1} contains an incident degree-2 vertex. Note that all edges in G_i are new and incident with v_i. If v_i has three neighbors in G_i, then v_i will disappear, and there are no more edges in G'_i. Hence, G'_i has at most two edges. Note that both edges are incident with v_i.

(C2) Vertex v has one neighbor in G'_{i-1}. If v has degree 2 in H_{i-1}, then v disappears in the next step and G'_i does not contain it. Since v_i has at most one edge that stays in G'_i, the number of edges in G'_i is not larger than in G'_{i-1}. If v has degree 1 in H_{i-1}, then, by construction, all other vertices in G_{i-1} have also degree at most 1 in H_{i-1}. Hence, $i = 3$, that is, v and its neighbor are the first two vertices that we introduced.

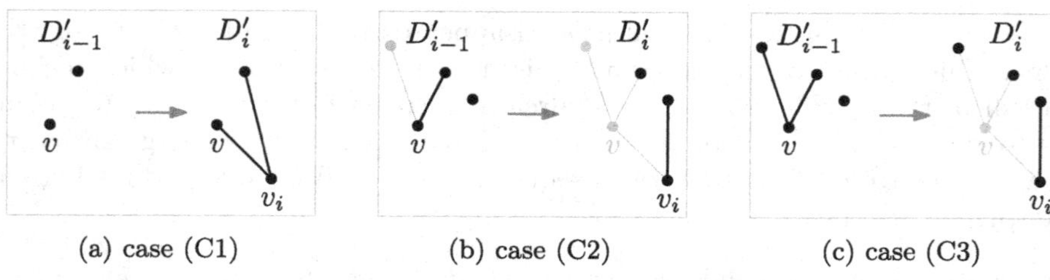

(a) case (C1) (b) case (C2) (c) case (C3)

Fig. 5. Cases considered in the proof of Theorem 4. In all of them, the number of edges in G'_i is maximized. Gray vertices and edges were visible in some previous steps.

(C3) Vertex v has two neighbors in G'_{i-1}. In this case, the two edges incident with v are the only edges in G'_{i-1}. Then v disappears as v_i is its last neighbor. Therefore, G'_i contains at most one edge that v_i may have introduced.

We have shown that, in each case, the number of visible edges in G'_i, for $4 \leq i \leq n-1$, is at most two. Note that, if there are two edges, then they share an endpoint. In the full version [12], we also show that we can always find a position of the vertices such that each frame is outerplanar and straight-line.

To see the linear runtime, note that we can choose v_i and update H_i in amortized constant time by using a suitable data structure [13]. The other steps of our construction require constant time for each vertex v_i. □

5 Forest Storyplans

Clearly, any triangle is an obstruction for a graph to admit a forest storyplan. Interestingly, for planar and subcubic graphs this is the only obstruction for the existence of a forest storyplan as we show now.

Theorem 5 ($\star$). *Every Δ-free subcubic graph admits a straight-line forest storyplan. Such a storyplan can be computed in linear time and has at most five edges per frame.*

Proof Sketch. We use the storyplan from the proof of Theorem 4. By construction, we never get a cycle since we consider triangle-free graphs. □

As a warm-up for our main result, we briefly show the following weaker result.

Observation 1. *Every Δ-free outerplanar graph admits a straight-line forest storyplan, and such a storyplan can be computed in linear time.*

Proof. Let G be a Δ-free outerplanar graph, and let Γ be an outerplanar straight-line drawing of G. Let $\sigma = \langle v_1, v_2, \ldots, v_n \rangle$ be the circular order of the vertices along the outer face of Γ (which can easily be determined in linear time [11]). We claim that σ yields a forest storyplan of G. (Note that the positions of the vertices in Γ will make this storyplan straight-line.)

To this end, we show that there is no frame where a complete face of Γ is visible. If this is true, then no frame contains a complete cycle. This is due to the fact that, in outerplanar graphs, the vertex set of every cycle contains the vertex set of at least one face. Let $F = \langle v_{i_1}, v_{i_2}, \ldots, v_{i_k} \rangle$ with $i_1 < i_2 < \cdots < i_k$ be a face of G. Since G is Δ-free, we have $k \geq 4$. Note that v_{i_1} and v_{i_3} as well as v_{i_2} and v_{i_4} are not adjacent. Since G is outerplanar, v_{i_2} may be adjacent only to vertices that appear in σ between (and including) v_{i_1} and v_{i_3}. Therefore, v_{i_2} disappears before v_{i_4} appears. Hence it is indeed not possible that all vertices of the same face appear in a frame. □

Now we improve upon the simple result above. Note, however, that we do not guarantee a linear running time any more.

Theorem 6. *Every Δ-free planar graph admits a straight-line forest storyplan, and such a storyplan can be computed in polynomial time.*

Proof. Let G be a Δ-free planar graph, and let Γ be a planar straight-line drawing of G. In the desired forest storyplan for G, we use the position of the vertices in Γ.

We first give a rough outline of our iterative algorithm and then describe the details. In each iteration (which spans one or more steps of the storyplan that we construct), we *pick* a vertex on the current outer face, which means that we add it and its neighbors (if they are not visible yet) to the storyplan one by one. In this way, after each iteration, at least one vertex disappears, namely the one we picked.

Let $G_1 = G$ and, for $i \in \{1, 2, \ldots\}$, let v_i be the vertex that we pick in iteration i, and let G_{i+1} be the subgraph of G_i that we obtain after removing the vertices (and the edges incident to them) that disappear in iteration i; see Fig. 7b. The algorithm terminates as soon as G_i is a forest and adds the remaining vertices in arbitrary order to the storyplan under construction. We call vertices and edges incident with the (current) outer face *outer*. The others are *inner*.

We always pick outer vertices. For this reason, only two types of vertices are problematic for avoiding cycles: the endpoints of *chords* (i.e., inner edges incident with two outer vertices) and the endpoints of *half-chords* (i.e., length-2 paths that connect two outer vertices via an inner vertex).

Let G'_i be the (embedded) subgraph of G_i (embedded according to Γ) that consists of all vertices and edges that lie on a simple cycle that bounds the outer face of G_i, plus every edge that connects two cycles, plus all chords and half-chords (and, thus, plus the inner vertices that lie on the half-chords) of G_i; see Figs. 6 and 7c. For example, the edges e and e' of G_2 in Fig. 7b are not part of G'_2. We say that a vertex of G'_i is *free* if it lies on the outer face and is not part of a chord or a half-chord.

Let H_i be the weak dual of G'_i (see Fig. 7c), i.e., the (embedded) multigraph that has a vertex for each inner face of G'_i and an edge for each pair of inner faces that are incident with a common edge of G'_i. Note that H_i is outerplane (since the inner vertices of G'_i form an independent set) and that H_i has no loops (since G'_i does not have leaves). We maintain the following invariants:

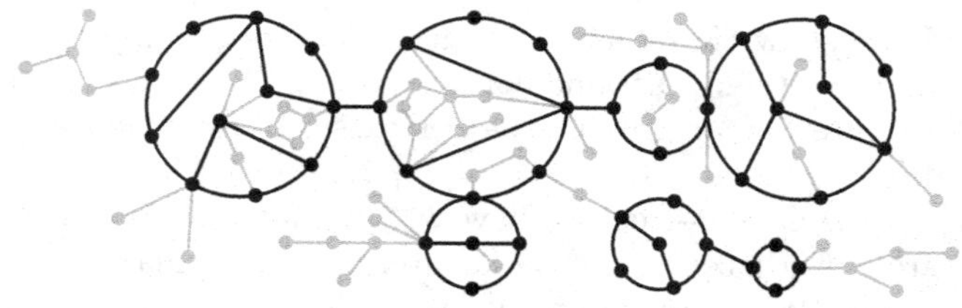

Fig. 6. From an embedded △-free planar graph G_i (black & gray), we obtain G_i' (black). Note that G_i' decomposes into seven simple cycles and two connected components. The outer edges and vertices of these connected components form cactus graphs. (Color figure online)

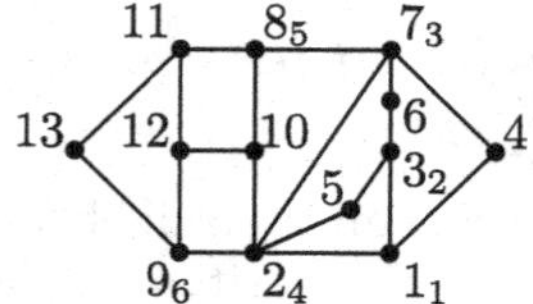

(a) △-free planar graph G with a forest storyplan

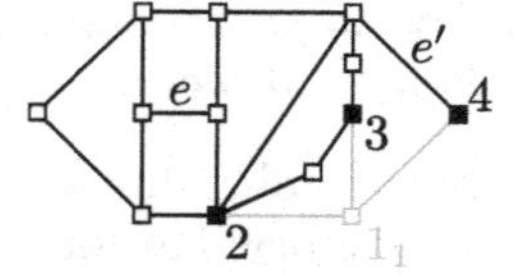

(b) iteration 2: subgraph G_2 (black) of $G_1 = G$

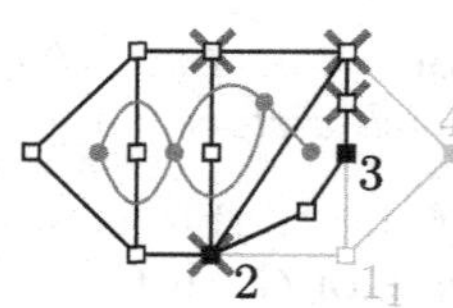

(c) subgraph G_2' of G_2 with weak dual H_2 (green)

Fig. 7. A △-free planar graph G where (a) shows a forest storyplan computed by our algorithm, (b) shows the result of the first iteration of the algorithm, and (c) shows the auxiliary graph for the second iteration. Subscripts refer to the iteration in which a vertex is picked. Red crosses mark vertices that may not be picked. (Color figure online)

(I1) At no point in time, the set of visible edges on the outer face forms a cycle.
(I2) *During* iteration i, the only inner vertices that may be visible are those that are adjacent to v_i and to no other visible vertex on the outer face.
(I3) *During* iteration i, the only inner edges that may be visible are those that are incident with v_i and to no other vertex on the outer face.
(I4) *At the end* of each iteration (after removing the vertices that are not visible any more and before picking a new one), only vertices and edges incident with the outer face are visible.

Obviously, if the invariants hold, the set of visible edges in each frame forms a forest. In order to guarantee that the invariants hold, we use the following rules that determine which vertices we may not pick; see Fig. 8. We call a vertex observing these rules *good*. Note that we always pick a good vertex on the outer face of G_i' – we will later argue that there always is one.

(R1) Do not pick a vertex v whose extended neighborhood $N[v] = \{v\} \cup \{u : uv \in E(G_i)\}$ contains all invisible vertices of the outer face of G_i'.
(R2) Do not pick an endpoint of a chord.
(R3) Do not pick a neighbor of an endpoint of a chord if the other endpoint of that chord is visible.
(R4) Do not pick an endpoint of a half-chord if the other endpoint is visible.

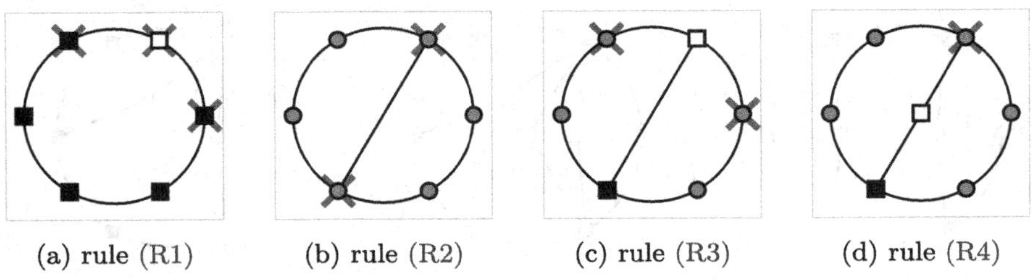

Fig. 8. Rules that determine which vertices may not be picked (marked by red crosses). Black squares represent visible vertices, white squares represent invisible vertices, and gray disks represent vertices that may be visible or invisible. (Color figure online)

Rule (R1) ensures that we do not close a cycle on the outer face, thus, invariant (I1) holds. Rule (R4) ensures that none of the visible inner vertices is adjacent to two visible vertices on the outer face (including the picked vertex), thus, invariant (I2) holds. Rules (R2) and (R3) ensure that no chords are visible. Together with rule (R4) and the fact that G is Δ-free, they ensure that the inner edges that are visible are incident with the picked vertex and no other vertex on the outer face. Thus, invariant (I3) holds. Invariant (I4) holds because we always pick a vertex on the outer face and remove it. As a result, the faces incident with the picked vertex become part of the outer face and the previously inner neighbors (if any) of the picked vertex become incident with the outer face.

It remains to prove that, as long as G_i is not a forest (and the algorithm terminates), there exists a vertex that can be picked without violating any of our rules. Our proof is constructive; we show how to find a vertex to pick.

We first show that H_i is a (collection of) cactus graph(s), that is, every edge of H_i lies on at most one cycle. Suppose that H_i contains an edge e_1 that lies on at least two simple cycles. If the interiors of the two cycles are disjoint, then e_1 is not incident to the outer face of H_i (contradicting H_i being outerplane). Otherwise, one of the cycles has at least one edge $e_2 \neq e_1$ in the interior of the other cycle, again contradicting H_i being outerplane.

We show in two steps that G_i (actually even G'_i) always contains a good vertex, which we pick. First, we show how to find a good vertex in the base case, that is, if the outer face of G'_i is a simple cycle. Then, we consider the general case where the outer face of G'_i is a (collection of) cactus graph(s). Here, we repeatedly apply the argument of the base case to find a good vertex. So, assume that the outer face of G'_i is a simple cycle and, hence, H_i is connected.

In the trivial case that the weak dual H_i is a single vertex, G'_i is a cycle of at least four free vertices. Due to invariant (I1), there is an invisible vertex $v \in G'_i$. Any non-neighbor of v in G'_i is a good vertex, which we can pick.

If H_i has a vertex of degree 1, which corresponds to a face f of G'_i, it means that f is incident with exactly one chord and to no half-chords. Since G is Δ-free, there are at least two free vertices in f. Note that at most one endpoint of the chord is visible (due to invariant (I3)). If one endpoint is indeed visible, then

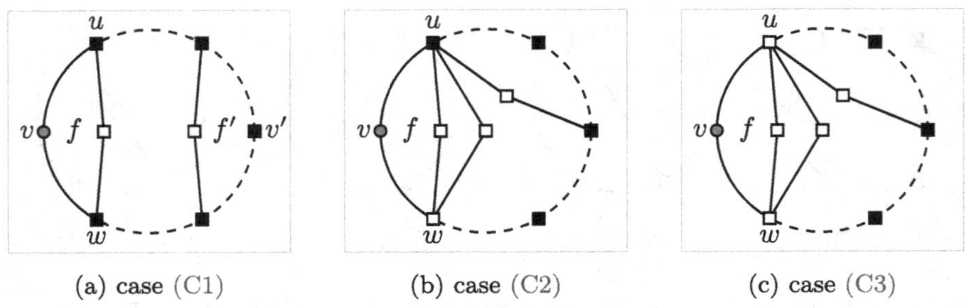

(a) case (C1) (b) case (C2) (c) case (C3)

Fig. 9. Cases when there is no chord in G'_i. We always find a good vertex.

its unique neighbor on the boundary of f that is not incident with the chord observes all rules and can be picked. If none of the endpoints of the chord is visible, then any free vertex of f can be picked.

Otherwise, all vertices of H_i have degree at least 2. Let F be the set of faces of G'_i that are incident with exactly one half-chord and to an arbitrary number of outer edges (but to no other inner edge). Note that in H_i, F corresponds to a set of vertices of degree 2. We now use the following two helpful claims. Their proofs are in the full version [12].

Claim 1 (⋆). The set F has cardinality at least 2.

Claim 2 (⋆). Let the edge e (or the edge pair $\{e_1, e_2\}$) be any chord (half-chord) of G'_i, let F_1 be the set of inner faces on the one side, and let F_2 be the set of inner faces on the other side of e (or $\{e_1, e_2\}$, resp.). Then, $F_1 \cap F \neq \emptyset$ and $F_2 \cap F \neq \emptyset$.

We continue to show that there is a good vertex on the outer face of G'_i, which we can pick. Assume first that G'_i does not have chords. Thus, all vertices of G'_i trivially observe rules (R2) and (R3). Let $f \in F$, and let u and w be the endpoints of the unique half-chord incident with f. If there is a free vertex v in f such that $N[v]$ does not contain the last invisible vertices of the outer face of G'_i, then we pick v. Rules (R1) and (R4) are observed by the definition of v. If, for every free vertex v in f, $N[v]$ contains all invisible vertices of the outer face of G'_i, consider the following three cases; see Fig. 9. The cases are ordered by priority; if we fulfill the conditions of multiple cases, the first case applies.

- (C1) Both u and w are visible. Then, consider a face $f' \in F$ different from f, which exists by Claim 1. Clearly, all of its vertices are visible, and we can pick any free vertex v' of f' without violating the rules.
- (C2) Exactly one of $\{u, w\}$ is visible. W.l.o.g., assume that u is visible and w is invisible. We claim that u observes all rules. Since w remains invisible after picking u, u observes rule (R1). If there was another half-chord incident with u, either it would again be incident with w, which does not violate rule (R4), or it would be incident with another vertex of G'_i, which is

visible. By Claim 2, however, there is another face $f' \in F$ on the other side of that half-chord. As all of the vertices of f' on the outer face are visible, we would be in case (C1) instead.

(C3) Both u and w are invisible. We claim that u observes the rules. Similar to case (C2), u observes rule (R1) (since w stays invisible) and rule (R4) (if there was another half-chord incident with u whose other endpoint is visible, we would be in case (C1) or in case (C2)).

Now assume that G'_i has one or more chords. Of course, each of these chords has at most one visible endpoint. The chords with exactly one visible endpoint divide G'_i into several subgraphs. Observe that at least one of these subgraphs contains no such chord in its interior and is bounded by only one of them (or no chord has a visible endpoint, then there is only one subgraph, namely G'_i). We call this subgraph $\hat{G}'_i$ and we let u and w denote the visible and invisible endpoints of the bounding chord e, respectively (or if there is only one subgraph, then u and w are just neighbors). By a case distinction on the facets incident to u and w, we can show that there is always a good vertex on the outer face of $\hat{G}'_i$, and hence on the outer face of G'_i. The details are in the full version [12].

Claim 3 (⋆). There is a good vertex on the outer face of $\hat{G}'_i$.

We have shown that there is always a good vertex on the outer face of G'_i if the outer face of G'_i is a simple cycle. Now assume that the outer face of G'_i is not just a simple cycle, but consists of one or multiple cactus graphs. If we have multiple cactus graphs, we can consider them individually. So, it suffices to consider the case where the outer face of G'_i is one (connected) cactus graph. Still, H_i may be disconnected. Let $C_1, C_2, \dots$ be the connected components of H_i, and let $\tilde{G}_1, \tilde{G}_2, \dots$ be the corresponding subgraphs of G'_i. Two subgraphs $\tilde{G}_j$ and $\tilde{G}_k$ may be connected by at most one common vertex or via a single edge. Otherwise, we consider them as non-connected (if they are connected by a path of length ≥ 2 in G_i, they are independent because the neighborhood of $\tilde{G}_j$ does not overlap $\tilde{G}_k$ and vice versa; these parts remain as a forest in the end). Let T be a graph with a vertex for each $\tilde{G}_1, \tilde{G}_2, \dots$ where two vertices are adjacent if and only if the corresponding subgraphs are connected. Since the outer face of G'_i is a cactus graph, T is a forest. Consider the subgraph $\tilde{G}_1$ and use the algorithm above to find a good vertex v. If v is a cut vertex, then check if it is also a good vertex in all subgraphs from $\{\tilde{G}_1, \tilde{G}_2, \dots\}$ where it is contained as well. Further, check for each neighbor w of v that is contained in a subgraph $\tilde{G}_j$ distinct from $\tilde{G}_1$ whether making w visible violates one of the invariants (note that this is a weaker criterion than checking if w is a good vertex and it implies that w and its neighbors in $\tilde{G}_j$ are not good vertices). If there is a subgraph $\tilde{G}_j$ where picking v breaks at least one rules (or making a neighbor of v visible breaks an invariant), then find a good vertex in $\tilde{G}_j$ (recall that there exists at least one good vertex) and proceed in the same way. Since T does not contain cycles, this procedure always terminates with a (globally) good vertex.

Concerning the running time, note that, if we maintain the outer face, we can find, for each vertex, its incident chords and half-chords in linear time. Further, our constructive proof can be turned into a polynomial-time algorithm as it includes only graph traversal and graph construction operations that can be executed in polynomial time. □

6 Open Problems

1. What is the complexity of deciding whether a given graph admits an outerplanar or a forest storyplan? We conjecture that recognition is NP-hard.
2. While we extended the existing planar storyplan problem into the direction of less powerful but easier-to-understand storyplans, one could also go into the opposite direction and investigate more powerful storyplans in order to be able to construct such storyplans for larger classes of graphs. For example, 1-planar storyplans would be a natural direction for future research.

References

1. Abdelaal, M., Lhuillier, A., Hlawatsch, M., Weiskopf, D.: Time-aligned edge plots for dynamic graph visualization. In: Banissi, E., et al. (eds.) Proceedings of 24th International Conference Information Visualisation (IV 2020), pp. 248–257 (2020). https://doi.org/10.1109/IV51561.2020.00048
2. Been, K., Nöllenburg, M., Poon, S.-H., Wolff, A.: Optimizing active ranges for consistent dynamic map labeling. Comput. Geom. **43**(3), 312–328 (2010). https://doi.org/10.1016/j.comgeo.2009.03.006
3. Binucci, C., et al.: Drawing trees in a streaming model. Inf. Process. Lett. **112**(11), 418–422 (2012). https://doi.org/10.1016/j.ipl.2012.02.011
4. Binucci, C., et al.: On the complexity of the storyplan problem. In: Angelini, P., von Hanxleden, R. (eds.) GD 2022. LNCS, vol. 13764, pp. 304–318. Springer, Cham (2023). https://doi.org/10.1007/978-3-031-22203-0_22. https://arxiv.org/abs/2209.00453
5. Borrazzo, M., Da Lozzo, G., Di Battista, G., Frati, F., Patrignani, M.: Graph stories in small area. J. Graph Algorithms Appl. **24**(3), 269–292 (2020). https://doi.org/10.7155/jgaa.00530
6. Burch, M.: The dynamic graph wall: visualizing evolving graphs with multiple visual metaphors. J. Vis. **20**(3), 461–469 (2017). https://doi.org/10.1007/s12650-016-0360-z
7. Da Lozzo, G., Rutter, I.: Planarity of streamed graphs. Theor. Comput. Sci. **799**, 1–21 (2019). https://doi.org/10.1016/j.tcs.2019.09.029
8. de Fraysseix, H., Pach, J., Pollack, R.: How to draw a planar graph on a grid. Combinatorica **10**(1), 41–51 (1990). https://doi.org/10.1007/BF02122694
9. Di Battista, G., et al.: Small point-sets supporting graph stories. In: Angelini, P., von Hanxleden, R. (eds.) GD 2022. LNCS, vol. 13764, pp. 289–303. Springer, Cham (2023). https://doi.org/10.1007/978-3-031-22203-0_21. https://arxiv.org/abs/2208.14126
10. Di Battista, G., Eades, P., Tamassia, R., Tollis, I.G.: Graph Drawing: Algorithms for the Visualization of Graphs. Prentice-Hall, Hoboken (1999)
11. Di Battista, G., Frati, F.: Small area drawings of outerplanar graphs. Algorithmica **54**(1), 25–53 (2009). https://doi.org/10.1007/11618058_9

12. Fiala, J., Firman, O., Liotta, G., Wolff, A., Zink, J.: Outerplanar and forest storyplans. arXiv report (2023). http://arxiv.org/abs/2311.13523
13. Matula, D.W., Beck, L.L.: Smallest-last ordering and clustering and graph coloring algorithms. J. ACM **30**(3), 417–427 (1983). https://doi.org/10.1145/2402.322385
14. Schnyder, W.: Embedding planar graphs on the grid. In: Proceedings of 1st ACM-SIAM Symposium Discrete Algorithms (SODA 1990), pp. 138–148 (1990). https://dl.acm.org/doi/10.5555/320176.320191

The Complexity of Cluster Vertex Splitting and Company

Alexander Firbas(✉), Alexander Dobler(✉), Fabian Holzer, Jakob Schafellner, Manuel Sorge(✉), Anaïs Villedieu(✉), and Monika Wißmann

TU Wien, Vienna, Austria
{alexander.firbas,alexander.dobler}@tuwien.ac.at,
{manuel.sorge,avilledieu}@ac.tuwien.ac.at

Abstract. Clustering a graph when the clusters can overlap can be seen from three different angles: We may look for cliques that cover the edges of the graph with bounded overlap, we may look to add or delete few edges to uncover the cluster structure, or we may split vertices to separate the clusters from each other. Splitting a vertex v means to remove it and to add two new copies of v and to make each previous neighbor of v adjacent with at least one of the copies. In this work, we study underlying computational problems regarding the three angles to overlapping clusterings, in particular when the overlap is small. We show that the above-mentioned covering problem is NP-complete. We then make structural observations that show that the covering viewpoint and the vertex-splitting viewpoint are equivalent, yielding NP-hardness for the vertex-splitting problem. On the positive side, we show that splitting at most k vertices to obtain a cluster graph has a problem kernel with $O(k)$ vertices. Finally, we observe that combining our hardness results with the so-called critical-clique lemma yields NP-hardness for Cluster Editing with Vertex Splitting, which was previously open (Abu-Khzam et al. [ISCO 2018]) and independently shown to be NP-hard by Arrighi et al. [IPEC 2023]. We observe that a previous version of the critical-clique lemma was flawed; a corrected version has appeared in the meantime on which our hardness result is based.

1 Introduction

In classical graph-clustering, we want to partition the input graph into clusters that are densely connected, while there are few connections between different clusters. However, in clusterings of real-world graphs the clusters often overlap [14]. We are interested here in exact algorithms for and complexity of such overlapping clustering problems. Without overlap, these are well-studied (see the survey [7]), but less so if we allow overlap [2,3,5,9]. In some applications, clusters may overlap but not very strongly. We focus mainly on this case.

A. Dobler—Supported by the Vienna Science and Technology Fund (WWTF) under grant 10.47379/ICT19035.
M. Sorge—Partly supported by the Alexander von Humboldt Foundation.

H. Fernau et al. (Eds.): SOFSEM 2024, LNCS 14519, pp. 226–239, 2024.
https://doi.org/10.1007/978-3-031-52113-3_16

To understand the complexity, a basic formulation of a clustering with small overlaps can focus on perfect clusterings, i.e., clusters are cliques and all edges of the input graph occur in a cluster. This leads to the SIGMA CLIQUE COVER (SCC) problem, where we seek a covering of the input graph by induced cliques and we want to minimize the total number of times the vertices are covered by the cliques (see Sect. 3 for a formal definition). SCC was previously studied in the context of displaying information in bioinformatics [11] and in combinatorics [8]. To our knowledge, its complexity was not known. We prove that SCC is NP-complete (Theorem 3.5).

An alternative view on overlapping clustering with small overlaps is that of splitting vertices: A vertex split is a graph operation that takes a vertex v and replaces it by two copies such that the union of the neighborhoods of the copies is equal to the neighborhood of the original vertex v. Given a graph and an integer k, we may then ask to perform at most k vertex-splitting operations in order to obtain a cluster graph (a disjoint union of cliques). The cliques in the obtained cluster graph then correspond to the clusters in the original graph. This yields the CLUSTER VERTEX SPLITTING (CVS) problem. In Sect. 4 we show that SCC and CVS are indeed equivalent, and thus both are NP-complete. On the positive side, we show that CVS is fixed-parameter tractable with respect to the number k of allowed splits, that is, it can be solved in $f(k) \cdot n^{O(1)}$ time where f is a computable function and n the number of vertices. Indeed, in Sect. 5 we show a stronger result, namely, that CVS admits an $O(k)$-vertex problem kernel, that is, we may produce with polynomial processing time an equivalent instance that contains $O(k)$ vertices (see Theorem 5.6). This result relies on an analysis of the structure of the so-called critical cliques of the input graph. Informally, a critical clique is an induced clique in the input graph with vertex set C such that all vertices in C have pairwise the same neighbors outside of C and such that there is no critical clique that strictly contains C.[1]

The CLUSTER EDITING WITH VERTEX SPLITTING (CEVS) problem [3] is closely related to the above two problems. The difference is that the underlying clustering model allows the clusters to be imperfect, that is, the clusters may miss a small number of edges and there may be a small number of edges that are not contained in any cluster. More precisely, in CEVS we are given a graph G and an integer k and we want to obtain a cluster graph from G by at most k modifications. As modifications we are allowed to split vertices and to add or delete edges. It was previously open whether CEVS is NP-hard [3] which has been independently and in parallel to our work been shown to be true [5]. Our impetus was to show NP-hardness of CEVS, too, and, indeed, combining our NP-hardness result for SCC with a so-called critical-clique lemma [3,4] yields NP-hardness of CEVS (see Sect. 7). We refrained from publishing this result at first, because the critical-clique lemma as stated by Abu-Khzam et al. [3,4] and used in references [5,6] is incorrect, see the counterexample in Sect. 6. Fortunately, after the appearance of our counterexample, a corrected variant of

[1] Alternatively, a critical clique is a maximal set of pairwise true twins.

the critical-clique lemma appeared [1], completing our alternative NP-hardness proof of CEVS.

Due to space constraints, statements marked with ★ are proved in the arXiv version of this paper [10].

2 Preliminaries

For a positive integer $n \in \mathbb{N}$ we use $[n]$ to denote $\{1, 2, \ldots, n\}$. For a set X, we denote by $\mathcal{P}(X)$ its power set. Moreover, for a family of sets $\mathcal{X}$, we write $\bigcup \mathcal{X}$ for the union of all sets member of $\mathcal{X}$, that is, $\bigcup_{X \in \mathcal{X}} X$. We denote disjoint unions by $\uplus$. Unless explicitly mentioned otherwise, all graphs are undirected and without parallel edges or self-loops. Given a graph G with vertex set $V(G)$ and edge set $E(G)$, we denote the neighborhood of a vertex $v \in V(G)$ by $N_G(v)$. If the graph G is clear from the context, we omit the subscript G. For $V' \subset V(G)$, we write $G[V']$ for the graph induced by the vertices V'. For $u, v \in V(G)$ we write uv as a shorthand for $\{u, v\}$, $G - v$ for $G[V \backslash \{v\}]$, and $d_G(v)$ for $|N_G(v)|$. The graph K_n is the complete graph on n vertices. We write $G \simeq H$ if a graph G is isomorphic to H. A *cluster graph* is a graph in which every connected component is a clique. Equivalently, a cluster graph does not contain a path P_3 with three vertices as an induced subgraph. A *vertex split* operation applied to a graph $G = (V, E)$ and $u \in V$ results in a graph $G' = (V', E')$ such that $V' = V \backslash \{u\} \cup \{v, w\}$ with $v, w \notin V$, and E' is obtained from E by making each vertex adjacent to u adjacent to at least one of v and w; that is, $N_{G'}(v) \cup N_{G'}(w) = N_G(u)$.

3 NP-Completeness of Sigma Clique Cover

To start, we will fix some notation. Leading up to the formulation of the sigma clique cover problem, we first define the notion of a sigma clique cover:

Definition 3.1. *Let G be a graph. Then, $\mathcal{C} \subseteq \mathcal{P}(V)$ is called a* sigma clique cover *of G if 1. $G[C]$ is a clique for all $C \in \mathcal{C}$ and 2. for each $e \in E(G)$, there is $C \in \mathcal{C}$ such that $e \in E(G[C])$, that is, all edges of G are "covered" by some clique of $\mathcal{C}$. The* weight *of a sigma clique cover $\mathcal{C}$ is* $\text{wgt}(\mathcal{C}) := \sum_{C \in \mathcal{C}} |C|$.

Now, we can formulate the associated decision problem:

Sigma Clique Cover (SCC)	
Input:	A tuple (G, s), where G is a graph and $s \in \mathbb{N}$.
Question:	Is there a sigma clique cover $\mathcal{C}$ of G with $\text{wgt}(\mathcal{C}) \leq s$?

Note that SCC is not equivalent to the well-studied Edge Clique Cover problem, whose optimization goal is to minimize $|\mathcal{C}|$ rather than $\text{wgt}(\mathcal{C})$. To show that SCC is NP-hard, we reduce from the Node Clique Cover problem. Analogous to the case of SCC, to define said problem formally, we first need introduce the notion of a node clique cover:

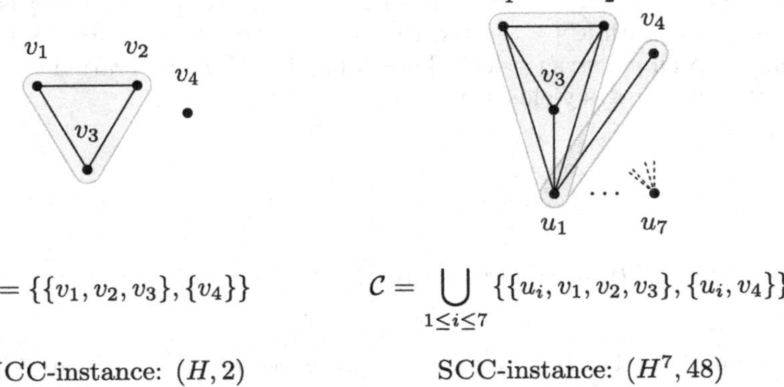

$\mathcal{C} = \{\{v_1, v_2, v_3\}, \{v_4\}\}$ $\qquad$ $\mathcal{C} = \bigcup_{1 \le i \le 7} \{\{u_i, v_1, v_2, v_3\}, \{u_i, v_4\}\}$

NCC-instance: $(H, 2)$ $\qquad$ SCC-instance: $(H^7, 48)$

Fig. 1. Example for our reduction from NCC to SCC. Covers are marked in gray.

Definition 3.2. *Let G be a graph. Then, $\mathcal{C} \subseteq \mathcal{P}(V)$ is called a* node clique cover *of G if 1. $G[C]$ is a clique for all $C \in \mathcal{C}$ and 2. for each $v \in V(G)$, there is $C \in \mathcal{C}$ such that $v \in V(G[C])$, that is, all vertices of G are "covered" by some clique $C \in \mathcal{C}$. The* size *of a node clique cover $\mathcal{C}$ is $|\mathcal{C}|$.*

With this, we can formulate the NP-hard [12] NODE CLIQUE COVER problem:

NODE CLIQUE COVER (NCC)	
Input:	A tuple (G, k), where G is a graph and $k \in \mathbb{N}$.
Question:	Is there a node clique cover $\mathcal{C}$ of G with $|\mathcal{C}| \le k$?

To formulate our reduction from NCC to SCC, we introduce notation to extend a graph with independent universal vertices.

Definition 3.3. *Let $G = (V, E)$ be a graph and $\ell \in \mathbb{N}$. Using a set $\{u_1, \ldots, u_\ell\}$ of ℓ new vertices called* universal vertices, *we construct a new graph G^ℓ with*

$$G^\ell := (V \cup \{u_1, \ldots, u_\ell\}, E \cup \{u_i v \mid 1 \le i \le \ell, v \in V\}).$$

Note that universal vertices themselves are not adjacent to each other. Informally, the main intuition behind our reduction from NCC is to add a sufficient number of universal vertices to the instances of NCC, such that concerning the derived instances of SCC, it will be "combinatorially favorable" to select cliques that contain a universal vertex. Refer to Fig. 1 for an example of the reduction.

Lemma 3.4. *Let $G = (V, E)$ be a graph and $\ell := 2|E| + 1$. Then, (G, s) is a positive instance of* NCC *if and only if $\left(G^\ell, \ell\,(|V| + s + 1) - 1\right)$ is a positive instance of* SCC.

Proof ($\Rightarrow$): Let $\mathcal{C}$ be a node clique cover of G with $|\mathcal{C}| \le s$. Without loss of generality, we assume that $\mathcal{C}$ is a partition of V—for otherwise if there are distinct $C', C'' \in \mathcal{C}$ with $C' \cap C'' \neq \emptyset$, then $\mathcal{C}' := (\mathcal{C}\setminus\{C'\}) \cup \{C'\setminus C''\}$ is a node clique cover of G with $|\mathcal{C}'| = |\mathcal{C}|$ and the number of nodes that are contained in more than one clique strictly less. Thus, applying this observation a sufficient number of times always yields a partition of V.

Let

$$\begin{aligned} \mathcal{A} &:= \{C \cup \{u_i\} \mid C \in \mathcal{C}, 1 \leq i \leq \ell\} \text{ and} \\ \mathcal{B} &:= \{\{v_1, v_2\} \mid v_1 v_2 \in E\}. \end{aligned}$$

We claim that $\mathcal{A} \uplus \mathcal{B}$ is a sigma clique cover of G^ℓ with $\operatorname{wgt}(\mathcal{A} \uplus \mathcal{B}) \leq \ell(|V| + s + 1) - 1$. First, we verify that $\mathcal{A} \uplus \mathcal{B}$ conforms to Definition 3.1, that is, it indeed is a sigma clique cover of G^ℓ. To that end, we begin by verifying that $G[C]$ is a clique for all $C \in \mathcal{A} \uplus \mathcal{B}$. By construction, we need to differentiate two cases: Firstly, let $C \cup \{u_i\} \in \mathcal{A}$. Since $G[C]$ is a clique, $E \subseteq E(G^\ell)$ and $\forall v \in V\colon u_i v \in E(G^\ell)$, it follows that $G^\ell[C \cup \{u_i\}]$ is also a clique. Secondly, let $\{v_1, v_2\} \in \mathcal{B}$. Similarly, since $G[\{v_1, v_2\}] \simeq K_2$ and $E \subseteq E(G^\ell)$, we have $G^\ell[\{v_1, v_2\}] \simeq K_2$.

Now, we prove that all edges of G^ℓ are "covered" by $\mathcal{A} \uplus \mathcal{B}$. Two cases need to be verified: Consider any $v_1 v_2 \in E$, i.e., those edges that are "inherited" from G to G^ℓ. We see that $\{v_1, v_2\} \subseteq B$ by definition. Furthermore, consider any $u_i v \in E(G^\ell)\setminus E$, i.e., those edges added to G in the construction of G^ℓ. Observe that since $\exists C \in \mathcal{C}$ with $v \in C$, we have $\{u_i, v\} \subseteq C \cup \{u_i\} \in \mathcal{A}$.

We conclude that $\mathcal{A} \uplus \mathcal{B}$ is a sigma clique cover of G^ℓ and proceed to verify that the claimed bound on the weight holds. By definition of $\mathcal{A}$, we obtain

$$\begin{aligned} \operatorname{wgt}(\mathcal{A}) &= \sum_{\substack{C \in \mathcal{C}, \\ 1 \leq i \leq \ell}} |C \cup \{u_i\}| \\ &= \sum_{\substack{C \in \mathcal{C}, \\ 1 \leq i \leq \ell}} |C| + 1 && \{u_1, \ldots, u_\ell\} \cap C = \emptyset \text{ for all } C \in \mathcal{C} \\ &= \ell \sum_{C \in \mathcal{C}} |C| + 1 \\ &= \ell \left(|\mathcal{C}| + \sum_{C \in \mathcal{C}} |C| \right) \\ &= \ell\,(|\mathcal{C}| + |V|). && \mathcal{C} \text{ is a partition of } V \end{aligned}$$

Clearly, $\operatorname{wgt}(\mathcal{B}) = 2|E| = \ell - 1$. Now, using $\mathcal{A} \cap \mathcal{B} = \emptyset$, we derive

$$\begin{aligned} \operatorname{wgt}(\mathcal{A} \uplus \mathcal{B}) &= \operatorname{wgt}(\mathcal{A}) + \operatorname{wgt}(\mathcal{B}) \\ &= \ell(|\mathcal{C}| + |V| + 1) - 1 \\ &\leq \ell(s + |V| + 1) - 1. && |\mathcal{C}| \leq s. \end{aligned}$$

Thus, the forward direction of the proof is established.

($\Leftarrow$): Let $\mathcal{S}$ be a sigma clique cover of G^ℓ with

$$\text{wgt}(\mathcal{S}) \leq \ell(|V| + s + 1) - 1,$$

and let

$$\begin{aligned} u^* &\in \operatorname*{argmin}_{u \in \{u_1,\dots,u_\ell\}} \text{wgt}\left(\{C \in \mathcal{S} \mid u \in C\}\right), \\ \mathcal{X} &:= \{C \in \mathcal{S} \mid u^* \in C\}, \text{ and} \\ \mathcal{N} &:= \{C \backslash \{u^*\} \mid C \in \mathcal{X}\}. \end{aligned}$$

We claim that $\mathcal{N}$ is a node clique cover of G with $|\mathcal{N}| \leq s$.

First, we verify that $\mathcal{N}$ conforms to Definition 3.2, i.e., it indeed is a node clique cover of G.

To accomplish this, we begin by showing that $G[C]$ is a clique for all $C \in \mathcal{N}$. Let $C\backslash\{u^*\} \in \mathcal{N}$. By construction, we have $C \in \mathcal{X}$, and since $\mathcal{X} \subseteq \mathcal{S}$ with the assumption that $\mathcal{S}$ is a sigma clique cover, we obtain that $G^\ell[C]$ is a clique. Because the clique-graph-property is hereditary, $G^\ell[C\backslash\{u^*\}]$ too is a clique. Observe that $\{u_1, \dots, u_\ell\} \cap V(G^\ell[C\backslash\{u^*\}]) = \emptyset$, for otherwise there would be an edge between two universal nodes of G^ℓ, and $G \leq G^\ell$, yielding $G^\ell[C\backslash\{u^*\}] = G[C\backslash\{u^*\}]$. Hence, $G[C\backslash\{u^*\}]$ is a clique.

Next, we verify that all vertices of G are "covered" by $\mathcal{N}$. Let $v \in V$. Since $\mathcal{S}$ is a sigma clique cover of G^ℓ and $vu^* \in E(G^\ell)$, there is some $C \in \mathcal{S}$ s.t. $\{v, u^*\} \subseteq C$. It immediately follows that $v \in C\backslash\{u^*\} \in \mathcal{N}$.

Finally, we establish that $|\mathcal{N}| \leq s$. To that end, first, we derive $\text{wgt}(\mathcal{X}) \leq |V| + s$. Towards a contradiction, suppose that $\text{wgt}(\mathcal{X}) \geq |V| + s + 1$. Observe that since no $C \in \mathcal{S}$ can contain two different universal nodes of G^ℓ we get

$$\begin{aligned} \text{wgt}(\mathcal{S}) &\geq \sum_{u \in \{u_1,\dots,u_\ell\}} \text{wgt}\left(\{C \in \mathcal{S} \mid u \in C\}\right) && \text{choice of } u^*, \mathcal{X} \\ &\geq \ell \cdot \text{wgt}(\mathcal{X}) && \text{wgt}(\mathcal{X}) \geq |V| + s + 1 \\ &\geq \ell(|V| + s + 1) \\ &= \ell(|V| + s + 1) - 1 + 1 && \ell(|V| + s + 1) - 1 \geq \text{wgt}(\mathcal{S}) \\ &\geq \text{wgt}(\mathcal{S}) + 1. \end{aligned}$$

In total, this yields $\text{wgt}(\mathcal{S}) \geq \text{wgt}(\mathcal{S}) + 1$, hence $\text{wgt}(\mathcal{X}) \leq |V| + s$.

Now, towards the final contradiction, suppose $|\mathcal{N}| \geq s+1$. We obtain

$$\begin{aligned}
\text{wgt}(\mathcal{X}) &= \sum_{\substack{C \in \mathcal{S}, \\ u^* \in C}} |C| && \textit{definition of } \mathcal{N} \\
&= \sum_{C \in \mathcal{N}} |C \cup \{u^*\}| && u^* \notin C \textit{ for all } C \in \mathcal{N} \\
&= |\mathcal{N}| + \sum_{C \in \mathcal{N}} |C| && |\mathcal{N}| \geq s+1 \\
&\geq s + 1 + \sum_{C \in \mathcal{N}} |C| && \textit{double counting principle} \\
&= s + 1 + \sum_{v \in V} |\{C \in \mathcal{N} \mid v \in C\}| && \mathcal{N} \textit{ covers } V \\
&\geq s + 1 + |V|.
\end{aligned}$$

Thus, we have derived both $\text{wgt}(\mathcal{X}) \geq |V| + s + 1$ and $\text{wgt}(\mathcal{X}) \leq |V| + s$, a contradiction. Hence, we conclude that $|\mathcal{N}| \leq s$. □

Using this preliminary work, the NP-completeness proof is straightforward:

Theorem 3.5 (★). Sigma Clique Cover *is* NP-*complete.*

4 NP-Completeness of Cluster Vertex Splitting

We use Theorem 3.5 to show that also CVS is NP-complete:

Cluster Vertex Splitting (CVS)	
Input:	A tuple (G, k), where G is a graph and $k \in \mathbb{N}$.
Question:	Is there a sequence of at most k vertex splits that transforms G into a cluster graph?

The basic idea is to show that, in an n-vertex graph, finding a sigma clique cover with weight $n + k$ is equivalent to finding k vertices to split such that the resulting graph is a cluster graph: Given such a sigma clique cover, we may look at each clique and its overlap to the rest of the graph. We can split each such clique off the rest of the graph by splitting all vertices in the overlap. This results in splitting each vertex a number of times equal to the number of times it is covered by a clique minus one, that is, overall k vertex splits. In the other direction, taking a cluster graph obtained by splitting and projecting it onto the original vertices in the input graph will yield a sigma clique cover. Its weight corresponds to the sum of the sizes of the clusters, which is exactly the number of copies we have created, that is, $n + k$. See the full version of this paper [10] for a formal proof.

Theorem 4.1 (★). Cluster Vertex Splitting *is* NP-*complete.*

5 A Linear Kernel for CLUSTER VERTEX SPLITTING

We first introduce some basic notions in Sect. 5.1. In Sect. 5.2, we establish the groundwork for the first data-reduction rule of the kernel, which allows us to reduce certain critical cliques in a SIGMA CLIQUE COVER instance. The second rule of the kernel is based on Sect. 5.3, where we determine that SIGMA CLIQUE COVER instances that have been exhaustively reduced using the previously explored mechanism and still contain more than $3k$ vertices are negative instances. We then give the kernel in Sect. 5.4.

5.1 The Notions of Valency and Critical Cliques

We will frequently have to prove lower bounds for the weight that a sigma clique cover needs to have at minimum. This we will do by observing that particular vertices must be covered by at least a certain number of cliques each. For this, *valency* of a vertex v with respect to a sigma clique cover $\mathcal{C}$ counts the number of cliques that contain v that is, $\text{val}_{\mathcal{C}}(v) := |\{C \in \mathcal{C} \mid v \in C\}|$.

With this notation, we can express the weight of a sigma clique cover in an alternative manner: Via the definition of $\text{wgt}(\cdot)$ (Definition 3.1) and the principle of double counting, we obtain $\text{wgt}(\mathcal{C}) = \sum_{C \in \mathcal{C}} |C| = \sum_{v \in V(G)} \text{val}_{\mathcal{C}}(v)$.

Another key tool that we will use in this section is the concept of critical cliques, coined by Lin et al. [13]. The *closed neighborhood* of a vertex v in a graph G is $N_G(v) \cup \{v\}$. This allows us to consider an equivalence relation, where vertices of a graph are in the same class if and only if their closed neighborhoods coincide. The equivalence classes under this relation are called the *critical cliques* of G. Consider a critical clique C of G. Observe that it is fully connected "internally", that is, $G[C]$ is a clique, and that $N_G(v) \backslash C = N_G(w) \backslash C$ for any $v, w \in C$, which means that the vertices of C share a common "external neighborhood".

If we delete all but one vertex from each critical clique, we obtain a graph isomorphic to what we will call the critical clique graph of G; we will use the shorthand $\text{CC}(G)$ to refer to it. Formally:

Definition 5.1. *Let G be a graph. Consider the equivalence relation $R_G \subseteq V(G) \times V(G)$ where $(v, w) \in R_G$ if and only if $N(v) \cup \{v\} = N(w) \cup \{w\}$. We use $[v]_G$ to denote the equivalence class generated by $v \in V(G)$ and R_G. The* critical clique graph *of G, referred to using $\text{CC}(G)$, is given by*

$$V(\text{CC}(G)) := \{[v]_G \mid v \in V(G)\} \text{ and}$$
$$E(\text{CC}(G)) := \{[v]_G[w]_G \mid vw \in E(G) \wedge [v]_G \neq [w]_G\}.$$

The main intuition we make use of here is that members of the same critical clique are essentially "clones" of one another. Thus, it seems reasonable that, provided certain conditions are met, we are allowed to "shrink" certain critical cliques without removing a significant amount of "computational complexity" when solving the combinatorial problems we are interested in.

5.2 Towards a Rule to Shrink Critical Cliques

Consider the critical clique graph $\mathrm{CC}(G)$ of a graph G. We distinguish between two kinds of critical cliques:

1. Critical cliques $[v]_G$ such that their neighborhood, that is, $N_{\mathrm{CC}(G)}([v]_G)$, forms a clique in $\mathrm{CC}(G)$, and
2. critical cliques $[v]_G$, where said neighborhood does not form a clique.

In this section, we show that, with respect to the sigma clique cover problem, critical cliques of the first kind consisting of at least two vertices, can either safely be reduced in size, or deleted altogether (Lemma 5.4). Correspondingly, we will refer to them as *reducible critical cliques.* The second kind of critical cliques we will call *irreducible critical cliques.*

To help prove Lemma 5.4, we first observe that in any minimum-weight sigma clique cover of a graph, a vertex member of a critical clique of the first kind is always covered by precisely one clique. Furthermore, this clique can be determined explicitly (Lemma 5.3). We start with a useful observation:

Lemma 5.2 (★). *Let G be a graph, $\mathcal{C}$ a sigma clique cover of G, and $v \in C \in \mathcal{C}$. Then, $C \subseteq N_G(v) \cup \{v\}$.*

Now, we are ready to prove our auxiliary lemma that offers insight into the structure of minimum-weight sigma clique covers:

Lemma 5.3 (★). *Let G be a graph without isolated vertices and let $[v]_G$ be a critical clique in G such that $\mathrm{CC}(G)[N_{\mathrm{CC}(G)}([v]_G)]$ is a clique. Furthermore, let $\mathcal{C}$ be a minimum-weight sigma clique cover of G and let $C^* := N_G(v) \cup \{v\}$. Then, C^* is contained in $\mathcal{C}$. Moreover, C^* is the only clique of $\mathcal{C}$ that covers v.*

The next lemma shows the correctness of reducing reducible critical cliques:

Lemma 5.4 (★). *Let G be a graph without isolated vertices and let $[v]_G$ be one of its critical cliques such that $|[v]_G| \geq 2$ and $\mathrm{CC}(G)[N_{\mathrm{CC}(G)}([v]_G)]$ is a clique. Then, $(G, |V(G)|+k)$ is a positive instance of* SCC *if and only if $(G-v, |V(G-v)|+k)$ is.*

5.3 Towards a Rule to Recognize Negative Instances

In the previous section, we laid the foundation for a rule that minimizes the sizes of reducible critical cliques. Consider an instance $(G, |V(G)|+k)$ of Sigma Clique Cover that has been exhaustively reduced using the aforementioned rule. We now observe that, if this instance has more than $3k$ vertices, then it is a negative instance. This will serve as the basis for Rule II defined in Theorem 5.6.

We proceed as follows: We assume that G has more than $3k$ vertices and consider an arbitrary sigma clique cover $\mathcal{C}$ of G. Then, we provide two separate lower bounds on $\mathrm{wgt}(\mathcal{C})$. One bound is based on reducible critical cliques, while the other bound is based on irreducible critical cliques. Each lower bound individually is too weak, but the maximum of both will be greater than $|V(G)|+k$ in all cases, yielding that $(G, |V(G)|+k)$ is a negative instance.

Lemma 5.5 (★). *Let G be a graph such that none of its connected components are cliques and $k \in \mathbb{N}$. We divide $V(\mathrm{CC}(G))$ into the partition $A \uplus B$ where $v \in A$ if and only if $\mathrm{CC}(G)[N_{\mathrm{CC}(G)}(v)]$ is a clique. Furthermore, we set $\overline{A} := \bigcup A$ and $\overline{B} := \bigcup B$, that is, the partition of $V(G)$ induced by $A \uplus B$. If $|A| = |\overline{A}|$ and $|V(G)| > 3k$, then $(G, |V(G)| + k)$ is a negative instance of* SCC.

5.4 Deriving the Kernel

In the two preceding sections, we essentially derived two data reduction rules for the sigma clique cover problem. It remains to compile our results into a polynomial kernelization procedure for Cluster Vertex Splitting. Essentially, we convert a given instance (G, k) of Cluster Vertex Splitting into an equivalent instance of Sigma Clique Cover, apply the two reduction rules exhaustively, until finally converting the reduced instance back to an instance of Cluster Vertex Splitting. Refer to Fig. 2 for an example.

Theorem 5.6. Cluster Vertex Splitting *admits a problem kernelization mapping an instance (G, k) to an equivalent instance (G', k') satisfying $|V(G')| \leq 3k + 3$ and $k' \leq k$.*

Proof. Let an instance of Cluster Vertex Splitting be given through (G, k) and let G_0 be obtained from G by removing all isolated vertices. Observe that (G, k) is equivalent to $(G_0, k =: k_0)$ with respect to CVS. By the equivalence of CVS and Sigma Clique Cover, (G_0, k_0) is a positive instance of CVS if and only if $(G_0, |V(G_0)| + k_0)$ is a positive instance of Sigma Clique Cover. Next, we construct the sequences $G_0, \dots$ and $k_0, \dots$ by exhaustively applying the following set of rules:

Rule I: If there is a critical clique $[v]_{G_i} \in V(\mathrm{CC}(G_i))$ such that $[v]_{G_i}$ contains at least two vertices and $\mathrm{CC}(G_i)[N_{\mathrm{CC}(G_i)}([v]_{G_i})]$ is a clique, then $G_{i+1} := (G_i - v) - I$ and $k_{i+1} := k_i$, where I is the set of isolated vertices in $G_i - v$.
Rule II: If Rule I is not applicable to G_i, Rule II has not been used so far, and $|V(G_i)| > 3k_i$, then $G_{i+1} := P_3$ and $k_{i+1} := 0$.

The running time and kernel size are proven in the full version of this paper [10]. We claim that Rule I and Rule II are *correct*, that is, the instances $(G_i, |V(G_i)| + k_i)$ and $(G_{i+1}, |V(G_{i+1})| + k_{i+1})$ are equivalent with respect to the SCC problem for all $i \in \{0, \dots, \ell - 1\}$. Let G_i such that G_{i+1} was obtained by applying Rule I, and let v as well as I as used in the definition of Rule I. First, consider the case when $I \neq \emptyset$. Let $w \in I$. We have that $d_{G_i}(w) \geq 1$, because w is not isolated in G_i. At the same time, we know that $d_{G_i}(w) < 2$, for otherwise w would not be isolated in $G_i - v$. Thus, $d_{G_i}(w) = 1$, which forces $|[v]_{G_i}| = 2$. Since $w \notin [v]_{G_i}$ would imply $d_{G_i}(w) \geq 2$, we conclude that $[v]_{G_i} = \{v, w\}$, that is, $G_i[\{v, w\}] \simeq K_2$ is a connected component of G_i. Now, it is easy to see that $(G_i, |V(G_i)| + k)$ is equivalent to $(G_{i+1}, |V(G_{i+1})| + k)$ with respect to the SCC problem. Otherwise, $I = \emptyset$. By construction, G_i is free of isolated vertices. Thus, applying Lemma 5.4 yields that $(G_i, |V(G_i)| + k)$ is equivalent

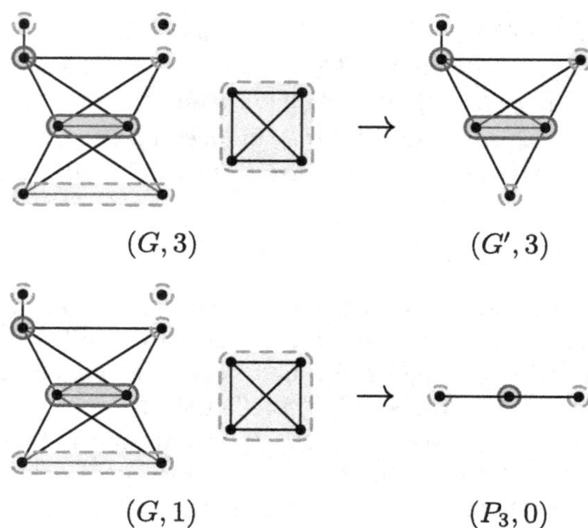

Fig. 2. Two instances of CVS and their corresponding kernel as given by Theorem 5.6. Reducible critical cliques are marked in green with dashed outlines, while irreducible critical cliques are marked in red with solid outlines. (Color figure online)

to $(G_i - v, |V(G_i - v)| + k) = (G_{i+1}, |V(G_{i+1})| + k)$ with respect to the SCC problem. Hence, Rule I is correct.

The correctness of Rule II essentially follows from Lemma 5.5; the proof is given in the full version [10]. In total, we have that $(G_\ell, |V(G_\ell)|+k_\ell)$ is a positive instance of SCC if and only if $(G_0, |V(G_0)| + k_0)$ is. By the equivalence between CVS and Sigma Clique Cover $(G_\ell, |V(G_\ell)| + k_\ell)$ is a positive instance of SCC if and only if (G_ℓ, k_ℓ) is a positive instance of CVS. Finally, we conclude that (G, k) is equivalent to (G_ℓ, k_ℓ) with respect to the CVS problem. □

6 The Critical-Clique Lemma

We now consider the critical-clique lemma for Cluster Editing With Vertex Splitting (CEVS) mentioned in the introduction. Let us first formally state the problem. In the following, by a *graph modification* we mean a vertex split, an edge addition, or an edge deletion.

Cluster Editing With Vertex Splitting	
Input:	A tuple (G, k), where G is a graph and $k \in \mathbb{N}$.
Question:	Is there a sequence of at most k graph modifications that transforms G into a cluster graph?

To state the critical-clique lemma, we first need an equivalence between the sequence of modifications in CEVS and a cover of the input graph by clusters.

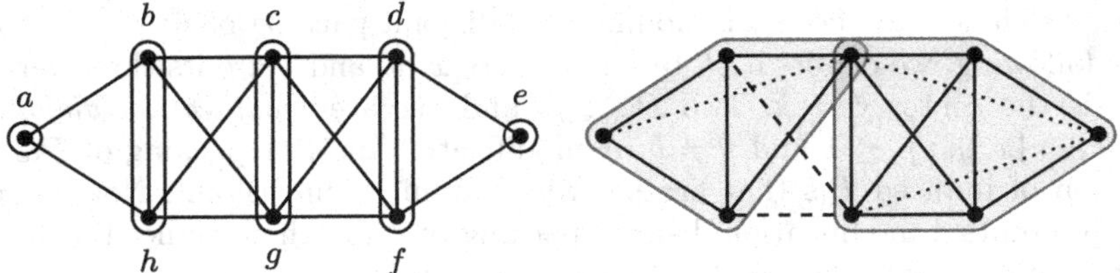

Fig. 3. Counterexample to the critical-clique lemma.

A *cover* of a graph G is a collection $\mathcal{C}$ of subsets of $V(G)$ such that $\bigcup_{C \in \mathcal{C}} C = V(G)$. The *cost* $\text{cst}_G(\mathcal{C})$ of a cover $\mathcal{C}$ is the number of non-edges contained in a set of $\mathcal{C}$ plus the number of edges not contained in any set of $\mathcal{C}$ plus the number of times each vertex is covered by a set beyond the first time. In formulas,

$$\text{cst}_G(\mathcal{C}) = \left|\left\{uv \in \binom{V}{2} \setminus E(G) \mid \exists C \in \mathcal{C} \colon \{u, v\} \subseteq C\right\}\right| + |\{uv \in E(G) \mid \forall C \in \mathcal{C} \colon \{u, v\} \nsubseteq C\}| + \left(\sum_{C \in \mathcal{C}} |C|\right) - |V(G)|.$$

Herein, $\binom{V}{2}$ denotes the set of all two-element subsets of V. If G is clear from the context, we omit the subscript G in cst_G.

The following lemma has been used implicitly by Abu-Khzam et al. [3] but we are not aware of a formal proof.

Lemma 6.1 (★). *Let G be a graph and k a positive integer. There is a sequence of at most k graph modifications to obtain from G a cluster graph if and only if G admits a cover of cost at most k.*

Recall the definition of critical cliques from Definition 5.1. The critical-clique lemma as stated by Abu-Khzam et al. [3] (see their Lemma 8) is as follows:

Lemma 6.2 (Incorrect). *Let G be a graph and $\mathcal{C}$ a cover of G of minimum cost. For each $C \in \mathcal{C}$ and each critical clique of G with vertex set K we have either $C \cap K = \emptyset$ or $K \subseteq C$.*

As far as we are aware, Lemma 6.2 is being used in references [3–6]. However, the example in Fig. 3 shows that Lemma 6.2 is incorrect: The left shows the input graph with marked critical cliques. The right shows a minimum-cost cover in which the left cover set contains the central critical clique only partially. Dashed edges are removed, dotted edges added, and vertices in both sets are split. The cost of the cover is 6.

Proposition 6.3. *The graph shown on the left in Fig. 3 needs at least 6 modifications to turn it into a cluster graph.*

Proof. We show that there is a modification-disjoint packing of six induced P_3s. In the following, we denote a P_3 by xyz, where x, y, and z are its three vertices and y is the center vertex. Two P_3s xyz and abc are *modification disjoint* if $|\{a, b, c\} \cap \{x, y, z\}| \leq 1$ and $y \neq b$. A modification-disjoint packing of P_3s is a collection of induced P_3s that are pairwise modification disjoint. Note that, if a graph admits a modification-disjoint packing of ℓ P_3s then we need at least ℓ modifications to turn the graph into a cluster graph.

Consider the following P_3s in the graph in Fig. 3: abc, cde, ahg, gfe, hcf, bgd. Note that they form a modification-disjoint packing. Thus we need at least 6 modifications to turn the graph into a cluster graph. □

There are other solutions of cost 6 that do not cut critical cliques. Thus, it is tempting to assume that, although not necessarily every optimal solution does not cut critical cliques, that there is always such an optimal solution. Indeed, after the appearance of our counterexample above, this has been proved to be true:

Lemma 6.4 (Abu-Khzam et al. [1]). *Let G be a graph and k a positive integer. If (G, k) admits a solution for* CEVS, *then there is a cover $\mathcal{C}$ of cost at most k such that for each critical clique K of G and each set $C \in \mathcal{C}$ we have either $K \subseteq C$ or $K \cap C = \emptyset$.*

7 The Complexity of Cluster Editing With Vertex Splitting

Based mainly on our NP-hardness proof of Cluster Vertex Splitting in conjunction with the corrected critical-clique lemma we obtain NP-hardness of Cluster Editing With Vertex Splitting:

Theorem 7.1 (★). *There is a polynomial-time many-one reduction from* CVS *to* CEVS, *showing that* CEVS *is* NP*-hard.*

8 Conclusion

We conclude with directions for future research. The constants in our kernelization for CVS (at most $3k + 3$ vertices, see Theorem 5.6) are already quite small, but it would be interesting to see whether they can be further improved. A problem kernel with a linear number of edges would also be interesting. A straightforward brute-force search on the kernel yields an algorithm solving CVS in $2^{O(k^2)} \cdot n^{O(1)}$ time, which can be improved to $2^{O(k \log k)} \cdot n^{O(1)}$ with further observations. Is it possible to obtain $2^{O(k)} \cdot n^{O(1)}$ time as well? Finally, we focused here on the case where the overlap between clusters is small. There are applications where the overlap is relatively large [14]. Thus, to get efficient algorithms in this case, it would be interesting to study parameterizations dual to k that measure the non-overlapping parts of the clustering.

References

1. Abu-Khzam, F.N., et al.: Cluster editing with vertex splitting. CoRR, abs/1901.00156v2 (2023). arxiv.org/abs/1901.00156v2
2. Abu-Khzam, F.N., Barr, J.R., Fakhereldine, A., Shaw, P.: A greedy heuristic for cluster editing with vertex splitting. In: Proceedings of the 4th International Conference on Artificial Intelligence for Industries (AI4I 2021), pp. 38–41. IEEE (2021). https://doi.org/10.1109/AI4I51902.2021.00017
3. Abu-Khzam, F.N., Egan, J., Gaspers, S., Shaw, A., Shaw, P.: Cluster editing with vertex splitting. In: Lee, J., Rinaldi, G., Mahjoub, A.R. (eds.) ISCO 2018. LNCS, vol. 10856, pp. 1–13. Springer, Cham (2018). https://doi.org/10.1007/978-3-319-96151-4_1
4. Abu-Khzam, F.N., Egan, J., Gaspers, S., Shaw, A., Shaw, P.: On the parameterized cluster editing with vertex splitting problem. CoRR, abs/1901.00156v1 (2019). arxiv.org/abs/1901.00156v1
5. Arrighi, E., Bentert, M., Drange, P.G., Sullivan, B., Wolf, P.: Cluster editing with overlapping communities. In: Proceedings of the 18th International Symposium on Parameterized and Exact Computation (IPEC 2023) (2023). Accepted for publication
6. Askeland, G.: Overlapping community detection using cluster editing with vertex splitting. Master's thesis, University of Bergen (2022)
7. Crespelle, C., Drange, P.G., Fomin, F.V., Golovach, P.A.: A survey of parameterized algorithms and the complexity of edge modification. Comput. Sci. Rev. **48**, 100556 (2023). https://doi.org/10.1016/j.cosrev.2023.100556
8. Davoodi, A., Javadi, R., Omoomi, B.: Edge clique covering sum of graphs. Acta Math. Hungar. **149**(1), 82–91 (2016). https://doi.org/10.1007/s10474-016-0586-1
9. Fellows, M.R., Guo, J., Komusiewicz, C., Niedermeier, R., Uhlmann, J.: Graph-based data clustering with overlaps. Discrete Optim. **8**(1), 2–17 (2011). https://doi.org/10.1016/j.disopt.2010.09.006
10. Firbas, A., et al.: The complexity of cluster vertex splitting and company. CoRR, abs/2309.00504 (2023). https://doi.org/10.48550/ARXIV.2309.00504
11. Gramm, J., Guo, J., Hüffner, F., Niedermeier, R., Piepho, H.-P., Schmid, R.: Algorithms for compact letter displays: comparison and evaluation. Comput. Stat. Data Anal. **52**(2), 725–736 (2007). https://doi.org/10.1016/j.csda.2006.09.035
12. Karp, R.M.: Reducibility among combinatorial problems. In: Miller, R.E., Thatcher, J.W. (eds.) Proceedings of a Symposium on the Complexity of Computer Computations. The IBM Research Symposia Series, pp. 85–103. Plenum Press, New York (1972). https://doi.org/10.1007/978-1-4684-2001-2_9
13. Lin, G.-H., Kearney, P.E., Jiang, T.: Phylogenetic k-root and steiner k-root. In: Goos, G., Hartmanis, J., van Leeuwen, J., Lee, D.T., Teng, S.-H. (eds.) ISAAC 2000. LNCS, vol. 1969, pp. 539–551. Springer, Heidelberg (2000). https://doi.org/10.1007/3-540-40996-3_46
14. Yang, J., Leskovec, J.: Structure and overlaps of ground-truth communities in networks. ACM Trans. Intell. Syst. Technol. **5**(2), 26:1–26:35 (2014). https://doi.org/10.1145/2594454

Morphing Graph Drawings in the Presence of Point Obstacles

Oksana Firman, Tim Hegemann, Boris Klemz, Felix Klesen, Marie Diana Sieper(✉), Alexander Wolff, and Johannes Zink

Institut für Informatik, Universität Würzburg, Würzburg, Germany
{oksana.firman,tim.hegemann,boris.klemz,felix.klesen,marie.sieper, johannes.zink}@uni-wuerzburg.de

Abstract. A crossing-free morph is a continuous deformation between two graph drawings that preserves straight-line pairwise noncrossing edges. Motivated by applications in 3D morphing problems, we initiate the study of morphing graph drawings in the plane in the presence of stationary point obstacles, which need to be avoided throughout the deformation. As our main result, we prove that it is NP-hard to decide whether such an obstacle-avoiding 2D morph between two given drawings of the same graph exists. This is in sharp contrast to the classical case without obstacles, where there is an efficiently verifiable (necessary and sufficient) criterion for the existence of a morph.

Keywords: Graph morphing · Point obstacles · NP-hard · Planar graph · Straight-line drawing

1 Introduction

In the field of Graph Drawing, a (*crossing-free*) *morph* between two straight-line drawings Γ_1 and Γ_2 of the same graph is a continuous deformation that transforms Γ_1 into Γ_2 while preserving straight-line pairwise noncrossing edges at all times. Morphing (beyond the above, strict definition in Graph Drawing) has applications in animation and computer graphics [12]. In this paper, we initiate the study of morphing graph drawings in the presence of stationary point obstacles, which need to be avoided throughout the motion.

Related Work. An obvious necessary condition for the existence of a crossing-free morph in $\mathbb{R}^2$ between two straight-line drawings Γ_1 and Γ_2 is that these drawings represent the same *plane graph* (i.e., a planar graph equipped with fixed combinatorial embedding and a distinguished outer face). It has been established long ago [7,16] that this (efficiently verifiable) criterion is also sufficient, i.e., a crossing-free morph in $\mathbb{R}^2$ between two straight-line drawings of the same plane graph always *exists*.

Work partially supported by DFG grants WO 758/9-1 and WO 758/11-1.

H. Fernau et al. (Eds.): SOFSEM 2024, LNCS 14519, pp. 240–254, 2024.
https://doi.org/10.1007/978-3-031-52113-3_17

More recent work [2,14] focuses on the efficient *computation* of such morphs. In particular, this involves producing a discrete description of the continuous motion. Typically, this is done in form of so-called piecewise linear morphs. In a *linear* morph, each vertex moves along a straight-line segment at a constant speed (which depends on the length of the segment) such that it arrives at its final destination at the end of the morph. The (unique) linear morph between Γ_1 and Γ_2 is denoted by $\langle \Gamma_1, \Gamma_2 \rangle$. A *piecewise linear* morph is created by concatenating several linear morphs, which are referred to as *(morphing) steps*. A piecewise linear morph consisting of k steps can be encoded as a sequence of $k+1$ drawings. Alamdari et al. [2] showed that two straight-line drawings of the same n-vertex plane graph always admit a crossing-free piecewise linear morph in $\mathbb{R}^2$ with $O(n)$ steps, which is best-possible. Their proof is constructive and corresponds to an $O(n^3)$-time algorithm, which was later sped up to $O(n^2 \log n)$ time by Klemz [14].

Other works are concerned with finding crossing-free morphs in $\mathbb{R}^2$ between two given drawings while preserving certain additional properties, such as convexity [3], upward-planarity [10], or edge lengths[1] [9], or with constructing crossing-free morphs in $\mathbb{R}^2$ that transform a given drawing to achieve certain properties, such as vertex visibilities [1] or convexity [13], while being in some sense monotonic, in order to preserve the so-called "mental map" [15] of the viewer.

Quite recent works [4–6] are concerned with transforming two drawings Γ_1, Γ_2 in the plane into each other by means of crossing-free morphs in the space $\mathbb{R}^3$. Such *2D–3D–2D* morphs are always possible [6]—even if Γ_1 and Γ_2 have different combinatorial embeddings—and they sometimes require fewer morphing steps than morphs that are restricted to the plane $\mathbb{R}^2$ [4,5]. Due to connections to the notoriously open UNKNOT problem, 3D–3D–3D morphs are not well understood and have, so far, only been considered for trees [4,5].

Our Model and Motivation. In this paper, we introduce and study a natural variant of the 2-dimensional morphing problem: given two crossing-free straight-line drawings Γ_1 and Γ_2 as well as a finite set of points P in $\mathbb{R}^2$, called *obstacles*, construct (or decide whether there exists) a crossing-free morph in $\mathbb{R}^2$ between Γ_1 and Γ_2 that *avoids* P, i.e., at no point in time throughout the deformation, the drawing is allowed to intersect any of the obstacles, which remain stationary. In particular, this problem naturally arises when constructing 2D–3D–2D morphs, where it is tempting to apply strategies for the classical 2-dimensional case on a subdrawing induced by the subset of the vertices contained in a plane π. Note that every edge between vertices on different sides of π intersects π in a point, which then acts as an obstacle for the 2-dimensional morph.

Contribution and Organization. We begin by stating some basic observations and preliminary results in Sect. 2. In particular, we observe that the necessary and sufficient condition for the classical case without obstacles is no longer sufficient

[1] In the fixed edge length scenario, the drawings are also known as linkages.

for our model (even when interpreting the obstacles as isolated vertices) and we present a stronger necessary condition (we say that the obstacles need to be "compatible" with the drawings), which is, however, still not sufficient. In fact, as our main result, we show that even if our condition is satisfied, it is NP-hard to decide whether an obstacle-avoiding morph exists (see Sect. 3):

Theorem 1. *Given a plane graph G, a set of obstacles P, and two crossing-free straight-line drawings Γ and Γ' in $\mathbb{R}^2$, it is NP-hard to decide whether there exists an obstacle-avoiding crossing-free morph in $\mathbb{R}^2$ between Γ and Γ'. The problem remains NP-hard when restricted to the case where G is connected, the drawings Γ and Γ' are identical except for the positions of four vertices, and the obstacles P are compatible with Γ and Γ'. (These statements hold regardless of whether the morph is required to be piecewise linear or not.)*

We remark that it is an essential part of the challenge to keep the edges straight-line during the morph – when dropping this requirement (i.e., when allowing edges to be deformed into arbitrary crossing-free curves or polylines), the problem can be solved efficiently [8]. The proof of Theorem 1 (by reduction from 3-SAT) is somewhat unusual from a Graph Drawing perspective: given a Boolean formula Φ, we describe the construction of a set of obstacles P and two (almost identical) drawings Γ_1, Γ_2 that exist irrespectively of the satisfiability of Φ, which instead corresponds to the existence of the obstacle-avoiding morph between the two drawings. In particular, we had to overcome the somewhat intricate challenge of designing gadgets that behave in a synchronous way. We conclude by discussing several open questions in Sect. 4. Claims marked with a clickable "$\star$" are proved in the full version of this article [11].

Conventions. In the remainder of the paper, we consider only morphs in the plane $\mathbb{R}^2$. We write "drawing" as a short-hand for "straight-line drawing in the plane $\mathbb{R}^2$" and, similarly, we write "(*planar*) morph" rather than "(crossing-free) morph in $\mathbb{R}^2$". For any positive integer n, we define $[n] = \{1, 2, \ldots, n\}$.

2 Preliminaries and Basic Observations

Let Γ_1 and Γ_2 be two drawings of the same plane graph G, and let P be a set of obstacles. We say that Γ_1 and Γ_2 are *blocked* by P if there is no planar morph between Γ_1 and Γ_2 that avoids P. Moreover, Γ_1 and Γ_2 are *blockable* if there exists a set of obstacles that blocks them. We start by observing that cycles are necessary to block drawings.

Proposition 1 ($\star$). *Let Γ_1 and Γ_2 be two drawings of the same plane forest F. Then Γ_1 and Γ_2 are not blockable.*

Proof (sketch). We describe how to construct a morph that avoids an arbitrary set of obstacles. Assume for now that F consists of a single tree, which we root at an arbitrary vertex r. We construct an obstacle-avoiding planar morph from Γ_1

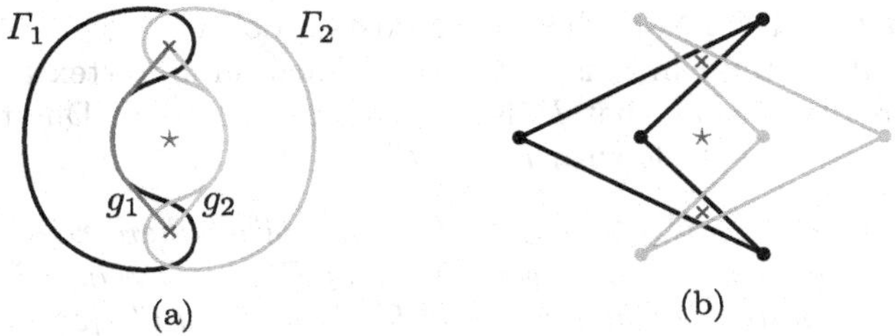

Fig. 1. The two simple closed curves Γ_1 and Γ_2 in (a) cannot be continuously deformed into each other without passing over one of the obstacles and while preserving simplicity. (The corresponding continuous deformation of the geodesic joining the two internal obstacles [blue crosses] within the closed curve would transform the curve g_1 into g_2 while keeping the endpoints fixed and without passing over the external obstacle [red star], which is impossible.) Consequently, the two drawings of a 4-cycle in (b) are blocked by the set of three obstacles; it is not compatible with the two drawings. (Color figure online)

to a drawing Γ_1' located in a disk that (i) is centered on the position of r in Γ_1, (ii) contains no obstacles, and (iii) whose radius is smaller than the distance between any pair of obstacles. This can be done by "contracting" the tree along its edges in a bottom-up fashion (i.e., starting from the leaves). We also morph Γ_2 into an analogously defined drawing Γ_2'. The drawings Γ_1' and Γ_2' can now be translated (far enough) away from the obstacles without intersecting them so that they can be transformed into each other by means of morphing techniques for the classical non-obstacle case [2,7,14,16].

If F contains multiple trees, it is easy to augment Γ_1 and Γ_2 to drawings of the same plane tree by inserting additional vertices and edges, thus reducing to the case of a single tree. □

We now turn our attention to the case when G may contain cycles. Recall that an obvious necessary condition for the existence of planar morph between two drawings is that they represent the same plane graph. Interpreting the obstacles in the set P as (isolated) vertices reveals that a planar morph between Γ_1 and Γ_2 that avoids P is possible only if each obstacle $p \in P$ is located in the same face in Γ_1 and Γ_2. However, as Fig. 1 shows, this condition is not sufficient. We say that P is *compatible* with Γ_1 and Γ_2 if there is a continuous deformation that transforms Γ_1 into Γ_2 while avoiding P and preserving pairwise noncrossing (*not necessarily straight-line*) edges at all times. The compatibility of P with Γ_1 and Γ_2 is obviously a necessary condition for the existence of a planar obstacle-avoiding morph. This condition can be checked efficiently [8]; note that it is violated in Fig. 1.

Compatibility is unfortunately still not sufficient for the existence of obstacle-avoiding morphs—even if the considered graph is just a cycle. In the following, we study and discuss this case in more detail. Let C_n denote the simple cycle with n vertices. Let Γ and Γ' be drawings of a plane C_n such that (i) Γ and Γ' are distinct (as mappings of C_n to the plane), but (ii) the closed curves realizing Γ and Γ' are identical, and (iii) the set of points of $\mathbb{R}^2$ used to represent vertices is the same in Γ and Γ'. Note that there exists an offset $o \in [n-1]$ such that, for every $i \in [n]$, vertex i in Γ is at the same location as vertex $i + o$ (modulo n) in Γ'. Therefore, we say that Γ' is a *shifted version* of Γ. Due to (ii), *every* set of obstacles is compatible with Γ and Γ'.

Proposition 2 (⋆). *Let $n \geq 6$ be an even integer. Then there exists a drawing Γ of C_n such that, for every shifted version Γ' of Γ, the drawings Γ and Γ' are blockable by seven obstacles that are compatible with Γ and Γ' (for an illustration, see Fig. 2).*

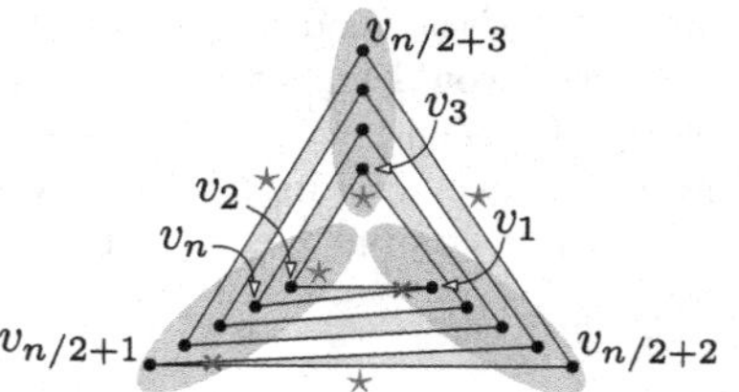

Fig. 2. A schematic drawing of C_n with vertices $v_1, \ldots, v_n$ (here $n = 14$) that cannot be morphed to a shifted version of it in a planar way while avoiding the two internal obstacles (blue crosses) and the five external obstacles (red stars). (Color figure online)

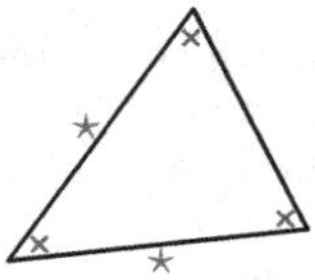

Fig. 3. Five obstacles suffice to block shifted versions of C_3.

It is plausible that Proposition 2 can be strengthened: even three obstacles seem to be sufficient for blocking two shifted drawings of an even-length cycle (we chose to use seven obstacles to simplify the proof). In contrast, mainly due to convexity, more obstacles are needed to block shifted drawings of C_3.

Proposition 3 (⋆). *Two drawings Γ_1 and Γ_2 of C_3 are not blockable by four obstacles that are compatible with Γ_1 and Γ_2.*

Proof (sketch). We perform a case distinction on the number of obstacles that are located in the interior of the cycle. Here, we only sketch the case with two inner obstacles $p_1 = (x_1, y_1)$ and $p_2 = (x_2, y_2)$. Assume without loss of generality that $x_1 = x_2$ and $y_1 > y_2$. There exists an $\varepsilon > 0$ such that the rectangle $R = [x_1 - \varepsilon, x_1 + \varepsilon] \times [y_2 - \varepsilon, y_1 + \varepsilon]$ lies in the interiors of Γ_1 and Γ_2 and, hence, does not contain any of the outer obstacles. Then, for any triangle $\Gamma \in \{\Gamma_1, \Gamma_2\}$, there is a planar morph that moves the vertices of Γ onto the boundary of R while avoiding the line segment $\overline{p_1p_2}$ and the region exterior to Γ. Finally, we show that two triangles with these properties can always be morphed into each other while staying in the closure of R and avoiding $\overline{p_1p_2}$. □

Our proof for (at most) two inner obstacles does not depend on the number of outer obstacles, which implies a partially stronger statement.

We have a tight upper bound for the number of obstacles needed to block drawings of C_3; see Fig. 3.

Proposition 4 (⋆). *Let Γ be a drawing of C_3, and let Γ' be a shifted version of Γ. Then Γ and Γ' are blockable by five obstacles compatible with Γ and Γ'.*

We now state a sufficient condition for the existence of planar obstacles-avoiding morphs between shifted drawings of a cycle. We call a degree-2 vertex in a drawing *free* if its two incident edges lie on a common line.

Proposition 5 (⋆). *Let Γ be a drawing of C_n, and let Γ' be a shifted version of Γ. If Γ contains a free vertex, then Γ and Γ' are not blockable by obstacles that are compatible with Γ and Γ'.*

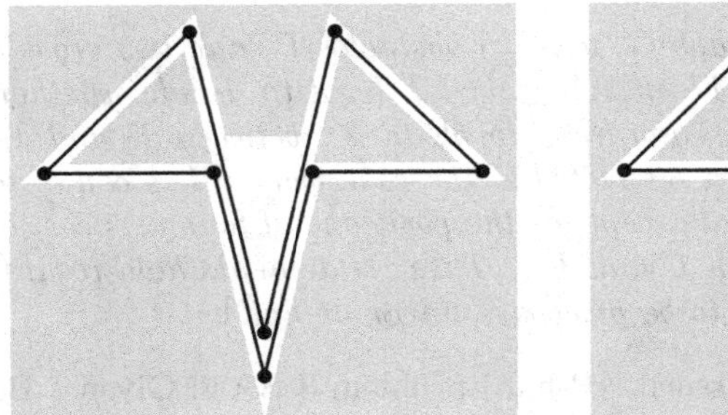
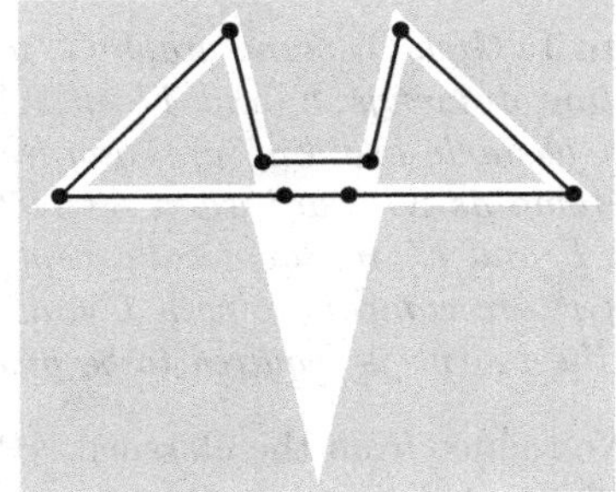

Fig. 4. Two drawings of C_8. If the shaded regions are densely filled with obstacles, the drawing on the left is essentially locked in place—it cannot be morphed planarly to a substantially different drawing without intersecting the obstacle regions. In particular, it cannot be morphed to the drawing on the right, even though this drawing contains two free vertices (and the obstacles are compatible with the two drawings).

Free vertices are helpful in other specific cases as well (in particular, they play a crucial role in our NP-hardness proof), but their usefulness is limited in general: their existence is not a sufficient condition for the existence of obstacle-avoiding morphs even when it comes to plane cycles; see Fig. 4.

Finally, we observe that two obstacles are not enough to block two drawings (with which the obstacles are compatible), regardless of the class of the represented graph.

Proposition 6 (⋆). *Let Γ_1 and Γ_2 be two drawings of the same plane graph G, and let P be a set of obstacles that are compatible with Γ_1 and Γ_2. If $|P| \leq 2$, then there exists a planar morph from Γ_1 to Γ_2 that avoids P.*

Proof (sketch). We interpret the (up to) two obstacles as isolated vertices of our plane graph (by compatibility, each obstacle belongs to the same face in Γ_1 and Γ_2). By known results [2,7,14,16], there is a morph M in which Γ_1 is deformed to Γ_2 without introducing crossings and without intersecting the obstacles. However, since the obstacles are treated as vertices, their positions might change over time. We can transform M into the desired morph by translating, rotating, and scaling the frame of reference as time goes on to ensure that the obstacles become fixpoints (one can think of this as moving, rotating, and zooming the camera in a suitable fashion). □

In view of the remark after Proposition 2, it seems that Proposition 6 is best-possible. In particular, the proof strategy does not generalize to three obstacles as affine transformations preserve the cyclic orientations of point triples.

3 Proof of Theorem 1

In this section, we show our main result.

Theorem 1. *Given a plane graph G, a set of obstacles P, and two crossing-free straight-line drawings Γ and Γ' in $\mathbb{R}^2$, it is NP-hard to decide whether there exists an obstacle-avoiding crossing-free morph in $\mathbb{R}^2$ between Γ and Γ'. The problem remains NP-hard when restricted to the case where G is connected, the drawings Γ and Γ' are identical except for the positions of four vertices, and the obstacles P are compatible with Γ and Γ'. (These statements hold regardless of whether the morph is required to be piecewise linear or not.)*

Proof. We reduce from the classical NP-hard problem 3-SAT. Given a Boolean formula $\Phi = \bigwedge_{i=1}^{m} c_i$ in conjuctive normal form over variables $x_1, x_2, \ldots, x_n$ whose clauses $c_1, c_2, \ldots, c_m$ consist of three literals each, we construct a plane graph G, two planar drawings Γ and Γ' of G, and a set P of obstacles that are compatible with Γ and Γ'. We show that there exists an obstacle-avoiding planar morph from Γ to Γ' if and only if Φ is satisfiable.

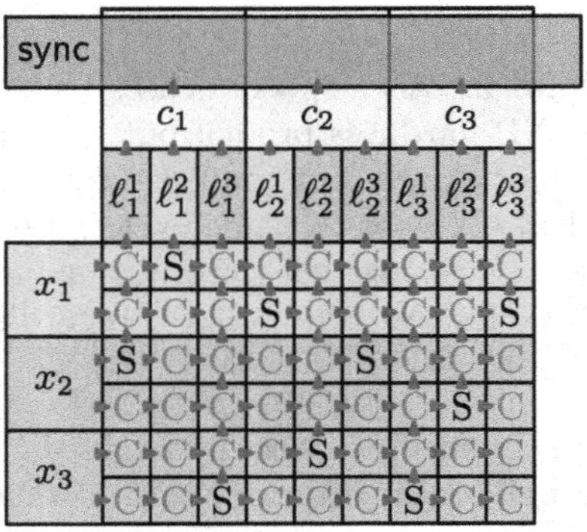

Fig. 5. General grid structure used in our NP-hardness reduction. Here, we use the formula $\Phi = c_1 \wedge c_2 \wedge c_3$, where $c_1 = (\ell_1^1 \vee \ell_1^2 \vee \ell_1^3) = (x_2 \vee x_1 \vee \neg x_3)$, $c_2 = (\ell_2^1 \vee \ell_2^2 \vee \ell_2^3) = (\neg x_1 \vee x_3 \vee x_2)$, and $c_3 = (\ell_3^1 \vee \ell_3^2 \vee \ell_3^3) = (\neg x_3 \vee \neg x_2 \vee \neg x_1)$. There are variable gadgets (left), clause and literal gadgets (top), split gadgets (S), crossing gadgets (C), and a synchronization gadget (sync) spanning over all clause gadgets. The gadgets have various states and orientations; dependencies are marked by triangular arrows.

Overview. In our reduction, we arrange obstacles, vertices, and edges such that we obtain a grid-like structure where we have two rows for each variable (one for each literal of the variable) and three columns for every clause (one for each literal in the clause); see Fig. 5. We then use several gadgets arranged within this grid-like structure. On the left side, the two rows of a variable terminate at a *variable gadget*. A variable gadget is in one of the states true, false, or unset. On the top side, the three columns of a clause are connected via three *literal gadgets* to a *clause gadget*. Each literal gadget and each clause gadget is in one of the states true or false. All clause gadgets are connected by a *synchronization gadget*. Within the column of each literal, we have a *split gadget* in one of the two rows of the corresponding variable x_i – in the upper row if the literal is x_i and in the lower row if the literal is $\neg x_i$. In all other grid cells, we have *crossing gadgets*. Every split gadget and every crossing gadget has a horizontal orientation (left/right) and a vertical orientation (bottom/top).

We first describe the general mechanism of our reduction. Independently of each other, every variable gadget can be morphed to reach any of its three states. In Γ and Γ', all crossing and split gadgets have the orientations left and bottom. A crossing or split gadget α can have orientation right only if the crossing/split gadget to the left of α has orientation right, too, or if α is adjacent to a variable gadget with state true (false) and α is in the upper (lower) row of the corresponding variable. Moreover, a crossing gadget can have orientation top only if the neighboring crossing/split gadget below has orientation top, too, and a split gadget can have orientation top only if it can also have orientation right. A literal gadget can be in the state true only if the gadget below it has orientation top. A clause gadget is in the state true if and only if at least one of its three literal gadgets is in the state true. Moreover, we can reach the final drawing Γ', where only the synchronization gadget differs from its drawing in Γ, if and only if all clause gadgets have state true simultaneously at some point in time. This ensures the correctness of our construction.

We now describe and visualize the geometric realization of our gadgets.

Forbidden Areas. In each gadget, we have *forbidden* areas where the vertices and edges cannot be drawn (henceforth drawn solid red). They are used to create a system of tunnels and cavities in which the edges of our drawings are placed and move; see, e.g., Fig. 6. We achieve this by densely filling the forbidden areas with obstacles, which are placed on a fine grid (as explained in more detail in the paragraph "Number and placement of obstacles" on page 12).

Variable Gadget. The variable gadget has a comparatively simple structure; see the three red boxes on the left side of Fig. 6. It has three vertices enclosed in a straight vertical tunnel of the forbidden area with one exit on the top right and one exit on the bottom right. As it is a straight tunnel, the middle vertex, which we call *decision vertex*, is a free (see Sect. 2 for the definition) vertex (see the thick blue–white vertex of x_3 in Fig. 6 for the arrangement in Γ and Γ'). We say that the variable gadget is in the state true (false) if this decision vertex

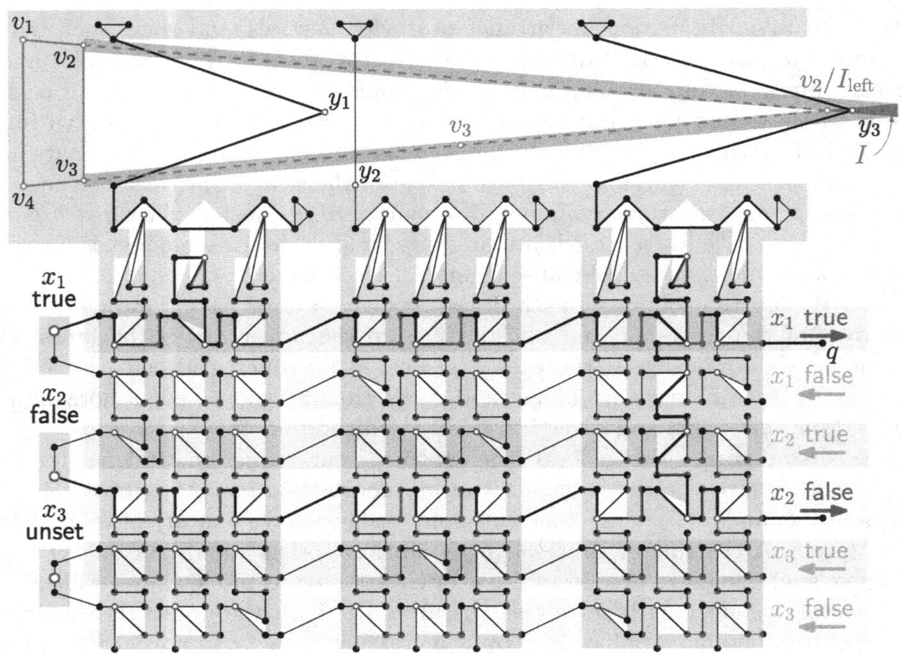

Fig. 6. Full construction for the instance $\Phi = (x_2 \vee x_1 \vee \neg x_3) \wedge (\neg x_1 \vee x_3 \vee x_2) \wedge (\neg x_3 \vee \neg x_2 \vee \neg x_1)$. Gadgets with orientation right/top/true use thicker strokes.

is moved to the top (bottom) position, and it is in the state unset otherwise. Consequently, we can move the top (bottom) vertex out of the tunnel if and only if we are in the state true (false). This, in turn, yields a free vertex for the adjacent row of crossing and split gadgets, which we describe next.

Split Gadget. A split gadget consists of a central vertex c of degree 3 together with paths to the left, right, and top; see Fig. 7. They are enclosed in a system of tunnels formed by the forbidden area. Figure 7a shows a split gadget σ in the base state as it appears in Γ and Γ'. If the crossing/split/variable gadget to the left of σ, which shares the vertex l with σ, has a free[2] vertex, then l can be moved to the next corner of the tunnel. Then, in turn, c can be moved to the bottom right corner of the white (obstacle-free) triangle in σ (see Fig. 7b), and the two other neighbors of c can be pushed one position to the right and one position up. In this case we say that the horizontal (vertical) orientation of σ is right (top). Otherwise, the horizontal (vertical) orientation of σ is left (bottom).

Crossing Gadget. A crossing gadget has a similar structure as a split gadget. However, we now have a central vertex c' of degree 4 with paths to the neigh-

[2] Here, this means that it is a crossing/split gadget that is oriented to the right or it is a variable gadget in its top row with state true or in its bottom row with state false.

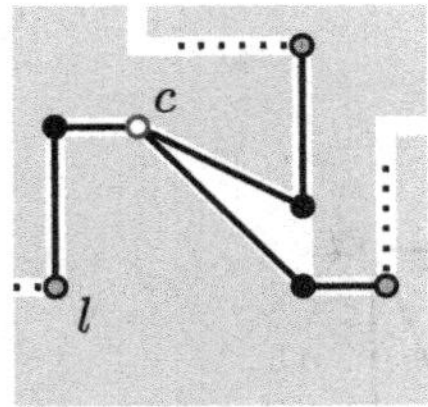

(a) Base state with horizontal and vertical orientation left and bottom.

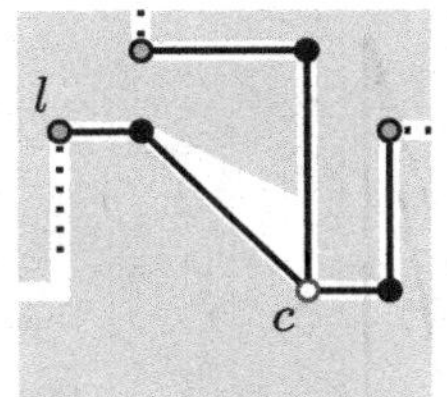

(b) If the gadget on the left has horizontal orientation right, vertices can be pushed one position further and we can reach the orientations right and top.

Fig. 7. Split gadget: from a horizontal row transporting a truth value, we "copy" the same truth value up to a vertical column.

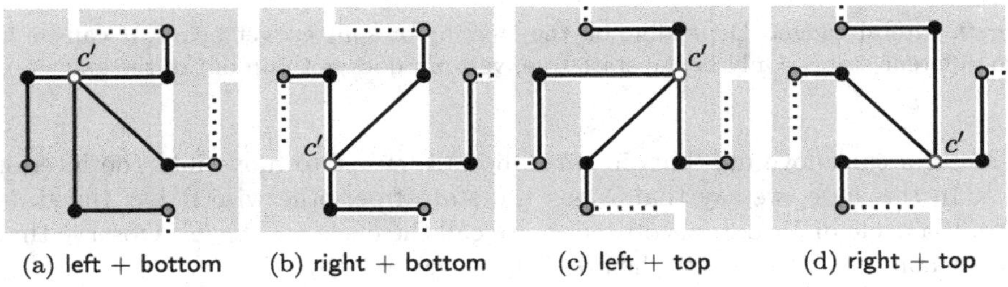

(a) left + bottom (b) right + bottom (c) left + top (d) right + top

Fig. 8. Crossing gadget: we transport truth values horizontally and vertically without influencing each other. The four possible combinations of orientations are illustrated.

boring gadgets to the left, right, top, and bottom. In the center of the crossing gadget χ in Fig. 8, there is a white (obstacle-free) square. In Γ and Γ', c' is placed in the top left corner of this square (see Fig. 8a). This is the base state of χ. If the gadget on the left of χ has a free vertex, we can push the adjacent vertices to the next corners of the tunnel such that c' can move to the bottom side of the square (see Figs. 8b and 8d). Only then, we can push the vertices of the path leaving the gadget on the right side to the next corner of the tunnel. In this case we say that the horizontal orientation of the crossing gadget is right; otherwise it is left. Symmetrically, if the gadget below χ has a free vertex, we can move c' to the right side of the square (see Figs. 8c and 8d). Only then we can push the vertices of the path leaving χ through the top side to the next corner of the tunnel. In this case we say that the vertical orientation of the crossing gadget is top; otherwise it is bottom. Observe that the states of the gadgets on the left and below χ *independently* determine the horizontal and vertical orientation of χ. This property assures that we can transport information along routes that cross each other, but do not influence each other.

Literal Gadget. Figure 9 shows a literal gadget λ in its base state. It consists of five vertices, one of which, r, is shared with the gadget β below; see Fig. 9. Only if the vertical orientation of β is top, vertex r can move to the original position

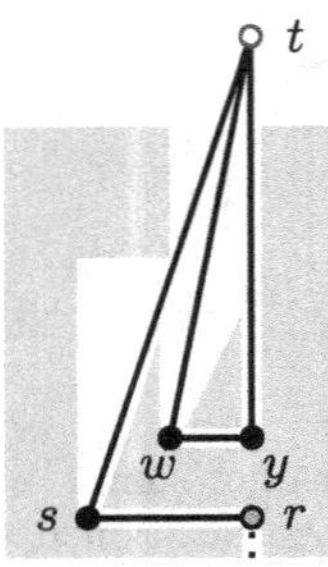

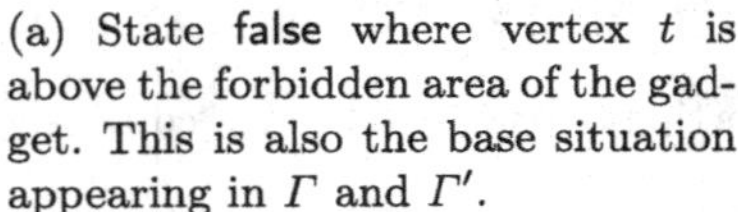
(a) State false where vertex t is above the forbidden area of the gadget. This is also the base situation appearing in Γ and Γ'.

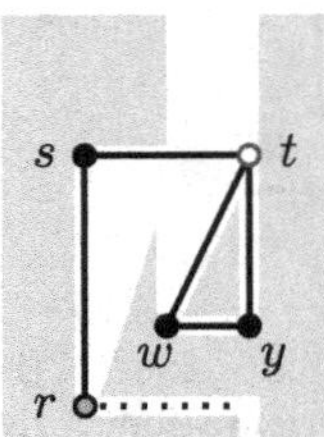

(b) State true where vertex t is inside the cavity formed by the forbidden area of the gadget. This state can only be reached if the split or crossing gadget below has orientation top.

Fig. 9. Literal gadget: Depending on the crossing or split gadget below, it can be in two different states. Only in the state true, vertex t does not pop out of the gadget.

of s and s can move up. This in turn allows vertex t to move into the interior of λ. In this case, we say that λ has the state true; otherwise it has the state false. If a cycle in Γ contains obstacles, we call the cycle an *anchor*. Observe that the anchor $\langle t, w, y\rangle$ restricts the area where we can move t.

Clause Gadget. Each clause c_i ($i \in [m]$) is represented by a clause gadget, which consists of a path of length 9 whose endpoints are anchored by two 3-cycles; see Fig. 10. In the base situation occurring in Γ and Γ', we have exactly one free vertex (denoted by y), which is at the bottom of a large rectangular obstacle-free region, which is also part of the synchronization gadget. Observe that, within this area, we cannot move y (up to a tiny bit) to the left or right due to its neighbors lying at (essentially) fixed positions (see Fig. 10a).

However, if one of the incident literal gadgets is in the state true, we get a second free vertex, due to which we can move z, which is a neighbor of y, onto the base position of y (see Fig. 10b). Now we can move y arbitrarily far to the right within this obstacle-free region (unless y is blocked by the edges of another clause gadget). Only when this is done for all clause gadgets simultaneously, the synchronization gadget (see below) can be morphed as desired. Thus, we say that a clause gadget is in the state true if at least one of its literal gadgets is in the state true; otherwise it is in the state false.

Synchronization Gadget. The synchronization gadget is a 4-cycle $\langle v_1, v_2, v_3, v_4\rangle$; see Fig. 6. In the base situation occurring in Γ, this cycle is drawn as an isosceles trapezoid T. In Γ', the 4-cycle is drawn as a shifted version (see Sect. 2 for the definition) of T. All sides of T except for the short parallel side are fixed by tunnels of the forbidden area. In particular, v_1 and v_4 can only be moved in an ε-region (for some small $\varepsilon > 0$) around their initial position, while v_2 and v_3 can potentially be moved to the right into the large obstacle-free rectangular

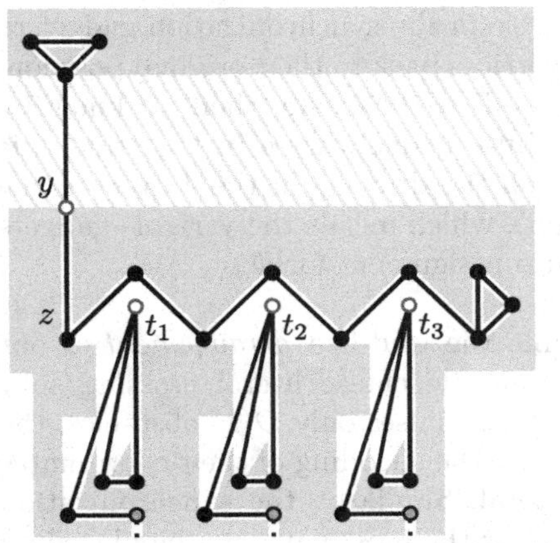

(a) The clause gadget is in the state false.

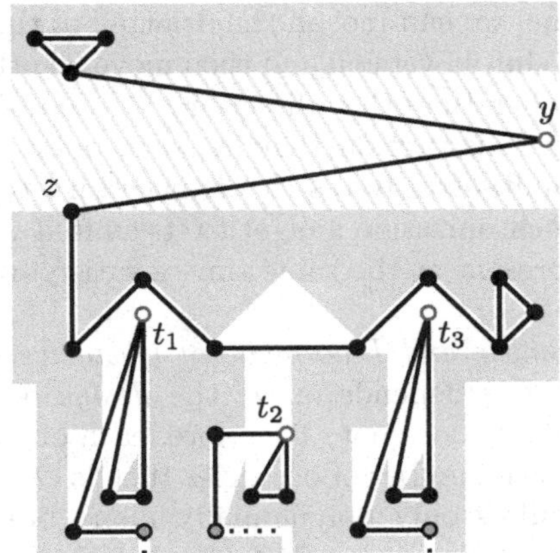

(b) The clause gadget is in the state true.

Fig. 10. Clause gadget together with three literal gadgets: If at least one of the literal gadgets is in the state true, the clause gadget is also in the state true. The obstacle-free region shared with the synchronization gadget is depicted in hatched green (Color figure online).

area (that is shared with all clause gadgets) within a small corridor (see the blue strips in Fig. 6). These corridors extend the legs of T, which have a specific angle depending on the number of clauses m such that the corridors intersect at a region I to the right side of all clause gadgets. Note that we cannot simply "rotate" T, since each of the ε-regions around the positions of v_1 and v_4 in Γ need to contain at least one vertex at any time. To reach a shifted version, there needs to be an intermediate drawing that has a third vertex within the tunnel between these ε-regions. Therefore, only one vertex remains to close the 4-cycle on the right side. That vertex needs to be placed inside I. Hence, we can reach the shifted version if and only if all of the clause gadgets are in the state true: in this case, for each clause gadget, we can move its vertex y into I so that the edges incident to y do not intersect the triangle $\triangle v_1 I_{\text{left}} v_4$, where I_{left} is the leftmost point of I. Thus, we can place v_2 onto I_{left} to make v_3 a free vertex and, therefore, reach the shifted drawing of the 4-cycle (cf. Proposition 5). In contrast, if a clause gadget is in the state false, at least one of the edges incident to its vertex y intersects $\triangle v_1 I_{\text{left}} v_4$ and, thus, v_2 and v_3 cannot reach the corridors of each other.

Correctness. In summary, if Φ admits a satisfying truth assignment, we can describe a (piecewise linear) morph from Γ to Γ' by moving the decision vertices in the variable gadgets according to the truth assignment, which allows to transport these truth values via the split and crossing gadgets to the literal and the clause gadgets. As all clause gadgets can reach the state true at the same time, we can morph the drawing of the 4-cycle in the synchronization gadget to its shifted version and then move all other vertices back to their original position.

For the other direction, if a morph from Γ to Γ' exists, then we know by our construction that at some point, all clause gadgets were in the state true simultaneously (as this is necessary to morph the drawing of the 4-cycle in the synchronization gadget to its shifted version), which means the variable gadgets represent at the same time a satisfying truth assignment for Φ.

Number and Placement of Obstacles. While the grid-like *arrangement* of our gadgets depends on Φ, the *design* of the variable/clause/literal/crossing/split gadgets does not. Therefore, each of these gadgets uses only $O(1)$ obstacles, the overall number of obstacles then is $O(nm)$, and the encoding of their coordinates requires only polynomially many bits in total. Similarly, the synchronization gadget also uses only $O(1)$ obstacles, which, for the most part, can be placed on the grid that is also used by the remaining gadget types. As an exception, two obstacles (in the top-right and bottom-right corner of the obstacles area interior to the 4-cycle) need to be placed on a refined grid to ensure that the area I lies to the right of all clause gadgets. Since vertices v_1 and v_2 have constant (horizontal) distance, we use an obstacle that lies $O(1/m)$ units below the highest possible position of v_1 and on the same x-coordinate as v_2 in Γ in order to bound the slope of the v_1v_2-tunnel from below. We bound the slope of the v_3v_4-tunnel symmetrically from above. As a result, the area I lies $\Omega(m)$ units to the right of the vertical line segment $\overline{v_2v_3}$. Since the width of the whole construction is $O(m)$, this suffices.

Thus, the encoding of the coordinates of the involved obstacles and the coordinates of the drawing of the 4-cycle requires only polynomially many bits.

Connectivity. So far, the graph in our reduction has $\Omega(m)$ connected components: a large connected component comprising all variable, split, crossing and literal gadgets, a connected component for all clause gadgets, as well as another connected component for the synchronization gadget. We can merge these components by adding edges without influencing the behavior of our gadgets: We add the edges $v_2y_1, y_1y_2, \ldots, y_{m-1}y_m$ together with a path from y_m to the large connected component of the other gadgets (e.g., in Fig. 6 from y_3 to q). □

4 Open Problems

1. In general, the considered decision problem is NP-hard (Theorem 1), but it can be solved efficiently for forests (Proposition 1). Are there other meaningful graph classes where this is the case? In particular, what about cycles or triangulations? It is conceivable that the latter case is actually easier since the placement of obstacles that are compatible with the two given drawings is quite limited. Regarding cycles, we emphasize that the existence of a free vertex is not a sufficient condition for the existence of a morph (cf. Fig. 4).
2. Does the problem lie in NP? Is it $\exists\mathbb{R}$-hard?

3. Does the problem become easier if there are only constantly many obstacles?
4. The drawings Γ and Γ' produced by our reduction are identical except for the position of four vertices. It seems quite plausible that our construction can be modified so that only three vertices change positions. Does the problem become easier when only up to two vertices may change positions?
5. It is easy to observe that if our reduction is applied to a satisfiable formula Φ with n variables and m clauses, there is a piecewise linear obstacle-avoiding morph between the produced drawings Γ and Γ' with $\Theta(n+m)$ steps, which is also necessary. Note that this number is not independent from the output size of the reduction. This motivates the following family of questions. Let k be a fixed arbitrary constant. Given two planar straight-line drawings Γ and Γ' of the same plane graph and a set of obstacles compatible with Γ and Γ', decide whether there exists a piecewise linear obstacle-avoiding morph from Γ to Γ' with at most k steps. For which values of k can this decision problem be answered efficiently?
6. Given two drawings of the same plane graph, how many compatible obstacles are necessary and sufficient to block them? Can this be computed efficiently?

References

1. Aichholzer, O.: Convexifying polygons without losing visibilities. In: Aloupis, G. Bremner, (eds.) CCCG, pp. 229–234 (2011). http://www.cccg.ca/proceedings/2011/papers/paper70.pdf
2. Alamdari, S., et al.: How to morph planar graph drawings. SIAM J. Comput. **46**(2), 824–852 (2017). https://doi.org/10.1137/16M1069171
3. Angelini, P., Da Lozzo, G., Frati, F., Lubiw, A., Patrignani, M., Roselli, V.: Optimal morphs of convex drawings. In: Arge, L., Pach, J., (eds.) SoCG. LIPIcs, vol. 34 pp. 126–140. Schloss Dagstuhl – Leibniz-Zentrum für Informatik (2015). https://doi.org/10.4230/LIPIcs.SOCG.2015.126
4. Arseneva, E., et al.: Pole dancing: 3D morphs for tree drawings. J. Graph Algorithms Appl. **23**(3), 579–602 (2019). https://doi.org/10.7155/jgaa.00503
5. Arseneva, E., Gangopadhyay, R., Istomina, A.: Morphing tree drawings in a small 3D grid. J. Graph Algorithms Appl. **27**(4), 241–279 (2023). https://doi.org/10.7155/jgaa.00623
6. Buchin, K., et al.: Morphing planar graph drawings through 3D. In: Gąsieniec, L. (ed.) SOFSEM 2023. LNCS, vol. 13878, pp. 80–95. Springer, Cham (2023). https://doi.org/10.1007/978-3-031-23101-8_6
7. Cairns, S.: Deformations of plane rectilinear complexes. Am. Math. Monthly **51**(5), 247–252 (1944)
8. Colin de Verdiére, É., de Mesmay, A.: Testing graph isotopy on surfaces. Discrete Comput. Geom. **51**, 171–206 (2014). https://doi.org/10.1007/s00454-013-9555-4
9. Connelly, R., Demaine, E.D., Rote, G.: Straightening polygonal arcs and convexifying polygonal cycles. Discrete Comput. Geom. **30**, 205–239 (2003). https://doi.org/10.1007/s00454-003-0006-7
10. Da Lozzo, G., Di Battista, G., Frati, F., Patrignani, M., Roselli, V.: Upward planar morphs. Algorithmica **82**(10), 2985–3017 (2020). https://doi.org/10.1007/s00453-020-00714-6

11. Firman, O., et al.: Morphing graph drawings in the presence of point obstacles. ArXiv report (2023). http://arxiv.org/abs/2311.14516
12. Gomes, J., Darsa, L., Costa, B., Velho, L.: Warping & Morphing of Graphical Objects. Morgan Kaufmann, Burlington (1999)
13. Kleist, L., Klemz, B., Lubiw, A., Schlipf, L., Staals, F., Strash, D.: Convexity-increasing morphs of planar graphs. Comput. Geom. **84**, 69–88 (2019). https://doi.org/10.1016/j.comgeo.2019.07.007
14. Klemz, B.: Convex drawings of hierarchical graphs in linear time, with applications to planar graph morphing. In: Mutzel, P., Pagh, R., Herman, G., (eds.) ESA, LIPIcs, vol. 204, pp. 57:1–57:15. Schloss Dagstuhl – Leibniz-Zentrum für Informatik (2021). https://doi.org/10.4230/LIPIcs.ESA.2021.57
15. Purchase, H.C., Hoggan, E., Görg, C.: How important is the "mental map"? – an empirical investigation of a dynamic graph layout algorithm. In: Kaufmann, M., Wagner, D. (eds.) GD 2006. LNCS, vol. 4372, pp. 184–195. Springer, Heidelberg (2007). https://doi.org/10.1007/978-3-540-70904-6_19
16. Thomassen, C.: Deformations of plane graphs. J. Combin. Theory Ser. B **34**(3), 244–257 (1983). https://doi.org/10.1016/0095-8956(83)90038-2

Word-Representable Graphs from a Word's Perspective

Pamela Fleischmann(✉), Lukas Haschke, Tim Löck, and Dirk Nowotka

Kiel University, Kiel, Germany
{fpa,lha,dn}@informatik.uni-kiel.de, stu229765@mail.uni-kiel.de

Abstract. Word-representable graphs were introduced in 2008 by Kitaev and Pyatkin in the context of semigroup theory. Graphs are called word-representable if there exists a word with the graph's nodes as letters such that the letters in the word alternate iff there is an edge between them in the graph. Until today numerous works investigated the word-representability of graphs but mostly from the graph perspective. In this work, we change the perspective to the words, i.e., we take classes of words and investigate the represented graphs. Our first subject of interest are the conjugates of words: we determine exactly which graphs are represented if we rotate the word. Afterwards, we look at k-local words introduced by Day et al. in 2017 in order to gain more insights into this class of words. Here, we investigate especially which graphs are represented by 1-local words. Lastly, we prove that the language of all words representing a graph is regular. We were also able to characterise k-representable graphs, solving an open problem.

1 Introduction

Word-representable graphs are graphs which can be encoded linearly by a word where the nodes of the graphs are the word's letters. For each two adjacent nodes in the graph, the associated letters must alternate in the word, and for each two nodes not connected by an edge, the associated letters must not alternate, e.g., the triangle graph, i.e. the complete graph with three nodes, is represented by the word `alfalfa` since all three letters alternate pairwise with each other witnessed by the projections `alala`, `afafa`, and `lflf`. Here, the projection of a word onto a set of letters is the word obtained by deleting all other letters, and two letters alternate if none of them occurs at consecutive positions in the appropriate projection of the word. The word `banana` also represents a graph with three nodes (it has three different letters), but only `a` and `n` are connected by an edge; `b` and `a` (and `b`, `n` resp.) are not adjacent since `aa` (and `nn` resp.) occurs in the projection `baaa` (and `bnn` resp.).

The theory of word-representable graphs was introduced in 2008 by Kitaev and Pyatkin [18] motivated as a tool for semigroup theory (cf. [19]). By now, word-representable graphs have applications in periodic scheduling [11,16], topology [22], and the power domination problem from physics [1]. From a theoretical point of view they are of interest since they generalise several important

H. Fernau et al. (Eds.): SOFSEM 2024, LNCS 14519, pp. 255–268, 2024.
https://doi.org/10.1007/978-3-031-52113-3_18

classes of graphs such as circle graphs, 3-colourable graphs, and comparability graphs. An introduction to word-representable graphs can be found in [14,16]. There are also several generalisations of word-representable graphs in the literature. Graphs representable by pattern-avoiding words have been heavily investigated in e.g. [12,15]. Other generalisations were presented in [2,13].

However, not all graphs are word-representable, witnessed by a pentagon with an additional node connected to all of the pentagon's nodes [18]. Thus, a line of research started to determine the graphs which are word-representable. A fundamental result was shown by Halldórsson et al. in [10], where they proved that a graph is word-representable iff it admits a semi-transitive orientation. For results about operations preserving word-representability, see [3,16] and the references therein. An asymptotic result about the number of word-representable graphs can be found in [4]. There is a lot of research about the word-representability of specific graph classes like grid graphs (cf. [9] and the references therein) and split graphs (cf. [17] and the references therein). From an algorithmic point of view it is important to note that deciding word-representability is NP-complete, and classical graph problems like vertex colouring remain NP-hard. An exception to this is the maximum clique problem, which is solvable in polynomial time (cf. [16]).

Still from the graphs perspective but with a focus on words, the minimal length representation of a graph is of interest. While Gaetz and Ji investigated this in the general case in [8], the research is often restricted to k-uniform words (each letter occurs exactly k times). Graphs that can be represented by a k-uniform word are called k-representable. For $k = 1$ and $k = 2$, k-representable graphs are well studied and coincide with the classes of complete graphs and circle graphs (cf. [16]). For higher values of k, no characterisation is known.

Our Contribution. First, we solve the open problem about a characterisation of k-representability by introducing the notion of *circle*-representation and investigating the cuts of edges. Afterwards, we investigate word-representable graphs from the perspective of combinatorics on words. Rotating a word gives the class of conjugate words, which is well studied, e.g., since the lexicographically smallest word in this class is the Lyndon word. We characterise, given a word, how the represented graphs change on rotating the word and present an efficient algorithm to compute them. Then, we show that palindromes present exactly a form of star graphs. Based on this result, we look in detail at the set of k-local words [5,7] and the graphs they represent. Here, we focus mainly on 1-local words since they are tightly related to palindromes, and we prove that every graph represented by a 1-local word is a comparability graph, respectively. Lastly, we show that all words representing the same graph form a regular language.

Structure of the Work. In Sect. 2 we introduce the basic definitions and notations. In Sect. 3 we present the characterisation for k-representability. The next two sections contain the results on the graphs represented by rotations of a given word and by k-local words. The proof of the regularity of the language of all words representing a graph can be found in Sect. 6.

2 Preliminaries

Let $\mathbb{N}$ denote the set of natural numbers $\{1, 2, 3, \dots\}$, and let $\mathbb{N}_0 := \mathbb{N} \cup \{0\}$. We define $[i, j] := \{n \in \mathbb{N} \mid i \leq n \leq j\}$ for $i, j \in \mathbb{N}$, and set $[n] := [1, n]$ for $n \in \mathbb{N}$ and $[n]_0 := [0, n]$. For a set M and $k \in \mathbb{N}_0$, let $\binom{M}{k}$ be the set of all of M's subsets of cardinality k. By $\dot{\cup}$ we denote the disjoint union of sets. We first introduce the notations from combinatorics on words and graph theory before we present the basic definitions from the domain of word-representable graphs.

An *alphabet* is a finite set $\Sigma = \{\mathtt{a}_1, \dots, \mathtt{a}_\ell\}$ of $\ell \in \mathbb{N}$ symbols, called *letters.* Σ^* denotes the set of all finite words over Σ, i.e., the free monoid over Σ. The empty word is denoted by ε and $\Sigma^+ = \Sigma^* \setminus \{\varepsilon\}$. The *length* of a word w is denoted by $|w|$. Define $\Sigma^k := \{w \in \Sigma^* \mid |w| = k\}$ for a $k \in \mathbb{N}$. The number of occurrences of a letter $\mathtt{a} \in \Sigma$ in a word $w \in \Sigma^*$ is denoted by $|w|_\mathtt{a}$, and w is called *k-uniform* for some $k \in \mathbb{N}$ if $|w|_\mathtt{a} = k$ for all $\mathtt{a} \in \Sigma$. Define the set of letters occurring in $w \in \Sigma^*$ by $\mathrm{alph}(w) = \{\mathtt{a} \in \Sigma \mid |w|_\mathtt{a} > 0\}$. The ith letter of a word w is given by $w[i]$ for $i \in [|w|]$. For a given word $w \in \Sigma^n$, the *reversal* of w is defined by $w^R = w[n]w[n-1]\cdots w[2]w[1]$. The powers of $w \in \Sigma^*$ are defined recursively by $w^0 = \varepsilon, w^n = ww^{n-1}$ for $n \in \mathbb{N}$. A word $u \in \Sigma^*$ is a *factor* of $w \in \Sigma^*$, if $w = xuy$ holds for some words $x, y \in \Sigma^*$. Here, u is a *suffix* of w if $y = \varepsilon$ holds. The set of factors of w is denoted by $\mathrm{Fact}(w)$. The factor $w[i]w[i+1]\cdots w[j]$ of w is denoted by $w[i..j]$ for $1 \leq i \leq j \leq |w|$. A *letter square* in w is a factor u of length 2 with $u[1] = u[2]$. A word w is a *palindrome* if $w = w^R$. A function $f : \Sigma^* \to \Sigma^*$ is called a *morphism*, if $f(xy) = f(x)f(y)$ for all $x, y \in \Sigma^*$ (notice that for a morphic function it suffices to have the images of all letters from Σ). We define the projective morphism $\pi_\Gamma : \Sigma^* \to \Sigma^*$ onto $\Gamma \subseteq \Sigma$ by $\pi_\Gamma(\mathtt{a}) = \mathtt{a}$ if $\mathtt{a} \in \Gamma$ and $\pi_\Gamma(\mathtt{a}) = \varepsilon$ otherwise for all $\mathtt{a} \in \Sigma$. Two words $w_1, w_2 \in \Sigma^*$ are called *conjugate* ($w_1 \sim w_2$) if there exist $x, y \in \Sigma^*$ such that $w_1 = xy$ and $w_2 = yx$. For further definitions from combinatorics on words, we refer to [20].

An *undirected graph* is a tuple $G = (V, E)$ with a finite set V of nodes and a set $E \subseteq \binom{V}{2}$ of edges. A graph $G = (V, E)$ is *directed*, if $E \subseteq \{(v_1, v_2) \in V^2 \mid v_1 \neq v_2\}$. The set $\binom{V}{2} \setminus E$ is called the set of *anti-edges* of a given undirected graph. The vertex set of a graph G is denoted by $V(G)$ and the edge set by $E(G)$. An *orientation* of an undirected graph is a directed graph that results from assigning a direction to each edge. An orientation of a graph is *transitive* if for all edges (u, v) and (v, z) there is also an edge (u, z). A graph is a *comparability graph* if it admits a transitive orientation. A *star graph* is a graph $(V \dot{\cup} \{x\}, \{\{x, v\} \mid v \in V\})$, i.e., there is a node x called the *centre* with edges to all other vertices, and there are no other edges in the graph. A graph $G = (V, E)$ is an *extended star graph*, if there exists $U \subseteq V$ such that (U, E) is a star graph, i.e., in addition to the star graph, we can have isolated (not incident with any edge) nodes. A *complete graph* is a graph without anti-edges. Given $k \in \mathbb{N}$, a *k-colouring* of a graph $G = (V, E)$ is a function $f : V \to [k]$ with $f(a) \neq f(b)$ for all $\{a, b\} \in E$. A graph is *k -colourable* if there is a k-colouring for it.

Definition 1. *Two letters* $\mathtt{a}, \mathtt{b} \in \Sigma$ alternate *in a word* $w \in \Sigma^*$ *if* $\pi_{\{\mathtt{a},\mathtt{b}\}}(w)$ *is* $(\mathtt{ab})^n$, $(\mathtt{ab})^n\mathtt{a}$, $(\mathtt{ba})^n$, *or* $(\mathtt{ba})^n\mathtt{b}$ *for some* $n \in \mathbb{N}_0$.

In other words, a and b alternate if no letter occurs at consecutive positions in the projection to a and b. Notice that a or b do not necessarily have to occur. For example, consider the word `rotator`. All letters (but a itself) alternate with a, but every other pair of letters does not, since we have, e.g., $\pi_{\{\mathtt{r},\mathtt{t}\}}(\mathtt{rotator}) = \mathtt{rttr}$.

Definition 2. *A graph* $G = (V, E)$ *is* represented by a word $w \in \Sigma^*$ *if* $\mathrm{alph}(w) = V$ *and for all* $\mathtt{a}, \mathtt{b} \in V$, a *and* b *alternate in* w *iff* $\{\mathtt{a}, \mathtt{b}\} \in E$. *A graph is* word-representable *iff there is a word representing it. A graph is* k-representable *for some* $k \in \mathbb{N}$ *iff there exists a* k*-uniform word representing it. The graph represented by a word* w *is denoted by* $G(w)$.

Since a alternates with every other letter and no other pair of letters alternates, the word `rotator` represents the graph depicted in Fig. 1.

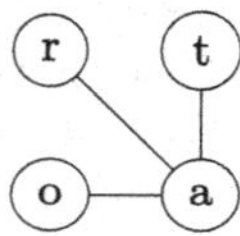

Fig. 1. Graph represented by the word `rotator`

The computational model we use is the standard unit-cost RAM with logarithmic word size.

3 k-Circle Representation

In this short section, we present a classical result on word-representable graphs. So far, there are well-known characterisations of k-representable graphs for $k \leq 2$ but none for $k \geq 3$. It is known that 2-representable graphs are exactly the 2-polygon-circle graphs, better known as circle graphs. However, this cannot be generalised to k-representable graphs being k-polygon-circle-graphs, as shown in [6]. In order to give a characterisation for $k \geq 3$, we take a different but similar approach to [6] and define the notion of a *k-circle representation*.

Definition 3. *For* $k \in \mathbb{N}$, *a* k*-circle representation of a graph* G *is a circle with an inscribed* k*-gon (i.e. a* k*-gon with all the vertices being on the circle) for every vertex of* G *such that the sides of any two* k*-gons intersect* $2k$ *times iff there is an edge between the related vertices in the graph (cf. Fig. 2).*

Lemma 4. *The sides of two* k*-gons inscribed into a circle cannot have more than* $2k$ *intersections.*

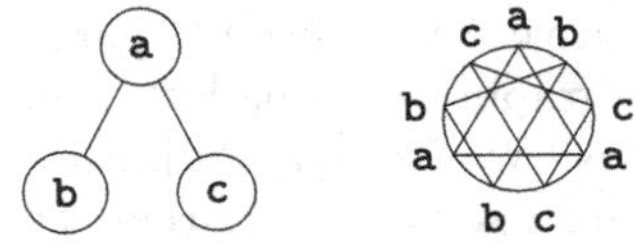

Fig. 2. A graph and a 3-circle-representation of it

By Lemma 4, we get a characterisation for k-representability for all $k \geq 3$.

Theorem 5. *For $k \geq 3$, a graph is k-representable iff there is a k-circle representation of it.*

Proof. Let $k \geq 3$. Let $G = (V, E)$ be k-representable by the k-uniform word $w \in V^*$. We choose $k \cdot |V|$ distinct points on the circle. Starting at any of them, we go around the circle clockwise and label each point with a letter in w such that the order of the word's letters is preserved. Let $\mathtt{a}_i$ be the ith point labelled with $\mathtt{a}$ for every $\mathtt{a} \in V$ and $i \in [k]$. Connecting every point $\mathtt{a}_i$ with $\mathtt{a}_{i+1}$ and $\mathtt{a}_k$ with $\mathtt{a}_1$ for $i \in [k-1]$ yields an inscribed k-gon for every vertex. Let $\mathtt{a}, \mathtt{b} \in V$ and let A, B be the polygons related to $\mathtt{a}$ and $\mathtt{b}$. First, consider $\{\mathtt{a}, \mathtt{b}\} \in E$, i.e., $\mathtt{a}$ and $\mathtt{b}$ alternate in w. Let $i \in [k]$ and $j = (i+1) \bmod k$. There is exactly one point $\mathtt{b}_l$ between $\mathtt{a}_i$ and $\mathtt{a}_j$, i.e., it was labelled after $\mathtt{a}_i$ and before $\mathtt{a}_j$. Thus, both of B's sides connected with $\mathtt{b}_l$ intersect the side between $\mathtt{a}_i$ and $\mathtt{a}_j$. Because there are k pairs $\mathtt{a}_i$ and $\mathtt{a}_j$, there are $2k$ intersections. Now, consider $\{\mathtt{a}, \mathtt{b}\} \notin E$, i.e., $\mathtt{a}$ and $\mathtt{b}$ do not alternate in w. There exists $i \in [k]$ with $j = (i+1) \bmod k$ such that there is no $\mathtt{b}_l$ between $\mathtt{a}_i$ and $\mathtt{a}_j$. This means the line from $\mathtt{a}_i$ to $\mathtt{a}_j$ does not intersect with any side of B. Because every of the k sides of A cannot have more than 2 intersections with B, A and B have less than $2k$ intersections. This concludes that we have a k-circle representation of G.

For the other direction assume $G = (V, E)$ has a k-circle representation. Now, we pick some point on the circle. We start with the empty word, go clockwise around the circle, and for any corner we pass, we append the graph's vertex related to the corner's polygon. Once we reach our starting point we have a k-uniform word w with $\text{alph}(w) = V$. For the same reason as above we have $2k$ intersections between the polygons A, B iff the related letters $\mathtt{a}$ and $\mathtt{b}$ alternate in w. Therefore, w represents G. □

By Theorem 5, we can construct a k-circle representation of $G(w)$ given a k-uniform word w by writing the letters on a circle and connecting all occurrences of the same letters as an inscribed polygon. Also, given a k-circle-representation we can determine a word representing the associated graph by reading the letters on the circle, starting with any letter. For example, the graph from Fig. 2 is represented by the word cabcacbab and all of its conjugates.

4 Graphs of Conjugate Words

Motivated by the k-circle representation in the last section, in this section we mainly investigate the graphs represented by the conjugates of a given word.

These conjugacy classes are a very well studied part of combinatorics on words since for instance the lexicographically smallest words in such a class are the famous *Lyndon words* [21]. All words in a conjugacy class can be obtained by writing the letters of the given word on a circle (cf. Sect. 3) and reading the words from each possible starting point. Thus, now our goal is to investigate the class of graphs represented by the conjugates of a given word w. Notice that for some $k \in \mathbb{N}$ two k-uniform conjugates always represent the same graph, cf. [18]. However in the general case, two conjugates do not necessarily represent the same graph, since by rotating the word, letter squares may appear or disappear. Consider for instance $w = \mathtt{abcba}$ representing a graph with the edges $\{\mathtt{a},\mathtt{c}\}$, $\{\mathtt{b},\mathtt{c}\}$. The conjugate $u = \mathtt{bcbaa}$ has still the edge $\{\mathtt{b},\mathtt{c}\}$ but not the edge $\{\mathtt{a},\mathtt{c}\}$.

Before we further investigate conjugates, we start with a straightforward result about a representative with all letter squares moved to the end.

Lemma 6. *Let $w \in \Sigma^*$ with $\mathtt{aa} \in \mathrm{Fact}(w)$ for a letter $\mathtt{a} \in \Sigma$. Then, w and $\pi_{\Sigma\setminus\{\mathtt{a}\}}(w)\mathtt{aa}$ represent the same graph.*

Lemma 6 guarantees that the following definition is well-defined, and the proposition afterwards shows that we can compute this representative in linear time in w's length if the alphabet's size is smaller than the word's length.

Definition 7. *Let $w \in \Sigma^*$ and $S = \{\mathtt{a} \in \Sigma \mid \mathtt{aa} \in \mathrm{Fact}(w)\}$. Set $w' = \prod_{s\in S} s^2$ as the concatenation of all letters from S squared obeying some fixed order on S. Define the* separated representative *of w by $\mathrm{sep}(w) = \pi_{\Sigma\setminus S}(w)w'$.*

Remark 8. By Lemma 6, $\mathrm{sep}(w)$ represents the same graph as w.

Proposition 9. *Given $w \in \Sigma^*$, $\mathrm{sep}(w)$ can be computed in time $O(|w| + |\Sigma|)$, and we have $|\mathrm{sep}(w)| \le |w| - \sum_{\mathtt{a}\in\Sigma,\mathtt{a}^2\in\mathrm{Fact}(w)}(|w|_{\mathtt{a}} - 2)$.*

The representative $\mathrm{sep}(w)$ makes the letter squares in a word explicit by stating them grouped at the end of the word. As mentioned above, conjugates of words may have letter squares while the word itself does not contain any. This can only occur if there is a conjugate yx of xy such that $x[1] = y[|y|]$ for some $x, y \in \Sigma^*$. To capture this property, we define the notion of a *circular square.*

Definition 10. *A pair $(\mathtt{a}, i)$ is a* circular square *in $w \in \Sigma^*$ with $|\mathrm{alph}(w)| > 1$ if $w[i] = w[1 + i \bmod |w|] = \mathtt{a}$.*

Consider again the word $w = \mathtt{rotator}$ having the circular square $(\mathtt{r}, 7)$. Besides w itself the conjugacy class contains the words `otatorr`, `tatorro`, `atorrot`, `torrota`, `orrotat`, `rrotato` all having only `r` as a square at different positions.

Lemma 11. *The number of circular squares in $w \in \Sigma^*$ is the same as in every conjugate of w.*

Remark 12. Given $w \in \Sigma^*$, $(w[|w|], |w|)$ is the only possibility for a circular square which is not a letter square. Note that for each letter square there exists exactly one conjugate of w such that the letter square becomes a circular square which is not a letter square: `lossless` has two letter squares, and the conjugates `slesslos` and `slossles` have circular squares which are not letter squares.

The following theorem characterises exactly for each given pair of letters $\mathtt{a}, \mathtt{b} \in \text{alph}(w)$ whether there are conjugates of w representing graphs with and without the edge $\{\mathtt{a}, \mathtt{b}\}$.

Theorem 13. *Let $w \in \Sigma^*$ and $\mathtt{a}, \mathtt{b} \in \text{alph}(w)$ with $\mathtt{a} \neq \mathtt{b}$. Set $u = \pi_{\{\mathtt{a},\mathtt{b}\}}(w)$. Let k be the number of circular squares in u.*

1. *There is a conjugate of w representing a graph with an edge $\{\mathtt{a}, \mathtt{b}\}$ iff $k \leq 1$.*
2. *There is a conjugate of w representing a graph without an edge $\{\mathtt{a}, \mathtt{b}\}$ iff $k \geq 1$.*

Proof. First, let w' be a conjugate of w representing a graph with an edge $\{\mathtt{a}, \mathtt{b}\}$. Choose $x, y \in \Sigma^*$ with $w' = xy$ and $w = yx$. Set $u' = \pi_{\{\mathtt{a},\mathtt{b}\}}(w')$, which is a conjugate of u. We know that a and b alternate in xy, which means there is no letter square and at most one circular square in u'. Therefore, we have $k \leq 1$. For the other direction, assume $k \leq 1$. If there is no letter square in u, a and b alternate and there is an edge $\{\mathtt{a}, \mathtt{b}\}$ in $G(w)$. Otherwise, we can assume aa is a factor of u w.l.o.g. There are $x, y \in \Sigma^*$ with $w = xy$ such that $(\mathtt{a}, |u'|)$ is a circular square in $u' = \pi_{\{\mathtt{a},\mathtt{b}\}}(yx)$. Since $k \leq 1$, there is no letter square in u', and a and b alternate in yx. It follows that $G(yx)$ has an edge $\{\mathtt{a}, \mathtt{b}\}$.

For the second claim, let w' be a conjugate of w representing a graph without the edge $\{\mathtt{a}, \mathtt{b}\}$. Define $u' = \pi_{\{\mathtt{a},\mathtt{b}\}}(w')$, which is a conjugate of u. We know that a and b do not alternate in w', which means there is a letter square and a circular square in u'. Therefore, we have $k \geq 1$. For the other direction, assume $k \geq 1$. If there is a letter square in u, the word w is a conjugate of itself, and there is no edge $\{\mathtt{a}, \mathtt{b}\}$ in G. Otherwise, we have $u[1] = u[|u|]$. There are $x, y \in \Sigma^*$ with $xy = w$ such that aa is factor of $u' = \pi_{\{\mathtt{a},\mathtt{b}\}}(yx)$ w.l.o.g. Hence, yx is a conjugate of w representing a graph without the edge $\{\mathtt{a}, \mathtt{b}\}$. □

In order to further investigate the set of graphs represented by the conjugates of a given word $w \in \Sigma^*$, we introduce the notion of *optional edges*. These edges are exactly the edges in G that can be removed or added by the conjugates of w.

Definition 14. *Let $w \in \Sigma^*$ and $\mathtt{a}, \mathtt{b} \in \text{alph}(w)$ with $\mathtt{a} \neq \mathtt{b}$. We say that $\{\mathtt{a}, \mathtt{b}\}$ is an* optional edge *of w if there is exactly one circular square in $\pi_{\{\mathtt{a},\mathtt{b}\}}(w)$.*

In Fig. 3 the graph with optional edges for the word decide is depicted. Notice that we are investigating conjugates of words and obtain their graphs. Thus, we are talking about optional edges in words since representatives of the same graph can have different optional edges, e.g., the words aab and aabb both represent the same graph, but the edge $\{\mathtt{a}, \mathtt{b}\}$ is only optional in aab.

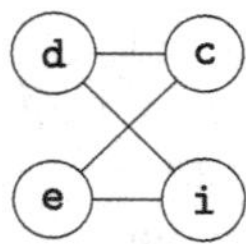

Fig. 3. Graph with optional edges in the word decide as edges

Lemma 15. *If* $(\mathtt{a}, i)$ *is the only circular square in a word* w *with* $\mathrm{alph}(w) = \{\mathtt{a}, \mathtt{b}\}$*, we have* $|w|_{\mathtt{a}} = |w|_{\mathtt{b}} + 1$*.*

Proposition 16. *Let* $w \in \Sigma^*$ *be a word representing the graph* $G = (\mathrm{alph}(w), E)$ *and* E' *be the set of optional edges of* w*. Then,* $(\mathrm{alph}(w), E')$ *is 2-colourable.*

Corollary 17. *For a graph* G*, there is a word* $w \in \Sigma^*$ *representing* G *with every edge and anti-edge of* G *being an optional edge of* w *iff* G *has at most two nodes.*

Before we present an algorithm that computes the graphs represented by the conjugates of a word, we investigate in more detail which conjugate contains a particular edge. For $w \in \Sigma^*$ and $\mathtt{a}, \mathtt{b} \in \Sigma$ with $\mathtt{a} \neq \mathtt{b}$, define the set of w's indices such that the rotation contains the edge $\{\mathtt{a}, \mathtt{b}\}$ by $I_w(\{\mathtt{a}, \mathtt{b}\}) = \{i \in [|w|] \mid \{\mathtt{a}, \mathtt{b}\} \in E(G(w[i+1..|w|]w[1..i]))\}$. For all $u, v \in \Sigma^*$ with $\mathtt{a}, \mathtt{b} \in \mathrm{alph}(uv)$ and $\mathtt{a} \neq \mathtt{b}$, we have immediately $\{\mathtt{a}, \mathtt{b}\} \in E(G(vu))$ iff $|u| \in I_{uv}(\{\mathtt{a}, \mathtt{b}\})$. On the other hand, given $w \in \Sigma^*$ and $G = (\mathrm{alph}(w), E)$, there is a conjugate of w representing G iff there is $i \in [|w|]$ with $\{a, b\} \in E$ iff $i \in I_w(\{\mathtt{a}, \mathtt{b}\})$. This leads to the following theorem.

Theorem 18. *Let* $w \in \Sigma^*$ *represent* $G = (\mathrm{alph}(w), E)$*,* $\mathtt{a}, \mathtt{b} \in \mathrm{alph}(w)$ *with* $\mathtt{a} \neq \mathtt{b}$*, and* $u = \pi_{\{\mathtt{a},\mathtt{b}\}}(w)$*. Let* k *be the number of circular squares in* u*.*

1. *If* $k = 0$*,* $I_w(\{\mathtt{a}, \mathtt{b}\}) = [|w|]$ *holds.*
2. *If* $k = 1$*, there exist* $n, m \in [|w|]$ *with* $w[n] = w[m]$ *and either* $\pi_{\{\mathtt{a},\mathtt{b}\}}(w[n..m]) = w[n]w[m]$ *or* $\pi_{\{\mathtt{a},\mathtt{b}\}}(w[n..|w|]w[1..m]) = w[n]w[m]$*. In the former case* $I_w(\{\mathtt{a}, \mathtt{b}\}) = [n, m-1]$ *holds, else* $I_w(\{\mathtt{a}, \mathtt{b}\}) = [n, |w|] \cup [1, m-1]$*.*
3. *If* $k > 1$*,* $I_w(\{\mathtt{a}, \mathtt{b}\}) = \emptyset$ *holds.*

Since the complete graph with two nodes has every edge and anti-edge as an optional edge, we investigate complete graphs in more detail from a word's perspective.

Lemma 19. *Every word* w *representing a complete graph with* n *nodes is of the form* $(w[1] \dots w[n])^k w[1] \dots w[|w| \bmod n]$ *for some* $k \in \mathbb{N}$*.*

By Lemma 19, we can present a full characterisation of the graphs that are represented by the conjugates of a word representing a complete graph.

Theorem 20. *Let* $w = uv$ *with* $w, u, v \in \Sigma^*$ *and* $|u| > 0$ *represent the complete graph* (V, E) *with* n *nodes. For* $m = |w| \bmod n$*,* vu *represents* $(V, E \setminus E')$ *with*

1. $E' = \{\{w[i], w[j]\} \mid i \in [|u|], j \in [m+1, n]\}$ *if* $|u| < m$,
2. $E' = \{\{w[i], w[j]\} \mid i \in [m], j \in [m+1, n]\}$ *if* $m \leq |u| \leq |w| - m$,
3. $E' = \{\{w[i], w[j]\} \mid i \in [|u| - |w| + m + 1, m], j \in [m+1, n]\}$ *if* $|w| - m < |u|$.

Due to Theorem 18, we can construct an efficient algorithm to compute I_w. It uses a matrix C to save indices of possible circular squares and an array first to save each letter's first occurrence. This is needed to detect circular squares in the projection that are not letter squares.

Proposition 21. *Let* $w \in \Sigma^*$. I_w *can be computed in* $O(|\operatorname{alph}(w)||w|)$ *time.*

We conclude this section by presenting an algorithm for efficient computation of the graphs represented by a word's conjugates.

Proposition 22. *Let* $w = \mathtt{a}w'$ *with* $\mathtt{a} \in \Sigma$ *and* $w' \in \Sigma^*$ *represent the graph* $G = (\operatorname{alph}(w), E)$. $G(w'\mathtt{a})$ *is computable in* $O(|\operatorname{alph}(w)|)$ *time given* G *and* I_w.

It suffices to calculate I_w from Proposition 22 once for all conjugates. However, after the computation of a conjugate's graph, I_w needs to be updated by reducing every entry by 1. This can be avoided by using an offset, which is applied to every entry. Since the computation of I_w takes $O(|\operatorname{alph}(w)||w|)$ time according to Proposition 21, we obtain the following theorem, which applies Proposition 22 $|w| - 1$ times for a given $w \in \Sigma^*$ using an offset to update I_w.

Theorem 23. *Given* $w \in \Sigma^*$ *and* $G(w)$, *all graphs represented by* w*'s conjugates can be computed in time* $O(|\operatorname{alph}(w)||w|)$.

5 Graphs Represented by *k*-Local Words

In this section, we investigate the graphs represented by k-local words. The notion of k-locality was introduced in [5] in the context of pattern matching and further investigated on words in [7]. Here, we connect 1-local words with comparability graphs. Before we introduce the notion of k-locality and show that each graph represented by a k-local word is $2k$-representable, we have a look at graphs represented by palindromes. Palindromes play an important role in 1-locality [5]. As a first result, we show that two letters in a palindrome can only alternate if one of them is in the palindrome's centre, which is at $\lceil \frac{1}{2}|w| \rceil$, implying that every non-empty palindrome represents an extended star graph.

Lemma 24. *Let* $w \in \Sigma^+$ *be a palindrome and* $k = \lceil \frac{1}{2}|w| \rceil$. *If* $\mathtt{a}, \mathtt{b} \in \Sigma$ *alternate in* w, *then* $w[k] \in \{\mathtt{a}, \mathtt{b}\}$.

Proposition 25. *Let* $w \in \Sigma^+$ *be a palindrome,* $U \subseteq \Sigma$ *be the set of letters alternating with* $w[k]$ *for* $k = \lceil \frac{1}{2}|w| \rceil$, *and let* $G = (\operatorname{alph}(w), E)$ *be the graph represented by* w. *Then,* $(U \dot\cup \{w[k]\}, E)$ *is a star graph with centre* $w[k]$.

The next proposition characterises the palindromes representing star graphs.

Proposition 26. *Let G be a star graph with centre* c *and vertex set $U\dot{\cup}\{\mathtt{c}\}$ for $|U| > 1$. The palindromes representing G are exactly the words*

$$sw_{n-1}\mathtt{c}...w_1\mathtt{c}w_0\mathtt{c}(w_0)^R(w_1\mathtt{c})^R...(w_{n-1}\mathtt{c})^R s^R$$

with $n \in \mathbb{N}$, $w_0, ..., w_n \in U^*$ 1-uniform, and $s \in (U\dot{\cup}\{\mathtt{c}\})^*$ being a suffix of $w_n\mathtt{c}$.

Applying Proposition 25, we can characterise when two palindromes represent the same graph. As expected, the centres of the palindromes play the key role.

Proposition 27. *Let $w_1, w_2 \in \Sigma^+$ be palindromes with $\mathrm{alph}(w_1) = \mathrm{alph}(w_2)$. Let U_1, U_2 be the sets of letters alternating with $w_1[k_1]$ and $w_2[k_2]$ for $k_1 = \lceil\frac{1}{2}|w_1|\rceil$ and $k_2 = \lceil\frac{1}{2}|w_2|\rceil$. The words w_1 and w_2 represent the same graph iff*

1. *$U_1 = U_2$ and $w_1[k_1] = w_2[k_2]$ or*
2. *$U_1 = \{w_2[k_2]\}$ and $U_2 = \{w_1[k_1]\}$ or*
3. *$U_1 = U_2 = \emptyset$.*

Theorem 28. *Non-empty palindromes represent exactly extended star graphs.*

Now, we are investigating the graphs represented by k-local words. Before we introduce these words formally, we give an example to give an understanding. Consider $w =$ banana and the enumeration $s = (\mathtt{b}, \mathtt{a}, \mathtt{n})$. Now, we mark letters: in a first step, we mark all occurrences of b and obtain $\overline{\mathtt{b}}\mathtt{anana}$. We call a consecutive factor of marked letters a block, and thus, we have one marked block. Now, we mark all occurrences of a, resulting in $\overline{\mathtt{ba}}\mathtt{n}\overline{\mathtt{a}}\mathtt{n}\overline{\mathtt{a}}$ having three marked blocks. In a last step, we mark the occurrences of n and get with $\overline{\mathtt{banana}}$ one marked block. Since the highest number of marked blocks we saw is 3, we say that w is 3-local. Notice that banana is also 2-local, witnessed by $(\mathtt{n}, \mathtt{a}, \mathtt{b})$. Now, we introduce these notions formally.

Definition 29. *Let $\overline{\Sigma} = \{\overline{x} \mid x \in \Sigma\}$ be the* set of marked letters. *For a word $w \in \Sigma^*$, a* marking sequence *of the letters occurring in w, is an enumeration $s = (x_1, x_2, ..., x_{|\mathrm{alph}(w)|})$ of $\mathrm{alph}(w)$. A letter x_i is called* marked at stage $k \in \mathbb{N}$ *if $i \leq k$. Moreover, we define w_k, the marked version of w at stage k, as the word obtained from w by replacing all x_i with $i \leq k$ by $\overline{x_i}$. A factor of w_k is a* marked block *if it only contains elements from $\overline{\Sigma}$ and the letters to the left and right (if existing) are from Σ. A word $w \in \Sigma^*$ is k-local for $k \in \mathbb{N}_0$ if there exists a marking sequence $(x_1, ..., x_{|\mathrm{alph}(w)|})$ of $\mathrm{alph}(w)$ such that for all $i \leq |\mathrm{alph}(w)|$, we have that w_i has at most k marked blocks. A word is called* strictly k-local *if it is k-local but not $(k-1)$-local.*

Before we present the results on the graphs represented by k-local words, we need a property of k-local words.

Lemma 30. *Let $w \in \Sigma^*$ be k-local with $|w|_{\mathtt{a}} > 2k$ for some $\mathtt{a} \in \Sigma$ and $k \in \mathbb{N}$. Then, $\mathtt{aa} \in \mathrm{Fact}(w)$.*

Our first results about k-local words connect k-locality and k-representability.

Proposition 31. *For a given $k \in \mathbb{N}$, a graph represented by a k-local word is $2k$-representable.*

Corollary 32. *There is no k such that every word-representable graph can be represented by a k-local word.*

Proposition 33. *Let $w \in \Sigma^*$ with $|\operatorname{alph}(w)| > 1$ be a strictly k-local word representing a graph G. There is a strictly $(k+1)$-local word representing G.*

Now, we have a more detailed look into the graphs represented by 1-local words. These words have a specific palindromic-like structure shown in [5]. While in [5] they concluded that it suffices to look at the condensed form of a word, we here need to keep letter squares in the middle of the word since it indicates that we do not have an edge in the corresponding graph. Therefore, we define the notion of the 1-*local normal form* of a word.

Definition 34. *A 1-local word $w \in \Sigma^*$ is in* normal form *(1l-NF) if we have either $w = \varepsilon$ or if there exist $\mathtt{a} \in \Sigma$, $n, m \in \{0, 1\}$, and $w' \in (\Sigma \backslash \{\mathtt{a}\})^*$ in 1l-NF such that $w = \mathtt{a}^n w' \mathtt{a}^m$.*

Remark 35. Let $w \in \Sigma^+$ be in 1l-NF with a marking sequence s starting with $\mathtt{x} \in \Sigma$ witnessing the locality, i.e., $w = w_1 \mathtt{x} w_2$ or $w = w_1 \mathtt{xx} w_2$ for some $w_1, w_2 \in (\Sigma \backslash \{\mathtt{x}\})^*$. From [7], we obtain inductively for all $\mathtt{y} \in \Sigma \backslash \operatorname{alph}(w)$, that $w_1 \mathtt{xyy} w_2$ or $w_1 \mathtt{xyyx} w_2$ are also in 1l-NF. Also based on [7], we can assume w_1 and w_2 to be condensed, thus 1-uniform on their respective alphabets.

The following results connect 1-local words and word-representability by words in 1l-NF, leading to the main theorem of this section.

Proposition 36. *Every graph representable by a 1-local word can be represented by a word in 1l-NF.*

Lemma 37. *Let $w = w_1 \mathtt{a} w_2 \mathtt{a} w_3 \in \Sigma^+$ with $\mathtt{a} \in \Sigma$, $w_1, w_2, w_3 \in (\Sigma \backslash \{\mathtt{a}\})^*$. If $w_1 w_3$ represents $G = (V, E)$ and w_1 and w_3 are 1-uniform, then $w_1 \mathtt{aa} w_3$ represents $(V \cup \{\mathtt{a}\}, E)$ and $w_1 \mathtt{a}$ and $\mathtt{a} w_3$ are 1-uniform.*

Lemma 38. *Let $w \in \Sigma^*$ be 1-local with $w = w_1 \mathtt{a} w_2 w_3$ or $w = w_1 w_2 \mathtt{a} w_3$ for a letter $\mathtt{a} \in \Sigma$ and $w_1, w_2, w_3 \in (\Sigma \backslash \{\mathtt{a}\})^*$. If $w_1 w_3$ represents $G = (V, E)$ and w_1 and w_3 are 1-uniform, then $w_1 \mathtt{a} w_3$ represents $(V \cup \{\mathtt{a}\}, E \cup \{\{\mathtt{a}, v\} \mid v \in V\})$ and $w_1 \mathtt{a}$ and $\mathtt{a} w_3$ are 1-uniform.*

We finish this section with a characterisation of the graphs represented by 1-local words. Therefore, we define the notion of 1-*local-representable graphs.*

Definition 39. *We inductively define* 1-local-representable graphs *as follows: the empty graph is 1-local-representable, and if (V, E) is 1-local-representable, then $(V \cup \{x\}, E)$ and $(V \cup \{x\}, E \cup \{\{x, v\} \mid v \in V\})$ are both 1-local-representable.*

Theorem 40. *1-local words represent exactly the 1-local-representable graphs.*

Proof. Let $w_1 \in \Sigma^*$ be a 1-local word representing the graph G. By Proposition 36, there is a word $w_2 \in (\operatorname{alph}(w_1))^*$ in 1l-NF representing G. Thus, there exist a letter $\mathtt{a} \in \mathrm{alph}(w_1)$ and a word $w_3 \in (\mathrm{alph}(w_1)\backslash\{\mathtt{a}\})^*$ in 1l-NF such that $w_2 \in \{\varepsilon, \mathtt{a}w_3\mathtt{a}, \mathtt{a}w_3, w_3\mathtt{a}\}$. Note that ε represents the empty graph, which is by definition 1-local-representable. Now, we can inductively apply Lemmas 37 and 38 and get that w_2 represents a 1-local-representable graph.

For the other direction, let G be a 1-local-representable graph. If G is empty, it is represented by the 1-local empty word. Otherwise, there are a 1-local-representable graph (V, E) and $\mathtt{a}$ such that either $G = (V\dot{\cup}\{\mathtt{a}\}, E)$ or $G = (V\dot{\cup}\{\mathtt{a}\}, E\dot{\cup}\{\{\mathtt{a}, v\} \mid v \in V\})$. We can assume by induction that there is a 1-local word $w \in V^*$ in 1l-NF that represents (V, E) and has a minimal marking sequence s starting with the letter $\mathtt{x}$. There are 1-uniform words $w_1, w_2 \in (V \setminus \{x\})^*$ such that $w = w_1\mathtt{x}w_2$ or $w = w_1\mathtt{xx}w_2$. Consider first $G = (V\dot{\cup}\{\mathtt{a}\}, E)$. If $w = w_1\mathtt{x}w_2$, then $w_1\mathtt{xaa}w_2$ represents G, since $\mathtt{a}$ does not alternate with any letter in this word. If $w = w_1xxw_2$, $w_1\mathtt{xaax}w_2$ represents G for the same reason. For both words $\mathtt{a} \circ s$ is a marking sequence witnessing the 1-locality. Secondly, consider $G = (V\dot{\cup}\{\mathtt{a}\}, E\dot{\cup}\{\{\mathtt{a}, v\} \mid v \in V\})$. If $w = w_1\mathtt{x}w_2$, then $w_1\mathtt{xa}w_2$ represents G, since w_1, w_2 are 1-uniform and $\mathtt{a}$ alternates with every letter in the word. Analogously, if $w = w_1\mathtt{xx}w_2$, then $w_1\mathtt{xax}w_2$ represents G. Again, $\mathtt{a} \circ s$ is a witness for the 1-locality. □

Corollary 41. *Every* 1*-local-representable graph is a comparability graph.*

Proof. If $G = (V, E)$ is 1-local-representable, then $(V \cup \{x\}, E \cup \{\{x, v\} \mid v \in V\})$ is also 1-local-representable and thus word-representable. The claim follows from [18]. □

6 The Language of a Graph

In this section, we show that the language of all words representing a given graph G is regular by constructing a deterministic finite automaton accepting this language. We firstly introduce the needed notions of finite automata.

A *deterministic finite automaton* is a 5-tuple $A = (Q, \Sigma, \delta, q_0, F)$ with a finite set of states Q, an alphabet Σ, a transition function $\delta : Q \times \Sigma \to Q$, an initial state $q_0 \in Q$, and a set of accepting states $F \subseteq Q$. A word $w \in \Sigma^*$ is *accepted* by A if there is a sequence of states $q_0, q_1, q_2, ..., q_{|w|}$ such that $\delta(q_{i-1}, w[i]) = q_i$ for all $i \in [|w|]$ and $q_{|w|} \in F$. The *language* of words accepted by A is denoted by $L(A)$. A language is *regular* if there is a deterministic finite automaton accepting it. We define $\delta^* : Q \times \Sigma^* \to Q$ by $\delta^*(q, \varepsilon) = q$ and $\delta^*(q, \mathtt{a}w') = \delta^*(\delta(q, \mathtt{a}), w')$ for $\mathtt{a} \in \Sigma$ and $w' \in \Sigma^*$.

In a first step, we construct a deterministic finite automaton $A_{\{\mathtt{a},\mathtt{b}\}}$ accepting exactly the language of all words over Σ where $\mathtt{a}, \mathtt{b} \in \Sigma$ alternate. In addition, we need a deterministic finite automaton $A_{\mathtt{a}}$ which accepts exactly the words containing an $\mathtt{a} \in \Sigma$.

Definition 42. *For* $\mathtt{a}, \mathtt{b} \in \Sigma$*, define* $A_{\{\mathtt{a},\mathtt{b}\}} = (Q, \Sigma, \delta, q_0, Q\backslash\{q_3\})$ *with* $Q = \{q_0, q_1, q_2, q_3\}$ *as given by Fig. 4 on the left hand side with additional transitions* $\delta(q, \mathtt{c}) = q$ *for all* $q \in Q$ *and* $\mathtt{c} \in \Sigma\backslash\{\mathtt{a}, \mathtt{b}\}$*. Moreover, define* $A_{\mathtt{a}} = (Q', \Sigma, \delta', q'_0, \{q'_1\})$ *with* $Q' = \{q'_0, q'_1\}$ *as given by Fig. 4 on the right hand side with additional transitions* $\delta'(q', \mathtt{c}) = q'$ *for all* $q' \in Q'$ *and* $\mathtt{c} \in \Sigma\backslash\{\mathtt{a}\}$*.*

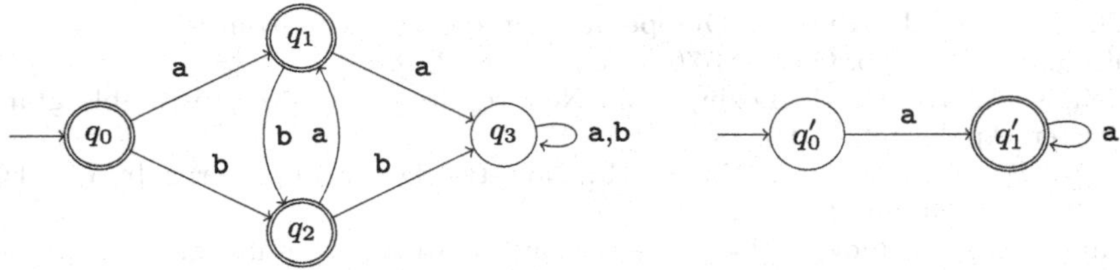

Fig. 4. Left hand side: $A_{\{a,b\}}$ and $\delta(q, c) = q$ for all $q \in Q$, $c \in \Sigma \backslash \{a, b\}$. Right hand side: A_a and $\delta(q, c) = q$ for all $q \in Q'$, $c \in \Sigma \backslash \{a\}$.

Lemma 43. *Let $w \in \Sigma^*$. Then, $w \in L(A_{\{a,b\}})$ iff* a *and* b *alternate in w, and $w \in L(A_a)$ iff* $a \in \text{alph}(w)$.

In the following theorem we combine these kinds of automata to construct an automaton accepting all words representing a given graph. Here, it is crucial that regular languages are closed under intersection and complement.

Theorem 44. *The language of words representing a given graph is regular.*

7 Conclusion

Our first result gives a geometrical characterisation of k-representability for $k \geq 3$. For this we introduced the notion of a k-circle representation of a graph. This notion led to the investigation of the graphs represented by the conjugates of a word. Here, we were able to fully characterise the graphs by defining circular squares. We also gave an efficient algorithm which calculates all associated graphs. In a next step, we investigated the graphs represented by palindromes. Based on these results, we had a look into k-local words. For us, interestingly, representability by k-local words and k-representability are closely related. Also, we were able to characterise the graphs represented by 1-local words, which are tightly related to palindromes. This raises the hope that k-local words for an arbitrary $k \in \mathbb{N}$ can also be characterised by a special form of graphs in order to understand k-local words better. Moreover, since it is proven to be NPC whether a word is k-local, fragments of these words may turn out to be determinable in P by using insights about graphs. We finished our work by showing that the language of words representing the same graph is regular. This might allow counting the number of words representing a given graph using automata theory. In future research several of our results could be transferred to generalisations of word-representable graphs.

References

1. Chandrasekaran, S., Sulthana, A.: k-Power domination of crown graph. IJAER **14**(13), 3066–3068 (2019)
2. Cheon, G.-S., Kim, J., Kim, M., Kitaev, S., Pyatkin, A.: On k-11-representable graphs. J. Comb. **10**(3) (2019)

3. Choi, I., Kim, J., Kim, M.: On operations preserving semi-transitive orientability of graphs. J. Comb. Optim. **37**(4), 1351–1366 (2018)
4. Collins, A., Kitaev, S., Lozin, V.V.: New results on word-representable graphs. Discret. Appl. Math. **216**, 136–141 (2017)
5. Day, J.D., Fleischmann, P., Manea, F., Nowotka, D.: Local patterns. In: FSTTCS. LIPIcs, vol. 93, pp. 24:1–24:14 (2017)
6. Enright, J.A., Kitaev, S.: Polygon-circle and word-representable graphs. Electron. Notes Discret. Math. **71**, 3–8 (2019)
7. Fleischmann, P., Haschke, L., Manea, F., Nowotka, D., Tsida, C.T., Wiedenbeck, J.: Blocksequences of k-local words. In: Bureš, T., et al. (eds.) SOFSEM 2021. LNCS, vol. 12607, pp. 119–134. Springer, Cham (2021). https://doi.org/10.1007/978-3-030-67731-2_9
8. Gaetz, M., Ji, C.: Enumeration and extensions of word-representants. Discret. Appl. Math. **284**, 423–433 (2020)
9. Glen, M.E.: Colourability and word-representability of near-triangulations. abs/1605.01688 (2018)
10. Halldórsson, M.M., Kitaev, S., Pyatkin, A.: On representable graphs, semi-transitive orientations, and the representation numbers. CoRR, abs/0810.0310 (2008)
11. Halldórsson, M.M., Kitaev, S., Pyatkin, A.: Graphs capturing alternations in words. In: Gao, Y., Lu, H., Seki, S., Yu, S. (eds.) DLT 2010. LNCS, vol. 6224, pp. 436–437. Springer, Heidelberg (2010). https://doi.org/10.1007/978-3-642-14455-4_41
12. Jones, M.E., Kitaev, S., Pyatkin, A.V., Remmel, J.B.: Representing graphs via pattern avoiding words. Electron. J. Comb. **22**(2), 2 (2015)
13. Kenkireth, B.G., Malhotra, A.S.: On word-representable and multi-word-representable graphs. In: Drewes, F., Volkov, M. (eds.) DLT 2023. LNCS, vol. 13911, pp. 156–167. Springer, Cham (2023). https://doi.org/10.1007/978-3-031-33264-7_13
14. Kitaev, S.: A comprehensive introduction to the theory of word-representable graphs. In: Charlier, É., Leroy, J., Rigo, M. (eds.) DLT 2017. LNCS, vol. 10396, pp. 36–67. Springer, Cham (2017). https://doi.org/10.1007/978-3-319-62809-7_2
15. Kitaev, S.: Existence of u-representation of graphs. J. Graph Theory **85**(3), 661–668 (2017)
16. Kitaev, S., Lozin, V.V.: Words and Graphs. Monographs in Theoretical Computer Science. An EATCS Series (2015)
17. Kitaev, S., Pyatkin, A.: On semi-transitive orientability of split graphs. CoRR, abs/2110.08834 (2021)
18. Kitaev, S., Pyatkin, A.V.: On representable graphs. J. Autom. Lang. Comb. **13**(1), 45–54 (2008)
19. Kitaev, S., Seif, S.: Word problem of the Perkins semigroup via directed acyclic graphs. Order **25**(3), 177–194 (2008)
20. Lothaire, M.: Combinatorics on Words. Cambridge Mathematical Library (1997)
21. Lyndon, R.C.: On Burnside's problem. Trans. Am. Math. Soc. **77**(2), 202–215 (1954)
22. Oliveros, D., Torres, A.: From word-representable graphs to altered Tverberg-type theorems. CoRR, abs/2111.10038 (2021)

The Complexity of Online Graph Games

Janosch Fuchs, Christoph Grüne(✉), and Tom Janßen

Department of Computer Science, RWTH Aachen University, Aachen, Germany
{fuchs,gruene,janssen}@algo.rwth-aachen.de

Abstract. Online computation is a concept to model uncertainty where not all information on a problem instance is known in advance. An online algorithm receives requests which reveal the instance piecewise and has to respond with irrevocable decisions. Often, an adversary is assumed that constructs the instance knowing the deterministic behavior of the algorithm. Thus, the adversary is able to tailor the input to any online algorithm. From a game theoretical point of view, the adversary and the online algorithm are players in an asymmetric two-player game.

To overcome this asymmetry, the online algorithm is equipped with an isomorphic copy of the graph, which is referred to as unlabeled map. By applying the game theoretical perspective on online graph problems, where the solution is a subset of the vertices, we analyze the complexity of these online vertex subset games. For this, we introduce a framework for reducing online vertex subset games from TQBF. This framework is based on gadget reductions from 3-Satisfiability to the corresponding offline problem. We further identify a set of rules for extending the 3-Satisfiability-reduction and provide schemes for additional gadgets which assure that these rules are fulfilled. By extending the gadget reduction of the vertex subset problem with these additional gadgets, we obtain a reduction for the corresponding online vertex subset game.

At last, we provide example reductions for online vertex subset games based on Vertex Cover, Independent Set, and Dominating Set, proving that they are *PSPACE*-complete. Thus, this paper establishes that the online version with a map of *NP*-complete vertex subset problems form a large class of *PSPACE*-complete problems.

Keywords: Online Algorithms · Computational Complexity · Online Algorithms Complexity · Two-Player Games · NP-complete Graph Problems · PSPACE-completeness · Gadget Reduction

1 Introduction

Online computation is an intuitive concept to model real time computation where the full instance is not known beforehand. In this setting, the instance

This work is funded by the Deutsche Forschungsgemeinschaft (DFG, German Research Foundation) – GRK 2236/1, WO 1451/2-1. The full version of this paper can be found on arXiv [6].

H. Fernau et al. (Eds.): SOFSEM 2024, LNCS 14519, pp. 269–282, 2024.
https://doi.org/10.1007/978-3-031-52113-3_19

is revealed piecewise to the online algorithm and each time a piece of information is revealed, an irrevocable decision by the online algorithm is required. To analyze the worst-case performance of an algorithm solving an online problem, a malicious adversary is assumed.

The adversary constructs the instance while the online algorithm has to react and compute a solution. This setting is highly asymmetric in favor of the adversary. Thus, for most decision problems, the adversary is able to abuse the imbalance of power to prevent the online algorithm from finding a solution that is close to the optimal one. To overcome the imbalance, there are different extensions of the online setting in which the online algorithm is equipped with some form of a priori knowledge about the instance. In this work, we analyze the influence of knowing an isomorphic copy of the input instance, which is also called an *unlabeled map*. With the unlabeled map, the algorithm is able to recognize unique structures while the online instance is revealed – like a vertex with unique degree – but it cannot distinguish isomorphic vertices or subgraphs.

The relation between the online algorithm and the adversary corresponds to players in an asymmetric two-player game, in which the algorithm wants to maximize its performance and the adversary's goal is to minimize it. The unlabeled map can be considered as the game board. One turn of the game consists of a move by the adversary followed by a move of the online algorithm. Thereby, the adversary reveals a vertex together with its neighbors and the online algorithm has to irrevocably decide whether to include this vertex in the solution or not. The problem is to evaluate whether the online algorithm has a *winning strategy*, that is, it is able to compute a solution of size smaller/greater or equal to the desired solution size k, for all possible adversary strategies.

Papadimitriou and Yannakakis make use of the connection between games and online algorithms for analyzing the canadian traveler problem in [13], which is an online problem where the task is to compute a shortest s-t-path in an a priori known graph in which certain edges can be removed by the adversary. They showed that the computational problem of devising a strategy that achieves a certain competitive ratio is *PSPACE*-complete by giving a reduction from TRUE QUANTIFIED BOOLEAN FORMULA, short TQBF.

Independently, Halldórsson [8] introduced the problems online coloring and online independent set on a priori known graphs, which is equivalent to having an unlabeled map. He studies how the competitive ratios improve compared to the model when the graph is a priori not known. Based on these results, Halldórsson et al. [9] continued the work on the online independent set problem without a priori knowing the graph. These results are then applied by Boyar et al. [3] to derive a lower bound for the advice complexity of the online independent set problem. Furthermore, they introduce the class of asymmetric online covering problems (AOC) containing ONLINE VERTEX COVER, ONLINE INDEPENDENT SET, ONLINE DOMINATING SET and others. Boyar et al. [4] analyze the complexity of these problems as graph property, namely the online vertex cover number, online independence number and online domination number, by showing their *NP*-hardness.

Moreover based on the work by Halldórsson [8], Kudahl [12] shows *PSPACE*-completeness of the decision problem Online Chromatic Number with Precoloring on an a priori known graph, which asks whether some online algorithm is able to color G with at most k colors for every possible order in which G is presented while having a precolored part in G. This approach is then improved by Böhm and Veselý [2] by showing that Online Chromatic Number is *PSPACE*-complete by giving a reduction from TQBF.

Our contribution is to analyze the computational complexity of a subclass of AOC problems that consider graph problems where the solution is a subset of the vertices. Similar to the problem Online Chromatic Number, we equip the online algorithm with an unlabeled map in order to apply and formalize the ideas of Böhm and Veselý. We call these problems *online vertex subset games* due to their relation to two-player games. While symmetrical combinatorial two-player games are typically *PSPACE*-complete [5], this principle does not apply to our asymmetrical setting. We are still able to prove *PSPACE*-completeness for the online vertex subset games based on Vertex Cover, Independent Set and Dominating Set by designing reductions such that the adversary's optimal strategy corresponds to the optimal strategy of the $\forall$-player in TQBF.

In order to derive reductions from TQBF to online vertex subset games, we identify properties describing the revelation or concealment of information to correctly simulate the $\forall$- and $\exists$-decisions as well as the evaluation of the quantified Boolean formula in the online vertex subset game. This simulation is modeled by disjoint and modular gadgets, which form a so-called gadget reduction – similar to already known reductions between NP-complete problems. Different forms of gadget reductions are described by Agrawal et al. [1] who formalize AC^0-gadget-reductions in the context of *NP*-completeness and by Trevisan et al. [15] who describe gadgets in reductions of problems that are formalized as linear programs. By formalizing gadgets capturing the above mentioned properties, we provide a framework to derive reductions for other online vertex subset games, which are based on problems that are gadget-reducible from 3-Satisfiability.

Paper Outline. First, we explain the online setting that we use throughout the paper and important terms, e.g., the online game, the problem class and the reveal model of our online problems. Secondly, we define the gadget reduction framework to reduce 3-Satisfiability to vertex subset problems. In the third section, we extend the framework to the online setting by identifying a set of important properties that must be fulfilled in the reduction. We also provide a scheme for gadgets that enforce these properties to generalize the framework to arbitrary vertex subset problems. In the fourth section, we detail the application of this framework to the problem Vertex Cover. Lastly, we apply the framework to the problems Independent Set and Dominating Set in the fifth section. At the end, we summarize the results and give a prospect on future possible work.

Neighborhood Reveal Model. Each request of the online problem reveals information about the instance for the online algorithm. The amount of information in each step is based on the reveal model. For an online problem with a map, the subgraph that arrives in one request is called *revelation subgraph.*

The neighborhood reveal model, which we use in this paper, was introduced by Harutyunyan et al. [10]. Within that model, the online algorithm gains information about the complete neighborhood of the revealed vertex. Nevertheless, the online algorithm has to make a decision on the current revealed vertex only but not on the exposed neighborhood vertices. All exposed but not yet revealed neighborhood vertices have to be revealed in the process of the online problem such that a decision can be made upon them. We denote the closed neighborhood of v with $N[v]$, that is, the set of v and all vertices adjacent to v.

Definition 1 (Neighborhood Reveal Model). *The neighborhood reveal model is defined by an ordering of graphs $(V_i, A_i, E_i)_{i \leq |V|}$. The reveal order of the adversary is defined by $adv \in S_{|V|}$, where $S_{|V|}$ is the symmetric group of size $|V|$. The graph G_i is defined by*

$$\begin{aligned} V_0 &= E_0 = \emptyset, \\ V_i &= V_{i-1} \cup N[v_{adv(i)}], && \text{for } 0 < i \leq |V| \\ E_i &= E_{i-1} \cup \{(v_{adv(i)}, w) \in E \mid w \in V_i\}, && \text{for } 0 < i \leq |V|. \end{aligned}$$

The revelation subgraph G' in the neighborhood reveal model is the subgraph of G_i defined by $G' = (V', E')$ with $V' = N[v_{adv(i)}]$ and $E' = \{\{v_{adv(i)}, w\} \in E\}$. The online algorithm has to decide whether $v_{adv(i)}$ is in the solution or not.

Online Vertex Subset Games. Throughout the paper, we consider a special class of combinatorial graph problems. The question is to find a vertex subset, whereby the size should be either smaller or equals, for minimization problems, or greater or equals, for maximization problems, some k, which is part of the input. Thereby, the vertex set needs to fulfill some constraints based on the specific problem. We call these problems vertex subset problems. Well-known problems like Vertex Cover, Independent Set and Dominating Set are among them.

We denote the online version with a map of a vertex subset problem P^{VS} with P^{VS}_o and define them as follows.

Definition 2 (Online Vertex Subset Game). *An online vertex subset game P^{VS}_o has a graph G and a $k \in \mathbb{N}$ as input. The question is, whether the online algorithm is able to find a vertex set of size smaller (resp. greater) or equals k, which fulfills the constraints of P^{VS} for all strategies of the adversary. Thereby, the online algorithm has access to an isomorphic copy of G and the adversary reveals the vertices according to the neighborhood reveal model.*

2 Gadget Reductions

Gadget reductions are a concept to reduce combinatorial problems in a modular and structured way. For the context of the paper, we define gadget reductions from 3-Satisfiability to vertex subset problems. The 3-Satisfiability instances $\varphi = (L, C)$ are defined by their literals L and their clauses C. We use a literal vertex v_ℓ for all $\ell \in L$ to represent a literal in the graph. There are implicit relations over the literals besides the explicit relation C, in that the reduction may be decomposed. For example, the relation between a literal and its negation, which is usually implicitly used to build up a *variable gadget*. These variable gadgets are connected by graph substructures that assemble the clauses as *clause gadgets*.

Definition 3 (Gadget Reduction from 3-Satisfiability to Vertex Subset Problems). *A gadget reduction $R_{gadget}(P^{VS})$ from a* 3-SATISFIABILITY *formula $\varphi = (L, C)$ to a vertex subset problem with graph $G_\varphi = (V, E)$ is a tuple containing functions from the literal set and all relations of the* 3-SATISFIABILITY *formula to the vertex set and all relations of the vertex subset problem. In the following, we denote the gadget based on element x to be $G_x := (V_x, E_x)$ with V_x being a set of vertices and E_x a set of edges, whereby the edges are potentially incident to vertices of a different gadget.*

The literal set of a 3-SATISFIABILITY *formula $\ell_1, \ell_2, \ldots, \ell_{|L|-1}, \ell_{|L|}$ is mapped to the vertex set of the graph problem. Thereby, each literal is mapped to exactly one vertex:*

$$R^{L\to V}_{gadget}(P^{VS}) : L \to V, \ell \mapsto G_\ell$$

The following relations on the literals are mapped as well.
(1) Literal - Negated Literal: $R^{L,\overline{L}}_{gadget}(P^{VS}) : R(L, \overline{L}) \to (V, E), (\ell, \overline{\ell}) \mapsto G_{\ell,\overline{\ell}}$
(2) Clause: $R^{C}_{gadget}(P^{VS}) : R(C) \to (V, E), C_j \mapsto G_{C_j}$
(3) Literal - Clause: $R^{L,C}_{gadget}(P^{VS}) : R(L, C) \to (V, E), (\ell, C_j) \mapsto G_{\ell,C_j}$
(4) Negated Literal - Clause: $R^{\overline{L},C}_{gadget}(P^{VS}) : R(\overline{L}, C) \to (V, E), (\overline{\ell}, C_j) \mapsto G_{\overline{\ell},C_j}$
Additionally, the following mapping allows for constant parts that do not change depending on the instance: $R^{const}_{gadget}(P^{VS}) : \emptyset \to (V, E), \emptyset \mapsto G_{const}$. Thereby, the vertices and edges of all gadgets are pairwise disjoint.

We use the more coarse grained view of *variable gadgets* as well. These combine the mappings $R^{L\to V}_{gadget}$ and $R^{L,\overline{L}}_{gadget}$ to R^{X}_{gadget}, and $R^{L,C}_{gadget}$ and $R^{\overline{L},C}_{gadget}$ to $R^{X,C}_{gadget}$, where X is the set of n variables.

The important function of the variable gadget is to ensure that only one of the literals of $\ell, \overline{\ell} \in L$ is chosen. On the other hand, the function of the clause gadget is to ensure together with the constraints of the vertex subset problem P^{VS} that the solution encoded on the literals fulfill the 3-SATISFIABILITY-formula if and only if the literals induce a correct solution. These functionalities are utilized in the correctness proof of the reduction by identifying the logical dependencies between the literal vertices v_ℓ for $\ell \in L$ and all other vertices based on the graph and the constraints of P^{VS} together with combinatorial arguments on the solution size. We denote these logical dependencies as *solution dependencies* as they are logical dependencies on the solutions of P^{VS}. Due to the asymmetric nature of the online problems, the adversary can reveal a solution dependent vertex before revealing the corresponding literal vertex. Thus, a decision on the solution dependent vertex is implicitly also a decision on the literal vertex, although it has not been revealed. We address this specific problem later in the description of the framework.

Definition 4 (Solution dependent vertices). *Given a gadget reduction, the following vertices of the reduction graph are solution dependent:*

1. *All literal vertices are solution dependent on their respective variable.*
2. *For a literal ℓ (resp. its negation $\overline{\ell}$), we denote the set of vertices that need to be part of the solution if v_ℓ (resp. $v_{\overline{\ell}}$) is part of the solution with V_ℓ (resp. $V_{\overline{\ell}}$). Then the vertices, that are in one but not both of these sets, i.e. $V_\ell \triangle V_{\overline{\ell}}$, are solution dependent on the corresponding variable.*

All vertices that are not solution dependent on any variable are called solution independent.

For example, in the reduction from 3-Satisfiability to Vertex Cover [7], the following solution dependencies apply: For each literal, the vertices v_ℓ and $v_{\overline{\ell}}$ are solution dependent on their respective variable. Furthermore, if a literal is part of the solution, all clause vertices representing its negation must also be part of the solution. Thus all clause vertices are solution dependent on their respective variable. Consequently, all vertices of the reduction graph for vertex cover are solution dependent.

3 A Reduction Framework for Online Vertex Subset Games

In this section, we present a general framework for reducing TQBF Game to an arbitrary online vertex subset game P_o^{VS} with neighborhood reveal model. The TQBF Game is played on a fully quantified Boolean formula, where one player decides the $\exists$-variables and the other decides the $\forall$-variables, in the order they are quantified. Deciding whether the $\exists$-player has a winning strategy is *PSPACE*-complete [14], and thus this reduction proves *PSPACE*-hardness for P_o^{VS}. We assume that the TQBF Game consists of clauses with at most three literals, which is also known to be *PSPACE*-complete [7].

Before we describe the reduction, we prove that the online game version of each vertex covering graph problem in *NP* is in *PSPACE*.

Theorem 1. *If P^{VS} is in NP, then P_o^{VS} is in PSPACE.*

This framework uses an (existing) gadget reduction of the vertex subset problem P^{VS} from 3-Satisfiability and extends it in order to give the online algorithm the ability to recognize the current revealed vertex. Due to the quantification of variables, we call the variable gadget of a $\forall$-variable a $\forall$-gadget (resp. $\exists$-gadget for an $\exists$-variable). Based on this, the online algorithm can use a one-to-one correspondence between the solution of the TQBF Game instance and the P_o^{VS} instance. The one-to-one correspondence between the $\forall$-variables and the $\forall$-gadgets is ensured by the knowledge of the adversary about the deterministic online algorithm. It simulates the response of the algorithm on the $\forall$-gadget.

Extension Gadgets. We extend the reduction graph G_φ of the offline problem with gadgets to a reduction graph for the online problem. These gadgets extend G_φ by connecting to a subset of its vertices. We denote these gadgets G_{ext} as *extension gadgets.*

Definition 5 (Graph Extension). *A graph extension of a graph $G = (V, E)$ by an extension gadget $G_{ext} = (V_{ext}, E_{ext}, E_{con})$ with the set of connecting edges $E_{con} \subseteq V \times V_{ext}$ is defined as $H = G \circ G_{ext}$, whereby*

$$V(H) = V \cup V_{ext},$$
$$E(H) = E \cup E_{ext} \cup E_{con}.$$

We further define $G \bigcirc_{i \in I} G^i_{ext} := (\ldots((G \circ G^{i_1}_{ext}) \circ G^{i_2}_{ext}) \circ \ldots) \circ G^{i_{|I|}}_{ext}$.

We also need the notion of *self-contained* gadgets. These do not influence the one-to-one correspondence between solutions of the online vertex subset game P_o^{VS} and TQBF GAME. In other words, optimal solutions on the graph and the extension gadget can be disjointly merged to obtain an optimal solution on the extended graph. Due to this independence, we are able to provide local information to the online algorithm via the map without changing the underlying formula. An example for self-contained extension gadgets is provided in Fig. 1. Note that, it can occur that self-containment depends on the extended graph.

Fig. 1. On the left, there is an example for an extension gadget that is self-contained w.r.t. the dominating set problem: No matter the solution on G, at least one vertex of G_{ext} has to be chosen. Additionally, choosing the black vertex of G_{ext} dominates all vertices attached to G, and thus any solution on G remains valid. On the right, there is an example for an extension gadget that is not self-contained w.r.t. the dominating set problem: If the solution on G contains the black vertices, it is also a solution for H, but the optimal solution on G_{ext} contains one vertex.

For our reduction framework, we introduce three types of self-contained extension gadgets: fake clause gadgets, dependency reveal gadgets and ID gadgets. The goal of these gadgets is that it is optimal for the adversary to reveal variables in the order of quantification, and that the online algorithm is able to assign the value of the $\exists$-variables, while the adversary is able to assign the value of the $\forall$-variables.

Fake Clause Gadgets. The number of occurrences of a certain literal in clauses is information that may allow the online algorithm to distinguish the literals of some $\forall$-variables, allowing the online algorithm to decide the assignment instead of the adversary. To avoid this information leak, we add gadgets for all possible non-existing clauses to the reduction graph. A *fake clause gadget* is only detectable if and only if a vertex, which is part of that clause gadget, is revealed by the adversary. The gadget needs to be self-contained such that the one-to-one correspondence between the solutions of the P_o^{VS} and TQBF GAME is not affected.

Definition 6 (Fake Clause Gadget). *A fake clause gadget $G_{fc}(C_j')$ for a non-existing clause $C_j' \notin C$ is an extension gadget that is self-contained. The fake clause gadgets are connected to the variable gadgets like the clause gadgets are to the variable gadgets according to the original gadget reduction, see Definition 3.*

All fake clause gadgets are pairwise disjoint. Let G_φ be the gadget reduction graph and

$$G_\varphi' := G_\varphi \bigcirc_{C_j' \notin C} G_{fc}(C_j').$$

After adding fake clause gadgets for all clauses $C_j' \notin C$ to G_φ, the revelation subgraphs of all vertices $v \in V(G_\varphi')$, which are part of a literal gadget, are pairwise isomorphic.

Dependency Reveal Gadgets. The two functions of the dependency reveal gadgets are that the adversary chooses the reveal order to be the order of quantification and the online algorithm knows the decision on the $\forall$-variables after the decision is made by the adversary. If the adversary deviates from the quantification order, the $\forall$-decision degenerates to an $\exists$-decision for the online algorithm. On the other hand, since the adversary forces the online algorithm to blindly choose the truth value of a $\forall$-quantified variable, the online algorithm does not know the chosen truth value. Thus, we need to reveal the truth value to the online algorithm whenever a solution dependent vertex of the next variables is revealed.

Definition 7 (Dependency Reveal Gadget). *A dependency reveal gadget $G_{dr}(x_i)$ for $\forall$-variable x_i is an extension gadget that is self-contained with the property: Let $\ell, \overline{\ell}$ be the literals of x_i. If a solution dependent vertex of x_j with $j \geq i$ is revealed to the online algorithm, the online algorithm is able to uniquely identify the vertices v_ℓ and $v_{\overline{\ell}}$.*

ID Gadgets. At last, the online algorithm needs information on the currently revealed vertex to identify it with the help of the map. For this, we introduce ID gadgets, which make the revelation subgraph of vertices distinguishable to a certain extent. Thus, the online algorithm is able to correctly encode the TQBF solution into the solution of the vertex subset game. The ID gadget is always connected exactly to the vertex it identifies, thus they are pairwise disjoint.

Definition 8 (ID Gadget). *An identification gadget $G_{id}(v)$ is a self-contained extension gadget connected to v such that the revelation subgraph of v is isomorphic to revelation subgraphs of vertices within a distinct vertex set $V' \subseteq V$.*

The General Reduction for Online Vertex Subset Games. With the gadget schemes defined above, we are able to construct a gadget reduction from TQBF GAME to P_o^{VS}. The idea of the reduction is to construct the optimal game strategy for the online algorithm to compute the solution to the TQBF GAME formula. Furthermore, encoding the solution to the TQBF GAME into the P_o^{VS} instance is a winning strategy by using the equivalence of the ∃- and ∀-gadgets to the ∃- and ∀-variables. At last, there is a one-to-one correspondence between the reduction graph solution of P^{VS} and the 3-SATISFIABILITY-solution.

A gadget reduction from 3-SATISFIABILITY to vertex subset problem P^{VS} can be extended such that P_o^{VS} is reducible from TQBF GAME as follows. Recall that G_φ is the gadget reduction graph of a fixed but arbitrary instance of P^{VS}.

1. Add fake clause gadgets for all clauses that are not in the TQBF GAME instance

$$G'_\varphi = G_\varphi \bigcirc_{c' \notin C} G_{fc}(c') \; .$$

2. Add dependency reveal gadgets for all ∀-variables x

$$G''_\varphi = G'_\varphi \bigcirc_{\substack{x \in X \\ x \text{ is } \forall}} G_{dr}(x) \; .$$

3. Add ID gadgets to all vertices

$$G'''_\varphi = G''_\varphi \bigcirc_{v \in V(G''_\varphi)} G_{id}(v) \; .$$

Then, if all gadgets can be constructed in polynomial time, G'''_φ is the corresponding reduction graph of P_o^{VS}. The gadget reduction also implies the following gadget properties, which individually have to be proven for a specific problem.

1. The fake clause gadgets are self-contained.
2. The dependency reveal gadgets are self-contained.
3. The ID gadgets are self-contained.
4. In G'''_φ, each solution dependent vertex which is not in a literal gadget of a ∀-variable has a unique revelation subgraph.
5. In G'''_φ, the two literal vertices of a ∀-variable have the same revelation subgraph, but different from vertices of any other gadget.
6. In G'''_φ, each vertex that is solution independent or part of an extension gadget has a revelation subgraph that allows for an optimal decision.

From the above construction, the following Lemmas 1 to 3, are fulfilled such that P_o^{VS} is proven to be *PSPACE*-hard in the following Theorem 2.

Lemma 1. *In the construction of the reduction, there is a one-to-one correspondence between the solution of the problem P_o^{VS} and* TQBF GAME*, if there is a one-to-one correspondence between the solutions in the gadget reduction from P^{VS} and* 3-SATISFIABILITY*. The equivalence is computable in PTIME.*

In the following, we show that the adversary has to reveal one literal vertex of each variable gadget before revealing vertices of other gadgets (except ID gadgets). Furthermore, the adversary has to adhere to the quantification order of the variables when revealing the first literal vertices of each gadget. If the adversary deviates from this strategy, it may allow the online algorithm to decide the truth assignment of ∀-variables. This may allow the algorithm to win a game based on an unsatisfiable formula. Thus, an optimal adversary strategy always follows the quantification order.

Lemma 2. *Every optimal game strategy for the adversary adheres to the reveal ordering*

$$G_{\ell_1} \text{ or } G_{\overline{\ell}_1} < G_{\ell_2} \text{ or } G_{\overline{\ell}_2} < \cdots < G_{\ell_n} \text{ or } G_{\overline{\ell}_n}, \tag{1}$$

$$G_{\ell} \text{ or } G_{\overline{\ell}} < G_c(C_j), \qquad \text{for all } \ell \in C_j \in C, \tag{2}$$

$$G_{\ell} \text{ or } G_{\overline{\ell}} < G_{fc}(C_j'), \qquad \text{for all } \ell \in C_j' \notin C, \tag{3}$$

$$G_{\ell} \text{ or } G_{\overline{\ell}} < G_{dr}(x), \qquad \text{for all } x \in X. \tag{4}$$

Lemma 3. *The vertex assignments of an ∃-variable gadget (resp. ∀-variable gadget) are equivalent to the decision of the ∃-player (resp. ∀-player) on an ∃-quantifier (resp. ∀-quantifier) in the* TQBF GAME. *The equivalence is computable in PTIME.*

Therefore, the solutions to the formula in the TQBF GAME and the solutions to the online vertex subset game are equivalent. Thus, the reduction graph G_φ''' is a valid reduction from TQBF GAME because the one-to-one correspondence between solutions is preserved, which concludes the proof of Theorem 2. At last, the online algorithm is able to win the game if and only if the TQBF GAME is winnable.

Theorem 2. *If P^{VS} is gadget reducible from* 3-SATISFIABILITY *and Lemmas 1 to 3 hold, then P_o^{VS} is PSPACE-complete.*

4 Vertex Cover

In this section, we use our reduction framework to show that the online vertex subset game based on the VERTEX COVER problem, the ONLINE VERTEX COVER GAME, is *PSPACE*-complete. VERTEX COVER was originally shown to be *NP*-complete by Karp [11] with a reduction from CLIQUE. However, since our reduction framework extends reductions from 3-SATISFIABILITY, we use an alternative reduction from Garey and Johnson [7].

Let φ be the 3-SATISFIABILITY-formula, let X be the set of n variables and let C be the set of m clauses of φ. We construct the following graph $G_\varphi = (V, E)$: For each variable x_i, introduce a variable gadget consisting of two vertices, connected by an edge. One of these vertices represents the positive literal, while the other represents the negative literal. Thus we refer to these vertices as literal vertices.

For each clause C_j, $J \in \{1, \ldots, m\}$, we construct a clause gadget, which is a triangle of vertices, where each vertex represents one of the literals in C_j. Finally, each vertex of a clause is connected to the literal it represents. An example of this construction is shown in Fig. 2.

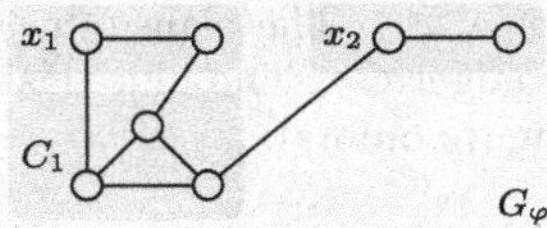

Fig. 2. The reduction graph for the reduction from 3-SATISFIABILITY to VERTEX COVER for instance $\varphi = (x_1 \vee \overline{x}_1 \vee x_2)$.

The dependencies in G_φ are of the type if a literal vertex is not contained in a solution, then all clause vertices representing the same literal must be contained in that solution. Therefore, all vertices in G_φ are solution dependent.

The ONLINE VERTEX COVER GAME has a graph G and a $k \in \mathbb{N}$ as input. It asks whether there is a winning strategy for the online algorithm, that is, it finds a vertex cover of size at most k for every reveal order while knowing an isomorphic copy of G.

Theorem 3. *The* ONLINE VERTEX COVER GAME *with the neighborhood reveal model and a map is PSPACE-complete.*

The containment of ONLINE VERTEX COVER GAME in *PSPACE* is already established by Theorem 1. To show hardness, we extend the above reduction for VERTEX COVER according to our framework. Therefore, we need to introduce fake clause gadgets, dependency reveal gadgets, and ID gadgets and prove that they fulfill the gadget properties, required by Lemmas 1 to 3.

An example for a fake clause gadget is shown in Fig. 3. Any optimal vertex cover on the fake clause gadget has size 3 and contains exactly the triangle representing the clause. In the neighborhood reveal model, fake clause gadgets can not be distinguished from real clause gadgets, as long as only vertices of variable gadgets are revealed by the adversary. However, as soon as a vertex of the fake clause gadget is revealed, it can be distinguished from a real clause gadget, as the vertex degrees are different.

The dependency reveal gadget reveals the solution dependencies to the online algorithm. An example is depicted in Fig. 4a. Since both the literal vertices and the vertices of clause gadgets are solution dependent, the online algorithm needs to be able to identify which variable they correspond to, and in the case of $\exists$-variables also which literal they correspond to. For that, we look at the degrees of all vertices in the graph G''_φ. The optimal solution for the dependency reveal gadget always contains exactly the center vertex of the star.

Finally, we define ID gadgets for literal vertices and all vertices in clause gadgets as they are solution dependent. An example is presented in Fig. 4b. The

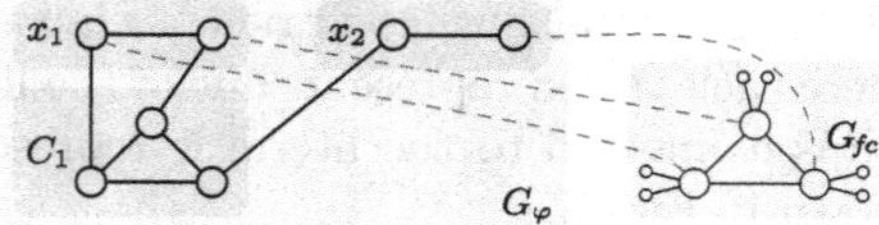

Fig. 3. The reduction graph for the reduction from 3-SATISFIABILITY to VERTEX COVER for instance $\varphi = (x_1 \vee \overline{x}_1 \vee x_2)$. The clause $(x_1 \vee \overline{x}_1 \vee \overline{x}_2)$ does not exist and is represented by a fake clause gadget G_{fc}. The blue dashed edges are the set E_{con} for the fake clause gadget. (Color figure online)

task of the ID gadget is to enable the algorithm to uniquely identify vertices of $\exists$-quantified literals and solution dependent vertices as well as to identify the $\forall$-quantified literals such that the literal vertices of one $\forall$-quantified variable are indistinguishable.

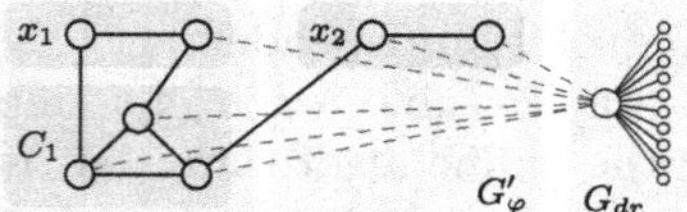

(a) The dependency reveal gadget for the $\forall$-variable x_1, with only one variable of higher index, is depicted. The fake clause gadgets of G'_φ are omitted.

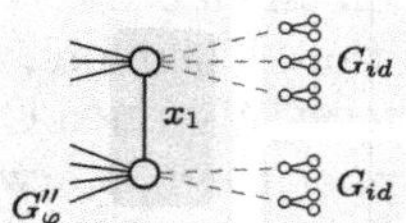

(b) The ID gadgets for the $\forall$-variable x_1 are shown. Both literal vertices have the same degree. For each gadget, the blue dashed edges are the set E_{con}.

Fig. 4. Dependency reveal gadget and ID gadget for VERTEX COVER.

Since all constructions are polynomial time computable, we established the requirements for Theorem 2, thus Theorem 3 is proven. The full construction of G'''_φ is shown in Fig. 5.

5 More Vertex Subset Problems

In this section, we apply Theorem 2 to the ONLINE INDEPENDENT SET GAME and ONLINE DOMINATING SET GAME. Like the ONLINE VERTEX COVER GAME, they take a graph G and a number $k \in \mathbb{N}$ as input. They ask whether there is a winning strategy for the online algorithm, that is, it finds an independent set (resp. dominating set) of size at least (resp. most) k for every reveal order while knowing an isomorphic copy of G.

Theorem 4. *The* ONLINE INDEPENDENT SET GAME *with the neighborhood reveal model and a map is PSPACE-complete.*

Theorem 5. *The* ONLINE DOMINATING SET GAME *with the neighborhood reveal model and a map is PSPACE-complete.*

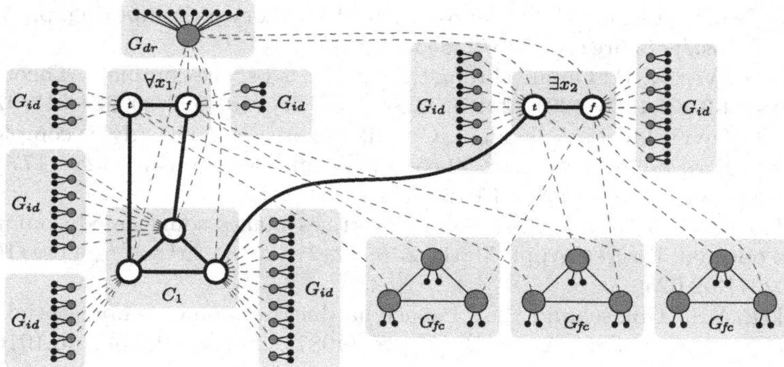

Fig. 5. Complete view on the reduction for the TQBF-instance $\forall x_1 \exists x_2\,(x_1 \vee \overline{x}_1 \vee x_2)$ to ONLINE VERTEX COVER GAME. The thick vertices and edges represent the original reduction. The blue dashed edges are the connecting edges of the extension gadgets. There are optimal solutions that contain all the gray vertices and none of the black vertices. Whether the white vertices are contained depends on the feasible solutions for the TQBF-formula.

6 Conclusion

We derived online games from the typical online setting in order to analyze their computational complexity. Furthermore, we developed a framework for online versions of vertex subset problems with neighborhood reveal model that allows reductions from TQBF GAME to show that these are *PSPACE*-complete. We showed particularly that the online versions VERTEX COVER, INDEPENDENT SET and DOMINATING SET with neighborhood reveal model are *PSPACE*-complete.

The gap between the complexity analysis that started with the *NP*-hardness for the problems from AOC under the vertex arrival model and our *PSPACE*-completeness results need to be closed. From our results the questions arises if the three problems VERTEX COVER, INDEPENDENT SET and DOMINATING SET are actually *PSPACE*-complete under the vertex arrival model. One way to show the *PSPACE*-completeness is to use our reduction framework and add a type of error correction gadget. However, the missing knowledge in the vertex arrival model might increase the asymmetry in favor of the adversary such that the complexity decreases and it remains *NP*-hard.

Additionally, the presented framework may be extended to more general subset problems from AOC where the solution is not a vertex subset.

References

1. Agrawal, M., Allender, E., Impagliazzo, R., Pitassi, T., Rudich, S.: Reducing the complexity of reductions. In: Leighton, F.T., Shor, P.W. (eds.) Proceedings of the

Twenty-Ninth Annual ACM Symposium on the Theory of Computing, pp. 730–738 (1997). https://doi.org/10.1145/258533.258671

2. Böhm, M., Veselý, P.: Online chromatic number is pspace-complete. Theory Comput. Syst. **62**(6), 1366–1391 (2018). https://doi.org/10.1007/s00224-017-9797-2
3. Boyar, J., Favrholdt, L.M., Kudahl, C., Mikkelsen, J.W.: The advice complexity of a class of hard online problems. Theory Comput. Syst. **61**(4), 1128–1177 (2017). https://doi.org/10.1007/s00224-016-9688-y
4. Boyar, J., Kudahl, C.: Adding isolated vertices makes some greedy online algorithms optimal. Discret. Appl. Math. **246**, 12–21 (2018). https://doi.org/10.1016/j.dam.2017.02.025
5. Fraenkel, A.S., Goldschmidt, E.: Pspace-hardness of some combinatorial games. J. Comb. Theory Ser. A **46**(1), 21–38 (1987). https://doi.org/10.1016/0097-3165(87)90074-4
6. Fuchs, J., Grüne, C., Janßen, T.: The complexity of online graph games (2023)
7. Garey, M.R., Johnson, D.S.: Computers and Intractability: A Guide to the Theory of NP-Completeness. W. H. Freeman (1979)
8. Halldórsson, M.M.: Online coloring known graphs. Electron. J. Comb. **7** (2000). https://doi.org/10.37236/1485
9. Halldórsson, M.M., Iwama, K., Miyazaki, S., Taketomi, S.: Online independent sets. Theor. Comput. Sci. **289**(2), 953–962 (2002). https://doi.org/10.1016/S0304-3975(01)00411-X
10. Harutyunyan, H.A., Pankratov, D., Racicot, J.: Online domination: the value of getting to know all your neighbors. In: Bonchi, F., Puglisi, S.J. (eds.) 46th International Symposium on Mathematical Foundations of Computer Science, MFCS, vol. 202, pp. 57:1–57:21 (2021). https://doi.org/10.4230/LIPIcs.MFCS.2021.57
11. Karp, R.M.: Reducibility among combinatorial problems. In: Miller, R.E., Thatcher, J.W. (eds.) Complexity of Computer Computations. IRSS, pp. 85–103. Springer, Cham (1972). https://doi.org/10.1007/978-1-4684-2001-2_9
12. Kudahl, C.: Deciding the on-line chromatic number of a graph with pre-coloring is PSPACE-complete. In: Paschos, V.T., Widmayer, P. (eds.) CIAC 2015. LNCS, vol. 9079, pp. 313–324. Springer, Cham (2015). https://doi.org/10.1007/978-3-319-18173-8_23
13. Papadimitriou, C.H., Yannakakis, M.: Shortest paths without a map. Theor. Comput. Sci. **84**(1), 127–150 (1991). https://doi.org/10.1016/0304-3975(91)90263-2
14. Stockmeyer, L.J., Meyer, A.R.: Word problems requiring exponential time: preliminary report. In: Aho, A.V., et al. (eds.) Proceedings of the 5th Annual ACM Symposium on Theory of Computing, pp. 1–9 (1973). https://doi.org/10.1145/800125.804029
15. Trevisan, L., Sorkin, G.B., Sudan, M., Williamson, D.P.: Gadgets, approximation, and linear programming (extended abstract). In: 37th Annual Symposium on Foundations of Computer Science, pp. 617–626 (1996). https://doi.org/10.1109/SFCS.1996.548521

Removable Online Knapsack with Bounded Size Items

Laurent Gourvès[1(✉)] and Aris Pagourtzis[2,3]

[1] Université Paris Dauphine-PSL, CNRS, LAMSADE, 75016 Paris, France
laurent.gourves@dauphine.fr

[2] National Technical University of Athens, 15780 Zografou, Greece
pagour@cs.ntua.gr

[3] Archimedes Research Unit, Athena RC, 15125 Marousi, Greece

Abstract. In the online unweighted knapsack problem, some items arrive in sequence and one has to decide to pack them or not into a knapsack of given capacity. The objective is to maximize the total size of packed items. In the traditional setting, decisions are irrevocable, and the problem cannot admit any ρ-competitive algorithm. The removable nature of the items allows to withdraw previously packed elements of the current solution. This feature makes the online knapsack problem amenable to competitive analysis under the ratio of $(\sqrt{5}-1)/2$, which is at the same time the best possible performance guarantee [12]. This article deals with refinements of the best possible competitive ratio of the online unweighted knapsack problem with removable items when either an upper or a lower bound on the size of the items is known.

Keywords: Online Algorithms · Knapsack · Competitive analysis

1 Introduction

This article deals with online computation where an instance is revealed over time and an irrevocable decision has to be made each time a portion of the input is known [1]. The problem under study in this work is *knapsack* [14]. In its online version, we are given the capacity C of a knapsack, and items are disclosed sequentially. Each item has a positive size and a positive weight. The goal is, as usual, to pack items into the knapsack so as to maximize their total weight, under the constraint that the total size of packed items does not exceed C. The online knapsack problem is appealing since it models many real life situations, and many articles have been devoted to it (see for example [2–5, 7, 12, 13, 15, 16]). Unfortunately, one can rapidly observe that no deterministic algorithm can exhibit a bounded competitive ratio, which is a worst case performance guarantee of an online algorithm against an ideal procedure which always makes

Aris Pagourtzis has been partially supported for this work by project MIS 5154714 of the National Recovery and Resilience Plan Greece 2.0 funded by the European Union under the NextGeneration EU Program.

H. Fernau et al. (Eds.): SOFSEM 2024, LNCS 14519, pp. 283–296, 2024.
https://doi.org/10.1007/978-3-031-52113-3_20

the optimal decision. This bad news holds even in the unweighted case (a.k.a. *subset sum*) where the weight of every item is equal to its size. Indeed, think of an unweighted instance where the first item is very small, say $\epsilon \cdot C$ with $1 \gg \epsilon > 0$. If the item is not packed, and no other item is disclosed, then the competitive ratio is $0/(\epsilon \cdot C) = 0$. If the item is packed, but a second item of size C arrives, then the first item prevents the second one from being packed. The competitive ratio is $(\epsilon \cdot C)/C = \epsilon$ because the right decision would be to pack the second item. In any case, the competitive ratio is at most ϵ, which can be arbitrarily small.

Then, how can we bypass this difficulty? Quoting the authors of [14], we need "*to find suitable restrictions on the pure online formulation which make sense from a real-world point of view and permit the construction and analysis of more successful algorithms.*" Many successful directions have been proposed for knapsack, including the notion of *removable items* introduced by Iwama and Taketomi [12]. An item is removable if its insertion in the knapsack is a revocable decision. In this setting, a deterministic $\frac{\sqrt{5}-1}{2}$-competitive online algorithm exists, along with a proof that $\frac{\sqrt{5}-1}{2}$ is the best possible competitive ratio [12].

In the present work, we aim at going further and combine item removability with an additional information on item sizes in order to devise a more accurate competitive analysis. The previously mentioned example comprises two "extreme" items whose size relative to the capacity is either very small or very big ($\epsilon \cdot C$ and C, respectively). Such an instance may poorly represent the full spectrum of concrete situations that require to solve an online knapsack problem. In practice, one can expect to deal with instances where the items' sizes are less heterogeneous. Moreover, especially when a decision maker has already faced several instances in the past, she or he may be knowledgeable of a bound on the size of every item. Therefore, we propose to exploit the existence and knowledge of such a bound towards a refined analysis of the best possible competitive ratio of deterministic algorithms for the online unweighted knapsack problem with removable items.

Setting and Contribution. As in [3,7,12,13], we consider the online unweighted knapsack (a.k.a. subset sum) problem. Consider a capacity C (a.k.a. budget) of 1 and some removable items disclosed online $o_1, o_2, \ldots$. Each item o_i has a size $|o_i| \in (0,1]$ and an identical weight of $|o_i|$. Assuming that $C = 1$ and $|o_i| \in (0,1]$ for all i is without loss of generality since dividing everything by C does not affect the competitive ratio. The number of items is not known in advance. Starting from an initial empty set S, an item o_i is revealed at every time step i (also called round), and one has to decide if o_i is packed in S or not. Elements inserted during previous rounds can be removed, but discarded items (i.e., items that were directly rejected or removed after their insertion) cannot be included afterwards. The value of S, denoted by $v(S)$ and defined as $\sum_{o_i \in S} |o_i|$, should never exceed the capacity. The objective is to maximize $v(S)$ when no more items are disclosed. The competitive ratio $\rho_{\mathcal{A}}$ of a deterministic online algorithm $\mathcal{A}$ is the worst case ratio between the value of the solution $S_{\mathcal{A}}$ built by $\mathcal{A}$ and the value of an optimal solution S^* which maximizes $v(S^*)$:

$\rho_{\mathcal{A}} := v(S_{\mathcal{A}})/v(S^*)$. The online unweighted knapsack problem with removable items admits a t-competitive online algorithm where

$$t := \frac{\sqrt{5}-1}{2} \approx 0.618 \tag{1}$$

is the golden ratio conjugate, and no deterministic online algorithm can be $(t+\epsilon)$-competitive for any positive ϵ [12].[1] Our contribution is to go beyond this result and propose refined bounds on the best competitive ratio based on known bounds on the size of the items. All the results of this article apply to *deterministic* algorithms for the removable online unweighted knapsack problem. In Sect. 2, some parameter $u \in (0, 1]$ such that any item's size is *upper bounded* by u is given, and we provide both lower and upper bounds on the best possible competitive ratio $\rho(u)$. In Sect. 3, a lower bound $\ell \in (0, 1]$ on the size of every item is known, and we characterize the best possible competitive ratio $\rho(\ell)$.

Due to space constraints, some technical elements are skipped and will be made available in an extended version of the article.

Related Work. The online knapsack problem was first studied by Marchetti-Spaccamela and Vercellis who made an average case analysis of the expected difference between the optimal and the approximate value [16]. A follow-up article, with a similar approach, is authored by Luecker [15].

Iwama and Taketomi introduced the notion of removable items and proved that the unweighted online knapsack problem admits a deterministic t-competitive algorithm, where t is the best possible performance guarantee [12]. Unfortunately, no competitive algorithm can exist for the online *weighted* knapsack with removable items [13]. Sometimes, removing a packed item comes with a cost which is equal to f times the item's size where $f > 1$ is a given *buyback factor*. The objective is to maximize the worth of packed items minus the cost paid for items which were removed after been packed. In this case, Han et al. studied the unweighted online knapsack problem [7] whereas Babaioff et al. considered a more general setting including the online weighted knapsack problem [2]. Their results include deterministic and randomized algorithms whose competitive ratio depends on f.

Other randomized algorithms for the online knapsack problem can be found in [5,8] when $f = 0$. For the weighted case (resp., unweighted case), the competitive ratio for removable items is between 0.5 and $\frac{e}{1+e} \approx 0.73$ (resp., between 0.7 and 0.8). When items cannot be removed, the best competitive ratio for the weighted and unweighted cases are 0 and 0.5, respectively.

In [13], the authors consider the online knapsack problem under the *resource augmentation* framework where the online knapsack is $R \geq 1$ times larger than its offline counterpart. Their results consist of bounds on the competitive ratio as a function of R for removable or weighted items. In the same vein, an alternative model supposes that the items can be placed in a buffer of size $K > 1$ before a selection of them is put in a knapsack of size 1 [10].

[1] Every competitive ratio of this article is in $[0, 1]$.

Sometimes, the items can be split and partially included in the knapsack. Han and Makino exploited this opportunity to derive a $k/(k+1)$-competitive online algorithm for the removable weighted case and a matching upper bound [11]. Here, k is the maximum number of times that an item can be cut.

In [4], Böckenhauer et al. explore the *advice complexity* of the online knapsack problem, where the goal is to evaluate the possible improvements on the competitive ratio provided by some additional information about the complete instance.

In a recent article by Böckenhauer et al. [3] on the online unweighted knapsack problem, the possibility to reserve an item (i.e., postpone the decision about it) is studied. Given $\alpha \in (0,1)$, reserving an item costs α times its value. This model is closely related to the one with removal cost [2,7] but, as opposed to "bought back" items, reserved items are not temporarily put into the knapsack. Thus, there is no hard capacity constraint. The authors characterize the best competitive ratio for all possible value of α [3].

The present work combines removable items with an additional information about the range of the items size. Having upper bounds on the worth of objects has already been done for analysing fair allocations of indivisible goods [6]. Concerning the online weighted knapsack problem, Babaioff et al. have used a parameter $\gamma \in (0,1]$ which restricts the size of any item to $\gamma \cdot C$. They gave a deterministic algorithm for the case $\gamma < 1/2$ whose competitive ratio tends to $1-2\gamma$ when the buyback factor goes to 0. The work of Chakrabarty et al. on the weighted knapsack problem also makes assumptions about both the size of objects and their weight-to-size ratio [17].

2 Upper Bounded Item Size

Given an upper bound $u \in (0,1]$ on the size of every item, we aim at determining the best possible competitive ratio $\rho(u)$ of deterministic online algorithms for the unweighted knapsack problem with removable items. Our results are depicted in Fig. 1. The lower (solid) and upper (dotted) bounds on $\rho(u)$ are non increasing functions u. One can observe that the competitive ratio is always above t, and the performance guarantee increases (and tends to 1) as the size of the items reduces. The curves of Fig. 1 meet for many values of u all over the interval $(0,1]$ but some gaps remain to be filled.

In this section, we begin with lower bounds on $\rho(u)$ obtained with a single parametric algorithm (cf. Sect. 2.1), followed by upper bounds on $\rho(u)$ induced by a family of instances (cf. Sect. 2.2).

2.1 Lower Bounds on the Competitive Ratio

Throughout our presentation we will say that an item is *rejected* if it is discarded immediately after its disclosure; if the item is first inserted in the solution and it is discarded at a later step then we say that it is *removed.* We consider a single

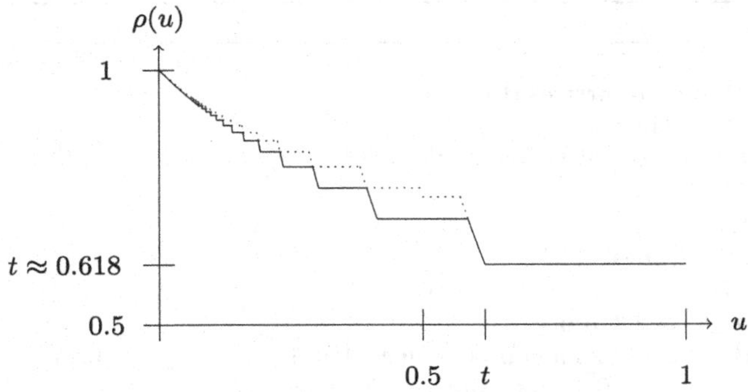

Fig. 1. Lower (solid) and upper (dotted) bounds on the competitive ratio ρ as a function of u.

parametric online algorithm described in Algorithm 1. For every positive integer k, the competitive ratio of Algorithm 1 is either a constant γ_k defined as

$$\gamma_k = \frac{k - 2 + \sqrt{k^2 + 4}}{2k}, \tag{2}$$

or a decreasing function of u equal to $\frac{1-u}{(k-1)u}$.

Note that γ_k belongs to $(0, 1]$ for all positive integers k. It increases with k, and $\gamma_1 = \frac{\sqrt{5}-1}{2} = t$. The rationale of γ_k originates from the following equality which will be interpreted and exploited later on.

$$(k-1)(1-\gamma_k) + \frac{1-\gamma_k}{\gamma_k} = \gamma_k \tag{3}$$

Other useful technical properties of γ_k (valid for all positive integers k) are the following four (their proof will appear in an extended version of this article).

$$0 < 1 - \gamma_k \leq \frac{1-\gamma_k}{\gamma_k} < 1/k \leq \frac{1-\gamma_k}{\gamma_k^2} \tag{4}$$

$$(k-1)\frac{1-\gamma_k}{\gamma_k^2} \leq 1 \tag{5}$$

$$(k+1)(1-\gamma_k) \geq \gamma_k \tag{6}$$

$$\frac{1 - \frac{1-\gamma_k}{\gamma_k^2}}{(k-1)\frac{1-\gamma_k}{\gamma_k^2}} = \gamma_k \tag{7}$$

Our analysis of the competitive ratio of Algorithm 1 is divided into two theorems for the sake of clarity. Theorems 1 and 2 correspond to intervals where the proposed lower bound on the competitive ratio is constant and decreasing with u, respectively. On Fig. 1, the i-th constant part of the solid curve, when counting from the right, corresponds to Theorem 1 when $i = k$ and $u \leq (1 -$

Algorithm 1

```
S ← ∅
while a new item o_i arrives do
    if v(S) ≥ γ_k then
        Reject o_i {S is not changed afterwards}
    else
        S ← S ∪ {o_i}
        if v(S) > 1 then
            Let B = {o ∈ S : |o| > 1 − γ_k}
            if v(B) > 1 then
                if B contains a subset Ŝ such that 1 ≥ v(Ŝ) ≥ γ_k then
                    S ← Ŝ {S is not changed afterwards}
                else
                    Remove the largest item of S
                end if
            else
                while v(S) > 1 do
                    Remove from S one element of S \ B {chosen arbitrarily}
                end while
            end if
        end if
    end if
end while
```

$\gamma_k)/\gamma_k^2$. The i-th decreasing part of the solid curve, still counting from the right, corresponds to Theorem 2 when $i = k-1$ and $u \leq 1/(k-1)$.

The solid curve depicted on Fig. 1 corresponds, for every possible value of $u \in (0,1]$, to the best (i.e., largest) lower bound offered by either Theorem 1 or Theorem 2, with an appropriate choice of the parameter k.

Theorem 1. *For all positive integers k, Algorithm 1 is γ_k-competitive when $u \in (0, \frac{1-\gamma_k}{\gamma_k^2}]$.*

Proof. Let B_i, $\hat{S}_i$, S_i, and S_i^* denote B, $\hat{S}$, S and an optimal solution at the end of round i of Algorithm 1 (i.e., when o_1 to o_i have been disclosed), respectively. We are going to prove that $v(S_i) \leq 1$ and $\frac{v(S_i)}{v(S_i^*)} \geq \gamma_k$ hold for all i.

An item o is said to be *small* if $0 < |o| \leq 1-\gamma_k$, *medium* if $1-\gamma_k < |o| \leq \frac{1-\gamma_k}{\gamma_k}$, *large* if $\frac{1-\gamma_k}{\gamma_k} < |o| \leq k^{-1}$, and *extra-large* if $k^{-1} < |o| \leq \frac{1-\gamma_k}{\gamma_k^2}$. Note that the extra-large category does not exist when $k = 1$. The categories lead to the following useful interpretation of Eq. (3): $k-1$ medium items plus one large or extra-large item constitute a desirable set because its total size is between $(k-1)(1-\gamma_k) + \frac{1-\gamma_k}{\gamma_k} = \gamma_k$ and $(k-1)\frac{1-\gamma_k}{\gamma_k} + \frac{1-\gamma_k}{\gamma_k^2} = 1$, i.e., it is a feasible set satisfying the guarantee γ_k.

The proof is by induction and we begin with the base case ($i = 1$). Since $S_1 = S_1^* = \{o_1\}$, we have that $v(S_1) \leq 1$ and $\frac{v(S_1)}{v(S_1^*)} = 1 \geq \gamma_k$. In order to prove

$v(S_i) \leq 1$ and $\frac{v(S_i)}{v(S_i^*)} \geq \gamma_k$ when $i > 1$, we make the induction hypotheses that both $v(S_{i-1}) \leq 1$ and $\frac{v(S_{i-1})}{v(S_{i-1}^*)} \geq \gamma_k$ hold.

If $v(S_{i-1}) \geq \gamma_k$ at line 4 of the algorithm, then $S_i = S_{i-1}$ (item o_i is rejected). Since $v(S_i^*) \leq 1$, the competitive ratio satisfies $\frac{v(S_i)}{v(S_i^*)} \geq \gamma_k$. The solution is not modified afterwards. Otherwise ($v(S_{i-1}) < \gamma_k$), the algorithm puts the new item o_i in the current solution (cf. line 6): $S_i = S_{i-1} \cup \{o_i\}$. If o_i fits (i.e., $v(S_i) \leq 1$ holds after the insertion of o_i), then $v(S_i) = v(S_{i-1}) + |o_i|$ and $v(S_i^*) \leq v(S_{i-1}^*) + |o_i|$. We get that $\frac{v(S_i)}{v(S_i^*)} \geq \frac{v(S_{i-1})+|o_i|}{v(S_{i-1}^*)+|o_i|} \geq \frac{v(S_{i-1})}{v(S_{i-1}^*)} \geq \gamma_k$ where the last inequality derives from the induction hypothesis.

From now on, suppose that $v(S_i) > 1$ holds after the insertion of o_i. We know from $v(S_{i-1}) < \gamma_k$ and $v(S_i) > 1$ that $|o_i| > 1 - \gamma_k$. In other words, o_i is not a small item. By construction, B_i is the subset of non-small items of S_i and it contains o_i. If $v(B_i) \leq 1$, then the algorithm executes the while loop containing line 17. The while loop starts with $v(S_i) > 1$ and removes small items of S_i until $v(S_i) \leq 1$. Since a small item has size at most $1 - \gamma_k$, we end up with a solution satisfying $\gamma_k \leq v(S_i) \leq 1$. Therefore, $\frac{v(S_i)}{v(S_i^*)} \geq \gamma_k$ and the solution is not modified afterwards. If $v(B_i) > 1$, then the algorithm tries to find in B_i a subset of non-small items $\hat{S}_i$ such that $1 \geq v(\hat{S}_i) \geq \gamma_k$, and sets S_i to $\hat{S}_i$ if $\hat{S}_i$ exists. In this case, the competitive ratio is reached and the solution is not modified afterwards.

Let us explain that verifying the existence of $\hat{S}_i$ is not difficult. B_i contains at most $k+1$ items for the following reasons: $|B_i| - 1$ non-small items of B_i were already present in S_{i-1}. Using $v(S_{i-1}) < \gamma_k$, and the fact that a non-small item has size at least $1 - \gamma_k$, we get that $|B_i| - 1 \leq k$ (indeed, Inequality (6) indicates that $k+1$ items of size $1 - \gamma_k$ have a total size of at least γ_k), which is equivalent to $|B_i| \leq k + 1$. Inequality (5) indicates that a set of $k - 1$ items of the largest possible size fits in the budget of 1. Together with $v(B_i) > 1$, we deduce that B_i cannot contain $k - 1$ (or less) non-small items. Thus, B_i contains k or $k + 1$ non-small items. Taking $k - 1$ items of B_i gives a feasible solution (cf. Inequality (5)). By taking the $k - 1$ biggest items of B_i, we can verify whether γ_k can be reached. If not, $\hat{S}_i$ possibly requires k items (when $|B_i| = k + 1$), which requires to test $k + 1$ possibilities. In all, at most $k + 2$ solutions are tested, where k is upper bounded by the number of disclosed items.

So far we have considered the cases where the algorithm was able to find a subset of items whose total size is between γ_k and 1. In these cases, the solution is not changed afterwards because the expected guarantee is reached, whichever item comes subsequently. Hereafter, we analyse the case where $\hat{S}_i \subseteq B_i$ such that $1 \geq v(\hat{S}_i) \geq \gamma_k$ does not exist. In this situation, the largest element is removed (cf. line 13). Note that line 13 is executed for round i, and line 13 was possibly executed during previous rounds. However, line 11 or the while loop were not previously executed because, so far, the guarantee has not been reached ($v(S_{i-1}) < \gamma_k$).

We have seen that B_i contains either k or $k + 1$ non-small items at line 8 of the algorithm. Let us make some observations which are valid for both cases.

(i) So far, the algorithm has not rejected or removed a small item. Only items of B_i were removed by the possible execution of line 13 (by construction, B_i is the set of non-small items).

(ii) The algorithm has not switched from one case to the other. If $|B_i| = k$, then it could not be $|B_j| = k + 1$ in a previous round $j < i$. Similarly, if $|B_i| = k + 1$, then it could not be $|B_j| = k$ in a previous round $j < i$. Indeed, suppose we have $|B_i| \neq |B_{i-1}|$.[2] Since we removed one item from B_{i-1}, and o_i was inserted afterwards to yield B_i, we get that $|B_i| = |B_{i-1}|$, contradiction.

(iii) We have $v(S_i) \leq 1$ after the largest element of B_i is removed (line 13). Indeed, S_i is built as follows: take S_{i-1}, add an item o_i, and remove its largest item. We get that $v(S_i) \leq v(S_{i-1})$ and $v(S_{i-1}) \leq 1$ holds by induction hypothesis.

(iv) An optimal (offline) solution contains at most $|B_i|-1$ non-small items. Since the algorithm keeps the smallest non-small items in B_i, and until now $v(B_i)$ is always strictly larger than 1, it is not possible to find $|B_i|$ non-small items (within the set of non-small items already disclosed) whose total size is at most 1.

It remains to prove $\frac{v(S_i)}{v(S_i^*)} \geq \gamma_k$ in both cases, namely $|B_i| \in \{k, k+1\}$ at line 8 of the algorithm.

Case $|B_i| = k$. The current solution S_i contains all the small items disclosed so far (cf. observation (i)). The total size of non-small items of S_i is at least $1 - \frac{1-\gamma_k}{\gamma_k^2}$ because we had $v(B_i) > 1$ and one item of size at most $\frac{1-\gamma_k}{\gamma_k^2}$ has been removed. Meanwhile, observation (iv) says that the optimum contains at most $|B_i| - 1 = k - 1$ non-small items, each of which has size at most $\frac{1-\gamma_k}{\gamma_k^2}$, and possibly all the small items disclosed so far. Therefore, the competitive ratio is lower bounded by

$$\frac{v(\{o : 0 < |o| \leq 1 - \gamma_k\}) + 1 - \frac{1-\gamma_k}{\gamma_k^2}}{v(\{o : 0 < |o| \leq 1 - \gamma_k\}) + (k-1)\frac{1-\gamma_k}{\gamma_k^2}} \geq \frac{1 - \frac{1-\gamma_k}{\gamma_k^2}}{(k-1)\frac{1-\gamma_k}{\gamma_k^2}} = \gamma_k$$

where the last equality is due to (7).

Case $|B_i| = k + 1$. Every item of B_i is medium. Indeed, exactly k items of B_i come from S_{i-1} and we know that $v(S_{i-1}) < \gamma_k$. If the non-small items of S_{i-1} were not exclusively medium items, then its total size would be at least $(k-1)(1-\gamma_k)+\frac{1-\gamma_k}{\gamma_k}$, which is equal to γ_k by (3), contradiction. The item of $B_i \setminus B_{i-1}$ (i.e., o_i) must be medium because the algorithm failed to find $\hat{S}_i$. Indeed, if o_i were large or extra-large, then one could combine it with $k - 1$ medium items of B_i and create $\hat{S}_i$ whose total size is between γ_k and 1 (cf. interpretation of (3)). No large or extra-large item was disclosed so far, because either this large or

[2] The assumption that the change in the size of B occurs between rounds $i - 1$ and i is made without loss of generality because we can apply the arguments to the round (possibly prior to i) during which the size of B is modified for the first time.

extra-large item could be used to produce $\hat{S}_i$ with $k-1$ medium items already present in the solution (contradiction with the non-existence of $\hat{S}_i$), or such a large or extra-large item was removed in a previous round j, but it corresponds to an incompatible situation where $|B_j| = k \neq |B_i|$, and we have seen that the algorithm does not switch from one case to the other (cf. observation (*ii*)).

The algorithm has kept all the small items disclosed so far and the k smallest medium items. The optimum S_i^* possibly contains all the small items disclosed so far, and at most k non-small items (cf. observation (*iv*)) which are medium because no large or extra-large item has been disclosed. The size of a medium item being between $1-\gamma_k$ and $\frac{1-\gamma_k}{\gamma_k}$, the competitive ratio can be lower bounded as follows.

$$\frac{v(\{o : 0 < |o| \leq 1-\gamma_k\}) + k(1-\gamma_k)}{v(\{o : 0 < |o| \leq 1-\gamma_k\}) + k\frac{1-\gamma_k}{\gamma_k}} \geq \frac{k(1-\gamma_k)}{k(\frac{1-\gamma_k}{\gamma_k})} = \gamma_k \qquad \square$$

When $k = 1$, Theorem 1 indicates that Algorithm 1 is t-competitive for all $u \in (0, 1]$ because $\gamma_1 = t$ and $\frac{1-\gamma_1}{\gamma_1^2} = 1$, thus generalizing the result of [12].

When $k \geq 2$, Inequality (5) indicates that $\frac{1-\gamma_k}{\gamma_k^2} \leq \frac{1}{k-1}$ so we can consider in the following theorem how Algorithm 1 performs when $u \in (\frac{1-\gamma_k}{\gamma_k^2}, \frac{1}{k-1}]$.

Theorem 2. *For all integers $k \geq 2$, Algorithm 1 is $\frac{1-u}{(k-1)u}$-competitive when $u \in (\frac{1-\gamma_k}{\gamma_k^2}, \frac{1}{k-1}]$.*

Proof. We keep the same notations as in the proof of Theorem 1. Only the notion of extra-large item changes: o is said to be extra-large if $k^{-1} < |o| \leq u$. All the cases of the proof of Theorem 1 remain unchanged and lead to a competitive ratio of γ_k except when $\hat{S}_i$ such that $\gamma_k \leq v(\hat{S}_i) \leq 1$ does not exist, and $|B_i| = k$. In this case, the current solution S_i contains all the small items disclosed so far. The total size of non-small items of S_i is at least $1-u$ because we had $v(B_i) > 1$ and one item of size at most u has been removed. Meanwhile, the optimum contains at most $|B_i|-1 = k-1$ non-small items, each of which has size at most u, and possibly all the small items disclosed so far. Note that $k-1$ items of size at most u fit in the budget since $u \leq \frac{1}{k-1}$ holds by assumption. The competitive ratio is lower bounded by

$$\frac{v(\{o : 0 < |o| \leq 1-\gamma_k\}) + 1-u}{v(\{o : 0 < |o| \leq 1-\gamma_k\}) + (k-1)u} \geq \frac{1-u}{(k-1)u}.$$

Using (7) and $\frac{1-\gamma_k}{\gamma_k^2} \leq u$, we get that $\gamma_k = \frac{1-(1-\gamma_k)/\gamma_k^2}{(k-1)(1-\gamma_k)/\gamma_k^2} \geq \frac{1-u}{(k-1)u}$. Therefore, the competitive ratio is $\frac{1-u}{(k-1)u}$ in the worst case. $\square$

To conclude this part, the combination of Theorems 1 and 2 gives the lower bounds on the competitive ratio depicted in solid on Fig. 1. Sometimes, the intervals of the theorems intersect for consecutive values of k. In this case, the best (i.e., largest) lower bound on the competitive ratio is retained.

2.2 Upper Bounds on the Competitive Ratio

Our upper bounds, depicted in dotted on Fig. 1, are obtained from instances showing that no deterministic online algorithm can have a competitive ratio larger than some specified value. We will often use the following simple observation: If the instances employed to show an upper bound of r on the competitive ratio only contain items of size at most x, then the competitive ratio is at most r for all $u \in [x, 1]$.

Let us first suppose that $u \in [\frac{4}{7}, 1]$, and consider the following instance.

Instance 1. *The first item o_1 has size u. If o_1 is not taken, then the competitive ratio is zero. The second item o_2 has size $1 - u + \epsilon$. Note that there exists an ϵ such that $1 - u + \epsilon \leq u$ because $u > 0.5$. Since $|o_1| + |o_2| > 1$, either o_1 or o_2 is kept, but not both. If o_2 replaces o_1, then no more item is disclosed. The competitive ratio is $\frac{1-u+\epsilon}{u}$ which tends to $\frac{1-u}{u}$ when ϵ goes to zero. Otherwise, o_1 is kept and a last item o_3 of size $u - \epsilon$ is disclosed. Again, the two items o_1 and o_3 do not fit. Since $|o_1| \geq |o_3|$, we can consider that o_1 is kept. The competitive ratio in this case is u because the optimum $\{o_2, o_3\}$ has value 1.*

Proposition 1. *Instance 1 gives an upper bound of $\frac{1-u}{u}$ on the competitive ratio when $u \in [\frac{4}{7}, t]$, and an upper bound of t when $u \in [t, 1]$.*

Proof. The upper bound is $\frac{1-u}{u}$ when $u \in [\frac{4}{7}, t]$ because $\frac{1-u}{u} \geq u$ in that interval. Since $t = \frac{\sqrt{5}-1}{2}$ is a root of $\frac{1-u}{u} = u$, an upper bound of t on the competitive ratio is derived from Instance 1 (fix u to t in the instance), and it is valid for all $u \in [t, 1]$ by the aforementioned observation. □

Note that the upper bound of t corresponds to one given in [12, Theorem 2]. Now suppose that $u \in (\frac{1}{2}, \frac{4}{7})$ and consider the following instance.

Instance 2. *The first 2 items o_1, o_2 have size $\frac{1}{4} - \epsilon$, where $\epsilon > 0$, and $\frac{1}{4}$, respectively. If any of them is not taken, then the competitive ratio tends to $\frac{1}{2}$. The next item o_3 has size $\frac{1}{2} + 2\epsilon$. If o_3 is kept, then one of o_1, o_2 must be removed. Then an item o_4 of size $\frac{1}{2} + \epsilon$ is disclosed, which does not fit in the current solution and can only replace o_3, leading to a solution of smaller value; hence, o_4 is not included, which gives a competitive ratio that tends to $\frac{3}{4}$ when ϵ goes to zero. Indeed, the total size of the current solution is at most $\frac{3}{4} + 2\epsilon$, while the optimal value is $\frac{1}{4} - \epsilon + \frac{1}{4} + \frac{1}{2} + \epsilon = 1$. If, on the other hand, o_3 is removed, then the instance is terminated, leading to a competitive ratio of $\frac{1/4-\epsilon+1/4}{1/4+1/2+2\epsilon} = \frac{1/2-\epsilon}{3/4+2\epsilon}$ that tends to $\frac{2}{3}$ as ϵ goes to zero.*

Proposition 2. *Instance 2 provides an upper bound of $\frac{3}{4}$ on the competitive ratio when $u \in (\frac{1}{2}, \frac{4}{7})$.*

Proof. The largest item of Instance 2 has size $1/2 + 2\epsilon$. The two cases lead to competitive ratios of $3/4 + 2\epsilon$ and $\frac{1/2-\epsilon}{3/4+2\epsilon}$, so we retain the largest one which tends to $3/4$ when ϵ goes to zero. By the above observation, the upper bound of $3/4$ is valid for $u > 1/2$. □

We finally consider a family of instances parameterized by an integer k such that $k \geq 2$, and the interval covered by each individual instance is $(\frac{1}{k+1}, \frac{1}{k}]$. Thus, the family allows us to cover the case $u \in (0, \frac{1}{2}]$.

Instance 3. *The first k items $o_1, \ldots, o_k$ have size $\frac{1-u}{k} + \epsilon$ each where ϵ is a small positive real satisfying*

$$0 < \epsilon \leq \frac{1}{k}\left(u - \frac{1}{k+1}\right). \tag{8}$$

If the first k items are not all taken, then the competitive ratio is at most $\frac{k-1}{k}$. The next item o_{k+1} has size u. Note that $u \geq \frac{1-u}{k} + \epsilon$ because of $u > \frac{1}{k+1}$ and (8). Since $\sum_{i=1}^{k+1} |o_i| > 1$, either $\{o_1, \ldots, o_k\}$ is kept (case A), or o_{k+1} replaces o_i for some $i \in \{1, \ldots, k\}$ (case B). If case A occurs, then keep disclosing items of size u until (i) either k items of size u have been disclosed but none of them entered the current solution, or (ii) one item of size u is put into the solution (hence, an item of size $\frac{1-u}{k} + \epsilon$ has been removed). If case A(ii) or case B occurs, then disclose an item of size $\frac{1-u}{k} + \epsilon$. One cannot improve the value of the current solution with this last item (it does not fit, and its size is smaller than the size of any other item of the current solution). The competitive ratio under case A(i) is $\frac{k(\frac{1-u}{k}+\epsilon)}{ku}$ which tends to $\frac{1-u}{ku}$ when ϵ goes to zero. The competitive ratio under cases A(ii) and B is $\frac{(k-1)(\frac{1-u}{k}+\epsilon)+u}{(k+1)(\frac{1-u}{k}+\epsilon)}$ which tends to $\frac{(k-1)(\frac{1-u}{k})+u}{(k+1)(\frac{1-u}{k})} = \frac{k-1+u}{(k+1)(1-u)}$ when ϵ goes to zero.

Proposition 3. *Instance 3 provides an upper bound of $\frac{1-u}{ku}$ when $u \in (\frac{1}{k+1}, c_k]$, and an upper bound of $\frac{1-c_k}{kc_k}$ for $u \geq c_k$, where*

$$c_k := \left(k^2 + k + 2 - \sqrt{(k^2 + k + 2)^2 - 4(k+1)}\right)/2.$$[3]

Proof. The upper bound derived from Instance 3 is $\max\left(\frac{k-1}{k}, \frac{1-u}{ku}, \frac{k-1+u}{(k+1)(1-u)}\right)$. Since $u \leq \frac{1}{k}$, we know that $\frac{k-1}{k} \leq \frac{1-u}{ku}$. Thus, the upper bound is $\max(\frac{1-u}{ku}, \frac{k-1+u}{(k+1)(1-u)})$ where $\frac{1-u}{ku}$ is a decreasing function of u while $\frac{k-1+u}{(k+1)(1-u)}$ is increasing. The cut point of these functions is $c_k \in (\frac{1}{k+1}, \frac{1}{k}]$. By the aforementioned observation, we have an upper bound on the competitive ratio of $\frac{1-c_k}{kc_k}$ when $u \geq c_k$. □

The combination of Propositions 1, 2 and 3 leads to the upper bounds depicted in dotted on Fig. 1.

3 Lower Bounded Item Size

Given a lower bound $\ell \in (0, 1]$ on the size of every item, we are going to show that the best possible competitive ratio $\rho(\ell)$ of deterministic online algorithms for the unweighted knapsack problem with removable items is as follows.

[3] Note that $\frac{1-c_k}{kc_k} = \gamma_{k+1}$ and $c_k = \frac{1-\gamma_{k+1}}{\gamma_{k+1}^2}$.

$$\rho(\ell) = \begin{cases} t, & \text{if } 0 \le \ell \le 1-t \\ \sqrt{\ell}, & \text{if } 1-t < \ell \le 1/2 \\ 1, & \text{if } 1/2 < \ell \le 1 \end{cases} \tag{9}$$

See (1) for the definition of t. One can observe that the competitive ratio is always above t. As for Sect. 2 where an upper bound was known, we begin with lower bounds on $\rho(\ell)$ followed by upper bounds on $\rho(\ell)$. Together, they constitute a characterization of $\rho(\ell)$ since there is no gap.

3.1 Lower Bounds on the Competitive Ratio

When $\ell \in [0, 1-t]$, the characterization of $\rho(\ell)$ given in (9) indicates a competitive ratio of t which can be obtained with either Algorithm 1 ($k = 1$), or with the algorithm of Iwama and Taketomi [12]. If $\ell > 1/2$, then there exists a simple algorithm with competitive ratio 1: maintain in the current solution the largest item encountered thus far. Since any solution contains at most one item, this simple algorithm is optimal. It remains to provide a $\sqrt{\ell}$-competitive algorithm for the case $\ell \in [1-t, 1/2]$, cf. Algorithm 2 and Theorem 3. The proof of Theorem 3 is deferred to an extended version of this article.

Algorithm 2. A $\sqrt{\ell}$-competitive algorithm for the case $1 - t \le \ell \le 1/2$

$S \leftarrow \emptyset$
while a new item o_i arrives **do**
    if $v(S) \ge \sqrt{\ell}$ **then**
        Reject o_i {*S is not changed afterwards*}
    else if $|o_i| \ge \sqrt{\ell}$ **then**
        $S \leftarrow \{o_i\}$ {*S is not changed afterwards*}
    else if $v(S) + |o_i| \le 1$ **then**
        $S \leftarrow S \cup \{o_i\}$
    else
        Let o be the unique item in S
        S gets the item of minimum size between o and o_i
    end if
end while

Theorem 3. *Algorithm 2 is $\sqrt{\ell}$-competitive when $\ell \in [1-t, 1/2]$.*

3.2 Upper Bounds on the Competitive Ratio

Let us begin with $\ell \in (0, 1-t]$.

Proposition 4. *The competitive ratio of deterministic online algorithms is at most t when $\ell \in (0, 1-t]$.*

Proof. We can reuse the bound provided by Iwama and Taketomi [12, Theorem 2]. Let us give the corresponding instance for the sake of readability. The first item o_1 has size $1 - t$. If o_1 is not taken, then the competitive ratio is zero. The second item o_2 has size $t + \epsilon$ where ϵ is a tiny positive real. Items o_1 and o_2 cannot be both taken without exceeding the budget. If o_2 does not replace o_1, then stop. The competitive ratio tends to $\frac{1-t}{t}$ when ϵ goes to zero. Otherwise o_2 replaces o_1, and a new item o_3 of size t is disclosed. The optimum $\{o_1, o_3\}$ has value 1 while the algorithm's solution has value at most $t + \epsilon$. The competitive ratio is either $\frac{1-t}{t}$ which is equal to t, or $t + \epsilon$, so we have an upper bound of t. In the instance, every item has size at least $1 - t$ because $t + \epsilon \geq t \geq 1 - t$, so the upper bound on the competitive ratio holds for all $\ell \in (0, 1 - t]$. □

The next step concerns the interval $(1 - t, \frac{1}{2}]$.

Proposition 5. *The competitive ratio of deterministic online algorithms is at most* $\sqrt{\ell}$ *when* $\ell \in (1 - t, \frac{1}{2}]$.

Proof. Consider the following instance. The first item o_1 has size ℓ. If o_1 is not taken, then the competitive ratio is zero. The second item o_2 has size $\sqrt{\ell} + \epsilon$. Since $\ell \leq 1$, we know that $\sqrt{\ell} + \epsilon \geq \ell$. Moreover, $\ell + \sqrt{\ell} + \epsilon > 1$ holds when $\ell \geq 1 - t$, so no feasible solution can contain both items. If o_2 does not replace o_1, then stop. The competitive ratio tends to $\ell/\sqrt{\ell} = \sqrt{\ell}$ when ϵ goes to zero. Otherwise o_2 replaces o_1, and a new item o_3 of size $1 - \ell$ is disclosed. The optimum $\{o_1, o_3\}$ has value 1 while the algorithm's solution has value at most $\sqrt{\ell} + \epsilon$ because $\sqrt{\ell} + \epsilon \geq 1 - \ell$ holds when $\ell > 1 - t$, leading to a competitive ratio which tends to $\sqrt{\ell}$ when ϵ goes to zero. Thus, both cases lead to the same upper bound of $\sqrt{\ell}$. □

No upper bound is needed when $\ell > 1/2$ because an optimal algorithm has been presented in the previous section.

4 Conclusion and Directions for Future Work

We have considered the removable online unweighted knapsack problem with bounded size items (denoted by u and ℓ for upper bound and lower bound, respectively). Our contribution consists of lower and upper bounds on the best competitive ratio for deterministic algorithms. The optimal ratio tends to 1 when the parameter u goes to 0. A direct extension to our work would be to close the gap for all possible values of u.

The reader may wonder what the situation is when the items are not removable but their size is bounded. Given an upper bound u (resp., lower bound ℓ) on the item sizes, the best possible competitive ratio is $1 - u$ (resp., ℓ) and simple deterministic algorithms can achieve these ratios.

For the future, we believe that it would be interesting to combine bounded item sizes with other approaches such as resource augmentation [13], buffering [10], reservation [3], advice [4], or item splitting [11]. Other possible extensions of the present work can be: exploring randomized online algorithms (as in [2,5,7,8]) with bounded size items, exploiting a possible prediction of the total number of disclosed items, or dealing weighted items restricted to convex functions of the size (as in [9]).

References

1. Albers, S.: Online algorithms: a survey. Math. Program. **97**(1–2), 3–26 (2003)
2. Babaioff, M., Hartline, J.D., Kleinberg, R.D.: Selling ad campaigns: online algorithms with cancellations. In: Chuang, J., Fortnow, L., Pu, P. (eds.) Proceedings 10th ACM Conference on Electronic Commerce (EC-2009), Stanford, California, USA, 6–10 July 2009, pp. 61–70. ACM (2009)
3. Böckenhauer, H., Burjons, E., Hromkovic, J., Lotze, H., Rossmanith, P.: Online simple knapsack with reservation costs. In: Bläser, M., Monmege, B. (eds.) STACS 2021, 16–19 March 2021, Saarbrücken, Germany (Virtual Conference). LIPIcs, vol. 187, pp. 16:1–16:18. Schloss Dagstuhl - Leibniz-Zentrum für Informatik (2021)
4. Böckenhauer, H., Komm, D., Královic, R., Rossmanith, P.: The online knapsack problem: advice and randomization. Theor. Comput. Sci. **527**, 61–72 (2014)
5. Cygan, M., Jez, L., Sgall, J.: Online knapsack revisited. Theory Comput. Syst. **58**(1), 153–190 (2016)
6. Demko, S., Hill, T.P.: Equitable distribution of indivisible objects. Math. Soc. Sci. **16**(2), 145–158 (1988)
7. Han, X., Kawase, Y., Makino, K.: Online unweighted knapsack problem with removal cost. Algorithmica **70**(1), 76–91 (2014)
8. Han, X., Kawase, Y., Makino, K.: Randomized algorithms for online knapsack problems. Theor. Comput. Sci. **562**, 395–405 (2015)
9. Han, X., Kawase, Y., Makino, K., Guo, H.: Online removable knapsack problem under convex function. Theor. Comput. Sci. **540**, 62–69 (2014)
10. Han, X., Kawase, Y., Makino, K., Yokomaku, H.: Online knapsack problems with a resource buffer. In: Lu, P., Zhang, G. (eds.) ISAAC 2019, 8–11 December 2019, Shanghai University of Finance and Economics, Shanghai, China. LIPIcs, vol. 149, pp. 28:1–28:14. Schloss Dagstuhl - Leibniz-Zentrum für Informatik (2019)
11. Han, X., Makino, K.: Online removable knapsack with limited cuts. Theor. Comput. Sci. **411**(44–46), 3956–3964 (2010)
12. Iwama, K., Taketomi, S.: Removable online knapsack problems. In: ICALP 2002, Malaga, Spain, 8–13 July 2002, Proceedings, pp. 293–305 (2002)
13. Iwama, K., Zhang, G.: Online knapsack with resource augmentation. Inf. Process. Lett. **110**(22), 1016–1020 (2010)
14. Kellerer, H., Pferschy, U., Pisinger, D.: Knapsack Problems. Springer, Berlin (2004). https://doi.org/10.1007/978-3-540-24777-7
15. Lueker, G.S.: Average-case analysis of off-line and on-line knapsack problems. J. Algorithms **29**(2), 277–305 (1998)
16. Marchetti-Spaccamela, A., Vercellis, C.: Stochastic on-line knapsack problems. Math. Program. **68**, 73–104 (1995)
17. Zhou, Y., Chakrabarty, D., Lukose, R.: Budget constrained bidding in keyword auctions and online knapsack problems. In: Papadimitriou, C., Zhang, S. (eds.) WINE 2008. LNCS, vol. 5385, pp. 566–576. Springer, Heidelberg (2008). https://doi.org/10.1007/978-3-540-92185-1_63

Faster Winner Determination Algorithms for (Colored) Arc Kayles

Tesshu Hanaka[1], Hironori Kiya[2], Michael Lampis[3], Hirotaka Ono[4], and Kanae Yoshiwatari[4(✉)]

[1] Department of Informatics, Faculty of Information Science and Electrical Engineering, Kyushu University, Fukuoka, Japan
hanaka@inf.kyushu-u.ac.jp

[2] Department of Informatics, Department of Core Informatics, Osaka Metropolitan University, Osaka, Japan
kiya@omu.ac.jp

[3] Université Paris-Dauphine, PSL University, CNRS, LAMSADE, 75016 Paris, France
michail.lampis@lamsade.dauphine.fr

[4] Department of Mathematical Informatics, Graduate School of Informatics, Nagoya University, Nagoya, Japan
yoshiwatari.kanae.w1@s.mail.nagoya-u.ac.jp, ono@nagoya-u.jp

Abstract. Arc Kayles and Colored Arc Kayles, two-player games on a graph, are generalized versions of well-studied combinatorial games Cram and Domineering, respectively. In Arc Kayles, players alternately choose an edge to remove with its adjacent edges, and the player who cannot move is the loser. Colored Arc Kayles is similarly played on a graph with edges colored in black, white, or gray, while the black (resp., white) player can choose only a gray or black (resp., white) edge. For Arc Kayles, the vertex cover number (i.e., the minimum size of a vertex cover) is an essential invariant because it is known that twice the vertex cover number upper bounds the number of turns of Arc Kayles, and for the winner determination of (Colored) Arc Kayles, $2^{O(\tau^2)}n^{O(1)}$-time algorithms are known, where τ is the vertex cover number and n is the number of vertices. In this paper, we first give a polynomial kernel for Colored Arc Kayles parameterized by τ, which leads to a faster $2^{O(\tau \log \tau)}n^{O(1)}$-time algorithm for Colored Arc Kayles. We then focus on Arc Kayles on trees, and propose a $2.2361^{\tau}n^{O(1)}$-time algorithm. Furthermore, we show that the winner determination Arc Kayles on a tree can be solved in $O(1.3831^n)$ time, which improves the best-known running time $O(1.4143^n)$. Finally, we show that Colored Arc Kayles is NP-hard, the first hardness result in the family of the above games.

Keywords: Arc Kayles · Combinatorial Game Theory · Exact Exponential-Time Algorithm · Vertex Cover

H. Fernau et al. (Eds.): SOFSEM 2024, LNCS 14519, pp. 297–310, 2024.
https://doi.org/10.1007/978-3-031-52113-3_21

1 Introduction

Arc Kayles is a combinatorial game played on a graph. In Arc Kayles, a player chooses an edge of an undirected graph G and then the selected edge and its neighboring edges are removed from G. In other words, a player chooses two adjacent vertices to occupy. The player who cannot choose adjacent two vertices loses the game.

Node Kayles, a vertex version of Arc Kayles, and Arc Kayles were introduced in 1978 by Schaefer [14]. The complexity of Node Kayles is shown to be PSPACE-complete, whereas the complexity of Arc Kayles is remains open. An important aspect of Arc Kayles is that it is a graph generalization of Cram, which is a well-studied combinatorial game introduced in [8]. Cram is a simple board game: two people alternately put a domino on a checkerboard, and the player who cannot place a domino will lose the game. Cram is interpreted as Arc Kayles, when a graph is a two-dimensional grid graph. Though Cram is quite more restricted than Arc Kayles, its complexity remains open. Since an algorithm for Arc Kayles also works for Cram, a study for Arc Kayles would help the study for Cram.

This paper presents new winner-determination algorithms together with elaborate running time analyses. The running time of our algorithms is parameterized by the vertex cover number of a graph. Note that the vertex cover number of a graph is strongly related to the number of turns of Arc Kayles, which is the total number of actions taken by two players, as seen below. Intuitively, the number of turns tends to reflect the complexity of a game because it is the depth of the game tree, and it is reasonable to focus on it when we design winner-determination algorithms.

The relation between the number of turns of Arc Kayles and the vertex cover number of the graphs it is played on is observed as follows. During a game of Arc Kayles, edges chosen by the players form a matching, and the player who completes a maximal matching wins; the number of turns in a game is the size of the corresponding maximal matching. Since the maximum matching size is at most twice the minimum maximal matching size, which is also at most twice the minimum vertex cover number, the number of turns of Arc Kayles is linearly upper and lower bounded by the vertex cover number.

1.1 Partisan Variants of Arc Kayles

In this paper, we also study partisan variants of Arc Kayles: Colored Arc Kayles and BW-Arc Kayles. In combinatorial game theory, a game is said to be *partisan* if some actions are available to one player and not to the other. Colored Arc Kayles, introduced in [19], is played on an edge-colored graph $G = (V, E_{\mathrm{B}} \cup E_{\mathrm{W}} \cup E_{\mathrm{G}})$, where $E_{\mathrm{B}}, E_{\mathrm{W}}, E_{\mathrm{G}}$ are disjoint. The subscripts B, W, and G of $E_{\mathrm{B}}, E_{\mathrm{W}}$, and E_{G} respectively stand for black, white, and gray. For every edge $e \in E_{\mathrm{B}} \cup E_{\mathrm{W}} \cup E_{\mathrm{G}}$, let $c(e)$ be the color of e, that is, B if $e \in E_{\mathrm{B}}$, W if $e \in E_{\mathrm{W}}$, and G if $e \in E_{\mathrm{G}}$. If $\{u, v\} \notin E_{\mathrm{B}} \cup E_{\mathrm{W}} \cup E_{\mathrm{G}}$, we set $c(\{u, v\}) = \emptyset$ for convenience. Since the first (black or B) player can choose black or gray edges, and the second (white or W) player can choose white or gray edges, COLORED ARC KAYLES is a partisan game. Note that COLORED ARC KAYLES with empty E_B and E_W is actually ARC KAYLES, which is no longer a partisan and is said to be *impartial*. We also name COLORED ARC KAYLES with empty E_G BW-ARC KAYLES, which is still partisan. This paper presents a fixed-parameter tractable (FPT) winner-determination algorithm also for COLORED ARC KAYLES, which is parameterized by vertex cover number.

Here, we introduce another combinatorial game called DOMINEERING. DOMINEERING is a partisan version of CRAM; one player can place a domino only vertically, and the other player can place one only horizontally. As ARC KAYLES is a graph generalization of CRAM, BW-ARC KAYLES is a graph generalization of DOMINEERING. Note that DOMINEERING is also a well-studied combinatorial game. In fact, several books of combinatorial game theory (e.g., [1]) use DOMINEERING as a common example of partisan games, though its time complexity is still unknown in general. Our algorithm mentioned above works for DOMINEERING.

1.2 Related Work

Node Kayles and Arc Kayles. As mentioned above, NODE KAYLES and ARC KAYLES were introduced in [14]. NODE KAYLES is the vertex version of ARC KAYLES; the action of a player in NODE KAYLES is to select a vertex instead of an edge, and then the selected vertex and its neighboring vertices are removed. The winner determination of NODE KAYLES is known to be PSPACE-complete in general [14], though it can be solved in polynomial time by using Sprague-Grundy theory [2] for graphs of bounded asteroidal number, such as comparability graphs and cographs. For general graphs, Bodlaender et al. propose an $O(1.6031^n)$-time algorithm [3]. Furthermore, they show that the winner of NODE KAYLES can be determined in time $O(1.4423^n)$ on trees. In [11], Kobayashi sophisticates the analysis of the algorithm in [3] from the perspective of the parameterized complexity and shows that it can be solved in time $O^*(1.6031^\mu)$, where μ is the modular width of an input graph[1]. He also gives an $O^*(3^\tau)$-time algorithm, where τ is the vertex cover number, and a linear kernel when parameterized by neighborhood diversity.

Different from NODE KAYLES, the complexity of ARC KAYLES has remained open for more than 45 years. Even for subclasses of trees, not much is known. For example, Huggans and Stevens study ARC KAYLES on subdivided stars with three paths [10]. To the best of our knowledge, until a few years ago no exponential-time algorithm for ARC KAYLES existed except for an $O^*(4^{\tau^2})$-time algorithm proposed in [13]. In [9,19], the authors showed that the winner of ARC KAYLES on trees can be determined in $O^*(2^{n/2}) = O(1.4143^n)$ time, which improves $O^*(3^{n/3})(= O(1.4423^n))$ by a direct adjustment of the analysis of Bodlaender et al.'s $O^*(3^{n/3})$-time algorithm for Node Kayles on trees.

[1] The $O^*(\cdot)$ notation suppresses polynomial factors in the input size.

BW-Arc Kayles and Colored Arc Kayles. BW-ARC KAYLES and COLORED ARC KAYLES were introduced in [9,19]. The paper presented an $O^*(1.4143^{\tau^2+3.17\tau})$-time algorithm for COLORED ARC KAYLES, where τ is the vertex cover number. The algorithm runs in time $O^*(1.3161^{\tau^2+4\tau})$ and $O^*(1.1893^{\tau^2+6.34\tau})$ for BW-ARC KAYLES and ARC KAYLES, respectively. This is faster than the previously known time complexity $O^*(4^{\tau^2})$ in [13]. They also give a bad instance for the proposed algorithm, which implies the running time analysis is asymptotically tight. Furthermore, they show that the winner of Arc Kayles can be determined in time $O^*((n/\nu+1)^\nu)$, where ν is the neighborhood diversity of an input graph. This analysis is also asymptotically tight.

Cram and Domineering. CRAM and DOMINEERING are well-studied in the field of combinatorial game theory. In [8], Gardner gives winning strategies for some simple cases. For CRAM on an $a\times b$ board, the second player can always win if both a and b are even, and the first player can always win if one of a and b is even and the other is odd. This can be easily shown by the so-called Tweedledum and Tweedledee strategy. For specific sizes of boards, computational studies have been conducted [17]. In [16], CRAM's endgame databases for all board sizes with at most 30 squares are constructed. As far as the authors know, the complexity to determine the winner for CRAM on general boards still remains open.

Finding the winning strategies of DOMINEERING for specific sizes of boards by using computer programs is well studied. For example, the cases of 8×8 and 10×10 are solved in 2000 [4] and 2002 [5], respectively. The first player wins in both cases. Currently, the status of boards up to 11×11 is known [15]. In [18], endgame databases for all single-component positions up to 15 squares for DOMINEERING are constructed. The complexity of DOMINEERING on general boards also remains open. Lachmann, Moore, and Rapaport show that the winner and a winning strategy of DOMINEERING on $m \times n$ board can be computed in polynomial time for $m \in \{1,2,3,4,5,7,9,11\}$ and all n [12].

1.3 Our Contribution

In this paper, we present FPT winner-determination algorithms with the minimum vertex cover number τ as a parameter, which is much faster than the existing ones. To this end, we show that COLORED ARC KAYLES has a polynomial kernel parameterized by τ, which leads to a $2^{O(\tau\log\tau)}n^{O(1)}$ time algorithm where n is the number of the vertices (Sect. 2); this improves the previous time complexity $2^{O(\tau^2)}n^{O(1)}$. For ARC KAYLES on trees, we show that the winner determination can be done in time $O^*(5^{\tau/2})(=O(2.2361^\tau))$ (Sect. 3), together with an elaborate analysis of time $O^*(7^{n/6})(=O(1.3831^n))$, which improves the previous bound $O^*(2^{n/2})(=O(1.4142^n))$ (Sect. 4). Finally, Sect. 5 shows that BW-ARC KAYLES is NP-hard, and thus so is COLORED ARC KAYLES. Note that this might be the first hardness result on the family of the combinatorial games shown in Sect. 1.2 except for NODE KAYLES.

1.4 Preliminaries

Let $G = (V, E)$ be an undirected graph. We denote $n = |V|$ and $m = |E|$, respectively. For an edge $e = \{u, v\} \in E$, we define $\Gamma(e) = \{e' \in E \mid e \cap e' \neq \emptyset\}$. For a graph $G = (V, E)$ and a vertex subset $V' \subseteq V$, we denote by $G[V']$ the subgraph induced by V'. For simplicity, we denote by $G - v$ instead of $G[V \setminus \{v\}]$ the graph obtained by deleting a vertex v, and by $G - V'$ instead of $G[V \setminus V']$ the graph obtained by deleting all vertices in V', respectively. For an edge subset E', we also denote by $G - E'$ the subgraph obtained from G by removing all edges in E' from G. A vertex set S is called a *vertex cover* if $e \cap S \neq \emptyset$ for every edge $e \in E$. Let τ denote the size of a minimum vertex cover of G, which is also called the *vertex cover number* of G. For the basic definitions of parameterized complexity such as fixed-parameter tractability, polynomial kernels, and graph parameters, we refer the reader to the standard textbook [7].

2 A Polynomial Kernel for Colored Arc Kayles

Our main result in this section is that COLORED ARC KAYLES admits a polynomial kernel when parameterized by the size τ of a given vertex cover. Since COLORED ARC KAYLES generalizes standard ARC KAYLES and our kernelization algorithm proceeds by deleting edges of the input graph, we obtain the same result for ARC KAYLES.

Before we proceed, let us give some intuition about the main idea. To make things simpler, let us first consider (standard) ARC KAYLES parameterized by the size of a vertex cover τ. One way in which we could hope to obtain a kernel could be via the following observation: if a vertex $x \in C$, where C is the vertex cover, has high degree (say, degree at least $\tau+1$), then we can guarantee that this vertex can always be played, or more precisely, that it is impossible to eliminate this vertex by playing on edges incident with its neighbors, since the game cannot last more than τ rounds. One could then be tempted to argue that, therefore, when a vertex has sufficiently high degree, we can delete one of its incident edges. If we thus bound the maximum degree of vertices of C, we obtain a polynomial kernel.

Unfortunately, there is a clear flaw in the above intuition: suppose that x is a high-degree vertex of C as before, xy an edge, and $x'y$ another edge of the graph, for $x' \in C$. If x' is a low-degree vertex, then deciding whether to play xy or another edge incident with x is consequential, as the player needs to decide whether the strategy is to eliminate x' by playing one of its incident edges, or by playing edges incident with its neighbors. We therefore need a property more subtle than simply a vertex that has high degree.

To avoid the flaw described in the previous paragraph, we therefore look for a dense sub-structure: a set of vertices $X \subseteq C$ such that there exists a set of vertices $Y \subseteq V \setminus C$ where all vertices of X have many neighbors in Y and at the same time vertices of Y have no neighbors outside X. In such a structure the initial intuition does apply: playing an edge xy with $x \in X$ and $y \in Y$ is equivalent to playing any other edge xy' with $y' \in Y$, because other vertices of

X "don't care" which vertices of Y have been eliminated (since vertices of X have high degree), while vertices outside of X "don't care" because they are not connected to Y. Our main technical tool is then to give a definition (Definition 1) which captures and generalizes this intuition to the colored version of the game: we are looking for (possibly non-disjoint) sets $X_W, X_B \subseteq C$ and $Y \subseteq V \setminus C$, such that each vertex of X_W and X_B has many edges playable by White or Black respectively with the other endpoint in Y, while edges incident with Y have their other endpoint in some appropriate part of $X_W \cup X_B$ (white edges in X_W, black edges in X_B, and gray edges in $X_W \cap X_B$). We show that if we can find such a structure, then we can safely remove an edge (Lemma 1) and then show that in polynomial time we can either find such a structure or guarantee that the size of the graph is bounded to obtain the main result (Theorem 1).

Definition 1. *Let $G = (V, E)$ be an instance of* Colored Arc Kayles, *with $E = E_W \cup E_B \cup E_G$, $C \subseteq V$ be a vertex cover of G of size τ, and $I = V \setminus C$. Then, for three sets of vertices X_W, X_B, Y we say that (X_W, X_B, Y) is a* dense triple *if we have the following: (i) $X_W, X_B \subseteq C$ and $Y \subseteq I$ (ii) for each $x \in X_W$ (respectively $x \in X_B$) there exist at least $\tau + 1$ edges in $E_W \cup E_G$ (respectively in $E_B \cup E_G$) incident with x with the other endpoint in Y (iii) for all $y \in Y$ all edges of E_G incident with y have their second endpoint in $X_W \cap X_B$ (iv) for all $y \in Y$ all edges of E_W (respectively of E_B) incident with y have their second endpoint in X_W (respectively in X_B).*

Lemma 1. *Let $G = (V, E)$ be an instance of* Colored Arc Kayles *and $C \subseteq V$ be a vertex cover of G of size τ. Then, for any edge $e \in E$ incident with a vertex of Y and any $r > 0$ we have the following: a player (Black or White) has a strategy to win Arc Kayles in G in at most r moves if and only if the same player has a strategy to win Arc Kayles in $G - e$ in at most r moves.*

Proof. We prove the lemma by induction on the size τ of C. For $\tau = 0$ the lemma holds, as there are no edges to delete. We therefore start with $\tau = 1$, so C contains a single vertex, say $C = \{x\}$. Suppose without loss of generality that Black is playing first (the other case is symmetric). If $X_B = \emptyset$ and $X_W = \emptyset$, then Y may only contain isolated vertices, so again there is no edge e that satisfies the conditions of the lemma, so the claim is vacuous. If $X_B = \emptyset$ and $X_W = \{x\}$, then any e that satisfies the conditions of the lemma must have $e \in E_W$. Clearly, deleting such an edge does not affect the answer, as this edge cannot be played. If on the other hand, $X_B = \{x\}$, then $|E_B| + |E_G| \geq 2$ (to satisfy condition (ii)), hence the current instance is a win for Black in one move, and removing any edge does not change this fact.

For the inductive step, suppose the lemma is true for all graphs with vertex cover at most $\tau - 1$. We must prove that optimal strategies are preserved in both directions. To be more precise, the optimal strategy of a player is defined as the strategy which guarantees that the player will win in the minimum number of rounds if the player has a winning strategy, or guarantees that the game will last as long as possible if the player has no winning strategy.

For the easy direction, suppose that the first player has an optimal strategy in $G - e$ which starts by playing an edge $f = ab$. This edge also exists in G, so we formulate a strategy in G that is at least as good for the first player by again initially playing f in G. Now, let $G_1 = G - \{a, b\}$ be the resulting graph, and $G_2 = G - e - \{a, b\}$ be the resulting graph when we play in $G - e$. If e has an endpoint in $\{a, b\}$, then G_1, G_2 are actually isomorphic, so clearly the first player's strategy in G is at least as good as her strategy in $G - e$ and we are done. Otherwise, $G_2 = G_1 - e$ and we claim that we can apply the inductive hypothesis to G_1 and G_2, proving that the two graphs have the same winner in the same number of moves and hence our strategy is winning for G. Indeed, G_1 has a vertex cover of size at most $\tau - 1$. Furthermore, if (X_W, X_B, Y) is a dense triple of G, then $(X_W \setminus \{a, b\}, X_B \setminus \{a, b\}, Y \setminus \{a, b\})$ is a dense triple of G_1, because Y contains at most one vertex from $\{a, b\}$, hence each vertex of $X_W \cup X_B$ has lost at most one edge connecting it to Y. Therefore, the inductive hypothesis applies, as $G_2 = G_1 - e$ and e is an edge incident with $Y \setminus \{a, b\}$.

For the more involved direction, suppose that the first player has an optimal strategy in G for which we consider several cases:

1. The optimal strategy in G initially plays an edge f that shares no endpoints with e.
2. The optimal strategy in G initially plays an edge f that shares exactly one endpoint with e.
3. The optimal strategy in G initially plays e.

For the first case, let G_1 be the graph resulting from playing f in G, and G_2 be the graph resulting from playing f in $G - e$. Again, as in the previous direction, we observe that we can apply the inductive hypothesis on G_1, G_2, and therefore playing f is an equally good strategy in $G - e$.

For the second case, it is even easier to see that playing f is an equally good strategy in $G - e$, as G_1, G_2 are now isomorphic (playing f in G removes the edge e that distinguishes G from $G - e$).

Finally, for the more involved case, suppose without loss of generality that Black is playing first in G and has an optimal strategy that begins by playing e, therefore $e \in E_B \cup E_G$. Let $e = xy$ with $x \in X_B$ and $y \in Y$. We will attempt to find an equally good strategy for Black in $G - e$. By condition (ii) of Definition 1, x has $\tau > 1$ other incident edges that Black can play, whose second endpoint is in Y. Let $e' = xy'$ be such an edge, with $y' \in Y$. Let $G_1 = G - \{x, y\}$ and $G_2 = G - \{x, y'\}$. It is sufficient to prove that G_1 and G_2 have the same winner in the same number of moves, if White plays first on both graphs. For this, we will again apply the inductive hypothesis, though this time it will be slightly more complicated, since G_1, G_2 may differ in many edges. We will work around this difficulty by *adding* (rather than removing) edges to both graphs, so that we eventually arrive at isomorphic graphs, without changing the winner.

Take G_1 and observe that $(X_W \setminus \{x\}, X_B \setminus \{x\}, Y \setminus \{y\})$ is a dense triple, as the vertex cover of G_1 has size at most $\tau - 1$, and each vertex of $X_B \cup X_W$ has lost at most one neighbor in Y. Add the vertex y to G_1 as an isolated vertex (this clearly does not affect the winner). Furthermore, $(X_W \setminus \{x\}, X_B \setminus \{x\}, Y)$ is a dense triple of the new graph. We now observe that adding a white edge from y to $X_W \setminus \{x\}$, or a black edge from y to $X_B \setminus \{x\}$, or a gray edge from y to $(X_W \cap X_B) \setminus \{x\}$ does not affect the fact that $(X_W \setminus \{x\}, X_B \setminus \{x\}, Y)$ is a dense triple. Hence, by inductive hypothesis, it does not affect the winner or the number of moves needed to win. Repeating this, we add to G_1 all the edges incident with y in G_2. We then take G_2, add to it y' as an isolated vertex, and then use the same argument to add to it all edges incident to y' in G_1 without changing the winner. We have thus arrived at two isomorphic graphs. □

Theorem 1. *There is a polynomial time algorithm which takes as input an instance G of* Colored Arc Kayles *and a vertex cover of G of size τ and outputs an instance G', such that G' has $O(\tau^3)$ edges, and for all $r > 0$ a player (Black or White) has a strategy to win in r moves in G if and only if the same player has a strategy to win in r moves in G'. Hence,* Colored Arc Kayles *admits a kernel with $O(\tau^3)$ edges.*

Proof. We describe an algorithm that finds a dense triple, if one exists, in the input graph $G = (V, E)$. If we find such a triple, we can invoke Lemma 1 to delete an edge from the graph, without changing the answer, and then repeat the process. Otherwise, we will argue that the G must already have the required number of edges. We assume that we are given a vertex cover C of G of size $\tau \geq 1$ and $I = V \setminus C$. If not, a 2-approximate vertex cover can be found in polynomial time using standard algorithms.

The algorithm executes the following rules exhaustively, until no rule can be applied, always preferring to apply lower-numbered rules.

1. If C contains an isolated vertex, delete it.
2. If there exists $x \in C$ such that x is incident with at most τ edges of $E_B \cup E_G$ and at most τ edges of $E_W \cup E_G$, then delete $N(x) \cap I$ from G.
3. If there exists $x \in C$ such that x is incident with at least 1 and at most τ edges of $E_B \cup E_G$, then for each $y \in I$ such that $xy \in E_B \cup E_G$, delete y from G.
4. If there exists $x \in C$ such that x is incident with at least 1 and at most τ edges of $E_W \cup E_G$, then for each $y \in I$ such that $xy \in E_W \cup E_G$, delete y from G.

The rules above can clearly be executed in polynomial time. We now first prove that the rules are safe via the following two claims.

Claim. If G contains a dense triple (X_W, X_B, Y), then applying any of the rules will result in a graph where (X_W, X_B, Y) is still a dense triple.

Proof. It is in fact sufficient to prove that the rules will never delete a vertex of $X_W \cup X_B \cup Y$, because if we only delete vertices outside a dense triple, the dense triple remains valid. Vertices removed by the first rule clearly cannot belong to $X_B \cup X_W$. For the second rule, we observe that if x satisfies the conditions of the rule, then $x \notin X_W \cup X_B$, as that would violate condition (ii) of Definition 1. Since $x \notin X_W \cup X_B$, for any $y \in I$ such that $xy \in E$, it must be the case that $y \notin Y$, therefore it is safe to delete such vertices. For the third rule, we observe that $x \notin X_B$, because that would violate condition (ii) of Definition 1. Therefore, if $y \in I$ such that $xy \in E_B \cup E_G$, we have $y \notin Y$ by conditions (iii) and (iv) of Definition 1, and it is safe to delete such vertices. The last rule is similar. □

Claim. If, after applying the rules exhaustively, the resulting graph is not edge-less, then we can construct a dense triple.

Proof. We can construct a dense triple (X_W, X_B, Y) as follows: place all remaining vertices of C which are still incident with an edge of E_W (respectively of E_B) into X_W (respectively into X_B), place all vertices of C still incident with an edge of E_G into both X_W and X_B, and place all remaining vertices of I into Y. The dense triple thus constructed is also a dense triple in the original graph.

We prove the claim by induction on the number of rule applications. Let $G_0 = G, G_1, G_2, \ldots, G_\ell$ be the sequence of graphs we obtain by executing the algorithm, where G_{i+1} is obtained from G_i by applying a rule. We first show that (X_W, X_B, Y) is a dense triple in the final graph G_ℓ. Consider a vertex $x \in X_W \setminus X_B$. By construction x is incident with an edge of E_W in G_ℓ but on no edge of $E_B \cup E_G$. We can see that x satisfies condition (ii) of Definition 1 because if it were incident with at most k edges of E_W, the second rule would have applied. Similarly, vertices of $X_B \setminus X_W$ satisfy condition (ii). For $x \in X_W \cap X_B$, by construction either x is incident with an edge of E_G or it is incident with edges from both E_W and E_B. Therefore, x is incident with at least 1 edge of $E_W \cup E_G$ and at least 1 edge of $E_B \cup E_G$. As a result, if x violated condition (ii), the third or fourth rules would have applied. Condition (iii) is satisfied because we placed all vertices of C incident with an edge of E_G into $X_W \cap X_B$. Condition (iv) is satisfied because we placed all vertices of C incident with an edge of E_W into X_W (similarly for E_B).

Having established the base case, suppose we have some $r < \ell$ such that (X_W, X_B, Y) is a dense triple in all of $G_{r+1}, \ldots, G_\ell$. We will show that (X_W, X_B, Y) is also a dense triple in G_r. If G_{r+1} is obtained from G_r by applying the first rule, this is easy to see, as adding an isolated vertex to G_{r+1} does not affect the validity of the dense triple. If on the other hand, we obtained G_{r+1} by applying one of the other rules, then we deleted from G_r some vertices of I. However, adding to G_{r+1} some vertices to I does not affect the validity of the dense triple, as the vertices of Y do not obtain new neighbors (hence conditions (iii) and (iv) remain satisfied), while condition (ii) is unaffected. We conclude that the constructed triple is valid in G. □

The last claim shows how to construct a dense triple in G if after applying the rules exhaustively the remaining graph is not edge-less. The kernelization algorithm is then the following: apply the rules exhaustively. When this is no longer possible, if the remaining graph is not edge-less, construct a dense triple and invoke Lemma 1 to remove an arbitrary edge of that triple. Run the kernelization algorithm on the remaining graph and return the result. Otherwise, if the graph obtained after applying all the rules is edge-less, we return the initial graph G.

What remains is to prove that when the kernelization algorithm ceases to make progress (that is, when applying all rules produces an edge-less graph), this implies that the given graph must have $O(\tau^3)$ edges. To see this, observe that to apply any rule, we need a vertex $x \in C$ which satisfies certain conditions. Once we apply that rule to x, the same rule cannot be applied to x a second time, because we delete an appropriate set of its neighbors. As a result, the algorithm will perform $O(\tau)$ rule applications. Each rule application deletes either an isolated vertex or at most $O(\tau)$ vertices of I. Each vertex of I is incident with $O(\tau)$ edges (since the other endpoint of each such edge must be in C). Therefore, each rule application removes $O(\tau^2)$ edges from the graph and after $O(\tau)$ rule applications we arrived at an edge-less graph. We conclude that the given graph contained $O(\tau^3)$ edges. □

Corollary 1. COLORED ARC KAYLES *can be solved in time* $\tau^{O(\tau)} + n^{O(1)}$ *on graphs on* n *vertices, where* τ *is the size of a minimum vertex cover of the input graph.*

Proof. Suppose that we have a vertex cover C of size τ (otherwise one can be found with standard FPT algorithms such as [6] in the time allowed). We first apply the algorithm of Theorem 1 in polynomial time to reduce the graph to $O(\tau^3)$ edges. Then, we apply the simple brute force algorithm which considers all possible edges to play for each move. Since the game cannot last for more than τ rounds (as each move decreases the size of the vertex cover), this results in a decision tree of size $\tau^{O(\tau)}$. □

Finally, a corollary of the above results is that ARC KAYLES also admits a polynomial kernel when parameterized by the number of rounds. This follows because the first player has a strategy to win in a small number of rounds only if the graph has a small vertex cover. Notice that this corollary cannot automatically apply to the colored version of the game, because if Black has a strategy to win in a small number of rounds, this only implies that the graph induced by the edge of $E_W \cup E_G$ (that is, the edges playable by White) has a small vertex cover.

Corollary 2. ARC KAYLES *admits a kernel of* $O(r^3)$ *edges and can be solved in time* $r^{O(r)} + n^{O(1)}$*, where the objective is to decide if the first player has a strategy to win in at most* r *rounds.*

Proof. Given an instance of ARC KAYLES G we first compute a maximal matching of G. If the matching contains at least $2r + 1$ edges, then we answer no, as the game will go on for at least $r + 1$ rounds, no matter which strategy the players follow. Otherwise, by taking both endpoints of all edges in the matching we obtain a vertex cover of size at most $4r$, and we can apply Theorem 1 and Corollary 1. □

3 Arc Kayles for Trees Parameterized by Vertex Cover Number

In [3], Bodlaender et al. showed that the winner of NODE KAYLES on trees can be determined in time $O^*(3^{n/3}) = O(1.4423^n)$. Based on the algorithm of Bodlaender et al., Hanaka et al. showed an $O^*(2^{n/2}) = O(1.4143^n)$ time algorithm to determine the winner of ARC KAYLES and NODE KAYLES on trees in [9]. This improvement is achieved by not considering the ordering of subtrees. Now, we show that the improved algorithm in [9] also runs in time $O^*(5^{\tau/2}) = O(2.2361^\tau)$, where τ is the vertex cover number.

We start with an introduction to the algorithm. The algorithm is based on the algorithm for NODE KAYLES of Bodlaender et al. [3], which uses the Sprague-Grundy theory. Any position of a game can be assigned a non-negative integer called nimber. 0 is assigned to a position P if and only if the second player wins in P in the game. Thus, in ARC KAYLES nimber of a graph G is 0 when G has no edge. When a graph has some edges, we calculate $mex(S)$. $mex(S)$ is the smallest non-negative integer which is not contained in S, where S is the set of non-negative integers. In a general game, for a position P where the winner is not trivial, S consists of nimbers of positions reachable from P in one move, and the nimber of P is $mex(S)$. Thus, in ARC KAYLES a nimber of a graph G with some edges is $mex(S)$, where S is the set of the nimbers of graphs which are reachable from G in one move. In addition, when the graph G is disconnected, the nimber of G can be obtained by computing bitwise XOR of the nimbers for each connected component.

The algorithm to determine the winner for ARC KAYLES on trees using Sprague-Grundy theory is as follows: Like a DFS, we calculate the nimber of input graph by calculating the nimbers of graphs which are reachable from input graph in one move, and so on. Once the position has been examined, the calculation result is held and is not calculated again. In memoization, each connected component of a tree is memorized and when for any vertex only the order of its children is different, it is regarded as the same tree.

The exponential part of the running time of the algorithm depends on the number of connected components that can be played in the game. When we play ARC KAYLES on a input graph T, which is a tree and the vertex cover number of T is τ, we claim that the number of connected components that can be played in the game is $O^*(5^{\tau/2}) = O(2.2361^\tau)$ (See appendix for details).

Theorem 2. *The winner of* ARC KAYLES *on a tree whose vertex cover number is τ can be determined in time $O^*(5^{\tau/2})(= O(2.2361^\tau))$.*

4 Arc Kayles for Trees

Continued from Sect. 3, we further analyze the winner determination algorithm in [9] for ARC KAYLES on trees. In [9], Hanaka et al. showed an $O^*(2^{n/2}) = O(1.4143^n)$-time algorithm to determine the winner of ARC KAYLES and NODE

KAYLES on trees, and we gave another running time of the algorithm of [9] with respect to vertex cover number in Sect. 3. Now, we improve the estimation of the running time of the algorithm and show that the winner of ARC KAYLES and NODE KAYLES on trees can be determined in time $O^*(7^{n/6})(= O(1.3831^n))$.

Theorem 3. *The winner of* ARC KAYLES *on a tree with n vertices can be determined in time* $O^*(7^{n/6}) = O(1.3831^n)$.

Theorem 4. *The winner of* NODE KAYLES *on a tree with n vertices can be determined in time* $O^*(7^{n/6})(= O(1.3831^n))$.

5 NP-Hardness of BW-Arc Kayles

The complexity to determine the winner of combinatorial games is expected to be PSPACE-hard, though no hardness results are known for (COLORED) ARC KAYLES so far. In this section, we prove that BW-ARC KAYLES is NP-hard.

Theorem 5. BW-ARC KAYLES *is NP-hard.*

Proof. We give a polynomial-time reduction from VERTEX COVER, which is the problem to decide whether G has a vertex cover of size at most τ. Let $G = (V, E)$ and τ be an instance of VERTEX COVER. Now we construct an edge-colored graph G' from G such that the black player has a winning strategy on G' playing first if and only if G has a vertex cover of size at most τ.

We construct G' as follows. The graph G' consists of three layers as shown in Fig. 1. The bottom layer corresponds to $G = (V, E)$; the vertex and edge sets are copies of V and E, which we call with the same name V and E. The edges in E are colored in white. The middle layer is a clique with size $2\tau - 1$, where the vertex set is $U = \{u_1, \dots, u_{2\tau-1}\}$ and all edges are colored in black. The top layer consists of two vertex sets $B = \{b_1, \dots, b_{2\tau-1}\}$ and $W = \{w_1, \dots, w_{2\tau-1}\}$, where they are independent. The bottom and middle layers are completely connected by black edge set $E_{V,U} = \{\{v, u\} \mid v \in V, u \in U\}$. The middle and top layers are connected by black edge set $E_{U,B} = \{\{u_i, b_i\} \mid i = 1, \dots, 2\tau - 1\}$ and white edge set $E_{U,W} = \{\{u_i, w_i\} \mid i = 1, \dots, 2\tau - 1\}$.

Let S be a vertex cover of G of size τ. For S, we define $E_{S,U} = \{\{v, u\} \in E_{V,U} \mid v \in S\}$. Note that the second (white) player can choose only edges in $E_{U,W}$ or E. The strategy of the first (black) player is as follows. In the first turn, the black player just chooses an edge in $E_{S,U}$. After that, the black player chooses an edge according to which edge the white player chooses right before the black turn. If the white player chooses an edge in $E_{U,W}$, let the black player choose an edge in $E_{S,U}$ in the next black turn. Otherwise, i.e., the white player chooses an edge in E, let the black player choose an edge in the middle layer in the next black turn. This is the strategy of the black player.

We now show that this is a winning strategy for the black player. If the black player following this strategy can choose an edge in every turn right after the white player's action, the black player is the winner. In fact, this procedure continues at most $2\tau - 1$ turns because exactly two vertices in U and at least

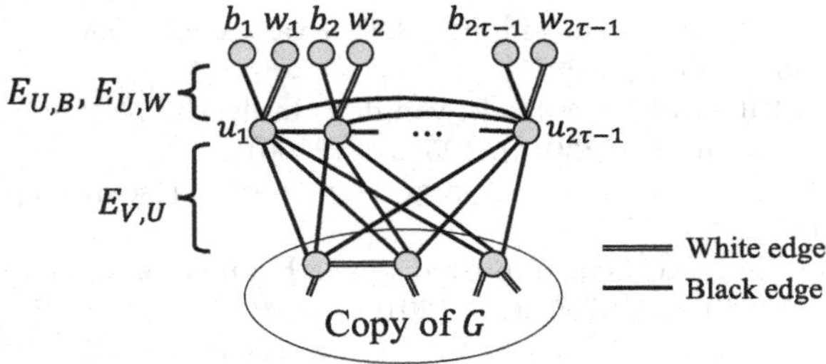

Fig. 1. The constructed graph G'.

one vertex in S are removed in every two turns (a white turn and the next black turn) under this strategy; after $2\tau - 1$ turns, no white edge is left and the next player is the white player. Thus what we need to show here is that the black player following this strategy can choose an edge in every turn right after the white player's action. Under this strategy, E can become empty before $2\tau - 1$ turns. In this case, the black player chooses an edge in $E_{U,B}$ instead of an edge in $E_{S,U}$ if the white player chooses an edge $E_{U,W}$. This makes that exactly two vertices in U are removed in every two turns. Then, the black player wins in the same way as above.

Next, we show that the white player has a winning strategy if G does not have a vertex cover of size τ, i.e. $|S| \geq \tau + 1$. The white player can win the game by selecting an edge in $E_{U,W}$ in every turn. Under this strategy, exactly one vertex in U is removed in white player's turn, and then the black player can play at most τ times because the size of U is $2\tau - 1$. An edge which black player can choose in his turn is $E_{U,B}$ or $E_{V,U}$, then the black player can remove vertices in S at most τ times. Therefore, after $2\tau - 1$ turns there are some vertices and white edges in the bottom layer and there is no black edge because U is empty. The winner is the white player. □

Now, we consider Colored Arc Kayles. Colored Arc Kayles is generalized of Arc Kayles and BW-Arc Kayles; edges are colored black, white and gray, and the black (resp., white) edges are selected by only the black (resp., white) player, while both the black and white players can select gray edges. Since Colored Arc Kayles includes BW-Arc Kayles, we also obtain the following corollary.

Corollary 3. Colored Arc Kayles *is NP-hard.*

References

1. Albert, M., Nowakowski, R., Wolfe, D.: Lessons in Play: An Introduction to Combinatorial Game Theory (2007)
2. Bodlaender, H.L., Kratsch, D.: Kayles and nimbers. J. Algorithms **43**(1), 106–119 (2002)

3. Bodlaender, H.L., Kratsch, D., Timmer, S.T.: Exact algorithms for kayles. Theoret. Comput. Sci. **562**, 165–176 (2015)
4. Breuker, D.M., Uiterwijk, J.W.H.M., van den Herik, H.J.: Solving 8×8 domineering. Theoret. Comput. Sci. **230**(1), 195–206 (2000)
5. Bullock, N.: Domineering: solving large combinatorial search spaces. ICGA J. **25**(2), 67–84 (2002)
6. Chen, J., Kanj, I.A., Xia, G.: Improved upper bounds for vertex cover. Theoret. Comput. Sci. **411**(40–42), 3736–3756 (2010)
7. Cygan, M., et al.: Parameterized Algorithms. Springer, Cham (2015). https://doi.org/10.1007/978-3-319-21275-3
8. Gardner, M.: Mathematical games: cram, crosscram and quadraphage: new games having elusive winning strategies. Sci. Am. **230**(2), 106–108 (1974)
9. Hanaka, T., Kiya, H., Ono, H., Yoshiwatari, K.: Winner determination algorithms for graph games with matching structures. Algorithmica (2023)
10. Huggan, M.A., Stevens, B.: Polynomial time graph families for arc kayles. Integers **16**, A86 (2016)
11. Kobayashi, Y.: On structural parameterizations of node kayles. In: Akiyama, J., Marcelo, R.M., Ruiz, M.-J.P., Uno, Y. (eds.) JCDCGGG 2018. LNCS, vol. 13034, pp. 96–105. Springer, Cham (2021). https://doi.org/10.1007/978-3-030-90048-9_8
12. Lachmann, M., Moore, C., Rapaport, I.: Who wins domineering on rectangular boards? arXiv preprint math/0006066 (2000)
13. Lampis, M., Mitsou, V.: The computational complexity of the game of set and its theoretical applications. In: Pardo, A., Viola, A. (eds.) LATIN 2014. LNCS, vol. 8392, pp. 24–34. Springer, Heidelberg (2014). https://doi.org/10.1007/978-3-642-54423-1_3
14. Schaefer, T.J.: On the complexity of some two-person perfect-information games. J. Comput. Syst. Sci. **16**(2), 185–225 (1978)
15. Uiterwijk, J.W.H.M.: 11 × 11 domineering is solved: the first player wins. In: Plaat, A., Kosters, W., van den Herik, J. (eds.) CG 2016. LNCS, vol. 10068, pp. 129–136. Springer, Cham (2016). https://doi.org/10.1007/978-3-319-50935-8_12
16. Uiterwijk, J.W.H.M.: Construction and investigation of cram endgame databases. ICGA J. **40**(4), 425–437 (2018)
17. Uiterwijk, J.W.H.M.: Solving cram using combinatorial game theory. In: Cazenave, T., van den Herik, J., Saffidine, A., Wu, I.C. (eds.) ACG 2019. LNCS, vol. 12516, pp. 91–105. Springer, Cham (2019). https://doi.org/10.1007/978-3-030-65883-0_8
18. Uiterwijk, J.W.H.M., Barton, M.: New results for domineering from combinatorial game theory endgame databases. Theoret. Comput. Sci. **592**, 72–86 (2015)
19. Yoshiwatari, K., Kiya, H., Hanaka, T., Ono, H.: Winner determination algorithms for graph games with matching structures. In: Bazgan, C., Fernau, H. (eds.) IWOCA 2022. LNCS, vol. 13270, pp. 509–522. Springer, Cham (2022). https://doi.org/10.1007/978-3-031-06678-8_37

Automata Classes Accepting Languages Whose Commutative Closure is Regular

Stefan Hoffmann(✉)

Universität Trier, Behringstraße 21, 54296 Trier, Germany
hoffmanns.tcs@gmail.com

Abstract. The commutative closure operation, which corresponds to the Parikh image, is a natural operation on formal languages occurring in verification and model-checking. Commutative closures of regular languages correspond to semilinear sets and, by Parikh's theorem, to the commutative closures of context-free languages. The commutative closure is not regularity-preserving on the class of regular languages, for example already the commutative closure of the simple language $(ab)^*$ is not regular. Here, we show that the commutative closure of a binary regular language accepted by a circular automaton yields a regular language. Then, we deduce a sufficient condition on the cycles in automata for regularity of the commutative closure. This yields this property, for example, for the following classes of automata: automata with threshold one transformation semigroups, automata with simple idempotents and almost-group automata. The fact that the commutative closure on group languages and polynomials of group languages is regularity-preserving is known in the literature. Polynomials of group languages correspond to level one-half of the group hierarchy. We also show that on the next level in this hierarchy, i.e., level one, this property is lost and the commutative closure is no longer regularity-preserving. Lastly, we give a binary circular automaton not contained in the largest proper positive variety $\mathcal{W}$ closed under shuffle and commutative closure.

Keywords: Automata Classes · Regularity-Preserving Operation · Regular Language · Commutative Closure · Commutation · Parikh Image

1 Introduction

Having applications in regular model checking [5,7,8], or arising naturally in the theory of traces [12,39] (one model for parallelism), the (partial) commutative closure has been extensively studied [5,7–9,11,13,14,16,17,19,22,28,29,36,37, 39].

The full commutative closure corresponds to the Parikh image, i.e., the image of the Parikh map [30] that counts the multiplicities of each letter in a word. Languages whose Parikh image equals the Parikh image of some regular language are called *slip languages* [15]. So, slip languages in a representation from

H. Fernau et al. (Eds.): SOFSEM 2024, LNCS 14519, pp. 311–325, 2024.
https://doi.org/10.1007/978-3-031-52113-3_22

which we can effectively obtain a regular language with the same Parikh image have decidable emptiness problem [35]. Examples of classes of slip languages with decidable emptiness problem are the context-free languages [30], reversal-bounded multicounter machines [24] or Parikh automata [25] and, due to decidability of emptiness, these languages are of interest in model checking and verification. Regularity of the commutative closure of regular languages is, in general, a decidable problem [13,16]. Here we investigate regularity of the commutative closure on regular languages, a question that has been investigated intensively before, see Sect. 3 or, for example [17] and the literature mentioned therein.

Contribution. First, we show that the commutative closure of a binary language accepted by a circular automaton yields a regular language. This result fails over larger alphabets. Indeed, the fact that results on the commutative closure might be sensitive to the alphabet size was known before in other instances. For example, Latteux [27] and Rigo [37] proved that the commutative closure of a binary slip language yields a context-free language, a result that is also false for larger alphabets. We then proceed to state general (and alphabet-independent) sufficient conditions for the regularity of the commutative closure of regular languages in terms of structural properties of accepting automata. The commutative closure of a polynomial of group languages, a language from the first half-level of the group hierarchy, yields a regular language [17]. We give a language showing that this is optimal in terms of the group hierarchy, i.e., a language on the next level with non-regular commutative closure. It is known, and in fact easily seen, that for the analogously defined Straubing-Thérien hierarchy [33,41,42], the commutative closure is regularity-preserving up to level three-half and this is optimal [23]. However, that level one-half is optimal in the group hierarchy does not appear to be stated in the literature and the witness language is not obvious. Lastly, we give a binary circular automaton accepting a language not in the positive variety $\mathcal{W}$, which is closed under commutative closure [17]. As binary circular automata are not contained in the other known classes of languages closed for commutative closure, this shows that languages accepted by binary circular automata yield a genuinely new class of regular languages with regular commutative closures.

2 Preliminaries

By Σ we denote a finite *alphabet* and by Σ^* the *set of finite sequences* with concatenation. The elements of Σ^* are also called *words*. Let $u \in \Sigma^*$ and $a \in \Sigma$. By $|u|$ we denote the *length* of u and by $|u|_a$ we denote the number of *occurrences of the letter* a in u. The *empty word*, i.e., the word of length zero, is denoted by ε. A *language* is a subset of Σ^*. Given $L \subseteq \Sigma^*$, the *Kleene plus* is $L^+ = \{u_1 \cdots u_n \mid n > 0 \land \forall i \in \{1, \ldots, n\} : u_i \in L\}$ and the *Kleene star* is $L^* = L^+ \cup \{\varepsilon\}$. Let Γ be an alphabet. A *morphism* is a map $\varphi \colon \Sigma^* \to \Gamma^*$ such that $\varphi(uv) = \varphi(u)\varphi(v)$ for $u, v \in \Sigma^*$. The *projection morphism (on Γ)* for $\Gamma \subseteq \Sigma$ is $\pi_\Gamma \colon \Sigma^* \to \Gamma^*$ given by $\pi_\Gamma(x) = \varepsilon$ for $x \in \Sigma \setminus \Gamma$, $\pi_\Gamma(x) = x$ for $x \in \Gamma$ and $\pi_\Gamma(x_1 \cdots x_n) = \pi_\Gamma(x_1)\pi_\Gamma(x_2) \cdots \pi_\Gamma(x_n)$ for $x_1, \ldots, x_n \in \Sigma$.

The *shuffle operation*, denoted by $\shuffle$, is defined by

$$u \shuffle v = \{w \in \Sigma^* \mid w = x_1 y_1 x_2 y_2 \cdots x_n y_n \text{ for some words } x_1, \ldots, x_n, y_1, \ldots, y_n \in \Sigma^* \text{ such that } u = x_1 x_2 \cdots x_n \text{ and } v = y_1 y_2 \cdots y_n\}$$

for $u, v \in \Sigma^*$ and $L_1 \shuffle L_2 := \bigcup_{x \in L_1, y \in L_2} (x \shuffle y)$ for $L_1, L_2 \subseteq \Sigma^*$. The *positive iterated shuffle* of L_1 is $L_1 \cup L_1 \shuffle L_1 \cup L_1 \shuffle L_1 \shuffle L_1 \cup \ldots$, the *iterated shuffle* adds $\{\varepsilon\}$ to it.

Let $u \in \Sigma^*$. The *commutation operation*, or *commutative closure*, is $[u] = \{v \in \Sigma^* \mid \forall a \in \Sigma : |u|_a = |v|_a\}$. This notation is taken from [17]. If $L \subseteq \Sigma^*$, then we set $[L] = \bigcup_{u \in L} [u]$. A language is called *commutative* if $[L] = L$.

A *finite, deterministic and complete automaton over* Σ *(or automaton for short)* is a quintuple $\mathcal{A} = (Q, \Sigma, \delta, q_0, F)$ such that Σ is the *input alphabet*, Q the finite *state set*, $\delta \colon Q \times \Sigma \to Q$ is a *total transition function*, q_0 the *start (or initial) state* and F the set of *final states*. We can extend δ to a function $\hat{\delta} \colon Q \times \Sigma^* \to Q$ in the usual way by setting $\hat{\delta}(q, \varepsilon) = q$ and $\hat{\delta}(q, ua) = \delta(\hat{\delta}(q, u), a)$ for $u \in \Sigma^*$, $a \in \Sigma$ and $q \in Q$. This function is also denoted by δ in the following. For $S \subseteq Q$ and $u \in \Sigma^*$, we set $\delta(S, u) = \{\delta(q, u) \mid q \in S\}$. We call $\mathcal{A}$ a *permutation automaton* if for every letter $a \in \Sigma$ the map $q \mapsto \delta(q, a)$ permutes the states.

By $L(\mathcal{A}) = \{u \in \Sigma^* \mid \delta(q_0, u) \in F\}$ we denote the language *recognized* (or *accepted*) by $\mathcal{A}$. A language $L \subseteq \Sigma^*$ is called *regular* if there exists an automaton $\mathcal{A}$ such that $L = L(\mathcal{A})$.

A *class of languages* $\mathcal{C}$ is a correspondence that associates with each alphabet Σ a set of languages $\mathcal{C}(\Sigma^*)$ over Σ.

We say a class $\mathcal{C}$ is *closed for some operation* if the operation, applied to languages from $\mathcal{C}(\Sigma^*)$, yields a language in $\mathcal{C}(\Sigma^*)$. We say $\mathcal{C}$ is a *class of regular languages* if $\mathcal{C}(\Sigma^*)$ is a set of regular languages for each Σ.

A *positive variety* [32] $\mathcal{V}$ is a class of regular languages such that, for all alphabets Σ and Γ,

1. $\mathcal{V}$ is a *lattice of languages*, i.e., closed under finite intersections and finite unions,
2. for every morphism $\varphi \colon \Sigma^* \to \Gamma^*$, if $L \in \mathcal{V}(\Gamma^*)$, then $\varphi^{-1}(L) \in \mathcal{V}(\Sigma^*)$,
3. for every $L \in \mathcal{V}(\Sigma^*)$ and $u \in \Sigma^*$ we have $u^{-1}L \in \mathcal{V}(\Sigma^*)$ and $Lu^{-1} \in \mathcal{V}(\Sigma^*)$.

As $\mathcal{V}$ is a lattice of languages we have $\{\emptyset, \Sigma^*\} \subseteq \mathcal{V}(\Sigma^*)$. We call $\mathcal{V}$ a *variety* if $\mathcal{V}(\Sigma^*)$ is also closed under complement.

Let $\mathcal{C}$ be a class languages. A *marked product* of languages over $\mathcal{C}(\Sigma^*)$ is a language of the form $L_0 a_0 L_1 a_1 \cdots a_k L_k$ with $k \geq 0$, $a_1, \ldots, a_k \in \Sigma$ and $L_1, \ldots, L_k \in \mathcal{C}(\Sigma^*)$. A *polynomial over languages in* $\mathcal{C}(\Sigma^*)$ is a finite union of marked products over $\mathcal{C}(\Sigma^*)$. By $\mathrm{Pol}(\mathcal{C})$ we denote the class of polynomials over $\mathcal{C}$, i.e., $L \in \mathrm{Pol}(\mathcal{C})(\Sigma^*)$ iff L is a polynomial over languages in $\mathcal{C}(\Sigma^*)$.

Example 1. Let Σ be an alphabet.

1. The class of all regular languages is a variety also closed under morphisms and shuffle.

2. The class of all regular commutative languages $\mathcal{C}om$ is a variety of languages.
3. The *variety of group languages* $\mathcal{G}$ is given by, for Σ,

$$\mathcal{G}(\Sigma^*) = \{L \subseteq \Sigma^* \mid L \text{ is accepted by a permutation automaton over } \Sigma\}.$$

4. Let $\mathcal{I}(\Sigma^*) = \{\emptyset, \Sigma^*\}$. Then $(\text{Pol}(\mathcal{I}))(\Sigma^*)$ are finite unions of languages of the form $\Sigma^* a_1 \Sigma^* \cdots a_n \Sigma^*$, $a_1, \ldots, a_n \in \Sigma$.
5. The class $\mathcal{J}$ of *piecewise testable languages* [40] is the Boolean closure of $\text{Pol}(\mathcal{I})$ and a variety.
6. The class $\mathcal{J}^-$ of complements of languages in $\text{Pol}(\mathcal{I})$, which is a positive variety.

We can build up a hierarchy of languages over $\mathcal{C}(\Sigma^*)$ by setting $\mathcal{C}_0(\Sigma^*) = \mathcal{C}(\Sigma^*)$ and, for $n \geq 0$, let $\mathcal{C}_{n+\frac{1}{2}}(\Sigma^*)$ be the polynomials over languages in $\mathcal{C}_n(\Sigma^*)$ and let $C_{n+1}(\Sigma^*)$ be the Boolean closure of languages from $\mathcal{C}_{n+\frac{1}{2}}(\Sigma^*)$.

By starting at level zero with the finite or cofinite[1] languages over Σ we obtain the *dot-depth hierarchy* [10], by starting with $\mathcal{C}_0(\Sigma^*) = \{\emptyset, \Sigma^*\}$ we obtain the *Straubing-Thérien hierarchy* [41,42] and by starting from $\mathcal{G}$ we obtain the *group hierarchy*. See [33] for more information on concatenation hierarchies.

We need the following lemma, which is an easy application of the pigeonhole principle.

Lemma 2.1. *Let $a_1, \ldots, a_n \in \mathbb{N}_0$. Then there exists a subset of $\{a_1, \ldots, a_n\}$ whose sum is divisible by n.*

3 Known Results

For the following classes of regular languages results for the commutation operation are known:

1. $\text{Pol}(\mathcal{G})$, $\text{Pol}(\mathcal{I})$, $\mathcal{J}$, $\mathcal{J}^-$, $\text{Pol}(\mathcal{J})$, $\text{Pol}(\mathcal{C}om)$,
2. the unique maximal positive variety $\mathcal{W}$ of languages that does not contain the language $(ab)^*$ for every $a \neq b$ [18],
3. the class $\mathcal{SE}(\mathcal{G})$, defined as the closure of $\mathcal{G}$ under shuffle, iterated shuffle, concatenation, union and Kleene star [20].

Languages from $\text{Pol}(\mathcal{J})$ can be written as finite unions of languages of the form $\Sigma_0^* a_1 \Sigma_1^* \cdots a_n \Sigma_n^*$ with $\Sigma_i \subseteq \Sigma$ and $a_1, \ldots, a_n \in \Sigma$ (see Arfi [1, Thm. 1.3]) and are also known as *Alphabetical Pattern Constraints* [5].

The next theorem summarizes results from Guaiana, Restivo and Salemi [19], Bouajjani, Muscholl and Touili [5], Cécé, Héam and Mainier [7,8], Cano, Guaiana and Pin [17] and Hoffmann [20].

Theorem 3.1. *The following properties hold:*

[1] A language is cofinite if its complement is finite.

1. *the classes* Pol($\mathcal{G}$), $\mathcal{W}$, Pol($\mathcal{I}$), $\mathcal{J}$, $\mathcal{J}^-$, Pol($\mathcal{J}$) *and* Pol($\mathcal{C}om$) *are all closed under commutation,*
2. *commutation on* $\mathcal{SE}(\mathcal{G})$ *is regularity-preserving.*

Actually, a couple of the results mentioned in Theorem 3.1 hold for more general partial commutation closures (sometimes also called trace closures [12]). However, as in this work we are only concerned with the (full) commutative closure, we do not introduce this more general notion and instead refer to the literature [12,17].

4 Circular Automata over a Binary Alphabet

The alphabet Σ is said to be a *binary alphabet* if $|\Sigma| = 2$. An automaton $\mathcal{A} = (Q, \Sigma, \delta, q_0, F)$ is called *circular* if there exists $b \in \Sigma$ cyclically permuting the states, i.e., $\delta(q, b^{|Q|}) = q$ and $Q = \{\delta(q, b^i) \mid i \geq 0\}$ for some (and hence all) $q \in Q$. So, the states can be arranged in a cycle traversed by the letter b.

Here, we show that for circular automata over a binary alphabet, the commutative closure yields a regular language. This result does not hold for larger alphabets, see Fig. 1. First, after introducing the positive variety $\mathcal{C}om^+$ and a characterization of it, we show a lemma about the regularity of the commutative closure of certain regular languages over a binary alphabet and another lemma on the commutative closure of languages accepted by circular automata over a binary alphabet. Then we prove the result as a consequence of both lemmas.

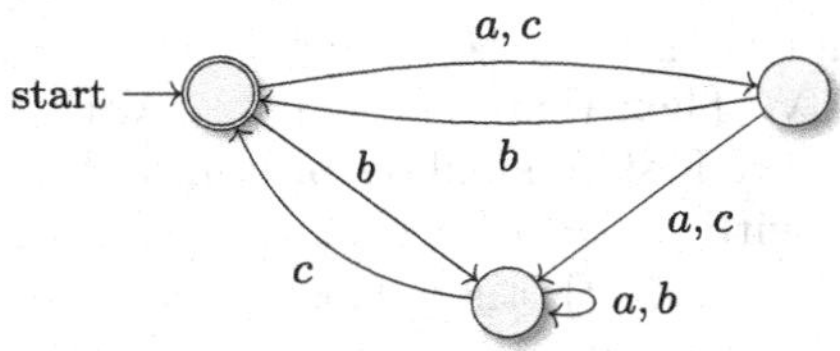

Fig. 1. A circular automaton $\mathcal{A}$ over $\Sigma = \{a, b, c\}$ such that the commutative closure of the recognized language is not regular, as $[L(\mathcal{A})] \cap \{a, b\}^* = \{u \in \{a, b\}^* : |u|_a = |u|_b\}$. The final state is marked by a double circle.

Pin [32] introduced the positive variety $\mathcal{C}om^+$ where $\mathcal{C}om^+(\Sigma^*)$ equals the lattice of languages generated by sets of the form $F(a, t) = \{u \in \Sigma^+ : |u|_a \geq t\}$ and $F(a, r, n) = \{u \in \Sigma^* : |u|_a \equiv r \pmod{n}\}$ for $a \in \Sigma$, $t \geq 0$ and $0 \leq r < n$. Every language in $\mathcal{C}om^+(\Sigma^*)$ is regular and the following characterization of those languages in $\mathcal{C}om^+(\Sigma^*)$ was obtained by Hoffmann [21, Thm. 31].

Theorem 4.1 (**[21, Thm. 31]**). *Let* $L \subseteq \Sigma^*$ *be commutative. Then we have* $L \in \mathcal{C}om^+(\Sigma^*)$ *if and only if there exists* $P > 0$ *such that for all* $a \in \Sigma$ *and* $u \in L$ *we have* $ua^P \in L$.

Observe that for commutative $L \subseteq \Sigma^*$ we have $ua^P \in L$ iff $[u] \sqcup\!\sqcup a^P \subseteq L$.

Lemma 4.2. *Let $L \subseteq \{a,b\}^*$ be a regular language. Then for each $n \geq 0$ the language $(b^n \sqcup\!\sqcup a^*) \cap [L] = \{u \in [L] : |u|_b = n\}$ is regular.*

Proof. We have $(b^n \sqcup\!\sqcup a^*) \cap [L] = b^n \sqcup\!\sqcup \{a^{|u|_a} : |u|_b = n \wedge u \in L\}$. As $\{a^{|u|_a} : |u|_b = n \wedge u \in L\} = \pi_{\{a\}}(L \cap (b^n \sqcup\!\sqcup a^*))$ and the regular languages are closed under morphisms, the shuffle operation and intersection, it follows that $(b^n \sqcup\!\sqcup a^*) \cap [L]$ is regular. □

Lemma 4.3. *Suppose $\mathcal{A} = (Q, \{a,b\}, \delta, q_0, F)$ is a circular automaton. Then there exists $N > 0$ such that*

$$\{u \in \Sigma^* : |u| \geq N \text{ and } [u] = [v] \text{ for some } v \in L(\mathcal{A})\}$$

is in $\mathcal{C}om^+(\{a,b\}^*)$.

Proof. Assume the letter b cyclically permutes the states. By finiteness, the letter a must induce at least one cycle on the states, i.e., there exist $q \in Q$ and $l > 0$ such that $\delta(q, a^l) = q$. In the following, we present the proof for the special case $l = 2$, which is perfectly general as it contains the basic proof idea while avoiding the notational burden of introducing additional indices. However, at the end we indicate how to adapt the proof for general $l > 0$.

Set $L = L(\mathcal{A})$, $n = |Q|$, $N = 2n^4$ and $P = 1 \cdot 2 \cdot 3 \cdots n$. Let $u \in [L]$. Then $[u] = [v]$ for some $v \in L$. As the letter b circularly permutes the states, it is $\delta(q_0, v) = \delta(q_0, vb^n)$. Hence, we have $vb^n \in L$ and so $[u] \sqcup\!\sqcup b^n = [vb^n] \subseteq [L]$. As P is a multiple of n we also get $[u] \sqcup\!\sqcup b^P \subseteq [L]$. In the remaining proof, we show that $u \in [L]$ with $|u| \geq N$ implies $u \sqcup\!\sqcup a^P \subseteq [L]$, which gives the claim by Theorem 4.1. Let $u \in [L]$ with $|u| \geq N$. Then there exists $v \in L$ with $[u] = [v]$ and we also have $|v| \geq N$. We divide the first N symbols of v into blocks of length n, i.e. we write $v = v_1 v_2 \cdots v_{2n^3} w$ with $|v_i| = n$ for each $i \in \{1, \ldots, 2n^3\}$ and $w \in \Sigma^*$. By finiteness, for each $i \in \{1, \ldots, 2n^3\}$, there exist $x_i, y_i, z_i \in \Sigma^*$ and $p_i \in Q$ such that $v_i = x_i y_i z_i$, $|y_i| > 0$, $p_i = \delta(q_0, v_1 \cdots v_{i-1} x_i)$ and $\delta(p_i, y_i) = p_i$.

First, assume $y_i = a^m$ for some $i \in \{1, \ldots, 2n^3\}$ and $0 < m \leq n$. As we can "pump up" the factor y_i and still end up at the same state, we find

$$\delta(q_0, v_1 \cdots v_{i-1} x_i y_i y_i^{P/m} z_i v_{i+1} \cdots v_{2n^3} w) = \delta(q_0, v) \in F.$$

Furthermore, $[v_1 \cdots v_{i-1} x_i y_i y_i^{P/m} z_i v_{i+1} \cdots v_{2n^3} w] = [v] \sqcup\!\sqcup a^P$ and so $[u] \sqcup\!\sqcup a^P = [v] \sqcup\!\sqcup a^P \subseteq [L]$.

Next, suppose $|y_i|_b > 0$ for all $i \in \{1, \ldots, 2n^3\}$. By the strong pigeonhole principle, at least $2n^2$ states among the states $p_1, \ldots, p_{2n^3}$ are equal. Without loss of generality, assume $p_1 = \ldots = p_{2n^2}$. As $|y_i| \leq n$ (and so $|y_i|_b \leq n$) for all $i \in \{1, \ldots, 2n^3\}$, using the strong pigeonhole principle again, we conclude that $2n$ factors among the factors $y_1, \ldots, y_{2n^2}$ have the same number of b's. Without loss of generality, assume $|y_1|_b = \ldots = |y_{2n}|_b$. Set

$$x = |y_1 \cdots y_n|_a \text{ and } y = |y_{n+1} \cdots y_{2n}|_a.$$

At least one of the numbers $x, y, x+y$ must be even, for if x and y are odd, then $x+y$ is even. Without loss of generality, assume x is even. Now, as, by the above choice, the factors $y_1, \ldots, y_n$ start and end at the same state, i.e., $p_1 = \ldots = p_n$ with $p_i = \delta(q_0, v_1 \cdots v_{i-1}x_i)$ and $\delta(p_i, y_i) = p_i$ for $i \in \{1, \ldots, n\}$, we can move them all into the first block and get a word in L (for the case that y is even we would use the same reasoning with $y_{n+1} \cdots y_{2n}$ and if $x+y$ is even we would use $y_1, \ldots, y_{2n}$ with the same reasoning in the following). More precisely,

$$\delta(q_0, x_1 y_1 y_2 \cdots y_n z_1 x_2 z_2 \cdots x_n z_n v_{n+1} \cdots v_{2n^3} w) = \delta(q_0, v) \in F.$$

Now, consider the state $q \in Q$ with $\delta(q, a^l) = q$ from above. As b cyclically permutes the states, we can choose $0 \leq M < n$ such that $\delta(q_0, x_1 b^M) = q$. Then, as x is even, and so $\delta(q, a^x) = q$, P is a multiple of l, $|y_1|_b > 0$ and b cyclically permutes the states, we have

$$\begin{aligned} \delta(q_0, x_1) &= \delta(q_0, x_1 b^{|y_1|_b \cdot n}) \\ &= \delta(q_0, x_1 b^{M + (|y_1|_b \cdot n - M)}) \\ &= \delta(q_0, x_1 b^M a^x b^{|y_1|_b \cdot n - M}) \\ &= \delta(q_0, x_1 b^M a^{x+P} b^{|y_1|_b \cdot n - M}). \end{aligned}$$

Hence, with $|y_1 \cdots y_n|_b = |y_1|_b \cdot n$ we get

$$\begin{aligned} &\delta(q_0, x_1 b^M a^{x+P} b^{|y_1 \cdots y_n|_b - M} z_1 x_2 z_2 \cdots x_n z_n v_{n+1} \cdots v_{2n^3} w) \\ &= \delta(q_0, x_1 z_1 x_2 z_2 \cdots x_n z_n v_{n+1} \cdots v_{2n^3} w) \\ &= \delta(q_0, x_1 y_1 y_2 \cdots y_n z_1 x_2 z_2 \cdots x_n z_n v_{n+1} \cdots v_{2n^3} w) \\ &= \delta(q_0, v) \in F. \end{aligned}$$

So, $x_1 b^M a^{x+P} b^{|y_1 \cdots y_n|_b - M} z_1 x_2 z_2 \cdots x_n z_n v_{n+1} \cdots v_{2n^3} w \in L$. Furthermore, it is $[b^M a^x b^{|y_1 \cdots y_n|_b - M}] = [y_1 \cdots y_n]$ and so

$$\begin{aligned} [x_1 b^M a^x b^{|y_1 \cdots y_n|_b - M} z_1 x_2 z_2 \cdots x_n z_n v_{n+1} \cdots v_{2n^3} w] &= [v], \\ [x_1 b^M a^{x+P} b^{|y_1 \cdots y_n|_b - M} z_1 x_2 z_2 \cdots x_n z_n v_{n+1} \cdots v_{2n^3} w] &= [v] \shuffle a^P. \end{aligned}$$

Hence, we find $[u] \shuffle a^P = [v] \shuffle a^P \subseteq [L]$.

Lastly, we indicate the argument for general $l > 0$, which uses the same idea as in the case $l = 2$. In this case, we consider $N = l \cdot n^4$ and for $u \in [L]$ with $|u| \geq N$ and $[u] = [v]$ for some $v \in L$ we find nonempty factors $y_1, \cdots, y_{ln}$ in $v \in L$ as above all having the same number of b's and looping at the same state in $\mathcal{A}$. If $|y_i|_b = 0$ for some $i \in \{1, \ldots, ln\}$, then we can pump up P times the letter a. Otherwise, we consider the l numbers $x_i = |y_{in+1} \cdots y_{in+n}|_a$ for $i \in \{0, 1, \ldots, l-1\}$ and all their sums. By Lemma 2.1 there exists a sum s that is a multiple of l. Then we argue as above using the y_i's corresponding to this sum by moving them to the front, rearrange the b's to reach the state $q \in Q$ with $\delta(q, a^l) = q$ and pump up $s + P$ times the letter a to obtain a word in L with the same number of a's and b's as va^P. □

Theorem 4.4. *Let $\Sigma = \{a, b\}$. If $\mathcal{A} = (Q, \Sigma, \delta, q_0, F)$ is a circular automaton, then the language $[L(\mathcal{A})]$ is regular.*

Proof. Set $L = L(\mathcal{A})$ an let $N > 0$ be the number according to Lemma 4.3. We have

$$[L] = \left(\bigcup_{i=0}^{N-1} ([L] \cap (b^i \sqcup\!\sqcup a^*)) \right) \cup ([L] \cap \{u \in \Sigma^* : |u|_b \geq N\}).$$

By Lemma 4.2 the languages $[L] \cap (b^i \sqcup\!\sqcup a^*)$ for $i \in \{0, 1, \ldots, N-1\}$ are regular. Furthermore, it is

$$[L] \cap \{u \in \Sigma^* : |u|_b \geq N\} = [L] \cap \{u \in \Sigma^* : |u| \geq N\} \cap b^N b^* \sqcup\!\sqcup a^*.$$

By Lemma 4.3 the language $[L] \cap \{u \in \Sigma^* : |u| \geq N\}$ is regular. So, we find that $[L]$ is regular. □

5 Structural Conditions on Automata

Here, we show that if each word accepted by a given automaton visits, for each letter $a \in \Sigma$ in the word, some state contained in a cycle labelled exclusively by a, then the automaton accepts a language whose commutative closure is regular. As a consequence we conclude that automata $\mathcal{A} = (Q, \Sigma, \delta, q_0, F)$ such that $\delta(Q, aa) = \delta(Q, a)$ for $a \in \Sigma$ accept languages whose commutative closure is regular. We then list a few automata families from the literature to which this result applies. Binary circular automata are not covered by the conditions mentioned in this section.

Hoffmann [21] introduced the positive variety $\mathcal{C}om_S^+$. By [21, Cor. 12] it follows that $L \subseteq \mathcal{C}om_S^+(\Sigma^*)$ if and only if L is a finite union of languages from $\{U \mid U \in \mathcal{C}om^+(\Gamma^*)$ for some $\Gamma \subseteq \Sigma.\}$. In [21, Thm. 33] the following characterization was obtained.

Theorem 5.1 ([21, Thm. 33]). *Let $L \subseteq \Sigma^*$ be commutative. Then we have $L \in \mathcal{C}om_S^+(\Sigma^*)$ if and only if there exists $P > 0$ such that for all $a \in \Sigma$ and $u \in L$ with $|u|_a > 0$ we have $ua^P \in L$.*

Theorem 5.2. *Let $\mathcal{A} = (Q, \Sigma, \delta, q_0, F)$ be an automaton with the property that for every word $u \in L(\mathcal{A})$ and letter $a \in \Sigma$ with $|u|_a > 0$, there exists a prefix $v \in \Sigma^*$ of u and $p > 0$ such that $\delta(\delta(q_0, v), a^p) = \delta(q_0, v)$, i.e., the state $\delta(q_0, v)$ is contained in a cycle whose transitions are solely labeled with the letter a. Then $[L(\mathcal{A})]$ is in $\mathcal{C}om_S^+(\Sigma^*)$.*

Proof. For each $a \in \Sigma$, let $P_a > 0$ be the least common multiple of the minimal a-cycle lengths in the automaton, i.e., set

$$P_a = \operatorname{lcm}\{p > 0 \mid \exists q \in Q : q = \delta(q, a^p) \wedge p = \min\{\overline{p} > 0 \mid \delta(q, a^{\overline{p}}) = q\}\},$$

Then set $P = \operatorname{lcm}\{P_a \mid a \in \Sigma\}$.

Let $u \in [L(\mathcal{A})]$ and $a \in \Sigma$ with $|u|_a > 0$. Then there exists $v \in [u]$ such that $\delta(q_0, v) \in F$. By assumption, we can write $v = ww'$ with $w, w' \in \Sigma^*$ and such that $\delta(\delta(q_0, w), a^P) = \delta(q_0, w)$. Hence $\delta(q_0, wa^P w') \in F$. So, as $wa^P w' \in L(\mathcal{A})$ and $[wa^P w'] = [u] \sqcup\!\sqcup a^P$, we can deduce $[u] \sqcup\!\sqcup a^P \subseteq [L(\mathcal{A})]$. Then Theorem 5.1 gives the claim. □

For a binary circular automata every state is in a cycle for the letter cyclically permuting the states. However, as the other letter can be chosen arbitrarily, it is easy to give examples of binary circular automata accepting infinitely many words where the condition from Theorem 5.2 does not apply. For example, when the other letter maps every state to the predecessor state for the cycle except one state that is mapped to itself and appropriately chosen start state and final states.

Theorem 5.3. *Let $\mathcal{A} = (Q, \Sigma, \delta, q_0, F)$ be an automaton such that $\delta(Q, aa) = \delta(Q, a)$ for each letter $a \in \Sigma$. Then the commutative closure $[L(\mathcal{A})]$ is a regular language[2] in $\mathcal{C}om_S^+(\Sigma^*)$.*

Proof. Let $u \in [L(\mathcal{A})]$. Then there exists $v \in L(\mathcal{A})$ with $v \in [u]$. We can suppose $u \neq \varepsilon$. Then let $a \in \Sigma$ be such that we can write $v = waw'$. Set $S = \delta(Q, a)$. Now $\delta(q_0, wa) \subseteq S$ and, as by assumption $\delta(S, a) = S$, the state set S is permuted by the letter $a \in \Sigma$. So, as each permutation can be partitioned into disjoint cycles [6], each state in S is contained in a cycle for the letter a , i.e., there exists $p > 0$ such that $\delta(\delta(q_0, wa), a^p) = \delta(q_0, wa)$. Hence, the commutative closure is regular by Theorem 5.2. □

The minimal automaton for the language $(ab)^*$ (this is the automaton from Fig. 1 but with the transitions for the letter c removed) fulfills $\delta(Q, aaa) = \delta(Q, aa)$ but $[(ab)^*] = \{u \in \{a, b\}^* : |u|_a = |u|_b\}$ is not regular.

Let $\mathcal{A} = (Q, \Sigma, \delta, q_0, F)$ be with n states. Then $\mathcal{A}$ is called an *almost-group automaton* [3], if there exists a unique letter $a \in \Sigma$ with $|\delta(Q, a)| = n - 1$ and $\delta(Q, aa) = \delta(Q, a)$ (or equivalently for precisely $n - 1$ states $q \in Q$ there exists $p > 0$ such that $q = \delta(q, a^p)$) and every other letter $b \in \Sigma \setminus \{a\}$ acts as a permutation, i.e., $\delta(Q, b) = Q$. A letter a is called a *simple idempotent letter*, if $|\delta(Q, a)| = |Q| - 1$ and $\delta(q, aa) = \delta(q, a)$ for each $q \in Q$. An automaton $\mathcal{A}$ is called an *automaton with simple idempotents* [38] if every letter $a \in \Sigma$ is either a simple idempotent or permutes the state set. Both types of automata arise in the study of synchronizing automata, see [43] for more information on synchronizing automata. We say $\mathcal{A}$ has a *threshold one transformation semigroup* [2] if for each $u \in \Sigma^*$ and $q \in Q$ there exists $p > 0$ such that $\delta(q, uu^p) = \delta(q, u)$, equivalently $\delta(Q, uu) = \delta(Q, u)$ for each $u \in \Sigma^*$.

By the previous results, the next statement follows immediately.

Corollary 5.4. *Let $\mathcal{A} = (Q, \Sigma, \delta, q_0, F)$. Then $[L(\mathcal{A})]$ is regular in the following cases:*

[2] We note that this result can be stated a little more general using so-called chains of simple semigroups and well-quasi order arguments due to Kunc [26] but which we leave out due to space.

1. *$\mathcal{A}$ has a threshold one transformation semigroup,*
2. *$\mathcal{A}$ is an automaton with simple idempotents,*
3. *$\mathcal{A}$ is an almost-group automaton,*
4. *$\mathcal{A}$ is a permutation automaton (already known, see [17, Thm. 5.3] & [28, Thm. 5]),*
5. $\forall q \in F\ \forall a \in \Sigma\ \exists p > 0 : \delta(q, a^p) = q$, *i.e., the final states are contained in cycles for the letters.*

6 The Group Hierarchy

The commutative closure is regularity-preserving on languages of level one-half in the group hierarchy, i.e., languages in $\mathrm{Pol}(\mathcal{G})$ [17]. The commutative closure in the related Straubing-Thérien hierarchy has been investigated in [19]. Here, by giving an example language, we show that this operation is not regularity-preserving on languages from the next level in the group hierarchy.[3]

Theorem 6.1. *Let* $\Sigma = \{a, b\}$ *and set* $L = \{u \in \Sigma^* : |u| \equiv 0 \pmod 2\}$*. Then the following is true.*

1. $\Sigma^* a L a \Sigma^*$ *is a polynomial of group languages.*
2. *The language* $\overline{\Sigma^* a L a \Sigma^*}$ *equals* $b^* \cup b^*\{ab^i \mid i \text{ odd}\}^* ab^* = b^* \cup b^*(a(bb)^*b)^*ab^*$.
3. $[\overline{\Sigma^* a L a \Sigma^*}] = \{u \in \{a, b\}^* : |u|_b \geq |u|_a - 1\}$ *is not regular.*

Proof. We omit (1) because it is easy to see.

(2) First, suppose $u \notin \Sigma^* a L a \Sigma^*$ and $u \notin b^* \cup b^* ab^*$. Then if $u \in \Sigma^* ab^i a \Sigma^*$, we must have that i is odd. This gives the inclusion $\overline{\Sigma^* a L a \Sigma^*} \subseteq b^* \cup b^*\{ab^i \mid i \text{ odd}\}^* ab^*$.

Conversely, let $u \in b^* \cup b^*\{ab^i \mid i \text{ odd}\}^* ab^*$. Obviously, $b^* \cup b^* ab^* \subseteq \overline{\Sigma^* a L a \Sigma^*}$. So, assume we can write $u = b^{i_0} ab^{i_1} a \cdots b^{i_{m-1}} ab^{i_m}$ with $m \geq 2$ and $i_1, \ldots, i_{m-1} > 0$ being odd numbers. Now, let $u = u_1 a u_2 a u_3$ for $u_1, u_2, u_3 \in \Sigma^*$. Then $u_2 = b^{i_s} ab^{i_{s+1}} \cdots ab^{i_t}$ for two numbers $0 < s \leq t < m$. So, we have

$$|u_2| = |u_2|_a + |u_2|_b = (t - s) + i_s + \ldots + i_t.$$

If $t - s$ is odd (resp. even), then $i_s + \ldots + i_t$ is a sum with an even (resp. odd) number of summands where each summands is an odd number, so this gives an even (resp. odd) number and hence the total sum as a sum of an odd and an even (resp. even and an odd) number is odd. So there exists no factorization $u_1 a u_2 a u_3$ such that $|u_2|$ is even and we have $u \in \overline{\Sigma^* a L a \Sigma^*}$.

(3) Let $u \in \Sigma^*$ with $|u|_b \geq |u|_a - 1$. We can suppose $|u|_a \geq 1$. Set $v = (ab)^{|u|_a - 1} ab^{|u|_b - |u|_a + 1}$. Then $[u] = [v]$ and $v \in \overline{\Sigma^* a L a \Sigma^*}$ by (2).

Conversely, let $u \in [\overline{\Sigma^* a L a \Sigma^*}]$. If $|u|_a \leq 1$, then clearly $|u|_b \geq |u|_a - 1$. Otherwise, let $v \in \overline{\Sigma^* a L a \Sigma^*}$ with $[u] = [v]$. By (2) we have $v = b^{i_0} ab^{i_1} \cdots ab^{i_m}$

[3] See Sect. 8 for a simpler example due to an anonymous reviewer.

with $m \geq 2$ and $i_1, \ldots, i_{m-1}$ being odd. In particular, we have $i_1, \ldots, i_{m-1} > 0$. This gives $|u|_b \geq m-1$ and as $|u|_a = m$ we have $|u|_b \geq |u|_a - 1$.

So, we have $[\overline{\Sigma^* aLa\Sigma^*}] = \{u \in \{a,b\}^* : |u|_b \geq |u|_a - 1\}$. This language is not regular, as can be seen by applying the pumping lemma to $a^n b^{n+1}$ with pumping constant n. □

The language $\overline{\Sigma^* aLa\Sigma^*}$ can also be characterized as the language of those words that either contain at most one a or where two instances of the letter a are at positions of the same parity. For example, $ababa \in \overline{\Sigma^* aLa\Sigma^*}$ or $bababbbaba \in \overline{\Sigma^* aLa\Sigma^*}$.

7 The Positive Variety $\mathcal{W}$

Denote by $\mathcal{W}$ the unique maximal positive variety that does not contain the language $(ab)^*$ [18]. Cano, Guaiana & Pin [17] have shown that $\mathcal{W}$ is closed under commutation, and as the commutation of $(ab)^*$ gives a non-regular language it follows that it is the largest positive variety of languages with regular commutative closures (and so these closures are in $\mathcal{W}$ again). Here, we give a binary circular automaton accepting a language not in $\mathcal{W}$.[4] So Theorem 4.4 is not implied by closure of $\mathcal{W}$ under commutation; hence languages accepted by binary circular automata yield a genuinely new class of regular languages with regular commutative closures not covered by the classes in the literature.

Gómez & Pin [18] introduced the positive variety $\mathcal{W}$ and obtained the following alternative characterizations:

1. $\mathcal{W}$ is the largest proper positive variety in the variety of regular languages closed under shuffle.
2. $\mathcal{W}$ is the largest proper positive variety closed under length preserving morphisms, i.e., mappings $\varphi \colon \Sigma^* \to \Gamma^*$ with $\varphi(uv) = \varphi(u)\varphi(v)$ and $|\varphi(u)| = |u|$ for $u, v \in \Sigma^*$.
3. A subset $I \subseteq M$ in a monoid[5] M is an *ideal* if $MIM \subseteq I$. A nonempty ideal is *minimal* if for every nonempty ideal J of M with $J \subseteq I$ we have $J = I$. Let $a, b \in M$. Then b is an *inverse of* a if $aba = a$ and $bab = b$. The *syntactic ordered monoid of* $L \subseteq \Sigma^*$ is $\Sigma^*/{\equiv_L}$ with the order $\leq_L /{\equiv_L}$ induced from the order $\leq_L$ given by $u \leq_L v$ if
$$\forall x, y \in \Sigma^* \colon xvy \in L \Rightarrow xuy \in L$$
and $u \equiv_L v$ if and only if $u \leq_L v$ and $v \leq_L u$ for $u, v \in \Sigma^*$. Then for a regular language $L \subseteq \Sigma^*$ we have $L \in \mathcal{W}(\Sigma^*)$ if and only if for any pair a, b of mutually inverse elements of its syntactic ordered monoid and any

[4] We note that it follows easily from results in the literature that languages accepted by binary circular automata are not in the other classes mentioned in this work closed for commutation.

[5] For the definition of monoids and semigroups and their relation to automata and formal language theory we refer to the literature, for example [32].

element z of the minimal ideal of the submonoid generated by a and b we have $(abzab)^\omega \leq_L ab$ where x^ω in a finite monoid M and $x \in M$ denotes the unique idempotent[6] element of the form x^i for some $i \geq 0$.

The latter condition is quite involved, but it implies that membership in $\mathcal{W}$ is decidable. The class $\mathcal{W}$ is also closed under concatenation and length-decreasing morphisms. As Pol(J), Pol($\mathcal{C}om$) and Pol($\mathcal{G}$) are positive varieties not containing $(ab)^*$ they are contained in $\mathcal{W}$.

Cano, Guaiana & Pin [17] gave $(ab)^*(a^* + b^*)$ as an example language not in $\mathcal{W}$ whose commutative closure equals $\{a, b\}^*$ and so is regular. Next, we show that languages accepted by circular binary automata are also, in general, not in $\mathcal{W}$.

Theorem 7.1. *There exists a circular binary automaton accepting a language not in* $\mathcal{W}$.

Proof. Let $\mathcal{A} = (Q, \{a, b\}, \delta, 0, F)$ with $Q = \{0, 1, 2\}$, $F = \{0\}$, $\delta(q, b) = (q + 1) \bmod 3$ for $q \in \{0, 1, 2\}$ and $\delta(0, a) = 2$, $\delta(1, a) = 0$, $\delta(2, a) = 2$. The syntactic ordered monoid equals the monoid generated by the transformations induced by a and b, and, for two transformations $\delta_u, \delta_v \colon Q \to Q$ induced by words u and v, respectively, the order is given by $\delta_u \leq \delta_v$ if

$$\forall q \in Q \colon \delta(q, v) \in F \Rightarrow \delta(q, u) \in F.$$

This is actually a canonical construction to obtain the syntactic ordered monoid outlined in [17, Sect. 1.4]. In the following, we also let each word w stand for the induces transformation δ_w on the states and we refer to the words as if they are transformations. Set $u = babbabb$. Then a and u are mutually inverse transformations on the states and they generate the submonoid $\{b^3, a, u, aa, au, ua\}$ which has $\{aa\}$ as its minimal ideal. Now $(uaaaua)^\omega = (aa)^\omega = aa$. However, as $aa \notin L(\mathcal{A})$ but $ua \in L(\mathcal{A})$, i.e., $\delta(0, aa) \notin F$ but $\delta(0, ua) \in F$, we have $aa \not\leq ua$. So, the characterization for languages in $\mathcal{W}$ via the syntactic ordered monoid is not fulfilled for $L(\mathcal{A})$. □

8 Conclusion

Languages in level one of the group hierarchy are accepted by so-called *block groups* [31]. A reviewer pointed out that the syntactic monoid of $(ab)^*$ is the six-element Brandt monoid and is a block group, which follows by any of the many characterizations of this class of semigroups [31] (we refer to the literature for the definitions of the mentioned notions). As the commutative closure of $(ab)^*$ is a non-regular language this gives an easier example than the one given in Sect. 6 of a language in level one of the group hierarchy with non-regular commutative closure.

[6] An element $y \in M$ is idempotent if $yy = y$.

The precise relation of the classes of languages given by the structural condition on automata from Sect. 5 to $\mathcal{W}$ remains a topic for future investigations. Also, relations to so-called basic varieties of languages [4] or conjunctive varieties of languages [34].

Acknowledgement. I thank the anonymous reviewers for careful reading and spotting typos. I also thank one reviewer for mentioning the simpler argument for the result from Sect. 6 (see the conclusion, Sect. 8).

References

1. Arfi, M.: Opérations polynomiales et hiérarchies de concaténation. Theoret. Comput. Sci. **91**(1), 71–84 (1991). https://doi.org/10.1016/0304-3975(91)90268-7
2. Beaudry, M.: Membership testing in threshold one transformation monoids. Inf. Comput. **113**(1), 1–25 (1994). https://doi.org/10.1006/inco.1994.1062
3. Berlinkov, M.V., Nicaud, C.: Synchronizing almost-group automata. Int. J. Found. Comput. Sci. **31**(8), 1091–1112 (2020). https://doi.org/10.1142/S0129054120420058
4. Birkmann, F., Milius, S., Urbat, H.: On language varieties without boolean operations. In: Leporati, A., Martín-Vide, C., Shapira, D., Zandron, C. (eds.) LATA 2021. LNCS, vol. 12638, pp. 3–15. Springer, Cham (2021). https://doi.org/10.1007/978-3-030-68195-1_1
5. Bouajjani, A., Muscholl, A., Touili, T.: Permutation rewriting and algorithmic verification. Inf. Comput. **205**(2), 199–224 (2007). https://doi.org/10.1016/j.ic.2005.11.007
6. Cameron, P.J.: Permutation Groups. London Mathematical Society Student Texts. Cambridge University Press, Cambridge (1999). https://doi.org/10.1017/CBO9780511623677
7. Cécé, G., Héam, P., Mainier, Y.: Clôtures transitives de semi-commutations et model-checking régulier. Technique et Science Informatiques **27**(1–2), 7–28 (2008). https://doi.org/10.3166/tsi.27.7-28
8. Cécé, G., Héam, P., Mainier, Y.: Efficiency of automata in semi-commutation verification techniques. RAIRO Theor. Inform. Appl. **42**(2), 197–215 (2008). https://doi.org/10.1051/ITA:2007029
9. Clerbout, M., Latteux, M.: Semi-commutations. Inf. Comput. **73**(1), 59–74 (1987). https://doi.org/10.1016/0890-5401(87)90040-X
10. Cohen, R.S., Brzozowski, J.A.: Dot-depth of star-free events. J. Comput. Syst. Sci. **5**(1), 1–16 (1971). https://doi.org/10.1016/S0022-0000(71)80003-X
11. Cori, R., Perrin, D.: Automates et commutations partielles. RAIRO Theor. Inform. Appl. **19**(1), 21–32 (1985). https://doi.org/10.1051/ita/1985190100211
12. Diekert, V., Rozenberg, G. (eds.): The Book of Traces. World Scientific, Singapore (1995). https://doi.org/10.1142/2563
13. Ginsburg, S., Spanier, E.H.: Bounded regular sets. Proc. Am. Math. Soc. **17**(5), 1043–1049 (1966). https://doi.org/10.2307/2036087
14. Ginsburg, S., Spanier, E.H.: Semigroups, Presburger formulas, and languages. Pac. J. Math. **16**(2), 285–296 (1966). https://doi.org/10.2140/pjm.1966.16.285
15. Ginsburg, S., Spanier, E.H.: AFL with the semilinear property. J. Comput. Syst. Sci. **5**(4), 365–396 (1971). https://doi.org/10.1016/S0022-0000(71)80024-7

16. Gohon, P.: An algorithm to decide whether a rational subset of N^k is recognizable. Theoret. Comput. Sci. **41**, 51–59 (1985). https://doi.org/10.1016/0304-3975(85)90059-3
17. Gómez, A.C., Guaiana, G., Pin, J.É.: Regular languages and partial commutations. Inf. Comput. **230**, 76–96 (2013). https://doi.org/10.1016/j.ic.2013.07.003
18. Gómez, A.C., Pin, J.: Shuffle on positive varieties of languages. Theoret. Comput. Sci. **312**(2–3), 433–461 (2004). https://doi.org/10.1016/j.tcs.2003.10.034
19. Guaiana, G., Restivo, A., Salemi, S.: On the trace product and some families of languages closed under partial commutations. J. Autom. Lang. Comb. **9**(1), 61–79 (2004). https://doi.org/10.25596/jalc-2004-061
20. Hoffmann, S.: The commutative closure of shuffle languages over group languages is regular. In: Maneth, S. (ed.) CIAA 2021. LNCS, vol. 12803, pp. 53–64. Springer, Cham (2021). https://doi.org/10.1007/978-3-030-79121-6_5
21. Hoffmann, S.: Regularity conditions for iterated shuffle on commutative regular languages. Int. J. Found. Comput. Sci. **34**(08), 923–957 (2023). https://doi.org/10.1142/S0129054123430037
22. Hoffmann, S.: State complexity bounds for the commutative closure of group languages. J. Autom. Lang. Comb. **28**(1–3), 27–57 (2023). https://doi.org/10.25596/JALC-2023-027
23. Hoffmann, S.: State complexity of permutation and the language inclusion problem up to parikh equivalence on alphabetical pattern constraints and partially ordered nfas. Int. J. Found. Comput. Sci. **34**(08), 959–986 (2023). https://doi.org/10.1142/S0129054123430025
24. Ibarra, O.H.: Reversal-bounded multicounter machines and their decision problems. J. ACM **25**(1), 116–133 (1978). https://doi.org/10.1145/322047.322058
25. Klaedtke, F., Rueß, H.: Monadic second-order logics with cardinalities. In: Baeten, J.C.M., Lenstra, J.K., Parrow, J., Woeginger, G.J. (eds.) ICALP 2003. LNCS, vol. 2719, pp. 681–696. Springer, Heidelberg (2003). https://doi.org/10.1007/3-540-45061-0_54
26. Kunc, M.: Regular solutions of language inequalities and well quasi-orders. Theoret. Comput. Sci. **348**(2–3), 277–293 (2005). https://doi.org/10.1016/j.tcs.2005.09.018
27. Latteux, M.: Cônes rationnels commutatifs. J. Comput. Syst. Sci. **18**(3), 307–333 (1979). https://doi.org/10.1016/0022-0000(79)90039-4
28. Commutative closures of regular semigroups of languages: L'vov, M. Cybern. Syst. Anal. (Cybern.) **9**, 247–252 (1973). https://doi.org/10.1007/BF01069078. translated (original in Russian) from Kibernetika (Kiev), No. 2, pp. 54–58, March-April, 1973
29. Muscholl, A., Petersen, H.: A note on the commutative closure of star-free languages. Inf. Process. Lett. **57**(2), 71–74 (1996). https://doi.org/10.1016/0020-0190(95)00187-5
30. Parikh, R.: On context-free languages. J. ACM **13**(4), 570–581 (1966). https://doi.org/10.1145/321356.321364
31. Pin, J.É.: PG = BG, a success story. In: Fountain, J. (ed.) NATO Advanced Study Institute, Semigroups, Formal Languages and Groups, pp. 33–47. Kluwer Academic Publishers (1995)
32. Pin, J.-E.: Syntactic semigroups. In: Rozenberg, G., Salomaa, A. (eds.) Handbook of Formal Languages, pp. 679–746. Springer, Heidelberg (1997). https://doi.org/10.1007/978-3-642-59136-5_10
33. Place, T., Zeitoun, M.: Generic results for concatenation hierarchies. Theory Comput. Syst. **63**(4), 849–901 (2019). https://doi.org/10.1007/s00224-018-9867-0

34. Polák, L.: A classification of rational languages by semilattice-ordered monoids. Archivum Mathematicum **040**(4), 395–406 (2004)
35. Rabin, M.O., Scott, D.S.: Finite automata and their decision problems. IBM J. Res. Dev. **3**(2), 114–125 (1959). https://doi.org/10.1147/rd.32.0114
36. Redko, V.N.: On the commutative closure of events. Doklady Akademija Nauk Ukrainskoj SSR (Kiev), also Dopovidi Akademij Nauk Ukrajnskoj RSR (= Reports of the Academy of Sciences of the Ukrainian SSR), pp. 1156–1159 (1963). (in Russian)
37. Rigo, M.: The commutative closure of a binary slip-language is context-free: a new proof. Discret. Appl. Math. **131**(3), 665–672 (2003). https://doi.org/10.1016/S0166-218X(03)00335-4
38. Rystsov, I.K.: Estimation of the length of reset words for automata with simple idempotents. Cybern. Syst. Anal. **36**(3), 339–344 (2000). https://doi.org/10.1007/BF02732984
39. Sakarovitch, J.: The "last" decision problem for rational trace languages. In: Simon, I. (ed.) LATIN 1992. LNCS, vol. 583, pp. 460–473. Springer, Heidelberg (1992). https://doi.org/10.1007/BFb0023848
40. Simon, I.: Piecewise testable events. In: Brakhage, H. (ed.) GI-Fachtagung 1975. LNCS, vol. 33, pp. 214–222. Springer, Heidelberg (1975). https://doi.org/10.1007/3-540-07407-4_23
41. Straubing, H.: A generalization of the Schützenberger product of finite monoids. Theoret. Comput. Sci. **13**, 137–150 (1981). https://doi.org/10.1016/0304-3975(81)90036-0
42. Thérien, D.: Classification of finite monoids: the language approach. Theoret. Comput. Sci. **14**, 195–208 (1981). https://doi.org/10.1016/0304-3975(81)90057-8
43. Volkov, M.V., Kari, J.: Černý's conjecture and the road colouring problem. In: Pin, J.É. (ed.) Handbook of Automata Theory, Volume I, pp. 525–565. European Mathematical Society Publishing House (2021). https://doi.org/10.4171/automata-1/15

Shortest Characteristic Factors of a Deterministic Finite Automaton and Computing Its Positive Position Run by Pattern Set Matching

Jan Janoušek and Štěpán Plachý(✉)

Faculty of Information Technology, Czech Technical University in Prague, Prague, Czech Republic
{Jan.Janousek,Stepan.Plachy}@fit.cvut.cz

Abstract. Given a deterministic finite automaton (DFA) A, we present a simple algorithm for constructing four deterministic finite automata that accept the shortest forbidden factors, the shortest forbidden suffixes, the shortest allowed suffixes, and the shortest forbidden prefixes. We refer to these automata as the shortest characteristic factors of automaton A. If the given automaton is local, and therefore the language it accepts is strictly locally testable, the sets of its shortest characteristic factors are finite, and these four automata are acyclic. This approach simplifies existing methods for the extraction of forbidden factors and also generalizes it for all classes of input DFAs. Furthermore, we demonstrate that this type of extraction can be used for a sublinear run of an automaton for certain inputs. We define a positive position run of a deterministic finite automaton, representing all positions in an input string where the automaton reaches a final state. Finally, we present an algorithm for computing the positive position run of the automaton, which utilizes pattern set matching of its shortest forbidden factors and its shortest allowed suffixes, provided that the sets are finite. We showcase the computation of the positive position run of a local automaton using backward pattern matching, which can achieve sublinear time.

Keywords: Finite automata · Local finite automata · Shortest characteristic factors · Strictly locally testable languages · Pattern matching · Positive position run of automata

1 Introduction

Finite automaton (FA) is a fundamental model of computation, widely studied in Computer Science and used in many applications [10,18]. The FA can also be used as a useful formalism in the process of creating a new algorithm for a regular

The authors acknowledge the support of the OP VVV MEYS funded project CZ.02.1.01/0.0/0.0/16_019/0000765 "Research Center for Informatics".

H. Fernau et al. (Eds.): SOFSEM 2024, LNCS 14519, pp. 326–339, 2024.
https://doi.org/10.1007/978-3-031-52113-3_23

language problem [2,13,14], especially because of its simplicity, efficiency and the availability of many operations on FA, such as determinisation, minimisation or construction for union, intersection, complement, product, iteration and reverse of regular languages. The outputs of the algorithm created as an FA and their positions during the reading of the input string are typically represented by the final states of the FA.

A strictly locally k-testable language (SLTL), $k \geq 1$, introduced in [12] and further studied in [3,6,21], is a language characterized by its factors of length k such that a string of length at least k is in the language if its prefix of length $k-1$, suffix of length $k-1$ and each factor of length k belong to a set of allowed prefixes, of allowed suffixes and of allowed factors, respectively. Such property of each part of the string can be equivalently defined using a forbidden set of factors having the same length. We note that in the definition there are no restrictions for strings shorter than k. For SLTL, since the length of the shortest such factors is bounded by k, the sets defining the language are finite. The class of SLTLs is accepted by local DFAs, where each string of length at least $k-1$ is synchronizing. The k is the smallest possible if the DFA is minimal.

A problem of extraction of the shortest forbidden factors, i.e. given a DFA finding the set of the shortest factors of a specific type (be it prefix, suffix, or inner factor) that guarantee non-acceptance when they are found in the input string, has been studied in the past and has its practical use-cases. Béal et al. [4] describe a quadratic-time algorithm to compute the set of minimal forbidden words of a factorial regular language. Rogers and Lambert [16,17] present algorithms for extracting the shortest forbidden factors and suffixes from a local DFA, which arose from a practical need in their research of analyzing stress patterns in human languages. They also mention the possibility of extracting the shortest forbidden prefixes by extracting suffixes from a DFA accepting reversed language.

Our improvement is based on the approach of Rogers and Lambert. Their algorithm works by constructing a powerset graph of the input DFA (i.e. a directed labeled graph where nodes are subsets of states and edges labeled with input symbols lead to a node containing destination states of any transition from a source state on the symbol) and looking for specific paths in the graph which correspond to the resulting words. This is done by traversing the graph from a specific state against the direction of edges. Since there can be multiple edges with the same symbol leading to the same state, each word can have multiple paths in such traversal. The traversal is therefore designed to go through all of these paths while also filtering them on the go to ensure we only get the shortest factors. A prerequisite of the traversal is also that the powerset graph is acyclic (on nodes of size at least 2), which is the case if and only if the DFA is local and as a result the shortest forbidden factors are finite.

We, however, show that no special traversal is needed if we modify the powerset construction in the right way. Our modification causes each word to have only one path, therefore we effectively create a DFA and its language is exactly the resulting set we want, or more precisely its reverse. A standard BFS or DFS is then sufficient to obtain the result if it is finite. With our approach, we can in a very straightforward way create DFAs for the shortest forbidden suffixes and factors, but also prefixes without the need to create a DFA for the reversed language, and we can also create a DFA for the shortest allowed suffixes. We will call these sets the shortest characteristic factors of the DFA.

Another advantage is that our extraction works for any DFA, not just local ones. If the shortest characteristic factors are finite then the DFAs are acyclic. Otherwise, they describe an infinite language. As mentioned, the language is the reverse of the result. If an unreversed result is needed, it is straightforward for DFAs with finite languages; otherwise, the standard construction of a DFA for the reversed language would be necessary.

Furthermore, we describe that extraction of the shortest characteristic factors can also be used to achieve a sublinear time of some FA-based algorithms by utilizing exact string pattern set matching. Pattern matching is a fundamental and widely studied problem [1,11]. Given a text and patterns, the pattern matching searches for all factors of the text that match the patterns. Among the most efficient pattern matching algorithms in practice are those based on backward pattern matching [7,8]. Although the time complexity of backward string pattern matching is generally $\mathcal{O}(n*m)$ in the worst case (for a text and a pattern of size n and m, respectively), due to the ability of the backward string pattern matching algorithms to skip parts of the text, they often perform sublinearly in practice [19,20].

We define a positive position run (PPR) of a deterministic finite automaton, which represents all positions in an input string in which the automaton reaches a final state. This notion commonly corresponds to the output of FA-based algorithms. Certain string pattern matching problems can for example be solved by FAs where positions of final states represent the occurrences of matches in the input string [9,14]. We then show the computation of PPR by matching sets of the shortest forbidden factors and the shortest allowed suffixes, if they are finite. The running time is then inherited from the pattern matching algorithm used, therefore with a backward pattern matching algorithm it can be sublinear for certain inputs. We demonstrate this using the Commentz-Walter algorithm. We are not aware of any algorithms computing runs of DFAs using pattern matching.

2 Basic Notions

An *alphabet* is a finite set of *symbols*. A *string* (or a *word*) w over an alphabet Σ is a sequence $w = a_1a_2\dots a_n$ such that $a_i \in \Sigma, n \geq i \geq 1$. $w[i]$ denotes the i-th symbol of w, $|w|$ is the length of w, w^R is the reverse of w, ε denotes the string of length 0 and Σ^* denotes the set of all string over Σ. A *language* over an alphabet Σ is a set $L \subseteq \Sigma^*$ and L^R denotes the reverse of the language. A *prefix* and a *suffix* of w is a string u such that $w = uv$ and $w = vu$, respectively, for some $v \in \Sigma^*$. The prefix and the suffix is *proper* if $|u| < |w|$. A factor of w is a string u such that $w = tuv$ for some $t, u, v \in \Sigma^*$ and it is *proper* if $|u| < |w|$.

A (partial) *deterministic finite automaton* (DFA) over an alphabet Σ is a 5-tuple $A = (Q, \Sigma, \delta, q_0, Q_f)$, where Q is a finite set of states, $\delta\colon Q \times \Sigma \to Q$ is a partial transition function, $q_0 \in Q$ is an initial state, and $Q_f \subseteq Q$ is a set of final states. The DFA is *complete* if δ is total. An *extended transition function*

$\delta^*: Q \times \Sigma^* \to Q$ is a partial function such that for each $q \in Q : \delta^*(q, \varepsilon) = q$ and if $\delta(q, a)$ and $\delta^*(\delta(q, a), w)$ exist, then $\delta^*(q, aw) = \delta^*(\delta(q, a), w)$, where $a \in \Sigma$ and $w \in \Sigma^*$. Otherwise $\delta^*(q, aw)$ is undefined. A DFA $A = (Q, \Sigma, \delta, q_0, Q_f)$ is said to *accept* a word $w \in \Sigma^*$ if $\delta^*(q_0, w) \in Q_f$. A *language of a state* $q \in Q$ is $L(q) = \{w \in \Sigma^* \mid \delta^*(q, w) \in Q_f\}$. A *language accepted by the* DFA A is $L(A) = L(q_0)$. A state $q \in Q$ is *reachable* if $\delta^*(q_0, w) = q$ for some $w \in \Sigma^*$. In this paper we will consider only DFAs without unreachable states. A is *minimal* if all states are reachable and $\forall q_1, q_2 \in Q : L(q_1) \neq L(q_2)$. A is *complete* if δ is complete. A *useless state* for A is any state $q \in Q$ such that $L(q) = \emptyset$. A minimal complete DFA has at most one useless state. A^R is a DFA such that $L(A^R) = L(A)^R$. A *run* of the DFA on word $w \in \Sigma^*$ is a sequence $q_0, \ldots, q_{|w|}$ such that $q_i = \delta^*(q_0, w_i)$, where w_i is a prefix of w and $|w_i| = i$.

For a DFA $A = (Q, \Sigma, \delta, q_0, Q_f)$ a word w is *synchronizing* to a set of states $S \subseteq Q$ if $\forall q \in Q : \delta^*(q, w) \in S$ if $\delta^*(q, w)$ exists. The DFA is k-*local* if each word of length at least k are synchronizing to a single state. If the DFA is k-local for some k then it is said to be *local*.

A *forbidden factor* of A is a word $w \in \Sigma^*$ such that $\forall q \in Q : \delta^*(q, w)$ is not defined or $L(\delta^*(q, w)) = \emptyset$. FF($A$) is the set of all forbidden factors of A and $w \in$ FF(A) is a *shortest forbidden factor* iff $u \notin$ FF(A) for all proper factors u of w. SFF(A) $\subseteq$ FF(A) is the set of all the shortest forbidden factors of A.

A *forbidden suffix* of A is a word $w \in \Sigma^*$ such that $\forall q \in Q : \delta^*(q, w) \notin Q_f$ or $\delta^*(q, w)$ is undefined. FS(A) is the set of all forbidden suffixes of A and w is a *shortest forbidden suffix* iff $w \notin$ FF(A) and $u \notin$ FS(A) for all proper suffixes u of w. SFS(A) $\subseteq$ FS(A) is the set of all the shortest forbidden suffixes of A.

An *allowed suffix* of A is a word $w \in \Sigma^*$ such that $w \notin$ FF(A) and $\forall q \in Q :$ $\delta^*(q, w) \in Q_f$ or $\delta^*(q, w)$ is not defined or $L(\delta^*(q, w)) = \emptyset$. AS($A$) is the set of all allowed suffixes of A and w is a *shortest allowed suffix* iff $u \notin$ AS(A) for all proper suffixes u of w. SAS(A) $\subseteq$ AS(A) is the set of all the shortest allowed suffixes of A.

A *forbidden prefix* of A is a word $w \in \Sigma^*$ such that $\delta^*(q_0, w)$ is undefined or $L(\delta^*(q_0, w)) = \emptyset$. FP($A$) is the set of all forbidden prefixes of A and w is a *shortest forbidden prefix* iff $w \notin$ FF(A) and $u \notin$ FP(A) for all proper prefixes u of w. SFP(A) $\subseteq$ FP(A) is the set of all the shortest forbidden prefixes of A.

If a DFA A is local then SFF(A), SFS(A), SAS(A) and SFP(A) are finite and $L(A)$ is strictly locally testable. We call these sets the *shortest characteristic factors* of A. An example of a local DFA and information on the sets of its characteristic factors can be found on Fig. 1.

We note that the shortest forbidden prefixes and suffixes are disjoint with the shortest forbidden factors since for characterization of the corresponding strictly locally testable language such prefix or suffix would be redundant. We also note that allowed suffixes are words that starting from any state always either end in a final state or fail (either by a missing transition or reaching a useless state). If however they only fail, then it's a forbidden factor instead. Therefore at least one path needs to end in a final state.

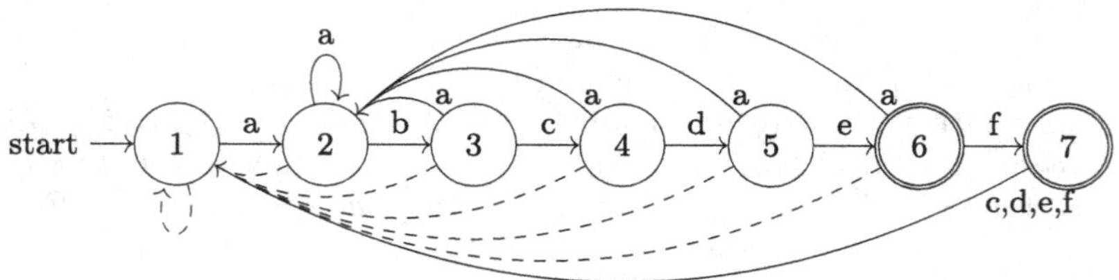

Fig. 1. Example of an incomplete 6-local DFA over alphabet $\{a, b, c, d, e, f\}$. Dashed transitions are for all symbols without explicit transition. The shortest characteristic factors are: $\text{SFP}(A) = \emptyset, \text{SFF}(A) = \{abcdefa, abcdefb\}$ and $\text{SAS}(A) = \{abcde, abcdef\}$. $\text{SFS}(A)$ contains 49 strings.

In Sect. 4 we introduce a pattern set matching approach to computing a version of an automaton run. As a consequence, the algorithm inherits properties of the pattern matching algorithm used. We demonstrate it using Commentz-Walter algorithm, which is a backwards pattern set matching algorithm that can skip over parts of the input string, which allows it to run in sublinear time for certain inputs and experiments indicate this is the case on average. The reversed pattern set is preprocessed into a trie structure with some additional information in each vertex, like functions shift1 and shift2, which are used by the algorithm to determine how far to shift in the input after a mismatch. We however strongly emphasize that any pattern set matching algorithm can be used and the choice is important only for the properties. For this reason, we do not describe the Commentz-Walter and all its notions in detail. For that see [8]. We consider a slight modification that matching certain patterns will cause the algorithm to terminate.

3 Extracting the Shortest Characteristic Factors with the Use of Finite Automaton

We first describe some properties of synchronizing strings and characteristic factors, which will serve as the basis of our constructions. The following lemmas describe how such strings behave in a DFA with a complete transition function. If the automaton's transition function is not complete, we can complete it by adding a new special useless state. Instead of all missing transitions, we add transitions to this useless state. We can equivalently state the synchronization property of a DFA after completing the transition function in the following way.

Lemma 1. *Let* $A = (Q, \Sigma, \delta, q_0, Q_f)$ *be a* DFA *without useless states and* $A' = (Q \cup U, \Sigma, \delta', q_0, Q_f)$ *be a complete* DFA *such that* U *is a set of useless states,* $\delta \subseteq \delta'$ *and* $L(A) = L(A')$*. A word* $w \in \Sigma^*$ *is synchronizing to a set of states* $S \subseteq Q$ *in* A *if and only if it is synchronizing to* $S \cup U$ *in* A'*.*

Consequently, the following lemmas hold.

Lemma 2. *Let $A = (Q, \Sigma, \delta, q_0, Q_f)$ be a complete* DFA *with $U \subseteq Q$ being a set of useless states. A word $w \in \Sigma^*$ is a forbidden factor of A if and only if w is synchronizing to U in A.*

Lemma 3. *Let $A = (Q, \Sigma, \delta, q_0, Q_f)$ be a complete* DFA. *A word $w \in \Sigma^*$ is a forbidden suffix of A if and only if w is synchronizing to $Q \setminus Q_f$.*

Lemma 4. *Let $A = (Q, \Sigma, \delta, q_0, Q_f)$ be a complete* DFA *with $U \subseteq Q$ being a set of useless states. A word $w \in \Sigma^*$ is an allowed suffix of A if and only if w is synchronizing to $Q_f \cup U$ and is not synchronizing to U.*

Corollary 1. *Let $A = (Q, \Sigma, \delta, q_0, Q_f)$ be a complete* DFA *with $U \subseteq Q$ being a set of useless states. A word $w \in \Sigma^*$ is an allowed suffix of A if and only if w is synchronizing to $Q_f \cup U$ in A and there exists a state $q \in Q$ such that $\delta^*(q, w) \in Q_f$.*

Lemma 5. *Let $A = (Q, \Sigma, \delta, q_0, Q_f)$ be a complete* DFA *with $U \subseteq Q$ being a set of useless states. A word $w \in \Sigma^*$ is a forbidden prefix of A if and only if $\delta^*(q_0, w) \in U$.*

3.1 The Shortest Characteristic Factors DFAs

Given a DFA A, we present new definitions for four simple DFAs that accept the shortest forbidden factors, the shortest forbidden suffixes, the shortest allowed suffixes and the shortest forbidden prefixes of the automaton A. If the DFA has these sets finite, which is guaranteed for local DFAs, the constructed DFAs are acyclic and extraction can then be done by some basic graph traversal algorithm, such as BFS or DFS.

The forbidden factors are all strings which, starting from all states, reach any useless state. For such a factor to be shortest, it cannot contain any other forbidden factor as a factor. If the factor contains an unnecessary suffix, then paths from all states will reach a useless state before reading the suffix. Therefore, at least one path must reach a useless state only at its end. Allowed suffixes are all strings that from all states end in either a final state or a useless state. However, at least one path must end in a final state, otherwise it would be a forbidden factor. Forbidden suffixes follow similar logic. Neither type of factor should contain an unnecessary prefix to be shortest. We note that the shortest characteristic suffix can contain another as a factor, as long as it's not a suffix.

Definition 1 (A shortest synchronizing template automaton). $\widehat{A}(A, \widehat{q_0}) = (\widehat{Q}, \Sigma, \widehat{\delta}, \widehat{q_0}, \widehat{Q}_f)$ *for some complete* DFA $A = (Q, \Sigma, \delta, q_0, Q_f)$ *with a set of useless states $U \subseteq Q$ and some $\widehat{q_0} \in \widehat{Q}$ such that:*

- $\widehat{Q} = \{\widehat{q} \mid \widehat{q} \colon Q' \to \{0, 1\}, Q' \subseteq Q\}$
- *For* $S \in \widehat{Q} : \mathrm{dom}(S) = \{q \mid (q, x) \in S, x \in \{0, 1\}\}$
- $\widehat{\delta}$ *is a partial function* $\widehat{Q} \times \Sigma \to \widehat{Q}$ *such that if* $\mathrm{dom}(S) \neq Q$ *then* $\widehat{\delta}(S, a) = \{(q, x) \mid \delta(q, a) \in \mathrm{dom}(S)\}$, *where* $x = \Big(\big(\delta\left(q, a\right), 1\big) \in S \wedge q \notin U\Big)$

- $\widehat{Q}_f = \{S \mid S \in \widehat{Q} \wedge \text{dom}(S) = Q \wedge (\exists q \in Q : (q,1) \in S)\}$

Based on the shortest synchronizing template automaton, we can define automata accepting the shortest forbidden factors, the shortest allowed suffixes, and the shortest forbidden suffixes and prove the correctness of their languages.

Definition 2. *For a complete* DFA $A = (Q, \Sigma, \delta, q_0, Q_f)$ *with a set of useless states* $U \subseteq Q$ *:*

- $\widehat{A}_{SFF}(A) = \widehat{A}(A, \widehat{q}_0)$*, where* $\widehat{q}_0 = U \times \{1\}$
- $\widehat{A}_{SAS}(A) = \widehat{A}(A, \widehat{q}_0)$*, where* $\widehat{q}_0 = (U \times \{0\}) \cup (Q_f \times \{1\})$
- $\widehat{A}_{SFS}(A) = \widehat{A}(A, \widehat{q}_0)$*, where* $\widehat{q}_0 = (U \times \{0\}) \cup (Q \setminus (Q_f \cup U) \times \{1\})$

Figures 3 and 4 show automata for the shortest forbidden factors and the shortest allowed suffixes, respectively, of DFA in Fig. 2.

Lemma 6. *Let* $A = (Q, \Sigma, \delta, q_0, Q_f)$ *be a complete* DFA *with a set of useless states* $U \subseteq Q$ *and* $\widehat{A}(A, \widehat{q}_0) = (\widehat{Q}, \Sigma, \widehat{\delta}, \widehat{q}_0, \widehat{Q}_f)$ *for some* $\widehat{q}_0 \in \widehat{Q}$*. Then* $\forall w \in \Sigma^*, S \in \widehat{Q} : \widehat{\delta}^*(\widehat{q}_0, w) = S$ *iff:*

1. $Q \neq \text{dom}\left(\widehat{\delta}(\widehat{q}_0, u)\right)$ *for all proper prefixes* u *of* w
2. $\text{dom}(S) = \{q \mid \delta^*(q, w^R) \in \text{dom}(\widehat{q}_0)\}$
3. $\forall (q,x) \in S : x = \left(\left(\delta^*(q, w^R), 1\right) \in \widehat{q}_0 \wedge \left(w = a'v \implies \delta^*(q, v^R) \notin U\right)\right)$ *for all* $a' \in \Sigma, v \in \Sigma^*$

Proof. We prove both implications inductively starting with the left-to-right. For $w = \varepsilon$ since $\widehat{\delta}(\widehat{q}_0, \varepsilon) = \widehat{q}_0$ all conditions trivially apply. For $w = w'a$ where $w' \in \Sigma^*, a \in \Sigma$ let $S' = \widehat{\delta}^*(\widehat{q}_0, w')$.

1. If S' exists then $Q \neq \text{dom}\left(\widehat{\delta}(\widehat{q}_0, u')\right)$ for all proper prefixes u' of w'. If $Q \neq \text{dom}(S')$ then $Q \neq \text{dom}\left(\widehat{\delta}(\widehat{q}_0, w')\right)$ as well. Otherwise, if $Q = \text{dom}(S')$ then by definition $\widehat{\delta}(S', a)$ does not exist. In such case or if S' does not exist then $\widehat{\delta}^*(\widehat{q}_0, w'a)$ does not exist either, which contradicts that $\widehat{\delta}^*(\widehat{q}_0, w) = S$.
2. $$\begin{aligned} \text{dom}(S) &= \{q \mid \delta(q,a) \in \text{dom}(S')\} \\ &= \{q \mid \delta^*(\delta(q,a), w'^R) \in \text{dom}(\widehat{q}_0)\} \\ &= \{q \mid \delta^*(q, aw'^R) \in \text{dom}(\widehat{q}_0)\} \end{aligned}$$
3. Let $v' \in \Sigma^*$. For each $(q,x) \in S$ applies:

$$\begin{aligned} x &= \left(\delta(q,a), 1\right) \in S' \wedge q \notin U \\ &= \left(\delta^*(\delta(q,a), w'^R), 1\right) \in \widehat{q}_0 \wedge (w' = a'v' \implies \delta^*(\delta(q,a), v'^R) \notin U) \wedge q \notin U \\ &= \left(\delta^*(q, aw'^R), 1\right) \in \widehat{q}_0 \wedge \left(w = a'v \implies \begin{cases} q \notin U & v = \varepsilon \\ \delta^*(q, av'^R) \notin U & v = v'a \end{cases}\right) \\ &= \left(\delta^*(q, w^R), 1\right) \in \widehat{q}_0 \wedge \left(w = a'v \implies \delta^*(q, v^R) \notin U\right) \end{aligned}$$

For the right-to-left implication for $w = \varepsilon$ all conditions trivially imply that $S = \widehat{q_0}$. For $w = w'a$ where $w' \in \Sigma^*, a \in \Sigma$ let $S' = \widehat{\delta}^*(\widehat{q_0}, w')$.

1. If $Q \neq \operatorname{dom}\left(\widehat{\delta}^*(\widehat{q_0}, u)\right)$ holds for all proper prefixes u of w then it also holds $Q \neq \operatorname{dom}\left(\widehat{\delta}^*(\widehat{q_0}, u')\right)$ and $Q \neq \operatorname{dom}\left(\widehat{\delta}^*(\widehat{q_0}, w')\right)$ for all proper prefixes u' of w'. That implies $Q \neq \operatorname{dom}(S')$.
2. If $\operatorname{dom}(S) = \{q \mid \delta^*(q, aw'^R) \in \operatorname{dom}(\widehat{q_0})\}$ then by reversing equations in proof of the left-to-right implication we obtain $\operatorname{dom}(S) = \{q \mid \delta(q, a) \in \operatorname{dom}(S')\}$.
3. Similarly from the 3rd condition we obtain $\forall (q, x) \in S : x = \Big(\big(\delta(q, a), 1\big) \in S' \wedge q \notin U\Big)$

All of these conditions by definition of $\widehat{\delta}$ imply that $\widehat{\delta}(S', a) = S$ which therefore means $\widehat{\delta}\big(\widehat{\delta}^*(\widehat{q_0}, w'), a\big) = \widehat{\delta}^*(q_0, w'a) = S$. □

Theorem 1. *Let* $A = (Q, \Sigma, \delta, q_0, Q_f)$ *be a complete* DFA *and* $\widehat{q_0} = (Q_0 \times 0) \cup (Q_1 \times 1)$ *for some* $Q_0, Q_1 \subseteq Q, Q_0 \cap Q_1 = \emptyset$. $L(\widehat{A}(A, \widehat{q_0}))^R$ *is a set of all words* $w \in \Sigma^*$ *such that* w *is a shortest word synchronizing to* $Q_0 \cup Q_1$ *in* A *and there exists a state* $q' \in Q$ *such that* $\delta^*(q', w) \in Q_1$ *and if* $w = ua$ *for some* $u \in \Sigma^*$ *and* $a \in \Sigma$ *then* $\delta^*(q', u) \notin U$.

Proof. Let $\widehat{A}(A, \widehat{q_0}) = (\widehat{Q}, \Sigma, \widehat{\delta}, \widehat{q_0}, \widehat{Q}_f)$. If a $\widehat{q}_f \in \widehat{Q}_f$ then $Q = \operatorname{dom}(\widehat{q}_f)$ and there exists at least one $q' \in Q$ such that $(q', 1) \in \widehat{q}_f$. Then Lemma 6 ensures that $\widehat{\delta}(\widehat{q_0}, w^R) = \widehat{q}_f$ applies for all words w satisfying that $\forall q \in Q : \delta^*(q, w) \in Q_0 \cup Q_1$, which implies that w is a synchronizing word to $Q_0 \cup Q_1$, and $Q \neq \operatorname{dom}(\widehat{\delta}(\widehat{q_0}, v^R))$ for all proper suffixes v of w, which implies w is the shortest synchronizing word since none of its proper suffixes synchronizes all states, and finally that $\delta^*(q', w) \in Q_1$ and if $w = ua$, then $\delta^*(q', u) \notin U$. □

Combining Theorem 1 with Lemma 2, Corollary 1 and Lemma 3 respectively we can infer the following theorems for characteristic factors DFAs.

Theorem 2. $L(\widehat{A}_{SFF}(A))^R = SFF(A)$ *for a complete* DFA A.

Theorem 3. $L(\widehat{A}_{SAS}(A))^R = SAS(A)$ *for a complete* DFA A.

Theorem 4. $L(\widehat{A}_{SFS}(A))^R = SFS(A)$ *for a complete* DFA A.

For extracting the shortest forbidden prefixes, the matching of which we do not use in the algorithm computing the positive position run of a DFA, a direct approach can also be used, which is achieved by just modifying the set of final states in the automaton for the shortest forbidden factors. For a prefix to be forbidden, the string must reach a useless state from the initial state according to Lemma 5. That is a weaker condition than the one for forbidden factors, where we require a useless state to be reached from all states. For the prefix to be shortest, it can't contain another as a prefix, so the path from the initial state should reach a useless state only at its end. It also can't be a forbidden factor. Therefore, we define the following automaton.

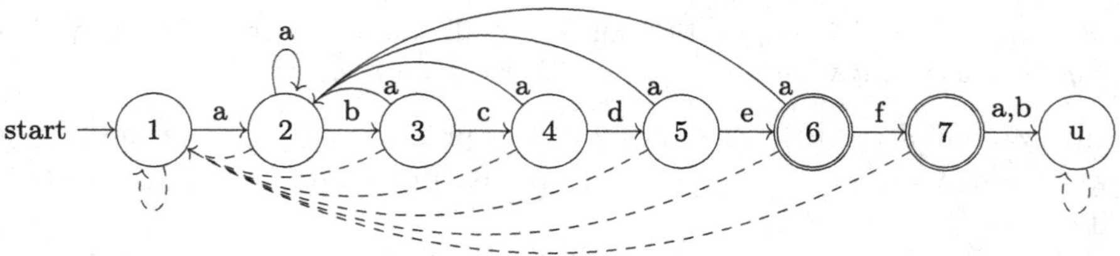

Fig. 2. The complete DFA over alphabet $\{a, b, c, d, e, f\}$ with a single useless state u, which is equivalent to the incomplete DFA in Fig. 1. Dashed transitions are for all symbols without explicit transition.

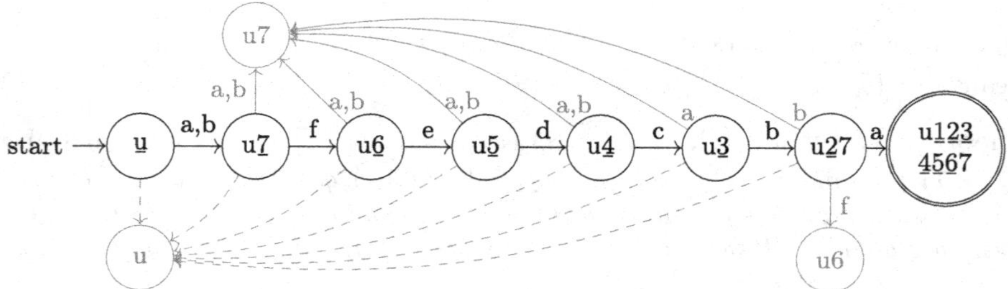

Fig. 3. DFA $\widehat{A}_{SFF}(A)$ of reverse of minimal forbidden factors of the DFA A in Fig. 2. Dashed transitions are for all symbols without explicit transition. Gray states are useless states and their transitions are omitted. Bottom dash within a state indicates that in a pair (q, x) x is 1. $\widehat{A}_{SFP}(A)$ would be identical except with no final states in this case.

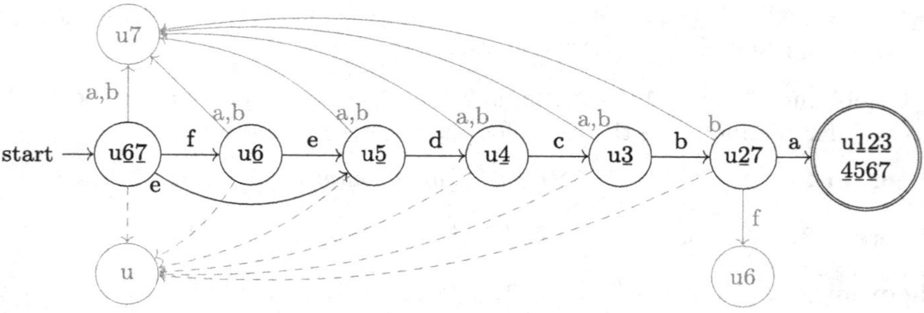

Fig. 4. DFA $\widehat{A}_{SAS}(A)$ of reverse of minimal allowed suffixes of the DFA A in Fig. 2. Dashed transitions are for all symbols without explicit transition. Gray states are useless states and their transitions are omitted. Bottom dash within a state indicates that in a pair (q, x) x is 1.

Definition 3. *Let* $\widehat{A}_{SFF}(A) = (\widehat{Q}, \Sigma, \widehat{\delta}, \widehat{q}_0, \widehat{Q}_f)$ *for a complete* DFA $A = (Q, \Sigma, \delta, q_0, Q_f)$. $\widehat{A}_{\mathrm{SFP}}(A) = (\widehat{Q}, \Sigma, \widehat{\delta}, \widehat{q}_0, \widehat{Q}'_f)$ *where* $\widehat{Q}'_f = \{S \in \widehat{Q}\}$ *such that* $\mathrm{dom}(S) \neq Q$ *and* $(q_0, 1) \in S$.

The following theorem can be inferred from Lemma 5 and Lemma 6.

Theorem 5. *$L(\widehat{A}_{SFP}(A))^R = SFP(A)$ for a complete DFA A.*

The shortest synchronizing template automaton is a variant of a subset construction, similar to Rogers and Lambert [17], and its size is therefore exponential in the worst case. More precisely, $|\widehat{Q}| \leq 3^{|Q|}$ for some $\widehat{A}(A, \widehat{q}_0) = (\widehat{Q}, \Sigma, \widehat{\delta}, \widehat{q}_0, \widehat{Q}_f)$ where $A = (Q, \Sigma, \delta, q_0, Q_f)$. Given the properties of its language, however, in each DFA for the shortest characteristic factors the part of the automaton with useful states is proportional to the size of the characteristic set, if the set is finite and, according to experiments of Rogers and Lambert [17], it is reasonable in practice.

4 Positive Position Run and Its Computation Using Pattern Matching of Characteristic Factors

We define a positive run and a positive position run of a DFA, which represents all configurations and all positions in an input string, respectively, in which the automaton reaches a final state. An example can be found in Table 1.

Definition 4. *Let $r = q_0, \ldots, q_{|w|}$ be a run of a* DFA *$A = (Q, \Sigma, \delta, q_0, Q_f)$ on a string $w \in \Sigma^*$. A* positive run *of the* DFA *on w is the set of all pairs (q_i, i) such that $q_i \in Q_f$. A* positive position run *of the* DFA *on w is the set of all indices i such that $q_i \in Q_f$.*

We describe how the positive position run of a DFA can be computed with the use of exact pattern set matching of forbidden factors and of allowed suffixes in an input string, provided the automaton has its shortest characteristic factors finite, which is guaranteed for local automata. We can use any existing matching algorithm for the matching of a finite set of patterns and the computation will share its properties. Because of certain pattern matching algorithms capable of skipping parts of input, the computation can perform sublinearly, which we demonstrate using the Commentz-Walter algorithm [8] with such a property. If the shortest characteristic sets are not finite, then the computation does not use pattern matching and the computation behaves as in the case of a standard implementation of the original automaton.

If a DFA has a finite set of the shortest characteristic factors, its language is strictly locally k-testable for some $k \geq 1$, and therefore if it is to accept an input string of length at least k, its prefix cannot be forbidden, there can't be any occurrence of a forbidden factor and the suffix must be allowed. Finding an occurrence of a forbidden factor therefore means that the run has failed, and any occurrence of an allowed suffix before a forbidden factor is found indicates that the automaton reached a final state. If the characteristic factors are finite it is possible to use some matching algorithm for a finite set of patterns for such task. The prefix is sufficient to check using the standard DFA run. This is covered in Algorithm 2. Figure 5 shows preprocessed trie of patterns for DFA in

Algorithm 1: Preprocessing of DFA

Require: Complete DFA $A = (Q, \Sigma, \delta, q_0, Q_f)$ with useless states $U \subseteq Q$

Ensure : k for which $L(A)$ is strictly locally $(k+1)$-testable (minimal if A is minimal DFA). If k is finite, then also SFF(A) and SAS(A)

$\widehat{A}_{SFF} \leftarrow \widehat{A}(A, U \times \{1\})$

$\widehat{A}_{SAS} \leftarrow \widehat{A}(A, (U \times \{0\}) \cup (Q_f \times \{1\}))$

Obtain $\widehat{A}_{SFP}$ from $\widehat{A}_{SFF}$

if *$\widehat{A}_{SFF}$ or $\widehat{A}_{SAS}$ contains cycles* **then return** $k \leftarrow \infty$

$\mathrm{SFF}(A) \leftarrow L(\widehat{A}_{SFF})^R$

$\mathrm{SAS}(A) \leftarrow L(\widehat{A}_{SAS})^R$

$k \leftarrow \max(\{|p| \mid p \in L(\widehat{A}_{SFP})\} \cup \{|p| - 1 \mid p \in SFF(A)\} \cup \{|p| \mid p \in SAS(A)\})$

Algorithm 2: Computing positive position run of DFA

Require: Complete DFA $A = (Q, \Sigma, \delta, q_0, Q_f)$ with useless states $U \subseteq Q$, input string w

Ensure : Positive position run PPR of A on s, i.e. set of indices where A reaches a final state in w

Using Algorithm 1 obtain k, $SFF(A)$ and $SAS(A)$.

$q \leftarrow q_0$

if $q_0 \in Q_f$ **then** $PPR \leftarrow \{0\}$

if $q_0 \in U$ **then return**

for $i \leftarrow 1$ **to** $\min(|w|, k)$ **do**

  $q \leftarrow \delta(q, w[i])$

  if $q \in Q_f$ **then** $PPR \leftarrow PPR \cup \{i\}$

  if $q \in U$ **then return**

if $|w| \leq k$ **then return** `// DFA for non-SLTL never gets past here`

Using a pattern set matching algorithm match $SFF(A)$ and $SAS(A)$ in w. If a forbidden factor is matched, then end. If an allowed suffix is matched, then add the end position index to PPR.

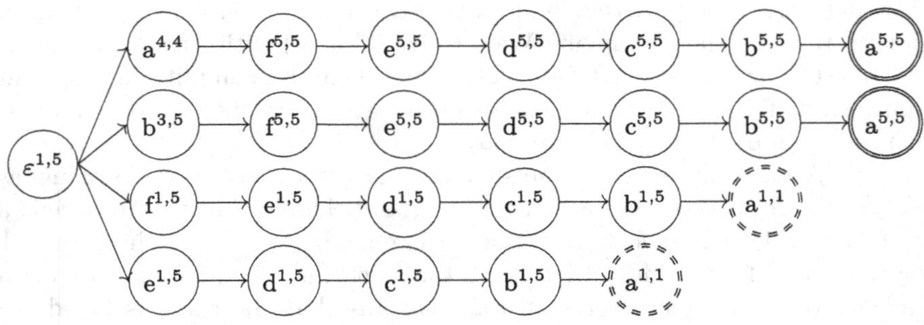

Fig. 5. Trie of the shortest forbidden factors and the shortest allowed suffixes of the DFA in Fig. 2. Solid final states represent forbidden patterns while dashed represent allowed patterns. Numbers in each node represent values of functions shift1 and shift2 as described in [8].

Table 1. Demonstration of running Algorithm 2 with the use of Commentz-Walter algorithm [8] for the input string in the second top row and the DFA in Fig. 2. In the second top row the orange cells mark separately handled prefix, the green cells mark positions of final states and the red cell marks match of a forbidden factor. The third row shows the run of the DFA on the input string. The other rows show the alignment of patterns stored in the trie in Fig. 5 during the Commentz-Walter algorithm. The green cells mark examined symbols that are matching and the red cells those where mismatch occurs. The positive position run for the DFA and the input string is $\{23, 24, 38, 39\}$. and the positive run is $\{(6, 23), (7, 24), (6, 38), (7, 39)\}$.

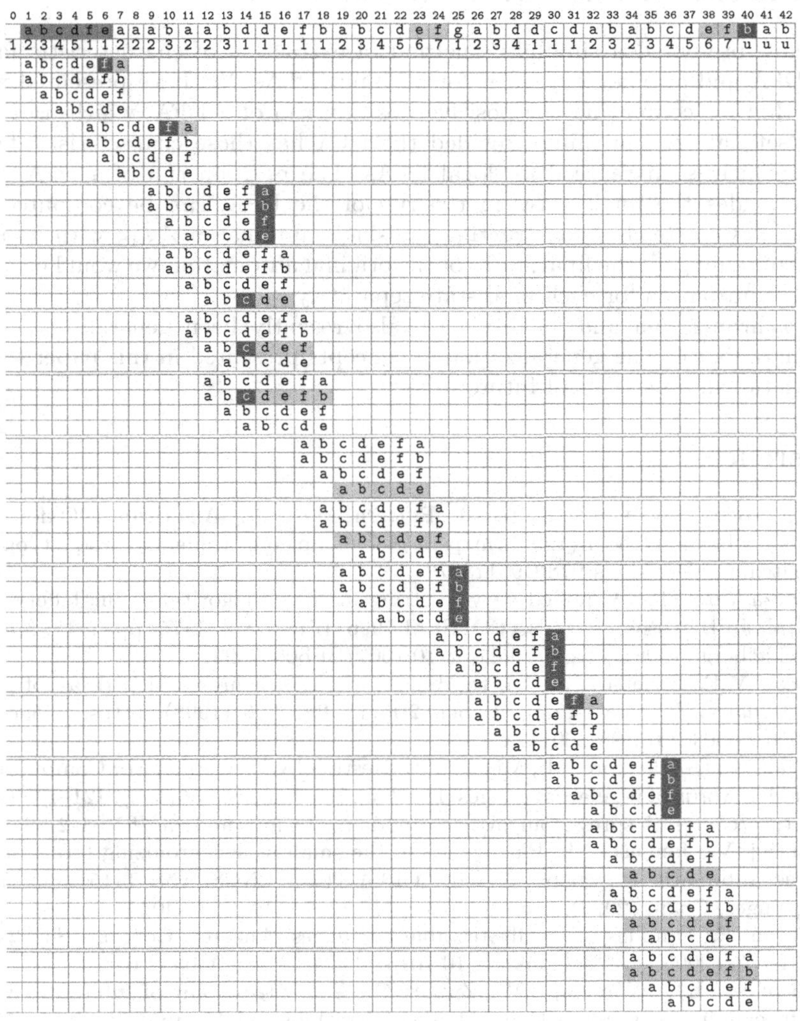

Fig. 2 and Fig. 1 shows the matching phase of Algorithm 2 with the use of the Commentz-Walter algorithm.

We note that for strings shorter than k there is no specific characterization given by the definition of SLTL. For such strings, it is a necessary but not sufficient condition that there is no occurrence of a forbidden characteristic factor. Similarly, the occurrence of an allowed suffix is not necessary. The same logic applies to cases when k is unbounded. In such case every input string falls into this category.

Theorem 6. *Given a complete* DFA A *and an input string* w *Algorithm 2 correctly computes the positive position run of* A *on* w.

5 Conclusion and Future Work

We have presented a simplified method of extraction of characteristic factors of a DFA and also generalized it for all input DFAs. We have presented a new algorithm for computing the positive position run of a DFA, which uses exact finite set pattern matching provided the DFA has those characteristic factors finite, which is guaranteed for local DFAs. The position run of a DFA, which also provides the final states in the run of the DFA, can be computed in a similar way: Instead of constructing the single automaton $\widehat{A}_{SAS}$ with its initial state containing all the final states of the original automaton, we would construct multiple automata for each final state separately.

Synchronization and locality have also recently been generalized for tree automata [5,15]. The generalization of concepts presented in this paper for tree automata can be a topic of future work.

References

1. Apostolico, A., Galil, Z. (eds.): Pattern Matching Algorithms. Oxford University Press, Oxford (1997). https://global.oup.com/academic/product/pattern-matching-algorithms-9780195113679
2. Baeza-Yates, R.: A unified view to string matching algorithms. In: Jeffery, K.G., Král, J., Bartošek, M. (eds.) SOFSEM 1996. LNCS, vol. 1175, pp. 1–15. Springer, Heidelberg (1996). https://doi.org/10.1007/BFb0037393
3. Béal, M.P., Senellart, J.: On the bound of the synchronization delay of a local automaton. Theoret. Comput. Sci. **205**(1), 297–306 (1998). https://doi.org/10.1016/S0304-3975(98)80011-X
4. Béal, M., Crochemore, M., Mignosi, F., Restivo, A., Sciortino, M.: Computing forbidden words of regular languages. Fundam. Informaticae **56**(1–2), 121–135 (2003). http://content.iospress.com/articles/fundamenta-informaticae/fi56-1-2-08
5. Blažej, V., Janoušek, J., Plachý, Š: On the smallest synchronizing terms of finite tree automata. In: Nagy, B. (ed.) CIAA 2023. LNCS, vol. 14151, pp. 79–90. Springer, Cham (2023). https://doi.org/10.1007/978-3-031-40247-0_5
6. Caron, P.: Families of locally testable languages. Theoret. Comput. Sci. **242**(1), 361–376 (2000). https://doi.org/10.1016/S0304-3975(98)00332-6
7. Cleophas, L.G., Watson, B.W., Zwaan, G.: A new taxonomy of sublinear right-to-left scanning keyword pattern matching algorithms. Sci. Comput. Program. **75**(11), 1095–1112 (2010). https://doi.org/10.1016/j.scico.2010.04.012
8. Commentz-Walter, B.: A string matching algorithm fast on the average. In: Maurer, H.A. (ed.) ICALP 1979. LNCS, vol. 71, pp. 118–132. Springer, Heidelberg (1979). https://doi.org/10.1007/3-540-09510-1_10

9. Crochemore, M., Hancart, C.: Automata for matching patterns. In: Rozenberg, G., Salomaa, A. (eds.) Handbook of Formal Languages, pp. 399–462. Springer, Heidelberg (1997). https://doi.org/10.1007/978-3-662-07675-0_9
10. Crochemore, M., Hancart, C.: Pattern matching in strings. In: Algorithms and Theory of Computation Handbook. Chapman & Hall/CRC Applied Algorithms and Data Structures Series. CRC Press (1999). https://doi.org/10.1201/9781420049503-c12
11. Crochemore, M., Rytter, W.: Text Algorithms. Oxford University Press, Oxford (1994). http://www-igm.univ-mlv.fr/%7Emac/REC/B1.html
12. McNaughton, R., Papert, S.: Counter-Free Automata. MIT Press, Cambridge (1971)
13. Melichar, B.: String matching with k differences by finite automata. In: ICPR 1996, Vienna, Austria, 25–19 August 1996, pp. 256–260. IEEE Computer Society (1996). https://doi.org/10.1109/ICPR.1996.546828
14. Melichar, B., Holub, J.: 6D classification of pattern matching problems. In: Proceedings of the Prague Stringology Club Workshop 1997, pp. 24–32 (1997)
15. Plachý, Š, Janoušek, J.: On synchronizing tree automata and their work–optimal parallel run, usable for parallel tree pattern matching. In: Chatzigeorgiou, A., et al. (eds.) SOFSEM 2020. LNCS, vol. 12011, pp. 576–586. Springer, Cham (2020). https://doi.org/10.1007/978-3-030-38919-2_47
16. Rogers, J., Lambert, D.: Extracting forbidden factors from regular stringsets. In: Proceedings of the 15th Meeting on the Mathematics of Language, London, UK, pp. 36–46. Association for Computational Linguistics (2017). https://doi.org/10.18653/v1/W17-3404
17. Rogers, J., Lambert, D.: Extracting subregular constraints from regular stringsets. J. Lang. Model. **7**, 143 (2019). https://doi.org/10.15398/jlm.v7i2.209
18. Sipser, M.: Introduction to the Theory of Computation, 3rd edn. Course Technology, Boston (2013)
19. Watson, B.W., Zwaan, G.: A taxonomy of sublinear multiple keyword pattern matching algorithms. Sci. Comput. Program. **27**(2), 85–118 (1996). https://doi.org/10.1016/0167-6423(96)00008-1
20. Watson, B.: Taxonomies and toolkits of regular language algorithms. Ph.D. thesis, Mathematics and Computer Science (1995). https://doi.org/10.6100/IR444299
21. Zalcstein, Y.: Locally testable languages. J. Comput. Syst. Sci. **6**(2), 151–167 (1972). https://doi.org/10.1016/S0022-0000(72)80020-5

Query Learning of Minimal Deterministic Symbolic Finite Automata Separating Regular Languages

Yoshito Kawasaki(✉), Diptarama Hendrian, Ryo Yoshinaka, and Ayumi Shinohara

Tohoku University, Sendai, Japan
yoshito.kawasaki.t6@dc.tohoku.ac.jp

Abstract. We propose a query learning algorithm for constructing a minimal DSFA M that separates given two regular languages L_+ and L_-, i.e., $L_+ \subseteq \mathcal{L}(M)$ and $L_- \cap \mathcal{L}(M) = \emptyset$. Our algorithm extends the algorithm for learning separating DFAs by Chen et al. (TACAS 2009) embedding the algorithm for learning DSFAs by Argyros and D'Antoni (CAV 2018). Since the problem of finding a minimal separating automaton is NP-hard, we also propose two heuristic methods to learn a separating DSFA which is not necessarily minimal. One runs faster and the other outputs smaller separating DSFAs. So, one of those can be chosen depending on the application requirement.

1 Introduction

Query learning is a prominent active learning model to obtain an unknown concept first proposed by Angluin [1]. In her setting, the learner's goal is to learn an unknown regular language by making queries to a *Minimally Adequate Teacher*, who answers two types of queries. A *membership query* asks whether an arbitrary string is in the target language, and an *equivalence query* asks whether the learner's hypothesis represents the target language and the teacher returns a counterexample if it does not. Her algorithm L^* terminates with a minimal Deterministic Finite Automaton (DFA) accepting the target language after posing a polynomial number of queries in the number of the states of the DFA and the maximum length of given counterexamples.

Since then, query learning has been extensively studied because of the wide variety of applications such as pattern recognition, model checking, program verification, etc. (see e.g. [7]). In practice, it is often the case that an answer to a membership query can be neither "yes" nor "no", but "don't care" (or "don't know"), due to the nature of the application or the teacher's computational limitations. Learning automata under such a situation has also been studied [10,12,14]. Among those, Chen et al. [4] tackled the problem of finding a minimal contextual assumption in an assume-guarantee rule in compositional verification. This requires to find an automaton M that separates two disjoint

D. Hendrian—He is currently working in Tokyo Medical and Dental University.

H. Fernau et al. (Eds.): SOFSEM 2024, LNCS 14519, pp. 340–354, 2024.
https://doi.org/10.1007/978-3-031-52113-3_24

regular languages L_+ and L_-; that is, M accepts every string in L_+ and no string in L_-. They proposed to use membership queries and containment queries instead of equivalence queries. Finding a minimal separating DFA is known to be an NP-hard problem [6]. They proposed an algorithm $\mathsf{L}^{\mathsf{Sep}}$ for exactly solving the problem and a heuristic variant of it for quickly finding a relatively small separating DFA.

On the other hand, recently, symbolic finite automata (SFAs), a generalization of classical automata, have been attracting much attention. In many applications like XML processing and program analysis involve huge or infinite alphabets. Symbolic automata have transition edges labeled with guard predicates, which concisely represent potentially infinite sets of concrete characters. Several query learning algorithms for different kinds of deterministic SFAs (DSFAs) have been proposed [2,3,5,11]. Among those, Argyros and D'Antoni's work [2] is prominent. They proposed an algorithm MAT^* that learns DSFAs using membership queries and equivalence queries with the aide of a predicate learner. It learns minimal DSFAs over any query learnable algebra.

In this paper, we propose an algorithm for finding a minimal DSFA that separates two disjoint languages over a huge or infinite alphabet. Our algorithm $\mathsf{MAT}^{\mathsf{sep}}$ extends $\mathsf{L}^{\mathsf{sep}}$ embedding MAT^*. Since the computational cost of the proposed algorithm is high, we also propose two heuristic methods to construct a relatively small separating DSFA. One runs faster and the other's outputs are smaller. So, one of those can be chosen depending on the application requirement.

2 Preliminaries

2.1 Boolean Algebras and Symbolic Automata

Symbolic automata have transitions that carry predicates over an *effective Boolean algebra* [2]. An effective Boolean algebra $\mathcal{A}$ is a tuple $(\mathfrak{D}, \Psi, [\![\_]\!])$ where $\mathfrak{D}$ is a *domain set*, Ψ is a set of *predicates* and $[\![\_]\!] : \Psi \to 2^{\mathfrak{D}}$ is a *denotation function*. The predicate set Ψ is closed under Boolean operations. For any predicates $\varphi_1, \varphi_2 \in \Psi$, one can find $\varphi_1 \wedge \varphi_2$, $\varphi_1 \vee \varphi_2$, and $\neg\varphi_1$ in Ψ such that $[\![\varphi_1 \vee \varphi_2]\!] = [\![\varphi_1]\!] \cup [\![\varphi_2]\!]$, $[\![\varphi_1 \wedge \varphi_2]\!] = [\![\varphi_1]\!] \cap [\![\varphi_2]\!]$, and $[\![\neg\varphi_1]\!] = \mathfrak{D} \setminus [\![\varphi_1]\!]$, respectively. There are special predicates $\bot, \top \in \Psi$ with $[\![\bot]\!] = \emptyset$ and $[\![\top]\!] = \mathfrak{D}$. It is decidable whether $[\![\varphi]\!] = \emptyset$ for any $\varphi \in \Psi$ and moreover there is an efficient procedure to find an element of $[\![\varphi]\!]$ if exists. We assume each predicate φ is assigned a positive integer $|\varphi|$, called the *size*.

Example 1 (Interval Algebra). The *interval algebra* over 32-bit non-negative integers $\mathfrak{D} = \{i \in \mathbb{N} \mid 0 \le i < 2^{32}\}$ is the Boolean closure of atomic predicates $l \in \mathfrak{D}$ which denotes $[\![l]\!] = \{x \in \mathfrak{D} \mid l \le x\}$. Examples of predicates in this algebra include $1 \wedge \neg 10 \vee 100 \wedge \neg 1000$, which denotes $[1, 10) \cup [100, 1000) = \{x \in \mathfrak{D} \mid 1 \le x < 10 \text{ or } 100 \le x < 1000\}$. Every predicate ψ has integers $l_1, r_1, \dots, l_k, r_k$ such that $0 \le l_1 < r_1 < \dots < l_k < r_k \le 2^{32}$ and $[\![\psi]\!] = \bigcup_{1 \le i \le k} [l_i, r_i)$.

Definition 1 (Three-valued symbolic finite automata). A *three-valued symbolic finite automata (3SFA) M* is a tuple $(\mathcal{A}, Q, q_0, Q_+, Q_-, Q_?, \Delta)$ where

$\mathcal{A} = (\mathfrak{D}, \Psi, [\![\_]\!])$ is an effective Boolean algebra; Q is a finite set of states; $q_0 \in Q$ is a state called the *initial state*; $Q_+, Q_-, Q_? \subseteq Q$ partition Q and they are respectively called *accepting*, *rejecting*, and *don't-care* states; and $\Delta\colon Q^2 \to \Psi$ is a *transition guard function.*

A two-valued SFA (i.e., a usual SFA) is a 3SFA with no don't-care states, which is often specified as a quintuple $(\mathcal{A}, Q, q_0, Q_+, \Delta)$, where $Q_- = Q \setminus Q_+$ and $Q_? = \emptyset$ are implicit.

A 3SFA M is *deterministic* if, for all $q \in Q$, $[\![\Delta(q, q_1)]\!] \cap [\![\Delta(q, q_2)]\!] = \emptyset$ for any distinct states q_1 and q_2. It is *total* if, for all $q \in Q$, $\bigcup_{q' \in Q} [\![\Delta(q, q')]\!] = \mathfrak{D}$. All 3SFAs throughout this paper are total and deterministic. When a 3SFA is total and deterministic, Δ induces a function δ from $Q \times \mathfrak{D}$ to Q defined by $\delta(q, a) = r$ iff $a \in [\![\Delta(q, r)]\!]$. The function δ is naturally extended to $Q \times \mathfrak{D}^* \to Q$ by $\delta(q, \varepsilon) = q$, where ε is the empty string, and $\delta(q, ua) = \delta(\delta(q, u), a)$ for $u \in \mathfrak{D}^*$ and $a \in \mathfrak{D}$. We furthermore extend δ to $2^Q \times \mathfrak{D}^* \to 2^Q$ by $\delta(Q', s) = \{\delta(q, s) \mid q \in Q'\}$. A word $u \in \mathfrak{D}^*$ is *accepted* and *rejected* by M if $\delta(q_0, u) \in Q_+$ and $\delta(q_0, u) \in Q_-$, and the languages accepted and rejected by M are $\mathcal{L}_+(M) = \{u \in \mathfrak{D}^* \mid \delta(q_0, u) \in Q_+\}$ and $\mathcal{L}_-(M) = \{u \in \mathfrak{D}^* \mid \delta(q_0, u) \in Q_-\}$, respectively. A language L is said to be *$\mathcal{A}$-regular* if L is accepted (equivalently, rejected) by some SFA M over $\mathcal{A}$. We say that a 3SFA M *separates* two disjoint languages L_+ and L_- if and only if $L_+ \subseteq \mathcal{L}_+(M)$ and $L_- \subseteq \mathcal{L}_-(M)$. If M separates $\mathcal{L}_+(M')$ and $\mathcal{L}_-(M')$ for a 3SFA M', we simply say M separates M'.

We give a characterization of two 3SFAs of which one separates the other. The following definitions and theorems are originally given for Mealy machines by Paull and Unger [15]. Here we translate their discussions for 3SFAs.

Definition 2 (Closed grouping, compatibles). Let M be a 3SFA and $\mathcal{G} = \{G_1, \dots G_m\}$ be a collection of sets of states of M. We say $\mathcal{G}$ is a *closed grouping* if (i) there exists $G \in \mathcal{G}$ such that $q_0 \in G$, (ii) for all $G \in \mathcal{G}$, $G \subseteq Q_+ \cup Q_?$ or $G \subseteq Q_- \cup Q_?$ and (iii) for all $(G, a) \in \mathcal{G} \times \mathfrak{D}$, there exists $G' \in \mathcal{G}$ such that $\delta(G, a) \subseteq G'$. A state set $G \subseteq Q$ is said to be a *compatible* if there exists a closed grouping $\mathcal{G}$ such that $G \in \mathcal{G}$. A compatible G is *maximal* if there exists no larger compatible that includes G.

Theorem 1. *A subset $G \subseteq Q$ is a compatible if and only if every pair of states in G forms a compatible.*

Theorem 2. *Suppose every state of 3SFA M is reachable from the initial state.*

- *If an SFA N separates M, there exists a closed grouping $\mathcal{G}$ of M such that $|\mathcal{G}| \leq |N|$.*
- *For any closed grouping $\mathcal{G}$ of M, there exists an SFA with $|\mathcal{G}|$ states that separates M.*

2.2 Equivalence Class and Representatives for 3SFA

Let us say that two characters $a, b \in \mathfrak{D}$ are *equivalent* in a 3SFA M if and only if for any predicate ψ used in M, either $\{a, b\} \subseteq [\![\psi]\!]$ or $\{a, b\} \cap [\![\psi]\!] = \emptyset$. A collection

D of characters that includes at least one element of each equivalence class is said to be a *representative set*. To decide the condition (iii) of the closedness of a grouping $\mathcal{G}$ (Definition 2), it suffices to check it for all $(G, a) \in \mathcal{G} \times D$ rather than all $(G, a) \in \mathcal{G} \times \mathfrak{D}$.

Proposition 1. *Every 3SFA M with n states has at most n^n equivalence classes, given by the following partition $\mathcal{P}$:*

$$\mathcal{P} = \{[\![\Delta(q_1, p_1) \wedge \ldots \wedge \Delta(q_n, p_n)]\!] \mid p_1, \ldots, p_n \in Q\} \setminus \{\emptyset\}. \quad (1)$$

There is a Boolean algebra $\mathcal{A}$ for which this upper bound is tight.

Example 2 (Representatives in Interval Algebra). Using Eq. (1) in Proposition 1, one can obtain the equivalence classes and a set of representatives. If a 3SFA M is over the interval algebra in Example 1, one can more efficiently find a set of representatives than naively implementing Eq. (1). Every guard predicate denotes a set $[l_1, r_1) \cup \cdots \cup [l_k, r_k)$, where $0 \le l_1 < r_1 < l_2 < \cdots < l_k < r_k \le 2^{32}$. Then all such l_i from all transition guards form a representative set. The size of the set is bounded by $\sum_{(q,q') \in Q^2} |\Delta(q, q')|$. The representative set obtained this way is not necessarily minimal, but minimization is easy.

2.3 Learning Model

This paper tackles the problem of finding a smallest SFA separating given $\mathcal{A}$-regular languages L_+ and L_-. Hereafter, we fix two $\mathcal{A}$-regular languages L_+ and L_- so that we do not have to mention L_+ and L_- explicitly.

Our algorithm $\mathsf{MAT}^{\mathsf{sep}}$ obtains a separating SFA using queries on the $\mathcal{A}$-regular languages L_+ and L_- and a query learning algorithm Λ for the Boolean algebra $\mathcal{A}$. The kinds of queries asked by the main algorithm $\mathsf{MAT}^{\mathsf{sep}}$ and those by the predicate learner Λ are different. The main algorithm $\mathsf{MAT}^{\mathsf{sep}}$ asks (language) membership, completeness, and soundness queries, which are always concerned with L_+ and L_-. On the other hand, the predicate learner Λ asks (predicate) membership and equivalence queries, which are about some character set represented by a predicate of the Boolean algebra.

A *(language) membership query* asks whether a string $u \in \mathfrak{D}^*$ belongs to either of the language L_+ or L_-. The answer to the query is $+$ if $u \in L_+$; $-$ if $u \in L_-$; and ? otherwise.

A *completeness query* asks whether M is *complete* with L_+ and L_-, i.e., $\mathcal{L}_+(M) \subseteq L_+$ and $\mathcal{L}_-(M) \subseteq L_-$. If not, the oracle gives a counterexample string in $(\mathcal{L}_+(M) \setminus L_+) \cup (\mathcal{L}_-(M) \setminus L_-)$.

A *soundness query* asks whether M is *sound* with L_+ and L_-, i.e., $L_+ \subseteq \mathcal{L}_+(M)$ and $L_- \subseteq \mathcal{L}_-(M)$. If not, the oracle gives a counterexample string in $(L_+ \setminus \mathcal{L}_+(M)) \cup (L_- \setminus \mathcal{L}_-(M))$.

Our goal is to find an SFA M on which the soundness query is answered "yes".

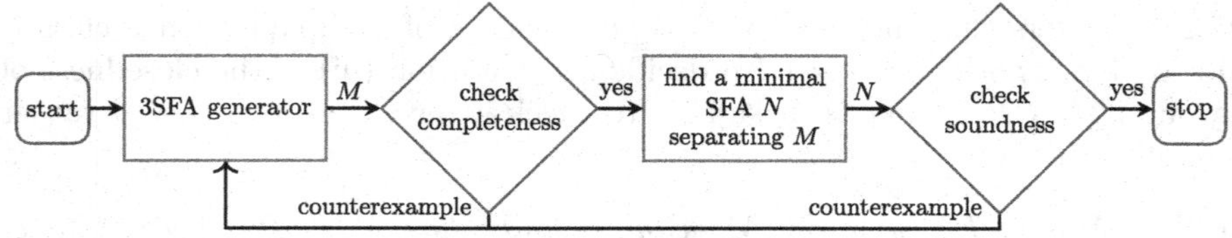

Fig. 1. Overview of $\mathsf{MAT}^{\mathsf{Sep}}$. The whole can be seen as a symbolic extension of $\mathsf{L}^{\mathsf{Sep}}$. The 3SFA generator computes a 3SFA extending (the hypothesis construction of) MAT^*.

A *(predicate) membership query* concerning a set $D \subseteq \mathfrak{D}$ asks whether a letter $a \in \mathfrak{D}$ belongs to D. The answer to the query is $+$ if $u \in D$, and $-$ otherwise.

An *equivalence query* asks whether $[\![\psi]\!] = D$ for an arbitrary predicate $\psi \in \Psi$. In the case where $[\![\psi]\!] \neq D$, the oracle gives a counterexample string in $([\![\psi]\!] \setminus D) \cup (D \setminus [\![\psi]\!])$.

The goal of a predicate learner Λ is, for any predicate $\varphi \in \Psi$, after some amount of queries, to guess a predicate $\psi \in \Psi$ such that $[\![\psi]\!] = [\![\varphi]\!]$.

3 Our Proposed Algorithm

Our learning algorithm $\mathsf{MAT}^{\mathsf{Sep}}$ can be viewed as a symbolic version of the algorithm $\mathsf{L}^{\mathsf{Sep}}$ by Chen et al. [4] for learning separating DFAs from 3DFAs. Their algorithm first "loosely" learns a 3DFA M which is complete, i.e., it holds $\mathcal{L}_+(M) \subseteq L_+$ and $\mathcal{L}_-(M) \subseteq L_-$. Since the goal is to obtain a DFA separating L_+ and L_-, exact identification of L_+ and L_- is not needed. Then, $\mathsf{L}^{\mathsf{Sep}}$ computes a minimal DFA that separates M, based on the technique by Paull and Unger [15]. The obtained DFA N separates $\mathcal{L}_+(M)$ and $\mathcal{L}_-(M)$ but not necessarily L_+ and L_-. If the 3DFA M is too "loose", it is quite likely that the DFA N does not separate L_+ and L_-, i.e., $L_+ \not\subseteq \mathcal{L}_+(N)$ or $L_- \not\subseteq \mathcal{L}_-(N)$ holds. In case N is not sound, we refine M using a counterexample in $(L_+ \setminus \mathcal{L}_+(N)) \cup (L_- \setminus \mathcal{L}_-(N))$. If N is sound, N is a minimal DFA separating L_+ and L_-. The correctness of the above algorithm is supported by the following lemma.

Lemma 1 ([4]). *Let M be a 3DFA complete with L_+ and L_-, $\hat{N}$ a minimal DFA separating L_+ and L_-, and N a minimal DFA separating $\mathcal{L}_+(M)$ and $\mathcal{L}_-(M)$. Then, N has no more states than $\hat{N}$.*

Our algorithm follows this outline and is illustrated in Fig. 1. Lemma 1 holds for symbolic automata as well. The important challenge is that we must perform the learning process over symbolic automata, manipulating predicates. We construct a complete 3SFA M by modifying Argyros and D'Antoni's SFA learning algorithm MAT^*. For computing a minimal separating SFA N, we design a symbolic version of Paull and Unger's algorithm.

3.1 3SFA Generator

Our 3SFA generator computes a 3SFA M using queries and a predicate learner Λ for the Boolean algebra $\mathcal{A}$. While $\mathsf{L}^{\mathsf{Sep}}$ extends Angluin's DFA learning algorithm

L*, ours extends Argyros and D'Antoni's algorithm MAT* [2], which is based on the TTT algorithm [9] rather than L*. Like most DFA learners do, the TTT algorithm finds strings u which have different Brzozowski derivatives $u^{-1}L = \{v \mid uv \in L\}$ of the learning target language L and uses them as states of the hypothesis automaton. Counterexamples are used for finding a new string whose Brzozowski derivative is different from any of those of the current state strings. The algorithm MAT* spawns instances $\Lambda^{(u,v)}$ of a predicate learner Λ for all pairs (u, v) of states and tries to let them learn the character set $\{a \in \mathfrak{D} \mid (ua)^{-1}L = v^{-1}L\}$, where the main algorithm tries to play the role of a teacher answering membership and equivalence queries asked by those predicate learner instances, though the answers to queries may be sometimes incorrect. The predicate in an equivalence query raised by $\Lambda^{(u,v)}$ is used as the label of the transition from u to v in the hypothesis 3SFA.

In our case, differently from MAT*, (i) we use completeness and soundness queries rather than equivalence queries, and (ii) we construct 3SFAs rather than SFAs. The first point introduces no new difficulties, since every counterexample to a completeness or soundness query is a counterexample to an equivalence query. Modifying the TTT algorithm to be a 3DFA learner is straightforward [8]. The state construction will be based on the Brzozowski derivatives of both L_+ and L_-. That is, we make two strings u and v two different states if $u^{-1}L_+ \neq v^{-1}L_+$ or $u^{-1}L_- \neq v^{-1}L_-$. We then spawn predicate learner instances $\Lambda^{(u,v)}$ for all pairs (u, v) of states and tries to let them learn the character set $\{a \in \mathfrak{D} \mid (ua)^{-1}L_+ = v^{-1}L_+ \text{ and } (ua)^{-1}L_- = v^{-1}L_-\}$. Other than those, our 3SFA generator is essentially the same as MAT*.

3.2 Finding a Minimal Separating SFA

Suppose we have obtained a 3SFA M which is complete with L_+ and L_-. Our next task is to find a minimal SFA N separating $\mathcal{L}_+(M)$ and $\mathcal{L}_-(M)$. By Theorem 2, a minimal SFA separating M can be obtained from a minimal closed grouping of M. Finding a minimal separating SFA, i.e., finding a closed grouping, is an NP-hard problem. We use a SAT solver for this task.

Finding a Minimal Closed Grouping with SAT Solver. Here we give our SAT formulation for finding a closed grouping of a 3SFA. It takes as inputs a 3SFA M, a set D of representative characters for M, and a size m of the output closed grouping $\mathcal{G} = \{G_1, \dots, G_m\}$. The constraints are formulated based on Definition 2 except that the domain is D instead of the universal set $\mathfrak{D}$. We try to find a solution to the following SAT formulation by increasing m from one by one. When we find a solution, it represents a minimal closed grouping through the interpretation presented below. The formula given below is not necessarily in the conjunctive normal form (CNF), but the standard technique transforms it into an equisatisfiable CNF formula of linear size. That is, if we have a subformula of the form $\bigvee_i \bigwedge_j F_{i,j}$, we introduce new variables f_i, replace the subformula with $\bigvee_i f_i$, and add the formula of the form $\bigwedge_j(\neg f_i \vee F_{i,j}) \wedge \bigvee_j(f_i \vee \neg F_{i,j})$. Applying this procedure recursively, we obtain a CNF formula.

Algorithm 1. Build an SFA from a closed grouping

Require: 3SFA $M = (\mathcal{A}, Q, q_0, Q_+, Q_-, Q_?, \Delta)$ and its closed grouping $\mathcal{G}$
Ensure: N is an SFA separating M

1: $G_0 \leftarrow$ some $G \in \mathcal{G}$ such that $q_0 \in G$
2: $F \leftarrow \{G \in \mathcal{G} \mid G \cap Q_+ \neq \emptyset\}$
3: $\hat{\Delta}(G, G') \leftarrow \bot$ for all $(G, G') \in \mathcal{G}^2$
4: **for** $G \in \mathcal{G}$ **do**
5: $\quad \psi \leftarrow \bot$
6: $\quad$ **for** $G' \in \mathcal{G}$ **do**
7: $\quad\quad \varphi \leftarrow \left(\bigwedge_{q \in G} \bigvee_{q' \in G'} \Delta(q, q')\right) \wedge \neg\psi$
8: $\quad\quad \hat{\Delta}(G, G') \leftarrow \varphi$
9: $\quad\quad \psi \leftarrow \psi \vee \varphi$
10: $N \leftarrow (\mathcal{A}, \mathcal{G}, G_0, F, \hat{\Delta})$

We prepare variables $S[q, i]$ meaning $q \in G_i$ and constants $T_0[q, a, q']$ meaning $\delta(q, a) = q'$ for $q, q' \in Q$, $a \in D$, and $i \in [m] = \{1, \ldots, m\}$. The SAT formula is given as follows.

$$S[q_0, 1] \tag{2}$$

$$\wedge \bigwedge_{i \in [m]} \bigwedge_{q \in Q_+} \bigwedge_{q' \in Q_-} \neg \left(S[q, i] \wedge S[q', i]\right) \tag{3}$$

$$\wedge \bigwedge_{i \in [m]} \bigwedge_{a \in D} \bigvee_{j \in [m]} \bigwedge_{q \in Q} \left(\neg S[q, i] \vee \bigvee_{q' \in Q} S[q', j] \wedge T_0[q, a, q']\right) \tag{4}$$

The restrictions (2), (3), and (4) correspond to (i), (ii), and (iii) of Definition 2, respectively, where $\bigwedge_{q \in Q} \left(\neg S[q, i] \vee \bigvee_{q' \in Q} S[q', j] \wedge T_0[q, a, q']\right)$ means $\delta(G_i, a) \subseteq G_j$. Letting $n = |Q|$, the size of the obtained formula is evaluated as $\mathcal{O}(1 + mn^2 + m^2n^2|D|) = \mathcal{O}(m^2n^2|D|)$. One may assume $|D| \in \mathcal{O}(n^n)$ by Proposition 1.

Constructing an SFA from a Closed Grouping. After finding a minimal closed grouping $\mathcal{G}$, we build an SFA N by Algorithm 1. A running example is shown in Fig. 2. We define $N = (\mathcal{A}, \mathcal{G}, G_0, F, \hat{\Delta})$, where G_0 is an arbitrary set in $\mathcal{G}$ that includes the initial state of M and $F = \{G \in \mathcal{G} \mid G \cap Q_+ \neq \emptyset\}$. In accordance with Paull and Unger's algorithm, we want $\hat{\delta}(G, a)$ to be any $G' \in \mathcal{G}$ such that $\delta(G, a) \subseteq G'$ for each $a \in \mathfrak{D}$. One might think defining $\hat{\Delta}$ by $\hat{\Delta}(G, G') = \bigwedge_{q \in G} \bigvee_{q' \in G'} \Delta(q, q')$ would be the symbolic counterpart of their construction. Actually, this yields a separating but non-deterministic automaton. We make $\hat{\Delta}$ deterministic using $\neg\psi$ at Line 7 of Algorithm 1, where ψ is the disjunction of the predicates on the edges from G determined so far. Then, the resulting SFA N is deterministic and separates the input 3SFA M.

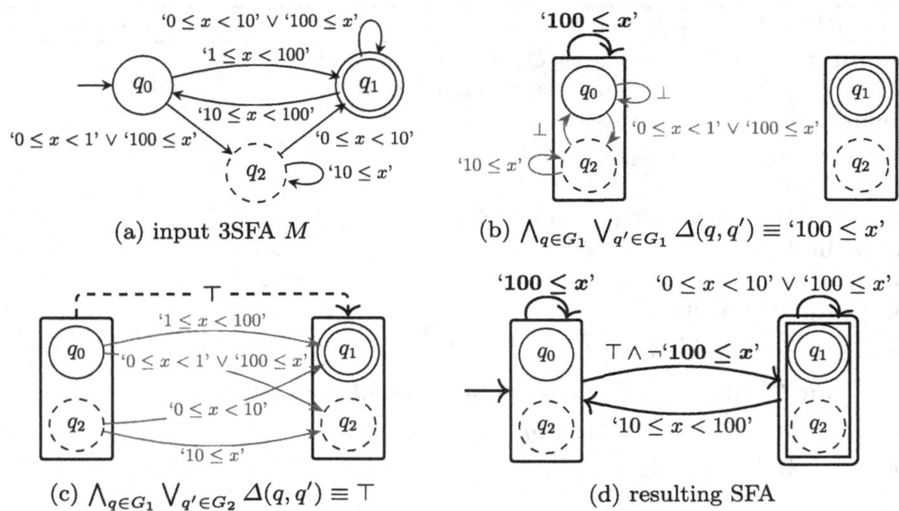

Fig. 2. A running example of Algorithm 1. (a) The input 3SFA M is over the interval algebra in Example 1. Here, the accepting, rejecting and don't-care states are represented by single, double and dashed circles, respectively. For ease of understanding, predicates are written in more intuitive forms, like '$1 \le x < 100$' instead of $1 \wedge \neg 100$. The 3SFA M has a minimal closed grouping $\{G_1, G_2\}$ with $G_1 = \{q_0, q_2\}$ and $G_2 = \{q_1, q_2\}$. (b) The label of the self-loop of G_1 is obtained by taking the disjunction of the predicates on edges from each state in G_1 and taking their conjunction. (c) The label of the edge from G_1 to G_2 is determined in the same way, but in addition we need to exclude the characters on the self-loop edge of G_1 to make the resulting SFA deterministic. (d) The other edge labels are determined in the same way and we obtain a separating SFA.

4 Correctness and Query Complexity

By Theorem 2 and Lemma 1, $\mathsf{MAT}^{\mathrm{Sep}}$ gives a minimal separating SFA when it terminates. The termination follows from the finiteness of the number of queries discussed below.

The query complexity of $\mathsf{MAT}^{\mathrm{Sep}}$ depends on that of the predicate learner Λ, just like MAT^* does. Actually the complexity can be evaluated exactly the same as MAT^*, since ternary classification is not more difficult than the binary case (e.g., [8]), and our learning process may stop before exactly identifying the ternary classification L_+, L_- and $\mathfrak{D}^* \setminus (L_+ \cup L_-)$. Let $\mathcal{M}^\Lambda(n)$ and $\mathcal{E}^\Lambda(n)$ denote the upper bounds on the numbers of MQs and EQs, respectively, raised by Λ for learning a target represented by a predicate of size at most n. Let $\mathcal{B}(M_*) = \max_{q \in Q} \max_{P \subseteq Q} \left|\bigvee_{p \in P} \Delta(q,p)\right|$ for a 3SFA $M_* = (\mathcal{A}, Q, q_0, Q_+, Q_-, Q_?, \Delta)$. This is the maximum size of a predicate that Λ may be expected to learn during the execution of the algorithm when $\mathcal{L}_+(M_*) = L_+$ and $\mathcal{L}_-(M_*) = L_-$.

Theorem 3. *Let $M_* = (\mathcal{A}, Q, q_0, Q_+, Q_-, Q_?, \Delta)$ be a minimal 3SFA such that $\mathcal{L}_+(M_*) = L_+$ and $\mathcal{L}_-(M_*) = L_-$. Then, $\mathsf{MAT}^{\mathrm{Sep}}$ will learn a minimal SFA sep-*

Algorithm 2. Greedily find a small closed grouping from maximal compatibles

Require: 3SFA $M = (\mathcal{A}, Q, q_0, Q_+, Q_-, Q_?, \Delta)$ and
the set of its maximal compatibles $\mathcal{G}_{\max}$

Ensure: $\hat{\mathcal{G}} \subseteq \mathcal{G}_{\max}$ is a closed grouping

$G_0 \leftarrow \arg\max_{G \in \{G \in \mathcal{G} \mid q_0 \in G\}} |G|$
$\hat{\mathcal{G}} \leftarrow \{G_0\}$
$V \leftarrow$ stack with a single element G_0
while V is not empty **do**
  $\psi \leftarrow \bot$
  for $G' \in \hat{\mathcal{G}}$ **do**
    $\psi \leftarrow \psi \vee \left(\bigwedge_{q \in G} \bigvee_{q' \in G'} \Delta(q, q')\right)$
  for G' **in** $\mathcal{G}_{\max} \setminus \hat{\mathcal{G}}$ in descending order of the size **do**
    $\varphi \leftarrow \left(\bigwedge_{q \in G} \bigvee_{q' \in G'} \Delta(q, q')\right) \wedge \neg\psi$
    if $[\![\varphi]\!] \neq \emptyset$ **then**
      $\hat{\mathcal{G}} \leftarrow \hat{\mathcal{G}} \cup \{G'\}$
      PUSH(V, G')
    $\psi \leftarrow \psi \vee \varphi$

arating L_+ and L_- using Λ with $\mathcal{O}\big(|Q|^2|\Delta|\mathcal{M}^\Lambda(k) + (|Q|^2 + |Q|\log m)|\Delta|\mathcal{E}^\Lambda(k)\big)$ membership queries[1] *and $\mathcal{O}\big(|Q||\Delta|\mathcal{E}^\Lambda(k)\big)$ completeness and soundness queries, where $k = \mathcal{B}(M_*)$, $|\Delta|$ denotes the number of state pairs $(q, q') \in Q^2$ with $\Delta(q, q') \neq \bot$, and m is the length of the longest counterexample given to* $\mathsf{MAT}^{\mathsf{Sep}}$.

5 Heuristics for Finding a Closed Grouping

In this section, we introduce a heuristic method for quickly computing a relatively small closed grouping. Although it is guaranteed that the output of $\mathsf{MAT}^{\mathsf{Sep}}$ is a minimal separating SFA, it is computationally expensive mainly to find a minimal closed grouping. Many applications require to quickly find a small separating SFA, which is not necessarily the smallest.

Greedy Heuristics. Chen et al. [4] proposed a greedy heuristic method to find a relatively small closed grouping of a 3DFA by considering only maximal compatibles. Note that Theorem 1 implies that compatibles correspond to cliques (therefore maximal compatibles correspond to maximal cliques) on the graph where the vertices are the states and the edges are the compatible pairs. We can identify incompatible pairs based on the following recursive nature of the incompatibility: (i) every pair of an accepting state and a rejecting state is incompatible and (ii) every pair that transitions to an incompatible pair by

[1] Argyros and D'Antoni's theorem claims a less tight bound $\mathcal{O}(|Q|^2|\Delta|\mathcal{M}^\Lambda(k) + |Q|^2 \log m|\Delta|\mathcal{E}^\Lambda(k))$ but they proved the bound presented here.

the same character is also incompatible. While the original algorithm [4,15] checks transitions for each character, our algorithm processes the predicates on the transition edges. That is, if a state pair $(p, q) \in Q \times Q$ has already been recognized to be incompatible and we have $[\![\Delta(p', p) \wedge \Delta(q', q)]\!] \neq \emptyset$, then the pair (p', q') is also incompatible.

Let $\mathcal{G}_{\max}$ be the set of maximal compatibles of a 3DFA M. Their algorithm first picks one of the largest compatibles G_0 from $\mathcal{G}_{\max}$ that includes the initial state q_0 of M and let $\hat{\mathcal{G}} = \{G_0\}$. Then, for each compatible $G \in \hat{\mathcal{G}}$ and each character a, it picks one of the largest compatibles G' from $\mathcal{G}_{\max}$ such that $\delta(G, a) \subseteq G'$ and puts G' into $\hat{\mathcal{G}}$, until $\hat{\mathcal{G}}$ converges. Then, the obtained set $\hat{\mathcal{G}}$ is a closed grouping.

Algorithm 2 is our symbolic extension, where we modified their algorithm so that a compatible in $\mathcal{G}_{\max}$ is added to $\hat{\mathcal{G}}$ only when we know that a current grouping $\hat{\mathcal{G}}$ is not closed for some characters. While their algorithm expands $\hat{\mathcal{G}}$ as soon as it finds that $\hat{\mathcal{G}}$ does not contain the largest compatible $G' \in \mathcal{G}_{\max}$ such that $\delta(G, a) \subseteq G'$ for some $G \in \hat{\mathcal{G}}$ and $a \in \mathfrak{D}$, our algorithm does not perform this if $\hat{\mathcal{G}}$ has some $G' \in \hat{\mathcal{G}}$ such that $\delta(G, a) \subseteq G'$. This modification sometimes reduces but never increases the size of a resulting closed grouping.

Hybrid Heuristics. We propose another heuristic method, which can be seen as a hybrid of the exact method and the above greedy heuristics. Limiting our candidates for components of a closed grouping to be only maximal compatibles like the greedy method, we pick the fewest possible compatibles to form a closed grouping using a SAT solver. The size of this formula is $\mathcal{O}(Nm + |D|m^2)$. This bound might appear worse than the size bound $\mathcal{O}(N^2n^2|D|)$ of the formulas used in the exact method, since m can be as big as $(n/3)^{n/3}$ [13]. However, in our experiments, m did not grow much compared to n, and thus the formulas used in this hybrid method were not too big.

6 Evaluation

In this section, we present some experimental results on our algorithm $\mathsf{MAT}^{\mathsf{Sep}}$. We consider SFAs over the interval algebra in Example 1. Any predicates consisting of k intervals can be learned with $\mathcal{O}(k \log |\mathfrak{D}|)$ membership queries and $\mathcal{O}(k)$ equivalence queries. Each time a counterexample is given, a learner can find two ends of some intervals by $\lceil \log |\mathfrak{D}| \rceil$ membership queries respectively.

We chose `MiniSat`[2] for the SAT solver used in the exact and the hybrid heuristic methods. All experiments were run on a machine with Intel®Core™ i9-7980XE CPU @ 2.60GHz and 64GB RAM, and were implemented in Dart based on the open source library `symbolicautomata`[3]. Section 6.1 compares the performances of the exact and the heuristic methods. Section 6.2 compares the numbers of queries raised by $\mathsf{MAT}^{\mathsf{Sep}}$ and $\mathsf{L}^{\mathsf{Sep}}$ for different alphabet sizes.

[2] http://minisat.se.
[3] https://github.com/lorisdanto/symbolicautomata.

6.1 Comparison in Separation: Exact vs Heuristic Methods

Here, we compare the performances of the exact and the heuristic methods for separation with respect to the total computation times and the sizes of the output SFAs. Input 3SFAs M over the interval algebra $\mathcal{A}$ in Example 1 are constructed, based on [4], from two randomly generated SFAs X and Y with two parameters: the number of states n and a partition $\mathcal{P} = \{\mathfrak{D}_1, \dots, \mathfrak{D}_m\}$ of the domain $\mathfrak{D}$. Those SFAs X and Y are constructed by the following procedure.

1. Let $Q = \{q_0, \dots, q_{n-1}\}$ and $\Delta(q_i, q_j) = \bot$;
2. Each state $q \in Q$ is added to F at 50% probability;
3. For each source state $q \in Q$,
   (a) each state $q' \in Q$ is put into a successor set S_q at 50% probability,
   (b) for each subdomain $\mathfrak{D}_j \in \mathcal{P}$, a successor state $q' \in S_q$ is randomly chosen, and update $\Delta(q, q')$ by $\Delta(q, q') \vee \varphi_j$ where $[\![\varphi_j]\!] = \mathfrak{D}_j$;
4. If the generated SFA $Z = (\mathcal{A}, Q, q_0, F, \Delta)$ is not a minimal one accepting $\mathcal{L}_+(Z)$, we restart the procedure from the beginning. Otherwise, Z is output.

The 3SFA M is defined so that $\mathcal{L}_+(M) = \mathcal{L}_+(X) \cap \mathcal{L}_+(Y)$ and $\mathcal{L}_-(M) = \mathcal{L}_-(X) \cap \mathcal{L}_-(Y)$. One can compute M by the standard technique of product automata. The parameters for generating X and Y can be different. We generated 3SFAs with parameters $(n_X, n_Y, \mathcal{P})$, where X and Y were generated with parameters $(n_X, \mathcal{P})$ and $(n_Y, \mathcal{P})$, respectively. It is often the case that M and minimal SFAs for $\mathcal{L}_+(M)$ and for $\mathcal{L}_-(M)$ have $n_X n_Y$ states but sometimes less. Therefore, it is guaranteed that a minimal separating SFA has at most $\min\{n_X, n_Y\}$ states, and finding such a small SFA is not trivial.

Our first experiment was conducted with the following parameters:

- $\mathcal{P} = \{[0, 10^0), [10^0, 10^1), \dots, [10^7, 10^8), [10^8, 2^{32})\}$ (fixed)
- $n_X = 5$ (fixed)
- $n_Y = 1, 2, \dots, 10$

We generated and tested on 100 instances for each parameter $(n_X, n_Y, \mathcal{P})$. The results are shown in Fig. 3a.

Next, we compared the performances by varying the number m of equivalence classes of characters. The parameters are set as follows:

- $n_X = n_Y = 5$ (fixed)
- $m = 2, 10, 20, 30, \dots, 100$
- $\mathcal{P}_m = \{\mathfrak{D}_j \mid 1 \leq j \leq m\}$ where $\mathfrak{D}_j = \left[\lfloor \frac{j-1}{m} d \rfloor, \lfloor \frac{j}{m} d \rfloor\right)$ for $d = |\mathfrak{D}| = 2^{32}$.

We generated and tested on 100 instances for each parameter $(m, n_X, n_Y, \mathcal{P}_m)$. The results are shown in Fig. 3b.

Figures 3a and 3b show that in both cases, the performance of the hybrid heuristic method is between the exact and the greedy method with respect to the computation time and the obtained SFA sizes. While the greedy method failed to find minimal SFAs in about 5% cases, it was only about 1% in the hybrid method. The greedy method runs roughly hundred times faster than the hybrid method, which runs roughly ten times faster than the exact method.

Thereby we used the greedy heuristic method in the following experiments.

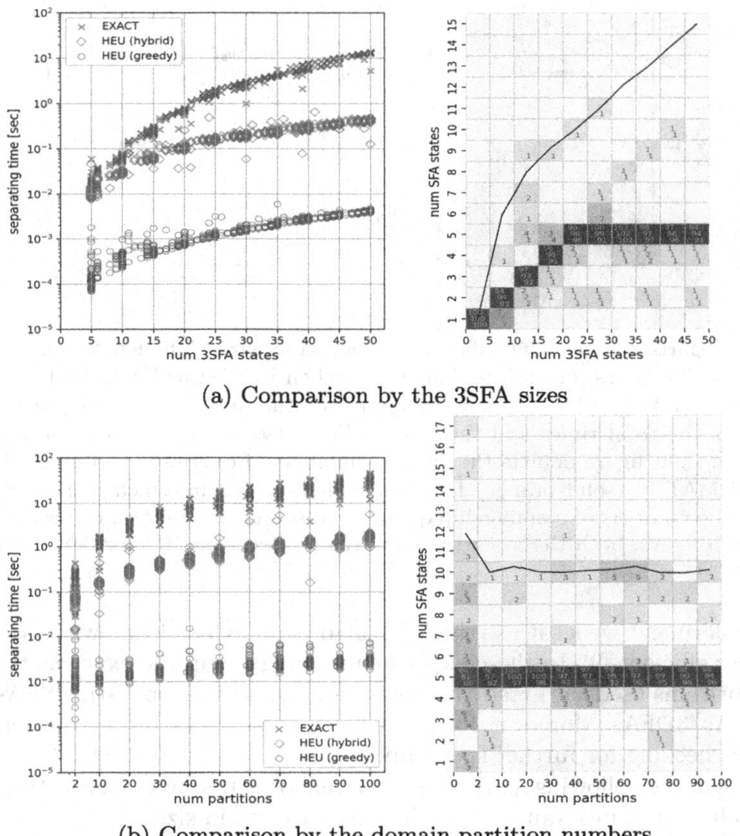

(a) Comparison by the 3SFA sizes

(b) Comparison by the domain partition numbers

Fig. 3. Comparison of the exact and the two heuristic methods. The scatter plots on the left figures compare the computation times of the three methods w.r.t. the input parameters. The heatmaps on the right figures depict the distributions of the sizes of the output SFAs, where 0's are omitted. For example, the cell at $((15, 20], 4)$ in the top-right table shows that the exact, hybrid, and greedy methods respectively obtained 99, 96, and 95 separating SFAs of size 4, when the size of the input 3SFA is more than 15 and at most 20. (For each cell, x-range is left-exclusive and right-inclusive.) The cell at $((60, 70], 4)$ in the bottom-right table shows that those three methods respectively obtained 3, 3, and zero separating SFAs of size 4, when the partition number m is in $(60, 70]$. While the greedy method failed to find minimal SFAs in 26 settings out of 1000 (a) and 69 out of 1000 (b), it was only in 11 (a) and 13 (b) settings by the hybrid method. The black curve shows the average numbers of maximal compatibles.

6.2 Comparison in Learning: SFA vs DFA

We compared the learning times and the numbers of queries of $\mathsf{MAT}^{\mathrm{Sep}}$ and the TTT version of $\mathsf{L}^{\mathrm{Sep}}$ under different sizes of the domain alphabet. While the original version of $\mathsf{L}^{\mathrm{Sep}}$ uses observation tables for computing complete 3DFAs,

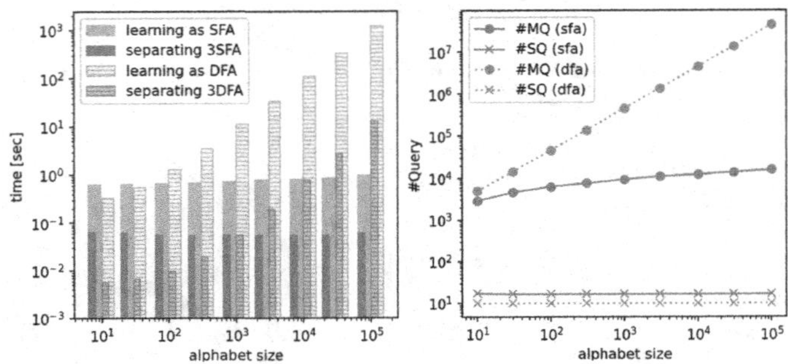

Fig. 4. Comparison of learning efficiency by alphabet sizes. The left figure depicts the average learning times for each alphabet size when learned as DFAs by $\mathsf{L}^{\mathsf{Sep}}$ (striped gray) and as SFAs by $\mathsf{MAT}^{\mathsf{Sep}}$ (solid orange) from the same teacher, in which the thin color shows the total times and the thick color shows the times spent on separating 3SFAs. The right figure depicts the average numbers of queries raised by $\mathsf{L}^{\mathsf{Sep}}$ (dotted gray) and $\mathsf{MAT}^{\mathsf{Sep}}$ (solid orange) for each alphabet size, in which o- and x-markers represent the numbers of membership queries and soundness queries, respectively. Note that in eath figure, both the x- and y-axes are on a log scale. (Color figure online)

in this experiment we let it use the TTT algorithm, which is known to be practically more efficient [9]. In these experiments, we gave up the exact method, and both algorithms used the greedy heuristics for computing separating SFAs/DFAs from 3SFAs/3DFAs. Moreover, following Chen et al. [4], we did not perform completeness checking for further speed up.

We used the following parameters to generate random 3SFAs M over the interval algebra $\mathcal{A}$ in Example 1 with different domain sizes d.

- $d = 10, 30, 100, 300, \ldots, 100000$ and let $\mathfrak{D} = \{i \in \mathbb{N} \mid 0 \leq i < d\}$
- $\mathcal{P}_d = \{\mathfrak{D}_j \mid 1 \leq j \leq 10\}$ where $\mathfrak{D}_j = \left[\frac{j-1}{10}d, \frac{j}{10}d\right)$
- $n_X = n_Y = 4$ (fixed)

We generated and tested on 100 instances for each parameter $(d, n_X, n_Y, \mathcal{P}_d)$.

The results are described in Fig. 4. One can observe that both the learning times and the numbers of queries by $\mathsf{MAT}^{\mathsf{Sep}}$ were almost independent of the domain size d, while those were linear in d for $\mathsf{L}^{\mathsf{Sep}}$.

7 Concluding Remarks

This paper has presented a learning algorithm $\mathsf{MAT}^{\mathsf{Sep}}$ to obtain a minimal separating SFA for given $\mathcal{A}$-regular languages L_+ and L_-, by combining the algorithm $\mathsf{L}^{\mathsf{Sep}}$ for learning separating DFAs by Chen et al. [4] and the one MAT^* for learning SFAs by Argyros and D'Antoni [2]. We have also proposed two heuristic algorithms. One is a slight improvement of the greedy heuristic algorithm by Chen et al. and the other is a hybrid of this and the exact method. The new heuristic algorithm shows an intermediate performance, so this can be used when one needs an algorithm that runs faster than the exact algorithm and that outputs a smaller SFA than the greedy algorithm. It is future work to demonstrate the performance of our algorithm using real data.

Acknowledgment. The authors are grateful to the anonymous reviewers for their helpful comments. This work is supported in part by JSPS KAKENHI Grant Numbers JP18K11150 (RY), JP20H05703 (RY), and JP21K11745 (AS).

References

1. Angluin, D.: Learning regular sets from queries and counterexamples. Inf. Comput. **75**(2), 87–106 (1987)
2. Argyros, G., D'Antoni, L.: The learnability of symbolic automata. In: Chockler, H., Weissenbacher, G. (eds.) CAV 2018. LNCS, vol. 10981, pp. 427–445. Springer, Cham (2018). https://doi.org/10.1007/978-3-319-96145-3_23
3. Argyros, G., Stais, I., Kiayias, A., Keromytis, A.D.: Back in black: towards formal, black box analysis of sanitizers and filters. In: IEEE Symposium on Security and Privacy, pp. 91–109 (2016)
4. Chen, Y.-F., Farzan, A., Clarke, E.M., Tsay, Y.-K., Wang, B.-Y.: Learning minimal separating DFA's for compositional verification. In: Kowalewski, S., Philippou, A. (eds.) TACAS 2009. LNCS, vol. 5505, pp. 31–45. Springer, Heidelberg (2009). https://doi.org/10.1007/978-3-642-00768-2_3
5. Drews, S., D'Antoni, L.: Learning symbolic automata. In: Legay, A., Margaria, T. (eds.) TACAS 2017. LNCS, vol. 10205, pp. 173–189. Springer, Heidelberg (2017). https://doi.org/10.1007/978-3-662-54577-5_10
6. Gold, E.M.: Complexity of automaton identification from given data. Inf. Control **37**(3), 302–320 (1978)
7. de la Higuera, C.: Grammatical Inference: Learning Automata and Grammars. Cambridge University Press, Cambridge (2010)
8. Isberner, M.: Foundations of active automata learning: an algorithmic perspective. Ph.D. thesis, Technical University of Dortmund (2015)
9. Isberner, M., Howar, F., Steffen, B.: The TTT algorithm: a redundancy-free approach to active automata learning. In: Bonakdarpour, B., Smolka, S.A. (eds.) RV 2014. LNCS, vol. 8734, pp. 307–322. Springer, Cham (2014). https://doi.org/10.1007/978-3-319-11164-3_26
10. Leucker, M., Neider, D.: Learning minimal deterministic automata from inexperienced teachers. In: Margaria, T., Steffen, B. (eds.) ISoLA 2012. LNCS, vol. 7609, pp. 524–538. Springer, Heidelberg (2012). https://doi.org/10.1007/978-3-642-34026-0_39
11. Maler, O., Mens, I.-E.: A generic algorithm for learning symbolic automata from membership queries. In: Aceto, L., Bacci, G., Bacci, G., Ingólfsdóttir, A., Legay, A., Mardare, R. (eds.) Models, Algorithms, Logics and Tools. LNCS, vol. 10460, pp. 146–169. Springer, Cham (2017). https://doi.org/10.1007/978-3-319-63121-9_8
12. Moeller, M., Wiener, T., Solko-Breslin, A., Koch, C., Foster, N., Silva, A.: Automata learning with an incomplete teacher (artifact). Dagstuhl Artifacts Ser. **9**(2), 21:1–21:3 (2023)
13. Moon, J., Moser, L.: On cliques in graphs. Israel J. Math. **3**, 23–28 (1965)

14. Neider, D.: Computing minimal separating DFAs and regular invariants using SAT and SMT solvers. In: Chakraborty, S., Mukund, M. (eds.) ATVA 2012. LNCS, pp. 354–369. Springer, Heidelberg (2012). https://doi.org/10.1007/978-3-642-33386-6_28
15. Paull, M.C., Unger, S.H.: Minimizing the number of states in incompletely specified sequential switching functions. IRE Trans. Electron. Comput. **EC–8**(3), 356–367 (1959)

Apportionment with Thresholds: Strategic Campaigns are Easy in the Top-Choice but Hard in the Second-Chance Mode

Christian Laußmann, Jörg Rothe, and Tessa Seeger(✉)

Heinrich-Heine-Universität Düsseldorf, MNF, Institut für Informatik, Düsseldorf, Germany
{christian.laussmann,rothe,tessa.seeger}@hhu.de

Abstract. In apportionment elections, a fixed number of seats in a parliament are distributed to parties according to their vote counts. Common procedures are divisor sequence methods like D'Hondt or Sainte-Laguë. In many countries, an electoral threshold is used to prevent very small parties from entering the parliament. Parties with fewer than a given number of votes are simply removed. We (experimentally) show that by exploiting this threshold, the effectiveness of strategic campaigns (where an external agent seeks to change the outcome by bribing voters) can be increased significantly, and prove that it is computationally easy to determine the required actions. To resolve this, we propose an alternative second-chance mode where voters of parties below the threshold receive a second chance to vote for another party. We establish complexity results showing that this makes elections more resistant to strategic campaigns.

Keywords: Apportionment · Bribery · Computational Complexity

1 Introduction

In parliamentary elections, votes are cast for parties which in turn compete for a fixed number of seats in parliament. An apportionment method is then used to apportion the seats to the parties according to their vote counts. Usually, such methods aim at apportioning the seats in a way that makes the parliament form a small but somehow proportional representation of the voters. Such a representative parliament can then efficiently discuss topics and decide laws in the name of the voters. In many countries, the basic procedure is extended by a so-called *legal electoral threshold* (simply threshold, for short)—a minimum number of votes a party must receive to participate in the apportionment process at all. For instance, in Germany, Poland, and Scotland a party must receive at least 5% of the total vote count to participate in the apportionment process.

Electoral thresholds are important for the government to quickly form and allow for effective decision-making by minimizing the effects of fragmentation of the parliament, i.e., by reducing the number of parties in it (see [16] for a study of how mechanical and psychological effects reduce fragmentation). Undoubtedly, with fewer parties in the parliament compromises can be made more efficiently. However, a disadvantage of the threshold is that voters supporting a party that did not make it above the threshold

H. Fernau et al. (Eds.): SOFSEM 2024, LNCS 14519, pp. 355–368, 2024.
https://doi.org/10.1007/978-3-031-52113-3_25

are not represented in the parliament at all because their votes are simply ignored. For example, more than 19% of the votes in the French election of the European Parliament in 2019 were lost due to a threshold of 5%, i.e., *"of five votes, just four become effective, and one is discarded as ineffective"* [15, p. 30].

Apart from these benefits and drawbacks of thresholds, we want to find out to what extent they can be exploited in *strategic campaigns*. In such scenarios, an external agent intends to change the election outcome in her favor by bribing voters within a certain budget to change their vote. That is, an external agent seeks to change a minimum number of votes in order to either ensure a party she supports receives at least ℓ seats in the parliament (constructive case), or to limit the influence of a party she despises by ensuring it receives no more than ℓ seats (destructive case).

Today's possibilities to process enormous amounts of data from social networks, search engines, etc., make it possible to predict the voting behavior of individuals and to target them with individualized (political) advertising. Because of this, strategic campaigns attract increasing attention in political elections. Given that these attempts are already being used in the real world, it is critical to understand the threat they pose. To assess these risks, it is essential to know how effective such campaigns can be and how easy it is to find optimal campaigns. Additionally, if there is a high risk, it is desirable to improve apportionment procedures to make them more resistant to such campaigns.

Related Work. The research line on *bribery*, a.k.a. *strategic campaigns*, was initiated by Faliszewski et al. [8]. Even earlier, Bartholdi et al. [3] introduced and studied *electoral control* where, in particular, either voters or candidates can be added to or deleted from an election by the election chair so as to make a preferred candidate win (see [12] for destructive control attacks where the goal is to preclude a despised candidate from winning). Bribery and control have been studied for a wide range of voting rules, as surveyed by Faliszewski and Rothe [11] and Baumeister and Rothe [5].

While both bribery and control have been mainly investigated for single-winner and multiwinner voting rules, Bredereck et al. [6] only recently initiated the study of bribery in apportionment elections. They show that an optimal strategic campaign for apportionment elections *without* a threshold can be computed in polynomial time. Their study is most closely related to our work. For the more general study of apportionment methods in mathematical and political context, we refer to the works by Pukelsheim [17] and Balinski and Young [1,2].

Our Contribution. Our first contribution is incremental to the work of Bredereck et al. [6]: We adapt their algorithm to run for apportionment elections *with* thresholds, and use binary search techniques for significantly accelerating the computation. Both improvements are important for the practical usability of the algorithms since most real-world (parliamentary) apportionment elections include a threshold, and the vote count is often in the order of 10 million, so algorithmic efficiency is of great importance. By testing the improved algorithm on real-world elections we observe that the campaigns can exploit the electoral threshold and significantly benefit from it. Further, we introduce the destructive variant of strategic campaigns. Our second and completely novel contribution is a simple extension of the usual apportionment procedure with electoral threshold: Voters who supported a party below the threshold can reuse their vote for one of the remaining parties above the threshold. These voters thus get a second chance. We provide complexity results showing that this modification renders the corresponding problems intractable, thus providing some protection of the election.

2 Preliminaries

We begin by introducing the following two notations. Throughout this paper, we denote the set $\mathscr{P} \setminus \{X\}$ by $\mathscr{P}_{-X}$, and we write $[x]$ as a shorthand for $\{1,\ldots,x\}$.

We now turn to the apportionment setting. There already exist simpler definitions of apportionment instances in the literature, but to treat the electoral threshold and the extension which we will propose in Sect. 5 conveniently, we propose the following definition that is close to the classical ones from single-winner and multiwinner voting and works in a two-stage process. An apportionment instance $I = (\mathscr{P}, \mathscr{V}, \tau, \kappa)$ consists of the set of m *parties* $\mathscr{P}$, a list of n *votes* $\mathscr{V}$ over the parties in $\mathscr{P}$, a *threshold* $\tau \in \mathbb{N} = \{0,1,2,\ldots\}$, and the *seat count* $\kappa \in \mathbb{N}$. Each vote in $\mathscr{V}$ is a strict ranking of the parties from most to least preferred, and we write $A \succ_v B$ if voter v prefers party A to B (where we omit the subscript v when it is clear from the context). We sometimes refer to the most preferred party of a voter v as v's *top choice*. We make the very natural assumption that we have more votes than we have both parties and seats. Note that in reality an electoral threshold is usually given as a relative threshold in percent (e.g., a 5% threshold). However, we can easily convert such a relative threshold into an absolute threshold, as required by our definition[1].

An apportionment instance will be processed in two steps: First, we compute a *support allocation* σ, then we compute a *seat allocation* α. The support allocation $\sigma : \mathscr{P} \to \mathbb{N}$ describes how many voters support each party. Depending on this support, the parties later receive a corresponding number of seats in the parliament. In classical apportionment settings (which we consider in Sects. 3 and 4), the support for each party is simply the number of top choices the party receives if the party receives at least τ top choices, otherwise, the support is 0. That is, votes for parties that receive less than τ top choices are ignored, and the voters have no opportunity to change their vote. We refer to this as the *top-choice mode*[2]. An alternative mode will be proposed in Sect. 5.

Example 1 (Support Allocation). Consider $\tau = 10$, $\mathscr{P} = \{A,B,C,D\}$, and

$$\mathscr{V} = (8\times A \succ B \succ C \succ D, \quad 12\times B \succ A \succ C \succ D, \quad 5\times B \succ D \succ C \succ A,$$
$$25\times C \succ A \succ D \succ B, \quad 10\times D \succ B \succ A \succ C).$$

In the top-choice mode, we obtain $\sigma(A) = 0$, $\sigma(B) = 17$, $\sigma(C) = 25$, and $\sigma(D) = 10$.

Given a support allocation, we can now determine the seat allocation by employing an *apportionment method*. As input, such a method takes the support allocation

[1] Note that in strategic campaigns, as we define them, the total number of voters never changes but only which party they vote for. Thus the threshold is also constant.

[2] Note that in the top-choice mode it would be sufficient to know the top choice of each voter. However, we need the full preference later in the second-chance mode. So for convenience, we assume complete rankings for both modes.

σ and the seat count κ, and computes the seat allocation $\alpha : \mathscr{P} \to \{0,\dots,\kappa\}$ satisfying $\sum_{A\in\mathscr{P}} \alpha(A) = \kappa$. Note that the threshold does not matter for apportionment methods because it was already applied in the computation of the support allocation. In this study, we focus on the class of *divisor sequence apportionment methods* including, for example, the D'Hondt method (also known as the Jefferson's method), and the Sainte-Laguë method (also known as the Webster method). A divisor sequence method is defined by a sequence $d = (d_1, d_2, \dots, d_\kappa) \in \mathbb{R}^\kappa$ with $d_i < d_j$ for all $i, j \in \{1,\dots,\kappa\}$ with $i < j$, and $d_1 \geq 1$. For each party $P \in \mathscr{P}$, we compute the *fraction list* $\left[\frac{\sigma(P)}{d_1}, \frac{\sigma(P)}{d_2}, \dots, \frac{\sigma(P)}{d_\kappa}\right]$. Then we go through the fraction lists of all parties to find the highest κ values (where ties are broken by some tie-breaking mechanism). Each party receives one seat for each of its list values that is among the κ highest values. D'Hondt is defined by the sequence $(1,2,3,\dots)$ and Sainte Laguë is defined by $(1,3,5,\dots)$.

Example 2 (D'Hondt). Suppose we allocate $\kappa = 6$ seats to the parties, party 1 has support 1104, party 2 has 363, party 3 has 355, and party 4 has 178. Then, the resulting D'Hondt fraction lists are:

$$\begin{aligned}
\text{party 1} &: [\mathbf{1104}, \ \mathbf{552}, \ \mathbf{368}, \ \mathbf{276}, \ 220.8, \ 184\ ],\\
\text{party 2} &: [\ \mathbf{363}, \ 181.5, \ 121, \ 90.8, \ 72.6, \ 60.5],\\
\text{party 3} &: [\ \mathbf{355}, \ 177.5, \ 118.3, \ 88.8, \ 71, \ 59.2],\\
\text{party 4} &: [\ 178, \ 89, \ 59.3, \ 44.5, \ 35.6, \ 29.7].
\end{aligned}$$

The $\kappa = 6$ highest values are highlighted in boldface. Party 1 thus receives four seats, parties 2 and 3 receive one seat each, and party 4 receives no seats at all.

Now we define strategic campaigns, modeled as a bribery scenario. We are given an apportionment instance, a budget K, and a number ℓ indicating the minimum number of seats we want to achieve for a distinguished party P^*. By bribing at most K voters to change their vote in our favor (i.e., we can alter their votes as we like), we seek to ensure that party P^* receives at least ℓ seats. To study whether finding successful campaigns (and checking whether there exist any at all) is tractable, we define the following decision problem (see [6,8]).

$\mathscr{R}$-THRESHOLD-APPORTIONMENT-BRIBERY

Given: An apportionment instance $(\mathscr{P}, \mathscr{V}, \tau, \kappa)$, a distinguished party $P^* \in \mathscr{P}$, and integers ℓ, $1 \leq \ell \leq \kappa$, and K, $0 \leq K \leq |\mathscr{V}|$.

Question: Is there a successful campaign, that is, is it possible to make P^* receive at least ℓ seats using apportionment method $\mathscr{R}$ by changing at most K votes in $\mathscr{V}$?

Note that since $|\mathscr{V}| \geq \kappa \geq \ell$ and $|\mathscr{V}| \geq K$, the encoding of κ, ℓ, and K does not matter for the complexity analysis. $\mathscr{R}$-DESTRUCTIVE-THRESHOLD-APPORTIONMENT-BRIBERY is defined analogously. This time, however, we ask whether it is possible that by changing at most K votes P^* receives *at most* ℓ seats, i.e., we want to limit the parliamentary influence of our target party. In both the constructive and the destructive cases, we assume tie-breaking to be to the advantage of P^*. That is, if P^* and another party P have the same value in their lists and only one seat is left for them, P^* will receive it.

3 Classical Top-Choice Mode

We start by analyzing the complexity of the two problems just defined in the classical top-choice mode of apportionment. We will see that for all divisor sequence methods both deciding whether a successful campaign exists and, if so, computing such a campaign can be done in polynomial time, in both the constructive and the destructive case.

Theorem 1. *Let $\mathscr{R}$ be a divisor sequence method. Then* $\mathscr{R}$-THRESHOLD-APPORTIONMENT-BRIBERY *and* $\mathscr{R}$-DESTRUCTIVE-THRESHOLD-APPORTIONMENT-BRIBERY *are in* P.

The proof of Theorem 1 is presented in the remainder of this section and relies on the lemmas we will present now. Algorithm 1 for $\mathscr{R}$-THRESHOLD-APPORTIONMENT-BRIBERY is given explicitly; we later describe how to adapt it for the destructive case. Note that both algorithms can also easily be adapted to compute an actual campaign (if one exists). The following lemma is crucial for the correctness of the algorithms.

Lemma 1. *The following two statements hold for all divisor sequence methods.*

1. *The maximum number of additional seats for party P^* by bribing at most K votes can always be achieved by convincing exactly K voters from parties in $\mathscr{P}_{-P^*}$ to vote for P^* instead.*
2. *The maximum number of seats we can remove from party P^* by bribing at most K votes can always be achieved by convincing exactly K voters from P^* to vote for parties in $\mathscr{P}_{-P^*}$ instead.*

Proof. Note that when a party receives additional support, the parties fraction list values increase, while they decrease when the support is decreased.

We begin with the first claim. Let's assume we found a way to convice voters to change their vote (in the following called *a bribery action*) such that P^* receives X additional seats. *Case 1:* There are parties other than P^* which receive additional votes. Let $P_i \neq P^*$ be such a party. Now consider that we move all votes that were moved to P_i to P^* instead. Clearly, P^*'s fraction list values increase, while those of P_i decrease. Thus P^* receives at least as many seats in this modified bribery action as in the original. *Case 2:* Votes are only moved from parties in $\mathscr{P}_{-P^*}$ to P^*. By moving (some) voters to party P_i instead of to P^*, the fraction list values of P^* would decrease, while those of P_i would increase. Therefore, P^* cannot receive more seats in this alternative bribery action compared to the original one. Finally, note that by the monotonicity of divisor sequence methods, moving more voters to P^* never makes P^* lose any seats. Thus we can spend the whole budget K on moving voters to P^*. This together with the two given cases implies that the best possible number of additional seats can always be achieved by moving K voters only from $\mathscr{P}_{-P^*}$ to P^* (although there might be other solutions which are equally good).

To prove the second claim, just swap the roles of P^* and the other parties. □

Lemma 1 is crucial for the correctness of the algorithms because it implies that we should exhaust the whole budget K for moving voters from other parties to P^* in the

Algorithm 1. Deciding whether Threshold-Apportionment-Bribery is possible.

```
Input: 𝒫, 𝒱, τ, κ, P*, K, ℓ
K ← min{n − σ(P*), K}
if σ(P*) + K < τ then
    return No
end if
compute γ {γ[P][x] is the minimum bribery budget needed to ensure that P receives exactly x seats before P* gets ℓ seats}

initialize table tab with κ − ℓ columns and m rows,
    where tab[0][0] ← 0 and the other entries are ∞.
let o : {1, ..., |𝒫_{−P*}|} → 𝒫_{−P*} be an ordering

for i ← 1 to |𝒫_{−P*}| do
    for s ← 0 to κ − ℓ do
        for (x, cost) ∈ γ[o(i)] do
            if s − x ≥ 0 then
                tmp ← tab[i − 1][s − x] + cost
                if tmp < tab[i][s] then
                    tab[i][s] = tmp
                end if
            end if
        end for
    end for
end for

for s ← 0 to κ − ℓ do
    if tab[|𝒫_{−P*}|][s] ≤ K then
        return Yes
    end if
end for
return No
```

constructive case, and for moving voters from P^* to other parties in the destructive case. That is, we do not need to consider moving voters within $\mathscr{P}_{-P^*}$.

Algorithm 1 decides whether a successful campaign exists in the constructive case. The algorithm works with every divisor sequence method. We now describe the algorithm intuitively. We first set K to the minimum of $n - \sigma(P^*)$ and K because this is the maximum number of votes we can move from other parties to P^*. Should it be impossible for P^* with this K to reach the threshold, we can already answer No, as P^* never receives any seat at all. The crucial part of the algorithm is computing the γ dictionary: As commented, $\gamma[P][x]$ gives the minimum number of votes that must be removed from party P so that P receives only x seats before P^* receives the ℓ-th seat, assuming P^* has exactly K additional votes in the end. Note that γ can be efficiently computed. We describe this in detail later. Note that we now define an order o over the parties. This can be any order; we just use it to identify each party with a row in the table which we now begin to fill. For each i, $1 \leq i \leq |\mathscr{P}_{-P^*}|$, and each s, $0 \leq s \leq \kappa - \ell$, the cell $\text{tab}[i][s]$ contains the minimum number of votes needed to be moved away from parties $o(1), \ldots, o(i)$ such that $o(1), \ldots, o(i)$ receive s seats in total before P^* is assigned its ℓ-th seat (again, assuming P^* has exactly K additional votes in the end). This table can also be efficiently computed with dynamic programming, as we describe later. Finally, we check if there exists a value of at most K in the last row of the table. If this holds, we answer Yes because there do exist bribes that do not exceed K and ensure that the other parties leave the ℓ-th seat for P^*.

Note that by tracing back through the table tab we can find the individual numbers of voters we need to move from each party to P^* for a successful campaign. This number does not sum up to K in many cases. If so, we can simply remove the remaining votes from arbitrary parties (except P^*).

Lemma 2. *Algorithm 1 decides* $\mathscr{R}$-THRESHOLD-APPORTIONMENT-BRIBERY *for every divisor sequence method* $\mathscr{R}$ *in polynomial time.*

Proof. We first prove that the algorithm indeed runs in polynomial time. For most of the algorithm this is easy to see: We essentially fill a table with $|\mathscr{P}|$ rows and at most κ columns. Since $\kappa \leq |\mathscr{V}|$, the table size is indeed polynomial in the input size. However, it is yet unclear how γ is computed. Computing γ works with a binary search for the jumping points of a function ϕ. Thereby, ϕ is defined as the number of seats a party with y votes receives before P^* receives ℓ seats, assuming P^* has exactly K additional votes in the end. Let q be the final vote count of P^* (i.e., with the K additional votes). Then, for a divisor sequence method with the sequence $d = (d_1, d_2, \ldots, d_\kappa)$ we have

$$\phi(y) = \begin{cases} 0 & \text{if } y \leq \tau \\ 0 & \text{if } y \leq q/d_\ell \\ \max\{z \in \{1, \ldots, \kappa\} \mid y/d_z > q/d_\ell\} & \text{otherwise} \end{cases} .$$

Finding the jumping points with binary search is in $\mathscr{O}(\kappa \cdot \log(K))$.

We now prove the correctness of the algorithm. Starting in the beginning, setting K to the minimum of $n - \sigma(P^*)$ and K is necessary to ensure that we never move more voters from parties in $\mathscr{P}_{-P^*}$ than allowed. Setting K higher than that would result in false positive results. For the remainder of this proof, we assume all K votes are moved from $\mathscr{P}_{-P^*}$ to P^*, i.e., P^* receives K additional votes in the end. This is optimal according to Lemma 1. The first if-statement returns NO if P^* cannot reach the threshold. This answer is correct since P^* can never get any seat as long as it is below the threshold, i.e., in this case the bribe is unsuccessful.

In the middle part of the algorithm, we fill a table. Recall that for each i, $1 \leq i \leq |\mathscr{P}_{-P^*}|$, and each s, $0 \leq s \leq \kappa - \ell$, the cell tab$[i][s]$ contains the minimum number of votes needed to be removed from parties $o(1), \ldots, o(i)$ such that $o(1), \ldots, o(i)$ receive s seats in total before P^* is assigned its ℓ-th seat. The values are computed dynamically from the previous row to the next row. This is possible because the seats that parties $o(1), \ldots, o(i)$ receive in total before P^* is assigned its ℓ-th seat are exactly the sum of the number of seats the parties receive individually before P^* receives its ℓ-th seat. Further, since this number can be computed directly by comparing the divisor list of the party with the divisor list of P^* (i.e., the ϕ function of each party is independent of other parties' support) the required bribery budget is also exactly the sum of the individual bribes. Thus the values in the list are indeed computed correctly.

Finally, if in the last row there exists a value of at most K, we correctly answer YES, by the following argument. Suppose we have a value of at most K in cell tab$[|\mathscr{P}_{-P^*}|][s]$. Then there are bribes that do not exceed K and ensure that the other parties receive at most s seats before P^* is assigned its ℓ-th seat. Since there are a total of κ seats available, and the other parties get $s \leq \kappa - \ell$ seats before P^* receives the ℓ-th seat, P^* will indeed receive its ℓ-th seat. However, if all cells of the last row contain a value greater than K, the given budget is too small to ensure that the other parties receive at most $\kappa - \ell$ seats before P^* receives its ℓ-th seat. Thus the other parties receive at least $\kappa - \ell + 1$ seats in this case, which leaves at most $\ell - 1$ seats for P^*, so we correctly answer NO. □

We can easily adapt Algorithm 1 for the destructive case. This time, we *remove* $\min\{K, \sigma(P^*)\}$ voters from party P^* and *add them to the other* parties. Of course, when P^* is pushed below the threshold, we immediately answer YES. For the destructive case, ϕ and γ need to be defined slightly different. Here, we define $\phi(y)$ as the number of seats a party with y votes receives before P^* is assigned its $(\ell+1)$-th seat (what we try to prevent). And $\gamma[P][x]$ is defined as the minimum number of votes we need to add to party P such that it receives at least x seats before P^* is assigned its $(\ell+1)$-th seat. Again, we fill the table with dynamic programming but this time, whenever we have filled a row completely, we check if it is possible for the parties corresponding to all yet filled rows to receive at least $\kappa - \ell$ seats before P^* receives its $(\ell+1)$-th seat by bribery. That is, we check whether we filled the last cell in the current row with a value of at most K or whether it would be possible to fill a cell beyond the last table cell in the current row with such a value. In that case we answer YES, since there are not enough seats left for P^* to be assigned its $(\ell+1)$-th seat with this bribery action. If this was never possible, we answer NO because the best we could do is to occupy at most $\kappa - \ell - 1$ seats with parties in $\mathscr{P}_{-P^*}$, which still leaves the $(\ell+1)$-th seat for P^*. This sketches the proof of the following lemma and completes the proof of Theorem 1.

Lemma 3. *For every divisor sequence method $\mathscr{R}$, Algorithm 1 (adapted as described) decides $\mathscr{R}$-*DESTRUCTIVE-THRESHOLD-APPORTIONMENT-BRIBERY *in polynomial time.*

Note that since we can decide in polynomial time whether there exists a successful campaign guaranteeing P^* at least ℓ seats by bribing at most K voters, we can also find the maximum number of seats we can guarantee for P^* with a budget of K (using a simple binary search) in polynomial time. Analogously, for the destructive variant, it is possible to efficiently determine the maximum number of seats we can steal from P^*. We do this in the experiments of the following section to test the effectiveness of the campaigns on real-world elections.

4 Experiment

As we showed in the previous section, computing successful and even optimal campaigns is computationally tractable. This indicates that such an attack would be relatively simple for a campaign manager to execute (at least from a computational standpoint). An immediate question that arises is whether the campaign is effective enough to be worth to be executed, i.e., how many seats can we actually gain for P^* in an optimal constructive campaign, and how many seats can we take away from P^* in an optimal destructive campaign?

In our experiment, we use three datasets from recent elections shown in Table 1. GREECE2023 is the *Greek parliamentary election 2023* with 300 seats to allocate

Table 1. The ten largest parties in datasets GREECE2023 (*Greek parliamentary election 2023*), IKE2022 (*Israel Knesset election 2022*), and BUL2023 (*2023 Bulgarian parliament election*).

GREECE2023	IKE2022	BUL2023
New Democracy (40.8%)	Likud (23.4%)	GERB-SDS (26.5%)
Syriza (20.1%)	Yesh Atid (17.8%)	PP-DB (24.6%)
PASOK (11.5%)	Religious Zionism (10.8%)	Revival (14.2%)
Communist (7.2%)	National Unity (9.1%)	Rights and Freedoms (13.8%)
Greek Solution (4.5%)	Shas (8.3%)	BSP (8.9%)
Victory (2.9%)	United Torah (5.9%)	Such a People (4.1%)
Freedom (2.9%)	Yisrael Beiteinu (4.5%)	Bulgarian Rise (3.1%)
MeRA25 (2.6%)	United Arab List (4.1%)	The Left! (2.2%)
Subversion (0.9%)	Hadash-Ta'al (3.8%)	Neutral Bulgaria (0.4%)
National Creation (0.8%)	Labor (3.7%)	Together (0.4%)

and a 3% threshold, IKE2022 is the *Israel Knesset election 2022* with 120 seats and a threshold of 3.25%, and BUL2023[3] is the *2023 Bulgarian parliament election* with 240 seats and a 4% threshold. The datasets were taken from the respective Wikipedia sites[4], with original language data available at https://votes25.bechirot.gov.il/nationalresults, https://ekloges.ypes.gr/current/v/home/parties/, and https://results.cik.bg/ns2023/rezultati/index.html.

We conducted our experiments as follows. In all three elections, we focus on a budget K equal to 0.25% of the total vote count. This is a relatively small fraction of the voters, and we find it plausible that a campaign manager could be able to influence that many voters. To show the effect of the threshold on the effectiveness of a campaign, we gradually raise the threshold in our experiment. As the distinguished party P^* we always choose the party with the highest voter support in the election, since it reaches all tested thresholds and is thus always present in the parliament. Lastly, we use D'Hondt in our experiments as a representative of the divisor sequence methods, since it is one of the most widely used in apportionment elections. We also conducted the same experiments with Sainte-Laguë with similar results.

Figure 1 illustrates the effectiveness of both the constructive (top row) and destructive (bottom row) campaigns run on the three real-world elections. One would expect some kind of proportionality, e.g., that 0.25% of the voters control approximately 0.25% of the seats. This is indeed what we observe for many values of the threshold. However, there are some spikes where with only 0.25% of voters one can make P^* gain sometimes 5% or even 10% of all seats on top in the constructive case. This is considerably more that one can expect from our small budget. Note that the spikes always occur at thresholds where a party is directly above the threshold. For instance, in BUL2023 we

[3] Here, we removed votes from the dataset which are labeled 'none of the above'.

[4] https://en.wikipedia.org/wiki/May_2023_Greek_legislative_election#Preliminary_results, https://en.wikipedia.org/wiki/2022_Israeli_legislative_election#Results, https://en.wikipedia.org/wiki/2023_Bulgarian_parliamentary_election#Results.

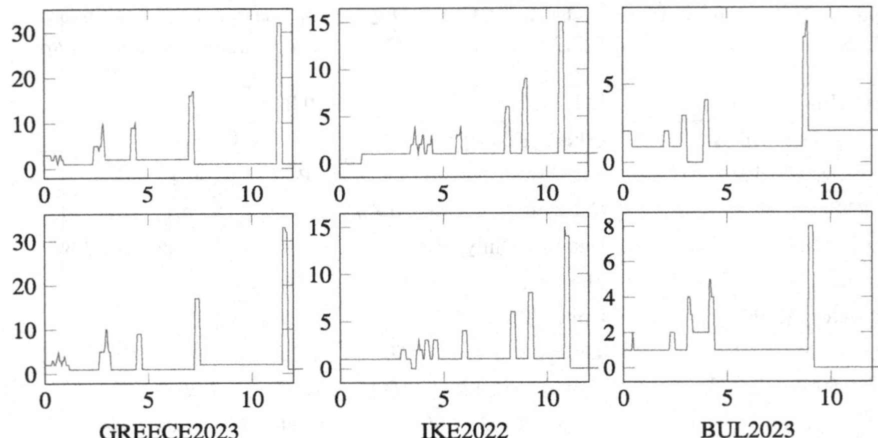

Fig. 1. The x-axis shows a variety of thresholds in percent of the number of voters n. The y-axis shows the maximally achievable number of additional seats (respectively, prevented seats) for the strongest party by D'HONDT-THRESHOLD-APPORTIONMENT-BRIBERY in the top row and by D'HONDT-DESTRUCTIVE-THRESHOLD-APPORTIONMENT-BRIBERY in the bottom row, each with a given budget of $K = 0.0025 \cdot n$.

see a peak at thresholds 2.0%, 2.9%, 3.9%, and 8.7%, which are exactly the values where *The Left!*, *Bulgarian Rise*, *Such a People*, and *BSP* are slightly above the threshold (see Table 1). This indicates that at these thresholds the campaign is focused on pushing a party below the threshold and free up its seats. For the destructive case, we can see similar peaks as in the constructive campaigns. However, this time the peaks are at thresholds where a party is just below the threshold. That is, the campaign is focused on raising a party above the threshold to steal seats from P^*.

Note that we also ran the experiments for other values for the bribe budget K and observed the following: For smaller budgets, we see narrower (and sometimes lower) peaks right at the thresholds where a party is just above it (respectively, just below it, in the destructive case), while for larger budgets, the peaks become wider (and sometimes also higher). Lastly, we again observe similar results when we choose the second- or third-strongest party instead of the strongest as our distinguished party P^*. Only with the third-strongest party as the distinguished party do we see a large spike when it is just below the threshold (or just above, in the destructive case), which, however, is in line with what one would expect.

5 The Second-Chance Mode

From the previous section we know that it is quite problematic if optimal campaigns are easy to compute, because it makes it very simple for a campaign manager to exert an enormous influence on the election outcome. Therefore, it would be of great advantage if there was a modification to the usual apportionment setting that makes the computation of optimal campaigns intractable. As mentioned in the introduction, another general problem of apportionment elections with thresholds is that voters for parties below the threshold are completely ignored. As a result, the parliament tends to be less representative. We now introduce the *second-chance mode* of voting in apportionment elections which will help resolve both of these problems at once. Unlike in the top-choice mode, in the second-chance mode voters for parties below the threshold get a second chance to vote. That is, we first determine the parties $\widehat{\mathscr{P}_\tau}$ that have at least τ top choices, i.e., the parties that make it above the threshold. Each voter now counts as a supporter for their most preferred party in $\widehat{\mathscr{P}_\tau}$. The second-chance voting process is reminiscent of the *single transferable vote (STV)* rule. However, it differs as STV is a single- or multiwinner voting rule and is not used for computing support allocations.

Example 3 (Second-Chance Voting). Consider $\tau = 2$, $\mathscr{P} = \{A,B,C,D\}$, and

$$\begin{aligned}\mathscr{V} = (A \succ B \succ C \succ D, \quad & A \succ B \succ C \succ D, \quad && A \succ B \succ C \succ D,\\ D \succ C \succ B \succ A, \quad & C \succ B \succ D \succ A, \quad && B \succ C \succ A \succ D,\\ B \succ D \succ A \succ C, \quad & B \succ A \succ C \succ D).\end{aligned}$$

Parties A and B receive three top choices each, whereas C and D both only receive one. Thus only A and B are above the threshold. Instead of ignoring the votes $C \succ B \succ D \succ A$ and $D \succ C \succ B \succ A$, in the second-chance mode we check which of A and B is more preferred in these votes. As in this example it is B in both votes, we add those two to the support of B, giving a total support of $\sigma(A) = 3$, $\sigma(B) = 5$, and $\sigma(C) = \sigma(D) = 0$.

Note that similar voting systems are already being used in Australia e.g. for the *House of Representative* and *Senate*[5]. In those elections, voters rank the candidates or parties from most to least preferred and votes for excluded choices are transferred according to the given ranking until the vote counts. In Sect. 3, we showed that in the classical apportionment setting, bribery can be solved efficiently for each divisor sequence method. To show that these problems are NP-hard in the second-chance mode, we provide reductions from the NP-complete HITTING SET problem [13].

HITTING SET

Given: A set $U = \{u_1, \dots, u_p\}$, a collection $S = \{S_1, \dots, S_q\}$ of nonempty subsets of U, and an integer K, $1 \leq K \leq \min\{p,q\}$.

Question: Is there a hitting set $U' \subseteq U$, $|U'| \leq K$, i.e., a set U' such that $U' \cap S_i \neq \emptyset$ for each $S_i \in S$?

Instead of just focusing on specific apportionment methods, in the following we generalize our results to a whole class of apportionment methods. We call an apportionment method *majority-consistent* if no party in $\mathscr{P}$ with less support than A receives more seats than A, where $A \in \mathscr{P}$ is a party with the highest support. Undoubtedly, this is a criterion every reasonable apportionment method should satisfy. Note that all divisor sequence methods and also the common Largest-Remainder-Method (LRM) (see, e.g., [6]) are majority-consistent. We now show that the second-chance mode of apportionment voting makes computing an optimal strategic campaign computationally intractable, and thus can prevent attempts of running such campaigns.

[5] https://www.aec.gov.au/learn/preferential-voting.htm.

Theorem 2. *For each majority-consistent apportionment method $\mathscr{R}$, $\mathscr{R}$-THRESHOLD-APPORTIONMENT-BRIBERY and $\mathscr{R}$-DESTRUCTIVE-THRESHOLD-APPORTIONMENT-BRIBERY are* NP*-hard in the second-chance mode. They are* NP*-complete if $\mathscr{R}$ is polynomial-time computable.*

Proof. Membership of both problems in NP is obvious whenever $\mathscr{R}$ is polynomial-time computable. We show NP-hardness of $\mathscr{R}$-THRESHOLD-APPORTIONMENT-BRIBERY by a reduction from HITTING SET. Let $(U,S,K) = (\{u_1,\dots,u_p\},\{S_1,\dots,S_q\},K)$ be an instance of HITTING SET with $q \geq 4$. In polynomial time, we construct an instance of $\mathscr{R}$-THRESHOLD-APPORTIONMENT-BRIBERY with parties $\mathscr{P} = \{c,c'\} \cup U$, a threshold $\tau = 2q+1$, $\ell = 1$ desired seat, $\kappa = 1$ available seat, and the votes

$$
\begin{aligned}
\mathscr{V} = (4q+2 \text{ votes } & c \succ \cdots, \\
4q+K+2 \text{ votes } & c' \succ \cdots, & (1)\\
\text{for each } j \in [q], 2 \text{ votes } & S_j \succ c' \succ \cdots, & (2)\\
\text{for each } i \in [p], q-\gamma_i \text{ votes } & u_i \succ c \succ \cdots, & (3)\\
\text{for each } i \in [p], q-\gamma_i \text{ votes } & u_i \succ c' \succ \cdots), & (4)
\end{aligned}
$$

where $S_j \succ c'$ denotes that each element in S_j is preferred to c', but we do not care about the exact order of the elements in S_j. Further, $2\gamma_i$ is the number of votes from group (2), in which u_i is at the first position. That is, it is guaranteed that each u_i receives exactly $2\gamma_i + (q-\gamma_i) + (q-\gamma_i) = 2q < \tau$ top choices, while c has $4q+2 \geq \tau$ and c' has $4q+K+2 \geq \tau$ top choices. Note that the voters in groups (2) and (4) use their second chance to vote for c', and those in group (3) use it to vote for c. It follows that c' currently receives exactly $2q+K$ more votes than c and thus wins the seat. We now show that we can make the distinguished party $P^* = c$ win the seat by bribing at most K voters if and only if there is a hitting set of size at most K.

($\Leftarrow$) Suppose there exists a hitting set $U' \subseteq U$ of size exactly K (if $|U'| < K$, it can be padded to size exactly K by adding arbitrary elements from U). For each $u_i \in U'$, we bribe one voter from group (1) to put u_i at their first position. These u_i now each receive the $2q+1$ top choices required by the threshold, i.e., they participate in the further apportionment process. Each u_i can receive a support of at most $4q+1$. Since the support of c is not affected by any bribes, no u_i can win the seat against c. Groups (3) and (4) do not change the support difference between c and c' and thus can be ignored. However, since U' is a hitting set, all $2q$ voters in group (2) now vote for a party in U' instead of c', reducing the difference between c and c' by $2q$. Further, we have bribed K voters from group (1) to not vote for c', which reduces the difference between c and c' by another K votes. Therefore, c and c' now have the same support, and since we assume tie-breaking to prefer c, party c wins the seat.

($\Rightarrow$) Suppose the smallest hitting set has size $K' > K$. That is, with only K elements of U we can hit at most $q-1$ sets from S. It follows that by bribing K voters from group (1) to vote for some $u_i \in U$ instead of c', we can only prevent up to $2(q-1)$ voters from group (2) to use their second chance to vote for c'. Thus we reduce the difference between c and c' by at most $2(q-1)+K$, which is not enough to make c win the seat. Now consider that we do not use the complete budget K on this strategy, i.e., to bribe voters of group (1), but only $K'' < K$. Note that by bringing only K'' parties from U above the threshold, we can only hit up to $2(q-1-(K-K''))$ sets from S. So the difference between c and c' is reduced by at most $2(q-1-(K-K''))+K''$ using this strategy. However, we now have a budget of $K-K''$ left to bribe voters, e.g., from group (2), to vote primarily for c without bringing any additional u_i above the threshold. It is easy to see that we will only reduce the difference between c and c' by at most $2(K-K'')$ with this strategy as, in the best case, c gains one supporter and c' loses one with a single bribery action. Thus we cannot reduce the difference between c and c' by more than $2(q-1-(K-K''))+K''+2(K-K'')=2(q-1)+K''$ with this mixed strategy. For each $K'' \leq K$, we have $2(q-1)+K'' < 2q+K$. Therefore, if there is no hitting set of size at most K, we cannot make the distinguished party c win against c'. The proof for the destructive variant works by swapping the roles of c and c'. □

6 Conclusions

We have studied strategic campaigns for apportionment elections with thresholds and introduced the second-chance mode of voting, where voters for parties below the threshold get a second chance to vote. The second-chance mode makes computing strategic campaigns intractable while they are easy to compute in the classical top-choice mode.

As future work, we propose to study other types of strategic campaigns (e.g., cloning of parties; see [7, 14, 18]). We already studied electoral control problems, in particular constructive and destructive control by adding or deleting parties or votes. It turns out that both, top-choice and second-chance mode are resistant to all four party control problems. For the proofs, it suffices to adapt the proofs of Bartholdi et al. [3] and Hemaspaandra et al. [12] showing that plurality voting is resistant to the corresponding control problems. For vote control in the top-choice mode, Algorithm 1 can be adapted showing that all four cases of vote control are in P for divisor sequence methods. However, this only works when the threshold is fixed, i.e., not given as percent of n. Regarding the second-chance mode, we have been able to show that all majority-consistent apportionment methods are resistant to vote control if the threshold is fixed. Another direction for future research is to study the complexity of these problems in restricted domains such as (nearly) single-peaked preferences [9, 10]. Also, studying the effectiveness of strategic campaigns in the second-chance mode using ILPs or approximation algorithms is an interesting direction for the future.

To make our strategic campaigns even more realistic, we propose to study more sophisticated cost functions such as *distance bribery* [4] where the cost of bribing a voter depends on how much we change the vote. We conjecture the problem to be harder under the assumption of distance bribery because of the observation that Lemma 1 no longer holds. That is, there are cases where it is more effective to move votes within $\mathscr{P}_{-P^*}$ than to move them to P^*. To illustrate this, suppose we have two seats, $\sigma(P^*)=7$, $\sigma(P_A)=4$, and $\sigma(P_B)=2$ with $\tau<2$. According to D'Hondt, P^* and P_A each receive one seat. Say $K=1$ but the cost for changing a vote from P_A to P^* is 2, and the cost for changing a vote from P_B to P^* is even higher. We thus cannot move a single voter to P^*, i.e., we cannot gain any seats for P^* by this strategy. However, if the cost for moving a voter from P_A to P_B is 1, we gain one seat for P^* by moving a voter from P_A to P_B.

While NP-hardness is desirable in the context of strategic campaigns, in the context of, e.g., margin of victory or robustness, the interpretations can be flipped, which can also be studied as future work. Finally, we suggest studying the extent to which voters' satisfaction with the parliament increases when the second-chance mode is used.

Acknowledgements. This work was supported in part by DFG grant RO-1202/21-1. We thank Niclas Boehmer, Robert Bredereck, and Martin Bullinger for their helpful comments during a seminar at Schloss Dagstuhl.

References

1. Balinski, M., Young, H.: The quota method of apportionment. Am. Math. Mon. **82**(7), 701–730 (1975)
2. Balinski, M., Young, H.: Fair Representation: Meeting the Ideal of One Man, One Vote. Yale University Press, New Haven (1982)
3. Bartholdi, J., III., Tovey, C., Trick, M.: How hard is it to control an election? Math. Comput. Model. **16**(8/9), 27–40 (1992)
4. Baumeister, D., Hogrebe, T., Rey, L.: Generalized distance bribery. In: Proceedings of the 33rd AAAI Conference on Artificial Intelligence, pp. 1764–1771. AAAI Press (2019)
5. Baumeister, D., Rothe, J.: Preference aggregation by voting. In: Rothe, J. (ed.) Economics and Computation. STBE, pp. 197–325. Springer, Heidelberg (2016). https://doi.org/10.1007/978-3-662-47904-9_4
6. Bredereck, R., Faliszewski, P., Furdyna, M., Kaczmarczyk, A., Lackner, M.: Strategic campaign management in apportionment elections. In: Proceedings of the 29th International Joint Conference on Artificial Intelligence, pp. 103–109. ijcai.org (2020)
7. Elkind, E., Faliszewski, P., Slinko, A.: Cloning in elections: finding the possible winners. J. Artif. Intell. Res. **42**, 529–573 (2011)
8. Faliszewski, P., Hemaspaandra, E., Hemaspaandra, L.: How hard is bribery in elections? J. Artif. Intell. Res. **35**, 485–532 (2009)
9. Faliszewski, P., Hemaspaandra, E., Hemaspaandra, L.: The complexity of manipulative attacks in nearly single-peaked electorates. In: Proceedings of the 13th Conference on Theoretical Aspects of Rationality and Knowledge, pp. 228–237. ACM Press (2011)
10. Faliszewski, P., Hemaspaandra, E., Hemaspaandra, L., Rothe, J.: The shield that never was: societies with single-peaked preferences are more open to manipulation and control. Inf. Comput. **209**(2), 89–107 (2011)
11. Faliszewski, P., Rothe, J.: Control and bribery in voting. In: Brandt, F., Conitzer, V., Endriss, U., Lang, J., Procaccia, A. (eds.) Handbook of Computational Social Choice, chap. 7, pp. 146–168. Cambridge University Press (2016)
12. Hemaspaandra, E., Hemaspaandra, L., Rothe, J.: Anyone but him: the complexity of precluding an alternative. Artif. Intell. **171**(5–6), 255–285 (2007)
13. Karp, R.: Reducibility among combinatorial problems. In: Miller, R., Thatcher, J. (eds.) Complexity of Computer Computations, pp. 85–103. Plenum Press (1972)
14. Neveling, M., Rothe, J.: The complexity of cloning candidates in multiwinner elections. In: Proceedings of the 19th International Conference on Autonomous Agents and Multiagent Systems, pp. 922–930. IFAAMAS (2020)
15. Oelbermann, K.F., Pukelsheim, F.: The European Elections of May 2019: electoral systems and outcomes (2020). https://doi.org/10.2861/129510. Study for the European Parliamentary Research Service
16. Pellicer, M., Wegner, E.: The mechanical and psychological effects of legal thresholds. Elect. Stud. **33**, 258–266 (2014)
17. Pukelsheim, F.: Proportional Representation. Springer, Heidelberg (2017)
18. Tideman, N.: Independence of clones as a criterion for voting rules. Soc. Choice Welfare **4**(3), 185–206 (1987)

Local Certification of Majority Dynamics

Diego Maldonado[1], Pedro Montealegre[2], Martín Ríos-Wilson[2](✉), and Guillaume Theyssier[3]

[1] Facultad de Ingeniería, Universidad Católica de la Santísima Concepción, Concepción, Chile
dmaldonado@ucsc.cl

[2] Facultad de Ingeniería y Ciencias, Universidad Adolfo Ibáñez, Santiago, Chile
{p.montealegre,martin.rios}@uai.cl

[3] Aix-Marseille Université, CNRS, I2M (UMR 7373), Marseille, France
guillaume.theyssier@cnrs.fr

Abstract. In majority voting dynamics, a group of n agents in a social network are asked for their preferred candidate in a future election between two possible choices. At each time step, a new poll is taken, and each agent adjusts their vote according to the majority opinion of their network neighbors. After T time steps, the candidate with the majority of votes is the leading contender in the election. In general, it is very hard to predict who will be the leading candidate after a large number of time-steps.

We study, from the perspective of local certification, the problem of predicting the leading candidate after a certain number of time-steps, which we call Election-Prediction. We show that in graphs with sub-exponential growth Election-Prediction admits a proof labeling scheme of size $\mathcal{O}(\log n)$. We also find non-trivial upper bounds for graphs with a bounded degree, in which the size of the certificates are sub-linear in n.

Furthermore, we explore lower bounds for the unrestricted case, showing that locally checkable proofs for Election-Prediction on arbitrary n-node graphs have certificates on $\Omega(n)$ bits. Finally, we show that our upper bounds are tight even for graphs of constant growth.

Keywords: Local Certification · Majority Dynamics · Proof Labeling Schemes

1 Introduction

Understanding social influence, including conformity, opinion formation, peer pressure, leadership, and other related phenomena, has long been a focus of

This research was partially supported by Centro de Modelamiento Matemático (CMM), FB210005, BASAL funds for centers of excellence from ANID-Chile (P.M), FONDECYT 1230599 (P.M.), Programa Regional STIC-AMSUD (CAMA) cod. 22-STIC-02 (P.M, M.R-W, G.T), ECOS project C19E02 (G.T., M.R-W.) and ANID FONDECYT Postdoctorado 3220205 (M.R-W).

H. Fernau et al. (Eds.): SOFSEM 2024, LNCS 14519, pp. 369–382, 2024.
https://doi.org/10.1007/978-3-031-52113-3_26

research in sociology [26,27]. With the advent of online social network platforms, researchers have increasingly turned to graph theory and network analysis to model social interactions [1]. In particular, opinion formation and evolution have been extensively studied in recent years [5].

One of the simplest and most widely studied models for opinion formation is the majority rule [6]. In this model, the opinion of an individual evolves based on the opinion of the majority of its neighbors. Specifically, consider an election with two candidates, labeled as 0 and 1, and let G be an undirected, connected, and finite graph representing the social network. Each node in the graph represents an individual with a preference for the candidate they will vote for. We call this preference *an opinion.* A particular assignment of opinions to each node is called a *configuration.* At the beginning, we consider that an initial configuration is given, representing the personal beliefs of each individual about their vote intentions. The configuration evolves in synchronous time-steps, where each individual updates its opinion according to the opinion of the majority of its neighbors. If the majority of its neighbors plan to vote for candidate 1, the node changes its opinion to 1. Conversely, if the majority of its neighbors prefer 0, the node switches to 0. In the event of a tie, where exactly half of the neighbors favor 1 and the other half favor 0, the individual retains their current opinion.

All graphs have configurations in which every node has the same opinion as the majority of its neighbors. These configurations are called *fixed points* since each node retains its opinion in subsequent time-steps. Interestingly, every graph admit non-trivial fixed points, where the local majority opinion of some nodes is different from the global majority. In general, the initial majority opinion among all nodes is not preserved when the opinions evolve in the majority dynamics. In fact, the majority opinion can switch between the two candidates in non-trivial ways, which depend on the graph properties as well as how the initial opinions are spread in the graph.

In fact, some initial configurations never converge to a fixed point. A very simple example that illustrate this point is the one of a network consisting of only two vertices connected by an edge, where one vertex initially has opinion 0 and the other has opinion 1, the two nodes exchange their opinions at each time-step, never reaching a fixed point. Configurations with this dynamical behavior are called limit cycles of period 2. Formally, a limit cycle of period 2 is a pair of configurations that mutually evolve into one another under the majority dynamics. In [23], it was shown that for any initial configuration on any graph, the majority dynamics either converge to a fixed point or a limit cycle of period 2. The results of [23] also imply that the described limit configuration (called *attractor*) is reached after a number of time-steps that is at most the number of edges in the input graph.

Therefore, even assuming that the global opinion of the society (represented by the network) evolves according to the majority rule, deciding *who wins the election* is a non-trivial task. During the last 25 years, there has been some effort in characterizing the computational complexity of this problem [22,32]. In this article we tackle this problem from the perspective of distributed algorithms and local decision.

Local Decision. Let $G = (V, E)$ be a simple connected n-node graph. A *distributed language* $\mathcal{L}$ is a (Turing-decidable) collection of tuples $(G, \mathsf{id}, \mathsf{In})$, called *network configurations*, where $\mathsf{In} : V \to \{0,1\}^*$ is called an *input function* and $\mathsf{id} : V \to [n^c]$ is an injective function that assigns to each vertex a unique identifier in $\{1, \ldots, n^c\}$ with $c > 1$. In this article, all our distributed languages are independent of the id assignments. In other words, if $(G, \mathsf{id}, \mathsf{In}) \in \mathcal{L}$ for some id, then $(G, \mathsf{id}', \mathsf{In}) \in \mathcal{L}$ for every other id'.

Given $t > 0$, a *local decision algorithm* for a distributed language $\mathcal{L}$ is an algorithm on instance $(G, \mathsf{id}, \mathsf{In})$, each node v of $V(G)$ receives the subgraph induced by all nodes at distance at most t from v (including their identifiers and inputs). The integer $t > 0$ depends only on the algorithm (not of the size of the input). Each node performs unbounded computation on the information received, and decides whether to accept or reject, with the following requirements:

- When $(G, \mathsf{id}, \mathsf{In}) \in \mathcal{L}$ then every node accepts.
- When $(G, \mathsf{id}, \mathsf{In}) \notin \mathcal{L}$ there is at least one vertex that rejects.

Distributed Languages for Majority Dynamics. Given a graph G and an initial configuration x. The *orbit* of x, denoted $\mathsf{Orbit}(x)$, is the sequence of configurations $\{x^t\}_{t>0}$ such that $x^0 = x$ and for every $t > 0$, x^t is the configuration obtained from x^{t-1} after updating the opinion of every node under the majority dynamics. We denote by ELECTION-PREDICTION the set of triplets (G, x, T), where the majority of the nodes vote 1 on time-step T starting from configuration x. Formally,

$$\textsc{Election-Prediction} = \left\{ (G,(x,T)) : \begin{array}{l} x \text{ is a configuration of } V(G), \\ T > 0, \\ \text{and } \sum_{v \in V(G)} x_v^T > \frac{|V(G)|}{2}. \end{array} \right\}$$

It is easy to see that there are no local decision algorithms for ELECTION-PREDICTION. That is, there are no algorithms in which every node of a network exchange information solely with nodes in its vicinity, and outputs which candidate wins the election. Intuitively, a local algorithm solving ELECTION-PREDICTION requires the nodes to count the states of other nodes in remote locations of the input graph. In fact, this condition holds even when there is no dynamic, i.e. $T = 0$. In that sense, the counting difficulty of problem ELECTION-PREDICTION hides the complexity of predicting the majority dynamics. For that reason, we also study the following problem:

$$\textsc{Reachability} = \left\{ (G,(x,y,T)) : \begin{array}{l} x \text{ is a configuration of } V(G), \\ T > 0, \text{ and } x^T = y. \end{array} \right\}.$$

Problem REACHABILITY is also hard from the point of view of local decision algorithms. Indeed, the opinion of a node v after t time-steps depends on the initial opinion of all the nodes at distance at most t from v. There are graphs for which the majority dynamics stabilizes in a number of time-steps proportional to the number of edges of the graph [23]. Therefore, no local algorithm will be able to even decide the opinion of a single vertex in the long term.

Local Certification. A locally checkable proof for a distributed language $\mathcal{L}$ is a prover-verifier pair where the prover is a non-trustable oracle assigning certificates to the nodes, and the verifier is a distributed algorithm enabling the nodes to check the correctness of the certificates by a certain number of communication rounds with their neighbors. Note that the certificates may not depend on the instance G only, but also on the identifiers id assigned to the nodes. In proof-labeling schemes, the information exchanged between the nodes during the verification phase is limited to the certificates. Instead, in locally checkable proofs, the nodes may exchange extra-information regarding their individual state (e.g., their inputs In or their identifiers, if not included in the certificates, which might be the case for certificates of sub-logarithmic size). The prover-verifier pair must satisfy the following two properties.

Completeness: Given $(G, \mathsf{In}) \in \mathcal{L}$, the non-trustable prover can assign certificates to the nodes such that the verifier accepts at all nodes;

Soundness: Given $(G, \mathsf{In}) \notin \mathcal{L}$, for every certificate assignment to the nodes by the non-trustable prover, the verifier rejects in at least one node.

The main complexity measure for both locally checkable proofs, and proof-labeling schemes is the size of the certificates assigned to the nodes by the prover. Another complexity measure is the number of communication rounds executed during the verification step. In this article, all our upper-bounds are valid for Proof Labeling Schemes with one-round certification, while all our lower-bounds apply to locally ceckable proofs with an arbitrary number of rounds of verification.

1.1 Our Results

We show that in several families of graphs there are efficient certification protocols for Reachability. More precisely, we show that there is a proof labeling scheme for Reachability with certificates on $\mathcal{O}(\log n)$ bits in n-node networks of sub-exponential growth.

A graph has sub-exponential growth if, for each node v, the cardinality of the set of vertices at distance at most r from v growths as a sub-exponential function in r, for every $r > 0$. Graphs of sub-exponential growth have bounded degree, and include several structured families of graphs such as the d-dimensional grid, for every $d > 0$. Nevertheless, not every class of graphs of bounded degree is of sub-exponential growth. For instance, a complete binary tree has exponential growth.

For graphs of bounded degree, we show that Reachability admits proof labeling schemes with certificates of sub-linear size. More precisely, we show that there is a proof labeling scheme for Reachability with certificates on $\mathcal{O}(\log^2 n)$ bits in n-node networks of maximum degree 3. Moreover, for each $\Delta > 3$ there exists $\epsilon > 0$ such that is a proof labeling scheme for Reachability with certificates on $\mathcal{O}(n^{1-\epsilon})$ bits in n-node networks of maximum degree $\Delta > 3$. Then, we show that all our upper-bounds are also valid for ELECTION-PREDICTION.

Then, we focus on lower-bounds. First, we show that in unrestricted families of graphs every, locally-checkable proof for the problem REACHABILITY as well as ELECTION-PREDICTION requires certificates of size $\Omega(n)$. We also show that even restricted to graphs of degree 2 and constant growth, every locally-checkable proof for REACHABILITY requires certificates of size $\Omega(n)$.

Our Techniques. Our upper bounds for the certification of REACHABILITY are based on an analysis of the maximum number of time-steps on which an individual may change its opinion during the majority dynamics. This quantity is in general unbounded. For instance, in an attractor which is a cycle of period two an oscillating node switches its state an infinite number of time-steps. However, when we look to two consecutive iterations of the majority dynamic, we obtain that the number of changes of a given node depends on the topology of the network. We show that the in the dynamic induced by two consecutive iterations of the majority dynamic (or, alternatively, looking one every two time-steps of the majority dynamics), the number of times that a node switches is state is constant on graphs of sub-exponential growth, logarithmic on graphs of maximum degree 3 and sublinear on graphs of bounded degree. The bound for graphs of sub-exponential growth was observed in [18], while the other two bounds can be obtained by a careful analysis of the techniques used in [18] (see Sect. 3 for further details). Roughly speaking, the idea consists in defining a function that assigns to each configuration a real value called the *energy* of the configuration. This energy function is strictly decreasing in the orbit of a configuration before reaching an attractor. In fact, through the definition of such function it can be shown that the majority dynamics reaches only fixed points or limit cycles of period two, in at most a polynomial number of time-steps [23]. In this article, we analyze a different energy function proposed in [18], from which we obtain the upper-bounds for the number of times that a node can switch states in two-step majority dynamic.

Our efficient proof labeling schemes are then defined by simply giving each vertex the list of time-steps on which it switches it state. From that information the nodes can reconstruct their orbit. We show that the nodes can use the certificates of the neighbors to verify that the recovered orbit corresponds to its real orbit under the majority dynamic. Our upper bounds for the certification of ELECTION-PREDICTION follow from the protocol used to certify REACHABILITY, and the use of classical techniques of local certification to count the total number of nodes in the graph, as well as the number of vertices that voted for each candidate.

Our lower-bounds are obtained using two different techniques. First, we show that in unrestricted families of graphs, every locally-checkable proof for the problem REACHABILITY or ELECTION-PREDICTION requires certificates of size $\Omega(n)$ by a reduction to the disjointedness problem in non-deterministic communication complexity. Then, we prove that restricted to graphs of degree 2 and constant growth, every locally-checkable proof for REACHABILITY requires certificates of size $\Omega(n)$ by using a locally-checkable proof for REACHABILITY to design a locally-checkable proof that accepts only if a given input graph is a cycle. In [24] it is shown that any locally checkable proof for the problem of distinguishing between a path of a cycle requires certificates of size $\Omega(\log n)$, implying that certifying REACHABILITY on graphs of degree at most 2 and constant growth (a cycle or a path) also requires $\Omega(\log n)$ certificates.

1.2 Related Work

Majority Dynamics for Modeling Social Influence. Numerous studies have been conducted on the majority dynamics. In [31], the authors studied how noise affects the formation of stable patterns in the majority dynamics. They found that the addition of noise can induce pattern formation in graphs that would otherwise not exhibit them. In [34] the authors explore opinion dynamics on complex social networks, finding that densely-connected networks tend to converge to a single consensus, while sparsely-connected networks can exhibit coexistence of different opinions and multiple steady states. Node degree influences the final state under different opinion evolution rules. Variations of the majority dynamics have been proposed and studied, such as the noisy majority dynamics [35], where agents have some probability of changing their opinion even when they are in the local majority, and the bounded confidence model [8,25], where agents only interact with others that have similar opinions.

Complexity of Majority Dynamics. Our results are in the line of a series of articles that aim to understand the computational complexity of the majority rule by studying different variants of the problem. In that context, two perspectives have been taken in order to show the P-Completeness. In [20] it is shown that the prediction problem for the majority rule is P-Complete, even when the topology is restricted to planar graphs where every node has an odd number of neighbors. The result is based in a crossing gadget that use a sort of *traffic lights*, that restrict the flow of information depending on the parity of the time-step. Then in [19] it is shown that the prediction problem for the majority rule is P-Complete when the topology is restricted to regular graphs of degree 3 (i.e. each node has exactly three neighbors). In [21], the authors study the majority rule in two dimensional grids where the edges have a *sign*. The *signed majority* consists in a modification of the majority rule, where the most represented state in a neighborhood is computed multiplying the state of each neighbor by the corresponding sign in the edge. The authors show that when the configuration of signs is the same on every site (i.e. we have an homogeneous cellular automata) then the dynamics and complexity of the signed majority is equivalent to the standard majority. Interestingly, when the configuration of signs may differ from site to site, the prediction problem is P-Complete.

Local Certification. Since the introduction of PLSs [29], different variants were introduced. As we mentioned, a stronger form of PLS are locally checkable proofs [24], where each node can send not only its certificates, but also its state and look up to a given radious. Other stronger forms of local certifications are t-PLS [12], where nodes perform communication at distance $t \geq 1$ before deciding. Authors have studied many other variants of PLSs, such as randomized PLSs [17], quantum PLSs [15], interactive protocols [7,28,33], zero-knowledge distributed certification [3], PLSs use global certificates in addition to the local ones [14], etc. On the other hand, some trade-offs between the size of the certificates and the number of rounds of the verification protocol have been exhibited [12]. Also, several hierarchies of certification mechanisms have been introduced, including games between a prover and a disprover [2,11].

PLSs have been shown to be effective for recognizing many graph classes. For example, there are compact PLSs (i.e. with logarithmic size certificates) for the recognition of acyclic graphs [29], planar graphs [13], graphs with bounded genus [9], H-minor-free graphs (as long as H has at most four vertices) [4], etc. In a recent breakthrough, Bousquet et al. [10] proved a "meta-theorem", stating that, there exists a PLS for deciding any monadic second-order logic property with $O(\log n)$-bit certificates on graphs of bounded *tree-depth*. This result has been extended by Fraigniaud et al [16] to the larger class of graphs with bounded *tree-width*, using certificates on $O(\log^2 n)$ bits.

Up to our knowledge, this is the first work that combines the study of majority dynamics and local certification.

2 Preliminaries

Let $G = (V, E)$ be a graph. We denote by $N_G(v)$ the set of neighbors of v, formally $N_G(v) = \{u \in V : \{u, v\} \in E\}$. The *degree* of v, denoted $d_G(v)$ is the cardinality of $N_G(v)$. The *maximum degree* of G, denoted Δ_G, is the maximum value of $d_G(v)$ taken over all $v \in V$. We say that two nodes $u, v \in V$ are *connected* if there exists a path in G joining them. In the following, we only consider connected graphs. The *distance* between u, v, denoted $d_G(u, v)$ is the minimum length (number of edges) of a path connecting them. The *diameter* of G is the maximum distance over every pair of vertices in G. For a node v, and $k \geq 0$, the *ball of radius k centered in v*, denoted by $B(v, k)$ is the set of nodes at distance at most k from v. Formally,

$$B_G(v, k) = \{u \in V : d_G(v, u) \leq k\}$$

We also denote by $\partial B_G(v, k) = B_G(v, k) \setminus B_G(v, k - 1)$ the border of $B_G(v, k)$. In the following, we omit the sub-indices when they are obvious by the context.

Let $G = (V, E)$ be a graph, $v \in V$ be an arbitrary node and $f : \mathbb{N} \to \mathbb{R}$ a function. We say that v has a f-bounded growth if there exist constants $c_1, c_2 > 0$ such that, for every $k > 0$, $c_1 f(k) \leq |\partial B(v, k)| \leq c_2 f(k)$. We also say that G has f-bounded growth if every node v has f-bounded growth. A family of graphs $\mathcal{G}$ has f- bounded growth if every graph in $\mathcal{G}$ has f-bounded growth. A family of graphs $\mathcal{G}$ has constant-growth (respectively linear, polynomial, sub-exponential, exponential)-growth if $\mathcal{G}$ has f-bounded growth, with f a constant (resp. linear, polynomial, sub-exponential, exponential) function. Observe that since $B(v, 1) = N(v)$, for every f-bounded graph G we have that $\Delta(G) \leq f(1)$.

2.1 Majority and Finite State Dynamics

Let $G = (V, E)$ be a connected graph. We assign to each node in G an initial opinion $v \mapsto x_v \in \{0, 1\}$. We call x a *configuration* for the network G. We call $x(t)$ the *configuration* of the network in time t. We define the majority dynamics in G by the following local rules for $u \in V$ and $t \geq 0$:

$$x_u^{t+1} = \begin{cases} 1 & \text{if } \sum_{v \in N(u)} x_v^t > \frac{d(v)}{2}, \\ x_u^t & \text{if } \sum_{v \in N(u)} x_v^t = \frac{d(v)}{2}, \\ 0 & \text{otherwise.} \end{cases}$$

where $x^0 = x$ is called an *initial configuration.* Notice that in the tie case (i.e. a node observe the same number of neighbors in each state), we consider that the node remains in its current state. Therefore, nodes of even degree may depend on their own state while nodes of odd degree do not. Therefore, we can also define the local rule of the majority dynamics as follows:

$$x_u^{t+1} = \text{sgn}\left(\sum_{v \in V} a_{uv} x_v^t - \frac{d(u)}{2}\right) \text{ where } a_{uv} = \begin{cases} 1 & \text{if } uv \in E, \\ 1 & \text{if } u = v \text{ and } d(u) \text{ is even,} \\ 0 & \text{otherwise.} \end{cases}$$

and $\text{sgn}(z)$ is the function that equals 1 when $z > 0$ and 0 otherwise.

Given a configuration x of a graph G and a vertex $v \in V(G)$, we define the *orbit* of x as the sequence of states that $\mathsf{Orbit}(x) = (x^0 = x, x^1, x^2, \ldots)$ that the majority dynamics visit when the initial configuration is x. We also define the *orbit of vertex* v as the sequence $\mathsf{Orbit}(x) = (x_v^0 = x_v, x_v^1, x_v^2, \ldots,)$.

Observe that the orbit of every configuration is finite and periodic. In other words, there exist non-negative integers $T = T(G, x)$ $p = p(G, x)$ such that $x^{T+p} = x^T$. Indeed, in an n-node graph it is possible to define exactly 2^n possible configurations. Therefore, in every orbit there is at least one configuration that is visited twice. The minimum T and p satisfying previous condition are denoted, respectively, the *transient length* and the *period* of configuration x.

Let x a configuration satisfying that $T(x, G) = 0$ is called an *attractor.* An attractor x satisfying $p(x, G) = 1$ is denoted a *fixed point.* Otherwise, it is denote a *limit-cycle* of period $p(x, G)$.

3 Certification Upper-Bounds

In this section, we give our certification upper bounds. Our analysis is based on the results of [18], where the authors aim to study the asymptotic behavior of the majority dynamics on infinite graphs. Our goal is to bound the number of changes in the two-step majority dynamics. We consider a variant of the energy operator. More precisely, for each $t > 0$ and $u \in V$, we aim to bound the number of time-steps on which the quantity $c_u^t(x) = |x_u^{t+1} - x_u^{t-1}|$ is non-zero.

Theorem 1. *Let G be a graph and x be an arbitrary configuration. Then, for every $r \in V$ and every $T > 0$, it holds:*

1. *If G is a graph of sub-exponential growth, then* $\sum_{t=1}^{T} c_r^t(x) = \mathcal{O}(1)$.
2. *If G is a graph of maximum degree 3, then* $\sum_{t=1}^{T} c_r^t(x) = \mathcal{O}(\log n)$.
3. *If G is a graph of maximum degree Δ, then* $\sum_{t=1}^{T} c_r^t(x) = \mathcal{O}(n^{1-\varepsilon})$*, where*

$$\varepsilon = \left(\frac{\log(\Delta+2)}{\log(\Delta)} - 1\right) > \frac{1}{\Delta\log(\Delta)}.$$

Proof. We show (1). The proofs of (2) and (3) can be found in the full version. We define an energy operator relative to r, giving weights to the edges of G that decrease exponentially with the distance from r. Formally, we denote by E_r^t the *energy operator centered in* r, defined as follows:

$$E_r^t(x) = \sum_{u,v\in V} \tilde{a}_{u,v}|x_u^{t+1} - x_v^t|$$

where $\tilde{a}_{u,v} = a_{u,v} \cdot \alpha^{\delta(u,v)}$, with $\alpha \in (0,1)$ a constant that we will fix later, and $\delta(u,v) = \min\{d(r,u), d(r,v)\}$. Then,

$$\begin{aligned} E_r^{t+1}(x) - E_r^t(x) &= \sum_{\{u,v\}\in E} \alpha^{\delta(u,v)}|x_u^{t+1} - x_v^t| - \sum_{\{u,v\}\in E} \alpha^{\delta(u,v)}|x_u^t - x_v^{t-1}| \\ &= \sum_{\{u,v\}\in E} \alpha^{\delta(u,v)}\left(|x_u^{t+1} - x_v^t| - |x_u^{t-1} - x_v^t|\right) \end{aligned}$$

We aim to upper bound $E_r^{t+1}(x) - E_r^t(x)$. Observe that for $u \in \partial B(r,i)$ and $v \in N(u)$, the value of $\delta(u,v)$ is either $i-1$ or i. Suppose that $c_u^t \neq 0$. We have that $\left(|x_u^{t+1} - x_v^t| - |x_u^{t-1} - x_v^t|\right)$ is maximized when almost half of the neighbors of u are in a different state than u in $t-1$, and exactly those neighbors are connected with edges of weight α^{i-1}. Therefore,

$$\begin{aligned}E_r^{t+1}(x) - E_r^t(x) =& \sum_{i=0}^{\infty} \sum_{u \in \partial B(r,i)} \sum_{v \in N(u)} \alpha^{\delta(u,v)} \left(|x_u^{t+1} - x_v^t| - |x_u^{t-1} - x_v^t| \right) \\ \leq & - c_r^t(x) \\ &+ \sum_{i=1}^{\infty} \sum_{\substack{u \in \partial B(r,i) \\ d(u) \text{ is even}}} c_u^t(x) \left(\left(\frac{d(u)}{2} \right) \alpha^{i-1} - \left(\frac{d(u)}{2} + 1 \right) \alpha^i \right) \\ &+ \sum_{i=1}^{\infty} \sum_{\substack{u \in \partial B(r,i) \\ d(u) \text{ is odd}}} c_u^t(x) \left(\left(\frac{d(u)-1}{2} \right) \alpha^{i-1} - \left(\frac{d(u)+1}{2} \right) \alpha^i \right)\end{aligned}$$

We now choose α. We impose that for each $u \in V \setminus \{r\}$ such that $d(u)$ is even,

$$\left(\frac{d(u)}{2} \right) \alpha^{i-1} - \left(\frac{d(u)}{2} + 1 \right) \alpha^i \leq 0 \Rightarrow \alpha \geq \frac{d(u)}{d(u)+2}; \tag{1}$$

and for each $u \in V \setminus \{r\}$ such that $d(u)$ is odd,

$$\left(\frac{d(u)-1}{2} \right) \alpha^{i-1} - \left(\frac{d(u)+1}{2} \right) \alpha^i \leq 0 \Rightarrow \alpha \geq \frac{d(u)-1}{d(u)+1}. \tag{2}$$

Picking $\alpha = \dfrac{\Delta}{\Delta+2}$ we obtain that conditions 1 and 2 are satisfied for every $u \in V \setminus \{r\}$. Then,

$$E_r^{t+1}(x) - E_r^t(x) \leq -c_r^t.$$

Using that $E_r^t(x) \geq 0$ for every $t > 0$, we obtain

$$\sum_{t=1}^{T} c_r^t(x) \leq \sum_{t=1}^{T} \left(E_r^t(x) - E_r^{t+1}(x) \right) = E_r^1(x) - E_r^T(x) \leq E_r^1(x)$$

To obtain our bound for $\sum_{t=1}^{T} c_r^t(x)$, we upper bound $E_r^1(x)$. Observe that

$$E_r^1(x) = \sum_{\{u,v\} \in E} \alpha^{\delta(u,v)} |x_u^{t+1} - x_v^t| = \sum_{i=0}^{\infty} \sum_{u \in \partial B(r,i)} \sum_{v \in N(u)} \alpha^{\delta(u,v)} |x_u^1 - x_v^0|$$

We have that the previous expression is maximized when, for each $u \in V$, almost half of the neighbors v of u satisfy $x_v^0 \neq x_u^1$, and the edge connecting such neighbors has the maximum possible weight. In that case, we obtain:

$$E_r^1(x) \leq \frac{\Delta}{2} \left(1 + \sum_{i=1}^{\infty} f(i) \alpha^{i-1} \right)$$

Since $f(i)$ is sub-exponential, we have that there exists a large enough $\ell > 0$ such that $f(\ell) \leq \left(\frac{\Delta+1}{\Delta}\right)^{\ell}$. Then,

$$\begin{aligned}
E_r^1(x) &\leq \frac{\Delta}{2}\left(1 + \sum_{i=1}^{\infty} f(i)\alpha^{i-1}\right) \\
&= \frac{\Delta}{2}\left(1 + \sum_{i=1}^{\ell} f(i)\alpha^{i-1} + \sum_{i=i^*+1}^{\infty} f(i)\alpha^{i-1}\right) \\
&\leq \frac{\Delta}{2} + \frac{\Delta+2}{2}\left(\left(\frac{\Delta+1}{\Delta}\right)^{\ell} \cdot \sum_{i=1}^{\ell}\left(\frac{\Delta}{\Delta+2}\right)^{i} + \sum_{i=\ell+1}^{\infty}\left(\frac{\Delta+1}{\Delta+2}\right)^{i}\right) \\
&= \frac{\Delta}{2} + \frac{\Delta(\Delta+2)}{4} \cdot \left(\frac{\Delta+1}{\Delta}\right)^{\ell} \cdot \left(1 - \left(\frac{\Delta}{\Delta+2}\right)^{\ell}\right) + \frac{(\Delta+2)^2}{2}\left(\frac{\Delta+1}{\Delta+2}\right)^{\ell+1}
\end{aligned}$$

Since $\Delta \leq f(1)$, we deduce that $\sum_{t=1}^{T} c_r^t(x) \leq E_r^1(x) = \mathcal{O}(1)$.

For a graph G let us define $\mathsf{Changes}(G) = \max_x \left(\max_v \sum_{t>0} c_v^t(x)\right)$. For a set of graphs $\mathcal{G}$ we define $\mathsf{Changes}(\mathcal{G}) = \max_{G\in\mathcal{G}} \mathsf{Changes}(G)$. Our proof labelling scheme for Reachability consists in giving each node u the sequence of time-steps t on which $c_u^t \neq 0$. From its own certificate and the certificate of its neighbors, a node v can compute its orbit.

Lemma 1. *For each n-node graph G there is a proof labeling scheme for problem* Reachability *with certificates of size $\mathcal{O}(\mathsf{Changes}(G) \cdot \log n)$.*

Theorem 1 pipelined with Lemma 1 gives the main result of this section.

Theorem 2.

1. *There is a 1-round proof labeling scheme for* Reachability *with certificates on $\mathcal{O}(\log n)$ bits in n-node networks of sub-exponential growth.*
2. *There is a 1-round proof labeling scheme for* Reachability *with certificates on $\mathcal{O}(\log^2 n)$ bits in n-node networks of maximum degree* 3.
3. *There is a 1-round proof labeling scheme for* Reachability *with certificates on $\mathcal{O}(n^{1-\epsilon}\log n)$ bits in n-node networks of maximum degree $\Delta > 2$, where $\varepsilon = 1/\Delta\log(\Delta)$.*

3.1 Upper-Bound for Election-Prediction

We now show the proof labeling schemes for Election-Prediction. Our bounds of the size of the certificates is obtained from Theorem 2 and the following result. Let us define Count-Ones as the problem of deciding, given a configuration x and a constant k, if in the graph there are exactly k nodes in state 1. In [29] it is shown that there is a PLS for Count-Ones with certificates of size $\mathcal{O}(\log n)$.

Lemma 2. *(see [29]) There is a proof labeling scheme for* COUNT-ONES *with certificates of size* $\mathcal{O}(\log n)$.

Roughly, the idea of the PLS of Lemma 2 the following: the certificate of a node v is a tuple $(\mathsf{root}, \mathsf{parent}, \mathsf{distance}, \mathsf{count})$ where root is the identifier of the root of a rooted spanning tree τ of G, parent is the identifier of the parent of v in τ, $\mathsf{distance}$ is the distance of v to the root and count is the number of nodes in state 1 in the subgraph of G induced by the descendants of v in τ. Then, every vertex checks the local coherence of the certificates, and the root also checks whether count equals k. The upper bounds for ELECTION-PREDICTION follow directly from Theorem 2 and Lemma 2.

Theorem 3.

1. *There is a 1-round proof labeling scheme for* ELECTION-PREDICTION *with certificates on* $\mathcal{O}(\log n)$ *bits in* n*-node networks of sub-exponential growth.*
2. *There is a 1-round proof labeling scheme for* ELECTION-PREDICTION *with certificates on* $\mathcal{O}(\log^2 n)$ *bits in* n*-node networks of maximum degree* 3.
3. *There is a 1-round proof labeling scheme for* ELECTION-PREDICTION *with certificates on* $\mathcal{O}(n^{1-\epsilon} \log n)$ *bits in* n*-node networks of maximum degree* $\Delta > 2$, *where* $\varepsilon = 1/\Delta \log(\Delta)$.

4 Lower-Bounds

We first prove that every locally checkable proof for problems REACHABILITY or ELECTION-PREDICTION on arbitrary n-node graphs require certificates of size $\Omega(n)$. The proof is a reduction from the disjointedness problem in non-deterministic communication complexity. In this problem, Alice receives a vector $a \in \{0,1\}^n$ and Bob a vector $b \in \{0,1\}^n$. The players can perform a series of communication rounds and have the task of deciding whether there exists a coordinate $i \in \{1,\dots,n\}$ such that $a_i = b_i$. In [30] it is shown that the non-deterministic communication complexity of this problem is $\Omega(n)$.

Theorem 4. *Every locally checkable proof certifying* ELECTION-PREDICTION *or* REACHABILITY *on arbitrary* n*-node graphs has certificates on* $\Omega(n)$ *bits.*

With respect to bounded degree graphs, we study the case in which $\mathcal{G}_2$ is the class of graphs with maximum degree at most 2. In other words, we focus in studying path graphs and cycle graphs. In this case, we show that REACHABILITY admits proof-label schemes of size $\Omega(\log n)$. We accomplish this task by a reduction to the task of verifying if G is a path or a cycle. More precisely, we define the problem

$$\text{CYCLE} = \{G \in \mathcal{G}_2 : G \text{ is a path graph.}\}$$

Lemma 3. *Let us suppose that* REACHABILITY *restricted to* $\mathcal{G}_2$ *admits a locally checkable proof with certificates of size* L. *Then,* CYCLE *admits a locally checkable proof with certificates of size* $2L$.

In [24] it is shown that every locally checkable proof for CYCLE has certificates of size $\Omega(\log n)$. We obtain the following theorem.

Theorem 5. *Every locally checkable proof for* REACHABILITY *on n-node graphs of maximum degree 2 has certificates on $\Omega(\log n)$ bits.*

References

1. Asuncion, A.U., Goodrich, M.T.: Turning privacy leaks into floods: surreptitious discovery of social network friendships and other sensitive binary attribute vectors. In: Proceedings of the 9th Annual ACM Workshop on Privacy in the Electronic Society, pp. 21–30 (2010)
2. Balliu, A., D'Angelo, G., Fraigniaud, P., Olivetti, D.: What can be verified locally? J. Comput. Syst. Sci. **97**, 106–120 (2018)
3. Bick, A., Kol, G., Oshman, R.: Distributed zero-knowledge proofs over networks. In: 33rd ACM-SIAM Symposium on Discrete Algorithms (SODA), pp. 2426–2458 (2022)
4. Bousquet, N., Feuilloley, L., Pierron, T.: Local certification of graph decompositions and applications to minor-free classes. In: 25th International Conference on Principles of Distributed Systems (OPODIS). LIPIcs, vol. 217, pp. 22:1–22:17. Schloss Dagstuhl - Leibniz-Zentrum für Informatik (2021)
5. Bredereck, R., Elkind, E.: Manipulating opinion diffusion in social networks. In: IJCAI International Joint Conference on Artificial Intelligence. International Joint Conferences on Artificial Intelligence (2017)
6. Castellano, C., Fortunato, S., Loreto, V.: Statistical physics of social dynamics. Rev. Mod. Phys. **81**(2), 591 (2009)
7. Crescenzi, P., Fraigniaud, P., Paz, A.: Trade-offs in distributed interactive proofs. In: 33rd International Symposium on Distributed Computing (DISC). LIPIcs, vol. 146, pp. 13:1–13:17. Schloss Dagstuhl - Leibniz-Zentrum für Informatik (2019)
8. Deffuant, G., Neau, D., Amblard, F., Weisbuch, G.: Mixing beliefs among interacting agents. Adv. Complex Syst. **3**(01n04), 87–98 (2000)
9. Esperet, L., Lévêque, B.: Local certification of graphs on surfaces. Theor. Comput. Sci. **909**, 68–75 (2022)
10. Feuilloley, L., Bousquet, N., Pierron, T.: What can be certified compactly? Compact local certification of MSO properties in tree-like graphs. In: Proceedings of the 2022 ACM Symposium on Principles of Distributed Computing, pp. 131–140 (2022)
11. Feuilloley, L., Fraigniaud, P., Hirvonen, J.: A hierarchy of local decision. Theor. Comput. Sci. **856**, 51–67 (2021)
12. Feuilloley, L., Fraigniaud, P., Hirvonen, J., Paz, A., Perry, M.: Redundancy in distributed proofs. Distrib. Comput. **34**(2), 113–132 (2021)
13. Feuilloley, L., Fraigniaud, P., Montealegre, P., Rapaport, I., Rémila, É., Todinca, I.: Compact distributed certification of planar graphs. Algorithmica 1–30 (2021)
14. Feuilloley, L., Hirvonen, J.: Local verification of global proofs. In: 32nd International Symposium on Distributed Computing. LIPIcs, vol. 121, pp. 25:1–25:17. Schloss Dagstuhl - Leibniz-Zentrum für Informatik (2018)
15. Fraigniaud, P., Gall, F.L., Nishimura, H., Paz, A.: Distributed quantum proofs for replicated data. In: 12th Innovations in Theoretical Computer Science Conference (ITCS). LIPIcs, vol. 185, pp. 28:1–28:20. Schloss Dagstuhl - Leibniz-Zentrum für Informatik (2021)

16. Fraigniaud, P., Montealegre, P., Rapaport, I., Todinca, I.: A meta-theorem for distributed certification. In: Parter, M. (ed.) SIROCCO 2022. LNCS, vol. 13298, pp. 116–134. Springer, Cham (2022). https://doi.org/10.1007/978-3-031-09993-9_7
17. Fraigniaud, P., Patt-Shamir, B., Perry, M.: Randomized proof-labeling schemes. Distrib. Comput. **32**(3), 217–234 (2019)
18. Ginosar, Y., Holzman, R.: The majority action on infinite graphs: strings and puppets. Discret. Math. **215**(1–3), 59–71 (2000)
19. Goles, E., Montealegre, P.: Computational complexity of threshold automata networks under different updating schemes. Theoret. Comput. Sci. **559**, 3–19 (2014)
20. Goles, E., Montealegre, P.: The complexity of the majority rule on planar graphs. Adv. Appl. Math. **64**, 111–123 (2015)
21. Goles, E., Montealegre, P., Perrot, K., Theyssier, G.: On the complexity of two-dimensional signed majority cellular automata. J. Comput. Syst. Sci. **91**, 1–32 (2018)
22. Goles, E., Montealegre, P., Salo, V., Törmä, I.: Pspace-completeness of majority automata networks. Theoret. Comput. Sci. **609**, 118–128 (2016)
23. Goles-Chacc, E., Fogelman-Soulié, F., Pellegrin, D.: Decreasing energy functions as a tool for studying threshold networks. Discret. Appl. Math. **12**(3), 261–277 (1985)
24. Göös, M., Suomela, J.: Locally checkable proofs in distributed computing. Theory Comput. **12**(1), 1–33 (2016)
25. Hegselmann, R., Krause, U.: Opinion dynamics and bounded confidence models, analysis and simulation. J. Artif. Soc. Soc. Simul. **5**(3) (2002)
26. Heider, F.: Attitudes and cognitive organization. J. Psychol. **21**(1), 107–112 (1946)
27. Kelman, H.C.: Compliance, identification, and internalization three processes of attitude change. J. Conflict Resolut. **2**(1), 51–60 (1958)
28. Kol, G., Oshman, R., Saxena, R.R.: Interactive distributed proofs. In: ACM Symposium on Principles of Distributed Computing, pp. 255–264. ACM (2018)
29. Korman, A., Kutten, S., Peleg, D.: Proof labeling schemes. Distrib. Comput. **22**(4), 215–233 (2010)
30. Kushilevitz, E.: Communication complexity. In: Advances in Computers, vol. 44. Elsevier (1997)
31. Mobilia, M., Redner, S.: Majority versus minority dynamics: phase transition in an interacting two-state spin system. Phys. Rev. E **68**, 046106 (2003)
32. Moore, C.: Majority-vote cellular automata, ising dynamics, and p-completeness. J. Stat. Phys. **88**, 795–805 (1997)
33. Naor, M., Parter, M., Yogev, E.: The power of distributed verifiers in interactive proofs. In: 31st ACM-SIAM Symposium on Discrete Algorithms (SODA), pp. 1096–115. SIAM (2020)
34. Nguyen, V.X., Xiao, G., Xu, X.J., Wu, Q., Xia, C.Y.: Dynamics of opinion formation under majority rules on complex social networks. Sci. Rep. **10**(1), 456 (2020)
35. Vieira, A.R., Crokidakis, N.: Phase transitions in the majority-vote model with two types of noises. Phys. A **450**, 30–36 (2016)

Complexity of Spherical Equations in Finite Groups

Caroline Mattes[1(✉)], Alexander Ushakov[2], and Armin Weiß[1]

[1] Universität Stuttgart, Institut für Formale Methoden der Informatik, Stuttgart, Germany
caroline.mattes@fmi.uni-stuttgart.de

[2] Stevens Institute of Technology, Hoboken, USA

Abstract. In this paper we investigate computational properties of the Diophantine problem for spherical equations in some classes of finite groups G. We classify the complexity of different variations of the problem, e.g., when G is fixed and when G is a part of the input.

When the group G is constant or given as multiplication table, we show that the problem can always be solved in polynomial time. On the other hand, for the permutation groups S_n (with n part of the input), the problem is NP-complete. The situation for matrix groups is quite involved: while we exhibit sequences of 2-by-2 matrices where the problem is NP-complete, in the full group $\mathrm{GL}(2, p)$ (p prime and part of the input) it can be solved in polynomial time. We also find a similar behaviour with subgroups of matrices of arbitrary dimension over a constant ring.

Keywords: Diophantine problem · finite groups · matrix groups · spherical equations · complexity · NP-completeness

2010 Mathematics Subject Classification. 20F10 · 68W30

1 Introduction

The study of equations has a long history in all of mathematics. Some of the first explicit general decidability results in group theory are due to Makanin [21] showing that equations over free groups are decidable. Let $F(Z)$ denote the free group on countably many generators $Z = \{z_i\}_{i=1}^{\infty}$. For a group G, an *equation over G with variables in Z* is a formal equality of the form $W = 1$, where

$$W = z_{i_1} g_1 \cdots z_{i_k} g_k \in F(Z) * G, \text{ with } z_{i_j} \in Z \text{ and } g_j \in G$$

and $*$ denotes the free product. We refer to $\{z_{i_1}, \ldots, z_{i_k}\}$ as the set of *variables* and to $\{g_1, \ldots, g_k\}$ as the set of *constants* (or *coefficients*) of $W = W(z_1, \ldots, z_k)$. A *solution* to an equation $W(z_1, \ldots, z_k) = 1$ over G is an assignment to the variables $z_1, \ldots, z_k$ that makes $W(z_1, \ldots, z_k) = 1$ true. The *Diophantine problem* (also called *equation satisfiability* or *polynomial satisfiability* problem) in a group G for a class of equations C is an algorithmic question to decide if a given equation $W = 1$ in C has a solution or not.

H. Fernau et al. (Eds.): SOFSEM 2024, LNCS 14519, pp. 383–397, 2024.
https://doi.org/10.1007/978-3-031-52113-3_27

One class of equations over groups that has generated much interest is the class of *quadratic* equations: equations where each variable z appears exactly twice (as either z or z^{-1}). It was observed in the early 80 s [8,25] that such equations have an affinity with the theory of compact surfaces (for instance, via their associated van Kampen diagrams). This geometric point of view sparked the initial interest in their study and has led to many interesting results, particularly in the realm of quadratic equations over free groups: solution sets were studied in [12], NP-completeness was proved in [9,16]. These results stimulated more interest in quadratic equations in various classes of (infinite) groups such as hyperbolic groups (solution sets described in [13], NP-complete by [17]), the first Grigorchuk group (decidability proved in [18], commutator width computed in [5]), free metabelian groups (NP-hard by [19], in NP for orientable equations by [20]), metabelian Baumslag–Solitar groups (NP-complete by [22]), etc.

Let us introduce some terminology: We say that quadratic equations $W = 1$ and $V = 1$ are *equivalent* if there is an automorphism $\phi \in \mathrm{Aut}(F(Z) * G)$ such that ϕ is the identity on G and $\phi(W) = V$. It is a well-known consequence of the classification of compact surfaces that any quadratic equation over G is equivalent, via an automorphism ϕ computable in time $O(|W|^2)$, to an equation in exactly one of the following three standard forms (see, [6,12]):

$$\prod_{j=1}^{k} z_j^{-1} c_j z_j = 1, \qquad (k \geq 1), \qquad (1)$$

$$\prod_{i=1}^{g} [x_i, y_i] \prod_{j=1}^{k} z_j^{-1} c_j z_j = 1, \qquad \prod_{i=1}^{g} x_i^2 \prod_{j=1}^{k} z_j^{-1} c_j z_j = 1, \qquad (g \geq 1, k \geq 0) \qquad (2)$$

with variables x_i, y_i and z_i. The number g is the *genus* of the equation, and both g and k (the number of constants) are invariants. The standard forms are called, respectively, *spherical* (1), *orientable of genus g* and *non-orientable of genus g* (2) according to the compact surfaces they correspond to.

In this paper we investigate spherical equations in finite groups. Let us remark that spherical equations naturally generalize fundamental (Dehn) problems of group theory, as the word and the conjugacy problem, if we allow the constants to be given as words over the generators. The complexity of solving equations in finite groups has been first studied by Goldmann and Russell [11] showing that the Diophantine problem in a fixed finite nilpotent group can be decided in polynomial time, while it is NP-complete in every finite non-solvable group.

Contribution. Here we study the complexity of solving spherical equations in finite groups both in the case that the group is fixed as well as that it is part of the input. In the latter case, we consider various different input models as well as different classes of groups. In particular, we show (here SPHEQ($\mathcal{G}$) denotes the Diophantine problem for the class $\mathcal{G}$ of groups):

- In a fixed group the satisfiable spherical equations form a regular language with commutative syntactic monoid – in particular, the Diophantine problem can be decided in linear time (Theorem 7).

- If the Cayley table is part of the input, we can solve the Diophantine problem in nondeterministic logarithmic space and, thus, in P (Proposition 8).
- For (symmetric) groups given as permutation groups the problem is NP-complete (Theorem 11).
- $\text{SPHEQ}((\text{ET}(2,n))_{n\in\mathbb{N}})$ is NP-complete (here, $\text{ET}(2,n)$ denotes the upper triangular matrices with diagonal entries in $\{\pm 1\}$, Corollary 15).
- $\text{SPHEQ}((\text{T}(2,p))_{p\in\mathbb{P}})$ is in P (here, $\text{T}(2,p)$ are the upper triangular matrices and $\mathbb{P}$ the set of primes, Theorem 16).
- $\text{SPHEQ}((\text{GL}(2,p))_{p\in\mathbb{P}})$ is in P (Theorem 22).
- $\text{SPHEQ}((\text{UT}(4,p))_{p\in\mathbb{P}})$ is in P where $\text{UT}(4,p)$ denotes the uni-triangular 4-by-4 matrices modulo p (Theorem 25).
- $\text{SPHEQ}((H_n^{(p)})_{(n,p)\in\mathbb{N}\times\mathbb{P}})$ is in P (here $H_n^{(p)}$ denotes the n-dimensional discrete Heisenberg group modulo p – note that the dimension and the prime field are both part of the input, Theorem 24).
- For every constant $m \geq 5$, we exhibit sequences of subgroups $G'_n \leq H_n \leq G_n \leq \text{GL}(n,m)$ such that $\text{SPHEQ}((G'_n)_{n\in\mathbb{N}})$ and $\text{SPHEQ}((G_n)_{n\in\mathbb{N}})$ are in P but $\text{SPHEQ}((H_n)_{n\in\mathbb{N}})$ is NP-complete (Corollary 27). In particular, NP-hardness does not transfer to super- nor sub-groups!

Due to space constraints many proofs are omitted or only short sketches are given. Full proofs can be found in the full version on arXiv [23].

2 Notation and Problem Description

By $\mathbb{P}$ we denote the set of all prime numbers. With $[k\mathinner{..}\ell]$ for $k,\ell\in\mathbb{Z}$ we denote the interval of integers $\{k,\ldots,\ell\}$.

We assume the reader to be familiar with basic group theory. Recall that each finite group embeds into some permutation or matrix group. A group can be presented by its *Cayley table* (or *multiplication table*). The Cayley table of a finite group G is a square matrix with rows and columns indexed by the elements of G. Row g and column h contains the product $gh \in G$. For a group G and $g_1,\ldots,g_\ell \in G$, $\langle g_1,\ldots,g_\ell\rangle$ denotes the subgroup generated by $g_1,\ldots,g_\ell$. If $g\in G$ we say that g is *conjugate* to h in G if $g^z := z^{-1}gz = h$ for some $z\in G$. Throughout, $\mathbb{Z}_n$ is the ring of integers modulo n or the corresponding additive group; we use C_2 to denote the multiplicative group of units $\{1,-1\}$ of $\mathbb{Z}$.

Permutation Groups. The symmetric group on n elements, denoted by S_n, is the group of order $n!$ of all bijective maps (aka. *permutations*) on the set $[1\mathinner{..}n]$. Here, A_n denotes the subgroup of S_n of all even permutations (see e.g. [3]). As a function, it maps $i_j \mapsto i_{j+1}$ and $i_k \mapsto i_1$. Let $\sigma,\sigma_1,\ldots,\sigma_k \in S_n$. Set $\text{mov}(\sigma) = \{\, i\in[1\mathinner{..}n] \mid \sigma(i)\neq i \,\}$ and $\text{smov}(\sigma_1,\ldots,\sigma_k) = \sum_{i=1}^{k} |\text{mov}(\sigma_i)|$. We say that $\sigma_1,\sigma_2\in S_n$ are disjoint if $\text{mov}(\sigma_1)\cap\text{mov}(\sigma_2)=\emptyset$.

Lemma 1 ([3], (15.2)). *Disjoint cycles in S_n commute. Moreover, every element in S_n can be written as the product of disjoint cycles in a unique way (up to the order of the cycles). This is called the disjoint cycle representation.*

We say that $\text{Cyc}(\sigma) = [j_1, \ldots, j_\ell]$ is the cycle structure of σ if it is the product of ℓ disjoint cycles of lengths $j_1, \ldots, j_\ell$ and denote by $|\text{Cyc}(\sigma)| = \ell$ the length of this cycle decomposition. The following lemma is an easy observation.

Lemma 2. *Let* $\sigma_1, \sigma_2 \in S_n$. *Then,* $|\text{mov}(\sigma_1\sigma_2)| \leq \text{smov}(\sigma_1, \sigma_2)$. *Moreover,* $\text{mov}(\sigma_1) \cap \text{mov}(\sigma_2) \neq \emptyset$ *if and only if* $|\text{mov}(\sigma_1\sigma_2)| < \text{smov}(\sigma_1, \sigma_2)$.

Lemma 3 ([3], **(15.3)**). *Let* $\sigma, \tau \in S_n$. *Then* σ *is conjugated to* τ *in* S_n *if and only if* $\text{Cyc}(\sigma) = \text{Cyc}(\tau)$ *(up to a permutation of the* j_i*).*

Matrix groups. The *general linear group* $\text{GL}(n, p)$ over the prime field $\mathbb{Z}_p$ is the group of all invertible n-by-n matrices. Further, let $\text{T}(n, p)$ be the subgroup of $\text{GL}(n, p)$ of upper triangular matrices. By $\text{ET}(n, p)$ we denote the subgroup of $\text{T}(n, p)$ of all matrices having only ± 1 on the diagonal (note that this is not a standard definition). Finally, let $\text{UT}(n, p)$ be the group of all matrices $A \in \text{T}(n, p)$ having only 1 on the diagonal. We denote the identity matrix by I or I_n.

Complexity and NP-Complete Problems. We assume the reader to be familiar with the complexity classes NL (nondeterministic logarithmic space), P and NP as well as NP-completeness (with respect to LOGSPACE reductions). It is well known that the following problems are NP-complete (see [10]):

3-PARTITION: Input: a list $S = (a_1, \ldots, a_{3k})$ of unary-encoded positive integers with $\frac{L}{4} < a_i < \frac{L}{2}$ for $L = \frac{1}{k}\sum_{i=1}^{3k} a_i$. Question: Is there a partition of $[1 \mathinner{..} 3k]$ into k (disjoint) sets $A_i = \{i_1, i_2, i_3\}$ with $a_{i_1} + a_{i_2} + a_{i_3} = L$ for all $i \in [1 \mathinner{..} k]$?

PARTITION: Input: a list $S = (a_1, \ldots, a_k)$ of binary-encoded positive integers. Question: Is there a subset $A \subseteq [1 \mathinner{..} k]$ such that $\sum_{i \in A} a_i = \sum_{i \notin A} a_i$?

3-EXACTSETCOVER: Input: a set A and subsets $A_1, \ldots, A_\ell \subseteq A$ with $|A_i| \leq 3$. Question: Is there a set $I \subseteq [1 \mathinner{..} \ell]$ with $\bigcup_{i \in I} A_i = A$ and $A_i \cap A_j = \emptyset$ for all $i \neq j$ in I? This is still NP-hard if each element occurs in at most 3 sets A_i.

Spherical Equations. Let G be a group. A *spherical equation* over G is an equation as defined in (1): $\prod_{i=1}^{k} z_i^{-1} c_i z_i = 1$ with $c_i \in G$ and z_i being variables for $i \in [1 \mathinner{..} k]$. We write:

$$\text{SPHEQ}(G) = \{(c_1, \ldots, c_k) \in G^k \mid k \in \mathbb{N}, \exists z_i \in G : \textstyle\prod_{i=1}^{k} z_i^{-1} c_i z_i = 1\}.$$

Note that this is a formal language over the alphabet G (we write $(c_1, \ldots, c_k)$ instead of $c_1 \cdots c_k$ as usual for formal languages to differentiate it from the product $c_1 \cdots c_k$ in G). Using this notation, we write $(c_1, \ldots, c_k) \in \text{SPHEQ}(G)$ if (1) has a solution. We identify $\text{SPHEQ}(G)$ with the computation problem

Input: $c_1, \ldots, c_k \in G$.
Question: Is $(c_1, \ldots, c_k) \in \text{SPHEQ}(G)$?

Note that we do not allow the individual c_i to be words over the generators as, for example, it is usual for the word and conjugacy problem; hence, these problems do not necessarily reduce to solving spherical equations.

Lemma 4. *Let* $\pi : \{1, \ldots, k\} \to \{1, \ldots, k\}$ *be a permutation. Then* $(c_1, \ldots, c_k) \in \text{SPHEQ}(G)$ *if and only if* $(c_{\pi(1)}, \ldots, c_{\pi(k)}) \in \text{SPHEQ}(G)$.

Proof. We have $x^{-1}cx\, y^{-1}dy = \tilde{y}^{-1}d\tilde{y}\, x^{-1}cx$ for $\tilde{y} = yx^{-1}c^{-1}x$. Thus, we can exchange c_i and c_{i+1}. By induction we can apply any permutation to the c_i. □

Let $c \in G$ and $\prod_{i=1}^{k} c_i^{z_i} = c$. This is indeed a spherical equation: Conjugating by z, multiplying by $z^{-1}cz$ and inserting zz^{-1} we obtain an equation as in (1). In the following we study this problem also with the group as part of the input. There are several ways how to represent a finite group as part of the input:

Input Models for Finite Groups. For a class/sequence of groups $\mathcal{G} = (G_n)_{n \in \mathbb{N}}$ we define the *satisfiability problem* (or *Diophantine problem*) for spherical equations for $\mathcal{G}$, denoted by $\text{SPHEQ}(\mathcal{G})$, as follows:

Input: $n \in \mathbb{N}$, a description of G_n and elements $c_1, \ldots, c_k \in G_n$.
Question: Is $(c_1, \ldots, c_k) \in \text{SPHEQ}(G_n)$?

We use the following representations of the input, which are frequently used in literature and computational algebra software [15]:

Constant. In this model, G is not part of the input but fixed. Thus, $\text{SPHEQ}(G)$ is defined as above.

Cayley Table. The problem $\text{SPHEQ}_{\text{CayT}}$ has the following input:

Input: The Cayley table of a group G and elements $c_1, \ldots, c_k$ in G.

Permutation groups. The problem $\text{SPHEQ}((S_n)_{n \in \mathbb{N}})$ has the following input:

Input: $n \in \mathbb{N}$ in unary and elements $c_1, \ldots, c_k \in S_n$ given as permutations.

Matrix Groups. For matrix groups we have two parameters, the dimension n and q for the ring $\mathbb{Z}_q$. Both can be either fixed or variable. Usually, the dimension n is given in unary and q as well as the entries of the matrices are given in binary. We describe the input for $\text{SPHEQ}((\text{GL}(n, q))_{q \in \mathbb{N}})$, but it is defined analogously for other matrix groups or if n is part of the input or q is fixed.

Input: $q \in \mathbb{N}$ in binary and matrices $c_1, \ldots, c_k$ in $\text{GL}(n, q)$.

Note that in the cases of $\text{SPHEQ}((\text{GL}(n, q))_{n,q \in \mathbb{N}})$ and $\text{SPHEQ}((S_n)_{n \in \mathbb{N}})$ it is natural to give n in unary since, if n is given in binary, we cannot even write down a single element of $\text{GL}(n, q)$ or S_n in polynomial space.

Subgroups. Let $\mathcal{G}$ be a sequence of groups (defined as above). By $\text{SPHEQ}(\text{SGR}(\mathcal{G}))$ we denote the problem to decide whether $(c_1, \ldots, c_k) \in \text{SPHEQ}(\langle g_1, \ldots, g_\ell \rangle)$ on input of a group $G \in \mathcal{G}$, elements $g_1, \ldots, g_\ell \in G$ and $c_1, \ldots, c_k \in \langle g_1, \ldots, g_\ell \rangle$. Note that this is a promise problem: we need that $c_1, \ldots, c_k \in \langle g_1, \ldots, g_\ell \rangle$.

A straightforward guess-and-check algorithm shows the following proposition. In the subgroup model we need to apply the reachability Lemma [4, Lemma 3.1].

Proposition 5. *Let* $\mathcal{G}$ *denote any class/sequence of groups given as matrix or permutation groups as above. Then* $\text{SPHEQ}(\mathcal{G})$ *and* $\text{SPHEQ}(\text{SGR}(\mathcal{G}))$ *are in* NP.

3 Fixed Finite Groups and Cayley Tables

Before we consider the complexity, let us mention an interesting property of spherical equations in finite simple groups:

Proposition 6. *If G is a finite non-abelian simple group, then every spherical equation of length (number of non-trivial c_i) at least $|G|^3 - |G| + 1$ has a solution.*

Next, let us turn to the complexity of spherical equations in a fixed finite group. The classes AC^0 and ACC^0 are defined in terms of circuit complexity, see [26, Def. 4.5, 4.34]. We have uniform $\mathsf{AC}^0 \subseteq$ uniform $\mathsf{ACC}^0 \subseteq \mathsf{LOGSPACE} \subseteq \mathsf{P}$.

Theorem 7. *Let G be a fixed finite group. Then the set of satisfiable spherical equations in G forms a regular language with commutative syntactic monoid. In particular,* $\textsc{SphEq}(G)$ *can be solved in linear time and is in uniform* ACC^0.

Note that Proposition 6 implies that $\textsc{SphEq}(G)$ is a regular language with an aperiodic syntactic monoid for every finite non-abelian simple group G; hence, it in AC^0 meaning that it is even easier than in the general case.

Proof. We view G as a finite alphabet. Then a spherical equation $\prod_{i=1}^{k} c_i^{z_i} = 1$ can be identified with the word $c_1 \cdots c_k \in G^*$. Let $\phi : G^* \to \mathcal{P}(G)$ (the power set of G, which is a monoid) denote the monoid homomorphism defined by $g \mapsto \{ x^{-1}gx \mid x \in G \}$. By Lemma 4, $\phi(G^*)$ is commutative. Further, $\prod_{i=1}^{k} c_i^{z_i} = 1$ is satisfiable if and only if $1 \in \phi(c_1 \cdots c_k)$ as outlined above. Thus, the set of satisfiable spherical equations is recognized by ϕ. By [24] all languages recognized by commutative (or solvable) monoids are in ACC^0. As spherical equations are a regular language ($\mathcal{P}(G)$ is finite), satisfiability can be decided in linear time. □

We can reduce satisfiability of a spherical equation $\prod_{i=1}^{k} z_i^{-1} c_i z_i = 1$ to reachability in the graph defined by $V = G \times [0\,..\,k]$ with edges from $(g, i-1)$ to $(gz^{-1}c_i z, i)$ for all $z \in G$ and $i \in [1\,..\,k]$. This yields the following result:

Proposition 8. $\textsc{SphEq}_{\mathrm{CayT}}$ *is in* NL *and it can be solved in time $k \cdot |G|^2$ on a random access machine.*

Note that, if G, H are finite groups and $g_i \in G$, $h_i \in H$ for $i \in [1\,..\,k]$, then $((g_1, h_1), \ldots, (g_k, h_k)) \in \textsc{SphEq}(G \times H)$ if and only if $(g_1, \ldots, g_k) \in \textsc{SphEq}(G)$ and $(h_1, \ldots, h_k) \in \textsc{SphEq}(H)$. Hence, we get the following corollary.

Corollary 9. *Let G be a fixed finite group, and let $\mathcal{G} = (G^n)_{n \in \mathbb{N}}$, n given in unary with $G^n = \underbrace{G \times \cdots \times G}_{n \text{ times}}$. Then,* $\textsc{SphEq}(\mathcal{G}) \in \mathsf{ACC}^0$.

Remark 10. By Proposition 8, $\textsc{SphEq}((G_n)_{n \in \mathbb{N}}) \in \mathsf{P}$ if $|G_n|$ is polynomial in the size of the description of G_n, the input model allows for a polynomial-time multiplication of group elements, and they are reasonably encoded: First, on input of the group G_n and a spherical equation, compute the Cayley table of G_n, then apply Proposition 8. On the other hand, by Corollary 9 there are classes/sequences of groups such that $|G_n|$ is exponential in n but $\textsc{SphEq}((G_n)_{n \in \mathbb{N}})$ is in ACC^0.

4 The Groups S_n and A_n

Theorem 11. SPHEQ$((S_n)_{n\in\mathbb{N}})$ *is* NP*-complete.*

Proof (sketch). By Proposition 5, SPHEQ$((S_n)_{n\in\mathbb{N}})$ is in NP. For the NP-hardness we reduce from 3-PARTITION (see Sect. 2): Let $A = (a_1, \ldots, a_{3k})$ be an instance of 3-PARTITION with $L = \frac{1}{k}\sum_{i=1}^{3k} a_i$, $\frac{L}{4} < a_i < \frac{L}{2}$. We define cycles $c_1, \ldots, c_{3k} \in S_n$ for $n = k(L+1)$ and a product of disjoint $L+1$-cycles c^*: $c_\ell = (1, \ldots, a_\ell + 1)$,

$$c^* = \prod_{i=1}^{k}((i-1)(L+1)+1, \ldots, i(L+1)). \tag{3}$$

We claim that the equation $\prod_{i=1}^{3k} c_i^{z_i} = c^*$ has a solution in S_n if and only if A is a positive instance of 3-PARTITION.

Let the sets $A_i = \{i_1, i_2, i_3\}$ be a partition of $[1\mathinner{..}3k]$ and $a_{i_1} + a_{i_2} + a_{i_3} = L$ (by a slight abuse of notation we use indices i_1, i_2, i_3 for elements of A_i). Cycles of the same length are conjugated (Lemma 3), so there are $x_i, y_i, v_i \in S_n$ such that

$$((i-1)(L+1)+1, \ldots, i(L+1)) = v_i^{-1} c_{i_1} v_i x_i^{-1} c_{i_2} x_i y_i^{-1} c_{i_3} y_i.$$

Hence, by (3) and Lemma 4 we have a solution.

Assume that a solution exists. Since $|\mathrm{Cyc}(c^*)| = k$, the cycles $c_i^{z_i}$ are not all disjoint. Starting with $\prod_{i=1}^{3k} c_i^{z_i}$ we multiply (at least $2k$) non-disjoint cycles and permute disjoint ones until we get c^* (Lemma 1). By Lemma 2, $\mathrm{smov}(c_1^{z_1}, \ldots, c_{3k}^{z_{3k}})$ decreases by at least $2k$. Since $|\mathrm{mov}(c^*)| = k(L+1)$, no more multiplications of non-disjoint cycles are possible. It follows that for each $i \in [1\mathinner{..}k]$ there exists $A_i = \{i_1, \ldots, i_{h_i}\}$ with $\prod_{\mu=1}^{h_i} c_{i_\mu}^{z_{i_\mu}} = ((i-1)(L+1)+1, \ldots, i(L+1))$. Thus, $\sum_{\mu=1}^{h_i} a_{i_\mu} \geq L$ and $a_i < \frac{L}{2}$ imply $h_i \geq 3$. So, $h_i = 3$ for all $i \in [1\mathinner{..}k]$. Since $kL = \sum_{i=1}^{3k} a_i$, we conclude $a_{i_1} + a_{i_2} + a_{i_3} = L$ for all i. □

Corollary 12. SPHEQ$((A_n)_{n\in\mathbb{N}})$ *is* NP*-complete.*

Thus, there is a sequence of simple groups $\mathcal{G}$ such that SPHEQ$(\mathcal{G})$ is NP-hard though Proposition 6 suggests that it might be easy for simple groups. Notice that $|A_n|^3 - |A_n| + 1$ is not polynomial in n.

5 Spherical Equations in Dihedral Groups

The dihedral group D_n of order $2n$ can be found as a subgroup of $\mathrm{GL}(2, n)$: Let $R = \begin{pmatrix} 1 & 1 \\ 0 & 1 \end{pmatrix}$, $S = \begin{pmatrix} 1 & 0 \\ 0 & -1 \end{pmatrix}$. Then, $D_n \cong \langle R, S \rangle$. Note that $\langle R \rangle \cong \mathbb{Z}_n$ and $\langle S \rangle \cong C_2$. We can present D_n as semidirect product: $D_n \cong \mathbb{Z}_n \rtimes C_2$ with C_2 acting on $\mathbb{Z}_n$ by $(k, \delta) \mapsto \delta k$. So, $D_n = \{\, (k, \delta) \mid k \in \mathbb{Z}_n,\ \delta = \pm 1 \,\}$ with multiplication defined by $(k_1, \delta_1)(k_2, \delta_2) = (k_1 + \delta_1 k_2, \delta_1 \cdot \delta_2)$. The neutral element is denoted by $(0, 1)$.

Let $(h_\ell, \gamma_\ell), (k_\ell, \delta_\ell) \in D_n$ for $\ell \in [1..m]$. Write $\Delta^m = \prod_{\ell=1}^m \delta_\ell$. By induction,

$$\prod_{\ell=1}^{m}(h_\ell, \gamma_\ell)(k_\ell, \delta_\ell)(h_\ell, \gamma_\ell)^{-1} = \Big(\sum_{\ell=1}^{m} \Delta^{\ell-1}\gamma_\ell k_\ell + \sum_{\ell=1}^{m} \Delta^{\ell-1}(1-\delta_\ell)h_\ell,\ \Delta^m\Big) \quad (4)$$

Theorem 13. *$\text{SPHEQ}((D_n)_{n\in\mathbb{N}})$, n given in binary, is* NP*-complete.*

Note that if we give n in unary notation, then $\text{SPHEQ}((D_n)_{n\in\mathbb{N}})$ can be solved in P. This follows from Proposition 8 because $|D_n|$ is $2n$ and, hence, polynomial.

Proof. By Proposition 5, $\text{SPHEQ}((D_n)_{n\in\mathbb{N}}) \in \text{NP}$. For the NP-hardness we show that $(a_1, \ldots, a_k) \in \mathbb{N}^k$ for some $k \in \mathbb{N}$ is a positive instance of PARTITION (see Sect. 2) if and only if $((a_1, 1), \ldots, (a_k, 1)) \in \text{SPHEQ}(D_n)$ with $n = 1 + \sum_{i=1}^k a_i$. Assume that there exists $A \subseteq [1..k]$ with $\sum_{i\in A} a_i = \sum_{i\notin A} a_i$. Let $z_i = (0, 1)$ if $i \in A$ and $z_i = (0, -1)$ if $i \notin A$. Then $\prod_{i=1}^k z_i(a_i, 1)z_i^{-1} = (0, 1)$ is a solution.

Now assume that $((a_1, 1), \ldots, (a_k, 1)) \in \text{SPHEQ}(D_n)$. By (4) there exist $\gamma_i \in \mathbb{Z}_n$ such that $\sum_{i=1}^k \gamma_i a_i = r \cdot n$ for some $r \in \mathbb{Z}$. Let $A = \{i \mid \gamma_i = 1\}$. Then

$$\sum_{i\in A} a_i - \sum_{i\notin A} a_i = r \cdot n = r \cdot \Big(1 + \sum_{i\in[1..k]} a_i\Big). \quad (5)$$

Thus, $r = 0$. Hence, by (5), $(a_1, \ldots, a_k)$ is a positive instance of PARTITION. □

Choosing n as a power of two, this reduction shows that $\text{SPHEQ}((D_{2^n})_{n\in\mathbb{N}})$ is NP-complete. Thus, there is a sequence of nilpotent groups such that $\text{SPHEQ}(\mathcal{G})$ is NP-hard (compared to [11] where it is in P for each fixed nilpotent group).

Remark 14. Let p_n be the n-th prime number. By Cramers conjecture [7], $p_{n+1} - p_n \in \mathcal{O}((\log p_n)^2)$. Assume that this conjecture is true and let $m = \sum a_i + 1$. Then we can find $p \in \mathbb{P}$, $p \in [m..m + (\log(m))^2]$ in polynomial time [2]. Choosing $n = p$ in the proof of Theorem 13 would yield that $\text{SPHEQ}((D_p)_{p\in\mathbb{P}})$ is NP-hard.

Let Q_m be the generalized Quaternion group of order $4m$. With the same reduction as for D_n we can show that $\text{SPHEQ}((Q_m)_{m\in\mathbb{N}})$ is NP-complete.

6 Spherical Equations for Two-by-Two Matrices

In this section we study 2-by-2 matrix groups with our main result Theorem 22 that $\text{SPHEQ}((\text{GL}(2,p))_{p\in\mathbb{P}}) \in \text{P}$. Let us first examine the case of upper triangular matrices. Using the embedding of D_n into $\text{ET}(2, n)$, we obtain:

Corollary 15. *$\text{SPHEQ}((\text{ET}(2, n))_{n\in\mathbb{N}})$ is* NP*-complete.*

It also follows that $\text{SPHEQ}(\text{SGR}((\text{GL}(m, q))_{q\in\mathbb{N}}))$ and $\text{SPHEQ}(\text{SGR}((\text{T}(m, q))_{q\in\mathbb{N}}))$ are NP-complete for every $m \in \mathbb{N} \setminus \{0, 1\}$.

On the first thought it seems that making the groups of interest bigger increases the complexity. However, this is not the case due to Theorem 16. Its proof is a reduction to solving linear equations. Be aware that we need $p \in \mathbb{P}$.

Theorem 16. *$\text{SPHEQ}((\text{T}(2, p))_{p\in\mathbb{P}})$ is in* P.

6.1 Spherical Equations in GL(2, p)

For a matrix $\left(\begin{smallmatrix} a & b \\ c & d \end{smallmatrix}\right) \in \mathrm{GL}(2,p)$ the *determinant* $\det(A)$ is $ad - bc$, its *trace* $\mathrm{tr}(A)$ is $a + d$. If λ_1, λ_2 are the eigenvalues of A, $\lambda_1, \lambda_2 \in \mathbb{Z}_p$ or $\lambda_1, \lambda_2 \in \mathbb{Z}_p[\sqrt{(\mathrm{tr}(A))^2 - 4\det(A)}] = \mathbb{Z}_p[\sqrt{\xi}]$ for any quadratic non-residue $\xi \in \mathbb{Z}_p$. Moreover, $\mathrm{tr}(A) = \lambda_1 + \lambda_2$ and $\det(A) = \lambda_1\lambda_2$. Let K be a field extension of $\mathbb{Z}_p$. Matrices $A, B \in \mathrm{GL}(2,K)$ are called *similar* if $A = X^{-1}BX$ for some $X \in \mathrm{GL}(2,\overline{K})$, where $\overline{K}$ is the algebraic closure of K. Then, $\det(A) = \det(B)$ and $\mathrm{tr}(A) = \mathrm{tr}(B)$. In our analysis we distinguish matrices of four types. We say that A is

- *scalar* if it is diagonal and $\lambda_1 = \lambda_2 \in \mathbb{Z}_p$.
- of type-I if $\lambda_1, \lambda_2 \in \mathbb{Z}_p$ and $\lambda_1 \neq \lambda_2$. In this case A is similar to its *standard form* $\left(\begin{smallmatrix} \lambda_1 & 0 \\ 0 & \lambda_2 \end{smallmatrix}\right)$.
- of type-II if $\lambda_1, \lambda_2 \notin \mathbb{Z}_p$ and $\lambda_1 \neq \lambda_2$. Then A is similar to its *standard form* $\left(\begin{smallmatrix} s & t \\ 1 & s \end{smallmatrix}\right)$ with $s, t \in \mathbb{Z}_p$, t not a quadratic residue, $\det(A) = s^2 - t$, $\mathrm{tr}(A) = 2s$.
- of type-III if A is not scalar, $\lambda_1, \lambda_2 \in \mathbb{Z}_p$ and $\lambda_1 = \lambda_2$. In this case A is similar to its *standard form* $\left(\begin{smallmatrix} \lambda_1 & 1 \\ 0 & \lambda_1 \end{smallmatrix}\right)$.

Every $A \in \mathrm{GL}(2,p)$ classifies as a matrix of one of these types. Note that A and A^{-1} have the same type. If A is non-scalar, its conjugacy class is uniquely defined by $\mathrm{tr}(A)$ and $\det(A)$ (however, scalar matrices can have the same trace and determinant as type-III matrices). For $S \subseteq \mathrm{GL}(2,p)$ define $\mathrm{tr}(S) = \{\, \mathrm{tr}(A) \mid A \in S \,\}$. Since the trace is invariant under conjugation, for any $A, B \in \mathrm{GL}(2,p)$,

$$\mathrm{tr}(\{\, A^Y B^Z \mid Y, Z \in \mathrm{GL}(2,p) \,\}) = \mathrm{tr}(\{\, AB^Z \mid Z \in \mathrm{GL}(2,p) \,\}).$$

For matrices A, B we write $T(A,B) = \mathrm{tr}(\{\, AB^Z \mid Z \in \mathrm{GL}(2,p) \,\})$.

Lemma 17. *The following problems are in polynomial time:*

(a) Given $A, B \in \mathrm{GL}(2,p)$, decide whether they are conjugate.
(b) Given $A \in \mathrm{GL}(2,p)$, determine its type.
(c) Given $A \in \mathrm{GL}(2,p)$ of type-II or type-III, compute its standard form.

The following problems are in random polynomial time:

(d) Given $A \in \mathrm{GL}(2,p)$ of type-I, compute its standard form.
(e) Given $A, B \in \mathrm{GL}(2,p)$, compute a matrix $Z \in \mathrm{GL}(2,p)$ with $A = B^Z$.

In the following lemmas we will investigate $T(A,B)$. We distinguish three cases according to the type of B. We will write $Z = \left(\begin{smallmatrix} v & x \\ y & z \end{smallmatrix}\right)$.

Lemma 18 (Type-I). *Let A be non-scalar and B of type-I. Then $T(A,B) = \mathbb{Z}_p$. Furthermore, given A and B as well as $k \in \mathbb{Z}_p$ one can find $Z \in \mathrm{GL}(2,p)$ satisfying $\mathrm{tr}(AB^Z) = k$ in polynomial time.*

Lemma 19 (Type-II). *For two matrices A and B of type-II we have $T(A,B) = \mathbb{Z}_p$. Furthermore, given A and B as well as $k \in \mathbb{Z}_p$ one can find a matrix $Z \in \mathrm{GL}(2,p)$ satisfying $\mathrm{tr}(AB^Z) = k$ in random polynomial time.*

Proof (sketch). Write $A = \left(\begin{smallmatrix} a & b \\ 1 & a \end{smallmatrix}\right)$ and $B = \left(\begin{smallmatrix} s & t \\ 1 & s \end{smallmatrix}\right)$ and $\delta = vz - xy$. We compute

$$\begin{aligned} T(A,B) &= \left\{ \tfrac{1}{\delta}(v^2b - x^2 - y^2bt + z^2t) + 2as \mid v,x,y,z \in \mathbb{Z}_p \text{ with } \delta \neq 0 \right\} \\ &\supseteq \left\{ \tfrac{1}{vz}(v^2b - x^2 + z^2t) + 2as \mid v,z \neq 0, x \in \mathbb{Z}_p \right\} && \text{(set } y = 0) \\ &= \left\{ \tfrac{v}{z}b + \tfrac{z}{v}t - \tfrac{z}{v}x^2 + 2as \mid v,z \neq 0, x \in \mathbb{Z}_p \right\} && \text{(replace } x \text{ by } zx) \end{aligned}$$

Note that t and b are quadratic non-residues. Using a counting argument, we can show that $T(A,B) = \mathbb{Z}_p$. Moreover, by applying an algorithm by Adleman, Estes, McCurley [1] a solution can be computed in random polynomial time. □

Lemma 20 (Type-III). *Let $A = \left(\begin{smallmatrix} a & b \\ c & d \end{smallmatrix}\right)$ be non-scalar and $B = \left(\begin{smallmatrix} s & 1 \\ 0 & s \end{smallmatrix}\right)$ of type-III. Then, $\mathbb{Z}_p \setminus \{s(a+d)\} \subseteq T(A,B)$ and it can be decided in polynomial time whether $s(a+d) \in T(A,B)$. Furthermore, a matrix $Z \in \mathrm{GL}(2,p)$ satisfying $\mathrm{tr}(AB^Z) = k$ for a given $k \in \mathbb{Z}_p$ can be found in random polynomial time.*

Proof (sketch). By matrix multiplication we obtain that

$$T(A,B) = \left\{ \tfrac{1}{\delta}\left(y^2(-b) + yz(a-d) + z^2c\right) + as + sd \mid y \neq 0 \text{ or } z \neq 0 \right\}.$$

If A is not scalar, then $y^2(-b) + yz(a-d) + z^2c \neq 0$ for some (not both trivial) y and z. Varying v and x we can get any nontrivial value for δ without changing $y^2(-b)+yz(a-d)+z^2c$ and attain all values in $\mathbb{Z}_p \setminus \{s(a+d)\}$. Finally, we obtain that $s(a+d) \in T(A,B)$ if and only if $(a-d)^2 + 4bc$ is a quadratic residue of p, which can be checked in polynomial time [14, Theorem 83]. □

Now, consider a special case when A and B are of type–III. In this case,

$$T(A,B) = \left\{ -\tfrac{y^2}{\delta} + 2as \mid y \in \mathbb{Z}_p,\ \delta \neq 0 \right\} = \mathbb{Z}_p$$

and, hence, if C is of type-I or type-II, then the equation $AB^Z = C^Y$ always has a solution because the conjugacy class is uniquely determined by $\det(C)$ and $\mathrm{tr}(C)$. Thus it remains to consider C of type-III.

Lemma 21 (3x Type-III). *Let $A, B, C \in \mathrm{GL}(2,p)$ be of type-III such that $\det(AB) = \det(C)$. Then $AB^{Z_2} = C^{Z_3}$ always has a solution for $Z_2, Z_3 \in \mathrm{GL}(2,p)$. Moreover, Z_2 and Z_3 can be found in random polynomial time.*

Theorem 22. $\textsc{SphEq}((\mathrm{GL}(2,p))_{p\in\mathbb{P}}) \in \mathsf{P}$. *Furthermore, a solution (if it exists) can be found in random polynomial time.*

The algorithm computing a solution might refuse an answer with a small probability. By repeated execution we obtain a polynomial time Las Vegas algorithm (always correct and expected polynomial running time).

Proof (sketch). By Theorem 7 we can assume $p \geq 5$. Let $C_1, \ldots, C_k \in \mathrm{GL}(2,p)$. We may assume that each C_i is non-scalar and that $\det(C_1 \cdots C_k) = 1$ and $k \geq 2$. If $k = 2$, we need to solve an instance of the conjugacy problem (Lemma 17).
If $k = 3$, write the equation as $C_1 C_2^{Z_2} = C_3^{Z_3}$. For C_3 not of type-III, $\det(C_3)$ and $\mathrm{tr}(C_3)$ uniquely define its conjugacy class. By Lemma 18–20, we can decide if $\mathrm{tr}(C_3) \in T(C_1, C_2)$ and, hence, if a solution exists. Otherwise, apply Lemma 21.
If $k \geq 4$, first multiply matrices until only three matrices C_1, C_2, C_3 remain. If none or all are of type-III, apply Lemma 18, 19 or 21. Otherwise, let C_3 be not of type-III. By Lemma 20, $T(C_1, C_2) \supseteq \mathbb{Z}_p \setminus \{\mathrm{tr}(C_1) \cdot \mathrm{tr}(C_2)/2\}$. If $\mathrm{tr}(C_3) \notin T(C_1, C_2)$, let $i \in \{1,2,3\}$ such that $C_i = DE$ for some $D, E \in \mathrm{GL}(2,p)$. Since $p \geq 5$, by Lemma 18–20 there is $U \in \mathrm{GL}(2,p)$ with $\mathrm{tr}(DE^U) \neq \mathrm{tr}(C_i)$ and $(\mathrm{tr}(DE^U))^2 \neq 4\det(DE^U)$. So DE^U is not of type-III or scalar and a solution exists. By Lemma 17–21 we can find a solution (if it exists) in random polynomial time. Note that we need to find the standard forms only for finding solutions. □

7 Matrix Groups in Higher Dimensions

In this section we are again in the *matrix group* input model considering the generalized Heisenberg groups (6) as well as the uni-triangular 4-by-4 matrices. Let us first consider matrices of arbitrary dimension in the subgroup model. Using the embedding of S_n into $\mathrm{GL}(n,q)$, we obtain the following:

Corollary 23. *For every constant* $q \in \mathbb{N} \setminus \{0,1\}$, $\textsc{SphEq}(\textsc{Sgr}((\mathrm{GL}(n,q))_{n\in\mathbb{N}}))$ *is* NP*-complete.*

Heisenberg Groups. The *generalized Heisenberg group* $H_n^{(p)}$ of dimension n over $\mathbb{Z}_p$ is the group

$$H_n^{(p)} = \left\{ \begin{pmatrix} 1 & \alpha_1 & a_2 \\ 0 & I & \alpha_3 \\ 0 & 0 & 1 \end{pmatrix} \;\middle|\; \alpha_1 \in \mathbb{Z}_p^{1\times(n-2)},\; a_2 \in \mathbb{Z}_p,\; \alpha_3 \in \mathbb{Z}_p^{(n-2)\times 1} \right\}. \tag{6}$$

Observe that $H_3^{(p)} = \mathrm{UT}(3,p)$ and $H_n^{(p)} \leq \mathrm{UT}(n,p)$ for each n and p.

Theorem 24. $\textsc{SphEq}((H_n^{(p)})_{(n,p)\in\mathbb{N}\times\mathbb{P}})$, *$p$ given in binary, is in* P.

The proof of Theorem 24 shows that spherical equations in $H_n^{(p)}$ can be reduced to linear equations. Notice that $\mathcal{H} = (H_n^{(p)})_{(n,p)\in\mathbb{N}\times\mathbb{P}}$ is a sequence of non-abelian matrix groups with $\textsc{SphEq}(\mathcal{H}) \in \mathsf{P}$ even if the dimension n and the prime field $\mathbb{Z}_p$ are both part of the input.

The Sequence $(\mathrm{UT}(4,p))_{p\in\mathbb{P}}$

Theorem 25. *$\textsc{SphEq}((\mathrm{UT}(4,p))_{p\in\mathbb{P}})$ is in* P.

Proof. Consider a spherical equation $\prod_{i=1}^{k} X_i C_i X_i^{-1} = 1$. Write $D_i = X_i C_i X_i^{-1}$, $A_i = \prod_{h=1}^{i} D_h$ and

$$X_i = \begin{pmatrix} 1 & x_i & w_i & u_i \\ 0 & 1 & z_i & v_i \\ 0 & 0 & 1 & y_i \\ 0 & 0 & 0 & 1 \end{pmatrix}, \qquad A_i = \begin{pmatrix} 1 & a_i^{(1)} & a_i^{(2)} & a_i^{(3)} \\ 0 & 1 & a_i^{(4)} & a_i^{(5)} \\ 0 & 0 & 1 & a_i^{(6)} \\ 0 & 0 & 0 & 1 \end{pmatrix}.$$

The entries of C_i and D_i are denoted in the same way as the ones of A_i. Then,

$$a_i^{(j)} = \sum_{h=1}^{i} c_h^{(j)} \text{ for } j \in \{1,4,6\}, \qquad a_i^{(2)} = \sum_{h=1}^{i} d_h^{(2)} + c_h^{(4)} a_{h-1}^{(1)},$$

$$a_i^{(5)} = \sum_{h=1}^{i} d_h^{(5)} + a_{h-1}^{(4)} c_h^{(6)}, \qquad a_i^{(3)} = \sum_{h=1}^{i} d_h^{(3)} + a_{h-1}^{(1)} d_h^{(5)} + a_{h-1}^{(2)} c_h^{(6)}.$$

The entries of X_i are variables. We have to check if the system of equations $a_k^{(j)} = 0$, $j \in [1..6]$ (in the following denoted by $(*)$) has a solution. Clearly, we need that $\sum_{i=1}^{k} c_i^{(\ell)} = 0$ for $\ell \in \{1,4,6\}$. We show that for each i both v_i and w_i appear only linearly and only in $a_k^{(3)}$ with coefficients $c_i^{(1)}$ or $c_i^{(6)}$. So if $c_i^{(1)} \neq 0$ or $c_i^{(6)} \neq 0$ for some i, then $(*)$ is solvable. Further, $(*)$ is linear if $c_i^{(4)} = 0$ for all i (hence, we can decide in polynomial time whether a solution exists). So, let $c_i^{(1)} = c_i^{(6)} = 0$ for all i. W.l.o.g. let $c_1^{(4)} \neq 0$. The equation $a_k^{(2)} = 0$ (resp. $a_k^{(5)} = 0$) has only x_i (resp. y_i) as variables. Thus, we can write x_1 (resp. y_1) as linear combination of the other x_i (resp. y_i) and insert this into $a_k^{(3)}$. We get

$$\sum_{i,j} \alpha_{ij} x_i y_j + \sum_i (\beta_i x_i + \delta_i y_i) + \zeta = 0. \tag{7}$$

with constants $\alpha_{ij}, \beta_i, \delta_i, \zeta$. Assume that there is $\alpha_{i_0 j_0} \neq 0$. Otherwise, (7) is linear. Setting $x_i = y_j = 0$ for $i \neq i_0$, $j \neq j_0$, (7) becomes $\alpha_{i_0 j_0} x_{i_0} y_{j_0} + \beta_{i_0} x_{i_0} + \delta_{i_0} y_{i_0} + \zeta = 0$. Choose y_{j_0} such that $\alpha_{i_0, j_0} y_{j_0} + \beta_{i_0} \neq 0$. Then, $x_{i_0} = (-\zeta - \delta_{i_0} y_{i_0})(b_{i_0 j_0} y_{j_0} + \beta_{i_0})^{-1}$. So also in this case $(*)$ is solvable. □

7.1 The Groups $\mathbb{Z}_m^k \rtimes C_2$

As before let $C_2 = \{\pm 1\}$. Then C_2 acts on $\mathbb{Z}_m$ by $(x,a) \mapsto x \cdot a$. The corresponding semidirect product $\mathbb{Z}_m \rtimes C_2$ is precisely the dihedral group D_m. We can define also the semidirect product $\mathbb{Z}_m^k \rtimes C_2 = \{(a,x) \mid a \in \mathbb{Z}_m^k, x = \pm 1\}$ with C_2 operating componentwise on $\mathbb{Z}_m^k$ meaning that the multiplication is given by

$$((a_1, \ldots, a_k), x) \cdot ((b_1, \ldots, b_k), y) = ((a_1 + x \cdot b_1, \ldots, a_k + x \cdot b_k), x \cdot y). \tag{8}$$

The neutral element is $((0, \ldots, 0), 1)$, which we also denote by $(0, 1)$.

Theorem 26. *Let $m = 3$ or $m \geq 5$ be fixed. Then* $\textsc{SphEq}((\mathbb{Z}_m^k \rtimes C_2)_{k \in \mathbb{N}})$ *is* NP*-complete.*

Before we prove Theorem 26 we explain one of its consequences. Consider the group $(\mathbb{Z}_m \rtimes C_2)^k = D_m^k$. We can embed $\mathbb{Z}_m^k \rtimes C_2$ into $(\mathbb{Z}_m \rtimes C_2)^k$ via

$$((a_1, \ldots, a_k), x) \mapsto ((a_1, x), \ldots, (a_k, x)).$$

As D_m embeds into $\mathrm{GL}(2, m)$, we also obtain an embedding of $\mathbb{Z}_m^k \rtimes C_2$ into $\mathrm{GL}(2k, m)$. Because m is fixed, $\mathbb{Z}_m \rtimes C_2$ and $\mathbb{Z}_m$ are fixed finite groups. Thus, according to Corollary 9, $\textsc{SphEq}(((\mathbb{Z}_m \rtimes C_2)^k)_{k \in \mathbb{N}})$ and $\textsc{SphEq}((\mathbb{Z}_m^k)_{k \in \mathbb{N}})$ are in P for k given in unary. So we have the following corollary:

Corollary 27. *There exist sequences of groups $\mathcal{G}' = (G'_n)_{n \in \mathbb{N}}$, $\mathcal{H} = (H_n)_{n \in \mathbb{N}}$ and $\mathcal{G} = (G_n)_{n \in \mathbb{N}}$ such that $G'_n \leq H_n \leq G_n$ for all $n \in \mathbb{N}$,* $\textsc{SphEq}(\mathcal{G}')$ *and* $\textsc{SphEq}(\mathcal{G})$ *are in* P *but* $\textsc{SphEq}(\mathcal{H})$ *is* NP*-complete.*

Proof (of Theorem 26). We reduce from 3-ExactSetCover (see Sect. 2). Let $A_1, \ldots, A_\ell$ be subsets of $[1 .. k]$. Let $H = \mathbb{Z}_m^{k+\ell} \rtimes C_2$ and write $a_{i,j}$ for the j-th entry of $a_i \in \mathbb{Z}_m^{k+\ell}$. We define 2ℓ elements $c_i = (a_i, 1) \in H$. For $i \in [1 .. \ell]$, $j \in [1 .. k]$ we set $a_{i,j} = a_{\ell+i,j} = 1$ if $j \in A_i$, $a_{\ell+i,k+i} = 1$ and $a_{i,j} = 0$ otherwise. We show that the equation

$$\prod_{i=1}^{2\ell} z_i c_i z_i^{-1} = ((\underbrace{2, \ldots, 2}_{k \text{ times}}, \underbrace{1, \ldots, 1}_{\ell \text{ times}}), 1) \tag{9}$$

has a solution if and only if $[1 .. k]$ together with $A_1, \ldots, A_\ell$ is a positive instance of 3-ExactSetCover. Let $\beta = (0, -1)$ and observe that

$$\beta c_i \beta^{-1} c_{\ell+i} = ((\underbrace{0, \ldots, 0}_{k+i-1 \text{ times}}, 1, 0, \ldots, 0), 1). \tag{10}$$

Let $I \subseteq [1 .. \ell]$ such that $\{A_i \mid i \in I\}$ is an exact covering. By the choice of the c_i, we have $a_{i,j} + a_{\ell+i,j} \in \{0, 2\}$ for all $i \in I$ and $j \in [1 .. k]$. So (10) implies

$$\prod_{i \in I} c_i \cdot c_{\ell+i} \cdot \prod_{i \notin I} \beta c_i \beta^{-1} \cdot c_{\ell+i} = ((\underbrace{2, \ldots, 2}_{k \text{ times}}, \underbrace{1, \ldots, 1}_{\ell \text{ times}}), 1).$$

Thus, according to Lemma 4 there exists a solution to (9).
If (9) has a solution, there is one with $z_i = (0, \pm 1)$ for all i. Then, $z_{\ell+j} = (0, 1)$ for $j \in [1 .. \ell]$. Otherwise we get a -1 in the last ℓ entries. Let $I = \{\, i \in [1 .. \ell] \mid z_i = (0, 1) \,\}$. By (9) and (10), $\sum_{i \in I}(a_{i,j} + a_{\ell+i,j}) = 2$ for $j \in [1 .. k]$ (in $\mathbb{Z}_m$). We have $a_{i,j} + a_{\ell+i,j} \in \{0, 2\}$ for $i \in [1 .. \ell]$, $j \in [1 .. k]$. For each j, $|\{i \in [1 .. \ell] \mid a_{i,j} + a_{\ell+i,j} = 2\}| \leq 3$. Since $m = 3$ or $m \geq 5$ for each j there is exactly one $i \in I$ with $a_{i,j} = 1$. So the A_i, $i \in I$ are an exact covering of $[1 .. k]$. □

8 Open Problems

- What is the complexity of $\text{SPHEQ}((\text{GL}(n,p))_{p\in\mathbb{P}})$ for $n \geq 3$? We conjecture that for every n there is a number f_n such that every spherical equation over $\text{GL}(n,p)$ consisting of at least f_n non-scalar matrices has a solution – so for each fixed n the problem would be still solvable in polynomial time. However, this might change if the dimension n is part of the input.
- What is the role of prime vs. arbitrary integer as modulus? In particular, is $\text{SPHEQ}((D_p)_{p\in\mathbb{P}})$ NP-hard? An affirmative answer to this question seems very unlikely without solving deep number-theoretic questions (see Remark 14). On the other hand, is it possible to extend the polynomial-time algorithm for $\text{SPHEQ}((\text{GL}(2,p))_{p\in\mathbb{P}})$ to $\text{SPHEQ}((\text{GL}(2,n))_{n\in\mathbb{N}})$? A possible obstacle here might be that one might have to factor n.
- What makes spherical equations difficult? Clearly the groups must be non-abelian – but on the other hand the example of $\text{GL}(2,p)$ shows that "too far from abelian" might be also easy.

Funding Information. Armin Weiß has been funded by DFG Grant WE 6835/1-2.

References

1. Adleman, L.M., Estes, D., McCurley, K.S.: Solving bivariate quadratic congruences in random polynomial time. Math. Comput. **48**, 17–28 (1987)
2. Agrawal, M., Kayal, N., Saxena, N.: Primes is in P. Ann. Math. **160**, 781–793 (2004)
3. Aschbacher, M.: Finite Group Theory. Cambridge Studies in Advanced Mathematics, 2nd edn. Cambridge University Press, Cambridge (2000)
4. Babai, L., Szemerédi, E.: On the complexity of matrix group problems i. In: IEEE Annual Symposium on Foundations of Computer Science (1984)
5. Bartholdi, L., Groth, T., Lysenok, I.: Commutator width in the first Grigorchuk group. Groups Geom. Dyn. **16**, 493–522 (2022)
6. Comerford, L.P., Edmunds, C.C.: Quadratic equations over free groups and free products. J. Algebra **68**, 276–297 (1981)
7. Cramér, H.: On the order of magnitude of the difference between consecutive prime numbers. Acta Arithmetica **2**(1), 23–46 (1936). http://eudml.org/doc/205441
8. Culler, M.: Using surfaces to solve equations in free groups. Topology **20**(2), 133–145 (1981)
9. Diekert, V., Robson, J.: Quadratic Word Equations, pp. 314–326. Springer, Heidelberg (1999). https://doi.org/10.1007/978-3-642-60207-8_28
10. Garey, M.R., Johnson, D.S.: Computers and Intractability: A Guide to the Theory of NP-Completeness. W. H. Freeman, New York (1979)
11. Goldmann, M., Russell, A.: The complexity of solving equations over finite groups. Inf. Comput. **178**(1), 253–262 (2002)
12. Grigorchuk, R., Kurchanov, P.: On quadratic equations in free groups. In: Proceedings of the International Conference on Algebra Dedicated to the Memory of A. I. Malcev. Contemporary Mathematics, vol. 131, pp. 159–171. AMS (1992)

13. Grigorchuk, R., Lysionok, I.: A description of solutions of quadratic equations in hyperbolic groups. Int. J. Algebra Comput. **2**(3), 237–274 (1992)
14. Hardy, G.H., Wright, E.M.: An Introduction to the Theory of Numbers, 4th edn. Oxford (1975)
15. Holt, D.F., Eick, B., O'Brien, E.A.: Handbook of Computational Group Theory. Chapman and Hall, Boca Raton (2005)
16. Kharlampovich, O., Lysenok, I., Myasnikov, A.G., Touikan, N.: The solvability problem for quadratic equations over free groups is NP-complete. Theor. Comput. Syst. **47**, 250–258 (2010)
17. Kharlampovich, O., Mohajeri, A., Taam, A., Vdovina, A.: Quadratic equations in hyperbolic groups are NP-complete. Trans. AMS **369**(9), 6207–6238 (2017)
18. Lysenok, I., Miasnikov, A., Ushakov, A.: Quadratic equations in the Grigorchuk group. Groups Geom. Dyn. **10**, 201–239 (2016)
19. Lysenok, I., Ushakov, A.: Spherical quadratic equations in free metabelian groups. Proc. Am. Math. Soc. **144**, 1383–1390 (2016)
20. Lysenok, I., Ushakov, A.: Orientable quadratic equations in free metabelian groups. J. Algebra **581**, 303–326 (2021)
21. Makanin, G.: Equations in a free group. Izvestiya AN SSSR, Ser. Mat. **46**, 1199–1273 (1982). (Russian, English translation in Math USSR Izvestiya, 21, 3 (1983))
22. Mandel, R., Ushakov, A.: Quadratic equations in metabelian Baumslag-Solitar groups. Int. J. Algebra Comput. (2023)
23. Mattes, C., Ushakov, A., Weiß, A.: Complexity of spherical equations in finite groups. arXiv eprints arxiv:2308.12841 (2023). https://doi.org/10.48550/arXiv.2308.12841
24. McKenzie, P., Péladeau, P., Thérien, D.: NC^1: the automata-theoretic viewpoint. Comput. Complex. **1**, 330–359 (1991)
25. Schupp, P.E.: Quadratic equations in groups, cancellation diagrams on compact surfaces, and automorphisms of surface groups. In: Word Problems, II (Conference on Decision Problems in Algebra, Oxford, 1976), Stud. Logic Foundations Math, North-Holland, Amsterdam, vol. 95, pp. 347–371 (1980)
26. Vollmer, H.: Introduction to Circuit Complexity: A Uniform Approach. Springer, Berlin (1999). https://doi.org/10.1007/978-3-662-03927-4

Positive Characteristic Sets for Relational Pattern Languages

S. Mahmoud Mousawi and Sandra Zilles(✉)

University of Regina, Regina, Canada
sandra.zilles@uregina.ca

Abstract. In the context of learning formal languages, data about an unknown target language L is given in terms of a set of *(word,label)* pairs, where a binary label indicates whether or not the given word belongs to L. A (polynomial-sized) characteristic set for L, with respect to a reference class $\mathcal{L}$ of languages, is a set of such pairs that satisfies certain conditions allowing a learning algorithm to (efficiently) identify L within $\mathcal{L}$. In this paper, we introduce the notion of *positive characteristic set*, referring to characteristic sets of only positive examples. These are of importance in the context of learning from positive examples only. We study this notion for classes of relational pattern languages, which are of relevance to various applications in string processing.

Keywords: Learning formal languages · Relational pattern languages · Characteristic sets

1 Introduction

Many applications in machine learning and database systems, such as, e.g., the analysis of protein data [3], the design of algorithms for program synthesis [12], or problems in pattern matching [4], deal with the problem of finding and describing patterns in sets of strings. From a formal language point of view, one way to address this problem is with the study for so-called pattern languages [1,14].

A pattern is a finite string of variables and terminal symbols. For instance, if the alphabet of terminal symbols is $\Sigma = \{a, b\}$, and $x_1, x_2, \ldots,$ denote variables, then $p = ax_1bax_1x_2$ is a pattern. The language generated by this pattern consists of all words that are obtained when replacing variables with finite words over Σ, where multiple occurrences of the same variable are replaced by the same word. Angluin [1] further required that no variable be replaced by the empty string, resulting in the notion of *non-erasing* pattern language, while Shinohara's *erasing* pattern languages allow erasing variables. For the pattern p above, the non-erasing language is $\{aw_1baw_1w_2 \mid w_1, w_2 \in \Sigma^+\}$, while the erasing language is $\{aw_1baw_1w_2 \mid w_1, w_2 \in \Sigma^*\}$. The latter obviously contains the former.

The repetition of the variable x_1 in p sets the words replaced for the two occurrences of x_1 in *equality* relation. Angluin already noted that relations other than equality could be studied as well— a thought that was later on explored by

H. Fernau et al. (Eds.): SOFSEM 2024, LNCS 14519, pp. 398–412, 2024.
https://doi.org/10.1007/978-3-031-52113-3_28

Geilke and Zilles [6]. For example, the *reversal* relation requires that the words replaced for the two occurrences of x_1 are the reverse of one another, which can be useful for modeling protein folding or behaviour of genes [6]. To the best of our knowledge, the first work to focus on specific relations aside from equality is that of Holte et al. [9], which studies the relations *reversal* as well as *equal length*, in the non-erasing setting. As the name suggests, *equal length* stipulates that the words replaced for the two occurrences of x_1 in p are of the same length.

Holte et al. focused on the question whether the equivalence of two patterns can be decided efficiently. While deciding equivalence is quite simple for non-erasing pattern languages with the equality relation [1], the situation is more complex for reversal and equal length relations. Holte et al.'s proposed decision procedure, which tests only a small number of words for membership in the two pattern languages under consideration, has applications beyond deciding equivalence. In particular, the sets of tested words serve as so-called teaching sets in the context of machine teaching [15] as well as characteristic sets in the context of grammatical inference from positive and negative data [8].

We continue this line of research in two ways: firstly, we study reversal and equal length relations for *erasing* pattern languages; secondly, we focus on the concept of characteristic set rather than the equivalence problem. From that perspective, there is a direct connection of our work to the theory of learning relational pattern languages. Our contributions are as follows:

(a) We introduce the notion of *positive characteristic sets*. Classical characteristic sets contain positive and negative examples, i.e., words over Σ together with a label indicating whether or not the word belongs to a target language to be learned. While characteristic sets are crucial for studying the (efficient) learnability of formal languages from positive and negative data [8], they are not suited to the study of learning from positive examples only. Positive characteristic sets are an adjustment to account for learning from only positive examples.

(b) Perhaps not surprisingly, we show that the existence of positive characteristic sets is equivalent to the existence of telltale sets, which are known to characterize classes of languages learnable from positive data in Gold's [7] model of learning in the limit [2]. Moreover, *small* positive characteristic sets yield *small* telltales. We thus make connections between grammatical inference from characteristic sets and learning in the limit using telltales.

(c) Holte et al. conjectured that the class of non-erasing pattern languages under the reversal relation possesses a system of small sets which, in our terminology, would be linear-size positive characteristic sets. We prove that, in the erasing case under the reversal relation, *no* system of positive characteristic sets exists (let alone a system of small such sets) when $|\Sigma| = 2$.

(d) For equal length, Holte et al. showed that, what we call a system of linear-size positive characteristic sets, exists for the class of non-erasing pattern languages, assuming $|\Sigma| \geq 3$. While their proof carries over to the erasing case as well, our main result states that even if $|\Sigma| = 2$, the class of erasing pattern languages under the equal length relation has linear-size positive characteristic sets (and thus also linear-size telltales). The proof techniques required for obtaining this result are vastly different from those used for the case $|\Sigma| \geq 3$.

2 Notation and Preliminary Results

Throughout this paper, we use Σ to denote an arbitrary finite alphabet. A language over Σ is simply a subset of Σ^*, and every string in Σ^* is called a word. We denote the empty string/word by ε. The reverse operation on a word $w = w_1 \dots w_l \in \Sigma^*$, with $l \in \mathbb{N}$ and $w_i \in \Sigma$, is defined by $w^{rev} := w_l \dots w_1$. For any set Z and any string $s \in Z^*$, we use $|s|$ to refer to the length of s.

Let L be a language over Σ. A set $S \subseteq \Sigma^* \times \{0, 1\}$ is consistent with L iff ($(s, 1) \in S$ implies $s \in L$) and ($(s, 0) \in S$ implies $s \notin L$). When learning a language L from positive and negative examples (without noise), a learning algorithm is presented with a set $S \subseteq \Sigma^* \times \{0, 1\}$ that is consistent with L; from this set it is supposed to identify L. We refer the reader to [8] for details of learning formal languages; the present paper will focus on structural properties of classes of formal languages that allow for efficient learning of formal languages from *only positive* examples; in particular, we consider sets $S \subseteq \Sigma^* \times \{1\}$.

2.1 Positive Characteristic Sets

While learning languages from positive examples is well-studied [2,7,11], we build a bridge between learnability from positive examples (via so-called *telltales* [2]) and (efficient) learnability from positive and negative examples (via (polynomial-size) *characteristic sets* [8]).

Definition 1 (adapted from [8]). *Let $\mathcal{L} = (L_i)_{i \in \mathbb{N}}$ be a family of languages over Σ. A family $(C_i)_{i \in \mathbb{N}}$ of finite subsets of $\Sigma^* \times \{0, 1\}$ is called a family of characteristic sets for $\mathcal{L}$ iff:*

1. *C_i is consistent with L_i for all $i \in \mathbb{N}$.*
2. *If $L_i \neq L_j$, and C_i is consistent with L_j, then C_j is not consistent with L_i.*

If, in addition, $C_i \subseteq \Sigma^* \times \{1\}$ for all $i \in \mathbb{N}$, then $(C_i)_{i \in \mathbb{N}}$ is called a family of positive characteristic sets for $\mathcal{L}$.

In the context of machine teaching, characteristic sets have also been called *non-clashing teaching sets* [5].

Every family $\mathcal{L} = (L_i)_{i \in \mathbb{N}}$ possesses a family of characteristic sets; it suffices to include in C_i, for each $j < i$ with $L_j \neq L_i$, one example from the symmetric difference of L_j and L_i, labelled according to L_i. In the literature, the *size* and *effective computability* of characteristic sets are therefore the main aspects of interest, and relate to notions of efficient learning of languages [8]. However, the mere *existence* of a family of positive characteristic sets is a non-trivial property. We will show below that it is equivalent to the existence of so-called telltales.

Definition 2 ([2]). *Let $\mathcal{L} = (L_i)_{i\in\mathbb{N}}$ be a family of languages over Σ. A family $(T_i)_{i\in\mathbb{N}}$ of finite subsets of Σ^* is called a family of telltales for $\mathcal{L}$ iff:*

1. *$T_i \subseteq L_i$ for all $i \in \mathbb{N}$.*
2. *If $i, j \in \mathbb{N}$, $L_i \neq L_j$, and $T_i \subseteq L_j$, then $L_j \not\subset L_i$.*

For families $\mathcal{L} = (L_i)_{i\in\mathbb{N}}$ with uniformly decidable membership, the existence of a family of telltales is equivalent to the learnability of $\mathcal{L}$ from positive examples only, in Gold's [7] model of learning in the limit [2]. We connect telltales to characteristic sets by observing that the existence of a telltale family is equivalent to the existence of a family of positive characteristic sets.

Proposition 1. *Let $\mathcal{L} = (L_i)_{i\in\mathbb{N}}$ be a family of languages over Σ. Then the following two statements are equivalent.*

1. *$\mathcal{L}$ possesses a family of positive characteristic sets.*
2. *$\mathcal{L}$ possesses a family of telltales.*

Proof. To show that Statement 1 implies Statement 2, let $(C_i)_{i\in\mathbb{N}}$ be a family of positive characteristic sets for $\mathcal{L}$. Define $C'_i = \{s \mid (s, 1) \in C_i\}$ for all $i \in \mathbb{N}$. We argue that $(C'_i)_{i\in\mathbb{N}}$ is a family of telltales for $\mathcal{L}$. To see this, suppose there are $i, j \in \mathbb{N}$ such that $L_i \neq L_j$, $C'_i \subseteq L_j$, and $L_j \subset L_i$. Then $C'_j \subset L_i$, so that C_j is consistent with L_i. Further, $C'_i \subseteq L_j$ implies that C_i is consistent with L_j. This is a contradiction to $(C_i)_{i\in\mathbb{N}}$ being a family of positive characteristic sets for $\mathcal{L}$. Hence no such $i, j \in \mathbb{N}$ exist, i.e., $(C'_i)_{i\in\mathbb{N}}$ is a family of telltales for $\mathcal{L}$.

To verify that Statement 2 implies Statement 1, let $(T_i)_{i\in\mathbb{N}}$ be a family of telltales for $\mathcal{L}$. The following (possibly non-effective) procedure describes the construction of a set C_i for any $i \in \mathbb{N}$, starting with $C_i = \emptyset$:

- For each $s \in T_i$, add $(s, 1)$ to C_i.
- For each $j < i$, if $L_i \not\subseteq L_j$, pick any $s_j \in L_i \setminus L_j$ and add $(s_j, 1)$ to C_i.

Obviously, C_i is finite. Now suppose there are i, j with $j < i$ and $L_i \neq L_j$, such that $C_i \cup C_j$ is consistent with L_i and L_j. If $L_i \not\subseteq L_j$, then, by construction, C_i contains $(s_j, 1)$ for some $s_j \notin L_j$, so that C_i is not consistent with L_i—a contradiction. So $L_i \subseteq L_j$. As C_j is consistent with L_i, we have $T_j \subseteq L_i \subseteq L_j$. By the telltale property, this yields $L_i = L_j$—again a contradiction. □

While the existence of a family of characteristic sets is trivially fulfilled, not every family $\mathcal{L} = (L_i)_{i\in\mathbb{N}}$ has a family of positive characteristic sets. For example, it was shown that the family containing Σ^* as well as each finite subset of Σ^* does not possess a family of telltales [2]. Therefore, by Proposition 1, it does not possess a family of positive characteristic sets.

In our study below, the size of positive characteristic sets (if any exist) will be of importance. The term *family of polynomial-sized positive characteristic sets/telltales* refers to a family of positive characteristic sets/telltales in which the size of the set C_i/T_i assigned to language L_i is bounded by a polynomial in the size of the underlying representation of L_i. The proof of Proposition 1 now has an additional consequence of interest in the context of efficient learning:

Corollary 1. *Let $\mathcal{L} = (L_i)_{i\in\mathbb{N}}$ be a family of languages over Σ. If $\mathcal{L}$ possesses a family of polynomial-sized positive characteristic sets then $\mathcal{L}$ possesses a family of polynomial-sized telltales.*

This corollary motivates us to focus our study on positive characteristic sets rather than on telltales—bounds on the size of the obtained positive characteristic sets then immediately carry over to telltales.

2.2 Relational Pattern Languages

Let $X = \{x_1, x_2, \ldots\}$ be a countably infinite set of variables, disjoint from Σ. Slightly deviating from Angluin's notation, a finite non-empty string p over $\Sigma \cup X$ is called a pattern, if no element from X occurs multiple times in p; the set of variables occurring in p is then denoted by $Var(p)$. For instance, for $\Sigma = \{a, b\}$, the string $p = x_1 a x_2 a b x_3 x_4 b$ is a pattern with $Var(p) = \{x_1, \ldots, x_4\}$. A pattern p generates words over Σ if we substitute each variable with a word, i.e., we apply a substitution φ to p. A substitution is a mapping $(\Sigma \cup X)^+ \to \Sigma^*$ that acts as a morphism with respect to concatenation and is the identity when restricted to Σ^+. The set of words generated from a pattern p by applying any substitution is a regular language, unless one constrains the substitutions. In the above example, p generates the regular language $\{w_1 a w_2 a b w_3 w_4 b \mid w_i \in \Sigma^* \text{ for } 1 \le i \le 4\}$.

One way of constraining substitutions is to force them to obey relations between variables. Angluin [1] allowed any subsets of variables to be in *equality* relation, requiring that all variables in equality relation must be substituted by identical strings. Furthermore, she constrained substitutions to be *non-erasing*, i.e., they map every variable to a non-empty string over Σ. For example, the pattern $p = x_1 a x_2 a b x_3 x_4 b$, under the constraint that x_3, x_4 are in equality relation and only non-erasing substitutions are allowed, generates the (non-regular!) language $\{w_1 a w_2 a b w_3 w_3 b \mid w_i \in \Sigma^+ \text{ for } 1 \le i \le 3\}$. Later, Shinohara removed the constraint of non-erasing substitutions, resulting in so-called erasing pattern languages [14]. Aside from equality, two relations that were studied in the context of non-erasing pattern languages are *reversal* and *equal length* [9]. We denote by *eq*, *rev*, and *len*, resp., the relations on Σ^* that correspond to equality, reversal, and equal length, resp. If $w, v \in \Sigma^*$, then (i) $(w, v) \in eq$ iff $w = v$; (ii) $(w, v) \in rev$ iff $w = v^{rev}$; (iii) $(w, v) \in len$ iff $|w| = |v|$. For instance, $(ab, bb) \in len \setminus (rev \cup eq)$, $(aab, baa) \in (rev \cup len) \setminus eq$, $(ab, ab) \in (eq \cup len) \setminus rev$, and $(aa, aa) \in eq \cap rev \cap len$. Clearly, $rev \cup eq \subseteq len$.

In general, a relational pattern is a pair (p, R), where p is a pattern and R is a binary relation over X. If r is a fixed binary relation over Σ^* and φ is any substitution, we say that φ is valid for R (or φ is an R-substitution), if $(x, y) \in R$ implies $(\varphi(x), \varphi(y)) \in r$. For example, if $r = len$, $\varphi(x) = bb$, $\varphi(y) = ab$, and $\varphi(z) = aaa$, then φ would be valid for $R = \{(x, y)\}$, but not for $R' = \{(x, z)\}$ or for $R'' = \{(x, y), (x, z)\}$.

If (p, R) is a relational pattern, then a *group* in (p, R) is any maximal set of variables in p that are pairwise related via the transitive closure of the symmetric closure of R over $Var(p)$. For example, when $p = a x_1 x_2 b a x_3 a b x_4 x_5 x_6$ and $R =$

$\{(x_1, x_3), (x_1, x_4), (x_2, x_6)\}$, then the groups are $\{x_1, x_3, x_4\}$, $\{x_2, x_6\}$, $\{x_5\}$. The group of a variable x is the group to which it belongs; we denote this group by $[x]$. A variable is *free* if its group is of size 1, like x_5 in our example.

3 Reversal Relation

In this section, we will fix $r = rev$. Note that a substitution that is valid for $\{(x, y), (y, z), (x, z)\}$, where x, y, z are three mutually distinct variables, must replace x, y, z all with the same string, and this string must be a palindrome. For convenience of the formal treatment of the *rev* relation, we avoid this special situation and consider only *reversal-friendly* relational patterns; a relational pattern (p, R) is reversal-friendly, if the transitive closure of the symmetric closure of R does not contain any subset of the form $\{(y_1, y_2), (y_2, y_3), \ldots, (y_n, y_1)\}$ for any odd $n \geq 3$.

Note that, for a reversal-friendly pattern (p, R), any group $[x]$ is partitioned into two subsets $[x]_{rev}$ and $[x]_{=}$. The former consists of the variables in $[x]$ for which any R-substitution φ is forced to substitute $\varphi(x)^{rev}$, while the latter contains variables in $[x]$ which φ replaces with $\varphi(x)$ itself.

If (p, R) is a reversal-friendly relational pattern, then $\varepsilon L_{rev}(p, R) := \{\varphi(p) \mid \varphi \text{ is an } R\text{-substitution}\}$. The class of (erasing) reversal pattern languages is now given by $\varepsilon\mathcal{L}_{rev} = \{\varepsilon L_{rev}(p, R) \mid (p, R) \text{ is a reversal-friendly relational pattern}\}$.

In the literature [9], so far only the non-erasing version of these languages was discussed: $L_{rev}(p, R) := \{\varphi(p) \mid \varphi \text{ is a non-erasing } R\text{-substitution}\}$. The class of non-erasing reversal pattern languages is then defined as $\mathcal{L}_{rev} = \{L_{rev}(p, R) \mid (p, R) \text{ is a reversal-friendly relational pattern}\}$. From [9], it is currently still open whether or not polynomially-sized positive characteristic sets exist. Using basic results on telltales, one can easily show that, for any alphabet size, the class $\mathcal{L}_{rev}$ has a family of positive characteristic sets.

Proposition 2. *Let Σ be any countable alphabet. Then $\mathcal{L}_{rev}$ has a family of positive characteristic sets.*

Proof. Let w be any word in $L_{rev}(p, R)$. Since non-erasing substitutions generate words at least as long as the underlying pattern, we have $|p'| \leq |w|$ for all p' for which there is some R' with $w \in L_{rev}(p', R')$. Thus, there are only finitely many languages $L \in \mathcal{L}_{rev}$ with $w \in L$. It is known that any family for which membership is uniformly decidable (like $\mathcal{L}_{rev}$), and which has the property that any word w belongs to at most finitely many members of the family, possesses a family of telltales [2]. By Proposition 1, thus $\mathcal{L}_{rev}$ has a family of positive characteristic sets. □

When allowing erasing substitutions, we obtain a contrasting result. In particular, we will show that, when $|\Sigma| = 2$, the class $\varepsilon\mathcal{L}_{rev}$ does not have a family of positive characteristic sets.

For convenience of notation, for a relational pattern (p, R), we will write $y = x^{rev}$ and use x^{rev} interchangeably with y, when $(x, y) \in R$, where x, y

are variables in p. Thus, we can rewrite a relational pattern (p, R) in alternate form $\overline{(p, R)}$. For example, we can write $\overline{(p, R)} = x_1 x_1^{rev} x_2 x_2^{rev} x_3 x_3^{rev}$ instead of $(p, R) = (x_1 y_1 x_2 y_2 x_3 y_3, \{(x_1, y_1), (x_2, y_2), (x_3, y_3)\})$. The reverse operation on a string of variables is defined as

$$(\nu_1 \nu_2 \ldots \nu_{l-1} \nu_l)^{rev} := (\nu_l)^{rev} (\nu_{l-1})^{rev} \ldots (\nu_2)^{rev} (\nu_1)^{rev},$$

where $\nu_i \in \{x_i, x_i^{rev}\}$ for all $1 \le i \le l$. Note that $(x_i^{rev})^{rev} := x_i$ and $(x_i)^{rev} := x_i^{rev}$. Now let $\bar{X} = X \cup \{x^{rev} \mid x \in X\}$.

Let R be any binary relation over X. A reversal-obedient morphism for R is a mapping $\phi : (\Sigma \cup \bar{X})^* \to (\Sigma \cup \bar{X})^*$ that acts as a morphism with respect to concatenation, leaves elements of Σ unchanged, and satisfies the following property for all $p \in (\Sigma \cup \bar{X})^*$: if $x, y \in \bar{X}$ occur in p and $(x, y) \in R$ then $\phi(x) = \phi(y)^{rev}$. For example, if $R = \{(x, y)\}$ and $a \in \Sigma$, such a mapping ϕ could be defined by $\phi(\sigma) = \sigma$ for all $\sigma \in \Sigma$, $\phi(x) = x_1 x_2$, $\phi(y) = x_2 x_1$, and $\phi(z) = a x_3$. This mapping ϕ would yield $\phi(axyaz) = \phi(axx^{rev}az) = ax_1x_2x_2x_1aax_3$.

The following lemma is helpful for establishing that positive characteristic sets do not in general exist for $\varepsilon\mathcal{L}_{rev}$. An analogous version for the equality relation was proven in [10]; our proof for the reversal relation is almost identical, and omitted due to space constraints.

Lemma 1. *Let $|\Sigma| \ge 2$. Suppose $(p, R), (p', R')$ are two arbitrary relational patterns where $p, p' \in X^+$. Then $\varepsilon L_{rev}(p, R) \subseteq \varepsilon L_{rev}(p', R')$ if and only if there exists a reversal-obedient morphism ϕ for R' such that $\phi(\overline{p', R'}) = \overline{(p, R)}$.*

Theorem 1. *Let $|\Sigma| = 2$. Then $\varepsilon\mathcal{L}_{rev}$ does not possess a family of positive characteristic sets.*

Proof. Fix $\Sigma = \{a, b\}$. We will show that $\varepsilon L_{rev}(x_1 x_1^{rev} x_2 x_2^{rev} x_3 x_3^{rev})$ does not have a positive characteristic set with respect to $\varepsilon\mathcal{L}_{rev}$. Our proof follows a construction used by Reidenbach [13] to show that $\varepsilon L_{eq}(x_1 x_1 x_2 x_2 x_3 x_3)$ does not have a telltale with respect to $\varepsilon\mathcal{L}_{eq}$. Let $(p, R) = x_1 x_1^{rev} x_2 x_2^{rev} x_3 x_3^{rev}$. In particular, we will show that for any finite set $T \subseteq \varepsilon L_{rev}(p, R)$ there exists a relational pattern (p', R') with $p' \in X^+$ such that $T \subseteq \varepsilon L_{rev}(p', R') \subset \varepsilon L_{rev}(p, R)$. This implies that $T \times \{1\}$ cannot be used as a positive characteristic set for $\varepsilon L_{rev}(p, R)$ with respect to $\varepsilon\mathcal{L}_{rev}$, which proves the theorem.

Fix $T = \{w_1, \ldots, w_n\} \subseteq \varepsilon L_{rev}(p, R)$. For each $w_i \in T$, define $\overleftarrow{\theta_i} : \Sigma^* \longrightarrow X^*$ by $\overleftarrow{\theta_i}(\sigma_1 \ldots \sigma_z) = \overleftarrow{\theta_i}(\sigma_1) \ldots \overleftarrow{\theta_i}(\sigma_z)$, for $\sigma_1, \ldots, \sigma_z \in \Sigma$, where, for $c \in \Sigma$, we set

$$\overleftarrow{\theta_i}(c) = \begin{cases} x_{2i-1}, & c = a \\ x_{2i}, & c = b \end{cases}$$

$T \subseteq \varepsilon L(p, R)$ implies for every $i \in \{1, \ldots, n\}$ the existence of a substitution θ_i (valid for R) such that $w_i = \theta_i(p)$. For each $w_i \in T$ we construct three strings of variables $\alpha_{i,1}, \alpha_{i,2}, \alpha_{i,3} \in X^*$, such that a concatenation of these strings in a specific way produces a pattern other than (p, R) that generates w_i. These $\alpha_{i,k}$ will be the building blocks for the desired pattern (p', R'). Consider two cases:

(*i*) Some $\sigma \in \Sigma$ appears in $\theta_i(x_3)$ exactly once while $\theta_i(x_1), \theta_i(x_2) \in \{\sigma'\}^*$ for $\sigma' \neq \sigma$, $\sigma' \in \Sigma$. Here we construct strings of variables as follows:
$\alpha_{i,1} := \overleftarrow{\theta_i}(\theta_i(x_1)\theta_i(x_2))$
$\alpha_{i,2} := \overleftarrow{\theta_i}(\theta_i(x_3))$
$\alpha_{i,3} := \varepsilon$

(*ii*) Not (*i*). In this case we simply set $\alpha_{i,k} := \overleftarrow{\theta_i}(\theta_i(x_k))$ where $1 \leq k \leq 3$.

In each case, $w_i \in \varepsilon L(\alpha_{i,1}\alpha_{i,1}^{rev}\alpha_{i,2}\alpha_{i,2}^{rev}\alpha_{i,3}\alpha_{i,3}^{rev})$.

Now define $(p', R') := y_1 y_1^{rev} y_2 y_2^{rev} y_3 y_3^{rev}$ where $y_k := \alpha_{1,k}\alpha_{2,k}\dots\alpha_{n,k} \in X^+$ for $1 \leq k \leq 3$. To conclude the proof, we show $T \subseteq \varepsilon L_{rev}(p', R') \subset \varepsilon L_{rev}(p, R)$.

To see that $T \subseteq \varepsilon L_{rev}(p', R')$, note $w_i \in \varepsilon L_{rev}(\alpha_{i,1}\alpha_{i,1}^{rev}\alpha_{i,2}\alpha_{i,2}^{rev}\alpha_{i,3}\alpha_{i,3}^{rev})$, and thus w_i can be generated from (p', R') by replacing all variables in $\alpha_{j,k}$, $j \neq i$, $1 \leq k \leq 3$, with the empty string.

To verify $\varepsilon L_{rev}(p', R') \subseteq \varepsilon L(p, R)$, it suffices to provide a reversal-obedient morphism $\Phi : X^* \longrightarrow X^*$ such that $\Phi(\overline{p, R}) = \overline{(p', R')}$. The existence of such Φ is obvious from our construction, namely $\Phi(x_k) = \alpha_{1,k}\alpha_{2,k}\dots\alpha_{n,k}$ for $1 \leq k \leq 3$.

Finally, we prove $\varepsilon L_{rev}(p', R') \subset \varepsilon L_{rev}(p, R)$. By way of contradiction, using Lemma 1, suppose there is a morphism $\Psi : X^* \longrightarrow X^*$ with $\Psi(\overline{p', R'}) = \overline{(p, R)}$.

By construction, (p', R') has no free variables, i.e., the group of each variable in p' has size at least 2. Hence we can decompose Ψ into two morphisms ψ and ψ' such that $\Psi(\overline{p', R'}) = \psi'(\psi(\overline{p', R'})) = \overline{(p, R)}$ and, for each $v_j \in Var(p')$: (i) $\psi(v_j) = \varepsilon$, if $|[v_j]| > 2$, (ii) $\psi(v_j) = v_j$, if $|[v_j]| = 2$. Since $\overline{(p', R')} = \Phi(\overline{p, R})$ and either $[\Phi(x_3)] = \emptyset$ or $|[\Phi(x_3)]| \geq 4$ (see cases (*i*) and (*ii*)), we conclude that $\psi(\Phi(x_3)) = \varepsilon$. Therefore,

$$\psi(\overline{p', R'}) = \overbrace{v_{j_1}v_{j_2}\dots v_{j_t}}^{\psi(\Phi(x_1))}\overbrace{v_{j_t}^{rev}v_{j_{t-1}}^{rev}\dots v_{j_1}^{rev}}^{\psi(\Phi(x_1^{rev}))}$$
$$\overbrace{v_{j_{t+1}}v_{j_{t+2}}\dots v_{j_{t+\ell}}}^{\psi(\Phi(x_2))}\overbrace{v_{j_{t+\ell}}^{rev}v_{j_{t+\ell-1}}^{rev}\dots v_{j_{t+1}}^{rev}}^{\psi(\Phi(x_2^{rev}))}\overbrace{\varepsilon}^{\psi(\Phi(x_3))}\overbrace{\varepsilon}^{\psi(\Phi(x_3^{rev}))}$$

where $t, \ell \geq 1$ and $v_{j_\zeta} \neq v_{j_\eta}$ for $\zeta \neq \eta$, $1 \leq \zeta, \eta \leq t + \ell$. To obtain a pattern with the same Parikh vector as of (p, R), ψ' must replace all except six variables with ε. By the structure of $\psi(\overline{p', R'})$, the only possible options up to renaming of variables are $\overline{(p_1, R_1)} := x_1x_2x_2^{rev}x_1^{rev}x_3x_3^{rev}$, $\overline{(p_2, R_2)} := x_1x_1^{rev}x_2x_3x_3^{rev}x_2^{rev}$ and $\overline{(p_3, R_3)} := x_1x_2x_3x_3^{rev}x_2^{rev}x_1^{rev}$. None of these are equivalent to (p, R). This is because there are words of the form $a^{2\kappa}bba^{2\xi}$ with $\kappa > \xi > 0$ that witness $\varepsilon L_{rev}(p, R) \not\subseteq \varepsilon L_{rev}(p_1, R_1)$, and $\varepsilon L_{rev}(p, R) \not\subseteq \varepsilon L_{rev}(p_3, R_3)$; with $0 < \kappa < \xi$ we obtain witnesses for $\varepsilon L_{rev}(p, R) \not\subseteq \varepsilon L_{rev}(p_2, R_2)$. Hence there is no morphism ψ' such that $(p, R) = \psi'(\psi((p', R')))$, and no Ψ such that $\Psi(\overline{(p', R')}) = \overline{(p, R)}$. This is a contradiction and therefore $\varepsilon L_{rev}(p', R') \subset \varepsilon L_{rev}(p, R)$. □

Example 1. Consider the set $T := \{w_1, w_2, w_3, w_4\}$ with

$$w_1 = \underbrace{a}_{\theta_1(x_1)}\underbrace{a}_{\theta_1(x_1^{rev})}\underbrace{a}_{\theta_1(x_2)}\underbrace{a}_{\theta_1(x_2^{rev})}\underbrace{b}_{\theta_1(x_3)}\underbrace{b}_{\theta_1(x_3^{rev})}$$

$$w_2 = \underbrace{ba}_{\theta_2(x_1)} \underbrace{ab}_{\theta_2(x_1^{rev})} \underbrace{a}_{\theta_2(x_2)} \underbrace{a}_{\theta_2(x_2^{rev})} \underbrace{ab}_{\theta_2(x_3)} \underbrace{ba}_{\theta_2(x_3^{rev})}$$

$$w_3 = \underbrace{b}_{\theta_3(x_1)} \underbrace{b}_{\theta_3(x_1^{rev})} \underbrace{bb}_{\theta_3(x_2)} \underbrace{bb}_{\theta_3(x_2^{rev})} \underbrace{bb}_{\theta_3(x_3)} \underbrace{bb}_{\theta_3(x_3^{rev})}$$

$$w_4 = \underbrace{aab}_{\theta_4(x_1)} \underbrace{baa}_{\theta_4(x_1^{rev})} \underbrace{b}_{\theta_4(x_2)} \underbrace{b}_{\theta_4(x_2^{rev})} \underbrace{bab}_{\theta_4(x_3)} \underbrace{bab}_{\theta_4(x_3^{rev})}$$

A pattern $\overline{(p', R')}$ generating these four strings and generating a proper subset of $\varepsilon L(x_1 x_1^{rev} x_2 x_2^{rev} x_3 x_3^{rev})$ is constructed as follows.

$$\alpha_1 = \underbrace{\overbrace{x_1 x_1}^{\alpha_{1,1}}}_{\overleftarrow{\theta_1}(a\ a)} \underbrace{\overbrace{x_1 x_1}^{\bar{\alpha}_{1,1}}}_{\overleftarrow{\theta_1}(a\ a)} \underbrace{\overbrace{x_2}^{\alpha_{1,2}}}_{\overleftarrow{\theta_1}(b)} \underbrace{\overbrace{x_2}^{\bar{\alpha}_{1,2}}}_{\overleftarrow{\theta_1}(b)} \overbrace{\varepsilon}^{\alpha_{1,3}} \overbrace{\varepsilon}^{\bar{\alpha}_{1,3}} \qquad \alpha_2 = \underbrace{\overbrace{x_4 x_3}^{\alpha_{2,1}}}_{\overleftarrow{\theta_2}(ba)} \underbrace{\overbrace{x_3 x_4}^{\bar{\alpha}_{2,1}}}_{\overleftarrow{\theta_2}(ab)} \underbrace{\overbrace{x_3}^{\alpha_{2,2}}}_{\overleftarrow{\theta_2}(a)} \underbrace{\overbrace{x_3}^{\bar{\alpha}_{2,2}}}_{\overleftarrow{\theta_2}(a)} \underbrace{\overbrace{x_3 x_4}^{\alpha_{2,3}}}_{\overleftarrow{\theta_2}(ab)} \underbrace{\overbrace{x_4 x_3}^{\bar{\alpha}_{2,3}}}_{\overleftarrow{\theta_2}(ba)}$$

$$\alpha_3 = \underbrace{\overbrace{x_6}^{\alpha_{3,1}}}_{\overleftarrow{\theta_3}(b)} \underbrace{\overbrace{x_6}^{\bar{\alpha}_{3,1}}}_{\overleftarrow{\theta_3}(b)} \underbrace{\overbrace{x_6 x_6}^{\alpha_{3,2}}}_{\overleftarrow{\theta_3}(bb)} \underbrace{\overbrace{x_6 x_6}^{\bar{\alpha}_{3,2}}}_{\overleftarrow{\theta_3}(bb)} \underbrace{\overbrace{x_6 x_6}^{\alpha_{3,3}}}_{\overleftarrow{\theta_3}(bb)} \underbrace{\overbrace{x_6 x_6}^{\bar{\alpha}_{3,3}}}_{\overleftarrow{\theta_3}(bb)} \qquad \alpha_4 = \underbrace{\overbrace{x_7 x_7 x_8}^{\alpha_{4,1}}}_{\overleftarrow{\theta_4}(aab)} \underbrace{\overbrace{x_8 x_7 x_7}^{\bar{\alpha}_{4,1}}}_{\overleftarrow{\theta_4}(baa)} \underbrace{\overbrace{x_8}^{\alpha_{4,2}}}_{\overleftarrow{\theta_4}(b)} \underbrace{\overbrace{x_8}^{\bar{\alpha}_{4,2}}}_{\overleftarrow{\theta_4}(b)} \underbrace{\overbrace{x_8 x_7 x_8}^{\alpha_{4,3}}}_{\overleftarrow{\theta_4}(bab)} \underbrace{\overbrace{x_8 x_7 x_8}^{\bar{\alpha}_{4,3}}}_{\overleftarrow{\theta_4}(bab)}$$

Hence, the pattern $\overline{(p', R')}$ is of the form:

$$\overline{(p', R')} = \underbrace{\overbrace{x_1 x_1}^{\alpha_{1,1}} \overbrace{x_4 x_3}^{\alpha_{2,1}} \overbrace{x_6}^{\alpha_{3,1}} \overbrace{x_7 x_7 x_8}^{\alpha_{4,1}}}_{\Phi(x_1)} \underbrace{\overbrace{x_8^{rev} x_7^{rev} x_7^{rev}}^{\alpha_{4,1}^{rev}} \overbrace{x_6^{rev}}^{\alpha_{3,1}^{rev}} \overbrace{x_3^{rev} x_4^{rev}}^{\alpha_{2,1}^{rev}} \overbrace{x_1^{rev} x_1^{rev}}^{\alpha_{1,1}^{rev}}}_{\Phi(x_1^{rev})}$$

$$\underbrace{\overbrace{x_2}^{\alpha_{1,2}} \overbrace{x_3}^{\alpha_{2,2}} \overbrace{x_6 x_6}^{\alpha_{3,2}} \overbrace{x_8}^{\alpha_{4,2}}}_{\Phi(x_2)} \underbrace{\overbrace{x_8^{rev}}^{\alpha_{4,2}^{rev}} \overbrace{x_6^{rev} x_6^{rev}}^{\alpha_{3,2}^{rev}} \overbrace{x_3^{rev}}^{\alpha_{2,2}^{rev}} \overbrace{x_2^{rev}}^{\alpha_{1,2}^{rev}}}_{\Phi(x_2^{rev})}$$

$$\underbrace{\overbrace{x_3 x_4}^{\alpha_{2,3}} \overbrace{x_6 x_6}^{\alpha_{3,3}} \overbrace{x_8 x_7 x_8}^{\alpha_{4,3}}}_{\Phi(x_3)} \underbrace{\overbrace{x_8^{rev} x_7^{rev} x_8^{rev}}^{\alpha_{4,3}^{rev}} \overbrace{x_6^{rev} x_6^{rev}}^{\alpha_{3,3}^{rev}} \overbrace{x_4^{rev} x_3^{rev}}^{\alpha_{2,3}^{rev}}}_{\Phi(x_3^{rev})}$$

Obviously, $S = \{s_1, s_2, s_3, s_4\} \subseteq \varepsilon L(p', R')$ and $\varepsilon L(p', R') \subseteq \varepsilon L(p, R)$. However, $aa\ bb\ aaaa \in \varepsilon L(p, R) \setminus \varepsilon L(p', R')$.

4 Equal-Length Relation

This section assumes $r = len$, i.e., variables in relation are replaced by words of equal length, independent of the actual symbols in those words. Recent work by Holte et al. [9] implies that, for non-erasing pattern languages, substitutions that replace each variable with a word of length at most 2 form positive characteristic sets, if the underlying alphabet has at least three symbols. In particular, without using the term "positive characteristic sets", Holte et al. showed that, if $|\Sigma| \geq 3$

and $\mathcal{S}_2(p,R) \subseteq L_{len}(p',R')$, then $L_{len}(p,R) \subseteq L_{len}(p',R')$. Here $L_{len}(p,R)$ refers to the non-erasing language generated by (p,R), i.e., the subset of $\varepsilon L_{len}(p,R)$ that results from applying only valid R-substitutions φ to p that do not erase any variable, i.e., $|\varphi(x)| \geq 1$ for all $x \in Var(p)$. Moreover, $\mathcal{S}_2(p,R)$ denotes the set of words in $L_{len}(p,R)$ that are generated by R-substitutions φ satisfying $\exists x \in Var(p)\ [\forall y \in [x]\ |\varphi(y)| = 2\ \wedge\ \forall y \in Var(p) \setminus [x]\ |\varphi(y)| = 1]$. Note that the positive characteristic sets $\mathcal{S}_2(p,R)$ used by Holte et al. are of size linear in $|p|$.

For learning-theoretic purposes, by contrast with $r = eq$ or $r = rev$, there is no real technical difference between the erasing and the non-erasing case when $r = len$. In particular, the result by Holte et al. carries over to the erasing case without any substantial change to the proof. Where substitutions replacing variables with strings of length either 1 or 2 were crucial in Holte et al.'s result, in the erasing case, it is sufficient to consider substitutions replacing variables with strings of length either 0 or 1. To formulate this result, we first define, for any relational pattern (p,R) and any $\ell \in \mathbb{N} \setminus \{0\}$:

$$\mathcal{S}_{\varepsilon,\ell}(p,R) = \{\varphi(p) \mid \exists x \in Var(p)\ [\forall y \in [x]\ |\varphi(y)| = \ell \wedge \forall y \notin [x]\ \varphi(y) = \varepsilon]\}.$$

Proposition 3. *Let $|\Sigma| \geq 3$. If $\mathcal{S}_{\varepsilon,1}(p,R) \subseteq \varepsilon L_{len}(p',R')$, then $\varepsilon L_{len}(p,R) \subseteq \varepsilon L_{len}(p',R')$. In particular, $\mathcal{S}_{\varepsilon,1}(p,R)$ is a linear-size positive characteristic set for $\varepsilon L_{len}(p,R)$ with respect to $\varepsilon\mathcal{L}_{len}$.*

Proof. Holte et al. [9] showed for the non-erasing case: If $\mathcal{S}_2(p,R) \subseteq L_{len}(p',R')$, then $L_{len}(p,R) \subseteq L_{len}(p',R')$. Their proof carries over directly to obtain: If $\mathcal{S}_{\varepsilon,1}(p,R) \subseteq \varepsilon L_{len}(p',R')$, then $\varepsilon L_{len}(p,R) \subseteq \varepsilon L_{len}(p',R')$.

Now suppose $\mathcal{S}_{\varepsilon,1}(p,R) \subseteq \varepsilon L_{len}(p',R')$ and $\mathcal{S}_{\varepsilon,1}(p',R') \subseteq \varepsilon L_{len}(p,R)$. Then we obtain $\varepsilon L_{len}(p,R) \subseteq \varepsilon L_{len}(p',R')$ and $\varepsilon L_{len}(p',R') \subseteq \varepsilon L_{len}(p,R)$, so that $\varepsilon L_{len}(p,R) = \varepsilon L_{len}(p',R')$. This implies that $\mathcal{S}_{\varepsilon,1}(p,R)$ is a positive characteristic set for $\varepsilon L_{len}(p,R)$ with respect to $\varepsilon\mathcal{L}_{len}$. □

The premise $|\Sigma| \geq 3$ in this result (both in our Proposition 3 and in Holte et al.'s result for the non-erasing case) plays a crucial role in the proof. This raises the question whether the result generalizes to binary alphabets. The main contribution of this section is to show that Proposition 3 holds for binary alphabets if we replace $\mathcal{S}_{\varepsilon,1}$ by $\mathcal{S}_{\varepsilon,2}$.

Theorem 2. *Let $|\Sigma| = 2$. If $\mathcal{S}_{\varepsilon,2}(p,R) \subseteq \varepsilon L_{len}(p',R')$, then $\varepsilon L_{len}(p,R) \subseteq \varepsilon L_{len}(p',R')$. In particular, $\mathcal{S}_{\varepsilon,2}(p,R)$ is a linear-size positive characteristic set for $\varepsilon L_{len}(p,R)$ with respect to $\varepsilon\mathcal{L}_{len}$.*

The remainder of this section is devoted to proving Theorem 2. We begin with the erasing version of [9, Lemma 8]; its proof is straightforward:

Lemma 2. *Suppose (p,R) is a relational pattern, where $p = q_1 x_i x_{i+1} q_2$ with $q_1, q_2 \in (\Sigma \cup X)^*$. Then $\varepsilon L_{len}(p,R) = \varepsilon L_{len}(p,R')$, where R' is obtained from R by swapping x_i with x_{i+1} in all relations.*

Using Lemma 2, we define a canonical way of representing relational pattern languages where the underlying relation is len.

Definition 3. *Suppose (p, R) is a pattern where p is of the general form $p = \vec{x}_1\omega_1\vec{x}_2\omega_2\ldots\vec{x}_n\omega_n\vec{x}_{n+1}$, with $n \geq 0$, $\vec{x}_1, \vec{x}_{n+1} \in X^*$, $\vec{x}_2, \ldots, \vec{x}_n \in X^+$, and $\omega_1, \ldots, \omega_n \in \Sigma^+$.*[1] *Suppose $[y_1], \ldots, [y_\kappa]$ are all the groups in (p, R), listed in a fixed order. The* equal-length normal form *of (p, R) is given by (p_{nf}, R) where*

$$p_{\text{nf}} = \vec{x}_1(1)\ldots\vec{x}_1(\kappa)\omega_1\ldots\omega_{n-1}\vec{x}_n(1)\ldots\vec{x}_n(\kappa)\omega_n\vec{x}_{n+1}(1)\ldots\vec{x}_n(\kappa),$$

where $\vec{x}_i(j)$ is the unique[2] *string of all variables in $Var(\vec{x}_i) \cap [y_j]$, written in increasing order of their indices in $X = \{x_1, x_2, \ldots\}$.*

By Lemma 2 we have $\varepsilon L_{len}(p, R) = \varepsilon L_{len}(p_{\text{nf}}, R)$.

Example 2. Let $p = x_1x_2\ a\ x_3\ b\ x_4x_5x_6\ bba\ x_7x_8x_9\ aa$, $R = \{(x_1, x_2), (x_1, x_8), (x_3, x_7), (x_5, x_8), (x_4, x_6)\}$. The groups in (p, R) are $\{x_1, x_2, x_5, x_8\}$, $\{x_3, x_7\}$, $\{x_4, x_6\}$, $\{x_9\}$. Then $p_{\text{nf}} = x_1x_2\ a\ x_3\ b\ x_5x_4x_6\ bba\ x_8x_7x_9\ aa$.

The sets $\mathcal{S}_{\varepsilon,\ell}(p, R)$ for $\ell = 1, 2$ turn out to be a core tool for deciding the equivalence of erasing relational pattern languages under the equal-length relation. Lemmas 3 through 7 are special cases of Theorem 2 which state that, for binary alphabets and for patterns of specific shapes, these sets form systems of positive characteristic sets.

Lemma 3. *Let $\Sigma = \{a, b\}$, $a \neq b$, $n_i \in \mathbb{N}$, $\vec{x}_1, \vec{x}'_1, \vec{x}_3, \vec{x}'_3 \in X^*$, $\vec{x}_2, \vec{x}'_2 \in X^+$.*

1. *Let $p = \vec{x}_1a^{n_1}b^{n_2+n_3}\vec{x}_2a\vec{x}_3$ and $p' = \vec{x}'_1a^{n_1}b^{n_2}\vec{x}'_2b^{n_3}a\vec{x}'_3$. Fix R, R'. Then $\mathcal{S}_{\varepsilon,1}(p, R) \subseteq \varepsilon L_{len}(p', R')$ implies $\varepsilon L_{len}(p, R) \subseteq \varepsilon L_{len}(p', R')$.*
2. *Let $p = \vec{x}_1ab^{n_1}\vec{x}_2b^{n_2}a^{n_3}\vec{x}_3$ and $p' = \vec{x}'_1a\vec{x}'_2b^{n_1+n_2}a^{n_3}\vec{x}'_3$. Fix R, R'. Then $\mathcal{S}_{\varepsilon,1}(p', R') \subseteq \varepsilon L_{len}(p, R)$ implies $\varepsilon L_{len}(p', R') \subseteq \varepsilon L_{len}(p, R)$.*

Proof. Let $c \in \{1, 2, 3\}$. If $x \in Var(p)$, then $d_c(x) := |[x]_R \cap Var(\vec{x}_c)|$ denotes the number of positions in $\vec{x}_c$ that correspond to variables in $[x]_R$. Likewise, for $x \in Var(p')$, we write $d'_c(x) := |[x]_{R'} \cap Var(\vec{x}'_c)|$.

We only prove the first statement; the second one is proved analogously. So suppose $\mathcal{S}_{\varepsilon,1}(p, R) \subseteq \varepsilon L_{len}(p', R')$. Without loss of generality, assume (p, R) is in equal-length normal form. First, we will establish the following claim.

Claim. Let $x \in Var(p)$, and let $\alpha, \beta \in \mathbb{N}$ satisfy $\alpha + \beta = d_2(x)$. Then there is some $t > 0$ and a family $(y_i)_{1 \leq i \leq t}$ of variables in p' such that the vector $\langle d_1(x), \alpha, d_3(x) + \beta\rangle$ is a linear combination of $\langle d'_1(y_i), d'_2(y_i), d'_3(y_i)\rangle$, $1 \leq i \leq t$.

Proof of Claim. Since (p, R) is in equal-length normal form, $\vec{x}_2$ contains a (possibly empty) substring of the form $z_1 \ldots z_{d_2(x)}$, where $[x] \cap Var(\vec{x}_2) = \{z_1, \ldots, z_{d_2(x)}\}$. Let ϕ be the substitution that erases every variable not contained in $[x]$, and replaces variables $y \in [x]$ as follows: if $y \in [x] \cap (Var(\vec{x}_1) \cup$

[1] Note that $n = 0$ results in a pattern in X^+.

[2] Uniqueness is given only modulo the fixed order $[y_1], \ldots, [y_\kappa]$ over all groups in p. There are canonical ways to define such an order, so that, for our purposes, we can consider the normal form of a pattern (p, R) to be unique.

$Var(\vec{x}_3)$) or if $y \in \{z_1, \ldots, z_\alpha\}$, then $\phi(y) = b$; if $y \in \{z_{\alpha+1}, \ldots, z_{d_2(x)}\}$, then $\phi(y) = a$. This substitution is valid for R; applied to p it generates the word

$$\phi(p) = b^{d_1(x)}\ a^{n_1}\ b^{n_2+n_3+\alpha}\ a^{\beta+1}\ b^{d_3(x)}\,.$$

By definition of ϕ, we have $\phi(p) \in \mathcal{S}_{\varepsilon,1}(p, R)$ and thus $\phi(p) \in \varepsilon L(p', R')$. Thus, there exists an R'-substitution ϕ' such that $\phi'(p') = \phi(p)$. Due to the shape of p', this implies $\phi'(\vec{x}_1') = b^{d_1(x)}$, $\phi'(\vec{x}_2') = b^{\alpha}$, and $\phi'(\vec{x}_3') = a^{\beta} b^{d_3(x)}$. For ϕ' to be valid for R', this entails the existence of a family $(y_i)_{1 \le i \le t}$ of variables in p' such that the vector $\langle d_1(x), \alpha, d_3(x) + \beta\rangle$ is a linear combination of the vectors $\langle d_1'(y_i), d_2'(y_i), d_3'(y_i)\rangle$, $1 \le i \le t$. □*(Claim.)*

With this claim in hand, we can prove $\varepsilon L(p, R) \subseteq \varepsilon L(p', R')$. To this end, let $w \in \varepsilon L(p, R)$. There are $w_1, w_2, w_3 \in \Sigma^*$ such that $w = w_1\ a^{n_1}\ b^{n_2+n_3}\ w_2\ a\ w_3$. Now define α and β as follows. If $w_2 \in \{b\}^*$, then $\alpha = |w_2|$. Otherwise, α is fixed such that $w_2 = b^{\alpha} a w_2'$ for some $w_2' \in \Sigma^*$. In either case, $\beta := |w_2| - \alpha = |w_2'| + 1$.

Since $w \in \varepsilon L(p, R)$, there is some $u > 0$ and a family $(z_j)_{1 \le j \le u}$ of variables in p such that the vector $\langle |w_1|, |w_2|, |w_3|\rangle$ is a linear combination of the vectors $\langle d_1(z_j), d_2(z_j), d_3(z_j)\rangle$, $1 \le j \le u$. Now choose $\alpha_1, \ldots, \alpha_u \ge 0$ such that $\alpha_j \le d_2(z_j)$ for all j, and let $\beta_j := d_2(z_j) - \alpha_j$. By the claim above, there is a family $(y_i^j)_{1 \le i \le t_j}$ of variables in p' such that the vector $\langle d_1(z_j), \alpha_j, d_3(z_j) + \beta_j\rangle$ is a linear combination of the vectors $\langle d_1'(y_i^j), d_2'(y_i^j), d_3'(y_i^j)\rangle$, $1 \le i \le t_j$. Thus, the vector $\langle |w_1|, \alpha, |w_3| + \beta\rangle$ is a linear combination of the vectors $\langle d_1'(y_i^j), d_2'(y_i^j), d_3'(y_i^j)\rangle$, $1 \le j \le u$, $1 \le i \le t_j$. Hence $w \in \varepsilon L(p', R')$, which completes the proof. □

Using similar techniques, we obtain the following series of lemmas, whose proofs are omitted due to space constraints.

Lemma 4. *Let* $\Sigma = \{a, b\}$, $a \neq b$, $n_1, n_2 \in \mathbb{N}$, $\vec{x}_1, \vec{x}_1', \vec{x}_3, \vec{x}_3' \in X^*$, $\vec{x}_2, \vec{x}_2' \in X^+$.

- *Let* $p = \vec{x}_1 a^{n_1} b^{n_2} \vec{x}_2 a^{n_3} \vec{x}_3$ *and* $p' = \vec{x}_1' a^{n_1} b^{n_2-1} \vec{x}_2' b a^{n_3} \vec{x}_3'$. *Fix* R, R'. *Then* $\mathcal{S}_{\varepsilon,1}(p, R) \subseteq \varepsilon L_{len}(p', R')$ *implies* $\varepsilon L_{len}(p, R) \subseteq \varepsilon L_{len}(p', R')$.
- *Let* $p = \vec{x}_1 a^{n_1} b \vec{x}_2 b^{n_2-1} a^{n_3} \vec{x}_3$, *and* $p' = \vec{x}_1' a^{n_1} \vec{x}_2' b^{n_2} a^{n_3} \vec{x}_3'$. *Fix* R, R'. *Then* $\mathcal{S}_{\varepsilon,1}(p', R') \subseteq \varepsilon L_{len}(p, R)$ *implies* $\varepsilon L_{len}(p', R') \subseteq \varepsilon L_{len}(p, R)$.

Lemma 5. *Let* $\Sigma = \{a, b\}$ *and* $a \neq b$. *Let* $p = \vec{x}_1 a^{n_1} b^{n_2} \vec{x}_2 a^{n_3} \vec{x}_3$ *and* $p' = \vec{x}_1' a^{n_1} \vec{x}_2' b^{n_2} a^{n_3} \vec{x}_3'$, *where* $n_1, n_2, n_3 \in \mathbb{N}$, $\vec{x}_1, \vec{x}_1', \vec{x}_3, \vec{x}_3' \in X^*$ *and* $\vec{x}_2, \vec{x}_2' \in X^+$. *Let* R *and* R' *be arbitrary relations over* X. *Then:*

1. $\mathcal{S}_{\varepsilon,2}(p, R) \subseteq \varepsilon L_{len}(p', R')$ *implies* $\varepsilon L_{len}(p, R) \subseteq \varepsilon L_{len}(p', R')$.
2. $\mathcal{S}_{\varepsilon,2}(p', R') \subseteq \varepsilon L_{len}(p, R)$ *implies* $\varepsilon L_{len}(p', R') \subseteq \varepsilon L_{len}(p, R)$.

Lemma 6. *Suppose* $p = \vec{x}_1 a^{n_1} b^{m_1} \vec{x}_2 a^{n_2} b^{m_2} \ldots \vec{x}_t a^{n_t} b^{m_t} \vec{x}_{t+1}$ *and*

$$p' = \vec{x}_1' a^{n_1} \vec{x}_2' b^{m_1} a^{n_2} \ldots \vec{x}_t' b^{m_{t-1}} a^{n_t} \vec{x}_{t+1}' b^{m_t} \vec{x}_{t+2}'\,,$$

where $n_i, m_j \in \mathbb{N}$, $\vec{x}_k, \vec{x}_{k'}' \in X^+$ *and* $\vec{x}_1, \vec{x}_1', \vec{x}_{t+1}, \vec{x}_{t+2}' \in X^*$, *for all* $i, j \in \{1, \ldots, t\}$, *all* $k \in \{2, \ldots, t\}$, *and all* $k' \in \{2, \ldots, t+1\}$. *Suppose* ($n_1 = 1$ *or* $m_1 = 1$), ($n_t = 1$ *or* $m_{t+1} = 1$), *and for each* $1 \le i \le (t-1)$, $m_i = 1$ *or* $n_{i+1} = 1$. *Then, for any two relations* R *and* R' *on* X, $\mathcal{S}_{\varepsilon,2}(p, R) \subseteq \varepsilon L_{len}(p', R') \subseteq \varepsilon L_{len}(p, R)$ *implies* $\varepsilon L_{len}(p, R) = \varepsilon L_{len}(p', R')$.

Lemma 6 motivates the following definition.

Definition 4. *Let* $t \geq 1$. *Let* $\pi = \vec{x}_1 a^{n_1} b^{m_1} \vec{x}_2 a^{n_2} b^{m_2} \ldots \vec{x}_t a^{n_t} b^{m_t} \vec{x}_{t+1}$ *and*

$$\pi' = \vec{x}'_1 a^{n_1} \vec{x}'_2 b^{m_1} a^{n_2} \ldots \vec{x}'_t b^{m_{t-1}} a^{n_t} \vec{x}'_{t+1} b^{m_t} \vec{x}'_{t+2},$$

where $m_1, n_i, m_j \in \mathbb{N}\backslash\{0\}$, $n_1, n_t, m_t \in \mathbb{N}$, $\vec{x}_k, \vec{x}'_{k'} \in X^+$, *and* $\vec{x}_1, \vec{x}'_1, \vec{x}_{t+1}, \vec{x}'_{t+2} \in X^*$, *for all* $i, j \in \{2, \ldots, t-1\}$, *all* $k \in \{2, \ldots, t\}$, *and all* $k' \in \{2, \ldots, t+1\}$.

Then π *and* π' *are called telltale conjugates provided that* ($n_1 = 1$ *or* $m_1 = 1$), ($n_t = 1$ *or* $m_t = 1$), *and* ($m_i = 1$ *or* $n_{i+1} = 1$) *for each* $1 \leq i \leq (t-1)$.

The main missing technical ingredient of the proof of Theorem 2 is the following lemma, which can be proven with a lengthy case analysis.

Lemma 7. *Let* (p, R) *and* (p', R') *be arbitrary patterns over* $\Sigma = \{a, b\}$. *Suppose there do not exist telltale conjugates* π, π' *such that* π *is a substring of* p *and* π' *is a substring of* p'. *Let* $\mathcal{S}_{\varepsilon,2}(p, R) \subseteq \varepsilon L_{len}(p', R') \subseteq \varepsilon L_{len}(p, R)$. *Then there exists an* $n \in \mathbb{N}$ *such that* $p, p' \in X^* \omega_1 X^+ \omega_2 \ldots X^+ \omega_n X^*$, *where* $\omega_1, \ldots, \omega_n \in \Sigma^+$. *In particular,* $L_{len}(p', R') = \varepsilon L_{len}(p, R)$.

Now, we can put everything together to prove Theorem 2.

Proof of Theorem 2. If p, p' do not contain any telltale conjugates as substrings then Theorem 2 holds by Lemma 7. If p, p' are telltale conjugates, then Theorem 2 holds by Lemma 6. Now, suppose p, p' contain telltale conjugates $\bar{p}$, $\bar{p'}$ as substrings. Let $p = \vec{x}_1 \overline{p} \vec{x}_2 \pi \vec{x}_3$ and $p' = \vec{x}'_1 \overline{p}' \vec{x}'_2 \pi' \vec{x}'_3$ where $\vec{x}_2, \vec{x}'_2 \in X^+$, $\vec{x}_1, \vec{x}'_1, \vec{x}_3, \vec{x}'_3 \in X^*$. Without loss of generality, the subpatterns $\overline{p}$, $\overline{p}'$, π, π' start and end with terminal letters. Theorem 2 holds, if π and π' are either identical or telltale conjugates. So suppose they are neither identical nor telltale conjugates.

Let $\pi = w_1 \vec{y}_1 w_2 \ldots \vec{y}_{n-1} w_n$ and $\pi' = w'_1 \vec{y}'_1 w'_2 \ldots \vec{y}'_{n'-1} w'_{n'}$ where $n, n' \geq 1$, $w_1, \ldots, w_n, w'_1, \ldots, w'_{n'} \in \Sigma^+$, and $\vec{y}_1, \ldots, \vec{y}_{n-1}, \vec{y}'_1, \ldots, \vec{y}'_{n'-1} \in X^+$. Now $w_1 \ldots w_n = w'_1 \ldots w'_{n'}$ (else the proof is completed as $\mathcal{S}_{\varepsilon,2}(p, R) \subseteq \varepsilon L_{len}(p', R') \subseteq \varepsilon L_{len}(p, R)$ is violated). Pick the minimum i such that $w_i \neq w'_i$. Without loss of generality, suppose $|w_i| < |w'_i|$. Select a variable $z \in Var(\vec{x}_i)$. Then, one can show that there is either an R-substitution θ such that $\theta(v) = \sigma \in \Sigma$ for $v \in [z]$ and $\theta(v') = \varepsilon$ for $v' \notin [z]$, or, an R-substitution ψ such that $\exists v_j \in [z]\ [\psi(v_j) \in \{\bar{\sigma}\sigma, \sigma\bar{\sigma}\} \wedge \forall v \in [z] \setminus \{v_j\}\ \psi(v) = \sigma^2]$, where $\bar{\sigma} \neq \sigma$, and $\psi(v') = \varepsilon$ for $v' \notin [z]$. Since $\theta(v_1 \ldots v_m) \in \{\sigma\}^+$ and $\psi(v_1 \ldots v_m) \in \{\sigma\}^+ \bar{\sigma} \{\sigma\}^+$, there is no other R'-substitution ϕ that can generate $\theta(p)$ or $\psi(p)$ by means of padding other variables. Since π and π' are not identical or telltale conjugates, then $\theta(p) \in \mathcal{S}_{\varepsilon,2}(p, R) \setminus \varepsilon L_{len}(p', R')$ or $\psi(p) \in \mathcal{S}_{\varepsilon,2}(p, R) \setminus \varepsilon L_{len}(p', R')$.

The case $p = \vec{x}_1 \overline{p} \vec{x}_2 \pi \vec{x}_3$ and $p' = \vec{x}'_1 \pi' \vec{x}'_2 \overline{p}' \vec{x}'_3$ and symmetric cases can be handled by complete analogy. This completes the proof. □

5 Conclusions

This paper provided insights into the connections between learning with characteristic sets and learning with telltales, via the notion of positive characteristic set. Its main contribution is to provide new (non-)learnability results for erasing relational pattern languages, again with a focus on positive characteristic sets.

To show that erasing pattern languages under the reversal relation have no positive characteristic sets (for binary alphabets), we used a construction that Reidenbach [13] devised to show the corresponding statement for the equality relation. An interesting open question from this construction is the following: Is it true that a relational pattern (p, R) has a telltale with respect to $\varepsilon\mathcal{L}_{rev}$ iff it has a telltale with respect to $\varepsilon\mathcal{L}_{eq}$? If yes, is there a constructive proof that shows how to transform telltales for either relation into telltales for the other?

Our main result states that, for binary alphabets, linear-size positive characteristic sets exist for erasing pattern languages under the equal length relation. A related open question is whether the same result holds true for the non-erasing case. Our proof makes use of the erasing property, and might need substantial adjustments for the non-erasing case.

Different real-world applications might give rise to different relations to study in the context of learning relational pattern languages. This paper analyzed only two of a multitude of possibilities; much future work is needed to gain a better understanding of the effects of various kinds of relations on learnability.

Acknowledgements. We thank R. Holte as well as the anonymous reviewers for helpful feedback. This work was supported by the Natural Sciences and Engineering Research Council of Canada (NSERC), both in the Canada Research Chairs program and in the Discovery Grants program. S. Zilles further acknowledges support through a Canada CIFAR AI Chair, held at the Alberta Machine Intelligence Institute (Amii).

References

1. Angluin, D.: Finding patterns common to a set of strings. J. Comput. Syst. Sci. **21**, 46–62 (1980)
2. Angluin, D.: Inductive inference of formal languages from positive data. Inf. Control **45**, 117–135 (1980)
3. Arikawa, S., Miyano, S., Shinohara, A., Kuhara, S., Mukouchi, Y., Shinohara, T.: A machine discovery from amino acid sequences by decision trees over regular patterns. New Gener. Comput. **11**, 361–375 (1993)
4. Clifford, R., Harrow, A.W., Popa, A., Sach, B.: Generalised matching. In: SPIRE, pp. 295–301 (2009)
5. Fallat, S., Kirkpatrick, D., Simon, H., Soltani, A., Zilles, S.: On batch teaching without collusion. J. Mach. Learn. Res. **24**, 1–33 (2023)
6. Geilke, M., Zilles, S.: Learning relational patterns. In: ALT, pp. 84–98 (2011)
7. Gold, E.: Language identification in the limit. Inf. Control **10**, 447–474 (1967)
8. de la Higuera, C.: Characteristic sets for polynomial grammatical inference. Mach. Learn. **27**(2), 125–138 (1997)
9. Holte, R.C., Mousawi, S.M., Zilles, S.: Distinguishing relational pattern languages with a small number of short strings. In: ALT, pp. 498–514 (2022)
10. Jiang, T., Salomaa, A., Salomaa, K., Yu, S.: Decision problems for patterns. J. Comput. Syst. Sci. **50**(1), 53–63 (1995)
11. Lange, S., Zeugmann, T., Zilles, S.: Learning indexed families of recursive languages from positive data: a survey. Theor. Comput. Sci. **397**(1–3), 194–232 (2008)

12. Nix, R.: Editing by example. ACM Trans. Program. Lang. Syst. **7**, 600–621 (1985)
13. Reidenbach, D.: A negative result on inductive inference of extended pattern languages. In: ALT, pp. 308–320 (2002)
14. Shinohara, T.: Polynomial time inference of extended regular pattern languages. In: Proceedings of RIMS Symposia on Software Science and Engineering, pp. 115–127 (1982)
15. Zhu, X., Singla, A., Zilles, S., Rafferty, A.: An overview of machine teaching. ArXiv arxiv:1801.05927 (2018)

Algorithms and Turing Kernels for Detecting and Counting Small Patterns in Unit Disk Graphs

Jesper Nederlof and Krisztina Szilágyi(✉)

Utrecht University, Utrecht, The Netherlands
{j.nederlof,k.szilagyi}@uu.nl

Abstract. In this paper we investigate the parameterized complexity of the task of counting and detecting occurrences of small patterns in unit disk graphs: Given an n-vertex unit disk graph G with an embedding of ply p (that is, the graph is represented as intersection graph with closed disks of unit size, and each point is contained in at most p disks) and a k-vertex unit disk graph P, count the number of (induced) copies of P in G.

For general patterns P, we give an $2^{O(pk/\log k)}n^{O(1)}$ time algorithm for counting pattern occurrences. We show this is tight, even for ply $p = 2$ and $k = n$: any $2^{o(n/\log n)}n^{O(1)}$ time algorithm violates the Exponential Time Hypothesis (ETH).

For most natural classes of patterns, such as connected graphs and independent sets we present the following results: First, we give an $(pk)^{O(\sqrt{pk})}n^{O(1)}$ time algorithm, which is nearly tight under the ETH for bounded ply and many patterns. Second, for $p = k^{O(1)}$ we provide a Turing kernelization (i.e. we give a polynomial time preprocessing algorithm to reduce the instance size to $k^{O(1)}$).

Our approach combines previous tools developed for planar subgraph isomorphism such as 'efficient inclusion-exclusion' from [Nederlof STOC'20], and 'isomorphisms checks' from [Bodlaender et al. ICALP'16] with a different separator hierarchy and a new bound on the number of non-isomorphic separations of small order tailored for unit disk graphs.

Keywords: Unit disk graphs · Subgraph isomorphism · Parameterized complexity

1 Introduction

A well-studied theme within the complexity of computational problems on graphs is how much structure within inputs allows faster algorithms. One of the most active research directions herein is to assume that input graphs are

Supported by the project CRACKNP that has received funding from the European Research Council (ERC) under the European Union's Horizon 2020 research and innovation programme (grant agreement No 853234).

H. Fernau et al. (Eds.): SOFSEM 2024, LNCS 14519, pp. 413–426, 2024.
https://doi.org/10.1007/978-3-031-52113-3_29

geometrically structured. The (arguably) two most natural and commonly studied variants of this are to assume that the graph can be drawn on $\mathbb{R}^2$ without crossings (i.e., it is planar) or it is the intersection graph of simple geometric objects. While this last assumption can amount to a variety of different models, a canonical and most simple model is that of *unit disk graphs*: Each vertex of the graph is represented by a disk of unit size and two vertices are adjacent if and only if the two associated disks intersect.

The computational complexity of problems on planar graphs has been a very fruitful subject of study: It led to the development of powerful tools such as *Bakers layering technique* and bidimensionality that gave rise to efficient approximation schemes and fast (parameterized) sub-exponential time algorithms for many NP-complete problems. One interesting example of such an NP-complete problem is *(induced) subgraph isomorphism*: Given a k-vertex pattern P and n-vertex host graph G, detect or count the number of (induced) copies of P inside G, denoted with $\text{sub}(P, G)$ (respectively, $\text{ind}(P, G)$). Here we think of k as being much smaller than n, and therefore it is very interesting to obtain running times that are only exponential in k (i.e. *Fixed Parameter Tractable time*). This problem is especially appealing since it generalizes many natural NP-complete problems (such as Independent Set, Longest Path and Hamiltonian Cycle) in a natural way, but its generality poses significant challenges for the bidimensionality theory: It does not give sub-exponential time algorithms for this problem.

Only recently, it was shown in a combination of papers ([4,7] and subsequently [13]) that, on planar graphs, subgraph isomorphism can be solved in $2^{O(k/\log k)}$ time[1] for general patterns, and in $2^{\tilde{O}(\sqrt{k})}n^{O(1)}$ time for many natural pattern classes, complementing the lower bound of $2^{o(n/\log n)}$ time from [4] based on the Exponential Time Hypothesis (ETH). It was shown in [13] that (induced) pattern occurrences can even be *counted* in sub-exponential parameterized (i.e. $2^{o(k)}n^{O(1)}$) time.

Unfortunately, most of these methods do not immediately work for (induced) subgraph isomorphism on geometric intersection graphs: Even unit disk graphs of bounded ply[2] are not H-minor free for any graph H (which significantly undermines the bidimensionality theory approach), and unit disk graphs of unbounded ply can even have arbitrary large cliques. This hardness is inherent to the graph class: Independent Set is $W[1]$-hard on unit disk graphs and, unless ETH fails, and it has no $f(k)n^{o(\sqrt{k})}$-time algorithm for any computable function f [6, Theorem 14.34]. On the positive side, it was shown in [14] that for bounded expansion graphs and fixed patterns, the subgraph isomorphism problem can be solved in linear time, which implies that subgraph isomorphism is fixed parameter tractable on unit disk graphs; however, their method relies on Courcelle's theorem and hence the dependence of k in their running time is very large and far from optimal.

[1] We let $\tilde{O}()$ omit polylogarithmic factors in its argument.

[2] The ply of an embedded unit disk graph is defined as the maximum number of times any point of the plane is contained in a disk.

Therefore, a popular research topic has been to design such fast (parameterized) sub-exponential time algorithms for specific problems such as Independent Set, Hamiltonian Cycle and Steiner Tree [2,3,10,16].

In this paper we continue this research line by studying the complexity of the decision and counting version of (induced) subgraph isomorphism. While some general methods such as contraction decompositions [15] and pattern covering [12] were already designed for graph classes that include (bounded ply) unit disk graphs, the fine-grained complexity of the subgraph problem itself restricted to unit disk graphs has not been studied and is still far from being understood.

Our Results. To facilitate the formal statements of our results, we need the following definitions: Given two graphs P and G, we define

$$\mathrm{ind}(P,G) = \{f : V(P) \to V(G) : f \text{ is injective}, uv \in E(P) \Leftrightarrow f(u)f(v) \in E(G)\}$$
$$\mathrm{sub}(P,G) = \{f : V(P) \to V(G) : f \text{ is injective}, uv \in E(P) \Rightarrow f(u)f(v) \in E(G)\}$$

Our main theorem reads as follows:

Theorem 1. *There is an algorithm that takes as input unit disk graphs P and G on k vertices and n vertices respectively, together with disk embeddings of ply p. It outputs $|\mathrm{sub}(P,G)|$ and $|\mathrm{ind}(P,G)|$ in $(pk)^{O(\sqrt{pk})}\sigma_{O(\sqrt{pk})}(P)^2 n^{O(1)}$ time.*

In this theorem, the parameter σ_s is a somewhat technical parameter of the pattern graph P that is defined as follows:

Definition 1 ([13]). *Given a graph P, we say that (A, B) is a* separation *of P of order s if $A \cup B = V(P)$, $|A \cap B| = s$ and there are no edges between $A \setminus B$ and $B \setminus A$. We say that (A, B) and (C, D) are* isomorphic separations *of P if there is a bijection $f : V(P) \to V(P)$ such that*

- *For any $u, v \in V(P)$, $uv \in E(P) \Leftrightarrow f(u)f(v) \in E(P)$*
- *$f(A) = C$, $f(B) = D$,*
- *For any $u \in A \cap B$, $f(u) = u$*

We denote by $\mathcal{S}_s(P)$ a maximal set of pairwise non-isomorphic separations of P of order at most s. We define $\sigma_s(P) = |\mathcal{S}_s(P)|$ as the number of non-isomorphic separations of P of order at most s.

Note that $\sigma_s(P) \geq \binom{k}{s}s!$. For ply $p = O(1)$ and many natural classes of patterns such as independent sets, cycles, or grids, it is easy to see that $\sigma_{O(\sqrt{k})}(P)$ is at most $2^{\tilde{O}(\sqrt{k})}$ and therefore Theorem 1 gives a $2^{\tilde{O}(\sqrt{k})}n^{O(1)}$ time algorithm for computing $|\mathrm{sub}(P,G)|$ and $|\mathrm{ind}(P,G)|$. We also give the following new non-trivial bounds on $\sigma_s(P)$ whenever P is a general (connected) unit disk graph[3]:

Theorem 2. *Let P be a k-vertex unit disk graph with given embedding of ply p, and s be an integer. Then: (a) $\sigma_s(P)$ is at most $2^{O(s\log k + pk/\log k)}$, and (b) If P is connected, then $\sigma_s(P) \leq 2^{O(s\log k)}$.*

[3] The proof of Theorem 2 can be found in the full version of the paper.

Using Theorem 1, this allows us to conclude the following result.

Corollary 1. *There is an algorithm that takes as input a k-vertex unit disk graph P and an n-vertex unit disk graph G, together with their unit disk embeddings of ply p, and outputs* $|\mathrm{sub}(P,G)|$ *and* $|\mathrm{ind}(P,G)|$ *in* $2^{O(pk/\log k)}n^{O(1)}$ *time.*

We show that this cannot be significantly improved even if $k = n$:

Theorem 3. *Assuming ETH, there is no algorithm to determine if* $\mathrm{sub}(P,G)$ *(*$\mathrm{ind}(P,G)$ *respectively) is nonempty for given n-vertex unit disk graphs P and G in $2^{o(n/\log n)}$ time, even when P and G have a given embedding of ply* 2.

Note that the ply of G is 1 if and only if G is an independent set so the assumption on the ply in the above statement is necessary. To our knowledge, this is the first lower bound based on the ETH excluding $2^{O(\sqrt{n})}$ time algorithms for problems on unit disks graphs (of bounded ply), in contrast to previous bounds that only exclude $2^{o(\sqrt{n})}$ time algorithms.

Clearly, a unit disk graph G of ply p has a clique of size p, i.e. its clique number $\omega(G)$ is at least p. On the other hand, it was shown in [8] that $p \geq \omega(G)/5$. In other words, parameterizations by ply and clique number are equivalent up to a constant factor.

1.1 Our Techniques

Our approach heavily builds on the previous works [4,7,13]: Theorem 1 is proved by using the dynamic programming technique from [4] that stores representatives of non-isomorphic separations to get the runtime dependence down from $2^{O(k)}$ to $\sigma_{O(\sqrt{pk})}(P)$, and the *Efficient Inclusion-Exclusion* technique from [13] to solve counting problems (on top of decision problems) as well.[4] We combine these techniques with a divide and conquer strategy that divides the unit disk graph in smaller graphs using horizontal and vertical lines as separators. As a first step in our proof, we give an interesting *Turing kernelization* for the counting problems that uses efficient inclusion-exclusion. Theorem 2 uses a proof strategy from [4] combined with a bound from [5] on the number of non-isomorphic unit disk graphs. Theorem 3 builds on a reduction from [4], although several alterations are needed to ensure the graph is a unit disk graph of bounded ply.

Organization. In Sect. 2 we provide additional notation and some preliminary lemmas. In Sect. 3 we provide a Turing kernel. In Sect. 4 we build on Sect. 3 to provide the proof of Theorem 1. In Sect. 5 we give an outline of the proof of Theorem 3. We finish with some concluding remarks in Sect. 6. The proofs that are omitted due to space restrictions (indicated with †) are provided in the full version on arXiv.

---

[4] Similar to what was discussed in [13], this technique seems to be needed even for simple special cases of Theorem 1 such as counting independent sets on subgraphs of (subdivided) grids.

2 Preliminaries

Notation. Given a graph G and a subset A of its vertices, we define $G[A]$ as the subgraph of G induced by A. Given a unit disk graph G, we say that G has *ply* p if there is an embedding of G such that every point in the plane is contained in at most p disks of G. Let P and G be unit disk graphs and let $|V(G)| = n$, $|V(P)| = k$.

We denote all vectors by bold letters, the all ones vector by $\mathbf{1}$ and the all zeros vector by $\mathbf{0}$. We use the Iverson bracket notation: for a statement P, we define $[P] = 1$ if P is true and $[P] = 0$ if P is false. We define $[k] = \{1, \ldots, k\}$. Given a function $f : A \to B$ and $C \subseteq A$, we define the restriction of f to C as $f|_C : C \to B$, $f|_C(c) = f(c)$ for all $c \in C$. If $g = f|_C$ for some C then we say that f *extends* g.

Definition 2. *For integers b, h, we say a unit disk graph G of ply p can be* drawn in a $(b \times h)$-box *with ply p if it has an embedding of ply p as unit disk graph in a $b \times h$ rectangle.*

Throughout this paper, we assume that the sides of the box are axis parallel, and the lower left corner is at $(0, 0)$. We assume that if a graph G can be drawn in a $(b \times h)$-box then we are given such an embedding.

Lemma 1 (†). *Given a unit disk graph G with a drawing in a $(b \times h)$-box with ply p, one can construct in polynomial time a path decomposition of G of width $4(\min\{b, h\} + 1)p$.*

Lemma 2 (Theorem 6.1 from [5]). *Let a non-decreasing bound $b = b(n)$ be given, and let $\mathcal{U}_n$ denote the set of unlabeled unit disk graphs on n vertices with maximum clique size at most b. Then $|\mathcal{U}_n| \leq 2^{12(b+1)n}$.*

Subroutines. The following lemma can be shown with standard dynamic programming over tree decompositions:

Lemma 3 (†). *Given P, G, a subset $P' \subseteq V(P)$, a path decomposition of G of width t, and a function $f : P' \to V(G)$, we can count $|\{g \in \mathrm{sub}(P, G) : g \text{ extends } f\}|$ and $|\{g \in \mathrm{ind}(P, G) : g \text{ extends } f\}|$ in time $\sigma_t(P) t^t n^{O(1)}$.*

The following lemma simply states that a long product of matrices can be evaluated quickly, but is nevertheless useful in a subroutine in the 'efficient inclusion-exclusion' technique.

Lemma 4 ([13]). *Given a set A, an integer h and a value $T[x, x'] \in \mathbb{Z}$ for every $x, x' \in A$, the value*

$$\sum_{x_1, \ldots, x_h \in A} \prod_{i=1}^{h-1} T[x_i, x_{i+1}] \tag{1}$$

can be computed in $O(h|A|^2)$ time.

Non-isomorphic Separations. In this paper we will work with non-isomorphic separations of small order, as defined in Definition 1. For separations $(C_1, D_1), (C_2, D_2) \in \mathcal{S}_s(P)$, we define

$$\mu((C_1, D_1), (C_2, D_2)) = |\{(C, D) : (C, D) \text{ is a separation of } P \text{ such that}$$
$$C \subseteq C_2 \text{ and } (C, D) \text{ isomorphic to } (C_1, D_1)\}|$$

Lemma 5 (†). *Given a graph P, one can compute $\mathcal{S}_s(P)$ and for each pair of separations $(C_1, D_1), (C_2, D_2) \in \mathcal{S}_s(P)$ the multiplicity $\mu((C_1, D_1), (C_2, D_2))$ in time $\sigma_s(P)n^{O(1)}$.*

3 Turing Kernel

We will now present a preprocessing algorithm for computing $|\text{sub}(P, G)|$ (the algorithm for $|\text{ind}(P, G)|$ is analogous) that allows us to assume that G can be drawn in a $(k \times k)$-box with ply p, i.e. that $|V(G)| \leq k^2 p$. This can be seen as a polynomial Turing kernel in case p and $\sigma_0(P)$ are polynomial in k. A *Turing kernel* of size f is an algorithm that solves the given problem in polynomial time, when given access to an oracle that solves instances of size at most $f(k)$ in a single step.

Lemma 6 describes how to reduce the width of G (and analogously the height of G). To prove it, we use the shifting technique. This general technique was first used by Baker [1] for covering and packing problems on planar graphs and by Hochbaum and Maass [9] for geometric problems stemming from VLSI design and image processing.

Intuitively, we draw the graph on a grid, and delete all the disks that intersect certain columns of the grid. After doing that, the remaining graph will consist of several small disconnected "building blocks". Each connected component of the pattern will be fully contained in one of the blocks, and since the blocks are small we can use the oracle to count the number of these occurrences. We take advantage of the fact that we can group together connected components that are isomorphic. We use Lemma 6 twice, to reduce the width and height of G to k.

Lemma 6. *Suppose we have access to an oracle that computes $|\text{sub}(P, G)|$ in constant time, where the host graph G can be drawn in a $(O(k) \times O(k))$-box with ply p. Then we can compute $|\text{sub}(P, G')|$ for host graphs G' that can be drawn in a box of height k with ply p in time $n \cdot \text{poly}(k) \cdot \sigma_0(P)^2$.*

Proof. For $i \in \{0, \ldots, k\}$ let $C_i = \{(x, y) \in \mathbb{R}^2 : x \equiv i \pmod{k+1}\}$. Informally, we draw a grid and select every $(k+1)$th vertical gridline. Let P_i be the set of all subgraphs of G that are isomorphic to P and are disjoint from C_i.

Note that every subgraph Q of G that is isomorphic to P is fully contained in at least one P_i. Indeed, every disk in Q can intersect at most one vertical grid line, so Q is disjoint from C_i for at least one value of i.

By the inclusion-exclusion principle, $\left|\bigcup_{i=0}^{k} P_i\right|$ equals

$$\sum_{\emptyset\subset C\subseteq\{0,\dots,k\}} (-1)^{|C|}\left|\bigcap_{i\in C} P_i\right| = \sum_{\ell=1}^{k+1}(-1)^{\ell} \sum_{0\le c_1<\dots<c_\ell\le k}\left|\bigcap_{j\in\{c_1,\dots,c_\ell\}} P_j\right|. \tag{2}$$

Let us show how we can compute $|\bigcap_{j\in\{c_1,\dots,c_\ell\}} P_j|$ quickly. For $a,b\in\{0,\dots,k\}$, we define $\mathcal{B}[a,b]$ as the subgraph of G contained by (open) stripes bounded by C_a and C_b. Formally, define $B[a,b]$ to be

$$\begin{cases}\{(x,y)\in\mathbb{R}^2:\ (\exists t\in\mathbb{N}_0)\ a+(k+1)t < x < b+(k+1)t\}, & \text{if } a\le b,\\ \{(x,y)\in\mathbb{R}^2:\ (\exists t\in\mathbb{N}_0)\ a+(k+1)(t-1) < x < b+(k+1)t\}, & \text{if } a> b.\end{cases}$$

We define $\mathcal{B}[a,b]$ as the induced subgraph of G such that all its disks are fully contained in $B[a,b]$. These sets are our "building blocks": after deleting $C_{c_1},\dots C_{c_\ell}$, the remaining graph is $\cup_{\alpha=1}^{\ell}\mathcal{B}[c_\alpha,c_{\alpha+1}]$, where we define $c_{\ell+1}=c_1$.

Let t be the number of non-isomorphic connected components of P and let $\mathcal{C}_0(P)=\{\mathcal{P}_1,\dots,\mathcal{P}_t\}$ be the set of representatives of all isomorphism classes of connected components of P. We can encode P as vector $\mathbf{p}=(p_1,\dots,p_t)$, where p_i is the size of the isomorphism class of $\mathcal{P}_i$.

Let $U=\{0,\dots,p_1\}\times\dots\times\{0,\dots,p_t\}$. For a t-dimensional vector $(v_1,\dots,v_t)\in U$ we define $P[(v_1,\dots,v_t)]$ as the subgraph of P that contains v_i copies of $\mathcal{P}_i$.

We would like to count in how many ways can we distribute the connected components of P to the building blocks. Equivalently, we can count the number of ways to assign a vector $\mathbf{v}^\alpha\in U$ to each block $\mathcal{B}[c_\alpha,c_{\alpha+1}]$ such that $\sum\mathbf{v}^\alpha=\mathbf{p}$.

Thus we have

$$\left|\bigcap_{j\in\{c_1,\dots,c_\ell\}} P_j\right| = \sum_{\mathbf{v}^1+\dots+\mathbf{v}^\ell=\mathbf{p}}\ \prod_{\alpha=1}^{\ell}|\mathrm{sub}(P[\mathbf{v}^\alpha],\mathcal{B}[c_\alpha,c_{\alpha+1}])|. \tag{3}$$

Note that $|U|=(p_1+1)\cdots(p_t+1)=\sigma_0(P)$: indeed, every vector $\mathbf{u}\in U$ corresponds to a unique separation $(V(P'),V(P-P'))$ of order 0, where P' consists of c_i copies of $\mathcal{P}_i$. Combining (2) and (3), we get that

$$\left|\bigcup_{i=0}^{k} P_i\right| = \sum_{\ell=1}^{k}(-1)^{\ell}T_\ell,$$

where

$$T_\ell = \sum_{\substack{0\le c_1<\dots<c_\ell\le k\\ \mathbf{v}^1+\dots+\mathbf{v}^\ell=\mathbf{p}}}\ \prod_{\alpha=1}^{\ell}|\mathrm{sub}(P[\mathbf{v}^\alpha],\mathcal{B}[c_\alpha,c_{\alpha+1}])|. \tag{4}$$

Suppose for now that we have computed $|\mathrm{sub}(P[\mathbf{v}^\alpha],\mathcal{B}[a,b])|$ for all $a,b\in\{0,\dots,k\}$, $\mathbf{v}^\alpha\in U$, and that we want to compute T_ℓ quickly.

To apply Lemma 4, we have to rewrite the sum (4) in such a way that the variables are pairwise independent. We replace the condition $c_i < c_{i+1}$ by multiplying with $[c_i < c_{i+1}]$. To replace the condition on the variables $\mathbf{v}^i$, we will re-index these variables by $\mathbf{u}^1, \ldots, \mathbf{u}^\ell$, where $\mathbf{u}^i = \sum_{j=1}^{i} \mathbf{v}^j$ for $i \in [\ell - 1]$ and $\mathbf{u}^\ell = \mathbf{p} - \mathbf{u}^{\ell-1}$, $\mathbf{u}^0 = \mathbf{0}$. Therefore, we have

$$T_\ell = \sum_{\substack{c_1,\ldots,c_\ell \in \{0,\ldots,k\} \\ \mathbf{u}^1,\ldots,\mathbf{u}^{\ell-1} \in U}} |\mathrm{sub}(P[\mathbf{p} - \mathbf{u}^{\ell-1}], \mathcal{B}[c_\ell, c_1])| \prod_{i=1}^{\ell-1} [c_i < c_{i+1}] \cdot [\mathbf{u}^{i-1} \leq \mathbf{u}^i]$$

$$\cdot |\mathrm{sub}(P[\mathbf{u}^i - \mathbf{u}^{i-1}], \mathcal{B}[c_i, c_{i+1}])|.$$

By Lemma 4, we can compute T_ℓ in time $\ell \cdot (k \cdot \sigma_0(P))^2$ if we are given $|\mathrm{sub}(P[\mathbf{u}], \mathcal{B}[a, b])|$ for all $\mathbf{u} \in U$, $a, b \in \{0, \ldots, k\}$.

It remains to show how we can compute $|\mathrm{sub}(P[\mathbf{u}], \mathcal{B}[a, b])|$ for given $\mathbf{u} \in U$, $a, b \in \{0, \ldots, k\}$. Let $C_1, \ldots, C_d$ be the connected components of $\mathcal{B}[a, b]$. Note that each C_i can be drawn in a $(k \times k)$-box with ply p, so we can use the oracle to compute $|\mathrm{sub}(P[\mathbf{w}], C_i)|$ for all $\mathbf{w} \in U$, $i \in [d]$. We would like to distribute the connected components of $P[\mathbf{u}]$ to $C_1, \ldots C_d$. We can do this by dynamic programming. For $i \in [d]$ and $\mathbf{w} \in U$, we define

$$T'[i, \mathbf{w}] = |\mathrm{sub}(P[\mathbf{w}], C_1 \cup \cdots \cup C_i)|$$

The recurrence is as follows:

$$T'[i, \mathbf{w}] = \sum_{\mathbf{w}' \leq \mathbf{w}} |\mathrm{sub}(P[\mathbf{w}'], C_i)| \cdot T[i - 1, \mathbf{w} - \mathbf{w}'],$$

where $\mathbf{w} \leq \mathbf{w}$**'** indicates that $\mathbf{w}$ is in each coordinate smaller than $\mathbf{w}$**'**. Thus we can compute $T'[i, \mathbf{u}]$ in time $d|U|^2 = d\sigma_0(P)^2 \leq (n/k)\sigma_0(P)^2$.

Therefore, we can compute $|\mathrm{sub}(P, G)|$ in time $n \cdot \mathrm{poly}(k) \cdot \sigma_0(P)^2$ for host graphs that can be drawn in a box of width k with ply p. □

Theorem 4. *For unit disk graphs P and G with given embedding of ply p, $|\mathrm{sub}(P, G)|$ can be computed in time $\sigma_0(P)^2 \cdot n \cdot \mathrm{poly}(k)$ when given access to an oracle that computes $|\mathrm{sub}(P, G)|$ where the host graph has size $O(k^2 p)$ in constant time. In particular, there is a Turing kernel for computing $|\mathrm{sub}(P, G)|$ when $\sigma_0(P)$ and p are polynomial in k.*

Proof. By Lemma 6, if G can be drawn in a box of width k with ply p, we can compute $|\mathrm{sub}(P, G)|$ in time $n \cdot \mathrm{poly}(k) \cdot \sigma_0(P)^2$. If not, we use the same approach as in the proof of Lemma 6. The obtained building blocks will be disjoint unions of subgraphs that can be drawn in a box of width k with ply p. Using dynamic programming and applying Lemma 6, we conclude that we can compute $|\mathrm{sub}(P, G)|$ in time $n \cdot \mathrm{poly}(k) \cdot \sigma_0(P)^2$. □

4 Proof of Theorem 1: The Algorithm

We present only the proof for $\text{sub}(P, G)$, since the proof for $\text{ind}(P, G)$ is analogous. Before we start with the proof, we need to give a number of definitions: Suppose that a unit disk embedding of G in a $(b \times h)$-box with ply p is given. For integers $0 \leq x \leq x' \leq b$, we define $G\langle x, x'\rangle$ as the induced subgraph of G whose vertex set consists are all vertices of G associated with disks in the unit disk embedding that are (partially) in between vertical lines x and x', i.e. the set of all disks that intersect the set $\{(a, b) \in \mathbb{R}^2 : x \leq a \leq x'\}$. We denote $G\langle x\rangle = G\langle x, x\rangle$.

Given functions $f_1 : D_1 \to R_1$ and $f_2 : D_2 \to R_2$, we say f_1 and f_2 are *compatible* if

- for all $u \in D_1 \cap D_2$, $f_1(u) = f_2(u)$, and
- for all $r \in R_1 \cap R_2$, we have $f_1^{-1}(r) = f_2^{-1}(r)$.

If f_1, f_2 are compatible, we define $f = f_1 \cup f_2$ as $f|_{D_1} = f_1$ and $f|_{D_2} = f_2$.

Note that in the above definition, the ranges of f_1 and f_2 matter. For example, the identity functions $f_1 : \{1\} \to \{1\}$ and $f_2 : \{2\} \to \{2\}$ are compatible, but the same functions are not compatible if we replace both ranges with $\{1, 2\}$.

Using Theorem 4, we can assume that G can be drawn in a $O(k) \times O(k)$ box with ply p. We will use dynamic programming. We will first define the sets of partial solutions that are counted in this dynamic programming algorithm. For variables

- integers $0 \leq x < x' \leq b$,
- separation (A, B) of P of order at most $2\sqrt{pk}$,
- injective $f : A \cap B \to G\langle x\rangle \cup G\langle x'\rangle$ such that $|f^{-1}(G\langle x\rangle)|, |f^{-1}(G\langle x'\rangle)| \leq \sqrt{pk}$

we define

$$T[x, x', (A, B), f] = \{g \in \text{sub}(P[A], G\langle x, x'\rangle) : g \text{ extends } f\}.$$

Note that T is indexed by *any* separation of P of order $2\sqrt{pk}$. We will later replace this with a set of non-isomorphic separations to obtain the claimed dependence $\sigma_{2\sqrt{pk}}(P)$ in the running time.

Informally, $T[x, x', (A, B), f]$ is the set of all patterns $P[A]$ in $G\langle x, x'\rangle$ such that f describes their behaviour on the "boundary" $G\langle x\rangle \cup G\langle x'\rangle$. We will now show how to compute the table entries. We consider two cases, depending on whether $x' - x$ is less than $\sqrt{k/p}$ or not.

Case 1: $x' - x \leq \sqrt{k/p}$
Note that in this case, the pathwidth of $G\langle x, x'\rangle$ is $O(\sqrt{pk})$ by Lemma 1. Using Lemma 3, we can compute $|\text{sub}(P, G)|$ in time $(pk)^{\sqrt{pk}}\sigma_{O(\sqrt{pk})}(P)n^{O(1)}$.

Case 2: $x' - x > \sqrt{k/p}$
Let $g \in T[x, x', (A, B), f]$, and let Q be the image of g. For $m \in \{x+1, \ldots, x'-1\}$, we say that Q is *sparse at* m if $|Q \cap G\langle m\rangle| \leq \sqrt{pk}$, i.e. the vertical line at m

intersects at most $\sqrt{pk}$ disks in Q. Since $|Q| \leq k$ and $x' - x > \sqrt{k/p}$, there is at least one m such that Q is sparse at m by the averaging principle. Therefore,

$$T[x, x', (A, B), f] = \bigcup_{m=x+1}^{x'-1} \{g \in T[x, x', (A, B), f] : g(A) \text{ is a sparse at } m\}.$$

By the inclusion-exclusion principle, $|T[x, x', (A, B), f]|$ is equal to

$$\sum_{\emptyset \subset X \subseteq \{x+1, \ldots, x'-1\}} (-1)^{|X|} \left|\{g \in T[x, x', (A, B), f] : g(A) \text{ is sparse at all } m \in X\}\right|.$$

Denoting $X = \{x_1, \ldots, x_\ell\}$, where $x_1 < \cdots < x_\ell$, we further rewrite this into

$$\sum_{\ell=1}^{x'-x-2} (-1)^\ell \sum_{x < x_1 < \cdots < x_\ell < x'} |\{g \in T[x, x', (A, B), f] : g(A) \text{ is sparse at } x_1, \ldots, x_\ell\}|.$$

Now we claim that, since $Q \cap G\langle m \rangle$ is a separator of $G[Q]$, $|T[x, x', (A, B), f]|$ can be further rewritten to express it recursively as follows:

Claim.

$$|T[x, x', (A, B), f]| = \sum_{\ell=1}^{x'-x-2} (-1)^\ell \sum_{(*)} \prod_{i=0}^{\ell} |T[x_i, x_{i+1}, (A_i, B_i), f_i]|,$$

where we let $x_0 = x$ and $x_{\ell+1} = x'$ for convenience and the sum $(*)$ goes over

- integers $x < x_1 < \cdots < x_\ell < x'$,
- separations (A_i, B_i) of P of order $2\sqrt{pk}$ for each $i = 0, \ldots, \ell$, such that
  - $\cup_{i=0}^{\ell} A_i = A$, and
  - $A_i \setminus B_i$ and $A_j \setminus B_j$ are disjoint for each $0 \leq i < j \leq \ell$,
- functions $f_i : A_i \cap B_i \to G\langle x_i \rangle \cup G\langle x_{i+1} \rangle$ for each $i = 0, \ldots, \ell$ such that $f, f_1, \ldots, f_\ell$ are pairwise compatible and $|f_i^{-1}(G\langle x_i \rangle)|, |f_i^{-1}(G\langle x_{i+1} \rangle)| \leq \sqrt{pk}$.

Proof of Claim. To prove this claim, consider first a function $g \in T[x, x', (A, B), f]$ such that $g(A)$ is sparse at $x_1, \ldots, x_\ell$. We describe how to find the separations (A_i, B_i) and functions f_i that correspond to g (for an example, see Fig. 1). Let $A_i = g^{-1}(G\langle x_i, x_{i+1} \rangle)$, $B_i = (P - A_i) \cup g^{-1}(G\langle x_i \rangle) \cup g^{-1}(G\langle x_{i+1} \rangle)$. Note that, since $g(A)$ is sparse at x_i and x_{i+1}, (A_i, B_i) is a separation of order at most $2\sqrt{pk}$. It is easy to see that $\cup A_i = g^{-1}(G\langle x, x' \rangle) = A$. Also, note that $A_i \setminus B_i = g^{-1}(G\langle x_i, x_{i+1} \rangle) \setminus (g^{-1}(G\langle x_i \rangle) \cup g^{-1}(G\langle x_{i+1} \rangle))$, so for any $i \neq j$, $A_i \setminus B_i$ and $A_j \setminus B_j$ are disjoint. We define $f_i : A_i \cap B_i \to G\langle x_i \rangle \cup G\langle x_{i+1} \rangle$ as $f_i = g|_{A_i \cap B_i}$. By construction, $f, f_1, \ldots, f_\ell$ are pairwise compatible.

Conversely, given pairwise compatible functions $g_0, \ldots, g_\ell$ such that $g_i \in T[x_i, x_{i+1}, (A_i, B_i), f_i]$, we show how to construct a function $g \in T[x, x', (A, B), f]$. Since g_i are compatible, we can define $g = g_0 \cup \cdots \cup g_\ell : A \to G\langle x, x' \rangle$. Since f, g_1, g_ℓ are pairwise compatible, g extends f. It is easy to see that this correspondence is one to one, which proves the claim.

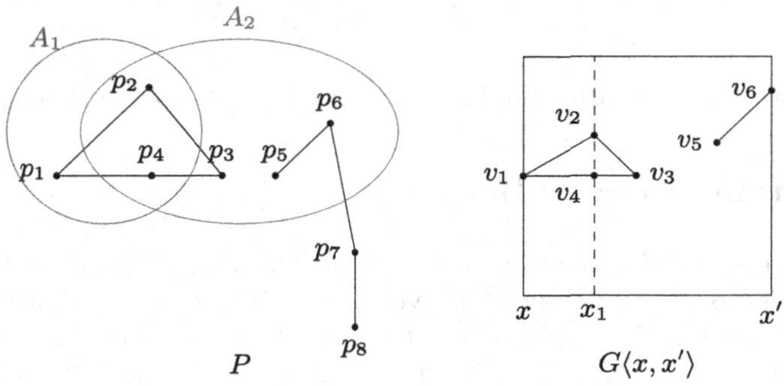

Fig. 1. The function $g : \{p_1, \ldots, p_6\} \to G\langle x, x'\rangle$ defined by $g(p_i) = v_i$, corresponds to functions $g_1 : A_1 \to G\langle x, x_1\rangle$ and $g_2 : A_2 \to G\langle x_1, x'\rangle$, where $g_1(p_i) = v_i$ and $g_2(p_i) = v_i$.

The next step is to rewrite the sum $(*)$ to match the form of Lemma 4. The only difference is that in (1) the summation is over variables that are pairwise independent.

Formally, let us define a square matrix M whose indices M_{ind} are of the form $(x_i, (A_i, B_i), f_i)$, where $x_i \in \{x, \ldots, x'\}$, $(A_i, B_i) \in \mathcal{S}_{2\sqrt{pk}}(P)$ and $f_i : A_i \cap B_i \to G\langle x_i\rangle \cup G\langle x_0\rangle$. Let $I_i = f_i^{-1}(G\langle x_i\rangle)$, $I_j = f_j^{-1}(G\langle x_j\rangle)$.

If $x_j \geq x_i$, f_i, f_j compatible and $|I_i|, |I_j| \leq \sqrt{pk}$ we define $M[(x_i, (A_i, B_i), f_i), (x_j, (A_j, B_j), f_j)]$ as

$$\mu((A_i, B_i), (A_j, B_j)) \cdot T[x_i, x_j, ((A_j \setminus A_i) \cup I_i, B_j \cup A_i), f_i|_{I_i} \cup f_j|_{I_j}],$$

and zero otherwise.

Intuitively, $M[(x_i, (A_i, B_i), f_i), (x_j, (A_j, B_j), f_j)]$ describes the number of ways to embed $P[A_j \setminus A_i]$ between lines x_i and x_j, where f_i and f_j describe the behaviour of these embeddings on lines x_i an x_j respectively. We observe that we can group isomorphic separations together, i.e. that instead of indexing by every separation, we can index by their representatives and take into account the multiplicities, which are described by μ.

Now we can rewrite the sum $(*)$ as

$$\sum_{(**)} M[(x_\ell, (A \setminus A_{\ell-1}, B \setminus B_{\ell-1}), f_{\ell-1}), (x', (A, B), f)]$$

$$\prod_{i=0}^{\ell-2} M[(x_i, (A_i, B_i), f_i), (x_{i+1}, (A_{i+1}, B_{i+1}, f_{i+1}))],$$

where the sum $(**)$ goes over $(x_0, (A_0, B_0), f_0), \ldots, (x_{\ell-1}, (A_{\ell-1}, B_{\ell-1}), f_{\ell-1}) \in M_{ind}$. Now by Lemma 4, we can compute the sum $(*)$ in time $\ell \cdot |M_{ind}|^2$. Let us bound the size of M_{ind}. Recall that $|\mathcal{S}_{2\sqrt{pk}}(P)| = \sigma_{2\sqrt{pk}}(P)$ and note that $G\langle x_i\rangle$ contains at most k^2p disks (since we can assume that G can be drawn in a $(O(k^2) \times O(k^2))$-box with ply p by Theorem 4). Thus we have

$$|M_{ind}| \leq k^2 \sigma_{2\sqrt{pk}}(P) \cdot (k^2 p)^{\sqrt{pk}},$$

Therefore, we can compute $|\text{sub}(P, G)|$ in time $k^{O(\sqrt{pk})} \cdot p^{O(\sqrt{pk})} \cdot \sigma_{O(\sqrt{pk})}(P)^2$.

5 Theorem 3: Lower Bound

In this section, we give a proof overview of Theorem 3, showing that under ETH there is no algorithm deciding whether $|\text{sub}(P, G)| > 0$ ($|\text{ind}(P, G)| > 0$ respectively) in time $2^{o(n/\log n)}$ even when the ply is two. The formal proof can be found in the full version. We will use a reduction from the String 3-Groups problem similar to the one in [4].

Definition 3. *The* String 3-Groups *problem is defined as follows. Given sets $A, B, C \subseteq \{0,1\}^{6\lceil \log n \rceil + 1}$ of size n, find n triples $(a, b, c) \in A \times B \times C$ such that for all i, $a_i + b_i + c_i \leq 1$ and each element of A, B, C occurs exactly once in a chosen triple.*

We call the elements of A, B, C strings. It was shown in [4] that, assuming the ETH, there is no algorithm that solves String 3-Groups in time $2^{o(n)}$. Given an instance (A, B, C) of String 3-Groups problem, we construct the corresponding host graph G and pattern P. Firstly, we modify slightly the strings in A, B, C to facilitate the construction of P and G. Let m be the length of the (modified) strings. For each $a \in A$, the connected component in G that corresponds to it consists of two paths $p_1 \dots p_m$ and $q_1 \dots q_m$, where p_i and q_i are connected by paths of length 3 if $a_i = 0$. For each $b \in B$, the connected component in P that corresponds to it consists of a path $t_1 \dots t_m$, where there is a path of length two attached to t_i if $b_i = 1$. The connected components corresponding to elements in C are constructed in a similar way. Finally, we add gadgets (triangles and 4-cycles) to each connected component in P and G to ensure we cannot "flip" the components in P. For an example, see Fig. 2.

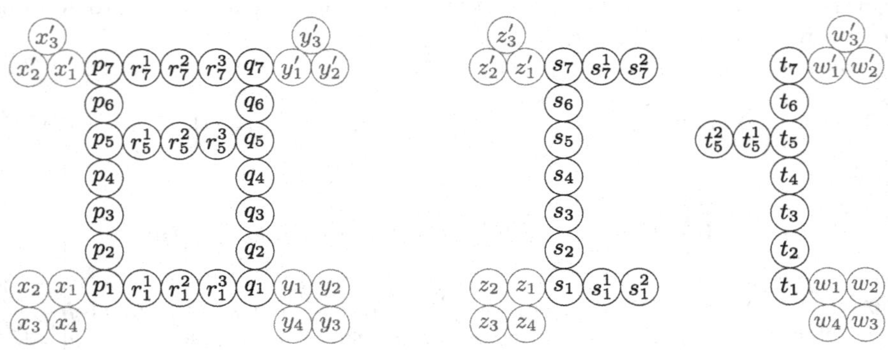

Fig. 2. Connected components corresponding to $a = 0111010 \in A$ (left), $b = 1000001 \in B$ (middle), $c = 0000100 \in C$ (right).

6 Concluding Remarks

We gave (mostly) sub-exponential parameterized time algorithms for computing $|\text{sub}(P, G)|$ and $|\text{ind}(P, G)|$ for unit disk graphs P and G. Since the fine-grained parameterized complexity of the subgraph isomorphism problem was only recently understood for planar graphs, we believe our continuation of the study for unit disk graphs is very natural, we hope it inspires further general results.

While our algorithms are tight in many regimes, they are not tight in all regimes. In particular, the (sub)-exponential dependence of the runtime in the ply is not always necessary: We believe the answer to this question may be quite complicated: For detecting some patterns, such as paths, $2^{O(\sqrt{k})}n^{O(1)}$ time algorithms are known [16], but it seems hard to extend it to the counting problem (and to all patterns with few non-isomorphic separations of small order).

For counting induced occurrences with bounded clique size our method can be easily adjusted to get a better dependence in the ply: I.e. our method can be used to a get a $(kp)^{O(\sqrt{k})}$ time algorithm for counting independent sets of size k in unit disk graphs of ply p (which is optimal under the ETH by [11]); is there such an improved independence on the ply for each pattern P?

Finally, we note it would be interesting to study the complexity of computing $|\text{sub}(P, G)|$ and $|\text{ind}(P, G)|$ for various pattern classes and various other geometric intersection graphs as well. Our results can be adapted to disk graphs where the ratio of the largest and smallest radius is constant (using a slight modification of Lemma 2). A possible direction for further research would be to determine for which patterns can one compute the above values on bounded ply disk graphs? Recent work [10] shows some problems admit algorithms running in sub-exponential time parameterized time.

References

1. Baker, B.S.: Approximation algorithms for np-complete problems on planar graphs. J. ACM (JACM) **41**(1), 153–180 (1994)
2. de Berg, M., Bodlaender, H.L., Kisfaludi-Bak, S., Marx, D., van der Zanden, T.C.: A framework for exponential-time-hypothesis-tight algorithms and lower bounds in geometric intersection graphs. SIAM J. Comput. **49**(6), 1291–1331 (2020). https://doi.org/10.1137/20M1320870
3. Bhore, S., Carmi, P., Kolay, S., Zehavi, M.: Parameterized study of steiner tree on unit disk graphs. Algorithmica **85**(1), 133–152 (2023). https://doi.org/10.1007/s00453-022-01020-z
4. Bodlaender, H.L., Nederlof, J., Zanden, T.C.V.D.: Subexponential time algorithms for embedding H-minor free graphs, vol. 55. Schloss Dagstuhl- Leibniz-Zentrum fur Informatik GmbH, Dagstuhl Publishing, August 2016. https://doi.org/10.4230/LIPIcs.ICALP.2016.9
5. Borgs, C., Chayes, J., Kahn, J., Lovász, L.: Left and right convergence of graphs with bounded degree. Random Struct. Algorithms **42**, 1–28 (2013). https://doi.org/10.1002/rsa.20414

6. Cygan, M., et al.: Parameterized Algorithms. Springer, Cham (2015). https://doi.org/10.1007/978-3-319-21275-3
7. Fomin, F.V., Lokshtanov, D., Marx, D., Pilipczuk, M., Pilipczuk, M., Saurabh, S.: Subexponential parameterized algorithms for planar and apex-minor-free graphs via low treewidth pattern covering. SIAM J. Comput. **51**(6), 1866–1930 (2022). https://doi.org/10.1137/19m1262504
8. Har-Peled, S., et al.: Stabbing pairwise intersecting disks by five points. Discret. Math. **344**(7), 112403 (2021)
9. Hochbaum, D.S., Maass, W.: Approximation schemes for covering and packing problems in image processing and VLSI. J. ACM (JACM) **32**(1), 130–136 (1985)
10. Lokshtanov, D., Panolan, F., Saurabh, S., Xue, J., Zehavi, M.: Subexponential parameterized algorithms on disk graphs (extended abstract). In: Naor, J.S., Buchbinder, N. (eds.) Proceedings of the 2022 ACM-SIAM Symposium on Discrete Algorithms, SODA 2022, Virtual Conference/Alexandria, VA, USA, 9–12 January 2022, pp. 2005–2031. SIAM (2022). https://doi.org/10.1137/1.9781611977073.80
11. Marx, D.: On the optimality of planar and geometric approximation schemes. In: 48th Annual IEEE Symposium on Foundations of Computer Science (FOCS 2007), 20–23 October 2007, Providence, RI, USA, Proceedings, pp. 338–348. IEEE Computer Society (2007). https://doi.org/10.1109/FOCS.2007.50
12. Marx, D., Pilipczuk, M.: Subexponential parameterized algorithms for graphs of polynomial growth. In: Pruhs, K., Sohler, C. (eds.) 25th Annual European Symposium on Algorithms, ESA 2017, 4–6 September 2017, Vienna, Austria. LIPIcs, vol. 87, pp. 59:1–59:15. Schloss Dagstuhl - Leibniz-Zentrum für Informatik (2017). https://doi.org/10.4230/LIPIcs.ESA.2017.59
13. Nederlof, J.: Detecting and counting small patterns in planar graphs in subexponential parameterized time. In: Makarychev, K., Makarychev, Y., Tulsiani, M., Kamath, G., Chuzhoy, J. (eds.) Proceedings of the 52nd Annual ACM SIGACT Symposium on Theory of Computing, STOC 2020, Chicago, IL, USA, 22–26 June 2020, pp. 1293–1306. ACM (2020). https://doi.org/10.1145/3357713.3384261
14. Nešetřil, J., De Mendez, P.O.: Sparsity: Graphs, Structures, and Algorithms, vol. 28. Springer Science & Business Media, Berlin, Heidelberg (2012). https://doi.org/10.1007/978-3-642-27875-4
15. Panolan, F., Saurabh, S., Zehavi, M.: Contraction decomposition in unit disk graphs and algorithmic applications in parameterized complexity. In: Chan, T.M. (ed.) Proceedings of the Thirtieth Annual ACM-SIAM Symposium on Discrete Algorithms, SODA 2019, San Diego, California, USA, 6–9 January 2019, pp. 1035–1054. SIAM (2019). https://doi.org/10.1137/1.9781611975482.64
16. Zehavi, M., Fomin, F.V., Lokshtanov, D., Panolan, F., Saurabh, S.: ETH-tight algorithms for long path and cycle on unit disk graphs. J. Comput. Geom. **12**(2), 126–148 (2021). https://doi.org/10.20382/jocg.v12i2a6

The Weighted HOM-Problem Over Fields

Andreea-Teodora Nász(✉)

Faculty of Mathematics and Computer Science, Universität Leipzig,
PO box 100 920, 04009 Leipzig, Germany
nasz@informatik.uni-leipzig.de

Abstract. The *HOM-problem*, which asks whether the image of a regular tree language under a tree homomorphism is again regular, is known to be decidable. In this paper, we prove the *weighted* HOM-problem for all fields decidable, provided that the tree homomorphism is *tetris-free* (a condition that generalizes injectivity). To this end, we reduce the problem to a property of the device representing the homomorphic image in question; to prove this property decidable, we then derive a pumping lemma for such devices from the well-known pumping lemma for regular tree series over fields, proved by Berstel and Reutenauer in 1982.

1 Introduction

The well-known model of finite-state automata has seen various extensions over the past decades. On the one hand, the qualitative evaluation of these acceptors was generalized to a quantitative one, leading to *weighted automata* [29]. Such devices assign a weight to each input word, and are thus suited to model numerical factors related to the input, such as costs, probabilities and consumption of resources or time. The research community focused on automata theory has studied weighted automata consistently and fruitfully [9,10,28]. Thereby, the favoured domains for weight calculations are often semirings [16,18], as they are both quite general and computationally efficient due to their distributivity.

Another dimension of generalization for finite-state automata targets their input, allowing them to handle more complex data structures such as infinite words [25], trees [5], graphs [4] and pictures [27]. In particular, *finite-state tree automata* and the *regular tree languages* they recognize were introduced independently in [7,31,32]. These devices find applications in a variety of areas like natural language processing [19], picture generation [8] and compiler construction [33]. Unsurprisingly, combining both types of generalizations leads to intricate yet fruitful research areas, and so several variants of *weighted tree automata* (WTA) and the *regular tree series* they recognize continue to be studied [11].

Tree homomorphisms are widely used in the context of term rewriting [13] and XML types [30]. A tree homomorphism is a structure-preserving transformation on trees which can duplicate subtrees, so the trees in the homomorphic image might have identical subtrees. Unfortunately, tree automata have limited memory, so they cannot ensure that certain subtrees are equal [12] (much like

H. Fernau et al. (Eds.): SOFSEM 2024, LNCS 14519, pp. 427–441, 2024.
https://doi.org/10.1007/978-3-031-52113-3_30

the classical (string) automata cannot ensure that the numbers of a's and of b's in a word are equal). Therefore, unlike in the word case, regular tree languages are not closed under tree homomorphisms. It was a long-standing open question if, given a regular tree language $\mathcal{L}$ and a tree homomorphism h as input, it is decidable whether $h(\mathcal{L})$ is again regular. This *HOM-problem* was finally solved in [6,14,15], with the help of a well-studied extension called *tree automata with constraints*; these devices can explicitely require certain subtrees to be equal, and can thus handle the duplications performed by h.

In the *weighted* HOM-problem, a regular tree series and a tree homomorphism are given as input. By its nature, this question requires a customized investigation for different semirings. Most recently, this problem was proved decidable for different scenarios [22,23], but in both cases, the semiring must be zero-sum free; this strong condition already excludes essential rings such as $\mathbb{Z}$. In this paper, we decide the weighted HOM-problem for all fields (and thus, all subspaces of fields), provided that the tree homomorphism is *tetris-free*, a property that generalizes injectivity.

The paper is structured as follows: In Sect. 2 we represent the homomorphic image of the input tree series by a WTA with constraints (WTAh). In Sect. 4 we show that, if the input tree homomorphism is tetris-free, then the weighted HOM-problem is equivalent to a certain decidable property of this WTAh. Proving said decidability relies on a pumping lemma for WTAh over fields, which we derive in Sect. 3 from the well-known pumping lemma for (regular) WTA over fields [1]. Finally, we present an example that illustrates why the approach is unsuited for non-tetris-free tree homomorphisms.

2 Preliminaries and Technical Background

We denote the set $\{0, 1, 2, \ldots\}$ by $\mathbb{N}$, and we let $[k] = \{1, \ldots, k\}$ for every $k \in \mathbb{N}$. Let A and B be sets. We write $|A|$ for the cardinality of A, and A^* for the set of finite strings over A. The empty string is ε and the length of a string w is $|w|$.

A *ranked alphabet* is a pair (Σ, rk) that consists of a finite set Σ and a rank mapping $\mathrm{rk}\colon \Sigma \to \mathbb{N}$. For every $k \geq 0$, we define $\Sigma_k = \mathrm{rk}^{-1}(k)$, and we sometimes write $\sigma^{(k)}$ to indicate that $\sigma \in \Sigma_k$. We often abbreviate (Σ, rk) by Σ, leaving rk implicit. Let Z be a set disjoint with Σ. The set of *Σ-trees over Z*, denoted by $T_\Sigma(Z)$, is the smallest set T that satisfies (i) $\Sigma_0 \cup Z \subseteq T$ and (ii) $\sigma(t_1, \ldots, t_k) \in T$ for every $k \in \mathbb{N}$, $\sigma \in \Sigma_k$ and $t_1, \ldots, t_k \in T$. We abbreviate $T_\Sigma(\emptyset)$ simply to T_Σ, and call any subset $L \subseteq T_\Sigma$ a *tree language*. Consider $t \in T_\Sigma(Z)$. The set $\mathrm{pos}(t) \subseteq \mathbb{N}^*$ of *positions of* t is defined by $\mathrm{pos}(t) = \{\varepsilon\}$ for every $t \in \Sigma_0 \cup Z$, and by $\mathrm{pos}\left(\sigma(t_1, \ldots, t_k)\right) = \{\varepsilon\} \cup \bigcup_{i \in [k]} \{ip \mid p \in \mathrm{pos}(t_i)\}$ for all $k \in \mathbb{N}$, $\sigma \in \Sigma_k$ and $t_1, \ldots, t_k \in T_\Sigma(Z)$. The set of positions of t inherits the lexicographic order $\leq_{\mathrm{lex}}$ from $\mathbb{N}^*$. The *size* $|t|$ of t is defined by $|t| = |\mathrm{pos}(t)|$ and the *height* $\mathrm{ht}(t)$ *of* t by $\mathrm{ht}(t) = \max_{p \in \mathrm{pos}(t)} |p|$. For $p \in \mathrm{pos}(t)$, the *label* $t(p)$ *of* t at p, the *subtree* $t|_p$ of t at p and the *substitution* $t[t']_p$ of t' *into* t at p are defined for $t \in \Sigma_0 \cup Z$ by $t(\varepsilon) = t|_\varepsilon = t$ and $t[t']_\varepsilon = t'$, and for $t = \sigma(t_1, \ldots, t_k)$ by $t(\varepsilon) = \sigma,\ t(ip') = t_i(p'),\ t|_\varepsilon = t,\ t|_{ip'} = t_i|_{p'},\ t[t']_\varepsilon = t'$, and finally $t[t']_{ip'} = \sigma(t_1, \ldots, t_{i-1}, t_i[t']_{p'}, t_{i+1}, \ldots, t_k)$ for every $k \in \mathbb{N}$, $\sigma \in \Sigma_k$, $t_1, \ldots, t_k \in T_\Sigma(Z)$, $i \in [k]$ and $p' \in \text{pos}(t_i)$. For every subset $S \subseteq \Sigma \cup Z$, we let $\text{pos}_S(t) = \{p \in \text{pos}(t) \mid t(p) \in S\}$ and we abbreviate $\text{pos}_{\{s\}}(t)$ by $\text{pos}_s(t)$ for every $s \in \Sigma \cup Z$. Let $X = \{x_1, x_2, \ldots\}$ be a fixed, countable set of formal variables. For $k \in \mathbb{N}$ we denote by X_k the subset $\{x_1, \ldots, x_k\}$. For any $t \in T_\Sigma(X)$ we let $\text{var}(t) = \{x \in X \mid \text{pos}_x(t) \neq \emptyset\}$. For $t \in T_\Sigma(Z)$, a subset $V \subseteq Z$ and a mapping $\theta\colon V \to T_\Sigma(Z)$, we define the *substitution* $t\theta$ *applied to* t by $v\theta = \theta(v)$ for $v \in V$, $z\theta = z$ for $z \in Z \setminus V$, and $\sigma(t_1, \ldots, t_k)\theta = \sigma\big(t_1\theta, \ldots, t_k\theta\big)$ for all $k \in \mathbb{N}$, $\sigma \in \Sigma_k$ and $t_1, \ldots, t_k \in T_\Sigma(Z)$. If $V = \{v_1, \ldots, v_n\}$, we write θ explicitly as $[v_1 \leftarrow \theta(v_1), \ldots, v_n \leftarrow \theta(v_n)]$, or simply as $[\theta(x_1), \ldots, \theta(x_n)]$ if $V = X_n$.

A *(commutative) semiring* [16,17] is a tuple $(\mathbb{S}, +, \cdot, 0, 1)$ that satisfies the following conditions: $(\mathbb{S}, +, 0)$ and $(\mathbb{S}, \cdot, 1)$ are commutative monoids, $\cdot$ distributes over $+$, and $0 \cdot s = 0$ for all $s \in \mathbb{S}$. Examples include $\mathbb{N} = \big(\mathbb{N}, +, \cdot, 0, 1\big)$, $\mathbb{Z} = \big(\mathbb{Z}, +, \cdot, 0, 1\big)$, $\mathbb{Q} = \big(\mathbb{Q}, +, \cdot, 0, 1\big)$, the Boolean semiring $\mathbb{B} = \big(\{0, 1\}, \vee, \wedge, 0, 1\big)$ and the arctic semiring $\mathbb{A} = \big(\mathbb{N} \cup \{-\infty\}, \max, +, -\infty, 0\big)$. When there is no risk of confusion, we refer to a semiring $(\mathbb{S}, +, \cdot, 0, 1)$ simply by its carrier set $\mathbb{S}$. A semiring is a *field* if it is (i) a *ring*, i.e. there exists $-1 \in S$ such that $1 + (-1) = 0$, and (ii) a *semifield*, i.e. for every $a \in \mathbb{S} \setminus \{0\}$ there exists a multiplicative inverse a^{-1} such that $a \cdot a^{-1} \in \mathbb{S}$. Let $\mathbb{F}$ be a field, then $\mathbb{S}$ is a *subsemiring* of $\mathbb{F}$ if $\mathbb{S} \subseteq \mathbb{F}$ and the operations of $\mathbb{S}$ are embeddable in $\mathbb{F}$, i.e., $(+_\mathbb{F})|_\mathbb{S} = +_\mathbb{S}$, $(\cdot_\mathbb{F})|_\mathbb{S} = \cdot_\mathbb{S}$, $0_\mathbb{S} = 0_\mathbb{F}$ and $1_\mathbb{S} = 1_\mathbb{F}$. The semirings $\mathbb{N}$ and $\mathbb{Z}$ are subsemirings of $\mathbb{Q}$, but not $\mathbb{B}$.

Let Σ be a ranked alphabet and Z a set. Any mapping $\varphi\colon T_\Sigma(Z) \to \mathbb{S}$ is called a *tree series* over $\mathbb{S}$, and its *support* is the set $\text{supp}(\varphi) = \{t \in T_\Sigma(Z) \mid \varphi(t) \neq 0\}$.

Given ranked alphabets Σ and Δ, let $h'\colon \Sigma \to T_\Delta(X)$ be a mapping such that for all $k \in \mathbb{N}$ and $\sigma \in \Sigma_k$, we have $h'(\sigma) \in T_\Delta(X_k)$. We extend h' to a mapping $h\colon T_\Sigma \to T_\Delta$ by $h(\alpha) = h'(\alpha) \in T_\Delta(X_0) = T_\Delta$ for all $\alpha \in \Sigma_0$, and by $h(\sigma(s_1, \ldots, s_k)) = h'(\sigma)[x_1 \leftarrow h(s_1), \ldots, x_k \leftarrow h(s_k)]$ for all $k \in \mathbb{N}$, $\sigma \in \Sigma_k$, and $s_1, \ldots, s_k \in T_\Sigma$. The mapping h is called the *tree homomorphism induced by* h', and we identify h' and its induced tree homomorphism h. We call h

- *nonerasing* if $h(\sigma) \notin X$ for all $\sigma \in \Sigma$,
- *nondeleting* if $\sigma \in \Sigma_k$ implies $\text{var}(h'(\sigma)) = X_k$ for all $k \in \mathbb{N}$,
- *input-finitary* if the preimage $h^{-1}(t)$ is finite for every $t \in T_\Delta$, and
- *tetris-free* if it is nondeleting, nonerasing and for $s, s' \in T_\Sigma$, $h(s) = h(s')$ implies (i) $\text{pos}(s) = \text{pos}(s')$ and (ii) $h\big(s(p)\big) = h\big(s'(p)\big)$ for all $p \in \text{pos}(s)$.

In other words, a nondeleting and nonerasing $h\colon T_\Sigma \to T_\Delta$ is tetris-free if we cannot combine the building blocks $h(\sigma)$, $\sigma \in \Sigma$ in different ways to build the same tree. Thus if we list all possible trees that can be generated from these building blocks, no tree will occur twice. This condition was introduced in [23] and generalizes injectivity: Intuitively, if a tree homomorphism h is tetris-free, then any non-injective behaviour of h is located entirely at the symbol level.

Example 1. Let $\Sigma = \{\alpha^{(0)}, \beta^{(0)}, \psi^{(2)}\}$ and $\Delta = \{a^{(0)}, f^{(3)}\}$. Consider the tree homomorphism $h\colon T_\Sigma \to T_\Delta$ that is induced by the mapping $h(\alpha) = h(\beta) = a$ and $h(\psi) = f(x_2, x_1, x_1)$. While h is not injective, it is tetris-free. However, the tree homomorphism $h'\colon T_\Sigma \to T_\Delta$ induced by $h'(\alpha) = a$, $h'(\beta) = f(a, a, a)$ and $h'(\psi) = f(x_2, x_1, a)$ is not: $\psi(\alpha, \alpha)$ and β violate the tetris-free condition.

If $h\colon T_\Sigma \to T_\Delta$ is nonerasing and nondeleting, then for every $s \in h^{-1}(t)$, we have $|s| \le |t|$. In particular, h is then input-finitary. Let $A\colon T_\Sigma \to \mathbb{S}$ be a tree series. Its *homomorphic image under* h is the tree series $h(A)\colon T_\Delta \to \mathbb{S}$ defined for every $t \in T_\Delta$ by $h(A)(t) = \sum_{s \in h^{-1}(t)} A(s)$. This relies on h to be input-finitary, otherwise the defining sum is not finite, so $h(A)(t)$ might not be well-defined. For this reason, we only consider nondeleting and nonerasing tree homomorphisms.

Recently it was shown [20,21] that such homomorphic images of regular tree languages can be represented efficiently using *weighted tree automata with hom-constraints* (WTAh) which were defined in [20], and first introduced for the Boolean case in [14]. All following concepts are illustrated in Example 5 below.

Definition 2 (cf. [21, Definition 1]). *Let $\mathbb{S}$ be a commutative semiring. A* weighted tree automaton over $\mathbb{S}$ with hom-constraints (WTAh) *is a tuple of the form $\mathcal{A} = (Q, \Sigma, F, R, \mathrm{wt})$ where Q is a finite set of states, Σ is a ranked alphabet, $F \subseteq Q$ is the set of final states, R is a finite set of rules of the form (ℓ, q, E) such that $\ell \in T_\Sigma(Q) \setminus Q$, $q \in Q$ and E is an equivalence relation on $\mathrm{pos}_Q(\ell)$, and $\mathrm{wt}\colon R \to \mathbb{S}$ assigns a weight to each rule.*

Rules of WTAh are typically depicted as $r = \ell \xrightarrow{E}_{\mathrm{wt}(r)} q$. The components of such a rule are the *left-hand side* ℓ, the *target state* q, the set E of *constraints* and the *weight* $\mathrm{wt}(r)$. A constraint $(p, p') \in E$ is listed as "$p = p'$", and if p is different from p', then p and p' are called *constrained positions.* The equivalence class of p in E is denoted $[p]_{\equiv_E}$. We generally omit the trivial constraints $(p, p) \in E$.

The WTAh is a *weighted tree grammar* (WTG) if $E = \emptyset$ (strictly speaking, E is the identity relation) for every rule $\ell \xrightarrow{E} q$, and a WTA in the classical sense [5] if additionally $\mathrm{pos}_\Sigma(\ell) = \{\varepsilon\}$. WTG and WTA are equally expressive, as WTG can be translated straightforwardly into WTA using additional states.

We are particularly interested in a specific subclass of WTAh, namely the *eq-restricted* WTAh [21]. In such a device, there is a designated *sink-state* whose sole purpose is to neutrally process copies of identical subtrees. More precisely, whenever subtrees are mutually constrained, there is one leading copy among them that can be processed as usual with arbitrary states and weights, while every other copy is handled exclusively by the weight-neutral sink-state.

Definition 3. *A WTAh $(Q, \Sigma, F, R, \mathrm{wt})$ is* eq-restricted *if it has a so-called* sink state $\bot \in Q \setminus F$ *such that (i) $\sigma(\bot, \dots, \bot) \to_1 \bot$ belongs to R for all $\sigma \in \Sigma$, and no other rules target $\bot$, and (ii) for every rule $\ell \xrightarrow{E} q$ with $q \neq \bot$, if $\mathrm{pos}_Q(\ell) = \{p_1, \dots, p_n\}$ and $q_i = \ell(p_i)$ for $i \in [n]$, the following conditions hold:*

1. *For each $i \in [n]$, the set $\{q_j \mid p_j \in [p_i]_{\equiv_E}\} \setminus \{\bot\}$ is a singleton.*
2. *There exists exactly one $p_j \in [p_i]_{\equiv_E}$ such that $q_j \neq \bot$.*

In other words, among each E" equivalence class there is only one occurrence of a state different from $\bot$, and every other E" related position is labelled by $\bot$. Moreover, $\bot$ processes every possible tree with weight 1. We denote the state sets of WTAh by $Q\dot{\cup}\{\bot\}$ instead of $Q \ni \bot$ to point out the sink-state.

Next, let us recall the semantics of WTAh from [21, Definitions 2 and 3].

Definition 4. *Let* $\mathcal{A} = \big(Q, \Sigma, F, R, \mathrm{wt}\big)$ *be a WTAh. A* run of $\mathcal{A}$ *is a tree over the ranked alphabet* $\Sigma \cup R$ *where the rank of a rule is* $\mathrm{rk}(\ell \xrightarrow{E} q) = \mathrm{rk}\big(\ell(\varepsilon)\big)$, *and it is defined inductively. Consider* $t_1, \ldots, t_n \in T_\Sigma$, $q_1, \ldots, q_n \in Q$ *and suppose that* ϱ_i *is a run of* $\mathcal{A}$ *for* t_i *to* q_i *with weight* $\mathrm{wt}(\varrho_i) = a_i$ *for each* $i \in [n]$. *Assume there exists* $\ell \xrightarrow{E}_a q$ *in* R *such that* $\ell = \sigma(\ell_1, \ldots, \ell_m)$, $\mathrm{pos}_Q(\ell) = \{p_1, \ldots, p_n\}$ *with* $\ell(p_i) = q_i$, *and that* $t_i = t_j$ *for all* $(p_i, p_j) \in E$. *Let* $t = \ell[t_1]_{p_1} \cdots [t_n]_{p_n}$, *then* $\varrho = \big(\ell \xrightarrow{E}_a q\big)(\ell_1, \ldots, \ell_m)[\varrho_1]_{p_1} \cdots [\varrho_n]_{p_n}$ *is a run of* $\mathcal{A}$ *for* t *to* q. *Its* weight $\mathrm{wt}(\varrho)$ *is computed as* $a \cdot \prod_{i \in [n]} a_i$. *If* $\mathrm{wt}(\varrho) \neq 0$, *then* ϱ *is* valid, *and if in addition,* $q \in F$ *for its* target state q, *then* ϱ *is* accepting. *The value* $\mathrm{wt}^q(t)$ *is the sum of all weights* $\mathrm{wt}(\varrho)$ *of runs of* $\mathcal{A}$ *for* t *to* q. *Finally, the tree series* $\|\mathcal{A}\| \colon T_\Sigma \to \mathbb{S}$ *recognized by* $\mathcal{A}$ *is defined simply by* $\|\mathcal{A}\| \colon t \mapsto \sum_{q \in F} \mathrm{wt}^q(t)$.

Since the weights of rules are multiplied, we assume wlog $\mathrm{wt}(r) \neq 0$ for all $r \in R$.

Example 5. Let $\Delta = \{a^{(0)}, g^{(2)}, f^{(3)}\}$ and $\mathcal{A}' = \big(Q\dot{\cup}\{\bot\}, \Delta, F', R', \mathrm{wt}\big)$ be the WTGh over $\mathbb{Z}$ with $Q = \{q, q_f\}$, $F' = \{q_f\}$ and the set of rules and weights

$$\begin{aligned} R' = \{\, & a \to_1 q\,, \quad g(a, q) \to_2 q\,, \quad f(q, q, \bot) \xrightarrow{2=3}_1 q_f\,, \\ & a \to_1 \bot\,, \quad g(\bot, \bot) \to_1 \bot\,, \quad f(\bot, \bot, \bot) \to_1 \bot\,\}. \end{aligned}$$

The constrained positions 2, 3 in the third rule satisfy (ii) from Definition 3, and the $\bot$" rules are as required in (i), so $\mathcal{A}'$ is eq-restricted. If we replace the third rule with $f(q, q, q) \xrightarrow{2=3}_1 q_f$, then the resulting WTAh is not eq-restricted any more. Let $t = f\big(a,\, g(a, a),\, g(a, a)\big) \in T_\Delta$. $\mathcal{A}'$ has a unique accepting run ϱ for t:

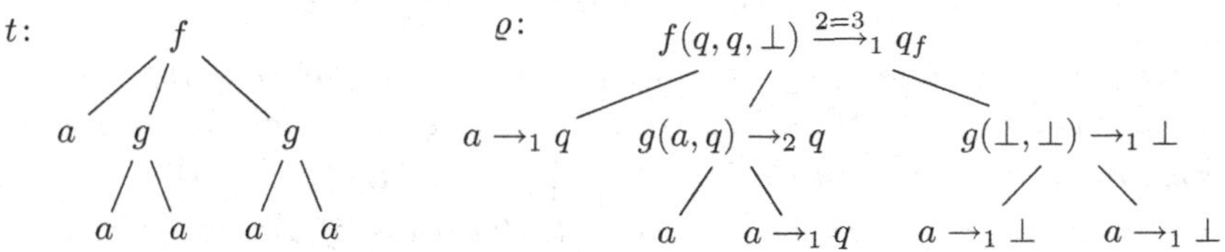

We have $\mathrm{wt}(\varrho) = 2$ despite $|\,\mathrm{pos}_g(t)| = 2$ because due to the eq-restriction, the duplicated subtree $t|_3$ is processed exclusively in the state $\bot$ with weight 1.

If a tree series is recognized by a WTA, it is called *regular*, and if it is recognized by an eq-restricted WTAh, then it is called *hom-regular*. This choice of name hints at the fact that eq-restricted WTAh are tailored to represent homomorphic images of regular tree series. The following example demonstrates this property.

Example 6. Consider $\Sigma = \{\alpha^{(0)}, \gamma^{(1)}, \psi^{(2)}\}$ and let $\mathcal{A} = (\{q, q_f\}, \Sigma, \{q_f\}, R, \mathrm{wt})$ be the WTA over $\mathbb{Z}$ with the following set of rules:

$$R = \{\alpha \to_1 q,\ \gamma(q) \to_2 q,\ \psi(q,q) \to_1 q_f\}.$$

It is $\mathrm{supp}(\|\mathcal{A}\|) = \{\psi(\gamma^n(\alpha), \gamma^m(\alpha)) \mid n, m \in \mathbb{N}\} = \{s \in T_\Sigma \mid \mathrm{pos}_\psi(s) = \{\varepsilon\}\}$ and $\|\mathcal{A}\| \colon \psi(\gamma^n(\alpha), \gamma^m(\alpha)) \mapsto 2^{n+m} = 2^{|\mathrm{pos}_\gamma(s)|}$. Let $\Delta = \{a^{(0)}, g^{(2)}, f^{(3)}\}$ and let $h \colon T_\Sigma \to T_\Delta$ be the tetris-free tree homomorphism induced by

$$h(\alpha) = a\,, \quad h(\gamma) = g(a, x_1) \ \text{ and } \ h(\psi) = f(x_2, x_1, x_1).$$

Then the eq-restricted WTAh $\mathcal{A}'$ from Example 5 recognizes $h(\|\mathcal{A}\|)$ defined by $\mathrm{supp}(h(\|\mathcal{A}\|)) = \{t \in T_\Delta \mid \mathrm{pos}_f(t) = \{\varepsilon\}\}$ and $h(\|\mathcal{A}\|) \colon t \mapsto 2^{|\mathrm{pos}_g(t) \setminus \mathrm{pos}_g(t|_3)|}$. The rules in R' are obtained from the rules in R by applying h to their left-hand sides, and the duplicated subtree at position 3 below f targets $\bot$ instead of q to avoid distorting the weight with an additional factor 2^n.

Formally, the following statement was shown in [20].

Lemma 7 (see [20, Theorem 19]). *Let $\mathcal{A} = (Q, \Sigma, F, R, \mathrm{wt})$ be a WTA over a commutative semiring $\mathbb{S}$ and $h \colon T_\Sigma \to T_\Delta$ a nondeleting and nonerasing tree homomorphism. There is an eq-restricted WTAh $\mathcal{A}'$ that recognizes $h(\|\mathcal{A}\|)$.*

As illustrated in Examples 5 and 6, the WTAh $\mathcal{A}'$ for the homomorphic image of a WTA $\mathcal{A}$ replaces each symbol σ in a rule of $\mathcal{A}$ by $h(\sigma)$, and preserves the original state behaviour, only adding $\bot$ along the duplicated subtrees. Thus, we can define a mapping that traces the runs of $\mathcal{A}$ to the runs of $\mathcal{A}'$.

Definition 8 (see [23, Definition 9]). *Let $\mathcal{A} = (Q, \Sigma, F, R, \mathrm{wt})$ be a WTA over a commutative semiring $\mathbb{S}$ and $h \colon T_\Sigma \to T_\Delta$ a nondeleting and nonerasing tree homomorphism. Let $\mathcal{A}'$ be the WTAh for $h(\|\mathcal{A}\|)$ provided by Lemma 7. Consider a rule $r = \sigma(q_1, \ldots, q_k) \to q$ of $\mathcal{A}$ and let $h(\sigma) = \delta(u_1, \ldots, u_n)$, then we set*

$$h^R(r) = \delta(u_1, \ldots, u_n)[\![q_1, \ldots, q_k]\!] \xrightarrow{E} q,$$

where the substitution $[\![q_1, \ldots, q_k]\!]$ replaces for every $i \in [k]$ only the $\leq_{lex}$" minimal occurrence of x_i in $\delta(u_1, \ldots, u_n)$ by q_i, and every other occurrence by $\bot$. The constraint set is defined as $E = \bigcup_{i \in [k]} \left[\mathrm{pos}_{x_i}(\delta(u_1, \ldots, u_n))\right]^2$.

The assignment h^R extends naturally to the runs of $\mathcal{A}$: For a run of the form $\vartheta = r = (\alpha \to q)$ with $\alpha \in \Sigma^0$, we set $h^R(\vartheta) = h^R(r)$. For $\vartheta = r(\vartheta_1, \ldots, \vartheta_k)$ with $r = \sigma(q_1, \ldots, q_k) \to q$ and $h(\sigma) = \delta(u_1, \ldots, u_n)$ we set

$$h^R(\vartheta) = (h^R(r))(u_1, \ldots, u_n)[\![h^R(\vartheta_1), \ldots, h^R(\vartheta_k)]\!];$$

here, the substitution $[\![h^R(\vartheta_1), \ldots, h^R(\vartheta_k)]\!]$ replaces for every $i \in [k]$ only the $\leq_{lex}$" minimal occurrence of x_i in $(h^R(r))(u_1, \ldots, u_n)$ by $h^R(\vartheta_i)$, and all other occurrences by the respective unique run to $\bot$ for the tree processed by ϑ_i.

Let us see how h^R acts on our example from above.

Example 9. Recall the WTA $\mathcal{A}$ and WTAh $\mathcal{A}'$ from Examples 5 and 6. We have

$$h^R: \quad \psi(q,q) \to q_f \quad \mapsto \quad f(q,q,\bot) \xrightarrow{2=3} q_f\,,$$

and for the unique run of $\mathcal{A}$ for the tree $\psi(\gamma(\alpha),\alpha)$, the image under h^R is

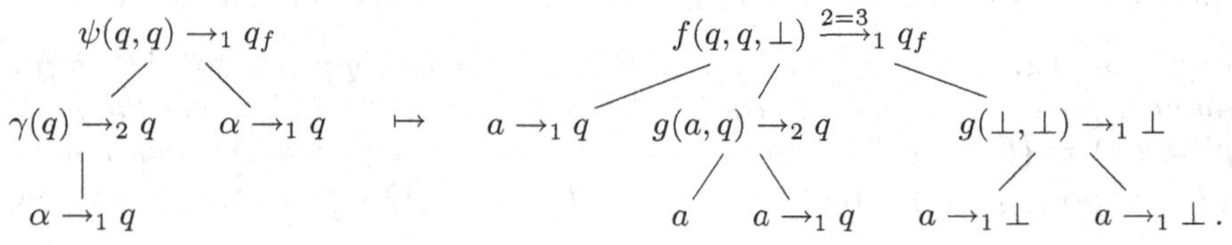

$$\begin{array}{c}\psi(q,q)\to_1 q_f\\ \gamma(q)\to_2 q \quad\quad \alpha\to_1 q\\ \alpha\to_1 q\end{array} \quad\mapsto\quad \begin{array}{c} f(q,q,\bot)\xrightarrow{2=3}_1 q_f\\ a\to_1 q \quad g(a,q)\to_2 q \quad g(\bot,\bot)\to_1 \bot\\ a \quad a\to_1 q \quad a\to_1\bot \quad a\to_1\bot\,.\end{array}$$

The following statement is a direct consequence of the proof of Lemma 7.

Lemma 10. *The mapping h^R from Definition 8 is well-defined on R, although not necessarily injective. Its image is $h^R(R) = \{r' \in R' \mid r' \text{ targets some } q \neq \bot\}$. If ϑ is a run of $\mathcal{A}$ for $s \in T_\Sigma$, then $h^R(\vartheta)$ is a run of $\mathcal{A}'$ for $h(s)$; conversely, for every run ϱ of $\mathcal{A}'$ for some $t \in T_\Delta$ to some $q \neq \bot$, there exists $s \in h^{-1}(t)$ and a run ϑ of $\mathcal{A}$ for s to q such that $h^R(\vartheta) = \varrho$, but $\mathrm{wt}(\vartheta)$ and $\mathrm{wt}'(\varrho)$ may differ.*

3 A Pumping Lemma Over Fields

The weighted HOM-problem takes a WTA $\mathcal{A}$ and a nondeleting, nonerasing tree homomorphism h as input, and asks whether $h(\|\mathcal{A}\|)$ is again regular. As mentioned earlier, the ℕ" variant of this problem was shown to be decidable in [22]. The proof presented there makes two assumptions on the semiring used for the weight calculations: First, it must be a subsemiring of a field, and second, it must be zero-sum free; the only common semiring that satisfies both conditions is ℕ. Remarkably, the strong condition of zero-sum freeness is only used to prove a pumping lemma for $h(\|\mathcal{A}\|)$. In this section, we derive an alternative pumping lemma over fields, provided that h is tetris-free. This way, we bypass the zero-sum freeness assumption, which allows us to lift the proof of [22] to the HOM-problem over fields, for tetris-free tree homomorphisms.

We begin by establishing a notation for the tree fragment read by a rule.

Definition 11. *Let $\mathcal{A}' = \big(Q\dot\cup\{\bot\}, \Delta, F, R, \mathrm{wt}\big)$ be an eq-restricted WTAh and let $r = \ell \xrightarrow{E} q$ be a rule of $\mathcal{A}'$ with some $q \neq \bot$. Let $\mathrm{pos}_{Q\setminus\{\bot\}}(\ell) = \{p_1,\ldots,p_k\}$. The Δ part of r is the tree $\widehat{\ell} = \ell[\bot]_{p_1}\cdots[\bot]_{p_k} \in T_\Delta(\{\bot\})$.*

The Δ part of a rule extracts the tree fragment from its left-hand side and overwrites every state label (for convenience simply with $\bot$). Note that ℓ can be easily recovered from $\widehat{\ell}$, E and the states $\ell(p_1),\ldots,\ell(p_k)$ in the correct order.

To prove the desired pumping lemma for our WTAh, we reduce it to the pumping lemma for WTA over fields proved by Berstel and Reutenauer in [1]. For this, we must construct a WTA related to the WTAh $\mathcal{A}'$ for $h(\|\mathcal{A}\|)$. The naive idea to simply use the input WTA $\mathcal{A}$ falls short: If h is not injective, there may be $s, s' \in \text{supp}(\|\mathcal{A}\|)$ with $h(s) = h(s') \notin \text{supp}(\|\mathcal{A}'\|)$ since in fields, different runs for $h(s)$ might cancel each other out, so we cannot lift the pumping lemma from $\mathcal{A}$ to $\mathcal{A}'$. Instead, we fabricate a new WTA that traces the behaviour of $\mathcal{A}'$ but ignores duplicated subtrees in order to remain regular. We will argue the well-definedness of this construction using some technical lemmas below.

Definition 12. *Let $\mathcal{A}' = (Q\dot{\cup}\{\bot\}, \Delta, F, R, \text{wt})$ be the eq-restricted WTAh from Lemma 7 for a WTA and a tetris-free tree homomorphism. Consider the ranked alphabet $\widehat{\Delta} = \{\widehat{\ell} \mid \ell$ is the left-hand side of some $r \in R\}$ with the rank function $\widehat{\text{rk}}(\widehat{\ell}) = |\text{pos}_{Q\setminus\{\bot\}}(\ell)|$. We define the WTA $\widehat{\mathcal{A}} = (Q \setminus \{\bot\}, \widehat{\Delta}, F, \widehat{R}, \widehat{\text{wt}})$ such that if $r = \ell \xrightarrow{E} q \in R$ with $q \neq \bot$ and $\text{pos}_{Q\setminus\{\bot\}}(\ell) = \{p_1, \ldots, p_k\}$ ordered lexicographically with $\ell(p_i) = q_i$ for all $i \in [k]$, then $\widehat{\ell}(q_1, \ldots, q_k) \to q \in \widehat{R}$ with weight $\text{wt}(r)$. No other rules are in $\widehat{R}$.*

The translation $\mathcal{A}' \mapsto \widehat{\mathcal{A}}$ induces a mapping $t \mapsto \widehat{t}$ defined inductively as follows: Consider $t \in T_\Delta$, a run ϱ of $\mathcal{A}'$ for t with $\varrho(\varepsilon) = \ell \xrightarrow{E} q$ and let $\text{pos}_{Q\setminus\{\bot\}}(\ell)$ be the set $\{p_1, \ldots, p_k\}$ in lexicographic order. Then $\widehat{t} = \widehat{\ell}(\widehat{t|_{p_1}}, \ldots, \widehat{t|_{p_2}}) \in T_{\widehat{\Delta}}$.

The WTA $\widehat{\mathcal{A}}$ reinterprets the trees $t \in T_\Delta$ as trees $\widehat{t} \in T_{\widehat{\Delta}}$ which, instead of symbols $\delta \in \Delta$, are now composed of the Δ parts of the rules of $\mathcal{A}'$. As the WTA $\widehat{\mathcal{A}}$, without the instrument of constraints at hand, cannot ensure equality of subtrees, all $\bot$" processed copies are discarded, and $\bot$ is not a state anymore.

Example 13. Recall the WTAh $\mathcal{A}'$ from Example 5. The ranked alphabet $\widehat{\Delta}$ is the set $\widehat{\Delta} = \{a^{(0)}, [g(a, \bot)]^{(1)}, [f(\bot, \bot, \bot)]^{(2)}\}$, and the WTA $\widehat{\mathcal{A}}$ is defined by $\widehat{\mathcal{A}} = (Q, \widehat{\Delta}, F', \widehat{R}, \widehat{\text{wt}})$ with the following set of rules and weights:

$$\widehat{R} = \{\, a \to_1 q\,, \quad [g(a, \bot)](q) \to_2 q\,, \quad [f(\bot, \bot, \bot)](q, q) \to_1 q_f \,\}\,.$$

For $t = f(a,\, g(a, a),\, g(a, a)) \in T_\Delta$ it is $\widehat{t} = [f(\bot, \bot, \bot)](a,\, [g(a, \bot)](a)) \in T_{\widehat{\Delta}}$:

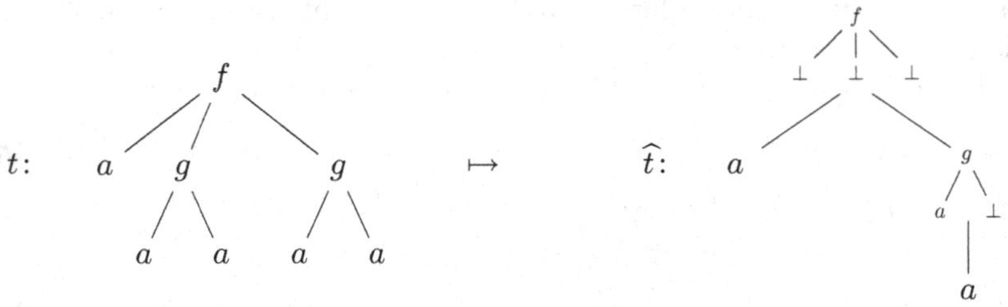

The following two lemmas are the basis for the correctness of our translation above. Unlike in a WTA where trees are read symbol-by-symbol, a rule of a WTAh processes an entire tree fragment; in general, there may be different ways to assemble a certain tree from these Δ parts of the rules of the WTAh, but by definition, tetris-free tree homomorphisms exclude this ambiguity.

Lemma 14. *Let* $\mathcal{A}' = (Q\dot{\cup}\{\bot\}, \Delta, F, R, \mathrm{wt})$ *be the eq-restricted WTAh from Lemma 7 for a WTA and a tetris-free tree homomorphism. For every* $t \in T_\Delta$, *the runs of* $\mathcal{A}'$ *for* t *differ only in the states they process, but neither in the* Δ *part of the rules they use, nor in their constraints. In particular, the set of positions related to any* $p'' \in \mathrm{pos}(t)$ *by the constraints of the rules used in a run coincides for all runs of* $\mathcal{A}'$ *for* t, *i.e. it is uniquely determined by* t.

Proof. Let ϱ and ϱ' be runs of $\mathcal{A}'$ for some $t \in T_\Delta$. By Lemma 10, there are two runs ϑ and ϑ' of $\mathcal{A}$ for some s and s', respectively, such that $h(s) = h(s') = t$, and $h^R(\vartheta) = \varrho$ and $h^R(\vartheta') = \varrho'$. Since h is tetris-free, it is $\mathrm{pos}(s) = \mathrm{pos}(s')$ and $h(s(p)) = h(s'(p))$ at every $p \in \mathrm{pos}(s)$. By the definition of h^R, these identical terms $h(s(p))$ and $h(s'(p))$ already determine the Δ parts of the rules used by ϱ and ϱ'. Moreover, the constraint sets are implicit to these terms, therefore ϱ and ϱ' can only differ in the states they process. □

The next lemma is again a consequence of the tetris-freeness. For details on the proof, we refer the reader to the full-length version [24].

Lemma 15. *Let* $\mathcal{A}'$ *be the eq-restricted WTAh from Lemma 7 for a WTA and a tetris-free tree homomorphism. If* $\mathcal{A}'$ *has two rules* r, r' *with the same* Δ *parts, then their constraint sets coincide as well.*

We are now ready prove that our translation $\mathcal{A}' \mapsto \widehat{\mathcal{A}}$ is correct:

Lemma 16. *The WTA* $\widehat{\mathcal{A}}$ *from Definition 12 is well-defined. The mapping* $t \mapsto \widehat{t}$ *induced by it is also well-defined and injective, and* $\|\mathcal{A}'\|(t) = \|\widehat{\mathcal{A}}\|(\widehat{t})$.

Proof. First, recall that ℓ can be recovered from $\widehat{\ell}$, E and the states $q_1, \ldots, q_k$. While $\widehat{\ell}$ and $q_1, \ldots, q_k$ are preserved in the rules of $\widehat{\mathcal{A}}$, E is uniquely determined by $\widehat{\ell}$ as stated in Lemma 15. Thus the weight function $\widehat{\mathrm{wt}}$ is well-defined.

Let $t \in T_\Delta$. By Lemma 14, all runs of $\mathcal{A}'$ for t have the same Δ parts, and these are precisely the alphabet symbols for $\widehat{t} \in T_{\widehat{\Delta}}$. Thus, the mapping $t \mapsto \widehat{t}$ is well-defined. Since E (and thus the positioning of every direct subtree) is uniquely determined by $\widehat{\ell}$ via Lemma 15, the mapping is also injective. Finally, $\widehat{\mathcal{A}}$ preserves the state behaviour and weights, so every run of $\mathcal{A}'$ for t to some $q \neq \bot$ corresponds to a run of $\widehat{\mathcal{A}}$ for $\widehat{t}$ to q, and vice versa, which proves the claim.

For illustration purposes, consider cases where Lemmas 14 and 15 do not hold.

Example 17. Recall the WTAh $\mathcal{A}'$ recognizing $h(\|\mathcal{A}\|)$ from Examples 5 and 6. Since h is tetris-free, $\mathcal{A}'$ satisfies Lemmas 14 and 15. If we add $\varphi^{(2)}$ to the input alphabet Σ, extend $\mathcal{A}$ to, say, $\mathcal{B}$ by adding the rule $\varphi(q, q) \to_{-2} q_f$, and extend h to $h_\star$ via $h_\star(\varphi) = f(x_1,\ g(a, x_2),\ g(a, x_1))$, then $h_\star$ is not tetris-free. The eq-restricted WTAh $\mathcal{B}'$ for $h_\star(\|\mathcal{B}\|)$ has the rule $f(q,\ g(a, q),\ g(a, \bot)) \xrightarrow{1=32}_{-2} q_f$, which allows an additional run $\varrho_\star$ for our tree $t = f(a,\ g(a, a),\ g(a, a))$:

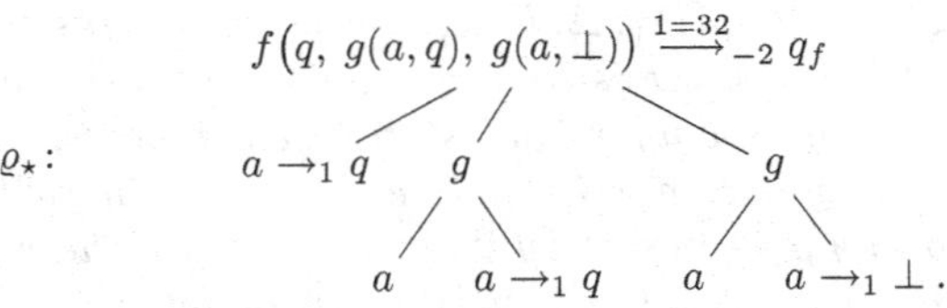

It is $\mathrm{wt}(\varrho) + \mathrm{wt}(\varrho_\star) = 0$, hence $t \notin \mathrm{supp}(\|\mathcal{B}'\|)$. The rules at $\varrho(\varepsilon)$ and $\varrho_\star(\varepsilon)$ have different Δ parts, so the statement in Lemma 14 does not hold. Indeed if we construct $\widehat{\mathcal{B}}$, we obtain the new symbol $\big[f\big(\bot, g(a,\bot), g(a,\bot)\big)\big]^{(2)} \in \widehat{\Delta}$ which provides a second tree $\widehat{t}_\star \in T_{\widehat{\Delta}}$ related to t:

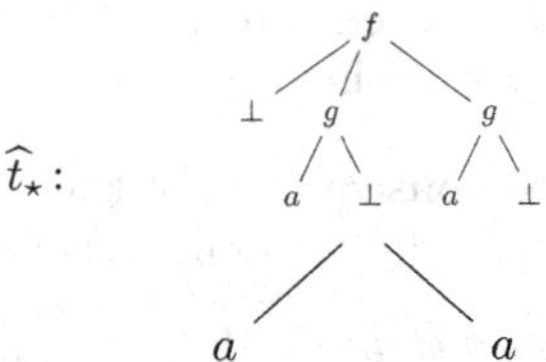

So, while the translation $t \mapsto \widehat{t}$ is still injective, it is not well-defined anymore. Moreover, it is $\widehat{t}, \widehat{t}_\star \in \mathrm{supp}(\|\widehat{\mathcal{B}}\|)$, despite $t \notin \mathrm{supp}(\|\mathcal{B}'\|)$.

On the other hand, instead of $\varphi^{(2)}$ let us add $\beta^{(0)}$ and $\kappa^{(2)}$ to Σ, and b to Δ. We extend $\mathcal{A}$ to, say, $\mathcal{C}$ by adding the rules $\beta \to_1 q$ and $\kappa(q,q) \to_1 q_f$, and h to $h^\star$ by setting $h^\star(\beta) = b$ and $h^\star(\kappa) = f(x_2, x_1, x_2)$. As before, $h^\star$ is not tetris-free. The WTAh $\mathcal{C}'$ has, compared to $\mathcal{A}'$, the additional rules $b \to_1 q$, $b \to_1 \bot$ and $f(q,q,\bot) \xrightarrow{1=3}_1 q_f$, so it does not satisfy Lemma 15. When constructing $\widehat{\mathcal{C}}$, we only add the symbol b to $\widehat{\Delta}$, but now there are two different rules whose Δ part is $f(\bot,\bot,\bot)$. It is $h^\star\big(\kappa(\alpha,\beta)\big) = f(b,a,b) \neq f(b,a,a) = h^\star\big(\psi(\alpha,\beta)\big)$; however, we have $\widehat{f(b,a,b)} = \widehat{f(b,a,a)} = \big[f(\bot,\bot,\bot)\big](b,a)$. Not only is it unclear which weight the rule $\big[f(\bot,\bot,\bot)\big](q,q) \to q_f$ should have in $\widehat{\mathcal{C}}$, but because the translation $t \mapsto \widehat{t}$ is not injective, we cannot recover $\|\mathcal{C}'\|$ from $\|\widehat{\mathcal{C}}\|$ anymore.

Next, we want to derive a pumping lemma for our WTAh $\mathcal{A}'$, which will be the foundation for deciding the weighted HOM-problem over fields. To this end, we apply the well-known pumping lemma for WTA proved by Berstel and Reutenauer [1] to the WTA $\widehat{\mathcal{A}}$. We require one more definition: that of a context.

Definition 18. *Let Δ be a ranked alphabet and $\Box \notin \Delta$. Any $C \in T_\Delta(\{\Box\}) \setminus T_\Delta$ is called a* multi-context. *If $|\mathrm{pos}_\Box(C)| = 1$, then C is a* context. *For a multi-context C with $\mathrm{pos}_\Box(C) = \{p_1, \ldots, p_n\}$ and $t_1, \ldots, t_n \in T_\Sigma(\{\Box\})$, we abbreviate $C[t_1]_{p_1} \cdots [t_n]_{p_n}$ to $C[t_1, \ldots, t_n]$, and if $t_1 = \ldots = t_n = t$, we simply write $C[t]$.*

Let us now recall the pumping lemma for WTA over fields with a slight adjustment, namely that the pumping takes place below a certain position.

Theorem 19 (cf. [1, Theorem 9.2]). *Let $\mathbb{F}$ be a field, Σ a ranked alphabet and $\mathcal{B}$ a WTA over $\mathbb{F}$ and Σ. There exists $N \in \mathbb{N}$ s.t. for every context C and $t_0 \in T_\Sigma$ such that $C[t_0] \in \text{supp}(\|\mathcal{B}\|)$ and $\text{ht}(t_0) \geq N$, there exists a sequence of pairwise distinct trees $t_1, t_2, \ldots$ such that $C[t_i] \in \text{supp}(\|\mathcal{B}\|)$ for all $i \in \mathbb{N}$.*

From this, we obtain the desired pumping lemma for the WTAh $\mathcal{A}'$.

Proposition 20 (Pumping Lemma). *Let $\mathbb{F}$ be a field and $\mathcal{A}'$ the eq-restricted WTAh from Lemma 7 for a WTA over $\mathbb{F}$ and Σ, and a tetris-free tree homomorphism $h\colon T_\Sigma \to T_\Delta$. There exists $N \in \mathbb{N}$ such that for every multi-context C and $t_0 \in T_\Delta$ such that $t := C[t_0] \in \text{supp}(\|\mathcal{A}'\|)$, $\text{ht}(t_0) \geq N$, and $\text{pos}_\square(C)$ is an equivalence class of mutually constrained positions in t, there exist infinitely many pairwise distinct trees $t_1, t_2, \ldots$ such that $C[t_i] \in \text{supp}(\|\mathcal{A}'\|)$ for all $i \in \mathbb{N}$.*

Proof. Recall from Lemma 14 that the equivalence relation of positions that are mutually constrained by a run of t, is uniquely determined by t. Let $\widehat{\mathcal{A}}$ be the WTA for $\mathcal{A}'$ from Definition 12 and let $\widehat{N}$ be the pumping constant for $\widehat{\mathcal{A}}$ from Theorem 19. We set $N = \widehat{N} \cdot \max_{\sigma \in \Sigma} \text{ht}\big(h(\sigma)\big)$. Let t be as in the statement, then $\widehat{t} \in \text{supp}(\widehat{\mathcal{A}})$ is of the form $\widehat{t} = \widehat{C}[\widehat{t_0}]$ with a context $\widehat{C}$ and $\text{ht}(\widehat{t_0}) \geq \widehat{N}$. Thus by Theorem 19, there is a sequence of trees $\widehat{t_1}, \widehat{t_2}, \ldots$ such that $\widehat{C}[\widehat{t_i}] \in \text{supp}(\|\widehat{\mathcal{A}}\|)$ for all $i \in \mathbb{N}$. In turn, each $\widehat{t_i}$ has a unique preimage t_i under the mapping $t \mapsto \widehat{t}$, and $\widehat{C}$ translates uniquely back to C. Thus we obtain t_i, $i \in \mathbb{N}$ with $\|\mathcal{A}'\|\big(C[t_i]\big) = \|\widehat{\mathcal{A}}\|\big(\widehat{C[t_i]}\big) = \|\widehat{\mathcal{A}}\|\big(\widehat{C}[\widehat{t_i}]\big) \neq 0$ for all $i \in \mathbb{N}$. □

4 The Tetris-Free Weighted HOM-Problem

In this section, we prove that the weighted HOM-problem over fields, restricted to tetris-free homomorphisms, is decidable. Formally, we show the following result.

Theorem 21. *Let $\mathbb{F}$ be a field, $\mathcal{A}$ a WTA over $\mathbb{F}$ and Σ, and $h\colon T_\Sigma \to T_\Delta$ a tetris-free tree homomorphism. It is decidable whether $h(\|\mathcal{A}\|)$ is regular.*

The approach to prove this result is quite natural: Nonregularity of $h\big(\|\mathcal{A}\|\big)$ is reduced to the following decidable property of the WTAh $\mathcal{A}'$ for $h(\|\mathcal{A}\|)$.

Definition 22 (see [22, Definition 10]). *Let $\mathcal{A}' = \big(Q\dot{\cup}\{\bot\}, \Delta, F, R, \text{wt}\big)$ be the eq-restricted WTAh from Lemma 7 for a WTA over a field and a tetris-free tree homomorphism, and let N be the pumping constant of $\mathcal{A}'$. We say that $\mathcal{A}'$ has the* large duplication property (LDP) *if there exists $t \in \text{supp}(\|\mathcal{A}'\|)$ with an accepting run ϱ, a position $p \in \text{pos}_R(\varrho)$ where $\varrho(p)$ has a nontrivial constraint set E, and a position p' that is constrained by E such that $\text{ht}(t|_{pp'}) \geq N$.*

A constraint that only acts on finitely many trees is expendable, since we can process these particular trees manually using additional states. If, however, $\mathcal{A}'$ has the LDP, then by our pumping lemma we obtain infinitely many trees to which a nontrivial constraint E applies, so we cannot bypass E. Thus, the LDP indicates that the constraints are indeed indispensable for representing $\|\mathcal{A}'\|$, and in turn these constraints cause nonregularity, as stated in Proposition 24.

The decision procedure of [22] for input $\mathcal{A}$ and h as above is now as follows.

1. Construct an eq-restricted WTAh $\mathcal{A}'$ recognizing $h(\|\mathcal{A}\|)$ via Lemma 7.
2. If $\mathcal{A}'$ has the LDP, then $h(\|\mathcal{A}\|)$ is not regular.
3. If $\mathcal{A}'$ does not have the LDP, then $h(\|\mathcal{A}\|)$ is regular.

For this procedure to be correct, the LDP must be (i) decidable and (ii) equivalent to the nonregularity of $\|\mathcal{A}'\|$. While proving (ii) only requires technical adaptations compared to [22], (i) presents new challenges since the pumping lemma for fields is weaker. We prove (i) indirectly by examining the WTA $\widehat{\mathcal{A}}$.

Proposition 23 (cf. [22, Lemma 11]). *Given as input a WTA $\mathcal{A}$ over a field, and a tetris-free tree homomorphism h, it is decidable whether the eq-restricted WTAh $\mathcal{A}'$ for $h(\|\mathcal{A}\|)$ from Lemma 7 has the LDP.*

Proof. Adopting the notation from Definition 22, let $t_0 = t|_{pp'}$. We will not decide the existence of a tree $t = C[t_0]$ in $\mathrm{supp}(\|\mathcal{A}'\|)$ as in the LDP directly, but instead decide whether its counterpart $\widehat{C[t_0]}$ exists in $\mathrm{supp}(\|\widehat{\mathcal{A}}\|)$. Consider thus the WTA $\widehat{\mathcal{A}}$ for $\mathcal{A}'$ constructed in Definition 12. We modify $\widehat{\mathcal{A}}$ by implementing a counter into its state set, which ensures that only trees of height less than N are attached to positions that are constrained in $\mathcal{A}'$. Then we check if any trees have been lost from $\mathrm{supp}(\|\widehat{\mathcal{A}}\|)$ in the process. If so, then the counterparts of these lost trees in $\mathcal{A}'$ confirm the LDP, otherwise $\mathcal{A}'$ does not have the LDP.

Formally, let $q \neq \bot$ and $\ell \xrightarrow{E} q$ a rule of $\mathcal{A}'$ with $\mathrm{pos}_{Q\setminus\{\bot\}}(\ell) = \{p_1, \ldots, p_k\}$ ordered lexicographically, and $\ell(p_i) = q_i$ for all $i \in [k]$. Suppose that $p_{i_1}, \ldots, p_{i_j}$ are the positions constrained by E. Then $\mathcal{A}'$ has the rule $\widehat{\ell}(q_1, \ldots, q_k) \to q$, and we replace it by the collection of all $\widehat{\ell}(\langle q_1, n_1\rangle, \ldots, \langle q_k, n_k\rangle) \to \langle q, n\rangle$ such that $n_1, \ldots, n_k, n \in [N]$, $n = \min\{\max_{i\in[k]}(n_i + |p_i|), N\}$ and $n_{i_1}, \ldots, n_{i_j} < N$. All these new rules have the same weight as $\widehat{\ell}(q_1, \ldots, q_k) \to q$. This operation is well-defined since by Lemma 15, the constraint E is uniquely determined by $\widehat{\ell}$. We proceed this way for every rule of $\widehat{\mathcal{A}}$ and denote the resulting WTA by $\widehat{\mathcal{B}}$.

Consider now the WTA recognizing $\|\widehat{\mathcal{A}}\| - \|\widehat{\mathcal{B}}\|$ defined by a disjoint union. Subtracting $\|\widehat{\mathcal{B}}\|$ removes all $\widehat{t}$ from $\mathrm{supp}(\|\widehat{\mathcal{A}}\|)$ s.t. all subtrees of $t \in \mathrm{supp}(\|\mathcal{A}'\|)$ pending from constrained positions are of height less than N; thus, the WTA for $\|\widehat{\mathcal{A}}\| - \|\widehat{\mathcal{B}}\|$ only accepts trees whose counterparts in $\mathcal{A}'$ satisfy the LDP. It remains to decide whether $\|\widehat{\mathcal{A}}\| - \|\widehat{\mathcal{B}}\|$ is the zero function by minimizing the WTA for it [2,3] and checking whether it has zero states. If indeed $\|\widehat{\mathcal{A}}\| = \|\widehat{\mathcal{B}}\|$, then $\mathcal{A}'$ has no tree that satisfies the condition of the LDP. If, however, there exists $\widehat{t} \in \mathrm{supp}(\|\widehat{\mathcal{A}}\|) \setminus \mathrm{supp}(\|\widehat{\mathcal{B}}\|)$, then its counterpart t satisfies the LDP. □

Finally, we can complete the proof of Theorem 21. The following statement, whose proof applies RAMSEY'S THEOREM [26], was first presented in [22] for the case of ℕ" weights (without the assumption of tetris-freeness). We have adapted the proof to our case and included it in the full-length version [24].

Proposition 24 (cf. [22, Prop. 13 and Thm. 17]). *Let $\mathcal{A}$ be a WTA over a field $\mathbb{F}$, h a tetris-free tree homomorphism, and $\mathcal{A}'$ the WTAh for $h(\|\mathcal{A}\|)$ constructed in Lemma 7. Then $h(\|\mathcal{A}\|)$ is regular iff $\mathcal{A}'$ does not have the LDP.*

Restrictihg the HOM-problem over fields to tetris-free tree homomorphisms is of essence: On the one hand, we use this assumption to construct a well-defined WTA $\widehat{\mathcal{A}}$ when proving that the LDP is decidable in Proposition 23. On the other hand, the statement of Proposition 24, which reduces the weighted HOM-problem to the LDP, also does not hold if h is not tetris-free:

Example 25. Consider $\mathcal{A}' = (\{q, q_f, q'_f, \bot\}, \{a^{(0)}, g^{(1)}, f^{(2)}\}, \{q_f, q'_f\}, R, \mathrm{wt})$ with

$$R = \{ \quad a \to_1 q, \qquad g(q) \to_1 q, \qquad f(q,q) \to_3 q_f, \\ f(q,\bot) \xrightarrow{1=2}_2 q_f, \qquad f(q,\bot) \xrightarrow{1=2}_{-2} q'_f \quad \} \cup R_\bot$$

where $R_\bot = \{a \to_1 \bot, g(\bot) \to_1 \bot, f(\bot,\bot) \to_1 \bot\}$.

The WTAh $\mathcal{A}'$ represents the image of a WTA under a suitable tree homomorphism, but not under any tetris-free one since $\mathcal{A}'$ does not satisfy Lemma 14. It is easy to see that $\mathcal{A}'$ has the LDP, e.g. with $C = f(\square, \square)$ and the sequence $t_i = g^i(a)$. However, the accepting runs for $C[t_i]$ that use constraints cancel each other out. Despite $\mathcal{A}'$ having the LDP, $\|\mathcal{A}'\|$ is the regular tree series with $\mathrm{supp}(\|\mathcal{A}'\|) = \{t \mid \mathrm{pos}_f(t) = \{\varepsilon\}\}$ and $\|\mathcal{A}'\|\colon f(g^i(a), g(^j(a)) \mapsto 3$ for all $i, j \in \mathbb{N}$. Thus, without the tetris-free assumption, Proposition 24 does not hold.

5 Conclusion

In this paper, we have proved that the weighted HOM-problem over fields for tetris-free tree homomorphisms is decidable. Formally, for a WTA $\mathcal{A}$ over a field, and a tetris-free tree homomorphism h as input, it is decidable whether $h(\|\mathcal{A}\|)$ is again regular. A tree homomorphism is tetris-free if its non-injective behaviour is located only at the symbol level, thus this property generalizes injectivity.

Our proof strategy is similar to [22]: We have reduced the HOM-problem to a decidable property of (the WTAh that recognizes) $h(\|\mathcal{A}\|)$. The homomorphism h has the ability to duplicate subtrees of its input trees, and we have shown that $h(\|\mathcal{A}\|)$ is regular iff h duplicates only finitely many subtrees of trees accepted by $\mathcal{A}$. This limited duplication is in turn decidable, and proving its decidability is our main contribution. For this, we presented a pumping lemma for the WTAh recognizing $h(\|\mathcal{A}\|)$, by translating it into a WTA and applying the pumping lemma for WTA over fields proved in [1].

Finally, we illustrate in our last example why our proof strategy for fields – unlike in the integer case [22] – requires tetris-freeness.

References

1. Berstel, J., Reutenauer, C.: Recognizable formal power series on trees. Theor. Comput. Sci. **18**(2), 115–148 (1982)
2. Bozapalidis, S.: Effective construction of the synthetic algebra of a recognizable series on trees. Acta Informatica **28**(4), 351–363 (1991)
3. Bozapalidis, S., Alexandrakis, A.: Représentations matricielles des séries d'arbre reconnaissables. RAIRO-Theor. Inform. Appl. **23**(4), 449–459 (1989)
4. Bozapalidis, S., Kalampakas, A.: Graph automata. Theor. Comput. Sci. **393**(1–3), 147–165 (2008). https://doi.org/10.1016/j.tcs.2007.11.022
5. Comon, H., et al.: Tree automata – Techniques and applications (2007)
6. Creus, C., Gascón, A., Godoy, G., Ramos, L.: The HOM problem is EXPTIME-complete. In: Proceedings of the 27th Annual IEEE Symposium on Logic in Computer Science, pp. 255–264. IEEE (2012). https://doi.org/10.1109/LICS.2012.36
7. Doner, J.: Tree acceptors and some of their applications. J. Comput. Syst. Sci. **4**(5), 406–451 (1970). https://doi.org/10.1016/S0022-0000(70)80041-1
8. Drewes, F.: Grammatical Picture Generation. Springer, Berlin, Heidelberg (2006). https://doi.org/10.1007/3-540-32507-7
9. Droste, M., Kuich, W., Vogler, H.: Handbook of Weighted Automata. Springer, Berlin, Heidelberg (2009). https://doi.org/10.1007/978-3-642-01492-5
10. Droste, M., Kuske, D.: Weighted automata (2021)
11. Fülöp, Z., Vogler, H.: Weighted tree automata and tree transducers. In: Handbook of Weighted Automata [9], chap. 9, pp. 313–403. https://doi.org/10.4171/Automata-1/4
12. Gécseg, F., Steinby, M.: Tree automata. Tech. Rep. 1509.06233, arXiv (2015)
13. Gilleron, R., Tison, S.: Regular tree languages and rewrite systems. Fund. Inform. **24**(1–2), 157–175 (1995)
14. Godoy, G., Giménez, O.: The HOM problem is decidable. J. ACM **60**(4), 1–44 (2013)
15. Godoy, G., Giménez, O., Ramos, L., Àlvarez, C.: The HOM problem is decidable. In: Proceedings of the 42nd ACM Symposium on Theory of Computing, pp. 485–494. ACM (2010). https://doi.org/10.1145/1806689.1806757
16. Golan, J.S.: Semirings and their Applications. Kluwer Academic, Dordrecht (1999). https://doi.org/10.1007/978-94-015-9333-5
17. Hebisch, U., Weinert, H.J.: Semirings. World Scientific (1998). https://doi.org/10.1142/3903
18. Hebisch, U., Weinert, H.J.: Semirings: algebraic theory and applications in computer science, vol. 5. World Scientific (1998)
19. Jurafsky, D., Martin, J.H.: Speech and Language Processing. Prentice Hall, Hoboken (2008)
20. Maletti, A., Nász, A.T.: Weighted tree automata with constraints. In: Diekert, V., Volkov, M. (eds.) Developments in Language Theory. DLT 2022. LNCS, vol. 13257, pp. 226–238. Springer, Cham (2022). https://doi.org/10.1007/978-3-031-05578-2_18
21. Maletti, A., Nász, A.T.: Weighted tree automata with constraints. Theory Comput. Syst. 1–28 (2023). https://arxiv.org/abs/2302.03434
22. Maletti, A., Nász, A., Paul, E.: Weighted HOM-problem for nonnegative integers (2023). https://doi.org/10.48550/arXiv.2305.04117
23. Nász, A.T.: Solving the weighted HOM-problem with the help of unambiguity. In: Gazdag, Z., Iván, S. (eds.) Proceedings of the 16th International Conference on Automata and Formal Languages, p. 200. Open Publishing Association (2023)

24. Nász, A.T.: The weighted HOM-problem over fields (2023). https://doi.org/10.48550/arXiv.2311.11067
25. Perrin, D.: Recent results on automata and infinite words. In: Chytil, M.P., Koubek, V. (eds.) MFCS 1984. LNCS, vol. 176, pp. 134–148. Springer, Heidelberg (1984). https://doi.org/10.1007/BFb0030294
26. Ramsey, F.P.: On a problem of formal logic. Proc. Lond. Math. Soc. **30**, 1–24 (1930)
27. Rosenfeld, A.: Picture Languages: Formal Models for Picture Recognition. Academic Press, Cambridge (2014)
28. Salomaa, A., Soittola, M.: Automata-Theoretic Aspects of Formal Power Series. Texts and Monographs in Computer Science. Springer, New York (1978). https://doi.org/10.1007/978-1-4612-6264-0
29. Schützenberger, M.P.: On the definition of a family of automata. Inform. Control **4**(2–3), 245–270 (1961). https://doi.org/10.1016/S0019-9958(61)80020-X
30. Schwentick, T.: Automata for xml-a survey. J. Comput. Syst. Sci. **73**(3), 289–315 (2007)
31. Thatcher, J.W.: Characterizing derivation trees of context-free grammars through a generalization of finite automata theory. J. Comput. Syst. Sci. **1**(4), 317–322 (1967). https://doi.org/10.1016/S0022-0000(67)80022-9
32. Thatcher, J.W., Wright, J.B.: Generalized finite automata theory with an application to a decision problem of second-order logic. Math. Syst. Theory **2**(1), 57–81 (1968). https://doi.org/10.1007/BF01691346
33. Wilhelm, R., Seidl, H., Hack, S.: Compiler Design – Syntactic and Semantic Analysis. Springer, Berlin, Heidelberg (2013). https://doi.org/10.1007/978-3-642-17540-4

Combinatorics of Block-Parallel Automata Networks

Kévin Perrot[1,2], Sylvain Sené[1,2], and Léah Tapin[2(✉)]

[1] Université publique, Marseille, France
[2] Aix Marseille Univ, CNRS, LIS, Marseille, France
leah.tapin@lis-lab.fr

Abstract. Automata networks are finite collections of entities (the automata), each automaton having its own set of possible states, which interact with each other over discrete time, interactions being defined as local functions allowing the automata to change their state according to the states of their neighbourhoods. Inspired by natural phenomena, the studies on this very abstract and expressive model of computation have underlined the very importance of the way (*i.e.* the schedule) according to which the automata update their states, namely the update modes which can be deterministic, periodic, fair, or not. Indeed, a given network may admit numerous underlying dynamics, these latter depending highly on the update modes under which we let the former evolve. In this paper, we focus on a new kind of deterministic, periodic and fair update mode family introduced recently in a modelling framework, called the block-parallel update modes by duality with the well-known and studied block-sequential update modes. We compare block-parallel to block-sequential update modes, then count them: (1) in absolute terms, (2) by keeping only representatives leading to distinct dynamics, and (3) by keeping only representatives giving rise to non-isomorphic limit dynamics. Put together, this paper constitutes a first theoretical analysis of these update modes and their impact on automata networks dynamics.

1 Introduction

Automata networks originated at the beginning of modern computer science in the 1940s, notably through the seminal works of McCulloch and Pitts on neural networks, and von Neumann on cellular automata, which have become since then widely studied models of computation. The former are classically embedded in a finite and heterogeneous structure (a graph) whereas the latter are embedded in an infinite but regular structure (a lattice). Whilst there exist deep differences between them, they both belong to the family of automata networks. This family groups together all distributed models of computation defined locally by means of automata which interact with each other over discrete time, so that the global computations they perform emerge from these local interactions governing them.

The end of the 1960s has underlined the prominent role of finite automata networks on which we focus in this paper, in the context of genetic regula-

H. Fernau et al. (Eds.): SOFSEM 2024, LNCS 14519, pp. 442–455, 2024.
https://doi.org/10.1007/978-3-031-52113-3_31

tion modelling [9,15]. The profiles of limit behaviours emerging from the system can represent for instance phenotypes, cellular types, or even biological rhythms [2,10]. The update modes have decisive effects on the dynamics of automata networks. Acquiring a better understanding of their influence has become a hot topic in the domain since Robert's seminal works on discrete iterations [14], leading to numerous further studies in the last two decades [1,11,12]. Works addressing the role of periodic update modes focused on block-sequential update modes, namely modes in which automata are partitioned into a list of subsets such that the automata of a same subset update their state all at once in parallel while the subsets are iterated sequentially.

We still do not know which "natural schedules" govern gene expression and regulation, although chromatin dynamics seems to play a key role [6,8]. In [4] an unexplored family, dual to the block-sequential one, is introduced and motivated, namely the family of block-parallel update modes. Rather than as as lists of sets, they are defined as sets of lists, or "partitioned orders", so that the automata of a list update their state sequentially according to the period of the list while the lists are triggered all in parallel at the initial time step. As highlighted by the authors, block-parallel update modes allow to capture endogenous biological timers/clocks of genetic or physiological origin, such as the aforementioned chromatin dynamics. Furthermore, they allow to break the property of fixed point set invariance (local update repetitions into a period are notably possible), letting automata networks have a richer range of dynamics. We give a first theoretical analysis of these modes, building a basis to further analyse their power of expressiveness.

Definitions are introduced in Sect. 2. Section 3 develops our main contributions and is divided into five parts. Section 3.1 characterises the update modes that are both block-sequential and block-parallel. We then address with closed formulas the counting of block-parallel modes: in absolute terms (Sect. 3.2), in terms of automata network dynamics (Sect. 3.3), up to isomorphic limit dynamics (Sect. 3.4). Numerical experiments are exposed in Sect. 3.5, suggesting that the number of block-parallel update modes to consider may be drastically reduced, when one is specifically interested in the asymptotic (limit) behaviour of dynamical systems. Perspectives are discussed in Sect. 4. Omitted and incomplete proofs can be found in the full version of this work [13].

2 Definitions

Let $[\![n]\!] = \{0, \ldots, n-1\}$, let $\mathbb{B} = \{0, 1\}$, let x_i denote the i-th component of vector $x \in \mathbb{B}^n$, let x_I denote the projection of x onto an element of $\mathbb{B}^{|I|}$ for some subset $I \subseteq [\![n]\!]$, let e_i be the i-th base vector, and $\forall x, y \in \mathbb{B}^n$, let $x + y$ denote the bitwise addition modulo two. Let $\sim$ denote the graph isomorphism, *i.e.* for $G = (V, A)$ and $G' = (V', A')$ we have $G \sim G'$ if and only if there is a bijection $\pi : V \to V'$ such that $(u, v) \in A \iff (\pi(u), \pi(v)) \in A'$.

Automata Networks. An automata network (AN) of size n is composed of a set of n *automata* $[\![n]\!]$, each holding a state from a finite alphabet X_i for $i \in [\![n]\!]$. A *configuration* is an element of $X = \prod_{i \in [\![n]\!]} X_i$. An *AN* is defined by a function $f : X \to X$, decomposed into n *local functions* $f_i : X \to X_i$ for $i \in [\![n]\!]$, where f_i is the i-th component of f. To obtain a discrete dynamical system on X, one must define when the automata update their state using their local function, which can be done in multiple ways, called *update modes.*

Block-Sequential Update Modes. A sequence $(W_\ell)_{\ell \in [\![p]\!]}$ with $W_\ell \subseteq [\![n]\!]$ for all $\ell \in [\![p]\!]$ is an *ordered partition* if and only if $\bigcup_{\ell \in [\![p]\!]} W_\ell = [\![n]\!]$ and $\forall i, j \in [\![p]\!], i \neq j \implies W_i \cap W_j = \emptyset$. An update mode $\mu = (W_\ell)_{\ell \in [\![p]\!]}$ is called *block-sequential* when μ is an ordered partition, and the W_ℓ are called *blocks.* The set of block-sequential update modes of size n is denoted BS_n. The update of f under $\mu \in \mathsf{BS}_n$ is given by $f_{(\mu)} : X \to X$ as follows:

$$f_{(\mu)}(x) = f_{(W_{p-1})} \circ \cdots \circ f_{(W_1)} \circ f_{(W_0)}(x),$$

$$\text{where } \forall i \in [\![n]\!],\ f_{(W_\ell)}(x)_i = \begin{cases} f_i(x) & \text{if } i \in W_\ell, \\ x_i & \text{otherwise.} \end{cases}$$

Block-Parallel Update Modes. In a block-sequential update mode, the automata in a block are updated simultaneously while the blocks are updated sequentially. A block-parallel update mode is based on the dual principle: the automata in a block are updated sequentially while the blocks are updated simultaneously. Instead of being defined as a sequence of unordered blocks, a block-parallel update mode will thus be defined as a set of ordered blocks. A set $\{S_k\}_{k \in [\![s]\!]}$ with $S_k = (i_0^k, \ldots, i_{n_k-1}^k)$ a sequence of $n_k > 0$ distinct elements of $[\![n]\!]$ for all $k \in [\![s]\!]$ is a *partitioned order* if and only if $\bigcup_{k \in [\![s]\!]} S_k = [\![n]\!]$ and $\forall i, j \in [\![s]\!], i \neq j \implies S_i \cap S_j = \emptyset$. An update mode $\mu = \{S_k\}_{k \in [\![s]\!]}$ is called *block-parallel* when μ is a partitioned order, and the sequences S_k are called *o-blocks* (for *ordered-blocks*). The set of block-parallel update modes of size n is denoted BP_n. With $p = \operatorname{lcm}(n_1, \ldots, n_s)$, the update of f under $\mu \in \mathsf{BP}_n$ is given by $f_{\{\mu\}} : X \to X$ as follows: $f_{\{\mu\}}(x) = f_{(W_{p-1})} \circ \cdots \circ f_{(W_1)} \circ f_{(W_0)}(x)$, where for all $\ell \in [\![p]\!]$ we define $W_\ell = \{i^k_{\ell \bmod n_k} \mid k \in [\![s]\!]\}$.

Basic Considerations. There is a natural way to convert a block-parallel update mode $\{S_k\}_{k \in [\![s]\!]}$ with $S_k = (i_0^k, \ldots, i_{n_k-1}^k)$ into a sequence of blocks of length $p = \operatorname{lcm}(n_1, \ldots, n_s)$. We define it as φ:

$$\varphi(\{S_k\}_{k \in [\![s]\!]}) = (W_\ell)_{\ell \in [\![p]\!]} \text{ with } W_\ell = \{i^k_{\ell \bmod n_k} \mid k \in [\![s]\!]\}.$$

In order to differentiate between sequences of blocks and sets of o-blocks, we denote by $f_{(\mu)}$ (resp. $f_{\{\mu\}}$) the dynamical system induced by f and μ when μ is a sequence of blocks (resp. a set of o-blocks), and simply f_μ when it is clear from the context. Moreover, abusing notations, we denote by $\varphi(\mathsf{BP}_n)$ the set of partitioned orders of $[\![n]\!]$ as sequences of blocks.

Block-sequential and block-parallel update modes are *periodic* (the same update procedure is repeated at each step), and *fair* (each automaton is updated

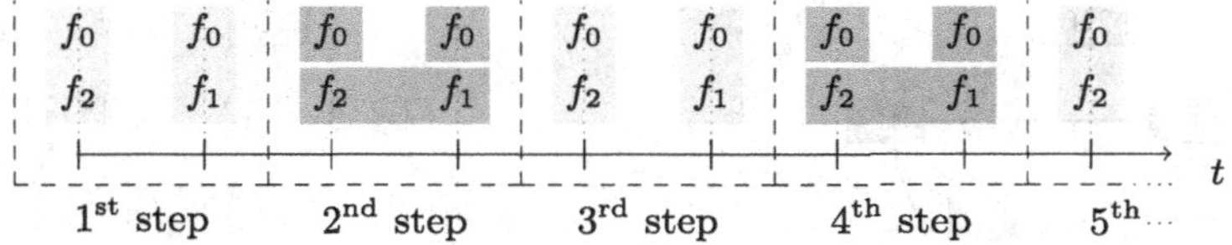

Fig. 1. Illustration of the execution along time of local transition functions according to block-parallel updating mode $\mu_{\mathsf{bp}} = \{(0), (2, 1)\}$.For the odd steps, we picture the blocks, and for the even steps, we picture the o-blocks.

at least once per step). We distinguish the concepts of *step* and *substep*. A step is the interval between x and $f_{(\mu)}(x)$ (or $f_{\{\mu\}}(x)$), and can be divided into $p = |\mu|$ (or $p = |\varphi(\mu)| = \text{lcm}(n_1, \ldots, n_s)$) substeps, corresponding to the elementary intervals in which only one block of automata is updated. The most basic update mode is the parallel μ_{par} which updates simultaneously all automata at each step. It is the element $([\![n]\!]) \in \mathsf{BS}_n$ and $\{(i) \mid i \in [\![n]\!]\} \in \mathsf{BP}_n$, with $\varphi(\{(i) \mid i \in [\![n]\!]\}) = ([\![n]\!])$.

Remark 1. Observe that in block-sequential update modes, each automaton is updated exactly once during a step, whereas in block-parallel update modes, some automata can be updated multiple times during a step. Update repetitions may have many consequences on the limit dynamics. For instance, the network of $n = 3$ automata such that $f_i(x) = x_{i-1 \mod n}$ under the update mode $\mu = (\{1, 2\}, \{0, 2\}, \{0, 1\})$, where each automaton is updated twice during a step, has 4 fixed points, among which 2, namely 010 and 101, cannot be obtained with block-sequential update modes (in this example, $\mu \notin \mathsf{BP}_n$).

Remark 2. Let $\mu = \{S_k\}_{k \in [\![s]\!]}$ be a block-parallel update mode. Each block of $\varphi(\mu)$ is of the same size, namely s, and furthermore each block of $\varphi(\mu)$ is unique.

Fixed Points, Limit Cycles and Attractors. Let f_μ be the dynamical system defined by an AN f of size n and an update mode μ. Let $p \geq 1$. A sequence of configurations $x^0, \ldots, x^{p-1} \in X$ is a *limit cycle* of f_μ if and only if $\forall i \in [\![p]\!], f_\mu(x^i) = x^{i+1 \mod p}$. A limit cycle of length $p = 1$ is a *fixed point*. The sequence of configurations $x^0, x^1, \ldots, x^{p-1} \in X$ is an *attractor* if and only if it is a limit cycle and there exist $x \in X$ and $i \in [\![p]\!]$ such that $f_\mu(x) = x^i$ but $x \notin \{x^0, \ldots, x^{p-1}\}$.

Example 1. Let $f : [\![3]\!] \times \mathbb{B} \times \mathbb{B} \to [\![3]\!] \times \mathbb{B} \times \mathbb{B}$ the automata network defined as:

$$f(x) = \begin{pmatrix} f_0(x) = \begin{cases} 0 & \text{if } ((x_0 = 0) \wedge (x_1 = x_2)) \vee (x_0 = x_1 = x_2 = 1) \\ 1 & \text{if } x_1 + x_2 \mod 2 = 1 \\ 2 & \text{otherwise} \end{cases} \\ f_1(x) = (x_0 \neq 0) \vee x_1 \vee x_2 \\ f_2(x) = ((x_0 = 1) \wedge x_1) \vee (x_0 = 2) \end{pmatrix}.$$

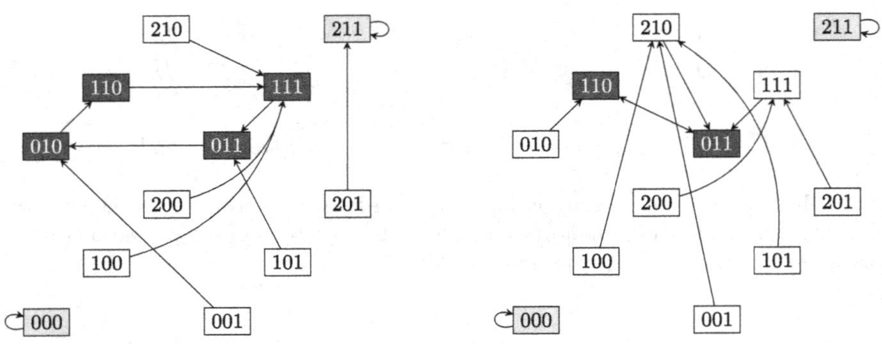

Fig. 2. The dynamics of $f_{(\mu_{\mathsf{bs}})}$ (left) and $f_{\{\mu_{\mathsf{bp}}\}}$ (right) from 1.

Let $\mu_{\mathsf{bs}} = (\{1\}, \{0, 2\})$ and $\mu_{\mathsf{bp}} = \{(0), (2, 1)\}$. The update mode μ_{bs} is block-sequential and μ_{bp} is block-parallel, with $\varphi(\mu_{\mathsf{bp}}) = (\{0, 2\}, \{0, 1\})$ as depicted in Fig. 1. Systems $f_{(\mu_{\mathsf{bs}})}$ and $f_{\{\mu_{\mathsf{bp}}\}}$ have different dynamics, as depicted in Fig. 2. They both have the same two fixed points and one limit cycle, but the similarities stop there. The limit cycle of $f_{(\mu_{\mathsf{bs}})}$ is of size 4, while that of $f_{\{\mu_{\mathsf{bp}}\}}$ is of size 2. Moreover, neither of the fixed points of $f_{\{\mu_{\mathsf{bp}}\}}$ is an attractor, while one of $f_{(\mu_{\mathsf{bs}})}$, namely 211, is. Both of these update modes' dynamics are unique in $\mathsf{BP}_3 \cup \mathsf{BS}_3$.

3 Counting Block-Parallel Update Modes

For the rest of this section, let $p(n)$ denote the number of integer partitions of n (multisets of integers summing to n), let $d(i)$ be the maximal part size in the i-th partition of n, let $m(i, j)$ be the multiplicity of the part of size j in the i-th partition of n. As an example, let $n = 31$ and assume the i-th partition is $(2, 2, 3, 3, 3, 3, 5, 5, 5)$, we have $d(i) = 5$ and $m(i, 1) = 0$, $m(i, 2) = 2$, $m(i, 3) = 4$, $m(i, 4) = 0$, $m(i, 5) = 3$. A partition will be the support of a partitioned order, where each part is an o-block. In our example, we can have:

$$\{(0,1),(2,3),(4,5,6),(7,8,9),(10,11,12),(13,14,15),\\(16,17,18,19,20),(21,22,23,24,25),(26,27,28,29,30)\},$$

and we picture it as the following *matrix-representation*:

$$\begin{pmatrix} 0 & 1 \\ 2 & 3 \end{pmatrix} \begin{pmatrix} 4 & 5 & 6 \\ 7 & 8 & 9 \\ 10 & 11 & 12 \\ 13 & 14 & 15 \end{pmatrix} \begin{pmatrix} 16 & 17 & 18 & 19 & 20 \\ 21 & 22 & 23 & 24 & 25 \\ 26 & 27 & 28 & 29 & 30 \end{pmatrix}.$$

We call *matrices* the elements of size $j \cdot m(i, j)$ and denote them $M_1, \ldots, M_{d(i)}$, where M_j has $m(i, j)$ *rows* and j *columns* (M_j is empty when $m(i, j) = 0$). The partition defines the dimensions of the matrices, and each row is an o-block.

For the comparison, the block-sequential update modes (ordered partitions of $[\![n]\!]$) are given by the ordered Bell numbers, sequence OEIS A000670. A closed formula for it is:

$$|\mathsf{BS}_n| = \sum_{i=1}^{p(n)} \frac{n!}{\prod_{j=1}^{d(i)} (j!)^{m(i,j)}} \cdot \frac{\left(\sum_{j=1}^{d(i)} m(i,j)\right)!}{\prod_{j=1}^{d(i)} m(i,j)!}.$$

Intuitively, an ordered partition of n gives a support to construct a block-sequential update mode: place the elements of $[\![n]\!]$ up to permutation within the blocks. This is the left fraction: $n!$ divided by $j!$ for each block of size j, taking into account multiplicities. The right fraction corrects the count because we sum on $p(n)$ the (unordered) partitions of n: each partition of n can give rise to different ordered partitions of n, by ordering all blocks (numerator, where the sum of multiplicities is the number of blocks) up to permutation within blocks of the same size which have no effect (denominator). The first ten terms are ($n = 1$ onward): 1, 3, 13, 75, 541, 4683, 47293, 545835, 7087261, 102247563.

3.1 Intersection of Block-Sequential and Block-Parallel Modes

In order to be able to compare block-sequential with block-parallel update modes, both of them will be written here in their sequence of blocks form (the usual form of block-sequential update modes and the rewritten form of block-parallel modes). First, $\varphi(\mathsf{BP}_n) \cap \mathsf{BS}_n \neq \emptyset$, since it contains at least $\mu_{\mathsf{par}} = ([\![n]\!]) = \varphi(\{(0),(1),\ldots,(n-1)\})$. However, neither $\mathsf{BS}_n \subseteq \varphi(\mathsf{BP}_n)$ nor $\varphi(\mathsf{BP}_n) \subseteq \mathsf{BS}_n$ are true. Indeed, $\mu_s = (\{0,1\},\{2\}) \in \mathsf{BS}_3$ but $\mu_s \notin \varphi(\mathsf{BP}_3)$ since a block-parallel cannot have blocks of different sizes in its sequential form. Symmetrically, $\mu_p = \varphi(\{(1,2),(0)\}) = (\{0,1\},\{0,2\}) \in \mathsf{BP}_3$ but $\mu_p \notin \mathsf{BS}_3$ since automaton 0 is updated twice. Nonetheless, we can precisely define their intersection.

Lemma 1. *$\mu \in (\mathsf{BS}_n \cap \varphi(\mathsf{BP}_n))$ if and only if μ is an ordered partition with p blocks of the same size s, if and only if there exists a partitioned order μ' with s o-blocks of the same size p such that $\varphi(\mu') = \mu$.*

Proof. Let $n \in \mathbb{N}$. We prove the first equivalence, the second follows directly.

($\Longrightarrow$) Let $\mu \in (\mathsf{BS}_n \cap \varphi(\mathsf{BP}_n))$. Since $\mu \in \mathsf{BS}_n$, μ is an ordered partition. Furthermore, $\mu \in \varphi(\mathsf{BP}_n)$ so all the μ's blocks are of the same size (Remark 2).

($\Longleftarrow$) Let $\mu = (W_\ell)_{\ell \in [\![p]\!]}$ be an ordered partition of $[\![n]\!]$ with all its blocks having the same size, denoted by s. Since μ is an ordered partition, $\mu \in \mathsf{BS}_n$. For each $\ell \in [\![p]\!]$, we can number arbitrarily the elements of W_ℓ from 0 to $s-1$ as $W_\ell = \{W_\ell^0, \ldots, W_\ell^{s-1}\}$. Now, let us define the set of sequences $\{S_k\}_{k \in [\![s]\!]}$ the following way: $\forall k \in [\![s]\!], S_k = \{W_\ell^k \mid \ell \in [\![p]\!]\}$. It is a partitioned order such that $\varphi(\{S_k\}_{k \in [\![s]\!]}) = \mu$, which means that $\mu \in \varphi(\mathsf{BP}_n)$. □

As a consequence of Lemma 1, given $n \in \mathbb{N}$, the set SEQ_n of sequential update modes such that every automaton is updated exactly once by step and only one automaton is updated by substep, is a subset of $(\mathsf{BS}_n \cap \varphi(\mathsf{BP}_n))$. Moreover, we can count the number of sequences of blocks in the intersection.

Proposition 1. *Given $n \in \mathbb{N}$, we have $|\mathsf{BS}_n \cap \varphi(\mathsf{BP}_n)| = \sum_{d|n} \frac{n!}{(\frac{n}{d}!)^d}$.*

Proof. The proof derives directly from the sequence OEIS A061095, which counts the *number of ways of dividing n labeled items into labeled boxes with an equal number of items in each box.* In our context, the "items" are the automata, and the "labeled boxes" are the blocks of the ordered partitions. □

3.2 Partitioned Orders

A block-parallel update mode is given as a partitioned order, *i.e.* an (unordered) set of (ordered) sequences. This concept is recorded as sequence OEIS A000262, described as the *number of "sets of lists"*. A nice closed formula for it is:

$$|\mathsf{BP}_n| = \sum_{i=1}^{p(n)} \frac{n!}{\prod_{j=1}^{d(i)} m(i,j)!}.$$

Intuitively, for each partition, fill all matrices ($n!$ ways to place the elements of $[\![n]\!]$) up to permutation of the rows within each matrix (matrix M_j has $m(i,j)$ rows). Another closed formula is presented in Proposition 2, which is used as the basis of implementations in Sect. 3.5. The first ten terms are ($n = 1$ onward): 1, 3, 13, 73, 501, 4051, 37633, 394353, 4596553, 58941091.

Proposition 2. *For any $n \geq 1$ we have:*

$$|\mathsf{BP}_n| = \sum_{i=1}^{p(n)} \prod_{j=1}^{d(i)} \binom{n - \sum_{k=1}^{j-1} k \cdot m(i,k)}{j \cdot m(i,j)} \cdot \frac{(j \cdot m(i,j))!}{m(i,j)!}.$$

Proof. Each partition is a support to generate different partitioned orders (sum on i), by considering all the combinations, for each matrix (product on j), of the ways to choose the $j \cdot m(i,j)$ elements of $[\![n]\!]$ it contains (binomial coefficient, chosen among the remaining elements), and all the ways to order them up to permutation of the rows (ratio of factorials). Observe that developing the binomial coefficients with $\binom{x}{y} = \frac{x!}{y! \cdot (x-y)!}$ gives

$$\prod_{j=1}^{d(i)} \binom{n - \sum_k}{j \cdot m(i,j)} \cdot (j \cdot m(i,j))! = \prod_{j=1}^{d(i)} \frac{(n - \sum_k)!}{(n - \sum_k - j \cdot m(i,j))!} = \frac{n!}{0!} = n!,$$

where $\sum_k$ is a shorthand for $\sum_{k=1}^{j-1} k \cdot m(i,k)$, which leads to retrieve the OEIS formula. □

3.3 Partitioned Orders up to Dynamical Equality

As for block-sequential update modes, given an AN f and two block-parallel update modes μ and μ', the dynamics of f under μ can be the same as that of f under μ'. To go further, in the framework of block-parallel update modes, there exist pairs of update modes μ, μ' such that for any AN f, the dynamics $f_{\{\mu\}}$ is the exact same as $f_{\{\mu'\}}$. As a consequence, in order to perform exhaustive searches among the possible dynamics, it is not necessary to generate all of them. We formalise this with the following equivalence relation.

Definition 1. *For $\mu, \mu' \in \mathsf{BP}_n$, we denote $\mu \equiv_0 \mu'$ when $\varphi(\mu) = \varphi(\mu')$.*

Example 2. Let $\mu_1 = \{(0,1),(2,3)\}$ and $\mu_2 = \{(2,1),(0,3)\}$. μ_1 and μ_2 are different partitioned orders, but $\varphi(\mu_1) = \varphi(\mu_2) = (\{0,2\},\{1,3\})$. Thus $\mu_1 \equiv_0 \mu_2$.

The following theorem shows that this equivalence relation is necessary and sufficient in the general case of ANs of size n, *i.e.* $\equiv_0$ captures the dynamical equivalence among block-parallel update modes.

Theorem 1. *For any $\mu, \mu' \in \mathsf{BP}_n$, $\mu \equiv_0 \mu' \iff \forall f : X \to X, f_{\{\mu\}} = f_{\{\mu'\}}$.*

Proof. Let μ and μ' be two block-parallel update modes of BP_n.

($\Longrightarrow$) Let us consider that $\mu \equiv_0 \mu'$, and let $f : X \to X$ be an AN. Then, we have $f_{\{\mu\}} = f_{(\varphi(\mu))} = f_{(\varphi(\mu'))} = f_{\{\mu'\}}$.

($\Longleftarrow$) Let us consider that $\forall f : X \to X, f_{\{\mu\}} = f_{\{\mu'\}}$. Let us assume for the sake of contradiction that $\varphi(\mu) \neq \varphi(\mu')$. For ease of reading, we will denote as $t_{\mu,i}$ the substep at which automaton i is updated for the first time with update mode μ. Then, there is a pair of automata (i,j) such that $t_{\mu,i} \leq t_{\mu,j}$, but $t_{\mu',i} > t_{\mu',j}$. Let $f : \mathbb{B}^n \to \mathbb{B}^n$ be a Boolean AN such that $f(x)_i = x_i \vee x_j$ and $f(x)_j = x_i$, and $x \in \mathbb{B}^n$ such that $x_i = 0$ and $x_j = 1$. We will compare $f_{\{\mu\}}(x)_i$ and $f_{\{\mu'\}}(x)_i$, in order to prove a contradiction. Let us apply $f_{\{\mu\}}$ to x. Before step $t_{\mu,i}$ the value of automaton i is still 0 and, most importantly, since $t_{\mu,i} \leq t_{\mu,j}$, the value of j is still 1. This means that right after step $t_{\mu,i}$, the value of automaton i is 1, and will not change afterwards. Thus, we have $f_{\{\mu\}}(x)_i = 1$. Let us now apply $f_{\{\mu'\}}$ to x. This time, $t_{\mu',i} > t_{\mu',j}$, which means that automaton j is updated first and takes the value of automaton i at the time, which is 0 since it has not been updated yet. Afterwards, neither automata will change value since $0 \vee 0$ is still 0. This means that $f_{\{\mu'\}}(x)_i = 0$. Thus, we have $f_{\{\mu\}} \neq f_{\{\mu'\}}$, which contradicts our earlier hypothesis. □

Let $\mathsf{BP}_n^0 = \mathsf{BP}_n / \equiv_0$ denote the corresponding quotient set, *i.e.* the set of block-parallel update modes to generate for exhaustive computer analysis of the possible dynamics in the general case of ANs of size n.

Theorem 2. *For any $n \geq 1$, we have:*

$$|\mathsf{BP}^0_n| = \sum_{i=1}^{p(n)} \frac{n!}{\prod_{j=1}^{d(i)} (m(i,j)!)^j} \tag{1}$$

$$= \sum_{i=1}^{p(n)} \prod_{j=1}^{d(i)} \prod_{\ell=1}^{j} \binom{n - \sum_{k=1}^{j-1} k \cdot m(i,k) - (\ell - 1) \cdot m(i,j)}{m(i,j)} \tag{2}$$

$$= \sum_{i=1}^{p(n)} \prod_{j=1}^{d(i)} \left(\binom{n - \sum_{k=1}^{j-1} k \cdot m(i,k)}{j \cdot m(i,j)} \cdot \prod_{\ell=1}^{j} \binom{(j - \ell + 1) \cdot m(i,j)}{m(i,j)} \right). \tag{3}$$

Proof. Formula 1 is a sum for each partition of n (sum on i), of all the ways to fill all matrices up to permutation within each column ($m(i,j)!$ for each of the j columns of M_j). Formula 2 is a sum for each partition of n (sum on i), of the product for each column of the matrices (products on j and ℓ), of the choice of elements (among the remaining ones) to fill the column (regardless of their order within the column). Formula 3 is a sum for each partition of n (sum on i), of the product for each matrix (product on j), of the choice of elements (among the remaining ones) to fill this matrix, multiplied by the number of ways to fill the columns of the matrix (product on ℓ) with these elements (regardless of their order within each column).

The equality of these three formulas is presented in [13] To prove that they count $|\mathsf{BP}^0_n|$, we now argue that for any pair $\mu, \mu' \in \mathsf{BP}_n$, we have $\mu \equiv_0 \mu'$ if and only if their matrix-representations are the same up to a permutation of the elements within columns (the number of equivalence classes is then counted by Formula 1). In the definition of φ, each block is a set constructed by taking one element from each o-block. Given that n_k in the definition of φ corresponds to j in the statement of the theorem, one matrix corresponds to all the o-blocks having the same size n_k. Hence, the $\ell \mod n_k$ operations in the definition of φ amounts to considering the elements of these o-blocks which are in a common column in the matrix representation. Since blocks are sets, the result follows. □

The first ten terms of the sequence $(|\mathsf{BP}^0_n|)_{n\geq 1}$ are: 1, 3, 13, 67, 471, 3591, 33573, 329043, 3919387, 47827093. They match the sequence OEIS A182666 (defined by its exponential generating function), and it is proven in [13] that they are indeed the same sequence.

3.4 Partitioned Orders up to Isomorphism on the Limit Dynamics

The following equivalence relation defined over block-parallel update modes turns out to capture exactly the notion of having isomorphic limit dynamics. It is analogous to $\equiv_0$, except that a circular shift of order i may be applied on the sequences of blocks.

Let σ^i denote the circular-shift of order $i \in \mathbb{Z}$ on sequences (shifting the element at position 0 towards position i).

Definition 2. *For $\mu, \mu' \in \mathsf{BP}_n$, we denote $\mu \equiv_\star \mu'$ when $\varphi(\mu) = \sigma^i(\varphi(\mu'))$ for some $i \in [\![|\varphi(\mu')|]\!]$ called the* shift. *Note that $\mu \equiv_0 \mu' \implies \mu \equiv_\star \mu'$.*

Notation 1. *Given $f_{\{\mu\}} : X \to X$, let $\Omega_{f_{\{\mu\}}} = \bigcap_{t \in \mathbb{N}} f^t{}_{\{\mu\}}(X)$ denote its* limit set *(abusing the notation of $f_{\{\mu\}}$ to sets of configurations), and $f^\Omega_{\{\mu\}} : \Omega_{f_{\{\mu\}}} \to \Omega_{f_{\{\mu\}}}$ its restriction to its limit set. Dynamics are deterministic, hence $f^\Omega_{\{\mu\}}$ is bijective.*

The next theorem shows that, if one is interested in the limit behaviour of ANs under block-parallel updates, then studying a representative from each equivalence class of the relation $\equiv_\star$ is necessary and sufficient to get the full spectrum of possible limit dynamics (recall that $\sim$ denotes graph isomorphism, which corresponds to the notion of conjugacy in dynamical system theory).

Theorem 3. *For any $\mu, \mu' \in \mathsf{BP}_n$, $\mu \equiv_\star \mu' \iff \forall f : X \to X, f^\Omega_{\{\mu\}} \sim f^\Omega_{\{\mu'\}}$.*

Proof (sketch). Let μ and μ' be two block-parallel update modes of BP_n.
($\Longrightarrow$) Let μ, μ' be such that $\mu \equiv_\star \mu'$ of shift $\hat{\imath} \in [\![p]\!]$, with $\varphi(\mu) = (W_\ell)_{\ell \in [\![p]\!]}$, $\varphi(\mu') = (W'_\ell)_{\ell \in [\![p]\!]}$ and $p = |\varphi(\mu)| = |\varphi(\mu')|$. It means that $\forall i \in [\![p]\!]$, we have $W'_i = W_{i+\hat{\imath} \bmod p}$, and for any AN f, we deduce that $\pi = f_{(W_0,\ldots,W_{\hat{\imath}-1})}$ is the desired isomorphism from $\Omega_{f_{\{\mu\}}}$ to $\Omega_{f_{\{\mu'\}}}$.
($\Longleftarrow$) We prove the contrapositive, from $\mu \not\equiv_\star \mu'$, by case analysis. In each case we build an AN f such that $f^\Omega_{\{\mu\}}$ is not isomorphic to $f^\Omega_{\{\mu'\}}$. In this sketch we detail only the simplest case.

(1) If in $\varphi(\mu)$ and $\varphi(\mu')$, there is an automaton $\hat{\imath}$ which is not updated the same number of times α and α' in μ and μ' respectively, then we assume without loss of generality that $\alpha > \alpha'$ and consider the AN f such that:
   - $X_{\hat{\imath}} = [\![\alpha]\!]$ and $X_i = \{0\}$ for all $i \neq \hat{\imath}$; and
   - $f_{\hat{\imath}}(x) = (x_{\hat{\imath}} + 1) \mod \alpha$ and $f_i(x) = x_i$ for all $i \neq \hat{\imath}$.

   It follows that $f^\Omega_{\{\mu\}}$ has only fixed points since $+1 \mod \alpha$ is applied α times, whereas $f^\Omega_{\{\mu'\}}$ has no fixed point because $\alpha' < \alpha$.
(2) If in $\varphi(\mu)$ and $\varphi(\mu')$, all the automata are updated the same number of times, then the transformation from μ to μ' is a permutation on $[\![n]\!]$ which preserves the matrices of their matrix representations. This case is harder and is fully presented in [13] through three subcases, in order to get extra hypotheses allowing to design specific ANs contradicting the isomorphism.

□

Let $\mathsf{BP}_n^\star = \mathsf{BP}_n / \equiv_\star$ denote the corresponding quotient set, *i.e.* the set of block-parallel update modes to generate for exhaustive computer analysis of the possible limit dynamics in the general case of ANs of size n.

Theorem 4. *Let $lcm(i) = lcm(\{j \in \{1, \ldots, d(i)\} \mid m(i,j) \geq 1\})$. For any $n \geq 1$, we have:*

$$|\mathsf{BP}_n^\star| = \sum_{i=1}^{p(n)} \frac{n!}{\prod_{j=1}^{d(i)} (m(i,j)!)^j} \cdot \frac{1}{lcm(i)}. \tag{4}$$

Proof. Let $\mu, \mu' \in \mathsf{BP}_n$ two update modes such that $\mu \equiv_\star \mu'$. Then their sequential forms are of the same length, and each automaton appears the same number of times in both of them. This means that, if an automaton is in an o-block of size k in μ's partitioned order form, then it is also in an o-block of the same size in μ''s. We deduce that two update modes of size n can only be equivalent as defined in Definition 2 if they are generated from the same partition of n.

Let $\mu \in \mathsf{BP}_n^0$, generated from partition i of n. Then $\varphi(\mu)$ is of length $\mathrm{lcm}(i)$. Since no two elements of BP_n^0 have the same block-sequential form, the equivalence class of μ in BP_n^0 contains exactly $\mathrm{lcm}(i)$ elements, all generated from the same partition i (all the blocks of $\varphi(\mu)$ are different). Thus, the number of elements of $\mathsf{BP}_n^\star$ generated from a partition i is the number of elements of BP_n^0 generated from partition i, divided by the number of elements in its equivalence class for $\mathsf{BP}_n^\star$, namely $\mathrm{lcm}(i)$. □

Remark 3. Formula 4 can actually be obtained from any formula in Theorem 2 by multiplying by $\frac{1}{\mathrm{lcm}(i)}$ inside the sum on partitions (from $i = 1$ to $p(n)$).

3.5 Implementations

Proof-of-concept Python implementations of three underlying enumeration algorithms for BP_n, BP_n^0 and $\mathsf{BP}_n^\star$ are available on the following repository: https://framagit.org/leah.tapin/blockpargen. We have conducted numerical experiments on a laptop, presented in Fig. 3. It shows the result of numerical experiments for n from 1 to 12.

Observe that the sizes of BP_n and BP_n^0 are comparable, whereas an order of magnitude is gained with BP_n^*, which may be significant for advanced numerical experiments regarding limit dynamics under block-parallel udpate modes. Enumerating BP_n, BP_n^0 and $\mathsf{BP}_n^\star$ up to $n = 8$ takes less than one second. For greater values, the time gain is significant when enumerating only the elements of $\mathsf{BP}_n^\star$.

n	BP_n	BP^0_n	$\mathsf{BP}^\star_n$
1	1 -	1 -	1 -
2	3 -	3 -	2 -
3	13 -	13 -	6 -
4	73 -	67 -	24 -
5	501 -	471 -	120 -
6	4051 -	3591 -	795 -
7	37633 -	33573 0.103s	5565 -
8	394353 0.523s	329043 0.996s	46060 0.161s
9	4596553 6.17s	3919387 12.2s	454860 1.51s
10	58941091 1min24s	47827093 2min40s	4727835 16.3s
11	824073141 21min12s	663429603 38min31s	54223785 3min13s
12	12470162233 5h27min38s	9764977399 9h49min26s	734932121 45min09s

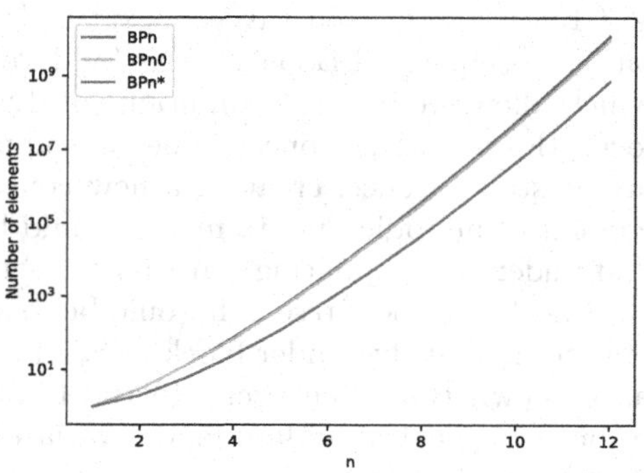

Fig. 3. Numerical experiments of our Python implementations on a standard laptop (processor Intel-Core™ i7 @ 2.80 GHz). For n from 1 to 12, the table (left) presents the size of BP_n, BP^0_n and $\mathsf{BP}^\star_n$ and running time to enumerate their elements (one representative of each equivalence class; a dash represents a time smaller than 0.1 second), and the graphics (right) depicts their respective sizes on a logarithmic scale.

4 Conclusion and Perspectives

In this article we settle the theoretical foundations to the study of block-parallel update modes in the AN setting. We first characterise their intersection with the classical block-sequential modes. Then, we provide closed formulas for counting: notably (1) a minimal set of representatives of block-parallel update modes that allow to generate the full spectrum of possible distinct dynamics, (2) a minimal set of representatives of block-parallel update modes that allow to generate the full spectrum of possible distinct limit dynamics up to isomorphism (*i.e.* the limit cycles lengths and distribution). Numerical experiments show that the computational gain is significant, in particular for the exhaustive study of how/when the fixed point invariance property is broken.

A major feature of block-parallel update modes is that they allow local update repetitions during a period. This is indeed the case for all block-parallel update modes which are not block-sequential (*i.e.* modes with at least two blocks of distinct sizes when defined as a partitioned order, cf. Lemma 1). Since we know that local update repetitions can break the fixed point invariance property which holds in block-sequential ANs (cf. the example given in Remark 1), it would be interesting to characterise the conditions relating these repetitions to the architecture of interactions (so called interaction graph) giving rise to the existence of new fixed points. More generally, as a complement to the results of Sect. 3.4, the following problem can be studied: given an AN f, to which extent is f block-parallel sensitive/robust? In [1], the authors addressed this question on block-sequential Boolean ANs by developing the concept of update digraphs which allows to capture conditions of dynamical equivalence at the syntactical level. However, this concept does not apply as soon as local update repetitions are at stake. Hence, creating a new concept of update digraphs in the general context of periodic update modes would be an essential step forward to explain and understand updating sensitivity/robustness of ANs.

Another track of research would be to understand how basic interaction cycles of automata evolve under block-parallel updates. For instance, the authors of [7] have shown that such cycles in the Boolean setting are somehow very robust to block-sequential update modes variations: the number of their limit cycles of length p is the same as that of a smaller cycle (of same sign) evolving in parallel. Together with the combinatorial analysis of [3], this provides a complete analysis of the asymptotic dynamics of Boolean interaction cycles. This gives rise to the following question: do interaction cycles behave similarly under block-parallel update modes variations? The local update repetitions should again play an essential role. In this respect, the present work sets the foundations for theoretical developments and computer experiments (Theorems 3 and 4). Such a study could constitute a first approach of the more general problem raised above, since it is well known that cycles are the behavioural complexity engines of ANs [14].

Eventually, since block-parallel schedules form a new family of update modes of which the field of investigation is still largely open today, we think that a promising perspective of our work would consist in dealing with the computational complexity of classical decision problems for ANs, in the lines of [5] about reaction systems. The general question to be addressed here is: do local update repetitions induced by block-parallel update modes make such decision problems take place at a higher level in the polynomial hierarchy, or even reach polynomial space completeness? We have early evidence of the latter.

Acknowledgments. The authors were funded mainly by their salaries as French State agents.This work has been secondarily supported by ANR-18-CE40-0002 FANs project (KP & SS), ECOS-Sud C19E02 SyDySy project (SS), and STIC AmSud 22-STIC-02 CAMA project (KP, LT & SS), and the MSCA-SE-101131549 ACANCOS (KP, LT & SS).

References

1. Aracena, J., Goles, E., Moreira, A., Salinas, L.: On the robustness of update schedules in Boolean networks. Biosystems **97**, 1–8 (2009)
2. Davidich, M.I., Bornholdt, S.: Boolean network model predicts cell cycle sequence of fission yeast. PLoS ONE **3**, e1672 (2008)
3. Demongeot, J., Noual, M., Sené, S.: Combinatorics of Boolean automata circuits dynamics. Disc. Appl. Math. **160**(4–5), 398–415 (2012)
4. Demongeot, J., Sené, S.: About block-parallel Boolean networks: a position paper. Nat. Comput. **19**, 5–13 (2020)
5. Dennunzio, A., Formenti, E., Manzoni, L., Porreca, A.E.: Complexity of the dynamics of reaction systems. Inf. Comput. **267**, 96–109 (2019)
6. Fierz, B., Poirier, M.G.: Biophysics of chromatin dynamics. Ann. Rev. Biophys. **48**, 321–345 (2019)
7. Goles, E., Noual, M.: Block-sequential update schedules and Boolean automata circuits. In: Proceedings of AUTOMATA 2010, pp. 41–50. DMTCS (2010)
8. Hübner, M.R., Spector, D.L.: Chromatin dynamics. Ann. Rev. Biophys. **39**, 471–489 (2010)
9. Kauffman, S.A.: Metabolic stability and epigenesis in randomly constructed genetic nets. J. Theor. Biol. **22**, 437–467 (1969)
10. Mendoza, L., Alvarez-Buylla, E.R.: Dynamics of the genetic regulatory network for Arabidopsis thaliana flower morphogenesis. J. Theor. Biol. **193**, 307–319 (1998)
11. Noual, M.: Updating automata networks. PhD thesis, École normale supérieure de Lyon (2012)
12. Paulevé, L., Sené, S.: Systems Biology Modelling and Analysis: Formal Bioinformatics Methods and Tools, Chapter Boolean Networks and their Dynamics: the Impact of Updates, pp. 173–250. Wiley, Hoboken (2022)
13. Perrot, K., Sené, S., Tapin, L.: On countings and enumerations of block-parallel automata networks. Preprint on arXiv:2304.09664 (2023)
14. Robert, F.: Discrete iterations: a metric study. Springer Series in Computational Mathematics, vol. 6. Springer, Heidelberg (1986). https://doi.org/10.1007/978-3-642-61607-5
15. Thomas, R.: Boolean formalization of genetic control circuits. J. Theor. Biol. **42**, 563–585 (1973)

On the Piecewise Complexity of Words and Periodic Words

M. Praveen[1,2], Ph. Schnoebelen[3(✉)], J. Veron[3], and I. Vialard[3]

[1] Chennai Mathematical Institute, Chennai, India
[2] CNRS, ReLaX, IRL 2000, Chennai, India
[3] Laboratoire Méthodes Formelles, Univ. Paris-Saclay, Gif-sur-Yvette, France
phs@lmf.cnrs.fr

Abstract. The piecewise complexity $h(u)$ of a word is the minimal length of subwords needed to exactly characterise u. Its piecewise minimality index $\rho(u)$ is the smallest length k such that u is minimal among its order-k class $[u]_k$ in Simon's congruence.

We study these two measures and provide efficient algorithms for computing $h(u)$ and $\rho(u)$. We also provide efficient algorithms for the case where u is a periodic word, of the form $u = v^n$.

1 Introduction

For two words u and v, we write $u \preccurlyeq v$ when u is a *subword*, i.e., a subsequence, of v. For example `SIMON` $\preccurlyeq$ `STIMULATION` while `HEBRARD` $\not\preccurlyeq$ `HAREBRAINED`. Subwords and subsequences play a prominent role in many areas of computer science. Our personal motivations come from descriptive complexity and the possibility of characterising words and languages via some short witnessing subwords.

Fifty years ago, and with similar motivations, I. Simon introduced *piecewise-testable* (PT) languages in his doctoral thesis (see [Sim72, Sim75, SS83]): a language L is PT if there is a *finite* set of words F such that the membership of a word u in L depends only on which words from F are subwords of u. PT languages have since played an important role in the algebraic and logical theory of first-order definable languages, see [Pin86, DGK08, Klí11] and the references therein. They also constitute an important class of simple regular languages with applications in learning theory [KCM08], databases [BSS12], linguistics [RHF+13], etc. The concept of PT languages has been extended to variant notions of "subwords" [Zet18], to trees [BSS12], infinite words [PP04, CP18], pictures [Mat98], or any combinatorial well-quasi-order [GS16].

When a PT language L can be characterised via a finite F where all words have length at most k, we say that L is piecewise-testable *of height* k, or k-PT. Equivalently, L is k-PT if it is closed under $\sim_k$, Simon's congruence of order k, defined via $u \sim_k v \overset{\text{def}}{\Leftrightarrow}$ u and v have the same subwords of length at most k. The

Work supported by IRL ReLaX. J. Veron supported by DIGICOSME ANR-11-LABX-0045.

H. Fernau et al. (Eds.): SOFSEM 2024, LNCS 14519, pp. 456–470, 2024.
https://doi.org/10.1007/978-3-031-52113-3_32

piecewise complexity of L, denoted $h(L)$ (for "height"), is the smallest k such that L is k-PT. It coincides with the minimum number of variables needed in any $\mathcal{B}\Sigma_1$ formula that defines L [DGK08].

The piecewise complexity of languages was studied by Karandikar and Schnoebelen in [KS19] where it is a central tool for establishing elementary upper bounds for the complexity of the FO^2 fragment of the logic of subwords.

In this paper we focus on the piecewise complexity of *individual words*. For $u \in A^*$, we write $h(u)$ for $h(\{u\})$, i.e., the smallest k s.t. $[u]_k = \{u\}$, where $[u]_k$ is the equivalence class of u w.r.t. $\sim_k$. We also introduce a new measure, $\rho(u)$, defined as the smallest k such that u is minimal in $[u]_k$ (wrt subwords).

We have two main motivations. Firstly it appeared in [KS19] that bounding $h(L)$ for a PT language L relies heavily on knowing $h(u)$ for specific words u in and out of L. For example, the piecewise complexity of a finite language L is exactly $\max_{u \in L} h(u)$ [KS19], and the tightness of many upper bounds in [KS19] relies on identifying a family of long words with small piecewise complexity. See also [HS19, Sect. 4]. Secondly the piecewise complexity of words raises challenging combinatorial or algorithmic questions. To begin with we do not yet have a practical and efficient algorithm that computes $h(u)$.

Our Contribution. Along $h(u)$, we introduce a new measure, $\rho(u)$, the *piecewise minimality index* of u, and initiate an investigation of the combinatorial and algorithmic properties of both measures. The new measure $\rho(u)$ is closely related to $h(u)$ but is easier to compute. Our main results are (1) theoretical results connecting h and ρ and bounding their values in contexts involving concatenation, (2) efficient algorithms for computing $h(u)$ and $\rho(u)$, and (3) an analysis of periodicities in the arch factorization of periodic words that leads to a simple and efficient algorithm computing $h(u^n)$ and $\rho(u^n)$ for periodic words u^n. Our motivation for computing $h(u^n)$ and $\rho(u^n)$ is that we see it as preparatory work for computing subword complexity measures on compressed data, see [SV23].

Related Work. In the literature, existing works on h mostly focus on $h(L)$ for L a PT-language, and provide general bounds (see, e.g., [KS19, HS19]). We are not aware of any practical algorithm computing $h(L)$ for L a PT-language given, e.g., via a deterministic finite-state automaton $\mathcal{A}$, and it is known that deciding whether $h(L(\mathcal{A})) \leq k$ is coNP-complete [MT15].

Regarding words, there is a rich literature on algorithms computing $\delta(u, v)$, the piecewise distance between two words: see [Sim03, FK18, BFH+20, GKK+21] and the references therein. Computing $h(u)$ and $\rho(u)$ amounts to maximising $\delta(u, v)$ over the set of all words v distinct from u —when computing $h(u)$—or a subset of these—for $\rho(u)$—.

Outline of the Paper. After recalling the necessary background in Sect. 2, we define the new measures $h(u)$ and $\rho(u)$ in Sect. 3 and prove some first elementary properties like monotonicity and convexity. In Sect. 4 we give efficient algorithms computing $h(u)$ and $\rho(u)$. In Sect. 5 we prove simple but new connections between Simon's side distance functions r, ℓ and Hébrard's arch factorization. This motivates the study of the arch factorization of periodic words and leads to a simple and efficient algorithm computing $h(u^n)$ and $\rho(u^n)$. For lack of space, most proofs are missing from this extended abstract: they can be found in the full version of the paper, available as `arXiv:2311.15431 [cs.FL]`.

2 Words, Subwords and Simon's Congruence

We consider finite words $u, v, \ldots$ over a finite alphabet A. The empty word is denoted with ϵ, the mirror (or reverse) of u with u^{R}, and we write $|u|$ for the length of u. We also write $|u|_a$ for the number of times the letter a appears in u. For a word $u = a_1 a_2 \cdots a_L$ of length L we write $\mathit{Cuts}(u) = \{0, 1, 2, \ldots, L\}$ for the set of positions between the letters of u. For $i \leq j \in \mathit{Cuts}(u)$, we write $u(i, j)$ for the factor $a_{i+1} a_{i+2} \cdots a_j$. Note that $u(0, L) = u$, $u(i_1, i_2) \cdot u(i_2, i_3) = u(i_1, i_3)$ and that $|u(i, j)| = j - i$. We write $u(i)$ as shorthand for $u(i-1, i)$, i.e., a_i, the i-th letter of u. With $\mathit{alph}(u)$ we denote the set of letters that occur in u. We often abuse notation and write "$a \in u$" instead of "$a \in \mathit{alph}(u)$" to say that a letter a occurs in a word u.

We say that $u = a_1 \cdots a_L$ is a *subword* of v, written $u \preccurlyeq v$, if v can be factored under the form $v = v_0 a_1 v_1 a_2 \cdots v_{L-1} a_L v_L$ where the v_i's can be any words (and can be empty). We write $\downarrow u$ for the set of all subwords of u: e.g., $\downarrow \mathtt{ABAA} = \{\epsilon, \mathtt{A}, \mathtt{B}, \mathtt{AA}, \mathtt{AB}, \mathtt{BA}, \mathtt{AAA}, \mathtt{ABA}, \mathtt{BAA}, \mathtt{ABAA}\}$.

Factors are a special case of subwords: u is a *factor* of v if $v = v' u v''$ for some v', v''. Furthermore, when $v = v' u v''$ we say that u is a *prefix* of v when $v' = \epsilon$, and is a *suffix* of v when $v'' = \epsilon$.

When $u \neq v$, a word s is a *distinguisher* (or a *separator*) if s is subword of exactly one word among u and v [Sim72].

For $k \in \mathbb{N}$ we write $A^{\leq k}$ for the set of words over A that have length at most k, and for any words $u, v \in A^*$, we let $u \sim_k v \overset{\text{def}}{\Leftrightarrow} \downarrow u \cap A^{\leq k} = \downarrow v \cap A^{\leq k}$. In other words, $u \sim_k v$ if u and v have the same subwords of length at most k. For example $\mathtt{ABAB} \sim_1 \mathtt{AABB}$ (both words use the same letters) but $\mathtt{ABAB} \not\sim_2 \mathtt{AABB}$ ($\mathtt{BA}$ is a subword of $\mathtt{ABAB}$, not of $\mathtt{AABB}$). The equivalence $\sim_k$, introduced in [Sim72, Sim75], is called Simon's congruence of order k. Note that $u \sim_0 v$ for any u, v, and $u \sim_k u$ for any k. Finally, $u \sim_{k+1} v$ implies $u \sim_k v$ for any k, and there is a refinement hierarchy $\sim_0 \supseteq \sim_1 \supseteq \sim_2 \cdots$ with $\bigcap_{k \in \mathbb{N}} \sim_k = \mathit{Id}_{A^*}$. We write $[u]_k$ for the equivalence class of $u \in A^*$ under $\sim_k$. Note that each $\sim_k$, for $k = 0, 1, 2, \ldots$, has finite index [Sim75, SS83, KKS15].

We further let $u \lesssim_k v \overset{\text{def}}{\Leftrightarrow} u \sim_k v \wedge u \preccurlyeq v$. Note that $\lesssim_k$ is stronger than $\sim_k$. Both relations are (pre)congruences: $u \sim_k v$ and $u' \sim_k v'$ imply $uu' \sim_k vv'$, while $u \lesssim_k v$ and $u' \lesssim_k v'$ imply $uu' \lesssim_k vv'$.

The following properties will be useful:

Lemma 2.1. *For all $u, v, v', w \in A^*$ and $a, b \in A$:*

1. *If $u \sim_k v$ and $u \preccurlyeq w \preccurlyeq v$ then $u \lesssim_k w \lesssim_k v$;*

2. *When $k > 0$, $u \sim_k uv$ if, and only if, there exists a factorization $u = u_1 u_2 \cdots u_k$ such that $alph(u_1) \supseteq alph(u_2) \supseteq \cdots \supseteq alph(u_k) \supseteq alph(v)$;*
3. *If $uav \sim_k ubv'$ and $a \neq b$ then $ubav \sim_k ubv'$ or $uabv' \sim_k uav$ (or both);*
4. *If $u \sim_k v$ then there exists $w \in A^*$ such that $u \lesssim_k w$ and $v \lesssim_k w$;*
5. *If $u \sim_k v$ and $|u| < |v|$ then there exists some v' with $|v'| = |u|$ and such that $u \sim_k v' \preccurlyeq v$;*
6. *If $uv \sim_k uav$ then $uv \sim_k ua^m v$ for all $m \in \mathbb{N}$;*
7. *Every equivalence class of $\sim_k$ is a singleton or is infinite.*

Proof. (1) is by combining $\downarrow u \subseteq \downarrow w \subseteq \downarrow v$ with the definition of $\sim_k$; (2–4) are Lemmas 3, 5, and 6 from [Sim75]; (5) is an immediate consequence of Theorem 4 from [Sim72, p. 91], showing that all minimal (wrt $\preccurlyeq$) words in $[u]_k$ have the same length —see also [SS83, Theorem 6.2.9] or [FK18]—; (6) is in the proof of Corollary 2.8 from [SS83]; (7) follows from (1), (4) and (6). □

The fundamental tools for reasoning about piecewise complexity were developed in Simon's thesis [Sim72]. First, there is the concept of "subword distance"[1] $\delta(u, v) \in \mathbb{N} \cup \{\infty\}$, defined for any $u, v \in A^*$, via

$$\delta(u,v) \stackrel{\text{def}}{=} \max\{k \mid u \sim_k v\} \tag{1}$$

$$= \begin{cases} \infty & \text{if } u = v, \\ |s| - 1 & \text{if } u \neq v \text{ and } s \text{ is a shortest distinguisher.} \end{cases} \tag{2}$$

Derived notions are the left and right distances [Sim72, p72], defined for any $u, t \in A^*$, via

$$r(u,t) \stackrel{\text{def}}{=} \delta(u, ut) = \max\{k \mid u \sim_k ut\}\,, \tag{3}$$

$$\ell(t,u) \stackrel{\text{def}}{=} \delta(tu, u) = \max\{k \mid tu \sim_k u\}\,. \tag{4}$$

Clearly r and ℓ are mirror notions. One usually proves properties of r only, and (often implicitly) deduce symmetrical conclusions for ℓ by the mirror reasoning.

Lemma 2.2 ([SS83, Lemma 6.2.13]). *For any words $u, v \in A^*$ and letter $a \in A$*

$$\delta(uv, uav) = \delta(u, ua) + \delta(av, v) = r(u, a) + \ell(a, v)\,. \tag{5}$$

3 The Piecewise Complexity of Words

In this section we define the complexity measures $h(u)$ and $\rho(u)$, give characterisations in terms of the side distance functions r and ℓ, compare the two measures and establish some first results on the measures of concatenations.

[1] In fact $\delta(u, v)$ is a measure of *similarity* and not of difference, between u and v. The associated distance is actually $d(u, v) \stackrel{\text{def}}{=} 2^{-\delta(u,v)}$ [SS83].

3.1 Defining Words via Their Subwords

The piecewise complexity of PT languages was defined in [KS16, KS19]. Formally, for a language L over A, $h(L)$ is the smallest index k such that L is $\sim_k$-saturated, i.e., closed under $\sim_k$. For a word $u \in A^*$, this becomes $h(u) \stackrel{\text{def}}{=} \min\{n \mid \forall v : u \sim_n v \implies u = v\}$: we call it the *piecewise complexity* of u.

Proposition 3.1. *For any $u \in A^*$,*

$$h(u) = \max_{\substack{u=u_1u_2\\ a\in A}} \delta(u, u_1 a u_2) + 1 \tag{6}$$

$$= \max_{\substack{u=u_1u_2\\ a\in A}} r(u_1, a) + \ell(a, u_2) + 1\,. \tag{H}$$

3.2 Reduced Words and the Minimality Index

Definition 3.2 ([Sim72, p. 70]). *Let $m > 0$, a word $u \in A^*$ is m-*reduced *if $u \not\sim_m u'$ for all strict subwords u' of u.*

In other words, u is m-reduced when it is a minimal word in $[u]_m$. This leads to a new piecewise-based measure for words, that we call the *minimality index*:

$$\rho(u) \stackrel{\text{def}}{=} \min\{m \mid u \text{ is } m\text{-reduced}\}\,. \tag{7}$$

Lemma 3.3 ([Sim72, p. 72]). *A non-empty word u is m-reduced iff $r(u_1, a) + \ell(a, u_2) < m$ for all factorizations $u = u_1 a u_2$ with $a \in A$ and $u_1, u_2 \in A^*$.*

This has an immediate corollary:

Proposition 3.4. *For any non-empty word $u \in A^*$*

$$\rho(u) = \max_{\substack{u=v_1av_2\\ a\in A}} r(v_1, a) + \ell(a, v_2) + 1\,. \tag{P}$$

Note the difference between Equations (H) and (P): $h(u)$ can be computed by looking at all ways one would insert a letter a inside u while for $\rho(u)$ one is looking at all ways one could remove some letter from u.

3.3 Fundamental Properties of Side Distances

The characterisations given in Propositions 3.1 and 3.4 suggest that computing $h(u)$ and $\rho(u)$ reduces to computing the r and ℓ side distance functions on prefixes and suffixes of u. This will be confirmed in Sect. 4.

For this reason we now prove some useful combinatorial results on r and ℓ. They will be essential for proving more general properties of h and ρ in the rest of this section, and in the analysis of algorithms in later sections.

Lemma 3.5. *For any word $u \in A^*$ and letters $a, b \in A$:*

$$r(ua, b) \leq 1 + r(u, a) . \tag{8}$$

The following useful lemma provides a recursive way of computing $r(u, t)$.

Lemma 3.6 ([Sim72, p. 71–72]). *For any $u, t \in A^*$ and $a \in A$:*

$$r(u, t) = \min\{r(u, a) \mid a \in alph(t)\} , \tag{R1}$$
$$r(u, a) = 0 \text{ if } a \text{ does not occur in } u , \tag{R2}$$
$$r(u, a) = 1 + r(u', au'') \text{ if } u = u'au'' \text{ with } a \notin alph(u'') . \tag{R3}$$

Corollary 3.7. *For any $u, v, t, t' \in A^*$ and $a \in A$*

$$alph(t) \subseteq alph(t') \implies r(u, t) \geq r(u, t') \text{ and } \ell(t, u) \geq \ell(t', u) , \tag{9}$$
$$a \notin alph(v) \implies r(uv, a) \leq r(u, a) \text{ and } \ell(a, vu) \leq \ell(a, u) . \tag{10}$$

Lemma 3.8. (Monotonicity of r and ℓ). *For all $u, v, t \in A^*$*

$$r(v, t) \leq r(uv, t) , \qquad \ell(t, u) \leq \ell(t, uv) . \tag{11}$$

Observe that $r(u, a)$ can be strictly larger than $r(uv, a)$, e.g., $r(aa, a) = 2 > r(aab, a) = 1$. However, inserting a letter in u cannot increase r or ℓ by more than one:

Lemma 3.9. *For any $u, v \in A^*$ and $a, b \in A$*

$$r(uav, b) \leq 1 + r(uv, b) , \qquad \ell(b, uav) \leq 1 + \ell(b, uv) . \tag{12}$$

3.4 Relating h and ρ

Theorem 3.10. *$h(u) \geq 1 + \rho(u)$ for any word u.*

The above inequality is an equality in the special case of binary words.

Theorem 3.11. *Assume $|A| = 2$. Then $h(u) = \rho(u) + 1$ for any $u \in A^*$.*

Remark 3.12. Theorem 3.11 cannot be generalised to words using 3 or more different letters. For example, with $A = \{\mathtt{A}, \mathtt{B}, \mathtt{C}\}$ and $u = \mathtt{CAACBABA}$, one has $h(u) = 5$ and $\rho(u) = 3$. Larger gaps are possible: $u = \mathtt{CBCBCBCBBCABBABABAAA}$ has $h(u) = 10$ and $\rho(u) = 6$. □

3.5 Subword Complexity and Concatenation

While the subwords of uv are obtained by concatenating the subwords of u and the subwords of v, there is no simple relation between $h(uv)$ or $\rho(uv)$ on one hand, and $h(u)$, $h(v)$, $\rho(u)$ and $\rho(v)$ on the other hand.

However, we can prove that h and ρ are monotonic and convex wrt concatenation.

We start with convexity.

Theorem 3.13 (Convexity). *For all* $u, v \in A^*$

$$\rho(uv) \leq \rho(u) + \rho(v), \qquad h(uv) \leq \max\{h(u) + \rho(v),\ \rho(u) + h(v)\}. \tag{13}$$

Note that the second inequality entails $h(uv) \leq h(u) + h(v) - 1$ and is in fact stronger.

Theorem 3.14 (Monotonicity). *For all* $u, v \in A^*$

$$h(u) \leq h(uv), \qquad h(v) \leq h(uv), \tag{14}$$

$$\rho(u) \leq \rho(uv), \qquad \rho(v) \leq \rho(uv). \tag{15}$$

4 Computing $h(u)$ and $\rho(u)$

Thanks to Propositions 3.1 and 3.4, computing h and ρ reduces to computing r and ℓ. For r and ℓ we may rephrase Lemma 3.6 in the following recursive form:

$$r(u, a) = \begin{cases} 0 & \text{if } a \notin u, \\ 1 + \min_{b \in a u_2} r(u_1, b) & \text{if } u = u_1 a u_2 \text{ and } a \notin u_2. \end{cases} \tag{16}$$

with a mirror formula for ℓ.

We now derive a reformulation that leads to more efficient algorithms.

Lemma 4.1. *For any word* $u \in A^*$ *and letters* $a, b \in A$

$$r(ub, a) = \begin{cases} 0 & \textit{if } a \notin ub, \\ 1 + r(u, a) & \textit{if } a = b, \\ \min \begin{Bmatrix} 1 + r(u_1, b) \\ r(u, a) \end{Bmatrix} & \textit{if } a \neq b \textit{ and } u = u_1 a u_2 \textit{ with } a \notin u_2. \end{cases} \tag{17}$$

The recursion in Eq. (16) involves the prefixes of u. We define the *r-table of u* as the rectangular matrix containing all the $r(u(0,i), a)$ for $i = 0, 1, \ldots, |u|$ and $a \in A$. In practice we write just $r(i, a)$ for $r(u(0,i), a)$.

Example 4.2. Let $u =$ `ABBACCBCCABAABC` over $A = \{\mathtt{A}, \mathtt{B}, \mathtt{C}\}$. The r-table of u is

i	0	1	2	3	4	5	6	7	8	9	10	11	12	13	14	15
w		A	B	B	A	C	C	B	C	C	A	B	A	A	B	C
$r(i, \mathtt{A})$	0	1	1	1	2	1	1	1	1	1	2	2	3	4	4	3
$r(i, \mathtt{B})$	0	0	1	2	2	1	1	2	2	2	2	3	3	3	4	3
$r(i, \mathtt{C})$	0	0	0	0	0	1	2	2	3	4	2	2	2	2	2	3

As an exercise in reading Eq. (17), let us check that the values in this r-table are correct. First $r(0, a) = 0$ for all letters $a \in A$ since a does not occur in $u(0,0) = \epsilon$ which is empty. Since $u(0,1) = \mathtt{A}$ does not contain B, we further have $r(1, \mathtt{B}) = 0$. And since $u(0,4) =$ `ABBA` does not contain C, we have $r(i, \mathtt{C}) = 0$ for all $i = 0, \ldots, 4$.

Let us now check, e.g., $r(6, a)$ for all $a \in A$. Since $u(0, 6) = \texttt{ABBACC}$ ends with $b = \texttt{C}$, the second case in Eq. (17) gives $r(6, \texttt{C}) = 1 + r(5, \texttt{C}) = 1 + 1 = 2$. For $a = \texttt{A}$, we find that the last occurrence of A in $u(0, 6)$ is at position 4. So $r(6, \texttt{A})$ is the minimum of $r(5, \texttt{A})$ and $1 + r(3, \texttt{C})$, which gives 1. For $a = \texttt{B}$, and since B last occurs at position 3 in $u(0, 6)$, $r(6, \texttt{B})$ is obtained as $\min\big(r(5, \texttt{B}), 1 + r(2, \texttt{C})\big) = \min(1, 1 + 0) = 1$. □

It is now clear that Eq. (17) directly leads to a $O(|A| \cdot |u|)$ algorithm for computing r-tables. The following code builds the table from left to right. While progressing through $i = 0, 1, 2, \ldots$, it maintains a table `locc` storing, for each $a \in A$, the position of its last occurrence so far.

```
'''Algorithm computing the r-table of u'''
# init locc & r[0,..]:
for a in A: locc[a]=0; r[0,a]=0
# fill rest of r & maintain locc[..]:
for i from 1 to |u|:
    b = u[i]; locc[b] = i;
    for a in A:
        if a == b:
            r[i,a] = 1 + r[i-1,a]
        else:
            r[i,a] = min(r[i-1,a], 1 + r[locc[a],b])
```

Corollary 4.3. *$h(u)$ can be computed in bilinear time $O(|A| \cdot |u|)$.*

Proof. The r-table and the ℓ-table of u are computed in bilinear time as we just explained. Then one finds $\max_{a \in A} \max_{i=0,\ldots,|u|} r(u(0, i), a) + \ell(a, u(i, |u|)) + 1$ by looping over these two tables. As stated in Proposition 3.1, this gives $h(u)$. □

For $\rho(u)$, we compute the r- and ℓ-vectors.

Definition 4.4 (r-vector, ℓ-vector, [Sim72, p. 73]). *The r-vector of $u = a_1 \cdots a_m$ is $\langle r_1, \ldots, r_m \rangle$ defined with $r_i = r(a_1 \cdots a_{i-1}, a_i)$ for all $i = 1, \ldots, m$. The ℓ-vector of u is $\langle \ell_1, \ldots, \ell_m \rangle$ defined with $\ell_i = \ell(a_i, a_{i+1} \cdots a_m)$ for all $i = 1, \ldots, m$.*

Remark 4.5. The *attribute* of u defined in [FK18, § 3] is exactly the juxtaposition of Simon's r- and ℓ-vectors with all values shifted by 1. □

Example 4.6. Let us continue with $u = \texttt{ABBACCBCCABAABC}$. Its r-vector and ℓ-vector are, respectively:

$$r\text{-vector: } \langle 0, 0, 1, 1, 0, 1, 1, 2, 3, 1, 2, 2, 3, 3, 2 \rangle$$
$$\ell\text{-vector: } \langle 3, 4, 3, 2, 4, 3, 2, 2, 1, 2, 1, 1, 0, 0, 0 \rangle$$

By summing the two vectors, looking for a maximum value, and adding 1, we quickly obtain

$$\max_{i=1,\ldots,|u|} r\big(u(0,i-1),u(i)\big) + \ell\big(u(i),u(i,|u|)\big) + 1 = 5\,,$$

which provides $\rho(u)$ as stated in Proposition 3.4. □

One could extract the r- and ℓ-vectors from the r- and ℓ-tables but there is a faster way.

The following algorithm that computes the r-vector of u is extracted from the algorithm in [BFH+20] that computes the canonical representative of u modulo $\sim_k$. We refer to [BFH+20] for its correctness. Its running time is $O(|A|+|u|)$ since there is a linear number of insertions in the stack L and all the positions read from L are removed except the last read.

```
'''Algorithm computing the r-vector of u'''
# init locc & L
for a in A: locc[a]=0;
L.push(0)
# fill rest of r & maintain locc[..]:
for i from 1 to |u|:
    a = u[i];
    while (head(L) >= locc[a]):
        j = L.pop()
    r[i] = 1+r[j] if j>0 else 0
    L.push(j)
    L.push(i)
    locc[a]=i
```

With a mirror algorithm, the ℓ-vector is also computed in linear time.

Corollary 4.7. *$\rho(u)$ can be computed in linear time $O(|A|+|u|)$.*

5 Arch Factorizations and the Case of Periodic Words

In this section we analyse periodicities in the arch decomposition of u^n and deduce an algorithm for $h(u^n)$ and $\rho(u^n)$ that runs in time $O(|A|^2|u|+\log n)$.

Let A be some alphabet. An A-arch (or more simply an "arch" when A is understood) is a word $s \in A^*$ such that s contains all letters of A while none of its strict prefixes does. In particular the last letter of s occurs only once in s. A *co-arch* is the mirror image of an arch. The arch factorization of a word $w \in A^*$ is the unique decomposition $w = s_1 \cdot s_2 \cdots s_m \cdot t$ such that $s_1,\ldots,s_m$ are arches and t, called the *rest* of w, is a suffix that does not contain all letters of A [Héb91]. If its rest is empty, we say that w is *fully arched.*

For example, ABBACCBCCABAABC used in Example 4.2 factorizes as ABBAC · CBCCA · BAABC · ϵ, with 3 arches and an empty rest. It is fully arched.

Reconsidering the r-table from Example 4.2 with the arch factorization perspective, we notice that, at the beginning of each arch, the value of $r(i, a)$ coincides with the arch number:

α

i	0	1	2	3	4	5	6	7	8	9	10	11	12	13	14	15
w		A	B	B	A	C	C	B	C	C	A	B	A	A	B	C
$r(i, \mathtt{A})$	0	1	1	1	2	1	1	1	1	1	2	2	3	4	4	3
$r(i, \mathtt{B})$	0	0	1	2	2	1	1	2	2	2	2	3	3	3	4	3
$r(i, \mathtt{C})$	0	0	0	0	0	1	2	2	3	4	2	2	2	2	2	3

There is in fact a more general phenomenon at work:

Lemma 5.1. *For any word $u \in A^*$, letter a in A, and A-arch s:*

$$r(s\,u, a) = 1 + r(u, a)\,. \tag{18}$$

Proof. By induction on the length of u. We consider two cases.
Case 1: If a does not occur in u then $r(u, a) = 0$ so the right hand side of (18) is 1. Since s is an arch it can be factored as $s = s_1 a s_2$ with a not occurring in s_2. Then $r(su, a) = 1 + \min_{b \in a s_2 u} r(s_1, b)$. Necessarily, the last letter of s, call it c, occurs in $a s_2 u$ and since the last letter of an arch occurs only once in the arch, c does not occur in s_1, i.e., $r(s_1, c) = 0$, entailing $\min_{b \in a s_2 u} r(s_1, b) = 0$ and thus $r(su, a) = 1$ as needed to establish (18).
Case 2: If a occurs in u, then

$$1 + r(u, a) = 1 + 1 + \min_{b \in a u_2} r(u_1, b)$$

for a factorization $u = u_1 a u_2$ of u

$$\stackrel{\text{i.h.}}{=} 1 + \min_{b \in a u_2} r(s u_1, b) = r(su, a)\,.$$

□

Corollary 5.2. *Let $v, u, t \in A^*$. If v is fully arched with k arches then*

$$r(v\,u, t) = k + r(u, t)\,, \qquad \ell(t, u\,v^{\mathtt{R}}) = k + \ell(t, u)\,. \tag{19}$$

5.1 Arch-Jumping Functions

Seeing how Simon's r and ℓ func‘to the arches and co-arches of a word, the arch-jumping functions from [SV23] will be helpful.

Definition 5.3 (α and β: arch-jumping functions). *Fix some alphabet A and some word w over A. For a position $i \in Cuts(w)$ we let $\alpha(i)$ be the smallest $j > i$ such that $w(i, j)$ is an arch. Note that $\alpha(i)$ is undefined if $w(i, |w|)$ does not contain all letters of A.*

Symmetrically, we let $\beta(i)$ be the largest $j < i$ such that $w(j, i)$ is a co-arch. This too is a partial function.

The following picture shows α and β on $w =$ ABBACCBCCABAABC from Example 4.2: $\alpha(2) = 5$, $\alpha(3) = 7$, α is undefined on $\{13, 14, 15\}$ and β on $\{0, 1, 2, 3, 4\}$.

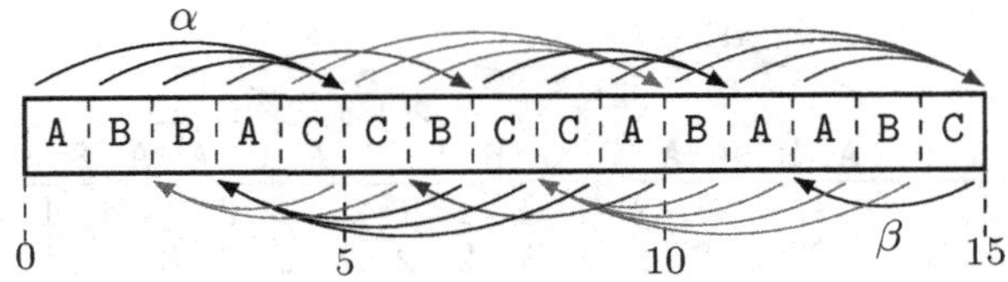

Lemma 5.4 ([SV23]). *When the values are defined, the following inequalities hold:*

$$i + |A| \leq \alpha(i) \,, \tag{20}$$

$$\alpha(i) \leq \alpha(i+1) \,, \tag{21}$$

$$i \leq \beta\alpha(i) \,, \tag{22}$$

$$\alpha(i) = \alpha\beta\alpha(i) \,, \tag{23}$$

$$i \leq \beta^n\alpha^n(i) \leq \beta^{n+1}\alpha^{n+1}(i) \leq \alpha(i) \,. \tag{24}$$

The arch factorization $w = s_1 \cdot s_2 \cdots s_m \cdot t$ of w can be defined in terms of α: m is the largest number such that $\alpha^m(0)$ is defined, each s_i is $w(\alpha^{i-1}(0), \alpha^i(0))$ and $r = w(\alpha^m(0), |w|)$ [SV23]. Co-arch factorizations can be defined similarly in terms of the β function.

5.2 Arch Factorization of Periodic Words

We now turn to *periodic words*, of the form $u \cdot u \cdots u$, i.e., u^n, where $n > 0$ is the number of times u is repeated. We let $L \stackrel{\text{def}}{=} |u|$ denote the length of u. Our first goal is to exhibit periodic patterns in the arch factorization of u^n.

Assume that $u \neq \epsilon$, with *alph*$(u) = A$. In order to study the arch factorization of u^n as a function of n, we set $w = u^\omega$ and consider the (infinite) arch factorization $u^\omega = s_1 \cdot s_2 \cdots s_m \cdots$: since w is infinite and since all the letters in A occur in u, α is defined everywhere over $\mathbb{N}$. For any $k \in \mathbb{N}$, we write λ_k for $\alpha^k(0)$, i.e., the cumulative length of w's first k arches.

Note that, over u^ω, $\alpha(i + L) = \alpha(i) + L$ since w is a periodic word. We say that $p \in \mathbb{N}$ *is an arch-period for* u^ω *starting at* i if there exists k such that $\alpha^k(i) \equiv \alpha^{k+p}(i) \mod L$. Such a period must exist for any $i \in \mathbb{N}$: the sequence $\alpha^0(i), \alpha^1(i), \ldots, \alpha^L(i)$ contains two values $\alpha^k(i)$ and $\alpha^{k'}(i)$ that are congruent modulo L and, assuming $k < k'$, one can pick $p = k' - k$.

Note that, since $\alpha(i+L) = \alpha(i)+L$ for any i, having $\alpha^k(i) \equiv \alpha^{k+p}(i) \mod L$ entails $\alpha^{k'}(i) \equiv \alpha^{k'+p}(i) \mod L$ for all $k' \geq k$. In fact, the *span* Δ, defined as $\alpha^{k+p}(i) - \alpha^k(i)$, does not depend on k once k is large enough.

Proposition 5.5. *There exists some integer* $p_u > 0$ *such that, for any* $i \in \mathbb{N}$, *the set of arch-periods starting at* i *are exactly the multiples of* p_u. *Consequently* p_u *is called* the arch-period *of* u.

Example 5.6. The following picture illustrates the case where u = AABBCC and $L = 6$. Starting at $i = 0$, one has $\alpha(0) = \lambda_1 = 5$ and $\alpha^4(0) = \lambda_4 = 17$, so $\alpha^1(0) \equiv \alpha^4(0) \mod L$ and $p_u = 3$ is the arch-period. □

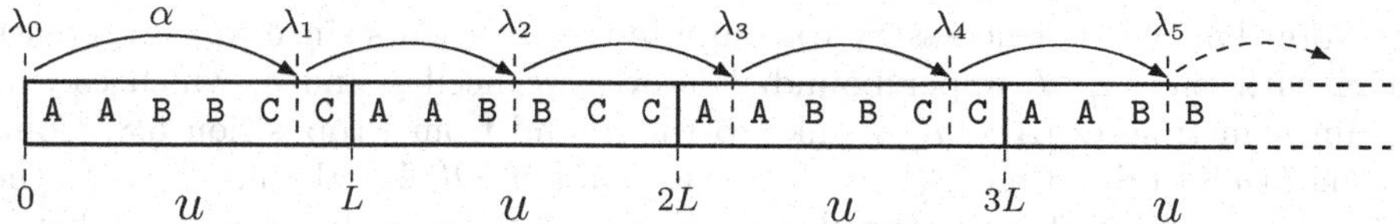

Recall that if u can be factored as $u = u_1u_2$ then u_2u_1 is a *conjugate* of u.

Proposition 5.7.
If v is a conjugate of u then u and v have the same arch-period.
If v is the mirror of u then u and v have the same arch-period.

Note that while p_u does not depend on the starting point i, the smallest k such that $\alpha^k(i) \equiv \alpha^{k+p_u}(i)$ usually does. In Example 5.6 one has $k = 1$ when starting from $i = 0$. But $\alpha^3(1) = 13 \equiv 1 = \alpha^0(1)$, so $k = 0$ works when starting from $i = 1$.

In the following, we shall always start from 0: the smallest k such that $\alpha^k(0) \equiv \alpha^{k+p_u}(0) \mod L$ is denoted by K_u and we further define $T_u = \lambda_{K_u} = \alpha^{K_u}(0)$ and $\Delta_u = \lambda_{K_u+p_u} - \lambda_{K_u} = \alpha^{K_u+p_u}(0) - \alpha^{K_u}(0)$. Note that Δ_u is a multiple of L, and we let $\Delta_u = \delta_u L$. Together, T_u and Δ_u are called the *transient* and the *span* of the periodic arch factorization. The *slope* σ_u is δ_u/p_u: after the transient part, moving forward by p_u arches in w is advancing through δ_u copies of u. In the above example, we have $p_u = 3$, $T = 5$ and $\Delta = 2L = 12$, hence $\sigma = \frac{2}{3}$ (here and below, we omit the u subscript when this does not cause ambiguities).

The reasoning proving the existence of an arch-period for u shows that at most L arches have to be passed before we find $\alpha^{k+p}(i) \equiv \alpha^k(i) \mod L$, so $k + p \leq L$, entailing $K_u + p_u \leq L$ and $\delta_u \leq L$. However, while $p_u = L$ is always possible, this does not lead to an L^2 upper bound for the span Δ_u. One can show the following:

Proposition 5.8 (Bounding span and transient). *For any $u \in A^*$, $T_u + \Delta_u \leq (|A| + 1) \cdot L$.*

5.3 Piecewise Complexity of Periodic Words

Theorem 5.9. *Assume alph(u) = A and write L for |u|. Further let T, Δ (= δL) and p be the transient, span and arch-period associated with u, and T′ be the transient associated with u^{R}.*
If $n \geq \frac{T+T'}{L}$ then $h(u^{n+\delta}) = h(u^n) + p$ and $\rho(u^{n+\delta}) = \rho(u^n) + p$.

Theorem 5.9 leads to a simple and efficient algorithm for computing $h(u^n)$ and $\rho(u^n)$ when n is large. We first compute p_u, Δ_u, T_u by factoring $u^{|A|}$ into arches. We obtain $T_{u^{\mathsf{R}}}$ in a similar way. We then find the largest m such that $(n-m)\delta_u L \geq T_u + T_{u^{\mathsf{R}}} + \Delta_u$. Writing n_0 for $n - mp_u\delta_u$, we then compute $h(u^{n_0})$ and $\rho(u^{n_0})$ using the algorithms from Sect. 4. Finally we use $h(u^n) = h(u^{n_0}) + mp$ and $\rho(u^n) = \rho(u^{n_0}) + mp$.

Note that it is not necessary to compute the transients since we can replace them with the $|A| \cdot L$ upper bound. However we need p and δ, which can be obtained in time $O(|A| \cdot |u|)$ thanks to the bound from Proposition 5.8. Computing $h(u^{n_0})$ takes time $O(|A|^2 \cdot |u|)$ since n_0 is in $O(|A|)$, thanks again to the bound on T and Δ. Finally the algorithm runs in time $O(|A|^2|u| + \log n)$, hence in linear time when A is fixed.

6 Conclusion

In this paper we focused on the piecewise complexity of individual words, as captured by the piecewise height $h(u)$ and the somewhat related minimality index $\rho(u)$, a new measure suggested by [Sim72] and that we introduce here.

These measures admit various characterisations, including Propositions 3.1 and 3.4 that can be leveraged into efficient algorithms running in bilinear time $O(|A| \cdot |u|)$ for $h(u)$ and linear time $O(|A| + |u|)$ for $\rho(u)$. Our analysis further allows to establish monotonicity and convexity properties for h and ρ, e.g., "$\rho(u) \leq \rho(uv) \leq \rho(u) + \rho(v)$", and to relate h and ρ.

In a second part we focus on computing h and ρ on periodic words of the form u^n. We obtain an elegant solution based on exhibiting periodicities in the arch factorization of u^n and as-yet-unnoticed connections between arch factorization and the side distance functions, and propose an algorithm that runs in polynomial time $O(|A|^2 \cdot |u| + \log n)$, hence in linear time in contexts where the alphabet A is fixed. This suggests that perhaps computing h and ρ on compressed data can be done efficiently, a question we intend to attack in future work.

References

[BFH+20] Barker, L., Fleischmann, P., Harwardt, K., Manea, F., Nowotka, D.: Scattered factor-universality of words. In: Jonoska, N., Savchuk, D. (eds.) DLT 2020. LNCS, vol. 12086, pp. 14–28. Springer, Cham (2020). https://doi.org/10.1007/978-3-030-48516-0_2

[BSS12] Bojańczyk, M., Segoufin, L., Straubing, H.: Piecewise testable tree languages. Logical Methods Comput. Sci. **8**(3), 1–32 (2012)

[CP18] Carton, O., Pouzet, M.: Simon's theorem for scattered words. In: Hoshi, M., Seki, S. (eds.) DLT 2018. LNCS, vol. 11088, pp. 182–193. Springer, Cham (2018). https://doi.org/10.1007/978-3-319-98654-8_15

[DGK08] Diekert, V., Gastin, P., Kufleitner, M.: A survey on small fragments of first-order logic over finite words. Int. J. Found. Comput. Sci. **19**(3), 513–548 (2008)

[FK18] Fleischer, L., Kufleitner, M.: Testing Simon's congruence. In: Proceedings of MFCS 2018, vol. 117 of Leibniz International Proceedings in Informatics, pp. 62:1–62:13. Leibniz-Zentrum für Informatik (2018)

[GKK+21] Gawrychowski, P., Kosche, M., Koß, T., Manea, F., Siemer, S.: Efficiently testing Simon's congruence. In: Proceedings of STACS 2021, vol. 187 of Leibniz International Proceedings in Informatics, pp. 34:1–34:18. Leibniz-Zentrum für Informatik (2021)

[GS16] Goubault-Larrecq, J., Schmitz, S.: Deciding piecewise testable separability for regular tree languages. In: Proceedings of ICALP 2016, vol. 55 of Leibniz International Proceedings in Informatics, pp. 97:1–97:15. Leibniz-Zentrum für Informatik (2016)

[Héb91] Hébrard, J.-J.: An algorithm for distinguishing efficiently bit-strings by their subsequences. Theor. Comput. Sci. **82**(1), 35–49 (1991)

[HS19] Halfon, S., Schnoebelen, Ph.: On shuffle products, acyclic automata and piecewise-testable languages. Inf. Process. Lett. **145**, 68–73 (2019)

[KCM08] Kontorovich, L., Cortes, C., Mohri, M.: Kernel methods for learning languages. Theor. Comput. Sci. **405**(3), 223–236 (2008)

[KKS15] Karandikar, P., Kufleitner, M., Schnoebelen, Ph.: On the index of Simon's congruence for piecewise testability. Inf. Process. Lett. **115**(4), 515–519 (2015)

[Klí11] Klíma, O.: Piecewise testable languages via combinatorics on words. Disc. Math. **311**(20), 2124–2127 (2011)

[KS16] Karandikar, P., Schnoebelen, Ph.: The height of piecewise-testable languages with applications in logical complexity. In Proceedings of CSL 2016, vol. 62 of Leibniz International Proceedings in Informatics, pp. 37:1–37:22. Leibniz-Zentrum für Informatik (2016)

[KS19] Karandikar, P., Schnoebelen, Ph.: The height of piecewise-testable languages and the complexity of the logic of subwords. Logical Methods Comput. Sci. **15**(2) (2019)

[Mat98] Matz, O.: On piecewise testable, starfree, and recognizable picture languages. In: Nivat, M. (ed.) FoSSaCS 1998. LNCS, vol. 1378, pp. 203–210. Springer, Heidelberg (1998). https://doi.org/10.1007/BFb0053551

[MT15] Masopust, T., Thomazo, M.: On the complexity of k-piecewise testability and the depth of automata. In: Potapov, I. (ed.) DLT 2015. LNCS, vol. 9168, pp. 364–376. Springer, Cham (2015). https://doi.org/10.1007/978-3-319-21500-6_29

[Pin86] Pin, J.É.: Varieties of Formal Languages. Plenum, New-York (1986)

[PP04] Perrin, D., Pin, J.-É.: Infinite words: Automata, Semigroups, Logic and Games, vol. 141 of Pure and Applied Mathematics Series. Elsevier (2004)

[RHF+13] Rogers, J., Heinz, J., Fero, M., Hurst, J., Lambert, D., Wibel, S.: Cognitive and sub-regular complexity. In: Morrill, G., Nederhof, M.-J. (eds.) FG 2012-2013. LNCS, vol. 8036, pp. 90–108. Springer, Heidelberg (2013). https://doi.org/10.1007/978-3-642-39998-5_6

[Sim72] Simon, I.: Hierarchies of Event with Dot-Depth One. PhD thesis, University of Waterloo, Waterloo, ON, Canada (1972)

[Sim75] Simon, I.: Piecewise testable events. In: Brakhage, H. (ed.) GI-Fachtagung 1975. LNCS, vol. 33, pp. 214–222. Springer, Heidelberg (1975). https://doi.org/10.1007/3-540-07407-4_23

[Sim03] Simon, I.: Words distinguished by their subwords. In: Proceedings of WORDS 2003 (2003)

[SS83] Sakarovitch, J., Simon, I.: Subwords. In: Lothaire, M. (ed.) Combinatorics on Words, vol. 17 of Encyclopedia of Mathematics and Its Applications, chap. 6, pp. 105–142. Cambridge University Press (1983)

[SV23] Schnoebelen, Ph., Veron, J.: On arch factorization and subword universality for words and compressed words. In: Proceedings of WORDS 20123, vol. 13899 of Lecture Notes in Computer Science, pp. 274–287. Springer, Heidelberg (2023). https://doi.org/10.1007/978-3-031-33180-0_21

[Zet18] Zetzsche, G.: Separability by piecewise testable languages and downward closures beyond subwords. In: Proceedings of LICS 2018, pp. 929–938. ACM Press (2018)

Distance Labeling for Families of Cycles

Arseny M. Shur[1(✉)] and Mikhail Rubinchik[2]

[1] Bar Ilan University, Ramat Gan, Israel
shur@datalab.cs.biu.ac.il
[2] SPGuide Online School in Algorithms, Karmiel, Israel

Abstract. For an arbitrary finite family of graphs, the distance labeling problem asks to assign labels to all nodes of every graph in the family in a way that allows one to recover the distance between any two nodes of any graph from their labels. The main goal is to minimize the number of unique labels used. We study this problem for the families $\mathcal{C}_n$ consisting of cycles of all lengths between 3 and n. We observe that the exact solution for directed cycles is straightforward and focus on the undirected case. We design a labeling scheme requiring $\frac{n\sqrt{n}}{\sqrt{6}} + O(n)$ labels, which is almost twice less than is required by the earlier known scheme. Using the computer search, we find an optimal labeling for each $n \leq 17$, showing that our scheme gives the results that are very close to the optimum.

Keywords: Distance labeling · Graph labeling · Cycle

1 Introduction

Graph labeling is an important and active area in the theory of computing. A typical problem involves a parametrized finite family $\mathcal{F}_n$ of graphs (e.g., all planar graphs with n nodes) and a natural function f on nodes (e.g., distance for *distance labeling* or adjacency for *adjacency labeling*). The problem is to assign labels to all nodes of every graph in $\mathcal{F}_n$ so that the function f can be computed solely from the labels of its arguments. Note that the algorithm computing f knows $\mathcal{F}_n$ but not a particular graph the nodes belong to. The main goal is to minimize the number of distinct labels or, equivalently, the maximum length of a label in bits. Additional goals include the time complexity of computing both f and the labeling function. In this paper, we focus solely on the main goal.

The area of graph labeling has a rather long history, which can be traced back at least to the papers [6,7]. The main academic interest in this area is in finding the limits of efficient representation of information. For example, the adjacency labeling of $\mathcal{F}_n$ with the minimum number of labels allows one to build the smallest "universal" graph, containing all graphs from $\mathcal{F}_n$ as induced subgraphs. Similarly, the optimal distance labeling of $\mathcal{F}_n$ gives the smallest "universal" matrix, containing the distance matrices of all graphs from $\mathcal{F}_n$ as principal minors.

The first author is supported by the grant MPM no. ERC 683064 under the EU's Horizon 2020 Research and Innovation Programme and by the State of Israel through the Center for Absorption in Science of the Ministry of Aliyah and Immigration.

H. Fernau et al. (Eds.): SOFSEM 2024, LNCS 14519, pp. 471–484, 2024.
https://doi.org/10.1007/978-3-031-52113-3_33

The distributed nature of labeling makes it also interesting for practical applications such as distributed data structures and search engines [2,8,12], routing protocols [9,20] and communication schemes [21].

The term *distance labeling* was coined by Peleg in 1999 [19], though some of the results are much older [15,21]. Let us briefly recall some remarkable achievements. For the family of all undirected graphs with n nodes it is known that the labels of length at least $\frac{n}{2}$ bits are necessary [16,18]. The first labeling scheme with $O(n)$-bit labels was obtained by Graham and Pollak [15]. The state-of-the-art labeling by Alstrup et al. [2] uses labels of length $\frac{\log 3}{2}n+o(n)$ bits[1] and allows one to compute the distances in $O(1)$ time (assuming the word-RAM model).

For planar graphs with n nodes, the lower bound of $\Omega(\sqrt[3]{n})$ bits per label and a scheme using $O(\sqrt{n}\log n)$ bits per label were presented in [13]. Recently, Gawrychowski and Uznanski [14] managed to shave the log factor from the upper bound. Some reasons why the gap between the lower and upper bounds is hard to close are discussed in [1].

For trees with n nodes, Peleg [19] presented a scheme with $\Theta(\log^2 n)$-bits labels. Gavoille et al. [13] proved that $\frac{1}{8}\log^2 n$ bits are required and $\approx 1.7\log^2 n$ bits suffice. Alstrup et al. [4] improved these bounds to $\frac{1}{4}\log^2 n$ and $\frac{1}{2}\log^2 n$ bits respectively. Finally, Freedman et al. [10] reached the upper bound $(\frac{1}{4} + o(1))\log^2 n$, finalizing the asymptotics up to lower order terms.

Further, some important graph families require only polynomially many labels and thus $O(\log n)$ bits per label. Examples of such families include interval graphs [11], permutation graphs [5], caterpillars and weighted paths [4].

Cycles are among the simplest graphs, so it may look surprising that graph labeling problems for the family $\mathcal{C}_n$ of all cycles up to length n are not settled yet. A recent result [3] states that $n+\sqrt[3]{n}$ labels are necessary and $n+\sqrt{n}$ labels are sufficient for the *adjacency* labeling of $\mathcal{C}_n$. Still, there is a gap between the lower and the upper bounds. (As in the case of planar graphs, this is the gap between $\sqrt[3]{n}$ and $\sqrt{n}$, though at a different level).

For the *distance* labeling of the family $\mathcal{C}_n$, the upper bound $O(n^{3/2})$ and the lower bound $\Omega(n^{4/3})$ on the minimal number of labels were proved in [17]. Apart from this paper, there are somewhat simpler folklore constructions leading to the same upper[2] and lower[3] bounds; note yet another gap between $\sqrt[3]{n}$ and $\sqrt{n}$.

In this paper we argue that the upper estimate is correct. We describe a distance labeling scheme for $\mathcal{C}_n$ that uses almost twice less labels than the folklore scheme. While this is a rather small improvement, we conjecture that our scheme produces labelings that are optimal up to an additive $O(n)$ term. To support this conjecture, we find the optimal number of labels for the families $\mathcal{C}_n$ up to $n = 17$. Then we compare the results of our scheme with an extrapolation of the optimal results, demonstrating that the difference is a linear term with a small constant. Finally, we describe several improvements to our scheme that further reduce this constant.

[1] Throughout the paper, log stands for the binary logarithm.

[2] E. Porat, private communication.

[3] S. Alstrup, private communication.

2 Preliminaries

Given two nodes u, v in a graph, the *distance* $d(u, v)$ is the length of the shortest (u, v)-path. Suppose we are given a finite family $\mathcal{F} = \{(V_1, E_1), \ldots, (V_t, E_t)\}$ of graphs and an arbitrary set $\mathcal{L}$, the elements of which are called *labels.* A *distance labeling* of $\mathcal{F}$ is a function $\phi : V_1 \cup \cdots \cup V_t \to \mathcal{L}$ such that there exists a function $d' : \mathcal{L}^2 \to \mathbb{Z}$ satisfying, for every i and each $u, v \in V_i$, the equality $d'(\phi(u), \phi(v)) = d(u, v)$. Since $d(u, u) = 0$, no label can appear in the same cycle twice. Thus for every single graph in $\mathcal{F}$ we view labels as unique names for nodes and identify each node with its label when speaking about paths, distances, etc.

In the rest of the paper, *labeling* always means distance labeling. A *labeling scheme* (or just *scheme*) for $\mathcal{F}$ is an algorithm that assigns labels to all nodes of all graphs in $\mathcal{F}$. A scheme is *valid* if it outputs a distance labeling.

We write C_n for the undirected cycle on n nodes and let $\mathcal{C}_n = \{C_3, \ldots, C_n\}$. The family $\mathcal{C}_n$ is the main object of study in this paper. We denote the minimum number of labels in a labeling of $\mathcal{C}_n$ by $\lambda(n)$.

2.1 Warm-Up: Labeling Directed Cycles

Consider a distance labeling of the family $\mathcal{C}_n^{\circlearrowleft} = \{C_3^{\circlearrowleft}, \ldots, C_n^{\circlearrowleft}\}$ of *directed* cycles. Here, the distance between two vertices is the length of the *unique* directed path between them. We write $\lambda_D(n)$ for the minimum number of labels needed to label $\mathcal{C}_n^{\circlearrowleft}$. A bit surprisingly, the exact formula for $\lambda_D(n)$ can be easily found.

Proposition 1. $\lambda_D(n) = \frac{n^2+2n+n \bmod 2}{4}$.

Proof. Let u and v be two labels in $C_i^{\circlearrowleft}$. Then $d(u, v) + d(v, u) = i$. Hence u and v cannot appear together is any other cycle. Thus every two cycles have at most one label in common. Then $C_{n-1}^{\circlearrowleft}$ contains at least $n - 2$ labels unused for $C_n^{\circlearrowleft}$, $C_{n-2}^{\circlearrowleft}$ contains at least $n - 4$ labels unused for both $C_n^{\circlearrowleft}$ and $C_{n-1}^{\circlearrowleft}$, and so on. This gives the total of at least $n + (n - 2) + (n - 4) + \cdots + n \bmod 2$ labels, which sums up exactly to the stated formula. To build a labeling with this number of labels, label cycles in decreasing order; labeling $C_i^{\circlearrowleft}$, reuse one label from each of the larger cycles such that neither label is reused twice. □

2.2 Basic Facts on Labeling Undirected Cycles

From now on, all cycles are undirected, so the distance between two nodes in a cycle is the length of the shortest of two paths between them. The maximal distance in C_i is $\lfloor i/2 \rfloor$. We often view a cycle as being drawn on a circumference, with equal distances between adjacent nodes, and appeal to geometric properties.

We say that a set $\{u, v, w\}$ of labels occurring in a cycle is a *triangle* if each of the numbers $d(u, v), d(u, w), d(v, w)$ is strictly smaller than the sum of the other two. Note that three nodes are labeled by a triangle if and only if they form an acute triangle on a circumference (see Fig. 1).

Lemma 1. *In a labeling of a family of cycles each triangle appears only once.*

Proof. If $\{u, v, w\}$ is a triangle in C_i, then $d(u, v) + d(u, w) + d(v, w) = i$. □

Lemma 1 implies the known lower bound on the number of labels for $\mathcal{C}_n$.

Proposition 2 ([17]). *Each labeling of* $\mathcal{C}_n$ *contains* $\Omega(n^{4/3})$ *distinct labels.*

Proof. As n tends to infinity, the probability that a random triple of labels of C_n forms a triangle approaches the probability that three random points on a circumference generate an acute triangle. The latter probability is 1/4: this is a textbook exercise in geometric probability. Thus, the set of labels of C_n contains $\Omega(n^3)$ triangles. By Lemma 1, the whole set of labels contains $\Omega(n^4)$ distinct triangles; they contain $\Omega(n^{4/3})$ unique labels. □

By *diameter* of a cycle C_i we mean not only the maximum length of a path between its nodes (i.e., $\lfloor i/2 \rfloor$) but also any path of this length. From the context it is always clear whether we speak of a path or of a number.

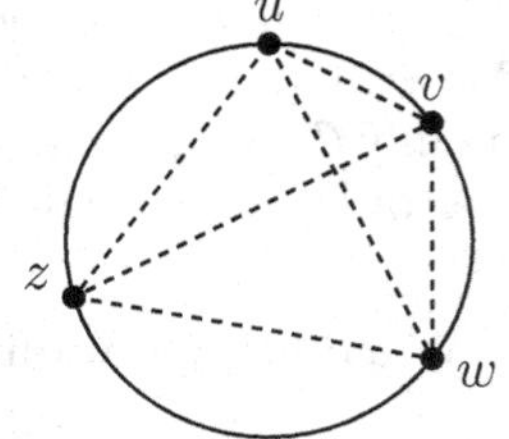

Fig. 1. Triangles in a labeled cycle. The sets $\{u, w, z\}$ and $\{v, w, z\}$ are triangles, while $\{u, v, w\}$ and $\{u, v, z\}$ are not.

Lemma 2. *For any labeling of two distinct cycles* C_i *and* C_j *there exist a diameter of* C_i *and a diameter of* C_j *such that every label appearing in both cycles belongs to both these diameters.*

Proof. Let $i > j$ and let $\mathcal{L}$ be the set of common labels of C_i and C_j. We assume $\#\mathcal{L} \geq 3$ as otherwise the diameters trivially exist. By Lemma 1, $\mathcal{L}$ contains no triangles. Hence for every triple of elements of $\mathcal{L}$ the maximum distance is the sum of the two other distances. Let u, v be labels with the maximum distance in $\mathcal{L}$. We have $d(u, v) \leq \lceil i/2 \rceil - 1$, because there are no larger distances in C_j. Then the shortest (u, v)-path in C_i is unique. Since $d(u, v) = d(u, w) + d(w, v)$ for any $w \in \mathcal{L}$ by the maximality of $d(u, v)$, all labels from $\mathcal{L}$ appear on this unique path, and hence on any diameter containing this path.

Though C_j may contain two shortest (u, v)-paths, all labels from $\mathcal{L}$ appear on one of them. Indeed, let $w, z \in \mathcal{L} \setminus \{u, v\}$. Considering the shortest (u, v)-path in C_i, we have w.l.o.g. $d(u, w) + d(w, z) + d(z, v) = d(u, v)$. This equality would be violated if w and z belong to different shortest (u, v)-paths in C_j. Thus, C_i has a diameter, containing the shortest (u, v)-path with all labels from $\mathcal{L}$ on it. □

The best known upper bound for cycle labeling is based on the following *folklore labeling scheme*, which is more economical than the scheme from [17]. For each $C_i \in \mathcal{C}_n$ we choose arbitrary adjacent nodes u and v and cover C_i by two disjoint paths: the path P_1 contains $\lceil i/2 \rceil$ nodes, including u, while P_2 contains $\lfloor i/2 \rfloor$ nodes, including v (see the example in Fig. 2). Each node from P_1 gets the label $(1, d_1, m_1)$, where d_1 is the distance to u and $m_1 = i \bmod \lceil \sqrt{n} \rceil$; each node from P_2 gets the label $(2, d_2, m_2)$, where d_2 is the distance to v, $m_2 = \lfloor i / \lceil \sqrt{n} \rceil \rfloor$.

Proposition 3 (folklore). *The folklore scheme is valid and uses* $\frac{3}{4}n\sqrt{n}+O(n)$ *labels to label* $\mathcal{C}_n$.

Proof. We prove the validity describing a procedure that derives the distance between two nodes from their labels (for illustration, see Fig. 2). Suppose that the labels (b,d,m) and (b',d',m') appear together in some (unknown) cycle. If $b=b'$, the two labels belong to the same path, so the distance between them is $|d-d'|$. If $b \neq b'$, we compute the length i of the cycle from the pair (m,m'). The two analysed labels are connected by a path of length $d+d'+1$ and by another path of length $i-d-d'-1$; comparing these lengths, we get the distance.

To compute the number of triples (b,d,m) used for labeling, note that there are two options for b, $\lfloor n/2 \rfloor+1$ options for d, and $\lceil\sqrt{n}\rceil$ options for m, for the total of $n\sqrt{n}+O(n)$ options. However, some triples are never used as labels. The label (b,d,m) is unused iff for every number $i \leq n$ compatible with m, the length of the path P_b in C_i is less than d. If $b=1$, the maximum i compatible with m is $i=(\lceil\sqrt{n}\rceil-1)\lceil\sqrt{n}\rceil+m$, which is $O(\sqrt{n})$ away from n. Therefore, each of $O(\sqrt{n})$ values of m gives $O(\sqrt{n})$ unused labels, for the total of $O(n)$. Let $b=2$. The maximum i compatible with m is $i=(m+1)\lceil\sqrt{n}\rceil-1$. The number of impossible values of d for this m is $(\lfloor n/2 \rfloor-1)-(\lfloor i/2 \rfloor-1)=\frac{\sqrt{n}-m}{2}\cdot\sqrt{n}+O(\sqrt{n})$. Summing up these numbers for $m=0,\ldots,\lceil\sqrt{n}\rceil-1$, we get $\frac{n\sqrt{n}}{4}+O(n)$ unused labels. In total we have $\frac{n\sqrt{n}}{4}+O(n)$ unused labels out of $n\sqrt{n}+O(n)$ possible; the difference gives us exactly the stated bound. □

3 More Efficient Labeling Scheme and Its Analysis

We start with more definitions related to labeled cycles. An *arc* is a labeled path, including the cases of one-node and empty paths. The labels on the arc form a string (up to reversal), and we often identify the arc with this string. In particular, we speak about *substrings* and *suffixes* of arcs. "To label a path P with an arc a" means to assign labels to the nodes of P to turn P into a copy of a.

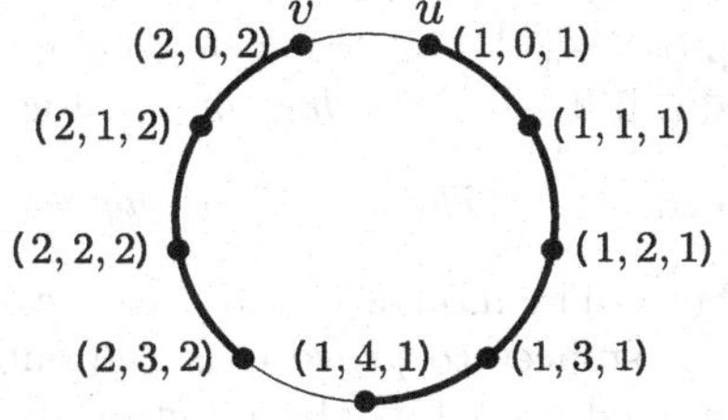

Fig. 2. Folklore scheme: labeling C_9 in the family $\mathcal{C}_{10}$. One has $\lceil\sqrt{10}\rceil=4$, $m_1=9 \bmod 4=1$, $m_2=\lfloor 9/4 \rfloor=2$.

By *intersection* of two labeled cycles we mean the labeled subgraph induced by all their common labels. Clearly, this subgraph is a collection of arcs. Lemma 2 says that the intersection is a subgraph of some diameter of each cycle. The intersections of two arcs and of a cycle and an arc are defined in the same way.

An *arc labeling* of a family of cycles is a labeling with the property that the intersection of any two cycles is an arc. Arc labelings are natural and easy to work with. Note that the folklore scheme produces an arc labeling: the intersection of two cycles is either empty or coincides with the path P_1 or the path P_2 of the smaller cycle. The schemes defined below produce arc labelings as well.

2-Arc Labeling Scheme. In this auxiliary scheme, cycles are labeled sequentially. The basic step is "label a cycle with two arcs". Informally, we partition the cycle into two paths of equal or almost equal length and label each of them with an arc (or with its substring of appropriate length). To specify details and reduce ambiguity, we define this step in the function $\mathsf{label}(a_1, a_2, C_j)$ below.

---

function $\mathsf{label}(a_1, a_2, C_j)$
if $|a_1| + |a_2| < j$ or $\min\{|a_1|, |a_2|\} < \lceil j/2 \rceil - 1$ **then**
    return error ▷ not enough labels for C_j
else
    label any path in C_j of length $\min\{|a_2|, \lfloor j/2 \rfloor + 1\}$ by a suffix of a_2
    label the remaining path in C_j by a suffix of a_1

The definition says that we label with suffixes (rather than arbitrary substrings) of arcs and use the longest possible suffix of the *second* arc. By default, we suppose that both suffixes can be read on the cycle in *the same direction* (i.e., the last labels from a_1 and a_2 are at the distance $\approx j/2$).

The 2-arc scheme starts with a set A of m pairwise disjoint arcs of sufficient length.

It calls the function $\mathsf{label}()$ for each pair of arcs in A. As a result, it produces a family of up to $\frac{m(m+1)}{2}$ labeled cycles. Lemma 3 below proves that the result is a labeling; we call it the *2-arc labeling.* In Fig. 3, such a labeling of the family $\mathcal{C}_5$ is shown.

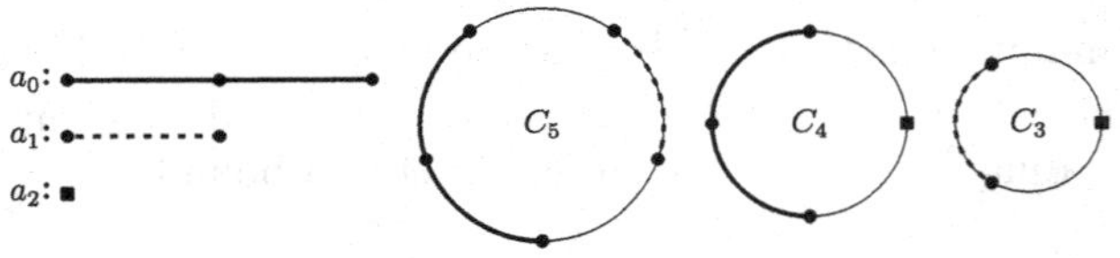

Fig. 3. Example: 2-arc labeling for the family $\mathcal{C}_5$.

Lemma 3. *The 2-arc labeling scheme is valid and produces arc labelings.*

Proof. The intersection of two cycles is an arc (possibly, empty) by construction, so it suffices to prove that the output of the 2-arc scheme is a labeling. Thus we need to define the function d' on labels. This is possible iff for every two labels u, v the distance $d(u, v)$ is the same for each cycle containing both u and v. Since the scheme uses each pair of arcs once, the labels u, v sharing several cycles belong to some arc $a \in A$. The intersection of a cycle C_i with a is at most the diameter of C_i. Hence $d(u, v)$ in C_i is the same as $d(u, v)$ in a, and this property holds for any cycle shared by u and v. Thus the scheme is valid. □

Remark 1. For a 2-arc labeling of the family $\mathcal{C}_n$ one can take $\sqrt{2n} + O(1)$ arcs of length $\frac{n}{2} + O(1)$ each, to the total of $\frac{n\sqrt{n}}{\sqrt{2}} + O(n)$ labels. So the 2-arc labeling beats the folklore labeling, which requires $\frac{3n\sqrt{n}}{4} + O(n)$ labels by Proposition 3.

Next, we develop the idea of 2-arc labeling to obtain a scheme that, as we believe, produces asymptotically optimal labelings for the families $\mathcal{C}_n$.

Chain Labeling Scheme. First we present the 2-arc labeling scheme for the family $\mathcal{C}_n$ as greedy Algorithm 1, which labels cycles in the order of decreasing length and proceeds in *phases* until all cycles are labeled. Each phase starts with creating a new arc with the function create($arc, length$); then the function label() is called in a cycle, each time using the new arc and one of the earlier created arcs. The length of the new arc is taken to barely pass the length condition for the first call to label(). In the preliminary phase 1 (lines 1–3), two arcs are created and the largest cycle is labeled. Other phases are iterations of the **while** cycle in lines 4–10. See Fig. 3 for the example.

Algorithm 1 : Greedy 2-arc labeling scheme for $\mathcal{C}_n$

```
1: create(a_0, ⌈n/2⌉); create(a_1, ⌊n/2⌋)
2: label(a_0, a_1, C_n)
3: i ← 2; j ← n − 1                    ▷ next arc to create; next cycle to label
4: while j > 2 do                                          ▷ start of ith phase
5:     create(a_i, ⌈j/2⌉ − 1)       ▷ minimum length of an arc needed to label C_j
6:     k ← 0                                                  ▷ next arc to use
7:     while k < i and j > 2 do
8:         label(a_k, a_i, C_j)
9:         j ← j − 1; k ← k + 1
10:    i ← i + 1
```

The *chain scheme* is a modification of Algorithm 1 that allows to use previously created arcs more efficiently. The difference is in the first parameter of the label function (line 8). During phase i, a *chain* is a path labeled by the concatenation $a_0a_1 \cdots a_{i-1}$ of the strings labeling all previously created arcs. Though formally the chain is an arc, the distance between labels in the chain may differ from the distance between the same labels in the already labeled cycles. For example, the string a_0a_1 labels both the cycle C_n with the diameter $\lfloor n/2 \rfloor + 1$ and a path of diameter $n - 1$ in the chain. However, with some precautions the chain can be used for labeling cycles. The chain scheme is presented below as Algorithm 2. The auxiliary function trim(c) deletes the suffix of the chain c that was used to label a cycle on the current iteration of the internal cycle.

Algorithm 2 : Chain labeling scheme for $\mathcal{C}_n$

```
1: create(a_0, ⌈n/2⌉); create(a_1, ⌊n/2⌋)
2: label(a_0, a_1, C_n)
3: i ← 2; j ← n − 1                    ▷ next arc to create; next cycle to label
4: while j > 2 do                                         ▷ start of phase i
5:     c ← a_0 a_1 ··· a_{i−1}                            ▷ chain for phase i
6:     create(a_i, ⌈j/2⌉ − 1)       ▷ minimum length of an arc needed to label C_j
7:     while |c| ≥ ⌈j/2⌉ − 1 and j > 2 do
8:         label(c, a_i, C_j)
9:         j ← j − 1; trim(c)        ▷ deleting the just used suffix from the chain
10:    i ← i + 1
```

To prove validity of the chain scheme, we need an auxiliary lemma.

Lemma 4. *Let j be the length of the longest unlabeled cycle at the beginning of i'th phase of Algorithm 2, $i \geq 2$. Then every substring of length $\leq \lfloor j/2 \rfloor + 1$ of the chain $c = a_0 a_1 \cdots a_{i-1}$ labels an arc in a cycle $C_{j'}$ for some $j' > j$.*

Proof. Note that if a string labels an arc in an already labeled cycle, then every its substring does the same. We proceed by induction on i. From line 2 we see that each substring of length $\lfloor n/2 \rfloor + 1$ of the string $a_0 a_1$ labels a diameter in C_n. Hence we have the base case $i = 2$. Since j decreases with each phase, the inductive hypothesis implies that at the start of $(i-1)$th phase each substring of length $\lfloor j/2 \rfloor + 1$ of $a_0 a_1 \cdots a_{i-2}$ labels an arc in some already labeled cycle. Let $C_{\hat{\jmath}}$ be the first cycle labeled at this phase. Then a diameter of $C_{\hat{\jmath}}$ is labeled with a suffix of $a_0 a_1 \cdots a_{i-2}$, say, a', and the remaining arc is labeled with the whole string a_{i-1}. Since $\hat{\jmath} > j$, both the prefix $a_0 a_1 \cdots a_{i-2}$ and the suffix $a' a_{i-1}$ of the chain c have the desired property: each substring of length $\leq \lfloor j/2 \rfloor + 1$ labels an arc in an already labeled cycle. As these prefix and suffix of c intersect by a substring a' of length $\geq \lfloor j/2 \rfloor + 1$, the whole chain c has this property. This proves the step case and the lemma. □

Lemma 5. *The chain labeling scheme is valid and builds an arc labeling.*

Proof. To prove that Algorithm 2 builds a labeling, it suffices to check the following property: in every cycle C_j, each pair of labels either does not appear in larger cycles or appears in some larger cycle at the same distance as in C_j. This property trivially holds for C_n, so let us consider a cycle C_j labeled at i'th phase, $i \geq 2$. The cycle C_j is labeled by two arcs: a substring of the chain $c = a_0 \cdots a_{i-1}$ and a suffix of the new arc a_i. We denote them by c' and a'_i respectively.

Suppose that a pair of labels (u, v) from C_j appeared in a larger cycle. Since all substrings of c used for labeling together with a_i are disjoint, u and v belong to the same arc (c' or a'_i). Then the shortest (u, v)-path in C_j is within this arc. If u, v are in a'_i, then $d(u, v)$ in C_j is the same as in the larger cycle containing a_i. If u, v are in c', then $d(u, v)$ in C_j is the same as in the larger cycle containing the arc c'; as $|c'| \leq \lceil j/2 \rceil - 1$, such a cycle exists by Lemma 4. Hence we proved that Algorithm 2 indeed builds a labeling; it is an arc labeling by construction. □

We call the labelings obtained by chain scheme *chain labelings*. Now we estimate the efficiency of the scheme.

Theorem 1. *A chain labeling of a family $\mathcal{C}_n$ of cycles uses $\frac{n\sqrt{n}}{\sqrt{6}} + O(n)$ labels.*

We first need a simpler estimate.

Lemma 6. *Algorithm 2 labels $\mathcal{C}_n$ using $O(\sqrt{n})$ phases and $O(n\sqrt{n})$ labels.*

Proof. Let us compare the runs of Algorithms 1 and 2 on the family $\mathcal{C}_n$. Suppose that ℓ_i is the length of a_i for Algorithm 2 and N_i is the number of cycles labeled by Algorithm 2 during the first i phases. For Algorithm 1, we denote the same parameters by ℓ'_i and N'_i. Note that $N_1 = N'_1 = 1$.

Let $i \geq 2$. If $N_{i-1} = N'_{i-1}$, then $\ell_i = \ell'_i$ as both algorithms begin the i'th phase with the same value of j. During this phase, Algorithm 1 labels i cycles, while Algorithm 2 labels at least i cycles (the length of the chain allows this), and possibly more. Hence $N_i \geq N'_i$. If $N_{i-1} > N'_{i-1}$, Algorithm 2 begins the i'th phase with smaller value of j compared to Algorithm 1. Then $\ell_i \leq \ell'_i$ and again, the length of the chain allows Algorithm 2 to label at least i cycles during the i'th phase. Hence $N_i > N'_i$ (or $N_i = N'_i = n - 2$ if both algorithms completed the labeling during this phase). Therefore, Algorithm 2 uses at most the same number of phases and at most the same number of labels as Algorithm 1. The latter uses $\sqrt{2n} + O(1)$ arcs and thus $O(n\sqrt{n})$ labels. The lemma is proved. □

Proof (of Theorem 1). The idea is to count distinct *pairs* of labels and infer the number of labels from the obtained result. First we count the pairs of labels that appear together in some cycle. We scan the cycles in the order of decreasing length and add "new" pairs (those not appearing in larger cycles) to the total. After applying Algorithm 2 to $\mathcal{C}_n$, each cycle C_j is labeled by two arcs of almost equal length: one of them is a substring of the chain c and the other one is a suffix of the current arc a_i. All pairs from c in C_j are not new by Lemma 4. All pairs between c and a_i are new by construction, and their number is $\frac{j^2}{4} + O(j)$. Over all cycles C_j, this gives $\frac{n^3}{12} + O(n^2)$ distinct pairs in total. All pairs from a_i are new if and only if C_j is the first cycle labeled in a phase. By Lemma 6, Algorithm 2 spends $O(\sqrt{n})$ phases. Hence there are $O(\sqrt{n})$ arcs a_i, each containing $O(n^2)$ pairs of labels, for the total of $O(n^{5/2})$ pairs. Therefore, the number of pairs that appear together in a cycle is $\frac{n^3}{12} + O(n^{5/2})$.

Next we count the pairs of labels that *do not* appear together. The labels in such a pair belong to different arcs. Let u, v be from $a_{i'}$ and a_i respectively, $i' < i$. If u and v *appear* together, then the largest cycle containing both u and v was labeled at phase i. Indeed, earlier phases have no access to v, and if u and v share a cycle labeled at a later phase, then Lemma 4 guarantees that they also appear in a larger cycle. Thus, to get the total number of pairs that do not appear together we count, for each phase i, the pairs (u, v) such that u is from the chain, v is from a_i, and neither of the cycles labeled during this phase contains both u and v.

There are three reasons why neither of the cycles labeled during phase i contains both u from the chain c and v from a_i. First, this can be the last phase, which is too short to use u. Since $|a_i| < n$ and $|c| = O(n\sqrt{n})$ by Lemma 6, the last phase affects $O(n^{5/2})$ pairs. Second, u can belong to a short prefix of c that remained unused during the phase due to the condition in line 7 of Algorithm 2. This prefix is shorter than a_i, so this situation affects less than $|a_i|^2$ pairs. As the number of phases is $O(\sqrt{n})$ (Lemma 6), the total number of such pairs is $O(n^{5/2})$. Third, v can belong to a prefix of a_i that was unused for the cycle containing u. The number of labels from a_i that were unused for *at least one* cycle during phase i does not exceed $|a_i| - |a_{i+1}|$, which gives $O(n)$ labels over all phases. Each such label v is responsible for $O(n\sqrt{n})$ pairs by Lemma 6, for the total of $O(n^{5/2})$ pairs. Thus there are $O(n^{5/2})$ pairs that do not appear together.

Putting everything together, we obtain that a chain labeling of $\mathcal{C}_n$ contains $p = \frac{n^3}{12} + O(n^{5/2})$ pairs of labels. Hence the number $ch(n)$ of labels is

$$ch(n) = \sqrt{2p} + O(1) = \frac{n\sqrt{n}}{\sqrt{6}} \cdot \sqrt{1 + O(n^{-1/2})} + O(1) = \frac{n\sqrt{n}}{\sqrt{6}} + O(n),$$

as required. □

4 Chain Labelings vs Optimal Labelings

The chain labeling beats the folklore labeling almost twice in the number of labels (Theorem 1 versus Proposition 3). However, it is not clear how good this new labeling is, given that the known lower bound (Proposition 2) looks rather weak. In this section we describe the results of an experimental study we conducted to justify Conjecture 1, stating that the chain labeling is asymptotically optimal.

Conjecture 1. $\lambda(n) = \frac{n\sqrt{n}}{\sqrt{6}} + O(n)$.

We proceed in three steps, which logically follows each other.

Step 1. Compute as many values of $\lambda(n)$ as possible and compare them to $\frac{n\sqrt{n}}{\sqrt{6}}$.

Outline of the Search Algorithm. To compute $\lambda(n)$, we run a recursive depth-first search, labeling cycles in the order of decreasing length. The upper bound *max* on the total number of labels is a global variable. The recursive function $\mathsf{labelCycle}(j, L, D)$ gets the length j of the cycle to label, the set L of existing labels, and the table D of known distances between them. The function runs an optimized search over all subsets of L. When it finds a subset X that is both

- *compatible*: all labels from X can be assigned to the nodes of C_j respecting the distances from D, and
- *large*: labeling C_j with $X \cup Y$, where the set Y of labels is disjoint with L, holds the total number of labels below the upper bound *max*,

it labels C_j with $X \cup Y$, adds Y to L to get some L', adds newly defined distances to D to get some D', and compares j to 3. If $j = 3$, the function reports (L', D'), sets $max = \#L'$, and returns; otherwise, it calls $\mathsf{labelCycle}(j-1, L', D')$. The value of *max* in the end of search is reported as $\lambda(n)$. See [22] for the C^{++} code.

Results. We managed to find $\lambda(n)$ for $n \leq 17$; for $n = 17$ the algorithm made over $5 \cdot 10^{12}$ recursive calls, which is 30 times bigger than for $n = 16$. Computing $\lambda(18)$ would probably require a cluster. The witness labelings can be found in the arXiv version (CoRR abs/2308.15242); the numbers $\lambda(n)$ fit well in between the bounds $\frac{n\sqrt{n}}{\sqrt{6}}$ and $\frac{n(\sqrt{n}+1)}{\sqrt{6}}$ (see Fig. 4). As a side result, we note that almost all optimal labelings we discovered are *arc labelings*.

If we view the "corridor" in Fig. 4 as an extrapolation for $\lambda(n)$ for big n, we have to refer $ch(n)$ to this corridor.

Step 2. Estimate the constant in the $O(n)$ term in Theorem 1, to compare $ch(n)$ to the results of step 1.

Results. We computed $ch(n)$ for many values of n in the range $[10^3..10^7]$. In all cases, $ch(n) \approx \frac{n(\sqrt{n}+1.5)}{\sqrt{6}}$, which is only $\frac{n}{2\sqrt{6}}$ away from the "corridor" in Fig. 4.

The natural next question is whether we can do better.

Step 3. Find resources to improve the chain scheme to come closer to $\lambda(n)$.

In order to reduce the amount of resources "wasted" by the chain scheme, we describe three improving tricks. An example of their use is an optimal labeling of C_{14} presented in Fig. 5.

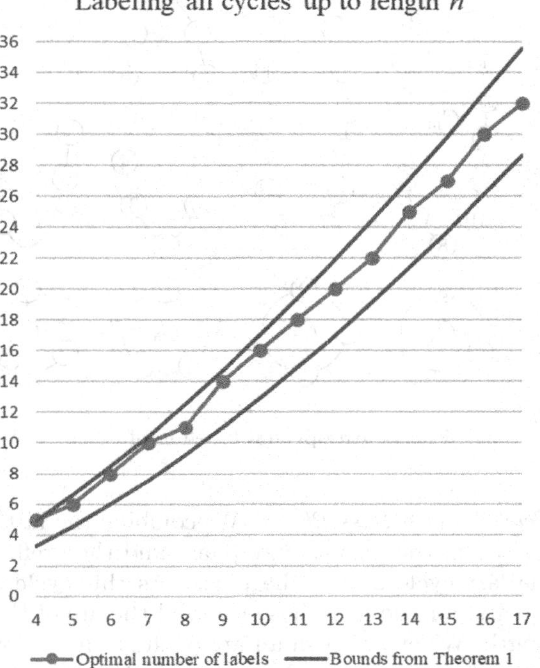

Fig. 4. Bound from Theorem 1 versus the optimal numbers $\lambda(n)$ of labels. The lower and upper bounds are $\frac{n\sqrt{n}}{\sqrt{6}}$ and $\frac{n(\sqrt{n}+1)}{\sqrt{6}}$ respectively.

Trick 1: Reusing Ends of Arcs. During a phase, if a cycle is labeled with the strings $a_1 \cdots a_j$ from the new arc and $c_x \cdots c_{x+j}$ or $c_x \cdots c_{x+j+1}$ from the chain, then it is correct to use for the next cycle the string $c_{x-j+1} \cdots c_x$ (resp., $c_{x-j} \cdots c_x$), thus reusing the label c_x; for example see C_{13} and C_{12} in Fig. 5.

The function label() is defined so that the above situation happens only in the beginning of the phase, so this trick saves 1 or 2 labels in the chain. Still, sometimes this leads to labeling an additional cycle during a phase.

Trick 2: Using Chain Remainders. In the end of a phase, we memorize the remainder c of the chain and the current arc a. Thus, at any moment we have the set S of such pairs of strings (initially empty). Now, before labeling a cycle we check whether S contains a pair (c, a) that can label this cycle. If yes, we extract (c, a) from S and run a "mini-phase", labeling successive cycles with c as the arc and a as the chain; when the mini-phase ends, with a' being the chain remainder, we add the pair (a', c) to S and proceed to the next cycle. Otherwise, we label the current cycle as usual. In Fig. 5, the pair $(12, abcdef)$ was added to S after phase 2. Later, the cycles C_6 and C_5 were labeled during a mini-phase with this pair; note that trick 1 was used to label C_5.

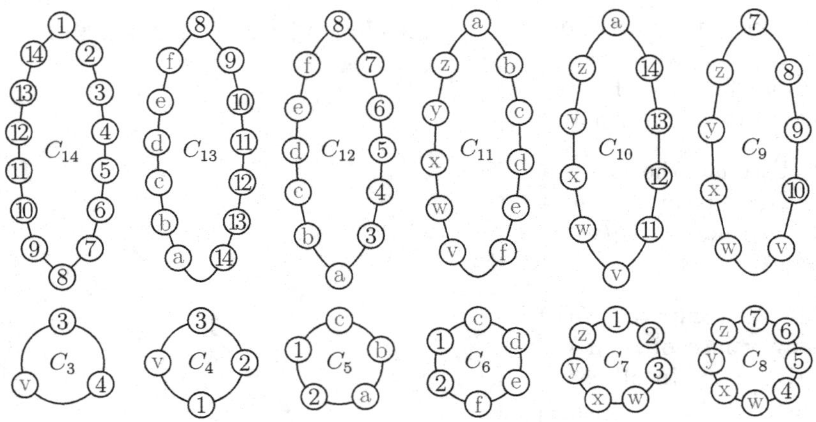

Fig. 5. An optimal labeling of $\mathcal{C}_{14}$ by the enhanced chain scheme.

Trick 3: Two-Pass Phase. We combine the last two phases as follows. Let a be the arc in the penultimate phase and the prefix a' of a was unused for labeling the last cycle during the phase. As this cycle consumed $|a| - |a'|$ labels from the arc, for the last phase we need the arc of the same (up to a small constant) length. We create such an arc of the form $\hat{a}a'$, where the labels from $\hat{a}$ are new (if $|a'| > |a|/2$, no new labels are needed). During the last phase, we reverse both the arc and the chain: the chain c is cut from the beginning, and the arc $\hat{a}a'$ is cut from the end. In this way, the labels from a' will not meet the labels from c for the second time, so the phase will finish correctly. An additional small trick can help sometimes: for a cycle, we use one less symbol from the arc than the maximum possible. This charges the chain by an additional label but this label can be reused from the previous cycle by employing trick 1. In Fig. 5, this was done for C_8. As a result, $a' = v$ did not meet the labels $1, \ldots, 6$ from the chain and we were able to label C_4 and C_3 without introducing an additional arc.

Results. We applied the *enhanced chain scheme*, which uses tricks 1–3, for labeling many families $\mathcal{C}_n$ for $n \in [10^3..10^7]$. In all cases, we get the number of labels $ch^+(n) \approx \frac{n(\sqrt{n}+1)}{\sqrt{6}}$, which is exactly the upper bound of the corridor in Fig. 4. In Table 1, we compare the results of chain schemes to the known optima, starting from the first nontrivial case $n = 7$.

Overall, the results gathered in Fig. 4 and Table 1 together with the behavior of $ch(n)$ and $ch^+(n)$ for big n give a substantial support to Conjecture 1.

5 Discussion and Future Work

The main open problem for distance labeling of the families $\mathcal{C}_n$ is the gap between the lower bound $\lambda(n) = \Omega(n\sqrt[3]{n})$ and the upper bound $\lambda(n) = O(n\sqrt{n})$. Our results suggest that the upper bound provides the correct asymptotics but improving the lower bound will probably need a new approach.

Table 1. Comparison of chain and optimal labelings for small n. The columns $ch(n)$, $ch^+(n)$, $\lambda(n)$ contain the number of labels for the chain scheme, enhanced chain scheme, and the optimal labeling, respectively; $ch^{++}(n)$ refers to the hybrid scheme: we perform the enhanced chain scheme till the two-pass phase, and replace this phase by a search for an optimal labeling of remaining cycles.

n	$ch(n)$	$ch^+(n)$	$ch^{++}(n)$	$\lambda(n)$
7	10	10	10	10
8	13	12	11	11
9	14	14	14	14
10	17	17	16	16
11	19	18	18	18
12	21	21	20	20

n	$ch(n)$	$ch^+(n)$	$ch^{++}(n)$	$\lambda(n)$
13	24	23	22	22
14	28	25	25	25
15	30	29	28	27
16	33	33	31	30
17	35	34	33	32
18	38	38	36	35?

This is pretty alike the situation with the distance labeling of planar graphs. Here, the gap (in terms of the length of a label in bits, i.e., in logarithmic scale) is between $\Omega(\sqrt[3]{n})$ and $O(\sqrt{n})$ and there is an evidence [1,14] that the upper bound gives the correct asymptotics but the existing approach does not allow to improve the lower bound. Another similar gap between the cubic root and the quadratic root appears in the adjacency labeling problem for $\mathcal{C}_n$ [3].

As a possible approach to the improvement of the lower bound for $\lambda(n)$ we propose to study the number $\lambda_k(n)$ of labels needed to label the family $\mathcal{C}_{n,k} = \{C_n, C_{n-1}, \dots, C_{n-k+1}\}$, starting from small k. Algorithm 1 and Lemma 2 imply $\lambda_2(n) = \lambda_3(n) = 1.5n + O(1)$ but already the next step is not completely trivial.

Acknowledgements. We are grateful to E. Porat for introducing the distance labeling problem to us. Our special thanks to A. Safronov for the assistance in computational experiments.

References

1. Abboud, A., Gawrychowski, P., Mozes, S., Weimann, O.: Near-optimal compression for the planar graph metric. In: Proceedings of the Twenty-Ninth Annual ACM-SIAM Symposium on Discrete Algorithms, SODA 2018, pp. 530–549. SIAM (2018)
2. Abiteboul, S., Alstrup, S., Kaplan, H., Milo, T., Rauhe, T.: Compact labeling scheme for ancestor queries. SIAM J. Comput. **35**(6), 1295–1309 (2006)
3. Abrahamsen, M., Alstrup, S., Holm, J., Knudsen, M.B.T., Stöckel, M.: Near-optimal induced universal graphs for cycles and paths. Discret. Appl. Math. **282**, 1–13 (2020)
4. Alstrup, S., Gørtz, I.L., Halvorsen, E.B., Porat, E.: Distance labeling schemes for trees. In: 43rd International Colloquium on Automata, Languages, and Programming, ICALP 2016. LIPIcs, vol. 55, pp. 132:1–132:16. Schloss Dagstuhl - Leibniz-Zentrum für Informatik (2016)
5. Bazzaro, F., Gavoille, C.: Localized and compact data-structure for comparability graphs. Discret. Math. **309**(11), 3465–3484 (2009)

6. Breuer, M.A.: Coding the vertexes of a graph. IEEE Trans. Inf. Theory **IT–12**, 148–153 (1966)
7. Breuer, M.A., Folkman, J.: An unexpected result on coding vertices of a graph. J. Math. Anal. Appl. **20**, 583–600 (1967)
8. Cohen, E., Kaplan, H., Milo, T.: Labeling dynamic XML trees. SIAM J. Comput. **39**(5), 2048–2074 (2010)
9. Eilam, T., Gavoille, C., Peleg, D.: Compact routing schemes with low stretch factor. J. Algorithms **46**(2), 97–114 (2003)
10. Freedman, O., Gawrychowski, P., Nicholson, P.K., Weimann, O.: Optimal distance labeling schemes for trees. In: Proceedings of the ACM Symposium on Principles of Distributed Computing, PODC 2017, pp. 185–194. ACM (2017)
11. Gavoille, C., Paul, C.: Optimal distance labeling for interval graphs and related graph families. SIAM J. Discret. Math. **22**(3), 1239–1258 (2008)
12. Gavoille, C., Peleg, D.: Compact and localized distributed data structures. Distrib. Comput. **16**(2–3), 111–120 (2003)
13. Gavoille, C., Peleg, D., Pérennes, S., Raz, R.: Distance labeling in graphs. J. Algorithms **53**(1), 85–112 (2004)
14. Gawrychowski, P., Uznanski, P.: Better distance labeling for unweighted planar graphs. Algorithmica **85**(6), 1805–1823 (2023)
15. Graham, R.L., Pollak, H.O.: On embedding graphs in squashed cubes. In: Alavi, Y., Lick, D.R., White, A.T. (eds.) Graph Theory and Applications. Lecture Notes in Mathematics, vol. 303, pp. 99–110. Springer, Heidelberg (1972). https://doi.org/10.1007/BFb0067362
16. Kannan, S., Naor, M., Rudich, S.: Implicit representation of graphs. SIAM J. Discrete Math. **5**(4), 596–603 (1992)
17. Korman, A., Peleg, D., Rodeh, Y.: Constructing labeling schemes through universal matrices. Algorithmica **57**, 641–652 (2010)
18. Moon, J.W.: On minimal n-universal graphs. Glasgow Math. J. **7**(1), 32–33 (1965)
19. Peleg, D.: Proximity-preserving labeling schemes and their applications. In: Widmayer, P., Neyer, G., Eidenbenz, S. (eds.) WG 1999. LNCS, vol. 1665, pp. 30–41. Springer, Heidelberg (1999). https://doi.org/10.1007/3-540-46784-X_5
20. Thorup, M., Zwick, U.: Compact routing schemes. In: Proceedings of the Thirteenth Annual ACM Symposium on Parallel Algorithms and Architectures, SPAA 2001, pp. 1–10. ACM (2001)
21. Winkler, P.M.: Proof of the squashed cube conjecture. Combinatorica **3**, 135–139 (1983)
22. Distance labeling for small families of cycles, source code (2023). https://tinyurl.com/tc8nd39s

On the Induced Problem for Fixed-Template CSPs

Rustem Takhanov[(✉)]

Nazarbayev University, 53 Kabanbay Batyr Ave, Astana, Kazakhstan
rustem.takhanov@nu.edu.kz

Abstract. The Constraint Satisfaction Problem (CSP) is a problem of computing a homomorphism $\mathbf{R} \to \mathbf{\Gamma}$ between two relational structures, where $\mathbf{R}$ is defined over a domain V and $\mathbf{\Gamma}$ is defined over a domain D. In a fixed template CSP, denoted CSP($\mathbf{\Gamma}$), the right side structure $\mathbf{\Gamma}$ is fixed and the left side structure $\mathbf{R}$ is unconstrained. In the last two decades it was discovered that the reasons that make fixed template CSPs polynomially solvable are of algebraic nature, namely, templates that are tractable should be preserved under certain polymorphisms. From this perspective the following problem looks natural: given a prespecified finite set of algebras $\mathscr{B}$ whose domain is D, is it possible to present the solution set of a given instance of CSP($\mathbf{\Gamma}$) as a subalgebra of $\mathbb{A}_1 \times ... \times \mathbb{A}_{|V|}$ where $\mathbb{A}_i \in \mathscr{B}$?

We study this problem and show that it can be reformulated as an instance of a certain fixed-template CSP over another template $\mathbf{\Gamma}^{\mathscr{B}}$.

We study conditions under which CSP($\mathbf{\Gamma}$) can be reduced to CSP($\mathbf{\Gamma}^{\mathscr{B}}$). This issue is connected with the so-called CSP with an input prototype, formulated in the following way: given a homomorphism from $\mathbf{R}$ to $\mathbf{\Gamma}^{\mathscr{B}}$ find a homomorphism from $\mathbf{R}$ to $\mathbf{\Gamma}$. We prove that if $\mathscr{B}$ contains only tractable algebras, then the latter CSP with an input prototype is tractable. We also prove that CSP($\mathbf{\Gamma}^{\mathscr{B}}$) can be reduced to CSP($\mathbf{\Gamma}$) if the set $\mathscr{B}$, treated as a relation over D, can be expressed as a primitive positive formula over $\mathbf{\Gamma}$.

Keywords: Lifted language · Lifted constraint instance · CSP with input prototype · Constraint satisfaction problem · Tractability

1 Introduction

The *constraint satisfaction problem (CSP)* can be formalized in the variable-value form as a problem of finding an assignment of values to a given set of variables, subject to constraints. There is also an equivalent formulation of it as a problem of finding a homomorphism $h : \mathbf{R} \to \mathbf{\Gamma}$ for two given finite relational structures $\mathbf{R}$ and $\mathbf{\Gamma}$.

A special case, when the second relational structure is some fixed $\mathbf{\Gamma}$ (called a template) and the domain of $\mathbf{\Gamma}$ is Boolean, was historically one of the first NP-hard problems [6], and its study has attracted some attention since the 70 s [8,16]. Feder and Vardi [7] formulated the so-called dichotomy conjecture

H. Fernau et al. (Eds.): SOFSEM 2024, LNCS 14519, pp. 485–499, 2024.
https://doi.org/10.1007/978-3-031-52113-3_34

that states that for a template $\mathbf{\Gamma}$ over an arbitrary finite domain, CSP($\mathbf{\Gamma}$) is either polynomially solvable or NP-hard. In [11] Jeavons showed that the complexity of CSP($\mathbf{\Gamma}$) is determined by the so-called polymorphisms of $\mathbf{\Gamma}$. A polymorphism of a relation $\varrho \subseteq D^n$ is defined as an m-ary function $p : D^m \to D$ such that ϱ is closed under the operation p that is applied component-wise to tuples from ϱ. A polymorphism of a template $\mathbf{\Gamma}$ is defined as a function that is a polymorphism of all relations in $\mathbf{\Gamma}$. Jeavons's result implies that if two languages $\mathbf{\Gamma}$ and $\mathbf{\Gamma}'$ have the same polymorphisms, then CSP($\mathbf{\Gamma}$) and CSP($\mathbf{\Gamma}'$) are log-space inter-reducible. Further development of this method [3,13,15,17] made it possible to precisely delineate the borderline between the polynomial-time and NP-complete templates of the CSP. Bulatov [4] and Zhuk [20] independently confirmed the Feder-Vardi conjecture.

Polymorphisms are interesting objects by themselves. Obviously, if a template $\mathbf{\Gamma}$ has a polymorphism p, then for any instance $\mathbf{R}$ the set of solutions is preserved by p. In other words, any such polymorphism induces an algebra on the solution set. Is there any simple generalization of that algebra that also takes into account the structure $\mathbf{R}$?

Inspired by this observation, we suggest generalizing algebras induced by polymorphisms of $\mathbf{\Gamma}$ in the following way. Suppose that we have a finite set $\mathcal{B}$ of similar algebras, i.e. every element $\mathbb{A} \in \mathcal{B}$ is a tuple $(D, o_1^{\mathbb{A}}, ..., o_k^{\mathbb{A}})$ where $o_i^{\mathbb{A}} : D^{n_i} \to D, i \in 1, \ldots, k$. Now, any instance of CSP($\mathbf{\Gamma}$) corresponds to another search problem, which we call *the induced problem*. More precisely, let (V, C) be an instance of CSP($\mathbf{\Gamma}$), where V is a set of variables and C is a set of constraints. The goal of the induced problem is to assign an algebra $\mathbb{A}_v \in \mathcal{B}$ to each variable $v \in V$ in such a way that for every constraint $\langle (v_1, ..., v_k), \varrho \rangle \in C$, the relation ϱ is a subalgebra of $\mathbb{A}_{v_1} \times ... \times \mathbb{A}_{v_p}$. If we are able to find such an assignment, it can be proved that the set of all solutions of the initial CSP instance (which is a subset of D^V) is a subalgebra of $\prod_{v \in V} \mathbb{A}_v$.

Motivation. Green and Cohen [10] considered the following computational problem which is a special case of the induced problem. Given a CSP over $D = \{1, \cdots, d\}$, they formulate a search problem: for any variable $v \in V$ find a permutation $\pi_v : D \to D$ such that if an assignment to $(v_1, ..., v_p)$ is constrained to be in $\varrho \subseteq D^p$ (in the initial CSP), then the resultant permuted relation $\varrho' = \{(\pi_{v_1}(x_1), \cdots, \pi_{v_p}(x_p)) | (x_1, \cdots, x_p) \in \varrho\}$ is max-closed (i.e. has a polymorphism $\max(x, y)$).

Indeed, suppose that we solved the described problem. Then, the initial CSP can be modified by a simple substitute $\varrho \to \varrho'$ in every constraint. This new CSP instance is called the permuted CSP. Then to any solution $h : V \to D$ of the permuted CSP (whose existence can be identified efficiently because the CSP is max-closed), we correspond a solution of the initial one by $h'(v) = \pi_v^{-1}(h(v))$.

To find such permutations $\pi_v, v \in V$ is computationally difficult in general. Green and Cohen's construction is a special case of ours, if we define a set of algebras

$$\mathcal{B} = \{\mathbb{A}_\pi | \pi \text{ is a permutation of } D\},$$

where for any permutation π of D we set

$$\mathbb{A}_\pi = \big(D, \max{}_\pi\big), \max{}_\pi(x, y) = \pi^{-1}\big(\max(\pi(x), \pi(y))\big).$$

The assignment of variables $v \longrightarrow \pi_v$ corresponds to the assignment $v \longrightarrow \mathbb{A}_{\pi_v}$ in our framework.

Our Results. First we prove that a search for an assignment $v \longrightarrow \mathbb{A}_v$ in the induced problem is equivalent to solving another fixed-template CSP($\mathbf{\Gamma}^{\mathcal{B}}$) and study the relationship between CSP($\mathbf{\Gamma}$) and CSP($\mathbf{\Gamma}^{\mathcal{B}}$). We prove that if the family $\mathcal{B}$ is tractable, i.e. all algebras in $\mathcal{B}$ are tractable (individually), and we are given a homomorphism $\mathbf{R} \longrightarrow \mathbf{\Gamma}^{\mathcal{B}}$, then a homomorphism $h : \mathbf{R} \longrightarrow \mathbf{\Gamma}$ can be found efficiently (Theorem 1). This result generalizes the desirable property of Green and Cohen's construction, in which, given appropriate permutations of domains, the initial CSP can be efficiently solved. As a corollary we obtain that if $\mathbf{\Gamma}$ maps homomorphically to $\mathbf{\Gamma}^{\mathcal{B}}$, then CSP($\mathbf{\Gamma}$) is reducible to CSP($\mathbf{\Gamma}^{\mathcal{B}}$) (Theorem 2). Further we prove that if the family $\mathcal{B}$ has a certain structure, namely, when a certain relation defined by $\mathcal{B}$ (called the trace of $\mathcal{B}$) is expressible as a primitive positive formula over $\mathbf{\Gamma}$, CSP($\mathbf{\Gamma}^{\mathcal{B}}$) can be reduced to CSP($\mathbf{\Gamma}$) (Theorems 3, 4 and Corollary 1).

Organization. In Sect. 2 we give all necessary definitions and state some basic facts that we need. In Subsect. 2.2 we describe the construction of *the "lifted language"*, taken from [14], and we introduce a novel framework of *CSPs with input prototype*. In Sect. 3 we introduce our main construction of the template $\mathbf{\Gamma}^{\mathcal{B}}$ and give examples of this construction. In Sect. 4 we prove the main result of the paper, that is an algorithm for CSP($\mathbf{\Gamma}$), given a homomorphism from an instance to $\mathbf{\Gamma}^{\mathcal{B}}$ (Theorem 1). Section 5 is dedicated to a Karp reduction of CSP($\mathbf{\Gamma}^{\mathcal{B}}$) to a non-uniform CSP that has basically two types of relations—those that are in Γ and a so called trace of $\mathcal{B}$ (Theorems 3 and 5). As a consequence we obtain that CSP($\mathbf{\Gamma}^{\mathcal{B}}$) can be reduced to CSP($\mathbf{\Gamma}$) if $\mathcal{B}$ is preserved under all polymorphisms of $\mathbf{\Gamma}$.

In Appendix one can find 7 examples demonstrating a usefullness of the induced problem (besides Green and Cohen's case). These examples show that for an appropriately chosen $\mathcal{B}$, the template $\mathbf{\Gamma}^{\mathcal{B}}$ is tractable by construction, for any $\mathbf{\Gamma}$. Any proof that is omitted in the main part of the paper can be found in Appendix.

2 Preliminaries

A problem is called *tractable* if it can be solved in polynomial time. We assume $\mathrm{P} \neq \mathrm{NP}$. Typically, a finite domain of CSP is denoted by D and a finite set of variables is denoted by V. We denote tuples in lowercase boldface such as $\mathbf{a} = (a_1, \ldots, a_k)$. Also for a mapping $h\colon A \longrightarrow B$ and a tuple $\mathbf{a} = (a_1, \ldots, a_k)$, where $a_j \in A$ for $j = 1, \ldots, k$, we will write $\mathbf{b} = (h(a_1), \ldots, h(a_k))$ simply as $\mathbf{b} = h(\mathbf{a})$. Let $\mathrm{ar}(\varrho)$, $\mathrm{ar}(\mathbf{a})$ stand for the arity of a relation ϱ and the size of a tuple **a**, respectively. A relational structure is a finite set and a tuple of relations of finite arity defined on that set. The set $\{1, ..., k\}$ is denoted by $[k]$.

Let us formulate the general CSP as a homomorphism problem.

Definition 1. *Let* $\mathbf{R} = (D^{\mathbf{R}}, r_1^{\mathbf{R}}, \cdots, r_s^{\mathbf{R}})$ *and* $\mathbf{R}' = (D^{\mathbf{R}'}, r_1^{\mathbf{R}'}, \cdots, r_s^{\mathbf{R}'})$ *be relational structures with a common signature (that is* $\mathrm{ar}(r_i^{\mathbf{R}}) = \mathrm{ar}(r_i^{\mathbf{R}'})$ *for every* $i \in [s]$*). A mapping* $h\colon D^{\mathbf{R}} \to D^{\mathbf{R}'}$ *is called a* homomorphism *from* **R** *to* **R**′ *if for every* $i \in [s]$ *and for any* $(x_1, \ldots, x_{\mathrm{ar}(r_i^{\mathbf{R}})}) \in r_i^{\mathbf{R}}$ *we have that* $((h(x_1), \ldots, h(x_{\mathrm{ar}(r_i^{\mathbf{R}})})) \in r_i^{\mathbf{R}'}$. *In this case, we write* $\mathbf{R} \xrightarrow{h} \mathbf{R}'$ *or just* $\mathbf{R} \to \mathbf{R}'$. *Also, we denote* $\mathrm{Hom}(\mathbf{R}, \mathbf{R}') = \{h \mid \mathbf{R} \xrightarrow{h} \mathbf{R}'\}$.

A finite relational structure $\mathbf{\Gamma} = (D, \varrho_1, \ldots, \varrho_s)$ over a fixed finite domain D is called a template.

Definition 2. *Let* D *be a finite set and* $\mathbf{\Gamma}$ *a template over* D. *Then the* **fixed template CSP** *for template* $\mathbf{\Gamma}$, *denoted* $\mathrm{CSP}(\mathbf{\Gamma})$, *is defined as follows: given a relational structure* **R** *of the same signature as* $\mathbf{\Gamma}$, *find a homorphism* $h : \mathbf{R} \to \mathbf{\Gamma}$.[1]

For $\mathbf{\Gamma} = (D, \varrho_1, \ldots, \varrho_s)$ we denote by Γ (without boldface) the set of relations $\{\varrho_1, \ldots, \varrho_s\}$ (which is called *the constraint language*) and by $\mathrm{CSP}(\Gamma)$ we denote $\mathrm{CSP}(\mathbf{\Gamma})$.

Definition 3. *A language* Γ *is said to be tractable if* $\mathrm{CSP}(\Gamma)$ *is tractable. Also,* Γ *is said to be NP-hard if* $\mathrm{CSP}(\Gamma)$ *is NP-hard.*

Any language Γ over a domain D can be associated with a set of operations on D, known as the polymorphisms of Γ [1], defined as follows.

Definition 4. *An operation* $g : D^m \to D$ *is a* polymorphism *of a relation* $\varrho \subseteq D^n$ *(or "*g preserves ϱ*", or "*ϱ is preserved by g*") if, for any* $m \times n$*-matrix whose columns are* $\overline{x}_1, \ldots, \overline{x}_n$ *and whose rows are all in* ϱ, *we have* $\big(g(\overline{x}_1), \ldots, g(\overline{x}_n)\big) \in \varrho$. *For any constraint language* Γ *over a set* D, *we denote by* $\mathrm{Pol}(\Gamma)$ *the set of all operations on* D *which are polymorphisms of every* $\varrho \in \Gamma$.

Let us denote the set of polymorphisms of Γ by $\mathrm{Pol}(\Gamma)$. Jeavons [11] showed that the complexity of $\mathrm{CSP}(\Gamma)$ is fully determined by $\mathrm{Pol}(\Gamma)$, which was the first step in developing the so-called algebraic approach to fixed template CSP. We will also use the notation $\mathrm{Pol}(\mathbf{\Gamma})$, meaning $\mathrm{Pol}(\Gamma)$.

2.1 Multi-sorted CSPs

For any finite collection of finite domains $\mathcal{D} = \{D_i | i \in I\}$, and any list of indices $(i_1, i_2, ..., i_m) \in I^m$, a subset ϱ of $D_{i_1} \times D_{i_2} \times \cdots \times D_{i_m}$, together with the list $(i_1, i_2, ..., i_m)$, is called a multi-sorted relation over $\mathcal{D}$ with arity m and signature $(i_1, i_2, ..., i_m)$. For any such relation ϱ, the signature of ϱ is denoted $\sigma(\varrho)$.

[1] Throughout the paper we define the CSP as a search problem, not as a decision problem, taking into account that both formulations are equivalent [3,5].

Definition 5. *Let* Γ *be a set of multi-sorted relations over a collection of sets* $\mathcal{D} = \{D_i | i \in I\}$*. The multi-sorted constraint satisfaction problem over* Γ*, denoted* MCSP(Γ)*, is defined to be the search problem with:*

Instance: *A triple* $(V; \delta; C)$ *where*

- V *is a set of variables;*
- δ *is a mapping from* V *to* I*, called the domain function;*
- C *is a set of constraints, where each constraint* $C \in C$ *is a pair* (s, ϱ)*, such that*
  - $s = (v_1, ..., v_{m_C})$ *is a tuple of variables of length* m_C*, called the constraint scope;*
  - ϱ *is an element of* Γ *with the arity* m_C *and signature* $(\delta(v_1), ..., \delta(v_{m_C}))$ *called the constraint relation.*

Question: *Find a solution (or, indicate its nonexistence), i.e., a function* ϕ*, from* V *to* $\cup_{i \in I} D_i$ *such that, for each variable* $v \in V$ *,* $\phi(v) \in D_{\delta(v)}$*, and for each constraint* $(s, \varrho) \in C$*, with* $s = (v_1, ..., v_m)$*, the tuple* $(\phi(v_1), ..., \phi(v_m))$ *belongs to* ϱ*?*

By construction a fixed template CSP, given in the form of a homomorphism problem, can be formulated as a multi-sorted CSP over a collection of domains $\mathcal{D} = \{D\}$. The problem of finding a homomorphism $h : \mathbf{R} \to \mathbf{\Gamma}$ where $\mathbf{R} = (V, r_1, \ldots, r_s)$ and $\mathbf{\Gamma} = (D, \varrho_1, \ldots, \varrho_s)$, is equivalent to the following set of constraints:

$$\{(\mathbf{v}, \varrho_i) | i \in [s], \mathbf{v} \in r_i\}. \tag{1}$$

Definition 6. *A set of multi-sorted relations,* Γ *is said to be tractable if* MCSP(Γ) *is tractable.*

Definition 7. *Let* $\mathcal{D}$ *be a collection of sets. An* m*-ary multi-sorted operation* t *on* $\mathcal{D}$ *is defined by a collection of interpretations* $\{t^D | D \in \mathcal{D}\}$*, where each* t^D *is an* m*-ary operation on the corresponding set* D*. A multi-sorted operation* t *on* $\mathcal{D}$ *is said to be a polymorphism of an* n*-ary multi-sorted relation* ϱ *over* $\mathcal{D}$ *with signature* $(\delta(1), ..., \delta(n))$ *if, for any* $m \times n$*-matrix* $[\overline{x}_1, \ldots, \overline{x}_n]$ *whose rows are all in* ϱ*, we have*

$$\left(t^{D_{\delta(1)}}(\overline{x}_1), \ldots, t^{D_{\delta(n)}}(\overline{x}_n)\right) \in \varrho. \tag{2}$$

For the set of multi-sorted relations Γ, MPol(Γ) denotes the set of all multi-sorted operations that are polymorphisms of each relation in Γ.

2.2 The Lifted Language

Let $\mathbf{\Gamma} = (D, \varrho_1, \ldots, \varrho_s)$ be a template and $\mathbf{R} = (V, r_1, \ldots, r_s)$ be a relational structure given as an input to CSP($\mathbf{\Gamma}$). The problem of finding a homomorphism $h : \mathbf{R} \to \mathbf{\Gamma}$ can be reformulated as an instance of the multi-sorted CSP in many different ways. We choose the most straightforward one as it gives an insight into the construction of the lifted language. We introduce for every variable $v \in V$

its unique domain $D_v = \{(v,a)|a \in D\}$. Thus, we get a collection of domains $\mathcal{D} = \{D_v|v \in V\}$.

For tuples $\mathbf{a} = (a_1, \ldots, a_p) \in D^p$ and $\mathbf{v} = (v_1, \ldots, v_p) \in V^p$ denote $d(\mathbf{v}, \mathbf{a}) = ((v_1, a_1), ..., (v_p, a_p))$. Now for a relation $\varrho \subseteq D^p$ and $\mathbf{v} = (v_1, \ldots, v_p) \in V^p$ we will define a multi-sorted relation $\varrho(\mathbf{v})$ over $\mathcal{D}$ with a signature $(v_1, \ldots, v_p)$ by

$$\varrho(\mathbf{v}) = \left\{d(\mathbf{v}, \mathbf{y})|\mathbf{y} \in \varrho\right\}. \tag{3}$$

The set of constraints $\{(\mathbf{v}, \varrho_i(\mathbf{v})) \; : \; i \in [s], \mathbf{v} \in r_i\} \cup \{(v, D_v) \; : \; v \in V\}$ defines an instance of the multi-sorted CSP whose solutions are in one-to-one correspondence with homomorphisms from $\mathbf{R}$ to $\mathbf{\Gamma}$. The correspondence between $h : V \to D$ and $h' : V \to \cup\mathcal{D}$ is established by the rule $h'(v) = (v, h(v))$.

Finally, we construct the language $\Gamma_\mathbf{R}$ (which is called the lifted language) that consists of multi-sorted relations over $\mathcal{D}$

$$\Gamma_\mathbf{R} = \{\varrho_i(\mathbf{v}) \; : \; i \in [s], \mathbf{v} \in r_i\} \cup \{D_v \; : \; v \in V\}. \tag{4}$$

Note that, if all relations in $\mathbf{R}$ are nonempty, the lifted language $\Gamma_\mathbf{R}$ contains all information about a pair $\mathbf{R}, \mathbf{\Gamma}$. After ordering its relations we get the template $\mathbf{\Gamma_R}$. A more general version of this language (formulated in terms of cost functions) first appeared in the context of *the hybrid CSPs* which is an extension of the fixed-template CSP framework (see Sect. 5.1 of [14]).

Note that this language defines $\mathrm{MCSP}(\mathbf{\Gamma_R})$, in which every variable is paired with its domain as in Definition 5. Sometimes the lifted language will be treated as a set of relations over a common domain $\cup_v D_v = V \times D$ (i.e. not multi-sorted), e.g. as in Theorem 4. Since all domains $D_v, v \in V$ are disjoint, $\mathrm{MCSP}(\mathbf{\Gamma_R})$ and $\mathrm{CSP}(\mathbf{\Gamma_R})$ are Karp reducible to each other.

The following lemma plays a key role in our paper. It shows that $\mathrm{MCSP}(\mathbf{\Gamma_R})$ is equivalent to another problem formulation called *the CSP with an input prototype* (see Sect. 6 of [18]).

Definition 8. *For a given template* $\mathbf{\Gamma}$ *and a relational structure* $\mathbf{P}$, **the CSP with an input prototype P** *is a problem, denoted* $\mathrm{CSP}^+_\mathbf{P}(\mathbf{\Gamma})$, *for which: a) an instance is a pair* $(\mathbf{R}, \chi)$ *where* $\mathbf{R}$ *is a relational structure and* $\chi : \mathbf{R} \to \mathbf{P}$ *is a homomorphism; b) the goal is to find a homomorphism* $h : \mathbf{R} \to \mathbf{\Gamma}$.

E.g., when $\mathbf{P} = ([4], \neq), \mathbf{\Gamma} = ([3], \neq)$, then $\mathrm{CSP}^+_\mathbf{P}(\mathbf{\Gamma})$ is a problem of finding a 3-coloring of a graph whose 4-coloring is given as part of input.

Lemma 1. $\mathrm{MCSP}(\mathbf{\Gamma_P})$ *and* $\mathrm{CSP}^+_\mathbf{P}(\mathbf{\Gamma})$ *are Karp reducible to each other in linear time.*

Proof. **Karp reduction of $\mathbf{MCSP(\Gamma_P)}$ to $\mathbf{CSP^+_P(\Gamma)}$.** Let $\mathbf{\Gamma} = \left(D, \varrho_1, ..., \varrho_s\right)$ and $\mathbf{P} = (V, r_1, \ldots, r_s)$ be given. Let I be an instance of $\mathrm{MCSP}(\mathbf{\Gamma_P})$ that consists of a set of variables W and a set of constraints C. For any $\varrho_i(\mathbf{v}) \in \Gamma_\mathbf{P}$ let us denote $f_i^\mathbf{v} = \{\mathbf{v}'|(\mathbf{v}', \varrho_i(\mathbf{v})) \in C\}$. Thus, $C = \left\{(\mathbf{v}', \varrho_i(\mathbf{v}))|i \in [s], \mathbf{v} \in r_i, \mathbf{v}' \in f_i^\mathbf{v}\right\}$.

Also, we are given an assignment $\delta : W \to V$, that assigns each variable $v \in W$ its domain $D_{\delta(v)}$. Denote $\mathbf{R} = (W, f_1, ..., f_s)$, where $f_i = \cup_{\mathbf{v} \in r_i} f_i^\mathbf{v}$.

According to Definition 5, $\varrho_i(\mathbf{v})$ is a relation with signature $\delta(\mathbf{v}')$, $\mathbf{v}' \in f_i^{\mathbf{v}}$. Therefore, for any $\mathbf{v}' \in f_i^{\mathbf{v}}$ its component-wise image $\delta(\mathbf{v}')$ is exactly the tuple $\mathbf{v}$. Since $\mathbf{v} \in r_i$, we conclude $\mathbf{R} \xrightarrow{\delta} \mathbf{P}$.

For $h : W \to D$, let us define $h^\delta : W \to V \times D$ by $h^\delta(v) = (\delta(v), h(v))$. Vice versa, to every assignment $g : W \to V \times D$ we will associate an assignment $g^f(x) = F(g(x))$ where F is a "forgetting" function, i.e. $F((v, a)) = a$. For any assignment $g : W \to V \times D$ that satisfies $g(v) \in D_{\delta(v)}$, by construction, we have $(g^f)^\delta = g$. By construction, g is a solution of our multi-sorted CSP if and only if $\mathbf{R} \xrightarrow{g^f} \mathbf{\Gamma}$. The latter is an instance of $\mathrm{CSP}^+_{\mathbf{P}}(\mathbf{\Gamma})$ with an input structure $\mathbf{R}$ and a homomorphism $\delta : \mathbf{R} \to \mathbf{P}$ given, and any solution h of it corresponds to a solution h^δ of $\mathcal{I}$. By construction, all computations are polynomial-time, so we proved that $\mathrm{MCSP}(\mathbf{\Gamma_P})$ can be polynomially reduced to $\mathrm{CSP}^+_{\mathbf{P}}(\mathbf{\Gamma})$.

Karp Reduction of $\mathrm{CSP}^+_{\mathbf{P}}(\mathbf{\Gamma})$ to $\mathrm{MCSP}(\mathbf{\Gamma_P})$. Let $\mathbf{\Gamma} = (D, \varrho_1, \ldots, \varrho_s)$, $\mathbf{P} = (V, r_1, \ldots, r_s)$. Suppose we are given an instance of $\mathrm{CSP}^+_{\mathbf{P}}(\mathbf{\Gamma})$ with an input structure $\mathbf{R} = (W, f_1, \ldots, f_s)$ and a homomorphism $\delta : \mathbf{R} \to \mathbf{P}$, i.e. our goal is to satisfy the set of constraints $\{(\mathbf{v}, \varrho_i) | i \in [s], \mathbf{v} \in f_i\}$. Let us construct an instance of $\mathrm{MCSP}(\mathbf{\Gamma_P})$:

$$\{(\mathbf{v}, \varrho_i(\delta(\mathbf{v}))) | i \in [s], \mathbf{v} \in f_i\}, \{(v, D_{\delta(v)}) | v \in W\}.$$

It is straightforward to check that if g is a solution for this instance then $h = g^f$ is a solution for $\mathrm{CSP}^+_{\mathbf{P}}(\mathbf{\Gamma})$ and visa versa. It remains to note that $\{\varrho_i(\delta(\mathbf{v})) | i \in [s], \mathbf{v} \in f_i\} \subseteq \mathbf{\Gamma_P}$. By construction both reductions take linear time on the size of input. □

3 The Construction

Suppose that we are given a list $o_1, \ldots, o_k$ of symbols with prescribed arities $n_1, \ldots, n_k$. This list is called the signature and denoted σ. An *algebra* with a signature σ is a tuple $\mathbb{A} = (D^{\mathbb{A}}, o_1^{\mathbb{A}}, o_2^{\mathbb{A}}, \ldots, o_k^{\mathbb{A}})$, where $D^{\mathbb{A}}$ denotes a finite domain of the algebra and $o_i^{\mathbb{A}} : (D^{\mathbb{A}})^{n_i} \to D^{\mathbb{A}}, i \in [k]$ denote its basic operations. Let us denote by $\mathcal{A}_D^\sigma$ the set of algebras with one fixed signature σ and over a single fixed domain D. Suppose we are given a collection $\mathcal{B} \subseteq \mathcal{A}_D^\sigma$ and a relational structure $\mathbf{\Gamma} = (D, \varrho_1, \ldots, \varrho_s)$ where ϱ_i is a relation over D.

Definition 9. *Let ϱ be an m-ary relation over D. Let us define*

$$\varrho^{\mathcal{B}} = \{(\mathbb{A}_1, \ldots, \mathbb{A}_m) \in \mathcal{B} \times \cdots \times \mathcal{B} \mid \varrho \text{ is a subalgebra of } \mathbb{A}_1 \times \cdots \times \mathbb{A}_m\}.$$

In other words, we define the relation $\varrho^{\mathcal{B}}$ as a subset of $\mathcal{B}^m$ that consists of tuples $(\mathbb{A}_1, \ldots, \mathbb{A}_m) \in \mathcal{B}^m$ such that for any $i \in [k]$, $(o_i^{\mathbb{A}_1}, o_i^{\mathbb{A}_2}, \ldots, o_i^{\mathbb{A}_m})$ is a component-wise polymorphism of ϱ.[2] *The last condition means that for any matrix $[\overline{x}_1, \ldots, \overline{x}_m] \in D^{n_i \times m}$ whose rows are all in ϱ, we have $(o_i^{\mathbb{A}_1}(\overline{x}_1), o_i^{\mathbb{A}_2}(\overline{x}_2), \ldots, o_i^{\mathbb{A}_m}(\overline{x}_m)) \in \varrho$.*

[2] $\varrho^{\mathcal{B}}$ can be empty.

Definition 10. *Given* Γ *and* $\mathcal{B}$*, we define* $\Gamma^{\mathcal{B}} = \{\varrho^{\mathcal{B}} | \varrho \in \Gamma\}$*. Analogously, if* $\mathbf{\Gamma} = (D, \varrho_1, ..., \varrho_s)$ *where* ϱ_i *is a relation over* D*, then we define* $\mathbf{\Gamma}^{\mathcal{B}} = (\mathcal{B}, \varrho_1^{\mathcal{B}}, \cdots, \varrho_s^{\mathcal{B}})$*.*

Now, given an instance $\mathbf{R}$ of CSP($\mathbf{\Gamma}$) we can consider $\mathbf{R}$ as an instance of CSP($\mathbf{\Gamma}^{\mathcal{B}}$). Let us decode Definitions 9 and 10. Any $h \in \text{Hom}(\mathbf{R}, \mathbf{\Gamma}^{\mathcal{B}})$ assigns to every variable $v \in V$ an algebra $h(v) \in \mathcal{B}$. For $j \in [s]$, $\mathbf{v} \in r_j$ our assignment satisfies $h(\mathbf{v}) \in \varrho_j^{\mathcal{B}}$, i.e. if $\mathbf{v} = (v_1, ..., v_p)$, then $(o_i^{h(v_1)}, ..., o_i^{h(v_p)})$ component-wise preserves ϱ_j. Suppose now that for any $v \in V$ we create a unique copy of the domain D, i.e. $D_v = \{(v, a) | a \in D\}$, and define $m_i^{D_v}$ as an interpretation of $o_i^{h(v)}$ on this copy D_v, i.e.

$$m_i^{D_v}((v, a_1), ..., (v, a_{n_i})) = (v, o_i^{h(v)}(a_1, ..., a_{n_i})).$$

Since for $j \in [s]$ and $\mathbf{v} = (v_1, ..., v_p) \in r_j$, $(o_i^{h(v_1)}, ..., o_i^{h(v_p)})$ component-wise preserves ϱ_j, operation m_i preserves the multi-sorted relation $\varrho_j(v_1, ..., v_p)$ (see Eq. (3)). In other words, m_i is a multi-sorted polymorphism of the lifted language $\Gamma_{\mathbf{R}}$. Thus, every assignment $h \in \text{Hom}(\mathbf{R}, \mathbf{\Gamma}^{\mathcal{B}})$ induces a system of multi-sorted polymorphisms $m_1, ..., m_k \in \text{MPol}(\Gamma_{\mathbf{R}})$.

For special cases of $\mathcal{B}$, the structure of the template $\mathbf{\Gamma}^{\mathcal{B}}$ has been studied in [19].

3.1 Example: Binary and Conservative Operations

This example is a direct generalization of Proposition 36 from [10]. Let us define $\mathcal{B}$ as the set of all algebras with a commutative and conservative binary operation over D, i.e.

$$\mathcal{B} = \{(D, b) | b(x, y) \in \{x, y\}, b(x, y) = b(y, x)\}.$$

It is a well-known fact that any commutative and conservative binary operation corresponds to a tournament on D, i.e. to a complete directed graph with a set of vertices D in which antiparallel arcs are not allowed (an identity $b(x, y) = y$ for distinct $x, y \in D$ corresponds to an arc from x to y in the tournament). Thus, $|\mathcal{B}| = 2^{\binom{|D|}{2}}$. We define a ternary operation m on the set $\mathcal{B}$ that acts as follows:

$$m((D, b_1), (D, b_2), (D, b_3)) = (D, b) \Leftrightarrow b(x, y) = b_1(b_2(x, y), b_3(x, y)).$$

By construction m outputs an element from $\mathcal{B}$, i.e. $m : \mathcal{B}^3 \to \mathcal{B}$. It is straightforward to check that the following identities hold for conservative and commutative operations:

$$b_1(b_1(x, y), b_2(x, y)) = b_1(b_2(x, y), b_1(x, y)) = b_2(b_1(x, y), b_1(x, y)) = b_1(x, y)$$

or $m(\mathbb{A}, \mathbb{A}, \mathbb{B}) = m(\mathbb{A}, \mathbb{B}, \mathbb{A}) = m(\mathbb{B}, \mathbb{A}, \mathbb{A}) = \mathbb{A}$, $\forall \mathbb{A}, \mathbb{B} \in \mathcal{B}$. Thus, m is a majority operation.

Now let us prove that for any $\varrho \subseteq D^k$, m is a polymorphism of $\varrho^{\mathcal{B}}$. Indeed, suppose that $(\mathbb{A}_1, \cdots, \mathbb{A}_k) \in \varrho^{\mathcal{B}}$, $(\mathbb{B}_1, \cdots, \mathbb{B}_k) \in \varrho^{\mathcal{B}}$ and $(\mathbb{C}_1, \cdots, \mathbb{C}_k) \in \varrho^{\mathcal{B}}$. We denote $\mathbb{A}_i = (D, a_i), \mathbb{B}_i = (D, b_i), \mathbb{C}_i = (D, c_i)$. By definition of $\varrho^{\mathcal{B}}$, we know that

$$\bar{x} = \begin{bmatrix} x_1 \\ \cdots \\ x_k \end{bmatrix}, \bar{y} = \begin{bmatrix} y_1 \\ \cdots \\ y_k \end{bmatrix} \in \varrho \Rightarrow \begin{bmatrix} b_1(x_1, y_1) \\ \cdots \\ b_k(x_k, y_k) \end{bmatrix}, \begin{bmatrix} c_1(x_1, y_1) \\ \cdots \\ c_k(x_k, y_k) \end{bmatrix} \in \varrho.$$

Therefore, component-wise application of $(a_1, \cdots, a_k)$ to the last two tuples also will result in a tuple from ϱ:

$$\begin{bmatrix} a_1(b_1(x_1, y_1), c_1(x_1, y_1)) \\ \cdots \\ a_k(b_k(x_k, y_k), c_k(x_k, y_k)) \end{bmatrix} \in \varrho.$$

The latter implies that $\big(m(\mathbb{A}_1, \mathbb{B}_1, \mathbb{C}_1), \cdots, m(\mathbb{A}_k, \mathbb{B}_k, \mathbb{C}_k)\big) \in \varrho^{\mathcal{B}}$.

Since, m is a majority polymorphism of any $\varrho^{\mathcal{B}}$, the problem $\mathrm{CSP}(\mathbf{\Gamma}^{\mathcal{B}})$ is tractable for any $\mathbf{\Gamma}$ [12]. Moreover, $\mathrm{CSP}(\mathbf{\Gamma}^{\mathcal{B}})$ can be solved by a local consistency checking algorithm. Note that if we define $\mathcal{B}^o \subset \mathcal{B}$ as the set of all tournament pairs that correspond to total orders on D, then $\mathrm{CSP}(\mathbf{\Gamma}^{\mathcal{B}^o})$ is NP-hard in general (see Proposition 38 from [10]).

This example shows that $\mathcal{B}$ can be such that $\mathrm{CSP}(\mathbf{\Gamma}^{\mathcal{B}})$ is tractable for any (possibly NP-hard) $\mathbf{\Gamma}$. Other examples of this kind can be found in Appendix.

4 The Complexity of $\mathbf{CSP}^{+}_{\mathbf{\Gamma}^{\mathcal{B}}}(\mathbf{\Gamma})$

Again, let Γ be a set of relations over D and $\mathcal{B} \subseteq \mathcal{A}^{\sigma}_{D}$. In the previous section we gave an example of a set of algebras $\mathcal{B}$ for which $\mathrm{CSP}(\mathbf{\Gamma}^{\mathcal{B}})$ is tractable for any $\mathbf{\Gamma}$ (possibly NP-hard). More such examples can be found in Appendix. Therefore, asking what can be achieved by substituting $\mathbf{\Gamma}$ with another template $\mathbf{\Gamma}'$ (e.g. $\mathbf{\Gamma}' = \mathbf{\Gamma}^{\mathcal{B}}$ for an appropriate $\mathcal{B}$), and the consequences of such substitutions, may be a promising research direction.

First we will study the following problem: if we managed to find a homomorphism from the input structure $\mathbf{R}$ to $\mathbf{\Gamma}^{\mathcal{B}}$, when can it help us to find $h : \mathbf{R} \to \mathbf{\Gamma}$? In Subsect. 2.2 we called this problem *the CSP with an input prototype* and denoted as $\mathrm{CSP}^{+}_{\mathbf{\Gamma}^{\mathcal{B}}}(\mathbf{\Gamma})$. It was prove to be equivalent to $\mathrm{CSP}(\mathbf{\Gamma}_{\mathbf{\Gamma}^{\mathcal{B}}})$. Thus, we will start with identifying conditions for the tractability of $\mathbf{\Gamma}_{\mathbf{\Gamma}^{\mathcal{B}}}$.

4.1 Conditions for the Tractability of $\mathbf{\Gamma}_{\mathbf{\Gamma}^{\mathcal{B}}}$

From the definition of lifted languages it is clear that for any relation $P \in \Gamma_{\mathbf{\Gamma}^{\mathcal{B}}}$ there exists a relation $\varrho \in \Gamma$ such that $P = \varrho(\mathbb{A}_1, \cdots, \mathbb{A}_p)$ where $(\mathbb{A}_1, \cdots, \mathbb{A}_p) \in \varrho^{\mathcal{B}}$. Since $\varrho^{\mathcal{B}} \subseteq \mathcal{B}^p$, we have $\mathbb{A}_1 \in \mathcal{B}, \cdots, \mathbb{A}_p \in \mathcal{B}$. From the definition of $\varrho(\mathbb{A}_1, \cdots, \mathbb{A}_p)$ (see (3)) we obtain that

$$\varrho(\mathbb{A}_1, \cdots, \mathbb{A}_p) \subseteq \{(\mathbb{A}_1, x_1) | x_1 \in D\} \times \cdots \times \{(\mathbb{A}_p, x_p) | x_p \in D\}.$$

For any algebra $\mathbb{A} \in \mathcal{B}$ let us denote by $\mathbb{A}^c$ a copy of $\mathbb{A}$, but with all operations redefined on its new unique domain $D^{\mathbb{A}^c} = \{(\mathbb{A}, x) | x \in D\}$ by

$$o_l^{\mathbb{A}^c}\big((\mathbb{A}, x_1), \cdots, (\mathbb{A}, x_{n_l})\big) = (\mathbb{A}, o_l^{\mathbb{A}}(x_1, \cdots, x_{n_l})).$$

If we introduce a new collection of domains $\{D^{\mathbb{A}^c} | \mathbb{A} \in \mathcal{B}\}$ parameterized by algebras from $\mathcal{B}$, then $\varrho(\mathbb{A}_1, \cdots, \mathbb{A}_p) \subseteq D^{\mathbb{A}_1^c} \times \cdots \times D^{\mathbb{A}_p^c}$ becomes a multi-sorted relation with signature $(\mathbb{A}_1, \cdots, \mathbb{A}_p)$. Thus, in the new notations, any relation from $\Gamma_{\mathbf{\Gamma}^{\mathcal{B}}}$ becomes multi-sorted over a collection of sets $\{D^{\mathbb{A}^c} | \mathbb{A} \in \mathcal{B}\}$. Also, denote $\mathcal{B}^c = \{\mathbb{A}^c | \mathbb{A} \in \mathcal{B}\}$.

Definition 11. *For a collection $\mathcal{B}$, we define* $\text{MINV}(\mathcal{B}^c)$ *as the set of all multi-sorted relations ϱ over a collection of sets $\{D^{\mathbb{A}^c} | \mathbb{A} \in \mathcal{B}\}$ such that for any $l \in [k]$, the n_l-ary multi-sorted operation $\{o_l^{\mathbb{A}^c} | \mathbb{A} \in \mathcal{B}\}$ is a polymorphism of ϱ.*

Lemma 2. *$\Gamma_{\mathbf{\Gamma}^{\mathcal{B}}}$ understood as a multi-sorted language over a collection of domains $\{D^{\mathbb{A}^c} | \mathbb{A} \in \mathcal{B}\}$ is a subset of* $\text{MINV}(\mathcal{B}^c)$.

Proof. Let us check that $\varrho_i(\mathbb{A}_1, ..., \mathbb{A}_p) \in \text{MINV}(\mathcal{B}^c)$ whenever $(\mathbb{A}_1, ..., \mathbb{A}_p) \in \varrho_i^{\mathcal{B}}$. The latter premise implies that for $h \in [k]$, $(o_h^{\mathbb{A}_1}, ..., o_h^{\mathbb{A}_p})$ is a component-wise polymorphism of ϱ_i, and therefore, $(o_h^{\mathbb{A}_1^c}, ..., o_h^{\mathbb{A}_p^c})$ is a component-wise polymorphism of $\varrho_i(\mathbb{A}_1, ..., \mathbb{A}_p)$. From the last we conclude that the multi-sorted operation $\{o_h^{\mathbb{A}^c} | \mathbb{A} \in \mathcal{B}\}$ is a polymorphism of $\varrho_i(\mathbb{A}_1, ..., \mathbb{A}_p)$, or $\varrho_i(\mathbb{A}_1, \cdots, \mathbb{A}_p) \in \text{MINV}(\mathcal{B}^c)$. □

The following definition is very natural.

Definition 12. *A collection $\mathcal{B}$ is called* **tractable** *if* $\text{MINV}(\mathcal{B}^c)$ *is a tractable constraint language.*

Thus, using Lemma 2 and Definition 12 we obtain the following key result.

Theorem 1. *If a collection $\mathcal{B}$ is tractable, then* $\text{MCSP}(\Gamma_{\mathbf{\Gamma}^{\mathcal{B}}})$ *is tractable.*

Remark 1. We gave a definition of the tractable collection $\mathcal{B}$ that serves our purposes. It can be shown that a collection $\mathcal{B}$ is tractable if and only if every algebra $\mathbb{A} \in \mathcal{B}$ is tractable (i.e. $\text{INV}(\{o_l^{\mathbb{A}} | l \in [k]\})$ is a tractable language). The fact is well-known in CSP studies, so we omit a proof of it.

Remark 2. In examples given in Appendix, sets of algebras $\mathcal{B}$, $\mathcal{B}_1$, $\mathcal{B}_{\text{com}}$, $\mathcal{B}_{\text{nu}}$, $\mathcal{B}_{\text{M}}$, $\mathcal{B}_{\text{wnu}}$ are tractable. Then, Theorem 1 and Lemma 1 give us the tractability of $\text{CSP}^+_{\mathbf{\Gamma}^{\mathcal{B}}}(\mathbf{\Gamma})$, $\text{CSP}^+_{\mathbf{\Gamma}^{\mathcal{B}_1}}(\mathbf{\Gamma})$, $\text{CSP}^+_{\mathbf{\Gamma}^{\mathcal{B}_{\text{com}}}}(\mathbf{\Gamma})$, $\text{CSP}^+_{\mathbf{\Gamma}^{\mathcal{B}_{\text{nu}}}}(\mathbf{\Gamma})$, $\text{CSP}^+_{\mathbf{\Gamma}^{\mathcal{B}_{\text{M}}}}(\mathbf{\Gamma})$, $\text{CSP}^+_{\mathbf{\Gamma}^{\mathcal{B}_{\text{wnu}}}}(\mathbf{\Gamma})$. The only problem that remains now is finding a homomorphism from an input structure to $\mathbf{\Gamma}^{\mathcal{B}}$, $\mathbf{\Gamma}^{\mathcal{B}_1}$, $\mathbf{\Gamma}^{\mathcal{B}_{\text{com}}}$, $\mathbf{\Gamma}^{\mathcal{B}_{\text{nu}}}$, $\mathbf{\Gamma}^{\mathcal{B}_{\text{M}}}$, $\mathbf{\Gamma}^{\mathcal{B}_{\text{wnu}}}$. In the examples we discussed that these tasks are all tractable.

Theorem 2. *If a collection $\mathcal{B}$ is tractable and $\mathbf{\Gamma} \to \mathbf{\Gamma}^{\mathcal{B}}$, then* $\text{CSP}(\mathbf{\Gamma})$ *is polynomial-time Turing reducible to* $\text{CSP}(\mathbf{\Gamma}^{\mathcal{B}})$.

Proof. For an instance $\mathbf{R}$ of $\text{CSP}(\mathbf{\Gamma})$, if $\mathbf{\Gamma} \to \mathbf{\Gamma}^{\mathcal{B}}$, then we can replace the right template $\mathbf{\Gamma}$ with $\mathbf{\Gamma}^{\mathcal{B}}$ and obtain a relaxed version of the initial CSP. Suppose that we are able to solve $\text{CSP}(\mathbf{\Gamma}^{\mathcal{B}})$. If the solution set of the relaxed problem is empty, then it all the more is empty for the initial one. But if we manage to find a single homomorphism from the input structure $\mathbf{R}$ to $\mathbf{\Gamma}^{\mathcal{B}}$, then the problem of finding a homomorphism $\mathbf{R} \to \mathbf{\Gamma}$ can be presented as an instance of $\mathrm{CSP}^+_{\mathbf{\Gamma}^{\mathcal{B}}}(\mathbf{\Gamma})$, or, by Lemma 1, of $\mathrm{MCSP}(\mathbf{\Gamma}_{\mathbf{\Gamma}^{\mathcal{B}}})$. Now, from the tractability of $\mathcal{B}$ and Theorem 1 we get that $\mathrm{MCSP}(\mathbf{\Gamma}_{\mathbf{\Gamma}^{\mathcal{B}}})$ is tractable and we efficiently find a homomorphism $h : \mathbf{R} \to \mathbf{\Gamma}$. □

Unfortunately, it is hard to satisfy the condition $\mathbf{\Gamma} \to \mathbf{\Gamma}^{\mathcal{B}}$ unless $\mathcal{B}$ contains constants. In examples from Appendix (and in an example from Sect. 3.1) it is not satisfied.

5 Reductions of $\mathrm{CSP}(\mathbf{\Gamma}^{\mathcal{B}})$ to $\mathrm{CSP}(\mathbf{\Gamma})$

Let us now find some conditions on $\mathcal{B}$ under which $\mathrm{CSP}(\mathbf{\Gamma}^{\mathcal{B}})$ is a fragment of $\mathrm{CSP}(\mathbf{\Gamma})$. Again, we are given $\mathbf{\Gamma} = (D, \varrho_1, ..., \varrho_s)$ where ϱ_l is a relation over D, $l \in [s]$ and $\mathcal{B} \subseteq \mathcal{A}^{\sigma}_D$. In this section we will show that under very natural conditions on $\mathcal{B}$, any instance of $\mathrm{CSP}(\mathbf{\Gamma}^{\mathcal{B}})$ can be turned into an instance of $\mathrm{CSP}(\mathbf{\Gamma})$. Let us introduce some natural definitions that will serve our purpose.[3] Let $D = [d]$. Given n, let $\alpha_n(1), \alpha_n(2), ..., \alpha_n(d^n)$ be a lexicographic ordering of D^n.

Definition 13. *The trace of $\mathcal{B}$, denoted $Tr(\mathcal{B})$, is the relation*

$$\left\{\left(o_1^{\mathbb{A}}(\alpha_{n_1}(1)), \cdots, o_1^{\mathbb{A}}(\alpha_{n_1}(d^{n_1})), \cdots, o_k^{\mathbb{A}}(\alpha_{n_k}(1)), \cdots, o_k^{\mathbb{A}}(\alpha_{n_k}(d^{n_k}))\right) \mid \mathbb{A} \in \mathcal{B}\right\}.$$

The arity of $Tr(\mathcal{B})$ is $\kappa(\mathcal{B}) = \sum_{s=1}^{k} d^{n_s}$.

Given a number n and an m-ary relation ϱ over D, let us denote ϱ^n the set of all tuples $(\overline{x}_1, ..., \overline{x}_m)$, where $\overline{x}_l \in D^n, l \in [m]$, such that all rows of the matrix $[\overline{x}_1, ..., \overline{x}_m]$ are in ϱ. Note that ϱ^n is a relation over D^n. According to the standard terminology of universal algebra, ϱ^n is a direct product $\prod_{i=1}^n \varrho$ of m-ary relations. It satisfies $|\varrho^n| = |\varrho|^n$. Also, let $\mathbf{\Gamma} \frown \mathcal{B} = (D, \varrho_1, \cdots, \varrho_s, Tr(\mathcal{B}))$.

Theorem 3. $\mathrm{CSP}(\mathbf{\Gamma}^{\mathcal{B}})$ *is Karp reducible to* $\mathrm{CSP}(\mathbf{\Gamma} \frown \mathcal{B})$.

Proof. Suppose that we are given an instance of $\mathrm{CSP}(\mathbf{\Gamma}^{\mathcal{B}})$, i.e. an input structure $\mathbf{R} = (V, r_1, \ldots, r_s)$. Our goal is to find a homomorphism $h : \mathbf{R} \to \mathbf{\Gamma}^{\mathcal{B}}$, i.e. to assign every variable $v \in V$ an algebra $h(v) \in \mathcal{B}$ in such a way that certain constraints are satisfied. Let us define an instance of $\mathrm{CSP}(\mathbf{\Gamma} \frown \mathcal{B})$ with a set of variables

$$W = \{\mathsf{a}[v, i, \alpha_{n_i}(j)] \mid v \in V, i \in [k], j \in [d^{n_i}]\}.$$

All variables take their values in D. A value assigned to the variable $\mathsf{a}[v, i, \alpha_{n_i}(j)]$ corresponds to $o_i^{h(v)}(\alpha_{n_i}(j))$ for $h : \mathbf{R} \to \mathbf{\Gamma}^{\mathcal{B}}$. For any $v \in V$, an assignment of a tuple of variables

$$T(v) = \left(\mathsf{a}[v, 1, \alpha_{n_1}(1)], ..., \mathsf{a}[v, 1, \alpha_{n_1}(d^{n_1})], ..., \mathsf{a}[v, k, \alpha_{n_k}(1)], ..., \mathsf{a}[v, k, \alpha_{n_k}(d^{n_k})]\right)$$

[3] The notion of the trace introduced below is used in a proof of Galois theory for functional and relational clones.

is constrained to be in $Tr(\mathcal{B})$. Any such constraint models an assignment of an algebra from $\mathcal{B}$ to v in the initial instance of CSP($\mathbf{\Gamma}^{\mathcal{B}}$). Indeed, an assignment of an algebra $h(v) \in \mathcal{A}_D^\sigma$ to v corresponds to assigning the tuple $T(v)$ the value

$$\left(o_1^{h(v)}(\alpha_{n_1}(1)), ..., o_1^{h(v)}(\alpha_{n_1}(d^{n_1})), ..., o_k^{h(v)}(\alpha_{n_k}(1)), ..., o_k^{h(v)}(\alpha_{n_k}(d^{n_k}))\right).$$

The latter tuple is in $Tr(\mathcal{B})$ if and only if $h(v) \in \mathcal{B}$.

The second type of constraints are

$$\langle(\mathsf{a}[v_1, j, \overline{x}_1], ..., \mathsf{a}[v_p, j, \overline{x}_p]), \varrho_l\rangle$$

for all $l \in [s]$, $(v_1, ..., v_p) \in r_l$, $j \in [k]$ and $(\overline{x}_1, ..., \overline{x}_p) \in (\varrho_l)^{n_j}$.

In the initial instance of CSP($\mathbf{\Gamma}^{\mathcal{B}}$) we have constraints of the following kind: assigned values for a tuple $(v_1, ..., v_p) \in r_l$ should be in $\varrho_l^{\mathcal{B}}$, i.e. $\left(o_j^{h(v_1)}, ..., o_j^{h(v_p)}\right)$ should component-wise preserve ϱ_l (for $j \in [k]$). This means that for any $(\overline{x}_1, ..., \overline{x}_p) \in (\varrho_l)^{n_j}$, we have $\left(o_j^{h(v_1)}(\overline{x}_1), ..., o_j^{h(v_p)}(\overline{x}_p)\right) \in \varrho_l$. Thus, for any $l \in [s], j \in [k]$, and any $(v_1, ..., v_p) \in r_l$, $(\overline{x}_1, ..., \overline{x}_p) \in (\varrho_l)^{n_j}$, we restrict $(\mathsf{a}[v_1, j, \overline{x}_1], ..., \mathsf{a}[v_p, j, \overline{x}_p])$ to take its values in ϱ_l, and a conjunction of all those constraints is equivalent to the initial constraints of CSP($\mathbf{\Gamma}^{\mathcal{B}}$).

Thus, we described, given an instance of CSP($\mathbf{\Gamma}^{\mathcal{B}}$), how to define a constraint satisfaction problem with relations in constraints taken either from $\{Tr(\mathcal{B})\}$, or from Γ. By construction, there is a one-to-one correspondence between solutions of the initial CSP and the constructed CSP. □

Let us denote by Γ^* the set of all relations over D that can be expressed as a primitive positive formula over Γ, i.e. as a syntactically correct formula of the form $\exists \mathbf{v} \varrho_1(\mathbf{v}_1) \wedge \cdots \wedge \varrho_l(\mathbf{v}_l)$ where $\varrho_i \in \Gamma, i \in [l]$ and $\mathbf{v}, \mathbf{v}_i, i \in [l]$ are lists of variables. From Theorem 3 and [11] we conclude:

Corollary 1. *If $Tr(\mathcal{B}) \in \Gamma^*$, then* CSP($\mathbf{\Gamma}^{\mathcal{B}}$) *is Karp reducible to* CSP($\mathbf{\Gamma}$).

Let us translate the premise of Corollary 1 to the language of polymorphisms. A well-known fact proved by Geiger [9] and by Bodnarchuk-Kalužnin-Kotov-Romov [2] is that $\varrho \in \Gamma^*$ if and only if ϱ is preserved by all polymorphisms from Pol(Γ). To reformulate Corollary 1 using this fact, we need to introduce some definitions.

Let f be an m-ary operation on D. An m-ary operation $f^{\mathcal{A}_D^\sigma}$ on $\mathcal{A}_D^\sigma$ is defined by the following rule: $f^{\mathcal{A}_D^\sigma}(\mathbb{A}_1, \cdots, \mathbb{A}_m) = \mathbb{A}$ if and only if for any $j \in [k]$,

$$o_j^{\mathbb{A}}(x_1, ..., x_{n_j}) = f\left(o_j^{\mathbb{A}_1}(x_1, ..., x_{n_j}), o_j^{\mathbb{A}_2}(x_1, ..., x_{n_j}), ..., o_j^{\mathbb{A}_m}(x_1, ..., x_{n_j})\right). \quad (5)$$

We say that $f^{\mathcal{A}_D^\sigma}$ preserves $\mathcal{B}$ if $f^{\mathcal{A}_D^\sigma}(\mathbb{A}_1, ..., \mathbb{A}_m) \in \mathcal{B}$ whenever $\mathbb{A}_1, ..., \mathbb{A}_m \in \mathcal{B}$. The following lemma directly follows from the definition of the trace. For completeness, its proof is given in Appendix.

Lemma 3. *An operation $p : D^m \to D$ is a polymorphism of $Tr(\mathcal{B})$ if and only if $p^{\mathcal{A}_D^\sigma}$ preserves $\mathcal{B}$.*

From Lemma 3 the following corollary is straightforward.

Corollary 2. *If* $p^{\mathcal{A}^\sigma_D}$ *preserves* $\mathcal{B}$ *for any* $p \in \mathrm{Pol}(\Gamma)$, *then* $\mathrm{CSP}(\mathbf{\Gamma}^{\mathcal{B}})$ *is polynomial-time reducible to* $\mathrm{CSP}(\mathbf{\Gamma})$.

Example 1. Let $D = \{0,1\}$ and $\Gamma = \{\{0\},\{1\}, xy \to z\}$. It is well-known that $\Gamma^* = \mathrm{Inv}(\{\wedge\})$ is a set of Horn predicates. Therefore, if $\wedge^{\mathcal{A}^\sigma_D}$ preserves $\mathcal{B}$, then $\mathrm{CSP}(\mathbf{\Gamma}^{\mathcal{B}})$ is polynomial-time solvable.

Theorem 3 can be slightly strengthened. In order to simplify our notation we will consider a case when the signature σ contains only one n-ary operation symbol o. Recall that $\varrho(\mathbf{v}) = \{d(\mathbf{v},\mathbf{y})|\mathbf{y} \in \varrho\}$ (see the Definition 3). Thus, $Tr(\mathcal{B})(\alpha_n(1),\cdots,\alpha_n(d^n))$ equals

$$\{((\alpha_n(1),y_1),\cdots,(\alpha_n(d^n),y_{d^n}))|(y_1,\cdots,y_{d^n}) \in Tr(\mathcal{B})\}.$$

Let us denote $\mathbf{\Gamma}^n = (D^n, \varrho_1^n,\cdots,\varrho_s^n)$ (the notation ϱ_i^n is introduced after Definition 13).

Theorem 4. $\mathrm{CSP}(\mathbf{\Gamma}^{\mathcal{B}})$ *is polynomial-time Karp reducible to*

$$\mathrm{MCSP}(\Gamma_{\mathbf{\Gamma}^n} \cup \{Tr(\mathcal{B})(\alpha_n(1),\cdots,\alpha_n(d^n))\}).$$

Proof. Let us return to the proof of Theorem 3 and to the CSP that we constructed in that proof. Since we have only one operation symbol in σ we will omit the second index in our variables. Recall that we had two types of constraints. Constraints of the first type require a tuple of variables

$$(\mathsf{a}[v,\alpha_n(1)],\cdots,\mathsf{a}[v,\alpha_n(d^n)])$$

to take its values in $Tr(\mathcal{B})$. Constraints of the second type are as follows: for any $l \in [s]$ and any $(v_1,...,v_p) \in r_l$, $(\overline{x}_1,...,\overline{x}_p) \in \varrho_l^n$, $(\mathsf{a}[v_1,\overline{x}_1],...,\mathsf{a}[v_p,\overline{x}_p])$ should take its values in ϱ_l. Thus, in the homomorphism reformulation of CSP, we have to find a homomorphism between a new pair of structures $\mathbf{R}' = (V', r_1',\ldots,r_s',\Xi)$ and $\mathbf{\Gamma} \frown \mathcal{B} = (D,\varrho_1,\cdots,\varrho_s,Tr(\mathcal{B}))$ where

$$V' = \{\mathsf{a}[v,\alpha_n(j)]|v \in V, j \in [d^n]\},$$
$$r_l' = \{(\mathsf{a}[v_1,\overline{x}_1],...,\mathsf{a}[v_p,\overline{x}_p])|(v_1,...,v_p) \in r_l, (\overline{x}_1,...,\overline{x}_p) \in \varrho_l^n\},$$
$$\Xi = \{(\mathsf{a}[v,\alpha_n(1)],\cdots,\mathsf{a}[v,\alpha_n(d^n)])|v \in V\}.$$

Let us define $\mathbf{\Gamma}^n_\xi = (D^n,\varrho_1^n,\cdots,\varrho_s^n,\xi)$ where $\xi = \{(\alpha_n(1),\cdots,\alpha_n(d^n))\}$. By construction a mapping $\delta : V' \to D^n$, where $\delta(\mathsf{a}[v,\overline{x}]) = \overline{x}$, is a homomorphism from $\mathbf{R}'$ to $\mathbf{\Gamma}^n_\xi$. Thus, we are given a homomorphism to $\mathbf{\Gamma}^n_\xi$ and our goal is to find a homomorphism to $\mathbf{\Gamma} \frown \mathcal{B}$ which is exactly the definition of $\mathrm{CSP}^+_{\mathbf{\Gamma}^n_\xi}(\mathbf{\Gamma} \frown \mathcal{B})$.

According to Lemma 1, $\mathrm{CSP}^+_{\mathbf{\Gamma}^n_\xi}(\mathbf{\Gamma} \frown \mathcal{B})$ is equivalent to $\mathrm{MCSP}((\mathbf{\Gamma} \frown \mathcal{B})_{\mathbf{\Gamma}^n_\xi})$. There are 2 types of relations in $(\mathbf{\Gamma} \frown \mathcal{B})_{\mathbf{\Gamma}^n_\xi}$: those that are in $\Gamma_{\mathbf{\Gamma}^n}$ and the relation $Tr(\mathcal{B})(\alpha_n(1),\cdots,\alpha_n(d^n))$. Therefore, we reduced $\mathrm{CSP}(\mathbf{\Gamma}^{\mathcal{B}})$ to $\mathrm{MCSP}(\Gamma_{\mathbf{\Gamma}^n} \cup \{Tr(\mathcal{B})(\alpha_n(1),\cdots,\alpha_n(d^n))\})$. □

To formulate a version of Theorem 4 for a general signature σ we need the notion of the disjoint union of relational structures. Given similar structures $\mathbf{T}_i = (A_i, \varrho_{i1}, ..., \varrho_{ik}), i \in [q]$, their disjoint union, denoted $\uplus_{i=1}^{q} \mathbf{T}_i$, is a structure $(B, \pi_1, ..., \pi_k)$ with the domain $B = \cup_{i=1}^{q} \{i\} \times A_i$ and relations $\pi_j = \cup_{i=1}^{q} \tau_{ij}$ where τ_{ij} is a reinterpretation of ϱ_{ij} as a relation over $\{i\} \times A_i$, i.e. $\tau_{ij} = \{((i, a_1), ..., (i, a_{\mathrm{ar}(\varrho_{ij})})) \mid (a_1, ..., a_{\mathrm{ar}(\varrho_{ij})}) \in \varrho_{ij}\}$. We denote B as $\uplus_{i=1}^{q} A_i$ and π_j as $\uplus_{i=1}^{q} \varrho_{ij}$. Let us denote $\mathbf{\Gamma}^{\sigma} = \uplus_{i=1}^{s} \mathbf{\Gamma}^{n_i}$. Let also $\gamma_1, \cdots, \gamma_N$ be the ordering of $\cup_{i=1}^{k} \{i\} \times D^{n_i}$ (the domain of $\mathbf{\Gamma}^{\sigma}$) in which first elements from $\{1\} \times D^{n_1}$ go (in lexicographic order), second $\{2\} \times D^{n_2}$ (in lexicographic order), etc. Then, a generalization of Theorem 4 is below. Its proof can be found in Appendix.

Theorem 5. $\mathrm{CSP}(\mathbf{\Gamma}^{\mathcal{B}})$ *is polynomial-time Karp reducible to*

$$\mathrm{M}CSP(\Gamma_{\mathbf{\Gamma}^{\sigma}} \cup \{Tr(\mathcal{B})(\gamma_1, \cdots, \gamma_N)\}).$$

Remark 3. $\mathrm{MCSP}(\Gamma_{\mathbf{\Gamma}^{\sigma}})$ is tractable, because $\mathbf{\Gamma}^{\sigma} \to \mathbf{\Gamma}$ and $\mathrm{CSP}^{+}_{\mathbf{\Gamma}^{\sigma}}(\mathbf{\Gamma})$ is a trivial problem (Lemma 1). Therefore, $\mathrm{CSP}(\Gamma_{\mathbf{\Gamma}^{\sigma}})$ is tractable. Moreover, by construction $\{Tr(\mathcal{B})(\gamma_1, \cdots, \gamma_N)\}$ is also tractable. Thus, Theorem 5 describes a reduction to an NP-hard language only if the union of those two tractable languages is NP-hard. We conducted some experimental studies with the latter constraint language using the Polyanna software which can be found in Appendix.

6 Conclusions

As examples in Appendix show, the induced problem $\mathrm{CSP}(\mathbf{\Gamma}^{\mathcal{B}})$ is often easier that the initial $\mathrm{CSP}(\mathbf{\Gamma})$. If $\mathcal{B}$ is tractable and one manages to find a homomorphism $\chi : \mathbf{R} \to \mathbf{\Gamma}^{\mathcal{B}}$, then finding $h : \mathbf{R} \to \mathbf{\Gamma}$ can be done efficiently. This inspires the whole family of algorithms based on reducing $\mathrm{CSP}(\mathbf{\Gamma})$ to $\mathrm{CSP}(\mathbf{\Gamma}^{\mathcal{B}})$. This generalizes Green and Cohen's reduction of CSPs to finding appropriate permutations of domains.

It is an open research problem to generalize the construction of the template $\mathbf{\Gamma}^{\mathcal{B}}$ to valued constraint languages (one such example can be found in Sect. 5 of [19]). Practical application of the reduction of $\mathrm{CSP}(\mathbf{\Gamma})$ to $\mathrm{CSP}(\mathbf{\Gamma}^{\mathcal{B}})$ is another topic of future research.

References

1. Barto, L., Krokhin, A., Willard, R.: Polymorphisms, and how to use them. In: Krokhin, A., Zivny, S. (eds.) The Constraint Satisfaction Problem: Complexity and Approximability, Dagstuhl Follow-Ups, vol. 7, pp. 1–44. Schloss Dagstuhl–Leibniz-Zentrum fuer Informatik, Dagstuhl (2017). https://doi.org/10.4230/DFU.Vol7.15301.1, http://drops.dagstuhl.de/opus/volltexte/2017/6959
2. Bodnarchuk, V., Kalužnin, L., Kotov, V., Romov, B.: Galois theory for post algebras. Cybernetics **5**(1-2), 243–252 (1969)
3. Bulatov, A., Krokhin, A., Jeavons, A.: Classifying the complexity of constraints using finite algebras. SIAM J. Comput. **34**(3), 720–742 (2005)

4. Bulatov, A.A.: A dichotomy theorem for nonuniform CSPS. CoRR abs/1703.03021 (2017). http://arxiv.org/abs/1703.03021
5. Cohen, D.A.: Tractable decision for a constraint language implies tractable search. Constraints **9**(3), 219–229 (2004). https://doi.org/10.1023/B:CONS.0000036045.82829.94
6. Cook, S.A.: The complexity of theorem-proving procedures. In: Proceedings of the Third Annual ACM Symposium on Theory of Computing, STOC 1971, pp. 151–158 (1971)
7. Feder, T., Vardi, M.Y.: The computational structure of monotone monadic SNP and constraint satisfaction: a study through datalog and group theory. SIAM J. Comput. **28**(1), 57–104 (1998)
8. Garey, M.R., Johnson, D.S.: Computers and Intractability; A Guide to the Theory of NP-Completeness. W. H. Freeman & Co., New York (1990)
9. Geiger, D.: Closed systems of functions and predicates. Pac. J. Math. **27**(1), 95–100 (1968)
10. Green, M.J., Cohen, D.A.: Domain permutation reduction for constraint satisfaction problems. Artif. Intell. **172**(8), 1094–1118 (2008)
11. Jeavons, P.: On the algebraic structure of combinatorial problems. Theor. Comput. Sci. **200**(1–2), 185–204 (1998)
12. Jeavons, P.G.: On the algebraic structure of combinatorial problems. Theoret. Comput. Sci. **200**(1–2), 185–204 (1998). https://doi.org/10.1016/S0304-3975(97)00230-2
13. Kearnes, K., Marković, P., McKenzie, R.: Optimal strong Mal'cev conditions for omitting type 1 in locally finite varieties. Algebra Univers. **72**(1), 91–100 (2014)
14. Kolmogorov, V., Rolínek, M., Takhanov, R.: Effectiveness of structural restrictions for hybrid CSPs. In: Elbassioni, K., Makino, K. (eds.) ISAAC 2015. LNCS, vol. 9472, pp. 566–577. Springer, Heidelberg (2015). https://doi.org/10.1007/978-3-662-48971-0_48
15. Maróti, M., McKenzie, R.: Existence theorems for weakly symmetric operations. Algebra Univers. **59**(3–4), 463–489 (2008)
16. Schaefer, T.J.: The complexity of satisfiability problems. In: Proceedings of the Tenth Annual ACM Symposium on Theory of Computing, STOC 1978, pp. 216–226 (1978)
17. Siggers, M.H.: A strong Mal'cev condition for locally finite varieties omitting the unary type. Algebra Univers. **64**(1–2), 15–20 (2010)
18. Takhanov, R.: Hybrid (V)CSPS and algebraic reductions. CoRR abs/1506.06540v1 (2015). https://arxiv.org/abs/1506.06540v1
19. Takhanov, R.: Hybrid VCSPs with crisp and valued conservative templates. In: Okamoto, Y., Tokuyama, T. (eds.) 28th International Symposium on Algorithms and Computation (ISAAC 2017). Leibniz International Proceedings in Informatics (LIPIcs), vol. 92, pp. 65:1–65:13. Schloss Dagstuhl–Leibniz-Zentrum fuer Informatik, Dagstuhl (2017). https://doi.org/10.4230/LIPIcs.ISAAC.2017.65, http://drops.dagstuhl.de/opus/volltexte/2017/8247
20. Zhuk, D.: The proof of CSP dichotomy conjecture. CoRR abs/1704.01914 (2017). http://arxiv.org/abs/1704.01914

Correction to: Parameterized Algorithms for Covering by Arithmetic Progressions

Ivan Bliznets, Jesper Nederlof, and Krisztina Szilágyi

Correction to:
Chapter 9 in: H. Fernau et al. (Eds.): *SOFSEM 2024: Theory and Practice of Computer Science*, LNCS 14519, https://doi.org/10.1007/978-3-031-52113-3_9

In the originally published version of chapter 9, there was a typo in the name of the author Jesper Nederlof. This has been corrected.

The updated version of this chapter can be found at
https://doi.org/10.1007/978-3-031-52113-3_9

H. Fernau et al. (Eds.): SOFSEM 2024, LNCS 14519, p. C1, 2024.
https://doi.org/10.1007/978-3-031-52113-3_35

Author Index

H. Fernau et al. (Eds.): SOFSEM 2024, LNCS 14519, pp. 501–502, 2024.
https://doi.org/10.1007/978-3-031-52113-3

Printed in the United States
by Baker & Taylor Publisher Services